INTRODUCTION TO

Spread-Spectrum Communications

INTRODUCTION TO
Spread-Spectrum Communications

Roger L. Peterson *Motorola Inc.*
Corporate Research Laboratories

Rodger E. Ziemer *University of Colorado*
—Colorado Springs

David E. Borth *Motorola Inc.*
Corporate Research Laboratories

PRENTICE HALL   Englewood Cliffs, NJ   07632

LIBRARY OF CONGRESS CATALOGING-IN-PUBLICATION DATA

Peterson, Roger L.
    Introduction to spread-spectrum communications / Roger L.
    Peterson, Rodger E. Ziemer, David E. Borth.
        p.   cm.
    Includes index.
    ISBN 0-02-431623-7
    1. Spread-spectrum communications.   I. Ziemer, Rodger E.
II. Borth, David E.   III. Title.
TK5103.45.P47   1995
621.382—dc20   94-27567
                                                    CIP

Acquisitions Editor: Linda Ratts
Production Supervisor: John Travis
Production Manager: Francesca Drago
Cover Designer: Jerry Votta
Buyer: Bill Scazzero
Editorial Assistant: Naomi Goldman

© 1995 by Prentice-Hall, Inc.
A Simon & Schuster Company
Englewood Cliffs, New Jersey 07632

The author and publisher of this book have used their best efforts in
perparing this book. These efforts include the development, research,
and testing of the theories and programs to determine their effectiveness.
The author and publisher shall not be liable in any event for incidental
or consequential damages in connection with, or arising out of, the
furnishing, performance, or use of these programs.

PRINTED IN THE UNITED STATES OF AMERICA

10 9 8 7 6 5 4 3 2 1

ISBN: 0-02-431623-7

Prentice-Hall International (UK) Limited, *London*
Prentice-Hall of Australia Pty. Limited, *Sydney*
Prentice-Hall Canada Inc., *Toronto*
Prentice-Hall Hispanoamericana, S.A., *Mexico*
Prentice-Hall of India Private Limited, *New Delhi*
Prentice-Hall of Japan, Inc., *Tokyo*
Simon & Schuster Asia Pte. Ltd., *Singapore*
Editora Prentice-Hall do Brasil, Ltda., *Rio de Janeiro*

CONTENTS

CHAPTER 5   **Initial Synchronization of the Receiver Spreading Code   221**

CHAPTER 6   **Performance of Spread-Spectrum Systems in Jamming Environments   319**

CHAPTER 9   Code-Division Multiple-Access Digital Cellular Systems   520

CHAPTER 10   Low-Probability-of-Intercept Methods   584

PREFACE

This book is an outgrowth of a book that the first two authors published in 1985 entitled *Digital Communications and Spread-Spectrum Systems.* After that book went out of print, we had many inquiries about how to obtain copies of it for the purposes of teaching courses in universities and industry on spread-spectrum communication systems. After thinking about it, we decided to add material to the original volume on areas involving spread spectrum that have become important since the first book was published. The most important body of knowledge included as a result is that on cellular mobile channels and systems, written by David Borth of Motorola, Inc. In addition, the material included in the first book has been expanded in many cases. We feel that the resulting book will fill a need for teaching the theory and applications of spread-spectrum communications at the senior–graduate level.

After a prologue, which sets the stage for the study of spread-spectrum communication systems, a review chapter on common digital communication methods begins the book. With coverage of this material, we feel that anyone having previous courses on basic communication theory, covering the application of Fourier theory to analog modulation systems and on probability and random processes, should, with diligence, be able to learn successfully about spread-spectrum systems from this book. All other necessary background material is self-contained in the book. It is in this sense that the book is termed an *introduction.*

In Chapter 2 we introduce the concept of spread-spectrum modulation and the reasons for its use. The most widely used types of spread-spectrum modulation are described, including *direct-sequence* (DS), *frequency-hopped* (FH), and *hybrid* DS/FH spread spectrum. The generation of pseudorandom digital sequences is important in any spread-spectrum system implementation. Chapter 3 provides a comprehensive introduction to the generation of pseudonoise (PN) sequences by means of linear feedback shift registers and the properties of PN sequences. Other types of sequences, such as Gold codes and nonlinear codes, are also described briefly.

An important function in any spread-spectrum system is synchronization of the locally generated despreading code with the spreading code generated at the transmitter. This synchronization process can be divided into two parts: initial synchronization (acquisition) and tracking. The former is the more complex of the two to analyze mathematically. Code tracking is therefore taken up first and is the subject of Chapter 4, with acquisition taken up in Chapter 5, even though code acquisition must chronologically precede tracking in implementation of a spread-spectrum receiver. The two main code tracking methods considered are referred to as *delay-locked-loop tracking* and *tau-dither tracking.* With suitable manipulations and definitions of the signal and noise processes within the loop, both techniques can be reduced to conventional phase-locked-loop implementations. Once this point is reached, the treatment of code tracking loops can make use of standard phase-locked-loop analysis techniques. Accordingly, Appendix A contains a summary of pertinent phase-locked-loop theory for those needing a brief introduction or refresher.

Initial synchronization of the spreading waveform is perhaps the most difficult spread-spectrum implementation problem. In Chapter 5 we treat this subject comprehensively. We begin with the simplest technique using swept serial search, progressing through a general analysis of stepped serial search, a discussion of multiple-dwell detection techniques, and finally to a detailed analysis of sequential detection techniques. The chapter ends with a short discussion of matched filter synchronization techniques.

An analysis of the performance of spread-spectrum systems in a jamming environment is the subject of Chapter 6. The chapter begins with a discussion of the system model, including barrage noise, partial band noise, pulsed noise, tone, multiple tone, and repeater jamming. Following this, the most commonly used digital modulation techniques are evaluated in most types of jamming. It is concluded that error correction coding is an essential component of any spread-spectrum system to provide adequate protection in jamming. Accordingly, Chapter 7 treats the performance of spread-spectrum systems that employ forward error correction. Some important coding schemes, including Reed–Solomon, Bose–Chaudhuri–Hocquenghem, and convolutional, are presented. A discussion of fundamental coding and decoding concepts, including Viterbi algorithm decoding of convolutional codes, is included for those not familiar with this subject.

Chapter 8, by David Borth, departs from spread-spectrum topics to introduce material on fading channels necessary for considering the performance of code-division multiple access (CDMA) systems used for cellular land-mobile communications, which is the topic of Chapter 9. A general fading channel model is specialized to the wide-sense-stationary uncorrelated scattering (WSSUS) fading channel model. Specializations of the WSSUS fading channel model to the more familiar time-selective, frequency-selective, and nondispersive fading models are then given. Modeling of the short-term characteristics of the land-mobile channel is done next in terms of the fading models discussed previously. Long-term behavior due to shadowing of the mobile channel in terms of lognormal fading is also discussed. The chapter ends with a discussion of coverage modeling, diversity, and diversity combining, including the Rake receiver.

Chapter 9, also by D. Borth, addresses the application of spread spectrum to CDMA in digital cellular systems. Fundamental concepts of cellular systems, including the concept of frequency reuse, are discussed first. Cochannel interference protection prediction by simple calculation and Monte Carlo simulation is considered next. The capacity of frequency division and CDMA cellular systems is then discussed, followed by a consideration of specific examples of CDMA digital cellular systems, including the North American DS-CDMA system, the Cooper and Nettleton system proposed in the late 1970s, the Bell Labs multilevel frequency-shift-keyed/FH system, the European GSM slow-FH system, and a hybrid slow-FH time-division multiple access/CDMA system for personal communications applications.

The book ends with a consideration of methods for uncooperatively detecting spread-spectrum signals and extracting their parameters, such as code clock rate. Both wideband energy detection of spread-spectrum signals and optimum receivers for detection of DS and FH spread-spectrum signals are considered. A detector for frequency-hopped or hybrid spread spectrum, which is a channelized receiver with a block moving window detector in each channel with combinations of these detections by a logical OR operation, is considered next as an approximation to the optimum detector. The chapter concludes with a consideration of delay-and-multiply receivers for extracting code clock rate, and optimization of a spread-spectrum signal to make this extraction as difficult as possible for the unintended interceptor. Chapter 10 was written by Rodger Ziemer.

The authors wish to recognize the encouragement provided by their colleagues to republish this material and suggestions for improvement of the book. One of us (Rodger Ziemer) gratefully acknowledges the research support provided by the Office of Naval Research over the past several years. This not only provided support during the summer when some of the material was written, but provided research funding for several interesting problem areas involving spread spectrum. Roger Peterson and David Borth thank Motorola Inc. for providing the technical environment in which we learned much of what we know about spread-spectrum communications. We also thank our associates for asking difficult questions and patiently awaiting answers which have provided the seeds for some of this material. All three authors thank Arthur Ross of Qualcomm, Inc. for providing Internet access to the computer programs referred to in Appendix F.

We are all grateful to our families for their consistent support throughout this project. Roger Peterson thanks his wife, Ann, for her patience, and dedicates his portions of this book to his daughter, Diane Ruth Peterson, for providing a source of energy and joy. Rodger Ziemer thanks his wife, Sandy, for her acceptance and patience of his infatuation with book writing. David Borth thanks his wife, Mary, and his children, Bridget, Jeffrey, and Brian, for their tolerance and compassion during the nights and weekends spent working on the book.

While the authors are not going to offer specific suggestions for the use of this book, we would like to point out that the material has been used successfully for formal courses and in independent study courses, as a self-study aid, and for short courses on spread-spectrum communications.

Roger L. Peterson
Rodger E. Ziemer
David E. Borth

ACKNOWLEDGMENTS

Grateful acknowledgment is extended to the following for permission to reprint copyrighted materials:

The Institute of Electrical and Electronics Engineers, Inc.:

Figure 1-15, reproduced from Figure 5, S. Chennakeshu and G. J. Saulner, "Differential Detection of $\pi/4$-Shifted DQPAK for Digital Cellular Radio," *IEEE Trans. on Veh. Tech.,* vol. 42, pp. 46–57, February 1993. Copyright © 1993 by IEEE. Reprinted by permission.

Figure 1-23, reproduced from Figures 4 and 6, P. J. Crepeau, "Uncoded and Coded Performance of MFSK & DPSK in Nakagami Fading Chennels," *IEEE Trans. on Commun. Tech.,* vol. COM-40, pp. 487–493, March 1992. Copyright © 1992 by IEEE. Reprinted by permission.

Figure A-16, reproduced from Figure 4, S. L. Goldman, "Second-Order Phase-Lock-Loop Acquisition Time in the Presence of Narrow-Ban Gaussian Noise," *IEEE Trans. on Commun. Tech.,* vol. COM-21, pp. 297–299, April 1973. Copyright © 1973 by IEEE. Reprinted by permission.

Figure 5-34, reproduced from Figure 2, R. B. Ward, "Acquisition of Pseudonoise Signals by Sequential Estimation," *IEEE Trans. on Commun. Tech.,* vol. COM-13, pp. 475–483, December 1965. Copyright © 1965 by IEEE. Reprinted by permission.

Figure 5-3, reproduced from Figure 4, W. R. Braun, "Performance Analysis for the Expanding Search PS Acq. Algorithm," *IEEE Trans. on Commun. Tech.,* vol. COM-30, pp. 424–435, March 1982. Copyright © 1982 by IEEE. Reprinted by permission.

Figure 5-11, reproduced from Figure 7, H. Urkowitz, "Energy Detection of Unknown Deterministic Signals," *Proc. IEEE,* vol. 55, pp. 523–531, April 1967. Copyright © 1967 by IEEE. Reprinted by permission.

Figures 5-37–5-39, reproduced from Figures 2, 3, 5, 6, 8, and 9, P. T. Nielson, "On the Acquisition Behavior of Binary Delay-Lock Loops," *IEEE Trans. Aerosp. & Elect. Syst.,* vol. AES-11, pp. 415–418, May 1975. Copyright © 1975 by IEEE. Reprinted by permission.

Macmillan Publishers:

Figure P-1, 1-2, 1-6, 1-7, 1-8, 1-9, 1-12, 1-16, 1-18, 1-19, 1-20, 4-5, and 7-13, reproduced from R. E. Ziemer and R. L. Peterson, *Introduction to Digital Communication,* New York, Macmillan Publishing, 1992. Reprinted by permission.

Figures 2-1–2-21, P2-6, P2-12, P2-13, P2-15, 3-1–3-9, 3-12–3-23, P3-8–P3-10, P3-17, 4-1–4-4, 4-6–4-28, P4-14, P4-16, P4-18, 5-1–5-11, 5-13–5-25, 5-34–5-39, 6-1–6-26, P6-7, 7-1, 7-3–7-12, 7-17, 7-18, 7-20, 7-21, 7-23, 7-25, 7-26, A-1–A-23, C-1, C-2, D-1–D-5, and E-1, reproduced from R. E. Ziemer and R. L. Peterson, *Digital Communications and Spread Spectrum Systems,* New York, Macmillan, 1985. Reprinted by permission.

McGraw-Hill, Inc.:

Figure reproduced from Figure 4-8, A. Viterbi, *Principles of Coherent Communications,* New York, McGraw-Hill, 1966. Reprinted by permission.

John Wiley & Sons, Inc.:

Figures A-17 and A-18, reproduced from Figures 2.2 and 2.3, W. F. Egan, *Frequency Synthesis by Phase Lock,* New York, John Wiley & Sons, 1981. Reprinted by permission.

Figures A-5, A-6, A-13a, and A-14, reproduced from Figures 2.3, 2.4, 5.2, and 5.5, F. M. Gardner, *Phaselock Techniques,* 2nd ed., New York, John Wiley & Sons, 1979. Reprinted by permission.

INTRODUCTION TO

Spread-Spectrum Communications

Prologue

The subject of this book is spread-spectrum communications. Spread-spectrum modulation refers to any modulation scheme that produces a spectrum for the transmitted signal much wider than the bandwidth of the information being transmitted *independently* of the information-bearing signal. Why would such an apparently wasteful approach to modulation be used? The reasons are several. Among them are:

1. To provide some degree of resistance to interference and jamming (called *jam resistance*—JR);
2. To provide a means for masking the transmitted signal in the background noise in order to lower the probability of intercept by an adversary (called *low probability of intercept*—LPI);
3. To provide resistance to signal interference from multiple transmission paths (i.e., *multipath*);
4. To permit the access of a common communication channel by more than one user (called *multiple access*);
5. To provide a means for measuring range, or distance between two points.

Spread-spectrum communications grew out of research efforts during World War II to provide secure means of communication in hostile environments. This work remained classified, for the most part, until the 1970s. In 1977, the first special issue of the *Institute of Electrical and Electronics Engineers Transactions in Communications* on spread spectrum appeared, to be followed by four more special issues [1–5] in the *Transactions* and the *IEEE Journal on Selected Areas on Communications*. In addition, other special issues in these journals and others on other topics [6–9] have devoted significant portions of their contents to spread-spectrum topics. Several books have also appeared that have dealt with spread spectrum and its various applications [10–14].

A block diagram of a digital communications system is shown in Figure P-1. This diagram is general in that it contains components that may not be found in all communications systems. The "standard" components that would appear in most communications systems include:

1. The source, including source encoder;
2. The data modulator;
3. Power amplification (and antenna if a radio propagation channel);

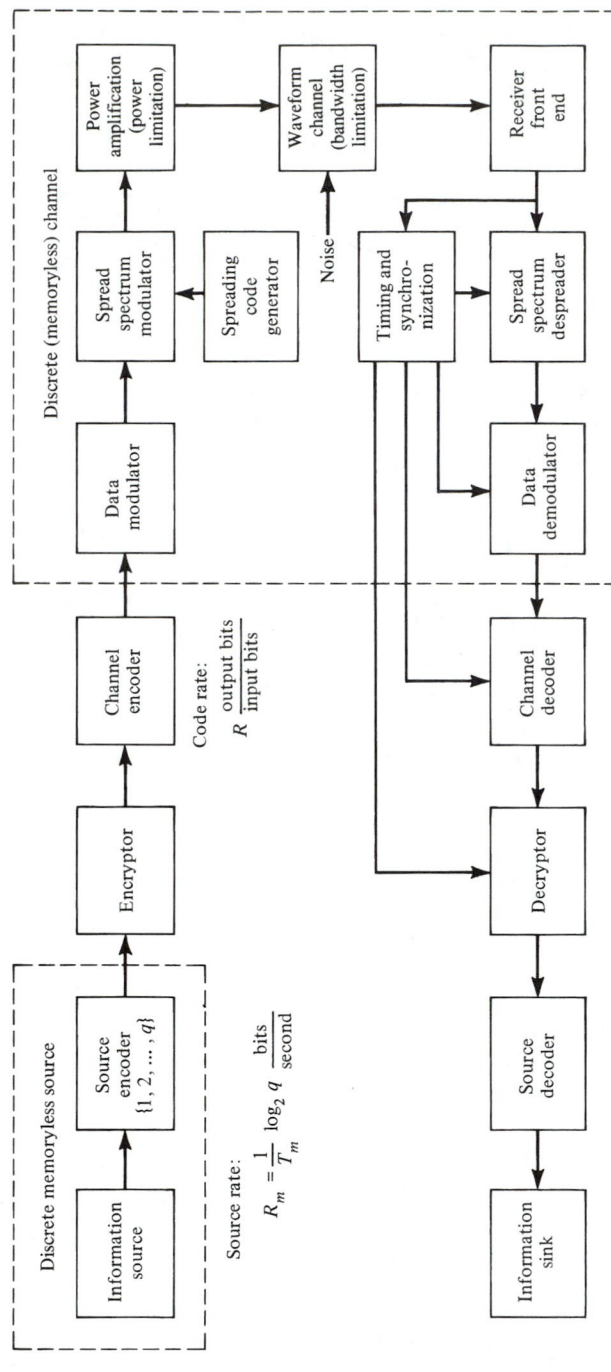

FIGURE P-1. Block diagram of a typical digital communication system.

4. Receiver front end (including the receiving antenna if a radio propagation channel);
5. Timing and synchronization;
6. Data demodulator;
7. Information sink, including source decoder.

The "nonstandard" components, not included in every communications system are:

1. Encryptor;
2. Channel encoder;
3. Spread-spectrum modulator;
4. Spread-spectrum despreader;
5. Channel decoder;
6. Decryptor.

Of these six nonstandard components, items 2 through 5 would be standard in a spread-spectrum communications system. It is obvious that the spread-spectrum modulator and despreader are necessary to make the system spread spectrum. What is less obvious is that the channel encoding and decoding functions (with the additional operations of interleaving and de-interleaving implied) are often necessary in a spread-spectrum system to regain performance lost to smart jammers. This and other topics, including the important issue of generation and synchronization of the spectrum spreading code, will be discussed in the chapters to follow in this book.

As mentioned previously, spread-spectrum communications systems were invented to provide secure communications in military environments, and these applications are enlarged upon in considerable detail in some of the references of this prologue [6, 8, 10–14]. One modern spread-spectrum communications system that was developed for military applications, but is now finding a host of civilian applications, is the Global Positioning System (GPS), also known by its military name NAVSTAR [11, 15]. These applications include geodesic survey, position location for civilian vehicles including navigation aids for travelers [16], position location for hunters and fishermen, and position location for commercial vehicles and ships. GPS makes use of 24 satellites in 12 hour orbits spaced uniformly around the earth whose positions are precisely known at all times (some of these are referred to as hot spares). Each satellite transmits two spread spectrum signals at 1575.42 MHz and 1227.60 MHz, referred to as the L1 and L2 signals, respectively. The L1 signal is spread by a short code to make acquisition easy.† It is called the Clear/Acquisition or C/A code, and has a coarse positioning accuracy relative to the code on the L2 signal which is called the Precise, or P, code (the difference in accuracy is about a factor of ten). Both the L1 and L2 signals also convey data about the satellite from which they originated at a fairly low bit rate. A potential user acquires the codes of at least four different satellites which are in view and, through simultaneous solution of the corresponding delay equations, is able to determine its position. With only the delay measurement, the position accuracy is of the order of tens of feet. With special processing techniques, known as differential GPS, the positioning accuracy is of the

---

† Some of the terms used in this discussion may seem foreign to the uninitiated reader. The purpose of this discussion is to merely provide an overview of some applications of spread spectrum. The details of spread spectrum synchronization, for example, is a subject of a later chapter.

order of a foot or less. Although GPS was developed as a military system, it has great civilian potential. These civilian applications are not without peril, however. For example, what happens in a time of war if all commercial airlines are using GPS for navigation and the military exercises its option to make the P code secure and unaccessible to civil applications [16]?

The GPS system was mentioned as an example of a spread-spectrum system that began as strictly military and ended up with a host of civilian applications that are growing daily. Another application of spread spectrum that is strictly civilian is cellular mobile radio. Although the first cellular mobile systems were narrowband and made use of frequency division multiple access for accommodating multiple users, standards have been written and pilot systems are being tested that make use of spread spectrum to accomplish the multiple access in future mobile radio systems. A whole chapter in this book is devoted to such systems.

A related system for which spread spectrum has been proposed as an accessing modulation is satellite-land-mobile communications. A number of such systems have been proposed and are in varying stages of development. Such systems use networks of satellites to provide world-wide communications between personal users with hand-held telephones in some cases and telephone booth-type facilities in other cases. Generally connections will take place through a network of multiple satellites. At least one of these proposed systems will use spread spectrum modulation for accessing communications satellites used as relays.

Another developing area in which spread-spectrum modulation is expected to play a significant role is personal communications systems (PCS) [17, 18]. This includes almost any application where a person can carry with them a communication system to communicate with their home or office. The concept here is that a person's telephone number will not be assigned to a *location* but rather to their *person*. The services provided will be not only voice, but data as well. The beginnings of PCS are evident in notebook computers with data and FAX modems, for example, but when fully developed, PCS will be much more pervasive in people's lives. In fact, the Federal Communications Commission has defined PCS as "a family of mobile or portable radio communications services which could provide services to individuals and businesses and be integrated with a variety of competing networks . . . The primary focus of PCS will be to meet communications requirements of people on the move.''[18]

This book is meant to provide a thorough coverage of the theory for spread-spectrum systems. It is termed an introduction because many of the applications discussed above are not dealt with. However, any course using this book as a text will provide the background necessary to further study spread-spectrum applications in depth.

References

[1] *IEEE Trans. on Commun.*, Vol. COM-25, Aug. 1977. Special issue on spread spectrum.

[2] *IEEE Trans. on Commun.*, Vol. COM-30, May 1982. Special issue on spread spectrum.

[3] *IEEE Journ. on Sel. Areas in Commun.,* Vol. 8, May 1990. Special issue on spread spectrum.

[4] *IEEE Journ. on Sel. Areas in Commun.,* Vol. 8, June 1990. Special issue on spread spectrum.

[5] *IEEE Journ. on Sel. Areas in Commun.,* Vol. 10, May 1992. Special issue on spread spectrum.

[6] *IEEE Trans. on Commun.,* Vol. COM-28, Sept. 1980. Special issue on military communications.

[7] *IEEE Journ. on Sel. Areas in Commun.,* Vol. SAC-2, July 1984. Special issue on mobile radio.

[8] *IEEE Journ. on Sel. Areas in Commun.,* Vol. SAC-3, Sept. 1985. Special issue on military communications.

[9] *IEEE Trans. on Veh. Tech.,* Vol. 40, May 1991. Special issue on mobile radio.

[10] M. K. SIMON, J. K. OMURA, R. A. SCHOLTZ, and B. A. LEVITT, *Spread Spectrum Communications Handbook* (New York: McGraw-Hill, 1994).

[11] R. E. ZIEMER and R. L. PETERSON, *Digital Communications and Spread Spectrum Systems* (New York: Macmillan, 1985).

[12] J. K. HOLMES, *Coherent Spread Spectrum Systems* (New York: John Wiley & Sons, 1982).

[13] D. L. NICHOLSON, *Spread Spectrum Signal Design: LPE and AJ Systems,* (Rockville, MD: Computer Science Press, 1988).

[14] R. C. DIXON, *Spread Spectrum Systems with Commercial Applications,* 3rd ed. (New York: John Wiley & Sons, 1994).

[15] *Global Positioning Systems,* Papers published in *Navigation,* Vols. I–III (Washington, D.C.: The Institute of Navigation, 1986).

[16] R. K. JURGEN, "Smart Cars and Highways Go Global," *IEEE Spectrum,* Vol. 28, pp. 26–36, May 1991.

[17] "Trends in Cordless and Cellular Communications," *IEEE Commun. Mag.,* Vol, 29, June 1991.

[18] B. Z. COBB, "Personal Communications Services," *IEEE Spectrum,* Vol. 30, pp. 20–25, June 1993.

CHAPTER 1

Basic Digital Communications Concepts

1-1 Introduction

The subject of this book is spread-spectrum communications. Chapters 2 through 10 address the fundamental concepts of spread-spectrum communications techniques and their applications. Spread-spectrum communications techniques are an extension of basic coherent and noncoherent digital modulation concepts. To provide a basis for the chapters that follow, many of these fundamental concepts are reviewed in this chapter. The techniques and concepts reviewed in this chapter include (1) matched-filter signal detection in additive white Gaussian noise (AWGN) backgrounds, (2) maximum-likelihood signal detection, (3) signal space concepts, (4) coherent and noncoherent modulation and demodulation/detection techniques, (5) bandwidth and power efficiency of various digital signaling techniques, and (6) the effects of flat fading channels on various digital signaling methods. This chapter acts as a summary of results with extensive references to provide additional information for the reader requiring more details. Those readers already familiar with the subject matter of this chapter may proceed directly to Chapter 2.

1-2 Detection of Binary Signals in Additive White Gaussian Noise

1-2.1 Coherent Modulation Schemes

The simplest possible digital communications system is one that transmits a sequence of binary symbols represented for convenience by $\{0,1\}$ from a transmitter to a receiver over a channel that degrades the transmitted signal with AWGN of two-sided spectral density $N_0/2$. The transmitted binary symbols are associated with two signaling waveforms, denoted as $s_1(t)$ and $s_2(t)$. These waveforms are defined to exist over the time interval $(0,T)$. One of these signals is transmitted each T seconds† so that the information transmission rate is $R_b = 1/T$ binary symbols (bits) per second. During signaling interval k, the transmitter associates a symbol, say a 1, with $s_1(t - kT)$ and the other symbol, say a 0, with $s_2(t - kT)$. The receiver is assumed to have perfect knowledge of both $s_1(t)$ and $s_2(t)$, including the precise time at which they could be received and the probability that they were transmitted (for simplicity, it is assumed in the sequel that they are equally likely), but does not

---

† The symbol T is used to denote *bit period*. Later, T_s is used to denote *symbol period* when signaling schemes that select from more than two transmitted signals are discussed.

know which signal was, in fact, transmitted. During each T-second signaling interval, the receiver observes the signaling waveform contaminated by AWGN and processes this information so as to minimize the probability of making an error.

It can be shown [1–4] that the minimum probability of error is achieved when the receiver guesses the transmitted signal to be that signal which, given the received signal plus noise waveform, was most likely to have been transmitted. Such a receiver is called a *maximum-likelihood receiver*. With these assumptions it can be shown [1] that the minimum probability of error is

$$P_E = Q[\sqrt{z(1 - R_{12})}] \tag{1-1}$$

where

$$Q(x) = \int_x^\infty \frac{e^{-u^2/2}}{\sqrt{2\pi}} \, du \tag{1-2}$$

is the Q-function, which is discussed and tabulated in Appendix B. The quantities z and R_{12} are defined as

$$z = \frac{(E_1 + E_2)/2}{N_0} = \frac{E_b}{N_0} \tag{1-3}$$

and

$$R_{12} = \frac{\sqrt{E_1 E_2}}{E} \rho_{12} \tag{1-4}$$

in which E_i, $i = 1, 2$, is the energy of signal i, defined as

$$E_i = \int_0^T |s_i(t)|^2 \, dt \tag{1-5}$$

and $E = (E_1 + E_2)/2$ is the average signal energy. The parameter ρ_{12} is the normalized correlation coefficient between signals, which is given by

$$\rho_{12} = \frac{1}{\sqrt{E_1 E_2}} \int_0^T s_1(t) s_2(t) \, dt \tag{1-6}$$

If $R_{12} = 0$, the signaling scheme is said to be *orthogonal*, while if $R_{12} = -1$, the signaling scheme is said to be *antipodal*. The probability of error is shown in Figure 1-1 as a function of $z = E_b/N_0$ in decibels for these two cases. For other values of R_{12}, probability-of-error results can be obtained by moving the curve of P_E for $R_{12} = 0$ by $10 \log_{10}(1 - R_{12})$ decibels either left or right, depending on the sign, and reading the corresponding probability of error from the ordinate.

The receiver for these binary signaling schemes can have one of two equivalently performing structures: a matched-filter implementation and a correlator implementation. A block diagram for the matched-filter receiver is shown in Figure 1-2a and is seen to consist of the matched filter followed by a sampler, which samples the output of the matched filter at the end of each T-second signaling interval, and a threshold comparator. For equally probable signals, the comparator threshold is set at

$$k = \tfrac{1}{2}[s_{o1}(T) + s_{o2}(T)] \tag{1-7a}$$

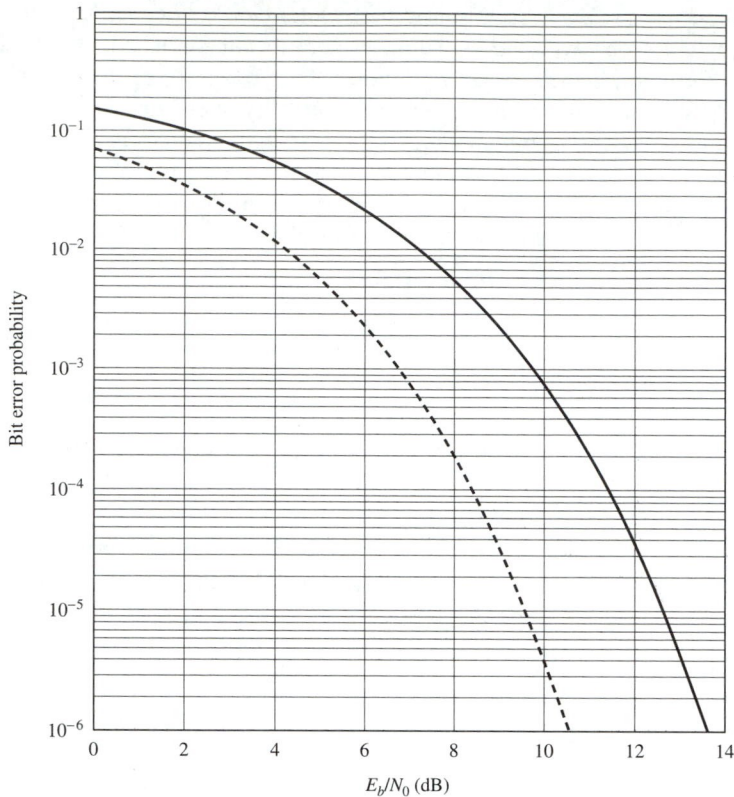

FIGURE 1-1. Probability of error for orthogonal (solid curve) and antipodal (dashed curve) signaling.

where $s_{o1}(T)$ and $s_{o2}(T)$ are the output signals from the matched filter at the sampling instant, corresponding to $s_1(t)$ and $s_2(t)$, respectively, at its input. The threshold can be shown to be [1]

$$k = \tfrac{1}{2}(E_2 - E_1) \tag{1-7b}$$

A matched filter for any signal has an impulse response that is the shifted time reverse of the signal (conjugate time reverse for a complex signal). Since we are dealing with two signals in this case, the matched filter is matched to the *difference* of the two signals and has an impulse response

$$h_0(t) = s_2(T - t) - s_1(T - t) \tag{1-8}$$

A block diagram for the correlator receiver is shown in Figure 1-2b and consists of a correlation operation with the difference of the two signals, followed by a sampler and threshold comparison. The correlator consists of a multiplier and integrator cascade. That the matched-filter and correlator implementations are equivalent may be demonstrated by writing down the signal plus noise outputs of each operation at the sampling instant and showing them to be equal.

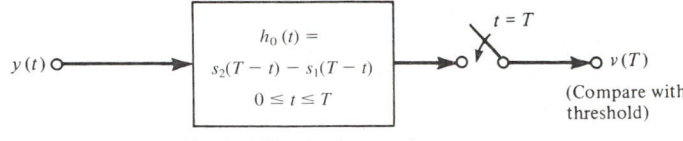

(a) Matched filter implementation

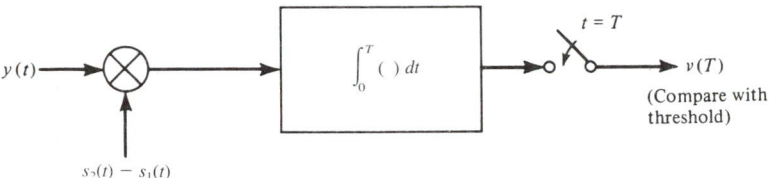

(b) Correlator implementation

FIGURE 1-2. (a) Matched-filter and (b) correlator implementations of the minimum-probability-of-error receiver for binary signal reception.

Several special cases are of interest in the binary signaling hierarchy. These are listed in Table 1-1 together with the thresholds and correlation coefficients in each case. Note that the results for the probability of error for the coherent signaling schemes of Table 1-1 do not account for errors in estimating the correct timing for the signaling interval or for establishing the carrier phase at the receiver. For example, if the carrier phase estimate at the receiver is in error by an amount ϕ, the probability of error for BPSK signaling is given by

$$P_E(\phi) = Q(\sqrt{2z \cos^2\phi}) \qquad |\phi| \leq \pi \tag{1-9}$$

If the phase error ϕ is a random variable with probability density function $p(\phi)$, the average probability of error is

$$\overline{P_E(\phi)} = \int_{-\pi}^{\pi} P_E(\phi)p(\phi)\,d\phi \tag{1-10}$$

For typical probability density functions for the phase error, it is usually necessary to integrate (1-10) numerically. For example, results are given in Table 1-2 for a Gaussian phase error probability density function of the form

$$p(\phi) = \frac{e^{-\phi^2/2\sigma_\phi^2}}{\sqrt{2\pi\sigma_\phi^2}} \tag{1-11}$$

As the signal-to-noise ratio, $z = E_b/N_0$, gets large, the probability of error approaches a nonreducible value. This is particularly apparent from the last column of Table 1-2.

The error probability as a function of E_b/N_0 for a given modulation scheme tells only half of the story, and is often referred to as a measure of its power efficiency. Also important is its bandwidth efficiency, defined to be the ratio of the bandwidth required to accept a given data rate divided by the data rate. For example, it is well

TABLE 1-1. Characteristics of Binary Digital Modulation Schemes[a]

Name	Signals	Threshold [eq. (1-7)]	Signal Correlation Coefficient, R_{12}
Antipodal baseband signaling	$s_{1,2}(t) = \pm\sqrt{\dfrac{E_b}{T}}, \quad 0 \le t \le T$	0	-1
Amplitude-shift keying (ASK)	$\left.\begin{array}{l} s_1(t) = 0 \\ s_2(t) = \sqrt{\dfrac{4E_b}{T}}\cos\omega_0 t \end{array}\right\} 0 \le t \le T$	$E_2/2$	0
(Binary) phase-shift keying (PSK)	$s_{1,2}(t) = \sqrt{\dfrac{2E_b}{T}}\sin(\omega_0 t \pm \cos^{-1}m), \quad 0 \le t \le T$ where $|m| \le 1$ is the modulation index	0	$2m^2 - 1$
Biphase-shift keying (BPSK)	$s_{1,2}(t) = \pm\sqrt{\dfrac{2E_b}{T}}\cos\omega_0 t, \quad 0 \le t \le T$	0	-1
Frequency-shift keying (FSK)	$\left.\begin{array}{l} s_1(t) = \sqrt{\dfrac{2E_b}{T}}\cos\omega_o t \\ s_2(t) = \sqrt{\dfrac{2E_b}{T}}\cos[(\omega_0 + \Delta\omega)t] \end{array}\right\} 0 \le t \le T$ where $\Delta\omega = \pi \times \text{integer}/T$	0	0

[a] In all cases, E_b is the average signal energy per bit, E_1 the energy of signal 1, and E_2 the energy of signal 2. All signaling schemes except antipodal baseband are referred to as coherent because the carrier phase must be known at the receiver to implement the matched or correlator detector.

TABLE 1-2. Error Probabilities for BPSK with Gaussian Phase Jitter

E/N_0 (dB)	P_E, $\sigma_\phi^2 = 0.01$ rad^2	P_E, $\sigma_\phi^2 = 0.05$ rad^2	P_E, $\sigma_\phi^2 = 0.1$ rad^2
9	3.68×10^{-5}	6.54×10^{-5}	2.42×10^{-4}
10	4.35×10^{-6}	1.05×10^{-5}	8.93×10^{-5}
11	3.04×10^{-7}	1.32×10^{-6}	3.78×10^{-5}
12	1.10×10^{-8}	1.59×10^{-7}	1.83×10^{-5}

known from Fourier theory that the spectrum of a rectangular pulse of duration T is

$$S(f) = AT \operatorname{sinc} fT \tag{1-12}$$

where

$$\operatorname{sinc} x = \frac{\sin \pi x}{\pi x} \tag{1-13}$$

When used to modulate a cosinusoid of frequency f_0, the spectrum of the rectangular pulse is centered around the carrier frequency, f_0.

$$S_m(f) = \frac{AT}{2} \{ \operatorname{sinc}[(f - f_0)T] + \operatorname{sinc}[(f + f_0)T] \} \tag{1-14}$$

The bandwidth of the main lobe of the magnitude of this spectrum is

$$B_{\text{RF}} = \frac{2}{T} \quad \text{Hz} \tag{1-15}$$

Since ASK and PSK involve square-pulse-modulated sinusoidal carriers, this is the null-to-null main-lobe bandwidth for these modulation schemes. For FSK, note that the minimum frequency spacing between cosinusoidal bursts at frequencies f_0 and $f_0 + \Delta f$ is $1/2T$ hertz to maintain orthogonality of the two signals. Each carrier must also have $1/T$ hertz on either side of it, giving a total bandwidth for FSK of

$$B_{\text{FSK}} = \frac{2.5}{T} \quad \text{Hz} \tag{1-16}$$

Since $1/T$ is the data rate, R_b, the bandwidth efficiencies of the various modulation schemes just considered are as given in Table 1-3.

TABLE 1-3. Bandwidth Efficiencies for Binary Modulation Schemes

Modulation Type	Bandwidth Efficiency (bps/Hz)
Rectangular pulse baseband	1
ASK, PSK, BPSK	0.5
FSK	0.4

1-2.2 Noncoherent Modulation Schemes

In situations where it is difficult to maintain phase stability, for example in fading channels, it is useful to employ modulation schemes that do not require the acquisition of a reference signal at the receiver which is in phase coherence with the received carrier. ASK and FSK are two modulation schemes that lend themselves well to noncoherent detection. Receivers for noncoherent detection of ASK and FSK are shown in Figure 1-3.

For noncoherent reception of binary ASK, the error probability for large signal-to-noise ratios is well approximated by

$$P_E \cong \tfrac{1}{2}e^{-z/2} \quad z \gg 1 \qquad \text{(noncoherent ASK)} \qquad (1\text{-}17)$$

For noncoherent detection of binary FSK, the probability of error is given exactly by

$$P_E = \tfrac{1}{2}e^{-z/2} \qquad \text{(noncoherent FSK)} \qquad (1\text{-}18)$$

Thus both perform the same for large signal-to-noise ratios. To compare this with coherent detection of FSK, the asympotic approximation for the Q-function given by

$$Q(x) = \frac{e^{-x^2/2}}{\sqrt{2\pi x}} \quad x \gg 1 \qquad (1\text{-}19)$$

is employed. Application of this to (1-1) with $R_{12} = 0$ gives

$$P_E = \frac{e^{-z/2}}{\sqrt{2\pi z}} \quad z \gg 1 \qquad \text{(coherent FSK)} \qquad (1\text{-}20)$$

Since the dominant behavior comes through the exponent in (1-20), it follows that coherent and noncoherent FSK have very nearly the same error probability perfor-

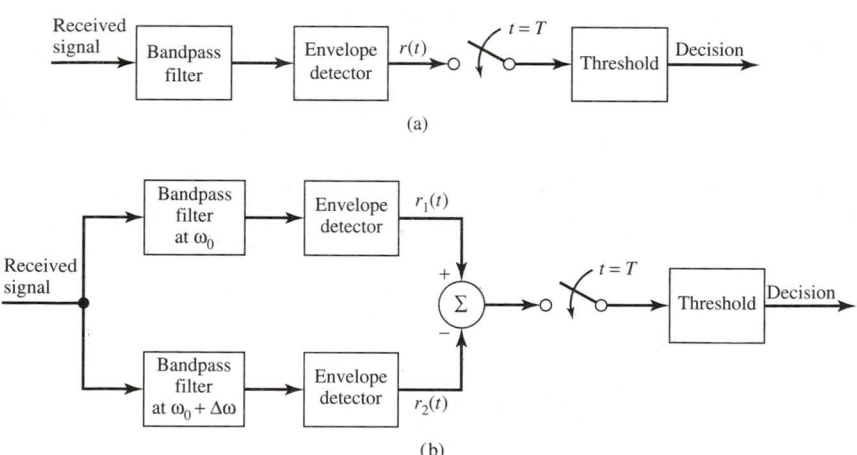

(a)

(b)

FIGURE 1-3. Receivers for the noncoherent detection of binary (a) ASK and (b) FSK.

TABLE 1-4. Example Illustrating the Differential Encoding Process

Message Sequence		1	0	0	1	1	1	0
Encoded Sequence	1	1	0	1	1	1	1	0
Transmitted Phase (rad)	0	0	π	0	0	0	0	π

mance at large signal-to-noise ratios, with coherent FSK slightly better due to the $z^{-1/2}$ in the denominator of (1-20).

There is one other binary modulation scheme which is, in a sense, noncoherent. It is differentially coherent PSK (DPSK), in which the phase of the preceding bit interval is used as a reference for the current bit interval. This technique depends on the channel being stable enough so that phase changes due to channel perturbations from a given bit interval to the succeeding one are inconsequential. It also depends on there being a known phase relationship from one bit interval to the next. This is ensured by differentially encoding the bits before phase modulation at the transmitter. Differential encoding is illustrated in Table 1-4. An arbitrary reference bit is chosen to start off the process. In Table 1-4, a 1 has been chosen. For each bit of the encoded sequence, the present bit is used as a reference for the following bit in the sequence. A 0 in the message sequence is encoded as a transition from the state of the reference bit to the opposite state in the encoded message sequence; a 1 is encoded as no change of state. Using these rules it is seen that the encoded sequence shown in Table 1-4 results.

Block diagrams of two possible receiver structures for DPSK are shown in Figure 1-4. The first is suboptimum but relatively simple to implement. The second is the

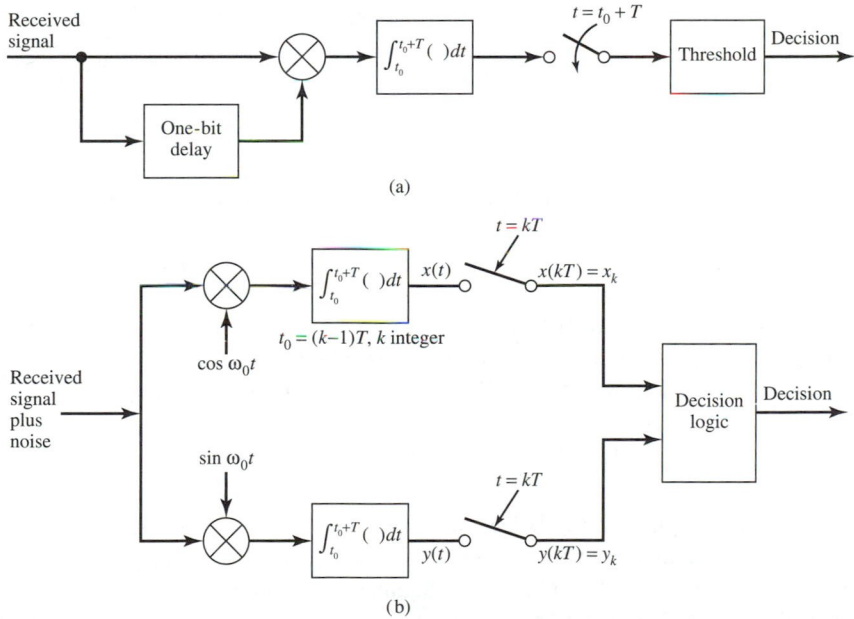

FIGURE 1-4. Receiver block diagrams for detection of DPSK: (a) suboptimum; (b) optimum.

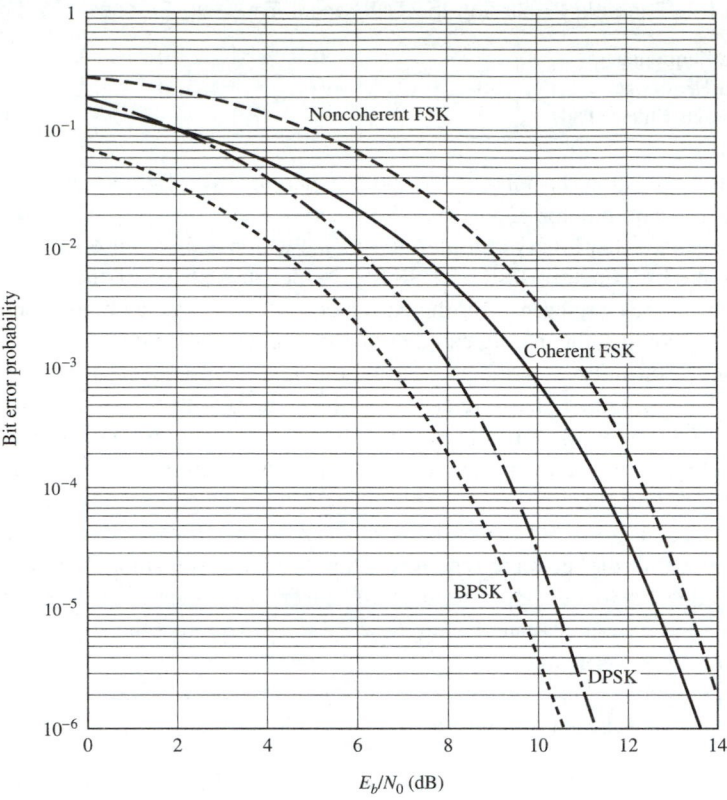

FIGURE 1-5. Comparison of error probabilities for BPSK, DPSK, coherent FSK, and noncoherent FSK.

optimum receiver for DPSK in AWGN. Its probability of error performance can be shown to be

$$P_E = \tfrac{1}{2}e^{-z} \qquad \text{(DPSK)} \qquad (1\text{-}21)$$

This can be compared with BPSK by again making use of the asymptotic approximation for the Q-function given by (1-19) in (1-1) with $R_{12} = -1$ to give the following approximate result for BPSK for large signal-to-noise ratios:

$$P_E = \frac{e^{-z}}{2\sqrt{\pi z}} \qquad z \gg 1 \qquad \text{(BPSK)} \qquad (1\text{-}22)$$

The exponential behavior for DPSK and BPSK is the same for large z; BPSK is slightly better due to the factor $z^{-1/2}$ in (1-22). Error probabilities for BPSK, DPSK, coherent FSK, and noncoherent FSK are compared in Figure 1-5. It is seen that less than 1 dB of degradation results in going from coherent to noncoherent detection at $P_E = 10^{-6}$.

1-3 Signal Detection in Geometric Terms

It is useful to view digital data transmission in geometric terms for a number of reasons. First, it provides a general framework that makes the analysis of several types of digital data transmission methods easier. Second, it provides an insight into the digital data transmission problem that allows one to see intuitively the power–bandwidth trade-offs possible. Third, it suggests ways to improve on standard modulation schemes. The mathematical basis for the geometric approach is known as signal space (Hilbert space in mathematical literature) theory. The first book to use this approach in the United States was that of Wozencraft and Jacobs [2], which was based, in part, on earlier work by the Russian Kotelnikov [5]. An early paper in the literature approaching signal detection from a geometric standpoint is Arthurs and Dym [6]. An overview of signal space concepts is given below.

1-3.1 Gram–Schmidt Procedure

Given a finite set of signals, denoted $s_i(t)$, $i = 1, 2, \ldots, M$ for $0 \leq t \leq T_s$, it is possible to find an orthonormal *basis set* for the space *spanned* by the signal set. Any signal in the space can then be represented as a linear combination of these basis functions.† The procedure, known as the Gram–Schmidt procedure, is easy to describe once some notation is defined. The *scalar product* of two signals, u and v, defined over the interval $[0, T_s]$ is defined as‡

$$(u,v) = \int_0^{T_s} u(t)v^*(t)\, dt \qquad (1\text{-}23)$$

and the *norm* of a signal is defined as

$$\|u\| = \sqrt{(u,u)} \qquad (1\text{-}24)$$

In terms of this notation, the Gram–Schmidt procedure is as follows:

1. Set $v_1(t) = s_1(t)$ and define the first orthonormal basis function as

$$\phi_1(t) = \frac{v_1(t)}{\|v_1\|} \qquad (1\text{-}25a)$$

2. Set $v_2(t) = s_2(t) - (s_2, \phi_1)\phi_1(t)$ and let the second orthonormal basis function be

$$\phi_2(t) = \frac{v_2(t)}{\|v_2\|} \qquad (1\text{-}25b)$$

---

† Because more than two possible signals can be sent during each signaling interval, the parameter T_s will now be used to denote the *signaling,* or *symbol, interval.*

‡ For the most part, signals will be real. However, it is sometimes convenient to represent signals as phasors or complex exponentials. This is discussed at the end of Chapter 2.

3. Set $v_3(t) = s_3(t) - (s_3, \phi_2)\phi_2(t) - (s_3, \phi_1)\phi_1(t)$ and let the next orthonormal basis function be

$$\phi_3(t) = \frac{v_3(t)}{\|v_3\|} \tag{1-25c}$$

4. Continue until all signals have been used. If one or more of the steps above yield $v_j(t)$'s for which $\|v_j(t)\| = 0$, omit these from consideration so that a set of $K \le M$ orthonormal functions is obtained. This is called a *basis set*.

Using the orthonormal basis set thus obtained, an arbitrary signal in the original set of signals† can be represented as

$$s_j(t) = \sum_{i=1}^{K} S_{ij}\phi_i(t) \qquad j = 1, 2, \ldots, M \tag{1-26}$$

where

$$S_{ij} = (s_j, \phi_i) = \int_0^{T_s} s_j(t)\phi_i^*(t) \, dt \tag{1-27}$$

With this procedure, any signal of the set can be represented as a point in a signal space [the coordinates of $s_j(t)$ are $S_{1j}, S_{2j}, \ldots, S_{Kj}$]. The representation of the signal in this space will be referred to as the signal vector. Thus the signal detection problem can be viewed geometrically. This is discussed in the following section.

1-3.2 Geometric View of Signal Detection

Given a set of M signals as discussed above, an M-ary digital communication system selects one of them with equal likelihood in each contiguous T_s-second interval and sends it through a channel in which white Gaussian noise of two-sided power spectral density $N_0/2$ is added. Letting the noise be represented by $n(t)$ and supposing that the jth signal is transmitted, the coordinates of the noisy received signal, here called the components of the data vector, are

$$Z_i = S_{ij} + N_j \quad i = 1, 2, \ldots, K \qquad \text{(signal } j \text{ transmitted)} \tag{1-28}$$

where

$$N_i = (n, \phi_i) = \int_0^{T_s} n(t)\phi_i^*(t) \, dt \tag{1-29}$$

Schematic diagrams of two receiver front ends that can be used to compute the coordinates of the data vector are shown in Figure 1-6. The first is called the

---

† In fact, it should be clear that any linear combination of the original set of signals can be thus represented.

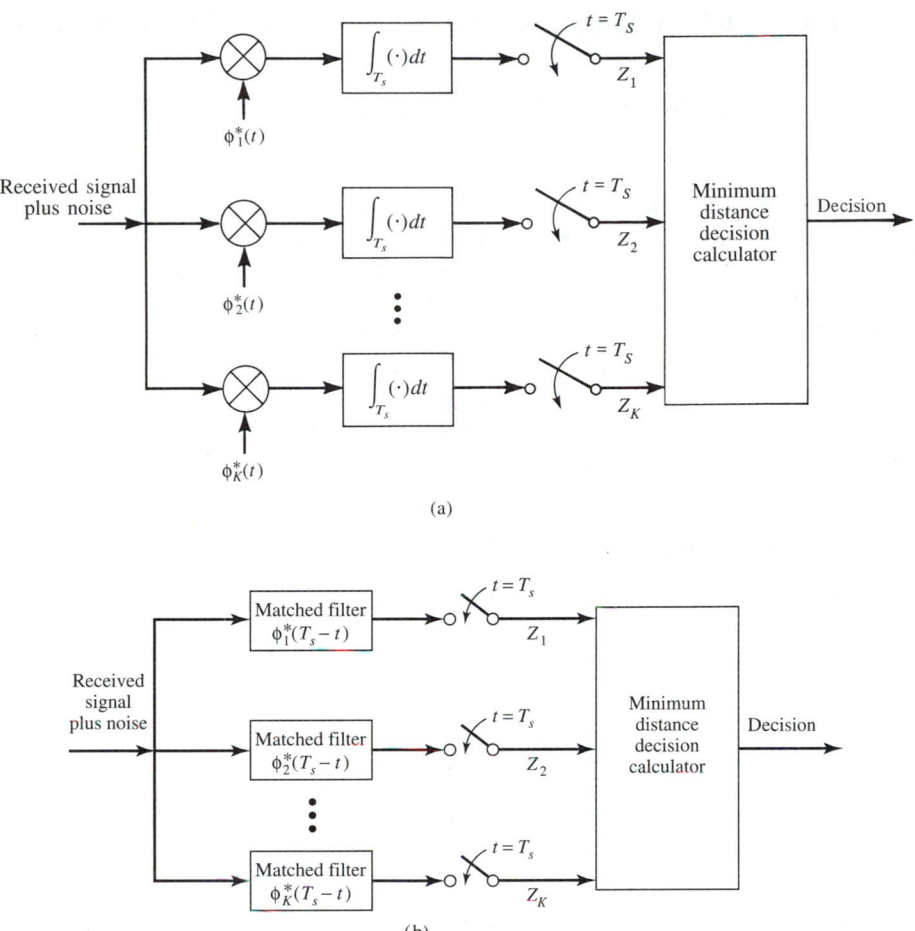

FIGURE 1-6. Receiver configurations for computing data vector components: (a) correlator realization; (b) matched-filter realization.

correlator realization, and the second is called the matched filter realization. Because the noise components are linear transformations of a Gaussian random process, they are also Gaussian and can be shown to have zero means and covariances:

$$\text{cov}(N_i, N_j) = E\{N_i N_j\} = \frac{N_0}{2} \delta_{ij} \tag{1-30}$$

where δ_{ij} is the Kronecker delta that is zero for the indices equal and zero otherwise. Consequently, the signal coordinates (1-28) are Gaussian with means S_{ij}, zero covariances, and variances $N_0/2$. Thus, given that signal $s_j(t)$ was sent, the joint conditional probability density function of the received data vector components is

$$p[z_1, z_2, \ldots, z_K|s_j(t)] = (\pi N_0)^{-K/2} \exp\left[-\frac{1}{N_0} \sum_{i=1}^{K} (z_i - S_{ij})^2\right]$$

$$j = 1, 2, \ldots, M \quad (1\text{-}31)$$

A reasonable strategy for deciding on the signal that was sent is to choose the most likely; that is, if each signal is transmitted with equal probability, maximize the conditional probability density function (1-31) by choosing the appropriate signal vector, which can also be shown equivalent to minimizing the average probability of error [2]. Given the form of (1-31), this is accomplished by minimizing its exponent. Minimizing the exponent is equivalent to minimizing the sum of the squares of the differences between the components of the received data vector and those of the signal vector (i.e., choosing the signal point that is closest in Euclidian distance to the received data point). This is illustrated in the following specific examples.

1-3.3 *M*-ary Phase-Shift Keying

Consider *M*-ary phase-shift keying (MPSK) for which the signal set is†

$$s_i(t) = \sqrt{\frac{2E_s}{T_s}} \cos\left[\omega_0 t + \frac{2\pi(i-1)}{M}\right] \qquad 0 \le t \le T_s$$

$$i = 1, 2, \ldots, M \quad (1\text{-}32)$$

where E_s is the signal energy, T_s the signal duration, and ω_0 the radian carrier frequency. The Gram–Schmidt procedure could be used to find the orthonormal basis set for expanding this signal set, but it is easier to expand (1-31) using trigonometric identities as

$$s_i(t) = \sqrt{E_s}\left[\cos\frac{2\pi(i-1)}{M}\sqrt{\frac{2}{T_s}}\cos\omega_0 t - \sin\frac{2\pi(i-1)}{M}\sqrt{\frac{2}{T_s}}\sin\omega_0 t\right]$$

$$= \sqrt{E_s}\left[\cos\frac{2\pi(i-1)}{M}\phi_1(t) - \sin\frac{2\pi(i-1)}{M}\phi_2(t)\right]$$

$$0 \le t \le T_s \qquad i = 1, 2, \ldots, M \quad (1\text{-}33)$$

where it follows that

$$\phi_1(t) = \sqrt{\frac{2}{T_s}}\cos\omega_0 t \qquad 0 \le t \le T_s$$

$$\phi_2(t) = \sqrt{\frac{2}{T_s}}\sin\omega_0 t \qquad 0 \le t \le T_s$$

$$(1\text{-}34)$$

Note that in this case $K = 2 \le M$. A signal space diagram is shown in Figure 1-7a

---

†The unknown phase that results from the signal propagating through the channel is usually estimated by a phase-locked-loop (PLL) circuit (see Appendix A). This operation is assumed perfect for purposes of analysis in this discussion. Hence this unknown phase at the receiver, estimated by the PLL, is taken as zero for convenience.

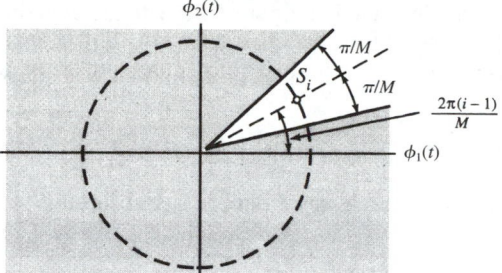

(a) Decision region (unshaded) for i^{th} transmitted phase

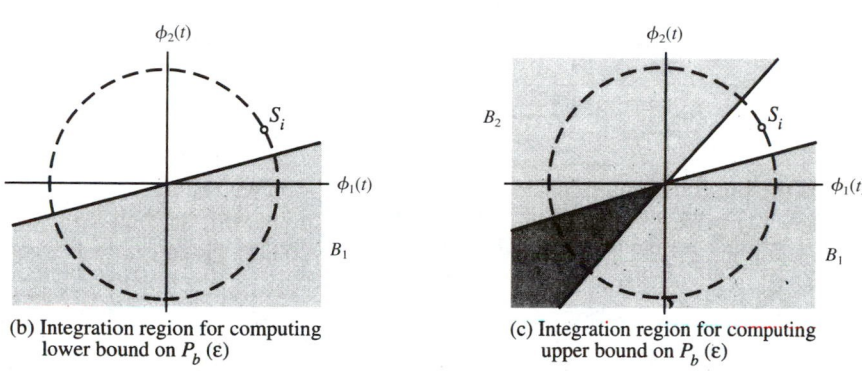

(b) Integration region for computing lower bound on $P_b\,(\varepsilon)$

(c) Integration region for computing upper bound on $P_b\,(\varepsilon)$

FIGURE 1-7. Signal space diagrams for MPSK: (a) diagram showing typical transmitted signal point with decision region; (b) and (c) half-planes for bounding the error probability. The probability of error is greater than the probability of the received data vector falling into one half-plane but less than the probability of it falling into either one of both half-planes.

(only the ith signal point is shown). The best decision strategy, as discussed above, chooses the signal point in signal space closest in Euclidian distance to the received data point. This is accomplished in Figure 1-7a by dividing the signal space up into decision regions associated with each signal point. If the received data point lands in a given region, the decision is made that the corresponding signal was transmitted. The probability of error is the probability that, given a certain signal was transmitted, the noise causes the data vector to land outside the corresponding decision region. The probability of *symbol* error, P_s, can be upper and lower bounded by [7]

$$Q\left(\sqrt{\frac{2E_s}{N_0}}\,\sin\frac{\pi}{M}\right) \leq P_s \leq 2Q\left(\sqrt{\frac{2E_s}{N_0}}\,\sin\frac{\pi}{M}\right) \qquad (1\text{-}35)$$

which is obtained by considering two half-planes above and below the wedge corresponding to the ith signal, as shown in Figure 1-7b and c. The upper bound is very tight† for M moderately large, a fact that follows by comparing the areas of the

---

† See P. J. Lee, "Computation of the Bit Error Rate of Coherent M-ary PSK with Gray Code Bit Mapping," *IEEE Trans. Commun.*, Vol. COM-34, pp. 488–491, May 1986, for an exact expression in terms of a single integral for the bit error probability.

plane with one wedge excluded in Figure 1-7a with the area of the two half-planes in Figure 1-7b and c. The probability of error curves will be given after conversion of symbol error probabilities to bit error probabilities is discussed.

1-3.4 Coherent *M*-ary Frequency-Shift Keying

The signal set for coherent *M*-ary frequency-shift keying (CMFSK) is given by

$$s_i(t) = \sqrt{\frac{2E_s}{T_s}} \cos[(\omega_0 + (i - 1)\Delta\omega)t] \qquad 0 \leq t \leq T_s$$

$$i = 1, 2, \ldots, M \quad (1\text{-}36)$$

The orthonormal basis function set for this modulation scheme is

$$\phi_i(t) = \sqrt{\frac{2}{T_s}} \cos[(\omega_0 + (i - 1)\Delta\omega)t] \qquad 0 \leq t \leq T_s$$

$$i = 1, 2, \ldots, M \quad (1\text{-}37)$$

Note that the signal space is *M*-dimensional as opposed to two-dimensional for the case of MPSK. A signal space diagram for $M = 3$ is shown in Figure 1-8. For moderately large *M*, the symbol error probability can be tightly upper bounded by [7]

$$P_s \leq (M - 1)Q\left(\sqrt{\frac{E_s}{N_0}}\right) \qquad (1\text{-}38)$$

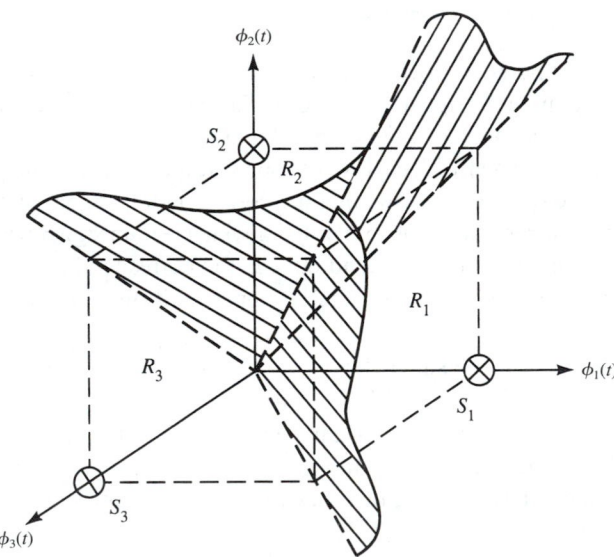

FIGURE 1-8. Signal space showing the possible transmitted signal points and decision boundaries for coherent 3-ary FSK.

Curves showing bit error probability as a function of signal-to-noise ratio will be presented later.

1-3.5 *M*-ary Quadrature-Amplitude-Shift Keying

M-ary quadrature-amplitude-shift keying (MQASK) uses the two-dimensional space of MPSK, but with multiple amplitudes. Many such two-dimensional configurations have been considered, but only a simple rectangular grid of signal points is considered here. The signal set can be expressed as

$$s_i(t) = \sqrt{\frac{2}{T_s}}(A_i \cos \omega_0 t + B_i \sin \omega_0 t) \qquad 0 \le t \le T_s \qquad (1\text{-}39)$$

where A_i and B_i are amplitudes taking on the values

$$A_i, B_i = \pm a, \pm 3a, \dots, \pm(\log_2 M - 1)a \qquad (1\text{-}40)$$

where M is assumed to be a power of 4. The parameter a can be related to the average signal energy by [7]

$$a = \sqrt{\frac{3E_s}{2(M-1)}} \qquad (1\text{-}41)$$

A signal space diagram is shown in Figure 1-9 for $M = 16$ with optimum partitioning of the decision regions. Each signal point is labeled with Roman numeral I, II,

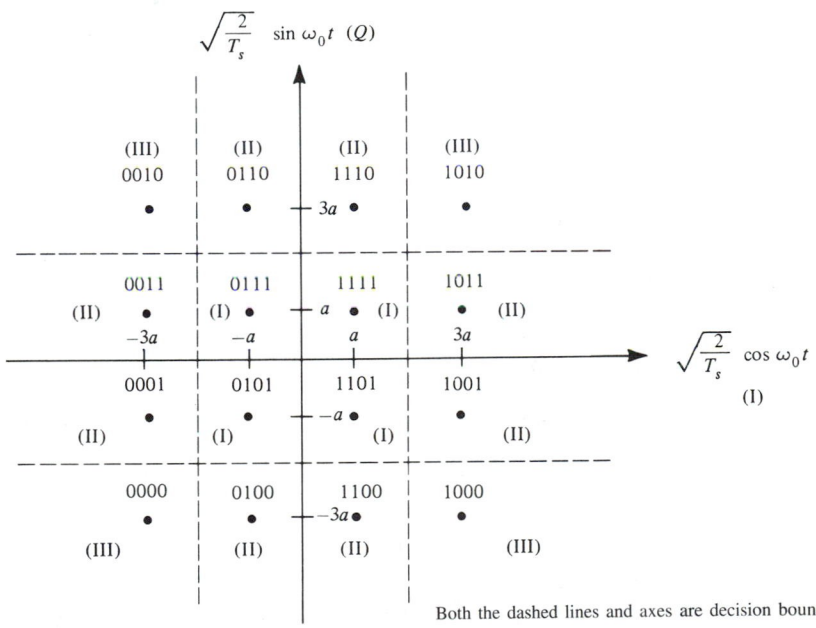

Both the dashed lines and axes are decision boundaries.

Roman numerals show decision region type.

FIGURE 1-9. Signal space diagram for 16-ary QASK.

or III. For the type I regions, the probability of correct reception is

$$P(C|\text{I}) = \left[1 - 2Q\left(\sqrt{\frac{2a^2}{N_0}} \right) \right]^2 \tag{1-42a}$$

For the type II regions, the probability of correct reception is

$$P(C|\text{II}) = \left[1 - 2Q\left(\sqrt{\frac{2a^2}{N_0}} \right) \right]\left[1 - Q\left(\sqrt{\frac{2a^2}{N_0}} \right) \right] \tag{1-42b}$$

and for the type III regions, it is

$$P(C|\text{III}) = \left[1 - Q\left(\sqrt{\frac{2a^2}{N_0}} \right) \right]^2 \tag{1-42c}$$

In terms of these probabilities, the probability of symbol error is [7]

$$P_s = 1 - \frac{1}{M}[(\sqrt{M} - 2)^2 P(C|\text{I}) + 4(\sqrt{M} - 2)P(C|\text{II}) + 4P(C|\text{III})] \tag{1-43}$$

Bit error probability plots will be provided later.

1-3.6 Differentially Coherent Phase-Shift Keying

Differentially coherent phase-shift keying (DPSK), first discussed in Section 1-2, can be generalized to more than two phases. Suppose that the transmitted carrier phase for symbol time $n - 1$ is α_{n-1} and the desired symbol phase for time n is β_n, which is assumed to take on a multiple of $2\pi/M$ radians. If it is desired to send the particular symbol (phase) $\beta_n = \Phi$, the *transmitted phase* at time n, α_n, is

$$\alpha_n = \alpha_{n-1} + \Phi \tag{1-44}$$

where α_{n-1} is the transmitted phase at time $n - 1$. Suppose that the phases detected corresponding to α_{n-1} and α_n are $\theta_{n-1} = \alpha_{n-1} + \gamma$ and $\theta_n = \alpha_n + \gamma$, respectively, where γ is the unknown phase shift introduced by the channel. The first stage of the receiver is a phase discriminator, which detects θ_n. From the previous decision interval, it is assumed that θ_{n-1} is available, so the receiver forms the difference

$$\begin{aligned} \theta_n - \theta_{n-1} &= (\alpha_n + \gamma) - (\alpha_{n-1} + \gamma) = \alpha_n - \alpha_{n-1} \\ &= \alpha_{n-1} + \Phi - \alpha_{n-1} = \Phi \end{aligned} \tag{1-45}$$

where no noise is assumed. In the presence of noise, the receiver must decide which $2\pi/M$ region $\theta_n - \theta_{n-1}$ falls into; hopefully, in this example, this is the region centered on Φ. Thus a correct decision will be made at the receiver when

$$\Phi - \frac{\pi}{M} < \theta_n - \theta_{n-1} \leq \Phi + \frac{\pi}{M} \tag{1-46}$$

A receiver block diagram implementing this decision strategy is shown in Figure 1-10. The error probability can be bounded by [8]

$$P_1 \leq P_s \leq 2P_1 \tag{1-47}$$

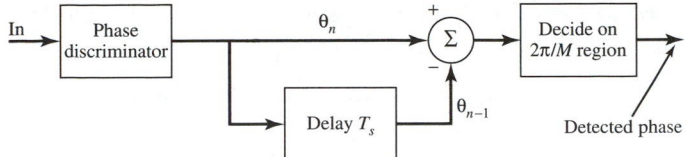

FIGURE 1-10. Block diagram for an *M*-ary differential PSK receiver.

where P_1 is upper and lower bounded by

$$P_1 \leq \frac{\pi}{2} \frac{\cos(\pi/2M)}{\sqrt{\cos(\pi/M)}} Q\left(2 \sqrt{\frac{E_s}{N_0}} \sin \frac{\pi}{2M}\right) \qquad (1\text{-}48a)$$

and

$$P_1 \geq \frac{1}{2} \frac{\cos(\pi/2M)}{\sqrt{\cos(\pi/M)}} \left[1 - 2Q\left(\pi \sqrt{\frac{E_s}{N_0}} \sin \frac{\pi}{M}\right)\right] Q\left(2 \sqrt{\frac{E_s}{N_0}} \sin \frac{\pi}{2M}\right) \qquad (1\text{-}48b)$$

1-3.7 Noncoherent *M*-ary FSK

The signal set for noncoherent MFSK can be expressed as

$$s_i(t) = \sqrt{\frac{2E_s}{T_s}} \cos(\omega_i t + \alpha) \qquad 0 \leq t \leq T_s \qquad i = 1, 2, \ldots, M \quad (1\text{-}49)$$

where E_s is the symbol energy, T_s the symbol duration, and α the unknown phase, which is modeled as a uniformly distributed random variable in $[0, 2\pi)$. The signal space is $2M$-dimensional and can be defined by the basis functions

$$\left.\begin{array}{l} \phi_{xi}(t) = \sqrt{\dfrac{2}{T_s}} \cos \omega_i t \\[2mm] \phi_{yi}(t) = \sqrt{\dfrac{2}{T_s}} \sin \omega_i t \end{array}\right\} \qquad 0 \leq t \leq T_s \qquad i = 1, 2, \ldots, M \qquad (1\text{-}50)$$

A fairly lengthy derivation [7] results in the symbol error probability expression

$$P_s = \sum_{k=1}^{M-1} \binom{M-1}{k} \frac{(-1)^k}{k+1} \exp\left(-\frac{k}{k+1} \frac{E_s}{N_0}\right) \qquad (1\text{-}51)$$

which reduces to the result for binary NFSK for $M = 1$. The optimum receiver is an extension of Figure 1-3 with a parallel set of inphase and quadrature filters and envelope detectors for each possible transmitted signal.

1-3.8 Hybrid Modulation Schemes

Quadrature-Multiplexed Modulation Schemes. The case of 4-ary, or quadriphase, PSK can be viewed as quadrature-multiplexed BPSK. Assume that

$m_1(t)$ and $m_2(t)$ are ± 1-valued binary signals, and consider the quadrature-multiplexed signal

$$
\begin{aligned}
s(t) &= A_1 m_1(t) \cos \omega_0 t - A_2 m_2(t) \sin \omega_0 t \\
&= A \cos[\omega_0 t + \theta(t)]
\end{aligned}
\tag{1-52}
$$

where, from trigonometric identities, it follows that

$$
\theta(t) = \tan^{-1} \frac{A_2 m_2(t)}{A_1 m_1(t)}
\tag{1-53a}
$$

and

$$
A = \sqrt{A_1^2 + A_2^2}
\tag{1-53b}
$$

Equation (1-53b) follows because of the binary-valued nature of $m_1(t)$ and $m_2(t)$. The quadrature components of (1-52) have powers

$$
P_1 = \frac{A_1^2}{2} \quad \text{and} \quad P_2 = \frac{A_2^2}{2}
\tag{1-54a}
$$

and the total power of the quadrature-multiplexed signal is

$$
P = \frac{A_1^2 + A_2^2}{2}
\tag{1-54b}
$$

In addition to representing a quadrature-multiplexed signal, it is seen that (1-52) can be viewed as a phase-modulated signal where the phase takes on the discrete steps

$$
\begin{array}{cc}
\tan^{-1} \dfrac{A_2}{A_1} & \tan^{-1} \dfrac{A_2}{A_1} + \pi \\[2ex]
-\tan^{-1} \dfrac{A_2}{A_1} & -\tan^{-1} \dfrac{A_2}{A_1} + \pi
\end{array}
\tag{1-55}
$$

A number of special cases can be considered for (1-52), depending on the nature of $m_1(t)$ and $m_2(t)$.

Unbalanced QPSK. If $m_1(t)$ and $m_2(t)$ are independent binary message sequences with differing symbol (bit) times, say T_1 and T_2, it may be desirable to maintain the same symbol energies in each quadrature component of (1-52). It follows from (1-54a) that this will be the case if

$$
P_1 T_1 = P_2 T_2 \quad \text{or} \quad \frac{A_1}{A_2} = \sqrt{\frac{T_2}{T_1}}
\tag{1-56}
$$

If this is the case, the bit error probabilities for correlator, or matched-filter detection, of the two quadrature symbol streams in AWGN will be the same. It follows from (1-55), however, that the phase steps of the modulated carrier will not be in 90° increments. This makes the implementation of the carrier acquisition process at the receiver more challenging [9].

Balanced QPSK. If $m_1(t)$ and $m_2(t)$ are derived from a binary bit stream, $d(t)$, with bit period T_b by associating the odd-indexed bits with $m_1(t)$ and the even-indexed bits with $m_2(t)$ with the switching times of $m_1(t)$ and $m_2(t)$ aligned, the result is quadrature phase-shift keying (QPSK). The symbol error probability is bounded by (1-35) but can be shown to be given exactly by [7]

$$P_s = 1 - \left[1 - Q\left(\sqrt{\frac{2E_s}{N_0}} \right) \right]^2 \cong 2Q\left(\sqrt{\frac{2E_s}{N_0}} \right) \qquad \frac{E_s}{N_0} \gg 1 \qquad (1\text{-}57)$$

Note that the approximate result shown in the rightmost expression of (1-57) is equal to the upper bound computed from (1-35) and results by neglecting the square of the Q-function in expanding the exact result. Also note that $m_1(t)$ and $m_2(t)$ need not be derived from a serial bit stream $d(t)$ but can, in fact, be separate data streams.

Offset Quadrature Phase-Shift Keying (OQPSK). To avoid the possibility of 180° phase shifts that may happen in QPSK when both $m_1(t)$ and $m_2(t)$ switch sign simultaneously, these quadrature data streams can be offset by one-half of a symbol period to produce OQPSK. In this case, only 90° phase shifts of the modulated carrier are possible. This gives less spectral regrowth of the sidelobes in situations where the modulated carrier is filtered and then passed through a nonlinear device such as a limiting amplifier [10]. The symbol error probability in AWGN is the same as for QPSK.

Minimum-Shift Keying (MSK). MSK results if the signaling pulses in the quadrature symbol streams of OQPSK are shaped by half-sinusoids [1,7,11,12].† Modulator and demodulator block diagrams are shown in Figure 1-11. It has the same bit error probability as BPSK as a function of E_b/N_0.

One form of MSK, referred to as *type-I MSK,* can be expressed as

$$\begin{aligned} s(t) &= A\left[d_1(t) \cos \frac{\pi t}{2T} \cos \omega_0 t - d_2(t) \sin \frac{\pi t}{2T} \sin \omega_0 t \right] \\ &= A \cos[\omega_0 t + \theta(t)] \end{aligned} \qquad (1\text{-}58)$$

where $d_1(t)$ and $d_2(t)$ are viewed as being derived from a single bit stream, $d(t)$, just as $m_1(t)$ and $m_2(t)$ were for balanced QPSK, and

$$\theta(t) = \tan^{-1}\left[\frac{d_2(t)}{d_1(t)} \tan \frac{\pi t}{2T} \right] \qquad (1\text{-}59)$$

For type-I MSK, it is noted that the cosine and sine weighting function in the in-phase and quadrature channels alternate in sign each half-cycle of the weighting periods, which are $T = 2T_s$ seconds in duration [i.e., the same durations as $d_1(t)$ and $d_2(t)$, which are staggered in their transitions just as for OQPSK]. If $d_1(t) = d_2(t)$ [i.e., successive bits of $d(t)$ are the same], it follows from (1-59) that

$$\theta(t) = \frac{\pi t}{2T} \qquad (1\text{-}60a)$$

---

† A trigonometric identity can be used to reduce (1-52) to its envelope-phase form in this case.

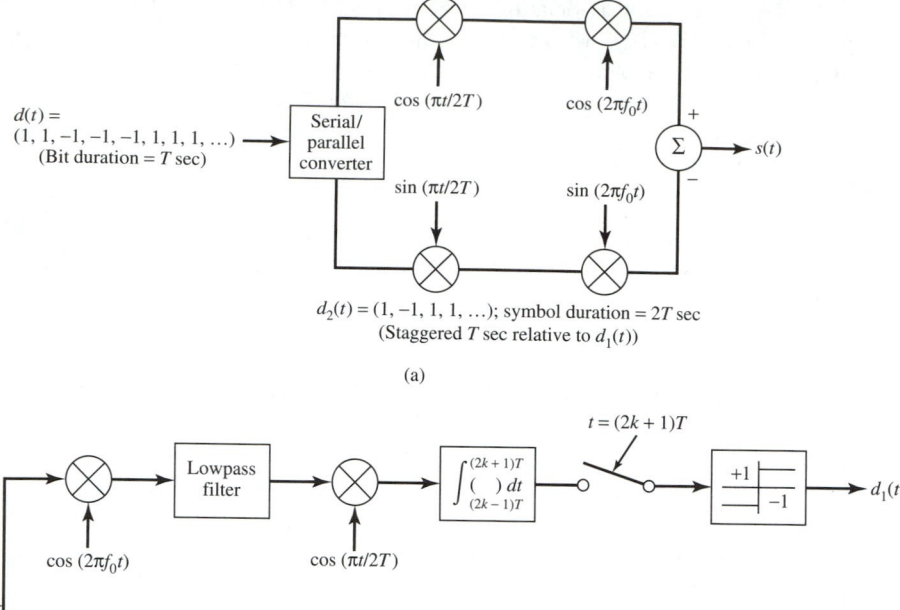

FIGURE 1-11. (a) MSK modulator and (b) demodulator.

and (1-58) becomes

$$s(t) = A \cos\left[2\pi\left(f_0 + \frac{1}{4T}\right)t + u_k\right] \qquad (1\text{-}60b)$$

where $u_k = 0$ or $k\pi$, k a nonzero integer, according to whether $d_1(t) = +1$ or -1, respectively. That is, the modulated carrier can be viewed as being increased in frequency from the carrier frequency f_0 by $1/4T$ hertz. [Note that the shift in frequency would have been a decrease had the sign of the second term in (1-58) been plus.] On the other hand, if $d_1(t) = -d_2(t)$ [i.e., successive bits of $d(t)$ are opposite], it follows from (1-59) that

$$\theta(t) = \frac{-\pi t}{2T} \qquad (1\text{-}61a)$$

and (1-59) becomes

$$s(t) = A \cos\left[2\pi\left(f_0 - \frac{1}{4T}\right)t + u_k\right] \qquad (1\text{-}61b)$$

where u_k is defined as previously. That is, the modulated carrier can be viewed as being decreased in frequency from the carrier frequency by $1/4T$ hertz. *This shows that type-I MSK can be viewed either as a phase modulation scheme or as a frequency modulation scheme.* Thus MSK can also be demodulated by means of a frequency discriminator and is sometimes referred to as continuous-phase frequency-shift keying with modulation index of $\frac{1}{2}$.

The unfolding of the phase of MSK with time is conveniently illustrated by means of a trellis diagram, which is shown for type-I modulation in Figure 1-12. All possible phase trajectories are shown in this diagram, assuming the modulation process started at time zero.

Another type of MSK, referred to as type-II, is obtained if the weightings on the quadrature symbol streams are always positive half-cosines or positive half-sines (depending on which quadrature symbol stream is being considered). The difference in the trellis diagram implied by this type of weighting is explored in the problems at the end of this chapter.

Another type of modulator for MSK, known as serial modulation, consists of shaping a BPSK signal with an appropriately chosen shaping filter. This will not be described here [1,11,13]. Since MSK can be viewed as an FSK signaling method, it can be also detected differentially and noncoherently [14,15]. Much insight into MSK modulation is provided in Ref. 16.

$\pi/4$**-Shifted-QPSK [17,18].** This scheme is a hybrid PSK modulation method. The signal constellation for $\pi/4$-shifted-QPSK, also known as symmetric differential phase exchange keying, is shown in Figure 1-13. It can be viewed as the superposition of two QPSK signal constellations offset by $45°$ relative to each other, resulting in eight possible phases. The solid lines in Figure 1-13 show the possible phase transitions. One advantage of $\pi/4$-shifted-QPSK is that it can be detected coherently, with a differential detector, or with a disciminator followed by an inte-

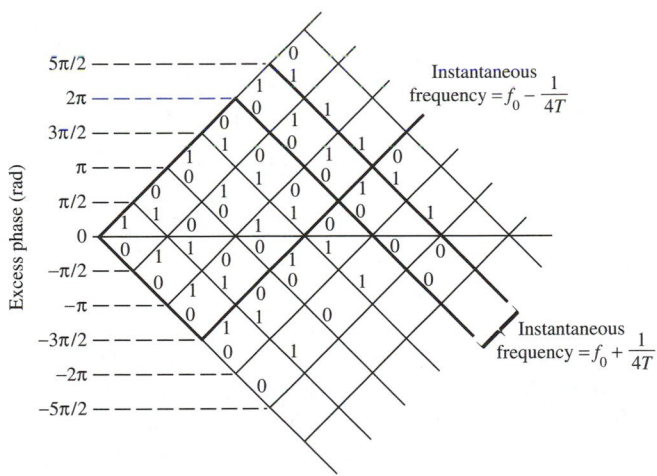

FIGURE 1-12. Trellis diagram for type-I MSK.

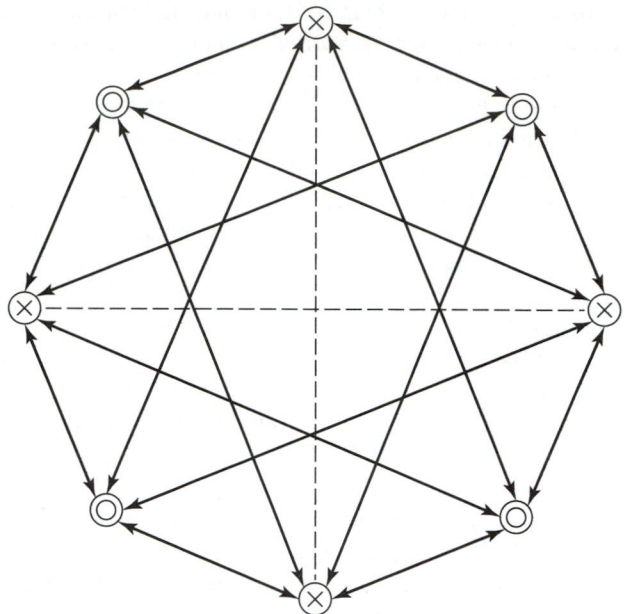

FIGURE 1-13. π/4-Shifted-QPSK signal constellation.

grate-and-dump filter. A second advantage of π/4-shifted-QPSK is that the modulated signal phasor does not pass through the origin, which gives it lower envelope deviation characteristics than BPSK or QPSK when filtered. When the phase transitions are differentially encoded, this modulation scheme is called π/4-shifted-differential-QPSK, or π/4-DQPSK. It simplifies the demodulation process in that a coherent carrier does not have to be acquired at the receiver and protects against loss of data due to phase slips. However, if a symbol error is made, a pair of symbol errors results. This modulation scheme is part of the North American digital cellular system standard [19]. It will be considered below.

Let the quadrature-channel symbols for the π/4-DQPSK signal in time slot k be represented by I_k and Q_k, respectively. They are given by

$$I_k = I_{k-1} \cos \Delta\phi_k - Q_{k-1} \sin \Delta\phi_k$$
$$Q_k = I_{k-1} \sin \Delta\phi_k + Q_{k-1} \cos \Delta\phi_k$$

$$(1\text{-}62)$$

where I_{k-1} and Q_{k-1} are the quadrature-channel symbols in time slot $k-1$. As for QPSK, quadrature-channel symbol streams, $d_1(t)$ and $d_2(t)$, are viewed as being derived from a single bit stream, $d(t)$, where odd-indexed bits are associated with $d_1(t)$ and even-indexed bits are associated with $d_2(t)$. To indicate a specific time slot, the subscript k is used. Bits d_{1k} and d_{2k} are paired (referred to as a di-bit) and mapped onto differentially encoded signal phases, $\Delta\phi_k$, using a Gray code as illustrated in Table 1-5. The π/4-DQPSK in-phase and quadrature signal components are then formed according to (1-62).

TABLE 1-5. Gray-Coded Phase Mapping for $\pi/4$-DQPSK

d_{1k} (LSB)	d_{2k} (MSB)	$\Delta\phi_k$
0	0	$\pi/4$
1	0	$3\pi/4$
1	1	$-3\pi/4$
0	1	$-\pi/4$

If ϕ_k represents the absolute phase angle for the kth symbol, trigonometric identities can be used to express (1-62) as

$$I_k = \cos\phi_k = \cos(\phi_{k-1} + \Delta\phi_k)$$
$$Q_k = \sin\phi_k = \sin(\phi_{k-1} + \Delta\phi_k) \tag{1-63}$$

The transmitted signal is then

$$s(t) = A[I_k \cos\omega_0 t - Q_k \sin\omega_0 t]$$
$$= A\cos(\omega_0 t + \phi_{k-1} + \Delta\phi_k) \tag{1-64}$$

In the North American cellular standard, raised cosine pulse shaping with a roll-off factor of 0.35 is specified for the transmitted signal [19].

One possible receiver structure for this type of modulation is shown in Figure 1-14. Its performance has been analyzed by Chennakeshu and Saulnier [18], with the result that the symbol error probability is given by

$$P_s = \int_{-\pi}^{\pi} \int_{\psi_2+\Delta\phi+\pi/4}^{\psi_2+\Delta\phi+7\pi/4} p(\psi_1 \mid \phi_1) p(\psi_2 \mid \phi_2)\, d\psi_1\, d\psi_2 \tag{1-65}$$

where

$$p(\psi_i \mid \phi_i) =$$
$$\frac{1}{2\pi} e^{-2z}[1 + \sqrt{8\pi z}\cos(\phi_i - \psi_i)e^{2z\cos^2(\phi_i - \psi_i)}\{1 - Q[2\sqrt{z}\cos(\phi_i - \psi_i)]\}] \tag{1-66}$$

for an additive white Gaussian noise channel. As previously, $z = E_b/N_0$. If, in addition, the channel introduces flat Rayleigh fading (described in detail in Section 1-5),

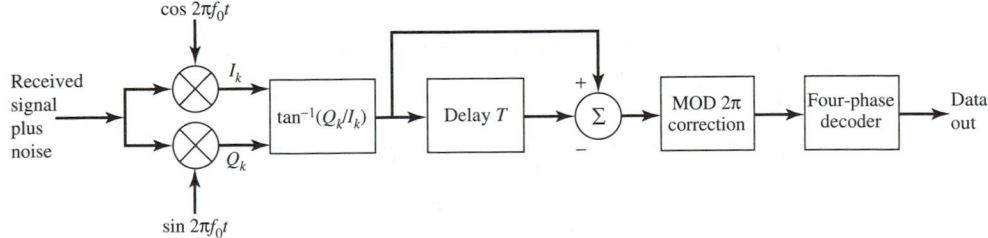

FIGURE 1-14. Receiver structure for $\pi/4$-DQPSK.

the conditional probability density function (1-66) is replaced by

$$p(\psi_i \mid \phi_i) =$$
$$\frac{1}{2\pi(1 + 2\bar{z})} + \frac{2}{\pi\bar{z}} \frac{\cos(\pi_i - \psi_i)}{a^3(\phi_i - \psi_i)} \left[\frac{\pi}{2} - \mu(\phi_i,\psi_i) - \frac{1}{2} \sin[2\mu(\phi_i,\psi_i)] \right] \quad (1\text{-}67a)$$

where

$$\mu(\phi_i,\psi_i) = \tan^{-1} \frac{-2\cos(\phi_i,\psi_i)}{a(\phi_i,\psi_i)} \quad (1\text{-}67b)$$

in which

$$a(\phi_i,\psi_i) = \sqrt{4\sin^2(\phi_i - \psi_i) + \frac{2}{z}} \quad (1\text{-}67c)$$

An expression for the bit error probability is

$$P_b = \int_{-\pi}^{+\pi} \int_{\psi_2 + \Delta\phi + \pi/4}^{\psi_2 + \Delta\phi + 5\pi/4} p(\psi_1 \mid \phi_1) p(\psi_2 \mid \phi_2)\, d\psi_1\, d\psi_2 \quad (1\text{-}68)$$

Performance results for both symbol and bit error probabilities in nonfading and Rayleigh-fading channels can be obtained numerically and are shown in Figure 1-15.

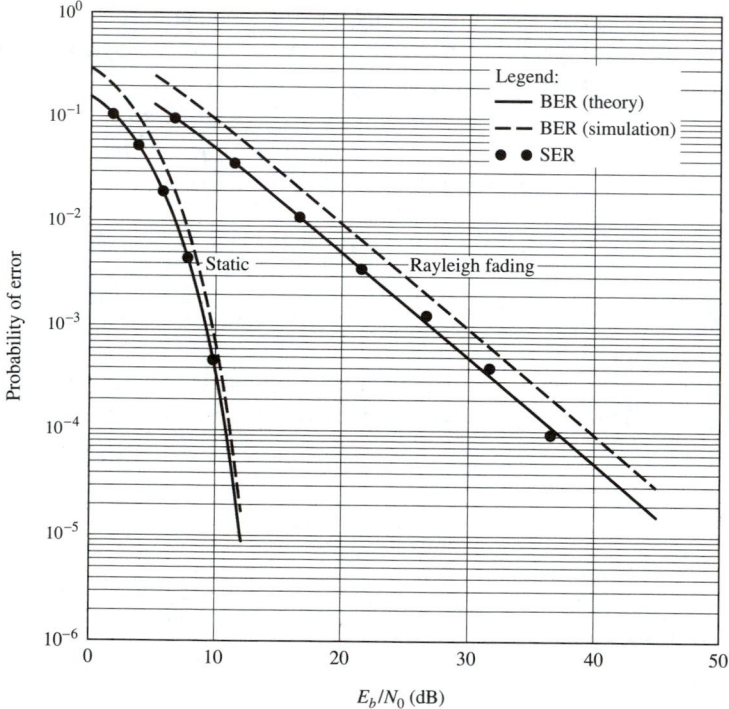

FIGURE 1-15. Symbol and bit error probabilities in nonfading and Rayleigh-fading channels for $\pi/4$-DQPSK. (From Ref. 18. Copyright © 1993 IEEE. Reprinted with permission.)

1-4 Comparison of Modulation Schemes

1-4.1 Bandwidth Efficiency

There are several ways to compare bandwidths of digital modulation schemes. One way is to compute out-of-band power as a function of bandwidth of an ideal brick wall filter. This requires integration of the power spectrum of the various modulation schemes being compared. The basis for bandwidth comparison used here will be the bandwidth required for the main lobe of the signal spectrum, which makes for somewhat simpler computation without undue loss of accuracy. For example, replacing T by the symbol duration, T_s, in (1-15), the radio-frequency bandwidth of the main lobe of a modulation scheme that uses a single modulated frequency to transmit the information, such as M-ary PSK, M-ary DPSK, or QASK, is

$$B_{\text{RF}} = \frac{2}{T_s} = 2R_s \qquad \text{Hz} \qquad (1\text{-}69)$$

where R_s is the symbol rate. However, for an M-ary modulation scheme, the symbol duration is related to the bit duration by

$$T_s = T_b \log_2 M = \frac{\log_2 M}{R_b} \qquad (1\text{-}70)$$

Thus the bandwidth in terms of bit rate for such modulation schemes is

$$B_{\text{RF}} = \frac{2R_b}{\log_2 M} \qquad \text{(MPSK, MDPSK, MQASK)} \qquad (1\text{-}71)$$

Now the ratio of bit rate to required bandwidth is called the bandwidth efficiency of a modulation scheme. In the case at hand, the bandwidth efficiency is

$$\text{bandwidth efficiency} = \frac{R_b}{B_{\text{RF}}} = 0.5 \log_2 M$$

$$\text{(MPSK, MDPSK, MQASK)} \quad (1\text{-}72)$$

For schemes using multiple frequencies to transmit information such as M-ary FSK, a more general approach is used. Each symbol is represented by a different frequency. For coherent M-ary FSK, the minimum separation per frequency required to maintain othogonality is $1/(2T_s)$ hertz. The two outside frequencies use $1/T_s$ hertz for the half of the main lobe on the left and right of the composite spectrum. The $M - 2$ interior frequencies require a minimum separation of $1/(2T_s)$ hertz (there are actually $M - 1$ interior slots), for a total RF bandwidth of

$$B_{\text{RF}} = \frac{1}{T_s} + \frac{M-1}{2T_s} + \frac{1}{T_s} = \frac{M+3}{2T_s} \qquad \text{(coherent MFSK)} \qquad (1\text{-}73)$$

Substitution of (1-70) gives a bandwidth efficiency of

$$\text{bandwidth efficiency} = \frac{R_b}{B_{\text{RF}}} = \frac{2 \log_2 M}{M+3} \qquad \text{(coherent MFSK)} \qquad (1\text{-}74)$$

For noncoherent MFSK, the minimum separation of the frequencies used to represent the symbols is taken as $2/T_s$ hertz for a total RF bandwidth of

$$B_{RF} = \frac{1}{T_s} + \frac{2(M-1)}{T_s} + \frac{1}{T_s} = \frac{2M}{T_s} \qquad \text{(noncoherent MFSK)} \qquad (1\text{-}75)$$

which has a bandwidth efficiency of

$$\text{bandwidth efficiency} = \frac{R_b}{B_{RF}} = \frac{\log_2 M}{2M} \qquad \text{(noncoherent MFSK)} \qquad (1\text{-}76)$$

A comparison of (1-72) with (1-74) and (1-76) shows that the bandwidth efficiency of MPSK and MQASK *increases* with M, whereas the bandwidth efficiency of MFSK *decreases* with M. This can be attributed to the dimensionality of the signal space staying constant with M for the former modulation schemes, whereas it increases with M for the latter. A comparison of bandwidth efficiency for the various modulation schemes considered in this chapter is given in Table 1-6.

1-4.2 Power Efficiency

The power efficiency of a modulation method is indicated by the value of E_b/N_0 required to yield a desired bit error probability, say 10^{-6}. This being the case, it is necessary to convert from symbol error probability to bit error probability, and from E_s/N_0 to E_b/N_0. The latter is straightforward since the difference between symbol energy and bit energy is symbol duration versus bit duration, which are related by

$$T_s = T_b \log_2 M \qquad (1\text{-}77)$$

giving

$$\frac{E_b}{N_0} = \frac{1}{\log_2 M} \frac{E_s}{N_0} \qquad (1\text{-}78)$$

TABLE 1-6. Comparison of Bandwidth Efficiencies in bps/Hz for Various Modulation Methods

M	MPSK, MDPSK, MQASK,[a] $\pi/4$-DQPSK[b]	Coherent MFSK	Noncoherent MFSK
2	0.5	0.400	0.250
4	1.0	0.571	0.250
8	1.5	0.545	0.188
16	2.0	0.421	0.125
32	2.5	0.286	0.078
64	3.0	0.179	0.047
128	3.5	0.107	0.027
256	4.0	0.062	0.066

[a]For MQASK, only values of M a power of 4 are applicable.

[b]$M = 4$ is applicable.

Conversion between symbol error probability and bit error probability is somewhat more complicated. First, considering two-dimensional modulation schemes such as MPSK and MQASK, it is assumed that the most probable errors are those in favor of adjacent signal points and that encoding is used which results in a single bit change in going from one signal point to an adjacent one (i.e., Gray encoding). Since there are $\log_2 M$ bits per symbol, the result is that bit error probability is related to symbol error probability approximately by

$$P_b \approx \frac{P_b}{\log_2 M} \qquad (1\text{-}79)$$

Finally, consider MFSK for which each symbol occupies a separate dimension in the signal space. Thus all symbol errors are equally probable, which means that each symbol error occurs with probability $P_s/(M-1)$. Suppose that for a given symbol error, k bits are in error. There are

$$\binom{\log_2 M}{k}$$

ways that this can happen, since each symbol represents $\log_2 M$ bits. This gives the average number of bit errors per symbol error as

$$\begin{array}{c}\text{average number of bit errors}\\ \text{per symbol error}\end{array} = \sum_{k=1}^{\log_2 M} k \binom{\log_2 M}{k} \frac{P_s}{M-1} = \frac{M \log_2 M}{2(M-1)} P_s \quad (1\text{-}80)$$

Since there are a total of $\log_2 M$ bits per symbol, the average bit error probability in terms of symbol error probability is

$$P_b = \frac{M}{2(M-1)} P_s \qquad (1\text{-}81)$$

The various modulation schemes considered in this chapter are compared on the basis of bit error probability versus E_b/N_0 in Figures 1-16 through 1-20.

1-5 Signaling Through Fading Channels

An important channel model is the flat-fading channel. Because the effects of flat fading on binary modulation performance are relatively easy to analyze, it will be considered here. Due to perturbations in the channel, the received signal is of the form

$$s_r(t) = R \cos[\omega_0 t + \theta_m(t) + \phi] \qquad (1\text{-}82)$$

where R and ϕ are random variables representing the effects of the fading. The modulation is represented by $\theta_m(t)$. *Flat fading* refers to the assumption that the fading is independent of frequency (i.e., the modulated signal is narrowband relative to channel filtering effects). Two important models for flat fading are: (1) R is Rayleigh distributed, and (2) R is Rician distributed. The former is a suitable model

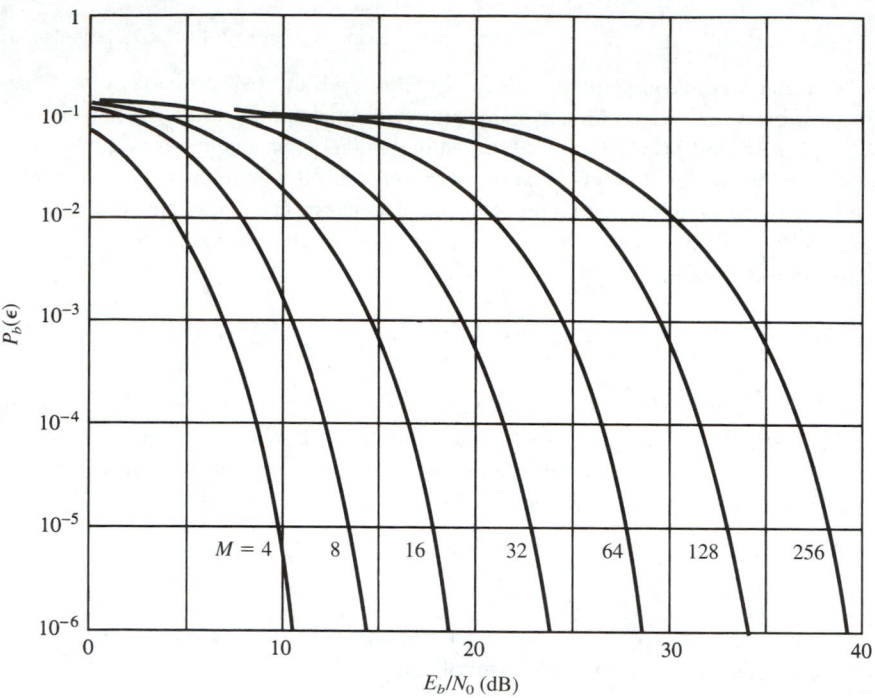

FIGURE 1-16. Bit error probability versus E_b/N_0 for coherent M-ary PSK.

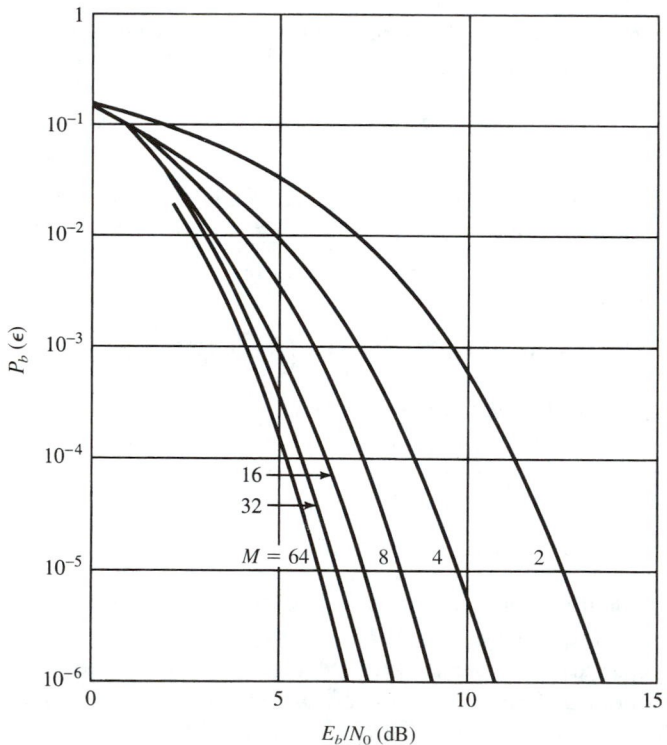

FIGURE 1-17. Bit error probability versus E_b/N_0 for coherent M-ary FSK.

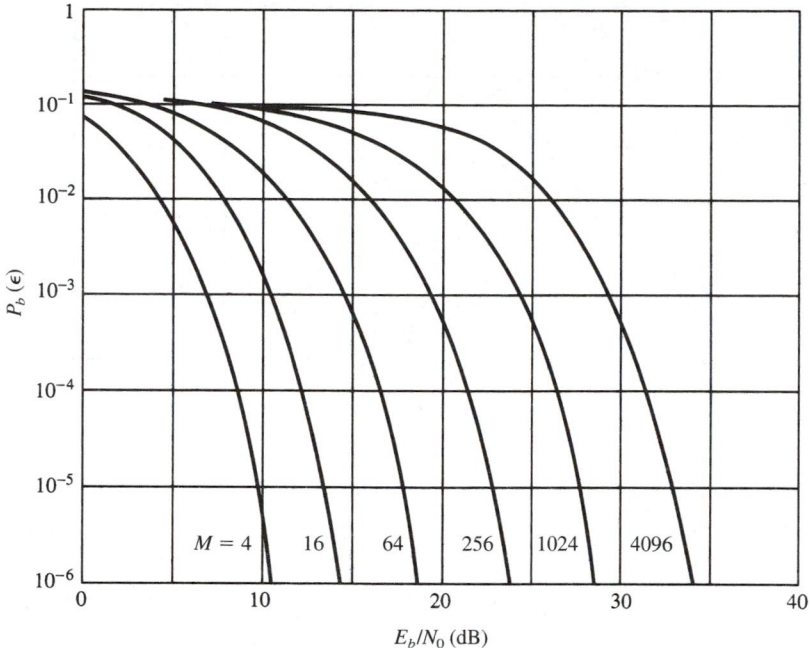

FIGURE 1-18. Bit error probability versus E_b/N_0 for coherent M-ary QASK.

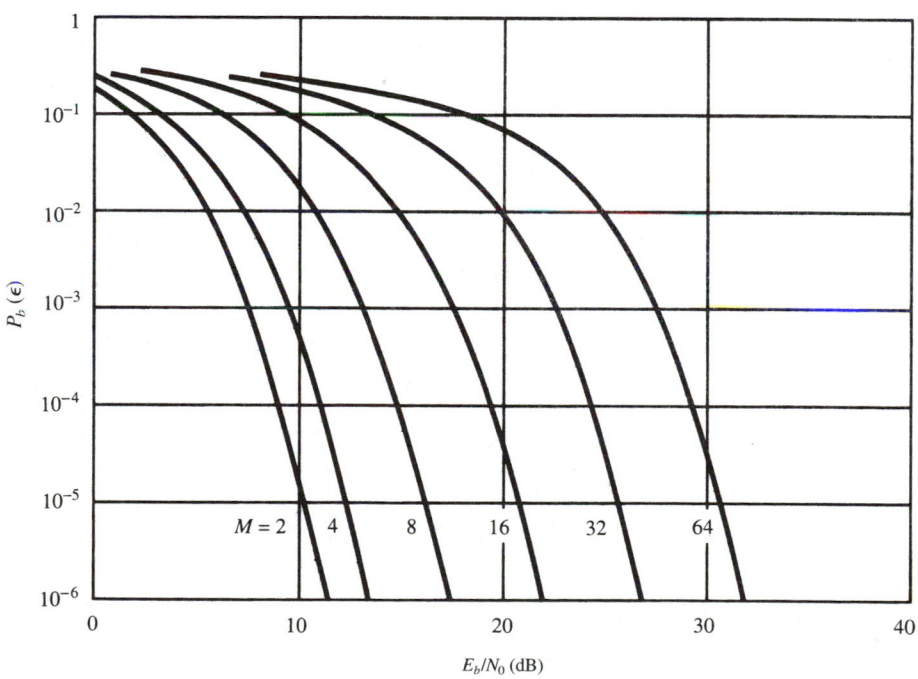

FIGURE 1-19. Bit error probability versus E_b/N_0 for M-ary DPSK.

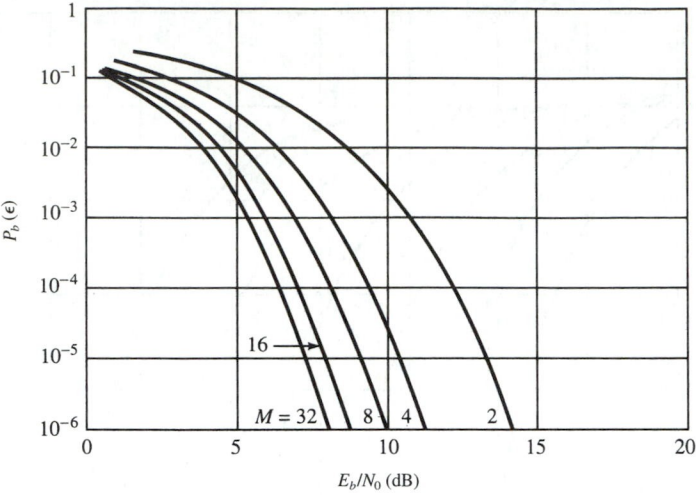

FIGURE 1-20. Bit error probability versus E_b/N_0 for noncoherent M-ary FSK.

for the case of several diffuse signal components of approximately the same powers being received, possibly due to multiple reflectors in the channel. The sum totals of the in-phase and quadrature signal components at the receiver are then approximately Gaussian by the central limit theorem, which means that the envelope, R, is Rayleigh and the phase, ϕ, is uniformly distributed in $(0, 2\pi)$. The Rician model is appropriate if the received signal consists of a direct, or specular, component plus several diffuse components of approximately equal powers. The Rayleigh model is the easiest one to analyze if it is assumed that for any coherent communication system the random phase is tracked exactly, an assumption that may not always be easily implemented. In the case of DPSK, it is only necessary that the phase be the same from one signaling interval to the next to allow the comparison of signals in adjacent bit intervals.

If these conditions are met, the probability of error expressions given previously are conditional on the received signal envelope, R, through the signal-to-noise ratio, z, which is

$$z = \frac{E_b}{N_0} = \frac{R^2 T}{2N_0} \tag{1-83}$$

Since R is assumed to be Rayleigh, it follows that z, being proportional to the square of R, is exponential with probability density function

$$f_z(z) = \frac{1}{\bar{z}} e^{-z/\bar{z}} \qquad z \geq 0 \tag{1-84}$$

where $\bar{z}$ is the average signal-to-noise ratio. The average probability of error for BPSK is then

$$\overline{P}_E = \int_0^\infty Q(\sqrt{2z}) \frac{1}{\bar{z}} e^{-z/\bar{z}} \, dz \tag{1-85}$$

This can be integrated exactly to give [4]

$$\overline{P}_E = \frac{1}{2}\left(1 - \sqrt{\frac{\overline{z}}{1+\overline{z}}}\right) \qquad \text{(BPSK)} \qquad (1\text{-}86)$$

A similar analysis for coherent FSK results in [4]

$$\overline{P}_E = \frac{1}{2}\left(1 - \sqrt{\frac{\overline{z}}{2+\overline{z}}}\right) \qquad \text{(coherent FSK)} \qquad (1\text{-}87)$$

Finally, for DPSK and noncoherent FSK, the results are [4]

$$\overline{P}_E = \frac{1}{2(1+\overline{z})} \qquad \text{(DPSK)} \qquad (1\text{-}88)$$

and

$$\overline{P}_E = \frac{1}{2+\overline{z}} \qquad \text{(noncoherent FSK)} \qquad (1\text{-}89)$$

respectively. These probability of error expressions are plotted in Figure 1-21 and compared with the corresponding nonfading results. Note that the penalty imposed by fading is severe.

For Rayleigh flat fading, the received signal consists of in-phase and quadrature components, which were Gaussian random variables. The resulting envelope of the received signal is then Rayleigh. A more realistic model in some cases consists of the received signal being composed of a steady, sinusoidal component along with Gaussian in-phase and quadrature components. Through standard transformation of variables techniques, the probability density function of the received signal envelope in this case can be shown to be *Rician,* which is of the form

$$p(r) = \frac{r}{\sigma^2}\exp\left(\frac{r^2 + a^2}{\sigma^2}\right)I_0\left(\frac{ar}{\sigma^2}\right) \qquad r \geq 0 \qquad (1\text{-}90)$$

where a is the amplitude of the sinusoidal component, σ^2 the variance of the Gaussian in-phase and quadrature components, and $I_0(x)$ the modified Bessel function of order zero. The sinusoidal component is sometimes called a direct or specular component, in analogy with a signal component reflected from an object or surface, and the quadrature Gaussian components, whose envelope is Rayleigh, is sometimes called the diffuse component in analogy with a component that is the sum total of a large number of contributions scattered from a rough surface or a large collection of objects.

The effect of Rician fading on noncoherent, noncoherent binary FSK can be analyzed with the result [7]

$$\overline{P}_E = \left(2 + \frac{\gamma^2\overline{z}}{1+\gamma^2}\right)^{-1}\exp\left[-\frac{\overline{z}}{2(1+\gamma^2) + \gamma^2\overline{z}}\right] \qquad (1\text{-}91)$$

where γ^2 is the ratio of powers in the diffuse to that in the specular received signal components. The error probability is plotted in Figure 1-22, where it is seen that the

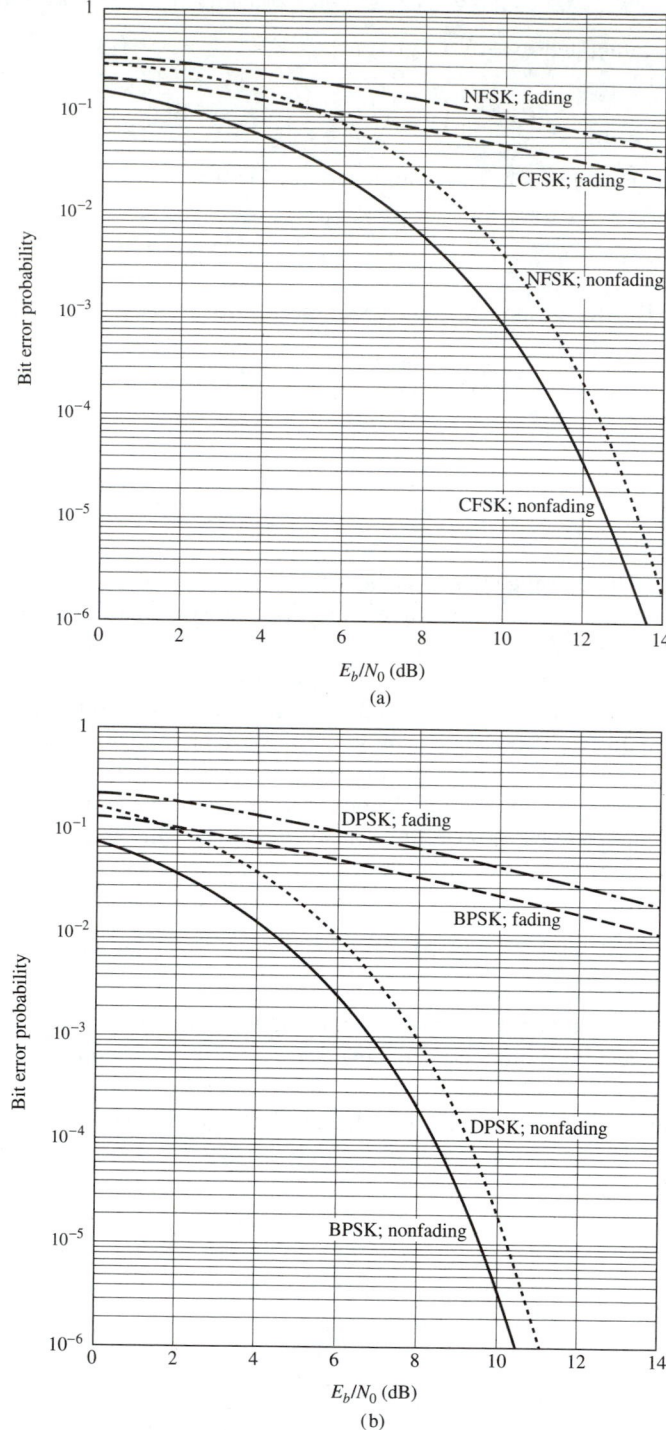

FIGURE 1-21. Error probabilities for various binary modulation schemes in flat-fading Rayleigh channels: (a) coherent and noncoherent FSK; (b) BPSK and DPSK.

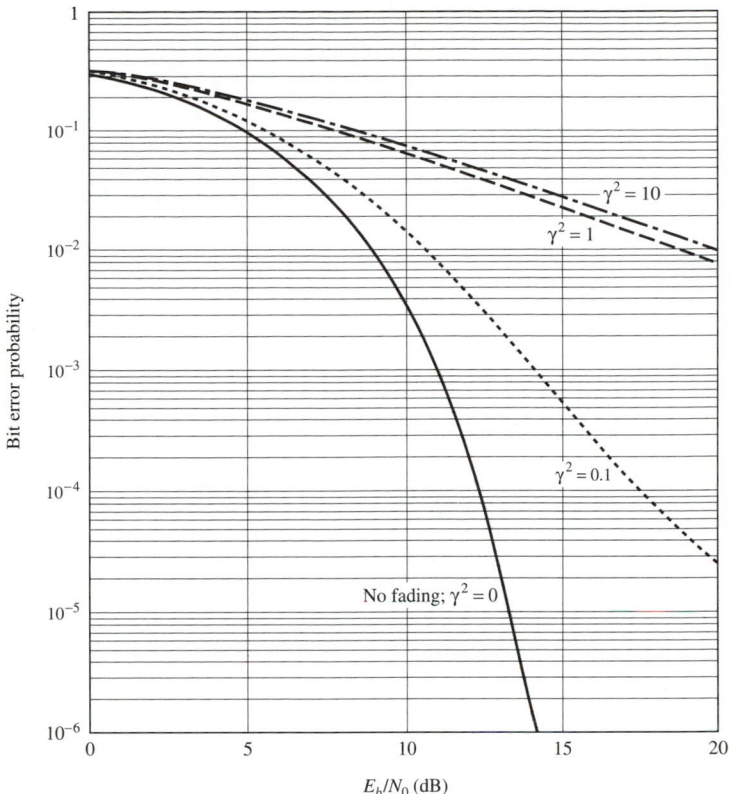

FIGURE 1-22. Error probability for noncoherent, binary FSK signaling in Rician flat fading.

presence of only a small amount of diffuse component has a severe effect on the probability of error. Sometimes the parameter $K = 1/\gamma^2$ is used and is referred to as the K factor.

Crepeau [20] has considered the effect of flat fading on M-ary DPSK and noncoherent FSK. To do so, he used another probabilistic model for the fading amplitude of the received signal called Nakagami-m, which is given by the probability density function

$$p(r) = \frac{2m^m r^{2m-1}}{\Gamma(m)\Omega^m} \exp\left(-\frac{mr^2}{\Omega}\right) \qquad r \geq 0 \qquad (1\text{-}92)$$

where Ω is the second moment of the random amplitude, R, $\Gamma(m)$ the gamma function, and m a parameter given by

$$m = \frac{\Omega^2}{\mathrm{var}(R^2)} \geq \frac{1}{2} \qquad (1\text{-}93)$$

The Nakagami-m probability density function is similar to the Rician density function in that for $m > \frac{1}{2}$ it is unimodal, with the width of the single peak becoming narrower as m gets larger. Similarly, the Rician pdf is unimodal, with the single peak becoming narrower as $K = 1/\gamma^2$ becomes larger. Crepeau found the bit error probability for noncoherent M-ary FSK in Nakagami-m fading to be

$$P_b = \frac{M}{2(M-1)} \sum_{i=1}^{M-1} \binom{M-1}{i} \frac{(-1)^{i+1}}{i+1} \left[\frac{m}{m + \frac{i}{i+1}(\log_2 M)\frac{E_b}{N_0}} \right]^i \tag{1-94}$$

Plots of this result are given in Figure 1-23 for several values of m. Again note that fading has a severe effect on the performance of the FSK communication system. Crepeau also considered the effect of coding and interleaving in the presence of

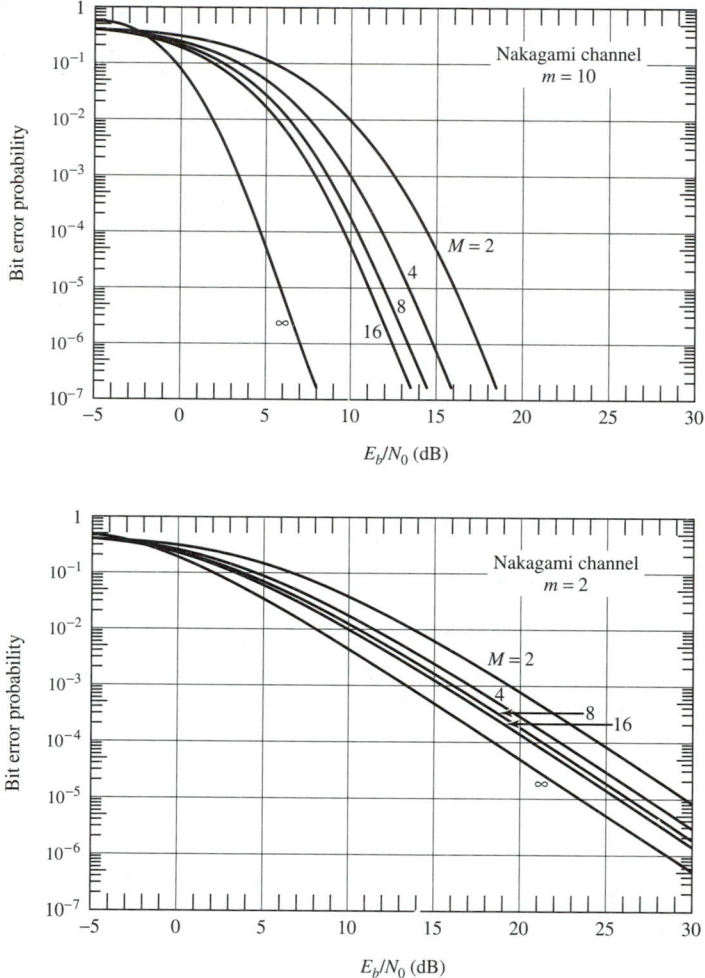

FIGURE 1-23. Error probability for noncoherent M-ary FSK signaling in Nakagami-m flat fading. (From Ref. 20. Copyright © 1992 IEEE. Reprinted with permission.)

Nakagami-m fading. The interleaving is assumed to randomize the errors from a fade completely. For a coding scheme that can correct t errors, the bit error probability varies as the inverse $m(t + 1)$th power of E_b/N_0 for large E_b/N_0.

1-6 Summary

In this chapter the performance of various digital modulation schemes has been considered from the standpoint of power and bandwidth efficiencies. Two types of channels were considered: additive white Gaussian noise and flat fading. Two types of modulation schemes were considered: those requiring the acquisition of a coherent reference at the receiver and those not requiring a coherent reference. The modulation methods considered were phase-shift keying (PSK), quadrature-amplitude-shift keying (QASK), differential phase-shift keying (DPSK), $\pi/4$-DQPSK, coherent frequency-shift keying (CFSK), and noncoherent frequency-shift keying (NFSK). With the exception of $\pi/4$-DQPSK, each of these modulation methods can be M-ary in that one of M possible signals can be transmitted during each signaling interval.

Two methods of comparison of M-ary modulation schemes are in terms of bandwidth efficiency, or the number of bits/second per hertz of bandwidth which can be transmitted, and in terms of power efficiency, or the E_b/N_0 required to achieve a specified bit error probability. The variation of these two measures with M depends on the dimensionality of the signal space for a given modulation method. For modulation schemes with increasing dimensionality versus M, such as for FSK, bandwidth efficiency decreases with M and power efficiency increases. For modulation schemes with constant dimensionality versus M, such as PSK, bandwidth efficiency increases with M and power efficiency decreases.

The effects of flat fading on digital signaling schemes are severe. One way to counteract facing is to use diversity, which spreads the transmitted signal power over many paths, or components, all of which are not expected to fade simultaneously with a high probability. Coding with interleaving is one such diversity scheme.

References

[1] R. E. ZIEMER and W. H. TRANTER, *Principles of Communications: Systems, Modulation, and Noise,* 4th ed. (Boston: Houghton Mifflin, 1995).

[2] J. M. WOZENCRAFT and I. M. JACOBS, *Principles of Communications Engineering* (New York: Wiley, 1965) (out of print, but available from Waveland Press, Prospect Heights, Illinois).

[3] R. E. BLAHUT, *Digital Transmission of Information* (Reading, Mass.: Addison-Wesley, 1990).

[4] J. G. PROAKIS, *Digital Communications,* 2nd ed. (New York: McGraw-Hill, 1989).

[5] V. A. KOTELNIKOV, *The Theory of Optimum Noise Immunity* (New York: Dover, 1960).

[6] A. ARTHURS and H. DYM, "On the Optimum Detection of Digital Signals in the Presence of White Gaussian Noise: A Geometric Interpretation of a Study of Three Basic Data Transmission Systems," *IRE Trans. Commun. Syst.*, Vol. CS-10, pp. 336–372, December 1962.

[7] R. E. ZIEMER and R. L. PETERSON, *Introduction to Digital Communication* (New York: Macmillan, 1992).

[8] V. K. PRABHU, "Error Rate Bounds for Differential PSK," *IEEE Trans. Commun.*, Vol. COM-30, pp. 2547–2550, December 1982.

[9] M. K. SIMON and W. K. ALEM, "Tracking Performance of Unbalanced QPSK Demodulators: Part I. Biphase Costas Loop with Passive Arm Filters," *IEEE Trans. Commun.*, Vol. COM-26, pp. 1147–1156, August 1978.

[10] R. M. GAGLIARDI, *Satellite Communications*, 2nd ed., (New York: Van Nostrand Reinhold, 1991).

[11] F. AMOROSO and J. A. KIVETT, "Simplified MSK Signaling Technique," *IEEE Trans. Commun.*, Vol. COM-25, pp. 433–441, April 1977.

[12] R. DE BUDA, "Coherent Demodulation of Frequency-Shift Keying with Low Deviation Ratio," *IEEE Trans. Commun.*, Vol. COM-20, pp. 429–435, June 1972.

[13] R. E. ZIEMER, C. R. RYAN, and J. H. STILWELL, "Conversion and Matched Filter Approximations for Serial Minimum-Shift Keying," *IEEE Trans. Commun.*, Vol. COM-30, pp. 495–509, March 1982.

[14] W. P. OSBORNE and M. B. LUNTZ, "Coherent and Noncoherent Detection of CPFSK," *IEEE Trans. Commun.*, Vol. COM-22, pp. 1023–1036, August 1974.

[15] M. K. SIMON and C. C. WANG, "Differential Versus Limiter-Discriminator Detection of Narrow-Band FM," *IEEE Trans. Commun.*, Vol. COM-31, pp. 1227–1234, November 1983.

[16] H. LIEB and S. PASUPATHY, "Error-Control Properties of Minimum Shift Keying," *IEEE Commun. Mag.*, Vol. 31, pp. 52–61, January 1993.

[17] F. G. JENKS, P. D. MORGAN, and C. S. WARREN, "Use of Four-Level Phase Modulation for Digital Mobile Radio," *IEEE Trans. Electromagn. Compat.*, Vol. EMC-14, pp. 113–130, November 1972.

[18] S. CHENNAKESHU and G. J. SAULNIER, "Differential Detection of $\pi/4$-DQPSK for Digital Cellular Radio," *IEEE Trans. Veh. Technol.*, Vol. VT-42, pp. 46–57, February 1993.

[19] EIA/TIA Interim Standard, *Cellular System Dual-Mode Mobile Station-Base Station Compatibility Standard*, IS-54-B, Telecommunications Industry Association, Washington, D.C.; April 1992.

[20] P. J. CREPEAU, "Uncoded and Coded Performance of MFSK and DPSK in Nakagami Fading Channels," *IEEE Trans. Commun.*, Vol. COM-40, pp. 487–493, March 1992.

Problems†

(1-1) Compare the E_b/N_0 in decibels required to yield a bit error probability of 10^{-6} for the following binary modulation schemes: **(a)** BPSK; **(b)** DPSK; **(c)** coherent FSK; **(d)** noncoherent FSK.

(1-2) Verify the entries in the right-hand columns of Table 1-1.

(1-3) Write a computer program to verify the numbers given in Table 1-2.

(1-4) Compute the ratio of the upper and lower bounds for P_1, given by (1-48a) and (1-48b), in the limits as E_s/N_0 approaches zero and infinity.

(1-5) Consider a channel that adds white Gaussian noise of spectral density of $N_0 = 10^{-8}$ W/Hz to a digitally modulated signal. A bit error probability of 10^{-5} is desired. Find the required average power for the received carrier for the following binary modulation schemes.
(a) BPSK at a data rate of 1 Mbps.
(b) Coherent FSK at a data rate of 100 kbps.
(c) ASK at a data rate of 3 kbps.
(d) What happens in each of the cases if the data rate is increased by a factor of 10?
(e) What happens in each of the cases if the noise power spectral density is made smaller by a factor of 10?

(1-6) Consider a 10-kHz bandpass channel with noise of spectral density of $N_0 = 10^{-9}$ W/Hz.
(a) For the coherent, binary modulation schemes of ASK, BPSK, and FSK, what is the maximum data rate that can be supported in each case?
(b) Find the received signal power required to give a bit error probability of 10^{-4} at each of the data rates found in part (a).

(1-7) Using the data provided in Table 1-2, estimate the degradation in E_b/N_0 in decibels for a bit error probability of 10^{-5} from ideal if the variance of the phase jitter is **(a)** 0.01; **(b)** 0.05; **(c)** 0.1 (the latter will require extension of the curve).

(1-8) Differentially encode the following binary sequences.
(a) 11100 11010 10000
(b) 11111 00000 11111
(c) 10101 01010 10101

---

†It is suggested that a computer mathematics package, such as Matlab, Mathcad, or Mathematica, be used in solving the problems with numerical answers.

(1-9) Consider a binary number $b_1b_2b_3 \cdots b_n$, with b_1 being the most significant bit (MSB) and b_n being the least significant bit (LSB). Let the corresponding Gray code bits be $g_1g_2g_3 \cdots g_n$. The Gray code for a given binary number is produced by the following algorithm:

$$g_1 = b_1$$

$$g_n = b_n \oplus b_{n-1} \qquad n \geq 2$$

where $\oplus$ represents modulo-2 addition (exclusive-OR or XOR) with truth table

		b_2	
		0	**1**
b_1	0	0	1
	1	1	0

Find the Gray encoded representation for the binary numbers 0 through 15_{10} and show that in going from a given number to an adjacent number, only one bit changes at a time, whereas this is not the case for the normal binary representation.

(1-10) Given the signal set

$$s_1(t) = e^{-t}u(t)$$

$$s_2(t) = e^{-2t}u(t)$$

$$s_3(t) = e^{-3t}u(t)$$

where $u(t)$ is the unit step, find an orthonormal basis set for the signal space spanned by these signals.

(1-11) Provide derivations for equations (1-42). Then, by considering the number of each type of region out of M, justify (1-43).

(1-12) Prove that (1-41) is true.

(1-13) Two data streams, one of 10 kbps and the other of 20 kbps, are to be transmitted on a common carrier using unbalanced QPSK.
  (a) What is the ratio of powers in the two quadrature channels such that both utilize the same energy per bit?
  (b) What are the transmitted phase values according to (1-55)?

(1-14) The baseband power spectrum for MSK can be shown to be

$$S_{\mathrm{MSK,\,BB}}(f) = \frac{16A^2T_b \cos^2(2\pi T_b f)}{\pi^2[1 - (4T_b f)^2]^2}$$

where A is the signal amplitude and T_b is the bit period. This spectrum is moved up in frequency and centered around f_c through the modulation process. The baseband power spectrum for QPSK (or OQPSK) is

$$S_{\text{QPSK, BB}}(f) = A^2 T_b \, \text{sinc}^2(2T_b f)$$

Show that the null-to-null main-lobe bandwidth of the modulated signal for MSK is $1.5/T_b$ versus $1/T_b$ hertz for QPSK.

(1-15) Plot a trellis diagram for MSK type II, clearly showing the differences from a trellis diagram for MSK type I.

(1-16) Show that the quadrature signal amplitudes, I_k and Q_k, for $\pi/4$-DQPSK take on the possible values 0, ± 1, and $\pm 2^{-1/2}$. Relate these to the values that ϕ_k takes on.

(1-17) Plot (1-66) for $z = 5, 7, 9$ dB versus $\psi_i - \phi_i$.

(1-18) Plot the bandwidth efficiencies of coherent M-ary PSK and FSK and noncoherent FSK versus M. Circle those points on the PSK curve that pertain to M-ary QASK.

(1-19) Sketch a trellis diagram for **(a)** QPSK; **(b)** OQPSK; **(c)** $\pi/4$-DQPSK.

(1-20) Take QPSK as a standard. Plot curves of the decibel increase or decrease (negative) required in E_b/N_0 over QPSK to produce a bit error probability of 10^{-6} for all the other M-ary modulation schemes considered in this chapter as a function of M. This will take one curve per modulation scheme. Note that QPSK requires about $E_b/N_0 = 10.54$ dB to give a bit error probability of 10^{-6}.

(1-21) A carrier component is sometimes included in BPSK to allow the acquisition of a reference coherent with the carrier at the receiver. Thus consider *phase-shift keying with carrier component,* which can be written as

$$s(t) = A \sin[\omega_0 t + d(t) \cos^{-1} a]$$

where $d(t)$ is a ± 1-valued binary message sequence with bit period T_b and a is a parameter called the modulation index.

(a) Show that $s(t)$ can be written as the sum of a carrier plus modulated-signal component as

$$s(t) = Aa \sin \omega_0 t + d(t)A\sqrt{1 - a^2} \cos \omega_0 t$$

(b) Show that the ratio of the powers in the carrier and modulation components of $s(t)$ is $a^2/(1 - a^2)$.

(c) Apply (1-1) to compute the bit error probability as a function of E_b/N_0 and a.

(1-22) Compute the increase in E_b/N_0 in decibels required to maintain a bit error probability of 10^{-3} for noncoherent FSK in a flat Rician fading channel over nonfading conditions for $\gamma^2 = 0.1, 1$, and 10. Compare the $\gamma^2 = 10$ case with the Rayleigh case for noncoherent FSK.

(1-23) Compute the increase in E_b/N_0 in decibels required to maintain a bit error probability of 10^{-3} in a flat Rayleigh-fading channel over nonfading conditions for the following binary modulation schemes: **(a)** BPSK; **(b)** DPSK; **(c)** coherent FSK; **(d)** noncoherent FSK.

(1-24) One scheme that has been proposed to combat fading is to use feedback from the receiver to the transmitter to let the transmitter know when the instantaneous signal-to-noise ratio is less than the average, or design, signal-to-noise ratio, $\bar{z}$, and simply not transmit information during those time intervals.

    (a) Using the exponential probability density function for the signal-to-noise ratio (1-84), compute the probability that $z < \bar{z}$.

    (b) Assuming that this represents the fraction of time that the channel is not usable, what is the effective data rate for a nominal 10-kbps system? (Note that there are practical limitations that must be overcome for such a system, including the need to maintain synchronization.)

(1-25) Write a computer program to verify the curves in Figure 1-23. What value of m appears to give the same fading effect as Rician fading with $\gamma^2 = 0.1$?

CHAPTER 2

Introduction to Spread-Spectrum Systems

2-1 Introduction

All of the modulation/demodulation techniques discussed so far have been designed to communicate digital information from one place to another as efficiently as possible in a stationary additive white Gaussian noise (AWGN) environment. The transmitted signals were selected to be relatively efficient in their use of the communication resources of power and bandwidth. The demodulators were designed to yield minimum bit error probability for the given transmitted signal in AWGN. Quantitative comparisons were made using the bandwidth and the E_b/N_0 required by the modem to achieve a specified bit error probability.

Although many real-world communication channels are accurately modeled as stationary AWGN channels, there are other important channels which do not fit this model. Consider, for example, a military communication system which might be jammed by a continuous wave (CW) tone near the modem's center frequency or by a distorted retransmission of the modem's own signal. The interference cannot be modeled as stationary AWGN in either of these cases. Another jammer may transmit pulsed AWGN which may not be stationary.

Another type of interference, which does not fit the stationary AWGN model, occurs when there are multiple propagation paths between the transmitter and receiver. The modem then interferes with itself via a delayed reception of its own signal. This phenomenon is called *multipath reception* and is a problem in line-of-sight microwave digital radios such as those used for long-haul telephone transmission and in urban mobile radio, among other places.

The remainder of this book is devoted to discussing a modulation and demodulation technique that can be used as an aid in mitigating the deleterious effects of the types of interference described above. This modulation and demodulation technique is called *spread spectrum* because the transmission bandwidth employed is much greater than the minimum bandwidth required to transmit the digital information. To be classified as a spread-spectrum system, the modem must have the following characteristcs:

1. The transmitted signal energy must occupy a bandwidth which is larger than the information bit rate (usually much larger) and which is approximately independent of the information bit rate.

2. Demodulation must be accomplished, in part, by correlation of the received signal with a replica of the signal used in the transmitter to spread the information signal.

A number of modulation techniques use a transmission bandwidth much larger than the minimum required for data transmission but are not spread-spectrum modulations. Low-rate coding, for example, results in increased transmission bandwidth but does not satisfy either of the conditions above. Wideband frequency modulation also results in a large transmission bandwidth but is not spread spectrum.

Spread-spectrum techniques can be very useful in solving a wide range of communications problems. The amount of performance improvement that is achieved through the use of spread spectrum is defined as the *processing gain* of the spread-spectrum system. That is, processing gain is the difference between system performance using spread-spectrum techniques and system performance not using spread-spectrum techniques, all else being equal. An often used approximation for processing gain is the ratio of the spread bandwidth to the information rate. In fact, some authors define processing gain as this or a similar bandwidth ratio. The particular definition chosen is of little consequence as long as it is always understood that real system performance improvement is the primary concern of the spread-spectrum system designer.

This chapter is intended to provide further motivation for the study of spread-spectrum systems and to introduce the most widely used types of spread-spectrum systems. Two important communication problems that can be partially solved using spread-spectrum techniques are described in Section 2-2. The two fundamental types of spread-spectrum systems, direct-sequence (DS) and frequency-hop (FH), are described in Sections 2-3 and 2-4, and hybrid forms are described in Section 2-5. The chapter concludes with a discussion of complex-envelope models used for spread spectrum.

2-2 Two Communications Problems

2-2.1 Pulse-Noise Jamming

Consider a coherent binary phase-shift-keyed (BPSK) communication system which is being used in the presence of a pulse-noise jammer. A *pulse-noise jammer* transmits pulses of bandlimited white Gaussian noise having total average power J referred to the receiver front end. The jammer may choose the center frequency and bandwidth of the noise to be identical to the receiver's center frequency and bandwidth. In addition, the jammer chooses its pulse duty factor ρ to cause maximum degradation to the communication link while maintaining constant average transmitted power J.

The bit error probability of a coherent BPSK system is given by (see Chapter 1)

$$P_E = Q\left(\sqrt{\frac{2E_b}{N_0}}\right) \tag{2-1}$$

where E_b represents the received energy per binary symbol and N_0 is the one-sided receiver front-end thermal noise power spectral density. When transmitting, the noise jammer increases the receiver noise power spectral density from N_0 to $N_0 + N_J/\rho$, where $N_J = J/W$ is the one-sided average jammer power spectral density and W is the one-sided transmission bandwidth. The jammer transmits using duty factor ρ, so that the average bit error probability is

$$\overline{P}_E = (1 - \rho)Q\left(\sqrt{\frac{2E_b}{N_0}}\right) + \rho Q\left(\sqrt{\frac{2E_b}{N_0 + N_J/\rho}}\right) \tag{2-2}$$

The jammer, given this formula, chooses ρ to maximize $\overline{P}_E$.

When a system is being designed to operate in a jamming environment, the maximum possible transmitter power is generally used and receiver front-end thermal noise can be safely neglected. In this case, the first term in (2-2) vanishes and $\overline{P}_E$ can be approximated by

$$\overline{P}_E \cong \rho Q\left(\sqrt{\frac{2E_b\rho}{N_J}}\right) \tag{2-3}$$

The Q-function can be bounded [1] by an exponential yielding

$$\overline{P}_E \leq \frac{\rho}{\sqrt{4\pi E_b\rho/N_J}} e^{-E_b\rho/N_J} \tag{2-4}$$

The maximum of this function over ρ can be found by taking the first derivative and setting it equal to zero. The maximizing ρ is found to be $\rho = N_J/2E_b$ and $\overline{P}_{E,\text{max}}$ is

$$\overline{P}_{E,\text{max}} = \frac{1}{\sqrt{2\pi e}} \frac{1}{2E_b/N_J} \tag{2-5}$$

Of course, the duty factor must be less than or equal to unity so that (2-5) applies only when $E_b/N_J \geq 0.5$. For $E_b/N_J < 0.5$, $\overline{P}_E$ is given by (2-3) with $\rho = 1.0$. Observe that the exponential dependence of bit error probability on signal-to-noise ratio of (2-1) has been replaced by an inverse linear relationship in (2-5). Equations (2-1) and (2-5) are plotted in Figure 2-1, where it can be seen that the optimized pulse noise jammer causes a degradation of approximately 31.5 dB relative to continuous jamming at a bit error probability of 10^{-5}.

The severe degradation in system performance caused by the pulse-noise jammer can be largely eliminated by using a combination of spread-spectrum techniques and forward error correction coding with appropriate interleaving. The effect of the spectrum spreading will be to change the abscissa of Figure 2-1 from E_b/N_J to E_bK/N_J, where K is a constant about equal to W/R, where R is the data rate of the spread-spectrum system. Error correction coding will be used to return from the inverse linear relation between error probability and signal-to-noise ratio to nearly the exponential relationship desired.

Finally, observe that in order to cause maximum degradation, the jammer must know the value of E_b/N_J at the receiver. This implies accurate knowledge of attenuation in both the transmitter-to-receiver path and jammer-to-receiver path. This knowledge would be difficult to obtain in a tactical environment, so that the results just described are worst-case. In addition, a real jammer would be limited in peak

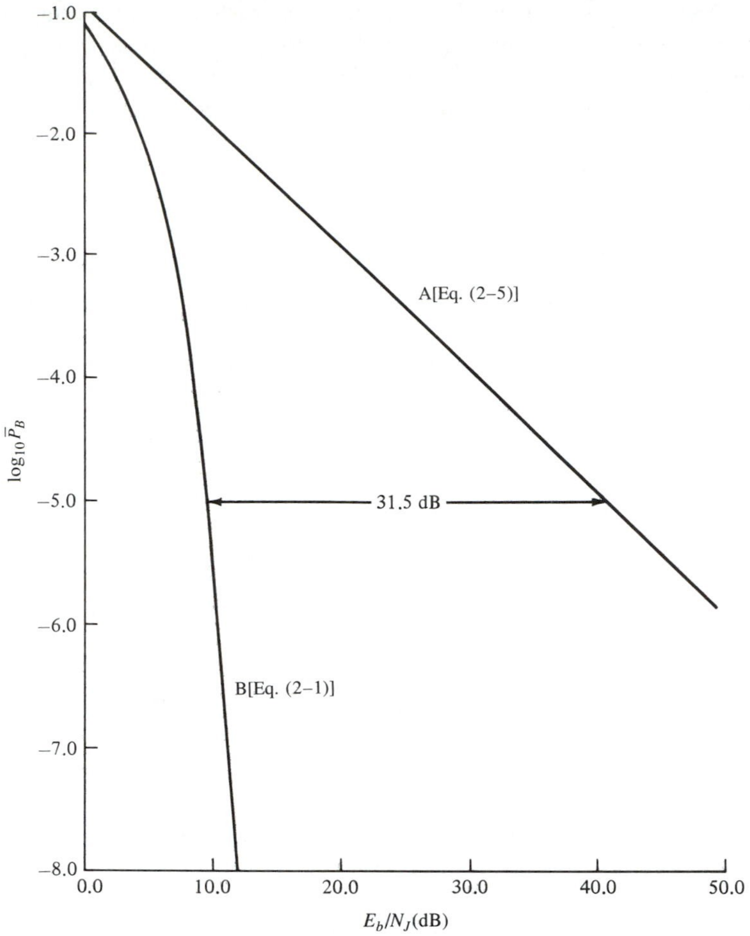

FIGURE 2-1. Bit error probability: A, worst-case pulse-noise jammer; B, continuous-noise jammer.

power output and would not be able to use an arbitrarily small duty factor. Despite these limitations, the pulse jammer is a serious threat to military communications systems.

2-2.2 Low Probability of Detection

Situations exist where it is desirable that a communication link be operated without knowledge of certain parties. *Low-probability-of-detection* (LPD) communication systems are designed to make their detection as difficult as possible by anyone but the intended receiver. This, of course, implies that the minimum signal power required to achieve a particular performance is used. The goal of the LPD system designer is to use a signaling scheme that results in the minimum probability of being detected within some time interval. Spread-spectrum techniques can significantly aid the system designer in achieving this goal.

Assume that the detector is using a radiometer. A radiometer detects energy received in a bandwidth W by filtering to this bandwidth, squaring the output of this filter, integrating the output of the squarer for time T, and comparing the output of the integrator at time T with a threshold as illustrated in Figure 2-2. If the integrator output is above a preset threshold at time T, the signal is declared present; otherwise, the signal is declared absent. The performance of the radiometer in detecting the desired communication signal is known if the probability density function (pdf) of the integrator output at time T is known. This probability density function is used to calculate the probability, P_d, of detecting the signal if it is indeed present, and probability of falsely declaring a detection when noise alone is present, P_{fa}.

Two approximations are often used for the integrator output pdf. The first, which is used when the *time–bandwidth product* TW is large relative to the received energy to noise power spectral density ratio E/N_0, is to approximate the pdf as Gaussian. In this case, P_d, P_{fa}, T, W, N_0, and $P = E/T$ are related by [2–5]

$$P_d = \Phi\left\{\left[\frac{P}{N_0}\sqrt{\frac{T}{W}} - \Phi^{-1}(1 - P_{fa})\right]\right\} \tag{2-6}$$

where†

$$\Phi(y) = \frac{1}{\sqrt{2\pi}}\int_{-\infty}^{y}\exp\left(-\frac{1}{2}\zeta^2\right)d\zeta \tag{2-7}$$

and $\Phi^{-1}(x)$ is equal to the variate y such that $\Phi(y) = x$. For a fixed P_{fa}, the probability of detection can be made smaller by reducing P/N_0 or increasing W. Integration time, T, is controlled by the detector, but W can be increased using spread-spectrum techniques. Thus another important application for spread spectrum is the reduction of signal detectability for fixed SNR, integration time, and detector false-alarm probability.

---

† The function $\Phi(y)$ is easily related to the Q-function, which can be calculated using the polynomial approximation given in Appendix B.

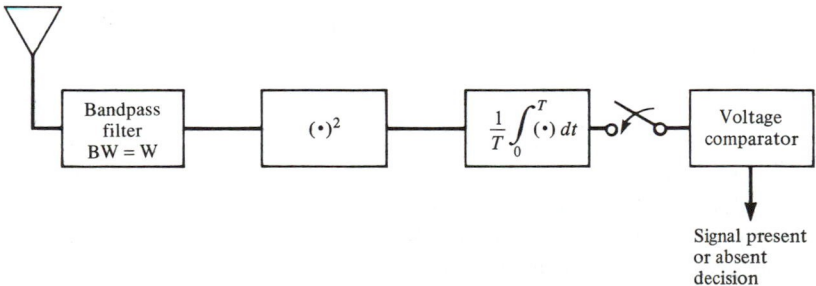

FIGURE 2-2. Energy detector or radiometer.

2-3 Direct-Sequence Spread Spectrum

One method of spreading the spectrum of a data-modulated signal is to modulate the signal a second time using a very wideband spreading signal. This second modulation is usually some form of digital phase modulation, although analog amplitude or phase modulation is conceptually possible. The spreading signal is chosen to have properties which facilitate demodulation of the transmitted signal by the intended receiver, and which make demodulation by an unintended receiver as difficult as possible. These same properties will also make it possible for the intended receiver to discriminate between the communication signal and jamming. If the bandwidth of the spreading signal is large relative to the data bandwidth, the spread-spectrum transmission bandwidth is dominated by the spreading signal and is nearly independent of the data signal.

Bandwidth spreading by direct modulation of a data-modulated carrier by a wideband spreading signal or code is called *direct-sequence* (DS) *spread spectrum.* Other types of spread-spectrum systems exist in which the spreading code is used to control the frequency or time of transmission of the data-modulated carrier, thus indirectly modulating the data-modulated carrier by a spreading code. These systems will be discussed later. The digital codes used for the spreading signal are discussed in detail in Chapter 3. The most common techniques used for direct-sequence spreading are discussed below.

2-3.1 BPSK Direct-Sequence Spread Spectrum

The simplest form of DS spread spectrum employs binary phase-shift keying (BPSK) as the spreading modulation. Ideal BPSK modulation results in instantaneous phase changes of the carrier by 180 degrees and can be mathematically represented as a multiplication of the carrier by a function $c(t)$ which takes on the values ± 1. Consider a constant-envelope data-modulated carrier having power P, radian frequency ω_0, and data phase modulation $\theta_d(t)$ defined by

$$s_d(t) = \sqrt{2P} \cos[\omega_0 t + \theta_d(t)] \tag{2-8}$$

This signal occupies a bandwidth typically between one-half and twice the data rate prior to DS spreading, depending on the details of the data modulation. BPSK spreading is accomplished by multiplying $s_d(t)$ by a function $c(t)$ representing the spreading waveform, as illustrated in Figure 2-3. The transmitted signal is

$$s_t(t) = \sqrt{2P} \, c(t) \cos[\omega_0 t + \theta_d(t)] \tag{2-9}$$

The signal of (2-9) is transmitted via a distortionless path having transmission delay T_d. The signal is received together with some type of interference and/or Gaussian noise. Demodulation is accomplished in part by remodulating with the spreading code appropriately delayed as shown in Figure 2-4. This remodulation or correlation of the received signal with the delayed spreading waveform is called *despreading* and is a critical function in all spread-spectrum systems. The signal

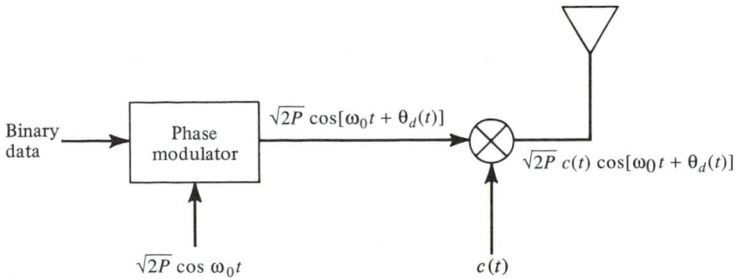

FIGURE 2-3. BPSK direct-sequence spread-spectrum transmitter.

component of the output of the despreading mixer is

$$\sqrt{2P}c(t - T_d)c(t - \hat{T}_d) \cos[\omega_0 t + \theta_d(t - T_d) + \phi] \qquad (2\text{-}10)$$

where $\hat{T}_d$ is the receiver's best estimate of the transmission delay. Since $c(t) = \pm 1$, the product $c(t - T_d) \times c(t - \hat{T}_d)$ will be unity if $\hat{T}_d = T_d$, that is, if the spreading code at the receiver is synchronized with the spreading code at the transmitter. When correctly synchronized, the signal component of the output of the receiver despreading mixer is equal to $s_d(t)$ except for a random phase ϕ, and $s_d(t)$ can be demodulated using a conventional coherent phase demodulator.

Observe that the data modulation above does not also have to be BPSK; no restrictions have been placed on the form of $\theta_d(t)$. However, it is common to use the same type of digital phase modulation for the data and the spreading code. When BPSK is used for both modulators, one phase modulator (mixer) can be eliminated. The double-modulation process is replaced by a single modulation by the modulo-2 sum of the data and the spreading code.

Figure 2-5 illustrates the direct-sequence spreading and despreading operation when the data modulation and the spreading modulation are BPSK. In this case, the data modulation is represented by a multiplication of the carrier by $d(t)$, where $d(t)$ takes on values of ± 1. Thus

$$s_d(t) = \sqrt{2P}\, d(t) \cos \omega_0 t \qquad (2\text{-}11)$$

$$s_t(t) = \sqrt{2P}\, d(t)c(t) \cos \omega_0 t \qquad (2\text{-}12)$$

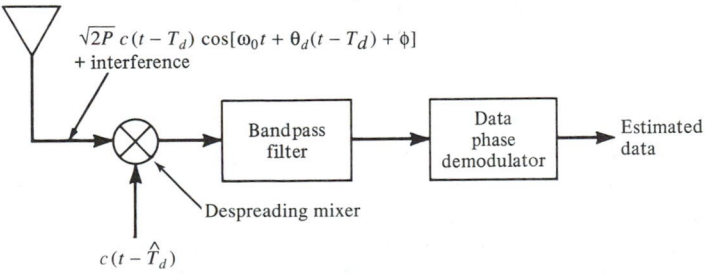

FIGURE 2-4. BPSK direct-sequence spread-spectrum receiver.

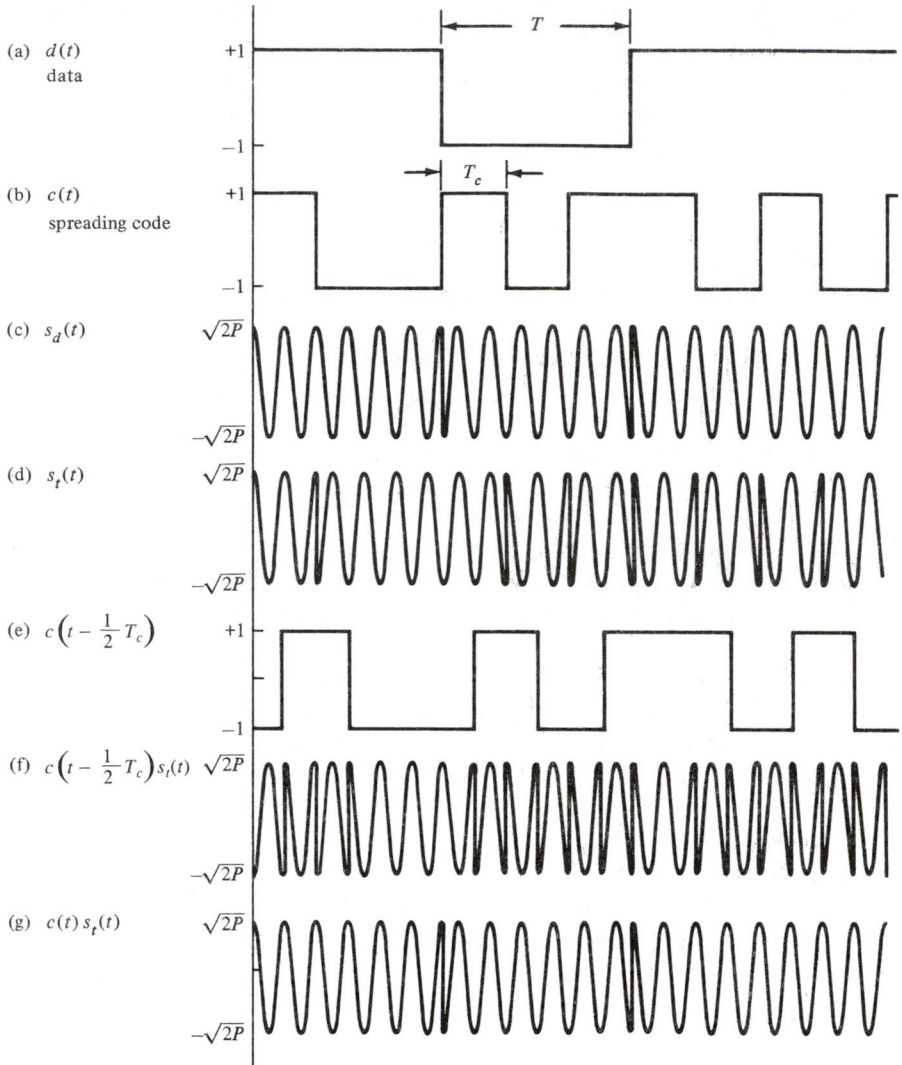

FIGURE 2-5. BPSK direct-sequence spreading and despreading.

The data and spreading waveforms are illustrated in Figure 2-5a and b, and $s_d(t)$ and $s_t(t)$ are illustrated in Figure 2-5c and d. Figure 2-5e represents an incorrectly phased input to the receiver despreading mixer assuming zero propagation delay, and Figure 2-5f shows the output of this mixer. Observe that Figure 2-5f is not equivalent to $s_d(t)$, illustrating that the receiver must be synchronized with the transmitter. Finally, Figure 2-5g shows the despreading mixer output when the despreading code is correctly phased. In this case $c(t)s_t(t) = s_d(t)$ and the data-modulated carrier has been recovered.

It is also instructive to consider the power spectra of the signals of Figure 2-5.

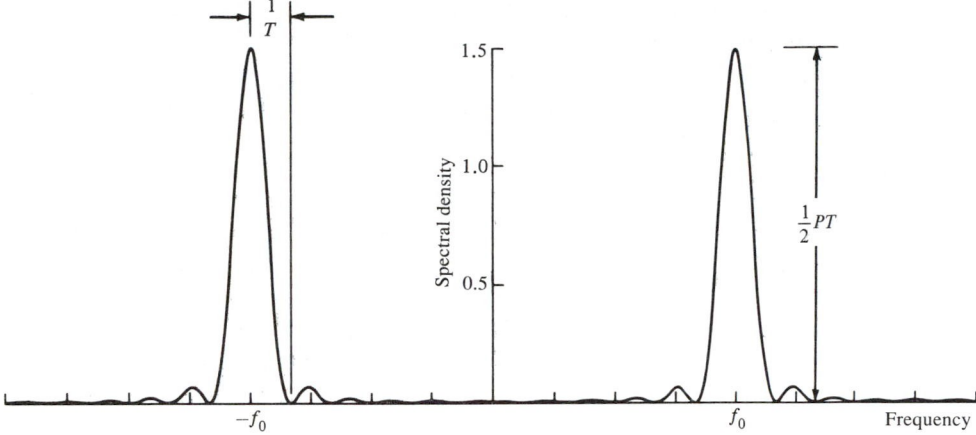

FIGURE 2-6. Power spectral density of data-modulated carrier.

The two-sided power spectral density in W/Hz of a binary phase-shift-keyed carrier is given by [6–9]

$$S_d(f) = \tfrac{1}{2}PT\{\text{sinc}^2[(f - f_0)T] + \text{sinc}^2[(f + f_0)T]\} \qquad (2\text{-}13)$$

which is plotted in Figure 2-6. Now observe that the signal $s_t(t)$ of Figure 2-5d is also a binary phase-shift-keyed carrier and therefore has a power spectral density (psd) which is given by (2-13) with T replaced by T_c, the duration of a spreading code symbol. The spreading code symbol duration T_c is often referred to as a spreading code *chip*. Figure 2-7 shows the power spectral density of $s_t(t)$ in the case where $T_c = T/3$. Observe that the effect of the modulation by the spreading code is to spread the bandwidth of the transmitted signal by a factor of 3, and that this spreading operation reduces the level of the psd by a factor of 3. In actual systems, this spreading factor is typically much larger than 3.

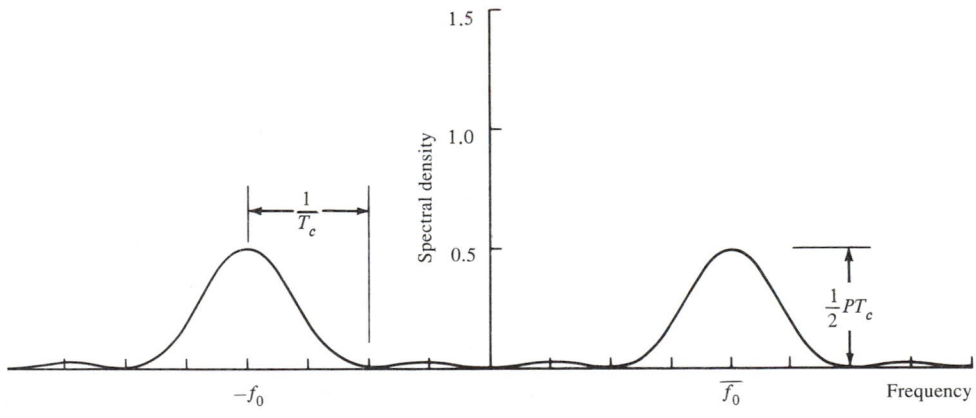

FIGURE 2-7. Power spectral density of data- and spreading code-modulated carrier.

Equation (2-13) applies only when both the data modulation and the spreading modulation are binary phase-shift keying, and when the data modulation and the spreading modulation are phase synchronous. In this case, since the data modulation is completely random, the signal $s_t(t)$ is a randomly biphase modulated signal and (2-13) applies. Consider again the case in which the data modulation is an arbitrary constant-envelope phase modulation. The data modulated carrier is represented by (2-8) and the transmitted signal is represented by (2-9). The power spectrum of the transmitted signal is calculated using the fact that the power spectrum and the autocorrelation function of a signal are a Fourier transform pair [1,10].

The data-modulated carrier is an ergodic random process, the spreading code is both deterministic and periodic, and their product, $s_t(t)$, is an ergodic random process. The signal $s_d(t)$ is independent of $c(t)$, so that the autocorrelation function $R_t(\tau)$ of the product $c(t)s_d(t)$ equals the product of the autocorrelation functions, that is,

$$R_t(\tau) = R_d(\tau)R_c(\tau) \tag{2-14}$$

Using the frequency convolution theorem [11] of Fourier transform theory, the power spectral density of $s_t(t)$, which is the Fourier transform of $R_t(T)$, is

$$S_t(f) = \int_{-\infty}^{\infty} S_d(f')S_c(f - f') \, df' \tag{2-15}$$

EXAMPLE 2-1

Calculate the power spectrum of the direct-sequence spread-spectrum transmitted signal when BPSK is used for both the data modulation and the spreading code modulation. Assume that the spreading code chip rate is 100 times the data rate, and that the period of the spreading code is infinite.

Solution: The power spectrum of the data-modulated carrier is given by (2-13). The power spectrum of the spreading code $c(t)$ is the Fourier transform of its autocorrelation function [1,10]

$$R_c(\tau) = \lim_{A \to \infty} \frac{1}{2A} \int_{-A}^{A} c(t')c(t' - \tau) \, dt'$$

When $\tau = 0$, this integral is equal to 1.0 since $c^2(t) = 1.0$. When $\tau \geq T_c$, the integral is zero since the code has been modeled as an infinite sequence of independent random binary digits. For $0 < \tau < T_c$ the integral is equal to the fraction of the chip time for which $c(t' - \tau) = c(t')$, as illustrated in Figure 2-8. Therefore,

$$R_c(\tau) = \begin{cases} 1 - \dfrac{|\tau|}{T_c} & |\tau| < T_c \\ \\ 0 & |\tau| \geq T_c \end{cases}$$

as illustrated in Figure 2-9. The Fourier transform of this triangular waveform is

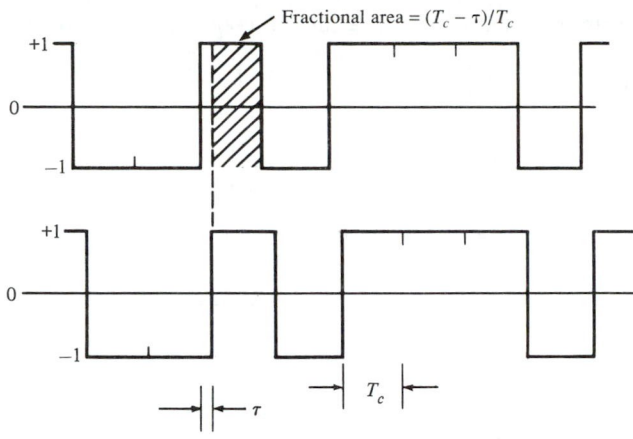

FIGURE 2-8. Calculation of autocorrelation function of an infinite sequence of random binary digits.

easily calculated. The result is

$$S_c(f) = T_c \, \text{sinc}^2 f T_c = \frac{T}{100} \, \text{sinc}^2 \frac{fT}{100}$$

The transmitted power spectrum is then

$$S_t(f) = \int_{-\infty}^{\infty} \frac{1}{2} PT \, \text{sinc}^2[(f' - f_0)T] \frac{T}{100} \, \text{sinc}^2 \left[(f - f') \frac{T}{100} \right] df'$$

$$+ \int_{-\infty}^{\infty} \frac{1}{2} PT \, \text{sinc}^2[(f' + f_0)T] \frac{T}{100} \, \text{sinc}^2 \left[(f - f') \frac{T}{100} \right] df'$$

Because the spreading chip rate is much larger than the data rate, the second sinc function in each integral is approximately constant over the range of significant

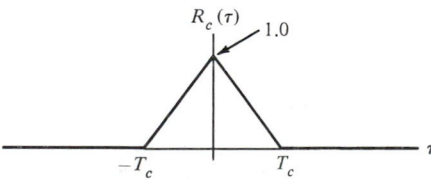

FIGURE 2-9. Autocorrelation function of an infinite sequence of random binary digits.

values of the first sinc function. Thus the convolution can be approximated by

$$S_t(f) \cong \frac{PT^2}{200} \text{sinc}^2\left[(f-f_0)\frac{T}{100}\right] \int_{-\infty}^{\infty} \text{sinc}^2[(f'-f_0)T] \, df'$$

$$+ \frac{PT^2}{200} \text{sinc}^2\left[(f+f_0)\frac{T}{100}\right] \int_{-\infty}^{\infty} \text{sinc}^2[(f'+f_0)T] \, df'$$

$$= \frac{1}{2}\frac{PT}{100}\left\{ \text{sinc}^2\left[(f-f_0)\frac{T}{100}\right] + \text{sinc}^2\left[(f+f_0)\frac{T}{100}\right] \right\}$$

It was claimed earlier that one of the advantages of using spread spectrum is that it will enable the receiver to reject deliberate interference or jamming. Interference rejection is accomplished by the receiver despreading mixer, which *spreads* the spectrum of the interference at the same time that the desired signal is *despread*. If the interference energy is spread over a bandwidth much larger than the data bandwidth, most of the energy will be rejected by the data filter.

Suppose that BPSK is used for both the data modulation and the spreading modulation and that the interference is a single tone having power J. Suppose that the jammer places the jamming tone directly in the center of the modem's transmission bandwidth. If no spectrum spreading were employed, the ratio of jamming power to signal power in the data bandwidth would be J/P. The power spectrum of the received signal is approximately

$$S_r(f) \cong \tfrac{1}{2}PT_c\{\text{sinc}^2[(f-f_0)T_c] + \text{sinc}^2[(f+f_0)T_c]\}$$

$$+ \tfrac{1}{2}J\{\delta(f-f_0) + \delta(f+f_0)\} \tag{2-16}$$

and the received signal is

$$r(t) = \sqrt{2P}\, d(t-T_d)c(t-T_d)\cos(\omega_0 t + \phi)$$

$$+ \sqrt{2J}\cos(\omega_0 t + \phi') \tag{2-17}$$

Assume that the receiver despreading code is correctly phased so that the output of the despreading mixer is

$$y(t) = \sqrt{2P}\, d(t-T_d)\cos(\omega_0 t + \phi)$$

$$+ \sqrt{2J}\, c(t-T_d)\cos(\omega_0 t + \phi') \tag{2-18}$$

and the power spectrum of $y(t)$ is

$$S_y(f) = \tfrac{1}{2}PT\{\text{sinc}^2[(f-f_0)T] + \text{sinc}^2[(f+f_0)T]\}$$

$$+ \tfrac{1}{2}JT_c\{\text{sinc}^2[(f-f_0)T_c] + \text{sinc}^2[(f+f_0)T_c]\} \tag{2-19}$$

Observe that the data signal has been despread to the data bandwidth, while the single-tone jammer has been spread over the full transmission bandwidth of the spread-spectrum system.

The power spectra of the signals discussed above are illustrated in Figure 2-10. The received power spectra are shown in Figure 2-10a, and the spectra after the

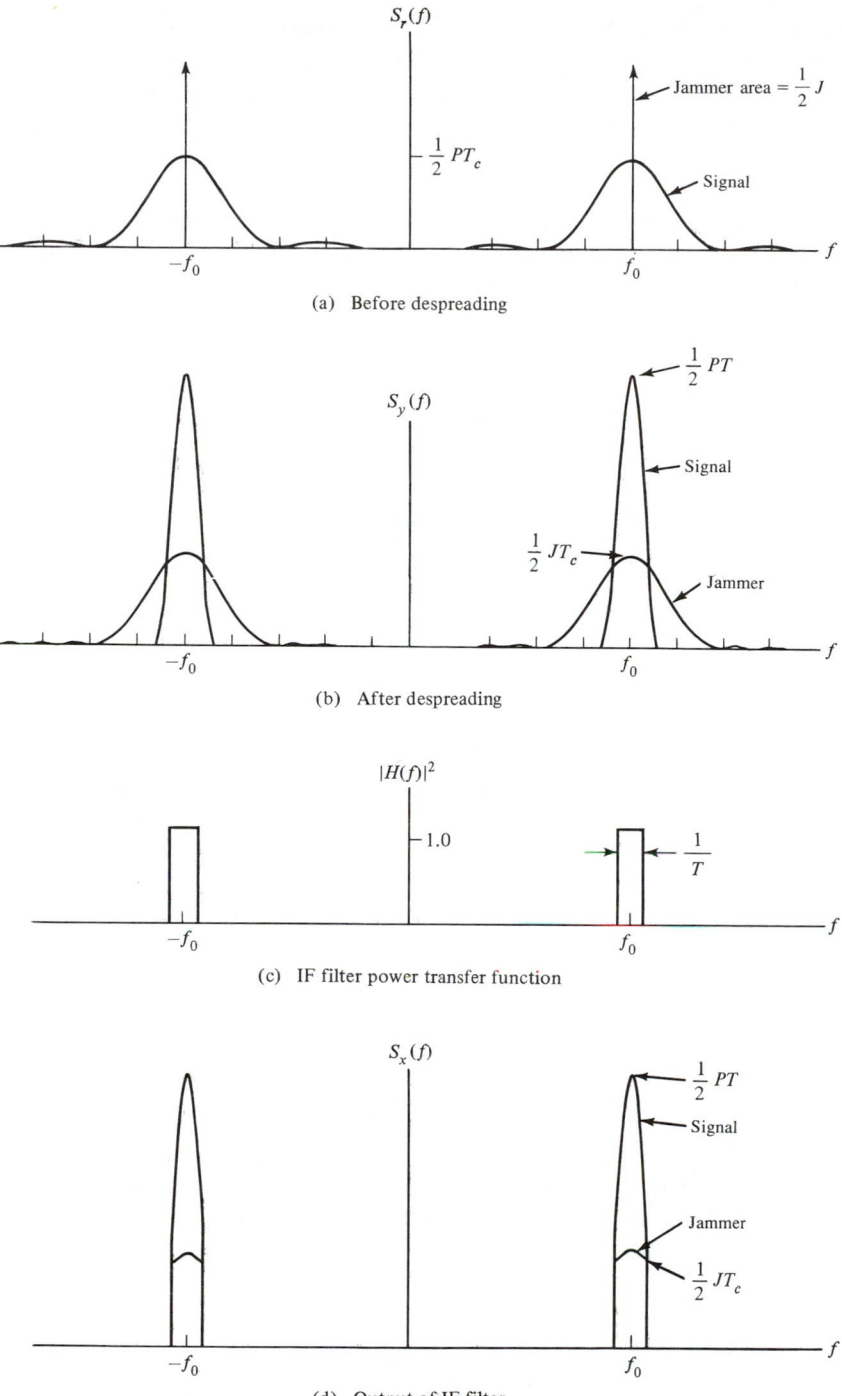

FIGURE 2-10. Receiver power spectral densities with tone jamming.

despreading mixer are shown in Figure 2-10b. The despreading operation in spread-spectrum receivers is followed by a filtering operation to limit the bandwidth at the input of the data demodulator to approximately the data bandwidth. The power transfer function of an ideal filter accomplishing this is shown in Figure 2-10c and the output of this filter is shown in Figure 2-10d. This ideal filter represents the noise equivalent bandwidth of an actual intermediate-frequency (IF) filter whose noise bandwidth is equal to the data rate. Nearly all of the signal power is passed by the IF filter. A large fraction of the spread jammer power, on the other hand, is rejected by this filter. The magnitude of the jammer power passed by the IF filter is

$$J_0 = \int_{-\infty}^{\infty} S_J(f)|H(f)|^2 \, df \tag{2-20}$$

where $S_j(f)$ is the power spectrum of the jammer after the despreading mixer. If an ideal bandpass IF filter as shown in Figure 2-10c is assumed, then

$$
\begin{aligned}
J_0 &= \int_{-f_0-1/2T}^{-f_0+1/2T} S_J(f) \, df + \int_{f_0-1/2T}^{f_0+1/2T} S_J(f) \, df \\
&\cong \frac{1}{2} JT_c \int_{-f_0-1/2T}^{-f_0+1/2T} \text{sinc}^2[(f+f_0)T_c] \, df \\
&\quad + \frac{1}{2} JT_c \int_{f_0-1/2T}^{f_0+1/2T} \text{sinc}^2[(f-f_0)T_c] \, df
\end{aligned}
\tag{2-21}
$$

For large ratios of data bandwidth to total spread bandwidth, that is, $T_c \ll T$, the sinc function is nearly constant over the range of the integration and

$$J_0 \cong J\frac{T_c}{T} \tag{2-22}$$

Thus the jamming power at the input to the data demodulator has been reduced by a factor T_c/T relative to its value without the use of spread spectrum. The processing gain of this very simple spread-spectrum system is equal to the inverse of this jammer power reduction factor, or

$$G_p = \frac{T}{T_c} \tag{2-23}$$

Other equivalent definitions of processing gain are possible and are often used. Throughout this book a consistent definition of processing gain as an improvement factor is used. This results in different formulas for G_p, depending on the particular system being considered.

2-3.2 QPSK Direct-Sequence Spread Spectrum

It is sometimes advantageous to transmit simultaneously on two carriers which are in phase quadrature. The principal reason for doing this is to conserve spectrum, since, for the same total transmitted power, the same bit error probability is achieved using one-half the transmission bandwidth [6]. Bandwidth efficiency is not usually of primary importance for antijam and low probability of detection applications of spread spectrum. However, quadrature modulations are used in spread-spectrum systems since they are more difficult to detect using feature detectors [5] in low-probability-of-detection applications and are less sensitive to some types of jamming. Spectrum efficiency is a primary concern for cellular radio applications of spread spectrum.

Both the data modulation and the spreading modulation can be placed on quadrature carriers using a number of techniques. If no restriction is placed on the data phase modulator, QPSK spreading modulation can be added using the system of Figure 2-11a. Observe that the power at either output of the quadrature hybrid is

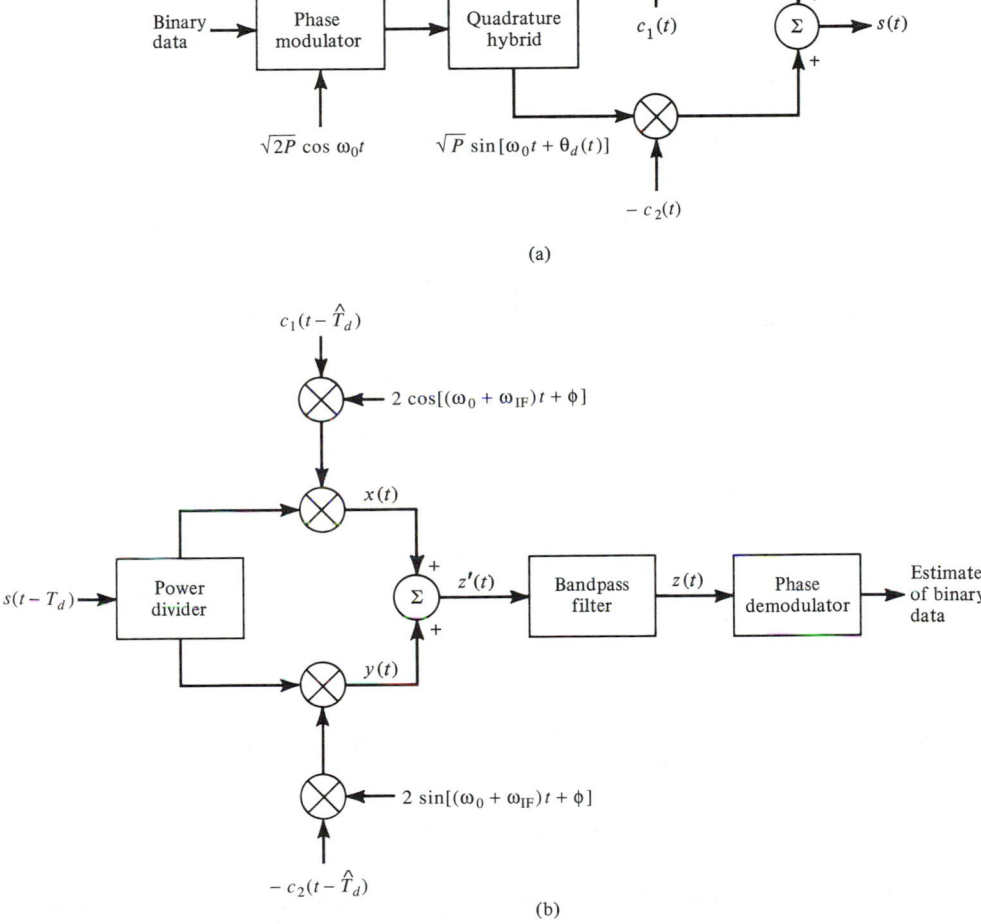

FIGURE 2-11. (a) QPSK spread-spectrum modulator with arbitrary data phase modulation; (b) QPSK spread-spectrum receiver for arbitrary data modulation.

one-half of the input power. The output of the QPSK modulator is

$$s(t) = \sqrt{P}\,c_1(t)\cos[\omega_0 t + \theta_d(t)] - \sqrt{P}\,c_2(t)\sin[\omega_0 t + \theta_d(t)]$$
$$\triangleq a(t) - b(t) \tag{2-24}$$

where $c_1(t)$ and $c_2(t)$ are the in-phase and quadrature spreading waveforms. When written this way both spreading waveforms are assumed to take on only values of ± 1. These spreading waveforms are assumed to be chip synchronous but otherwise totally independent of one another. This type of modulation is called *balanced QPSK modulation* [12] since the data modulation is balanced between the in-phase and quadrature-phase spreading channels.

The power spectrum of the QPSK spread-spectrum signal of (2-24) can be calculated by observing that both terms of this equation are identical, except for amplitude and a possible phase shift, to (2-9) for BPSK spread spectrum. Thus, since the two signals are orthogonal, the power spectrum of the sum signal equals the algebraic sum of the two power spectra. That this is true is most conveniently illustrated by calculating the autocorrelation function of $s(t)$, which is

$$\begin{aligned} R_s(\tau) &= E[s(t)s(t + \tau)] \\ &= E[a(t)a(t + \tau)] + E[b(t)b(t + \tau)] \\ &\quad - E[a(t)b(t + \tau)] - E[b(t)a(t + \tau)] \\ &= R_a(\tau) + R_b(\tau) - E[a(t)b(t + \tau)] - E[b(t)a(t + \tau)] \end{aligned} \tag{2-25}$$

If the functions $a(t)$ and $b(t)$ are orthogonal [8], the last two terms of (2-25) are equal to zero. This condition is satisfied in the present case since $c_1(t)$ and $c_2(t)$ are independent code waveforms.

The receiver for the transmitted signal of (2-24) is shown in Figure 2-11b. In this figure the bandpass filter is centered at frequency ω_{IF} and has a bandwidth sufficiently wide to pass the data-modulated carrier with negligible distortion. Using straightforward trigonometric identities, it can be shown that the components of $x(t)$ and $y(t)$ near the IF are

$$\begin{aligned} x(t) &= \sqrt{\frac{P}{2}}\,c_1(t - T_d)c_1(t - \hat{T}_d)\cos[\omega_{\mathrm{IF}}t - \theta_d(t)] \\ &\quad + \sqrt{\frac{P}{2}}\,c_2(t - T_d)c_1(t - \hat{T}_d)\sin[\omega_{\mathrm{IF}}t - \theta_d(t)] \end{aligned} \tag{2-26}$$

$$\begin{aligned} y(t) &= -\sqrt{\frac{P}{2}}\,c_1(t - T_d)c_2(t - \hat{T}_d)\sin[\omega_{\mathrm{IF}}t - \theta_d(t)] \\ &\quad + \sqrt{\frac{P}{2}}\,c_2(t - T_d)c_2(t - \hat{T}_d)\cos[\omega_{\mathrm{IF}}t - \theta_d(t)] \end{aligned} \tag{2-27}$$

If the receiver-generated replicas of the spreading codes are correctly phased, then

$$c_1(t - T_d)c_1(t - \hat{T}_d) = c_2(t - T_d)c_2(t - \hat{T}_d) = 1.0 \tag{2-28}$$

and the desired signals have been despread. These despread signals will pass through the bandpass filter. The undesired terms of (2-26) and (2-27) cancel, so that

$$z(t) = \sqrt{2P} \cos[\omega_{IF}t - \theta_d(t)] \tag{2-29}$$

In deriving the results above, perfect receiver carrier phase tracking has been assumed. Observe in (2-29) that the data-modulated carrier has been recovered. That is, the QPSK spreading modulation added by the transmitter has been removed by the receiver despreading operation. The signal $z(t)$ is the input to a conventional phase demodulator where data is recovered. Other forms of the receiver are possible. The particular placement of the mixers and filter shown, however, is typical of an arrangement that might be found in actual hardware.

Another configuration for a QPSK spread-spectrum modem is shown in Figure 2-12. In this case the in-phase and quadrature-phase QPSK channels are BPSK data

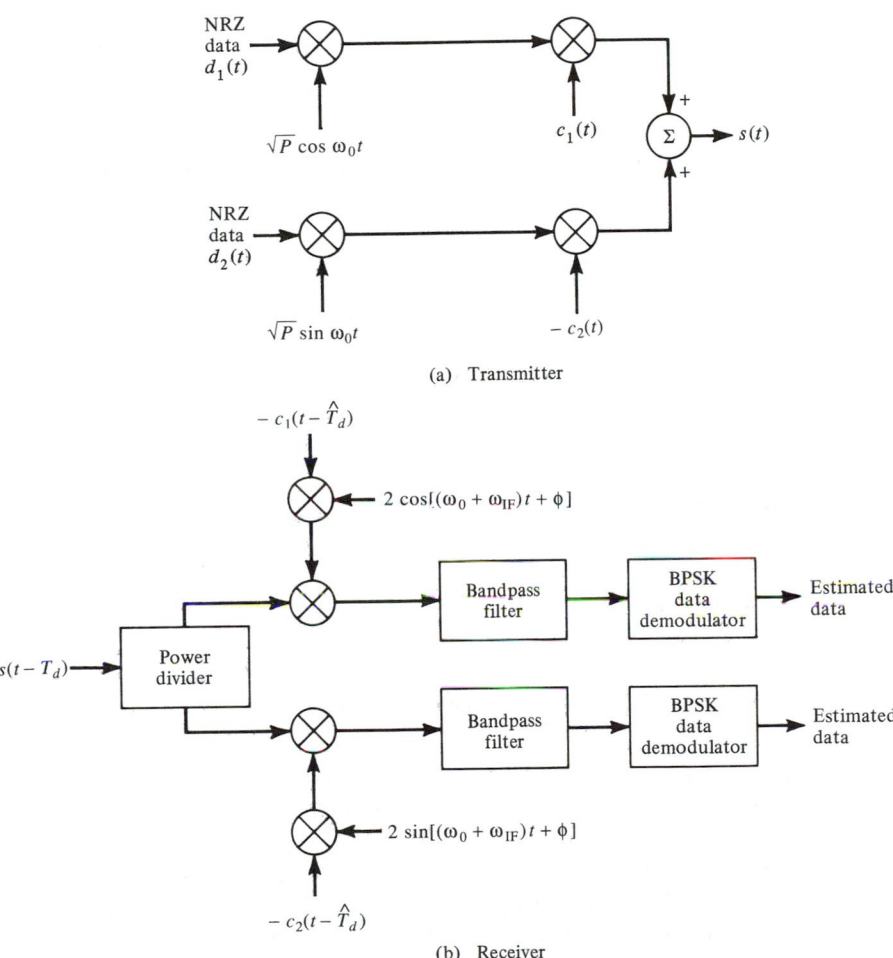

(a) Transmitter

(b) Receiver

FIGURE 2-12. Dual-channel QPSK direct-sequence spread-spectrum modem.

modulated using different BPSK data modulators. Thus different data are transmitted on the two QPSK channels. This modulation is called *dual-channel QPSK* [12]. The transmitted waveform for dual-channel QPSK is

$$s(t) = \sqrt{P}\, d_1(t)c_1(t) \cos \omega_0 t - \sqrt{P}\, d_2(t)c_2(t) \sin \omega_0 t \qquad (2\text{-}30)$$

which has total power P. The receiver for this waveform is shown in Figure 2-12b and is similar in operation to the balanced QPSK modem described above. A small but important variation on the signal of (2-30) yields one of the spread-spectrum signals used for the Tracking and Data Relay Satellite System (TDRSS). This variation is simply to permit the in-phase and quadrature channels to have unequal power. Thus the transmitted signal is

$$s(t) = \sqrt{2P_I}\, d_1(t)c_1(t) \cos \omega_0 t - \sqrt{2P_Q}\, d_2(t)c_2(t) \sin \omega_0 t \qquad (2\text{-}31)$$

Another variation applicable to all the QPSK spread-spectrum modems is to use offset QPSK for the spreading modulation. This variation is also employed on TDRSS.

The QPSK spread-spectrum modems discussed above are the principal types of QPSK modems either currently in use or widely discussed in the literature. Other variations, especially in the details of the implementations, are possible and may be more efficient under some conditions.

2-3.3 MSK Direct-Sequence Spread Spectrum

Another practical direct-sequence spread-spectrum modulation scheme is minimum shift keying (MSK). Although conventional schemes for generating MSK signals are more complex than the schemes for generating QPSK, the serial approach to MSK modulation is applicable to spread-spectrum systems and results in a system with the theoretical benefits of an in-phase and quadrature system together with hardware only slightly more complex than a BPSK system. The theory of both parallel and serial MSK implementations may be found in the literature [13–15] and will not be reviewed here.

A conventional MSK spread-spectrum modem is illustrated in Figure 2-13. The transmitter output signal is

$$s(t)$$
$$= \sqrt{2P}\, d(t)\left[c_1\!\left(t - \frac{T_c}{2}\right) \cos\!\left(\frac{\pi}{T_c}t\right) \cos \omega_0 t - c_2(t) \sin\!\left(\frac{\pi}{T_c}t\right) \sin \omega_0 t \right] \qquad (2\text{-}32)$$

It is left as an exercise to show that the MSK modulation is removed by the receiver of Figure 2-13b when both the carrier tracking and code tracking loops are operating perfectly.

Although the modem of Figure 2-13 ideally provides proper MSK spreading and despreading modulation, the hardware is relatively complex. The serial technique for MSK modulation can be used to simplify this hardware significantly. Two serial MSK spread-spectrum modems are illustrated in Figure 2-14. In both the transmitters and the receiver, the MSK conversion filter is a passive, linear, time-invariant filter. Observe that a single spreading code is used in these modems rather than the

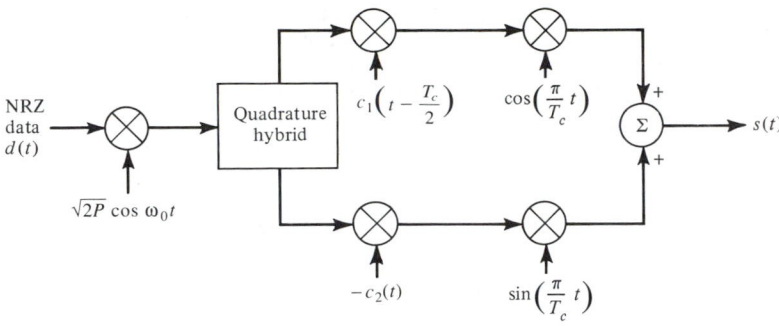

(a) Transmitter

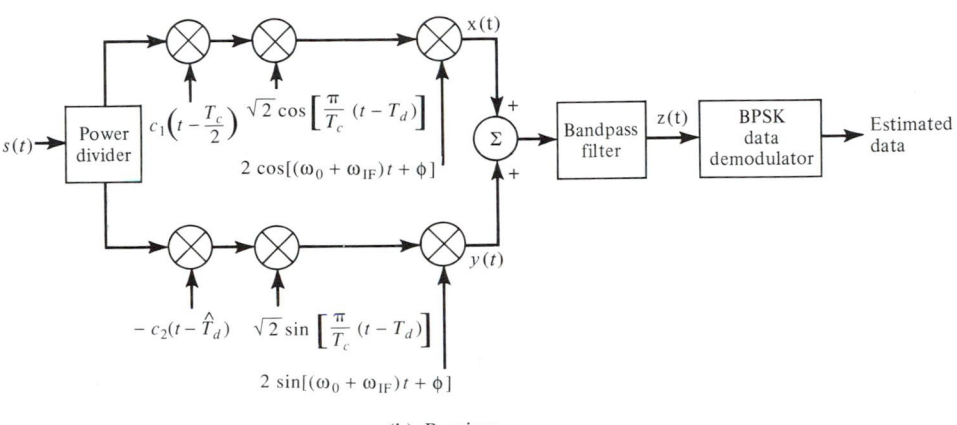

(b) Receiver

FIGURE 2-13. Minimum shift-keying spread-spectrum modem.

two separate spreading codes that are required in the conventional MSK modem. For the same performance, however, the serial MSK spreading code is required to operate at twice the rate of the codes in the conventional modulator, and therefore may be more difficult to implement. The two modems differ only in the placement of the data modulator.

A convenient tool for understanding MSK modulation is the excess phase trellis [14]. The excess phase trellis presents a graphical picture of the instantaneous phase difference between the transmitted signal and a carrier at the center frequency of the MSK spectrum. This excess phase is also, by definition, the phase of the complex envelope of the MSK signal. The excess phase trellis can also be used to describe the transmitted phase in a spread-spectrum system using MSK spreading together with BPSK data modulation. Assume that the data modulation is phase synchronous with the spreading modulation. Figure 2-15 is an example excess phase trellis for both modems of Figure 2-14 when the spreading code rate is three times the data rate.

Observe that the data modulation in transmitter 1 generates abrupt $\pi/2$ phase transitions in the output signal. This is illustrated in Figure 2-15c, which is derived

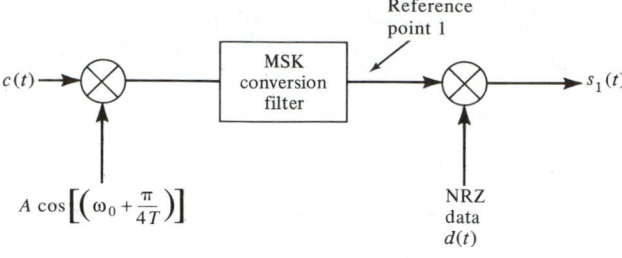

(a) Transmitter 1

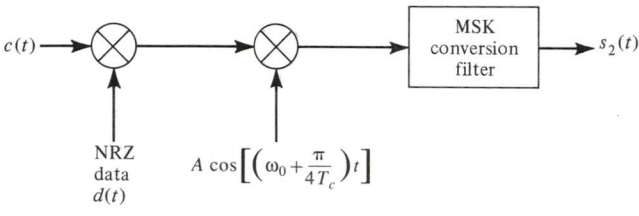

(b) Transmitter 2

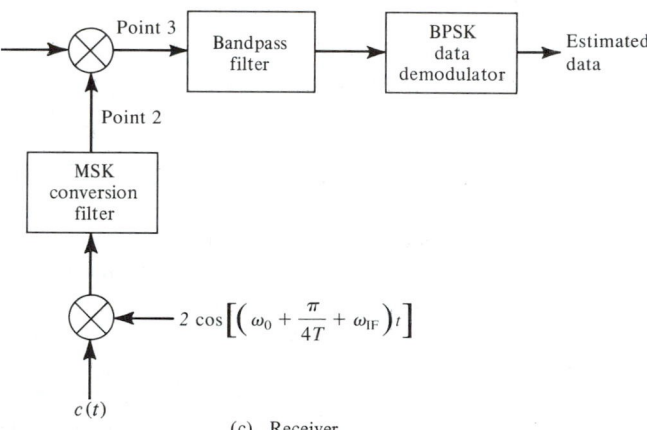

(c) Receiver

FIGURE 2-14. Serial MSK spread-spectrum modem.

from Figure 2-15d by adding a $\pi/2$ phase shift whenever the data is a "mark." In the receiver, the received signal is despread by remodulating with a replica of the spreading code. The excess phase of the despreading mixer output is found by subtracting the excess phase of the reference from the excess phase of the received signal. The result of this operation is illustrated in Figure 2-15e for transmitter 1. In this case, the BPSK data-modulated carrier has been recovered exactly.

In transmitter 2 of Figure 2-14b, the data are added modulo-2 to the spreading code prior to MSK modulation. The transmitted signal no longer has abrupt $\pi/2$

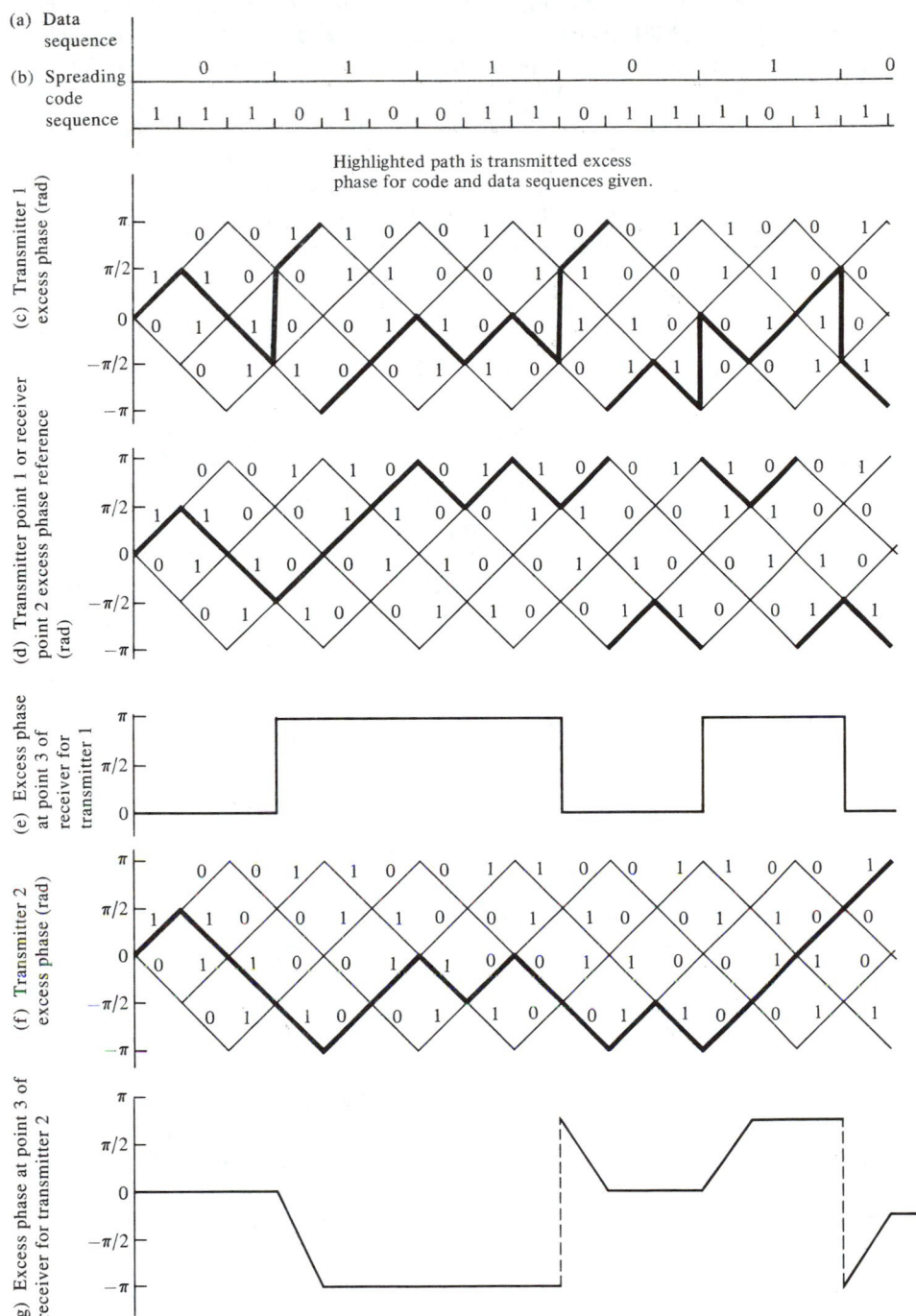

FIGURE 2-15. MSK spread-spectrum phase relationships.

phase transitions as illustrated in Figure 2-15f, and the transmitted power spectrum is precisely that of a MSK modulated carrier at the spreading code rate. The excess phase of the despreading mixer output in the receiver in this case is shown in Figure 2-15g, where it is seen that the BPSK data-modulated carrier is only approximately recovered. The data phase transitions have been slowed by the transmitter conversion filter. In this slowing down of the data phase transitions, the direction that the phasor takes between zero and $\pm\pi$ also becomes apparent. Although modem 2 recovers the data modulation only approximately, for high processing gains the approximation is very good and this circuit is extremely practical.

2-4 Frequency-Hop Spread Spectrum

A second method for widening the spectrum of a data-modulated carrier is to change the frequency of the carrier periodically. Typically, each carrier frequency is chosen from a set of 2^k frequencies which are spaced approximately the width of the data modulation bandwidth apart. The spreading code in this case does not directly modulate the data-modulated carrier but is instead used to control the sequence of carrier frequencies. Because the transmitted signal appears as a data-modulated carrier which is hopping from one frequency to the next, this type of spread spectrum is called *frequency-hop* (FH) *spread spectrum*. In the receiver, the frequency hopping is removed by mixing (down-converting) with a local oscillator signal which is hopping synchronously with the received signal.

2-4.1 Coherent Slow-Frequency-Hop Spread Spectrum

Although in most cases the frequency hopping is done noncoherently, a fully coherent frequency-hop system is theoretically possible and is of interest. Consider, for example, the FH system shown in Figure 2-16. The frequency synthesizer output is a sequence of tones of duration T_c, so $h_T(t)$ can be written

$$h_T(t) = \sum_{n=-\infty}^{\infty} 2p(t - nT_c) \cos(\omega_n t + \phi_n) \tag{2-33}$$

where $p(t)$ is a unit amplitude pulse of duration T_c starting at time zero, and ω_n and ϕ_n are the radian frequency and phase during the nth frequency-hop interval. The frequency ω_n is taken from a set of 2^k frequencies. In contrast to the DS system, where the spreading code was used one bit at a time, the spreading code here is used k bits at a time. The transmitted signal is the data-modulated carrier up-converted to a new frequency $(\omega_0 + \omega_n)$ for each FH chip,

$$s_t(t) = \left[s_d(t) \sum_{n=-\infty}^{+\infty} 2p(t - nT_c) \cos(\omega_n t + \phi_n) \right]_{\substack{\text{sum freq.}\\\text{components}}} \tag{2-34}$$

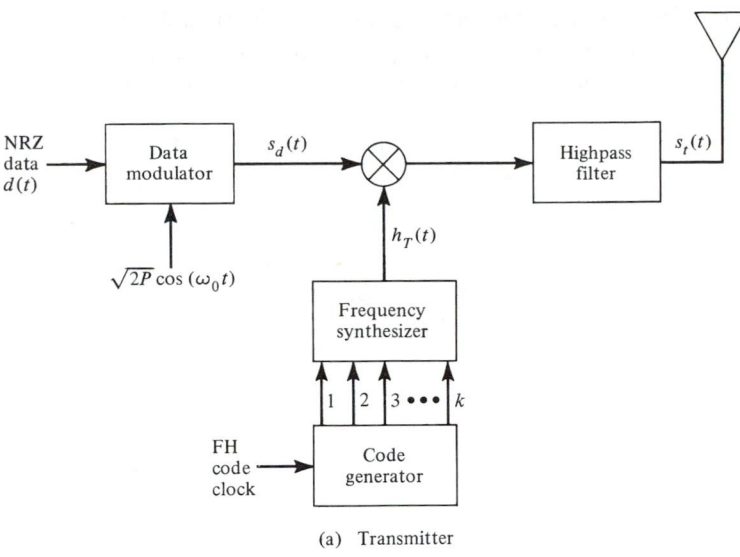

(a) Transmitter

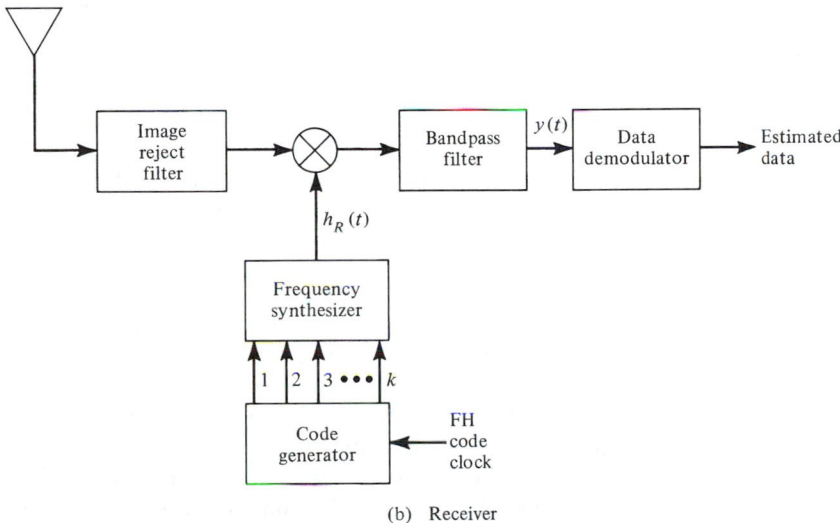

(b) Receiver

FIGURE 2-16. Coherent frequency-hop spread-spectrum modem.

Calculation of the transmitted power spectrum is accomplished using the frequency convolution theorem of Fourier transform theory. Define $S_d(f)$ to be the power spectral density of the data-modulated carrier and $S_h(f)$ to be the power spectral density of the hop carrier $h_T(t)$. These two signals are independent, so that the power spectrum of the transmitted signal is the sum frequency term of the convolution of $S_d(f)$ with $S_h(f)$.

The signal $h_T(t)$ may or may not be periodic. In most cases, if $h_T(t)$ were periodic, its period would be sufficiently long that little error would be made in considering the period infinite. This assumption is made in the following. Thus $h_T(t)$ is considered a purely random sequence of frequencies. For the coherent frequency-hop system being considered, the same phase ϕ_m is used each time $h_T(t)$ returns to frequency ω_m, that is, $\phi_n \in \{\phi_m, m = 1, 2, \ldots, 2^k\}$. With these assumptions, $S_h(f)$ is given by [8,9]

$$S_h(f) = \frac{1}{T_c^2} \sum_{n=-\infty}^{\infty} \left| \sum_{m=1}^{2^k} p_m G_m\left(\frac{n}{T_c}\right) \right|^2 \delta\left(f - \frac{n}{T_c}\right)$$

$$+ \frac{1}{T_c} \sum_{m=1}^{2^k} p_m(1 - p_m)|G_m(f)|^2 \qquad (2\text{-}35)$$

$$- \frac{2}{T_c} \sum_{m=1}^{2^k} \sum_{\substack{m'=1 \\ m'\neq m \\ m'>m}}^{2^k} p_m p_{m'} \text{Re}[G_m(f)G_{m'}^*(f)]$$

where p_m is the probability that frequency m is selected, and $G_m(f)$ is the Fourier transform of $g_m(t)$, where

$$g_m(t) = \begin{cases} 2p(t) \cos(\omega_m t + \phi_m) & 0 \le t \le T_c \\ 0 & \text{elsewhere} \end{cases} \qquad (2\text{-}36)$$

Observe that this psd has discrete components due to the assumption that the same phase is used each time $h_T(t)$ returns to frequency ω_m. The Fourier transform $G_m(f)$ is

$$G_m(f) = T_c \exp\{-j[\pi(f - f_m)T_c - \phi_m]\} \operatorname{sinc}[(f - f_m)T_c]$$

$$+ T_c \exp\{-j[\pi(f + f_m)T_c + \phi_m]\} \operatorname{sinc}[(f + f_m)T_c] \qquad (2\text{-}37)$$

Calculation of $S_h(f)$ can be simplified if the assumption is made that $G_m(f)$ and $G_{m'}(f)$ are nonoverlapping for $m \neq m'$. In this case, $G_m(f)G_{m'}^*(f) = 0$ and the third term and the cross products of the first term of (2-35) vanish. This assumption is very good whenever $1/T_c$ is small with respect to the minimum frequency spacing. Assuming also that all frequencies ω_m are equally likely leads to

$$S_h(f) \cong \frac{1}{(T_c 2^k)^2} \sum_{n=-\infty}^{\infty} \sum_{m=1}^{2^k} \left| G_m\left(\frac{n}{T_c}\right) \right|^2 \delta\left(f - \frac{n}{T_c}\right)$$

$$+ \frac{1}{T_c} \frac{1}{2^k}\left(1 - \frac{1}{2^k}\right) \sum_{m=1}^{2^k} |G_m(f)|^2 \qquad (2\text{-}38)$$

The discrete components of $S_h(f)$ are negligible only when $T_c 2^k \gg 1$, which may or may not be true in systems of interest.

EXAMPLE 2-2

Calculate the transmitted power spectral density for a coherent frequency hop system. The system employs BPSK data modulation with a data rate of 1 Mbps. The hop rate is 100×10^3 hops per second, the frequency spacing equals the data rate, and four frequencies are employed.

Solution: Since the frequency spacing is considerably larger than the hop rate, $G_m(f)$ and $G_{m'}(f)$ are nearly nonoverlapping (orthogonal) and (2-38) applies. With $T_c = 10^{-5}$ and $2^k = 4$ and $G_m(f)$ given by (2-37), $S_h(f)$ is approximately

$$S_h(f) \cong \frac{1}{2^{2k}} \sum_{n=-\infty}^{\infty} \sum_{m=1}^{2^k} \{\text{sinc}^2[(n - f_m T_c)] + \text{sinc}^2[(n + f_m T_c)]\} \, \delta\left(f - \frac{n}{T_c}\right)$$

$$+ \frac{T_c}{2^k}\left(1 - \frac{1}{2^k}\right) \sum_{m=1}^{2^k} \{\text{sinc}^2[(f - f_m)T_c] + \text{sinc}^2[(f + f_m)T_c]\}$$

The psd of the BPSK data-modulated carrier is given by (2-13).

Since the frequency spacing equals 1.0 MHz and the hop rate is 100×10^3, $f_m T_c$ is an integer so that the sinc function of the first term of $S_h(f)$ is sampled only at integer values and

$$S_h(f) \cong \frac{1}{2^{2k}} \sum_{m=1}^{2^k} [\delta(f - f_m) + \delta(f + f_m)]$$

$$+ \frac{T_c}{2^k}\left(1 - \frac{1}{2^k}\right) \sum_{m=1}^{2^k} \{\text{sinc}^2[(f - f_m)T_c] + \text{sinc}^2[(f + f_m)T_c]\}$$

The convolution of $S_d(f)$ and $S_h(f)$ is the desired result. This convolution may be simplified by observing that one of the sinc functions within the convolution integral varies much more slowly than the other. Thus an approximate result is obtained by considering one sinc function a constant. After some manipulations the final approximate result is

$$S_t(f) \cong \frac{PT}{2 \cdot 2^{2k}} \sum_{m=1}^{2^k} \{\text{sinc}^2[(f - f_m - f_0)T] + \text{sinc}^2[(f + f_m + f_0)T]\}$$

$$+ \left(1 - \frac{1}{2^k}\right)\frac{PT}{2 \cdot 2^k} \sum_{m=1}^{2^k} \{\text{sinc}^2[(f - f_m - f_0)T]$$

$$+ \text{sinc}^2[(f + f_m + f_0)T]\}$$

$$= \frac{1}{2} PT \frac{1}{2^k} \sum_{m=1}^{2^k} \{\text{sinc}^2[(f - f_m - f_0)T] + \text{sinc}^2[(f + f_m + f_0)T]\}$$

which is plotted in Figure 2-17.

> This final expression for the transmitted power spectral density is usually written down by inspection since it is simply the sum of the data-modulated carrier psd translated to all hop frequencies and weighted by the probability of transmitting that frequency. The rather lengthy derivation of this result here is intended to provide an understanding of the approximations used in arriving at this result.

The result of (2-35) is based on the fact that the same phase is being used each time the synthesizer of Figure 2-16 returns to frequency ω_m. If the frequency synthesizer phase is random for each successive time interval, the power spectral density of $h_T(t)$ is

$$S_h(f) = \frac{T_c}{2^k} \sum_{m=1}^{2^k} \{\text{sinc}^2[(f - f_m)T_c] + \text{sinc}^2[(f + f_m)T_c]\} \qquad (2\text{-}39)$$

which is derived in Appendix C. When this result is convolved with $S_d(f)$, a transmitted power spectral density identical to that calculated in Example 2-2 is obtained.

Returning now to the problem of transmitting data with a coherent FH spread-spectrum modem, the received signal $s_t(t - T_d)$ for the transmitter of Figure 2-16 is

$$s_t(t - T_d) = \sqrt{2P} \sum_{n=-\infty}^{\infty} p(t - T_d - nT_c)$$

$$\times \cos[(\omega_0 + \omega_n)t + \phi_n + \theta_d(t - T_d) - (\omega_0 + \omega_n)T_d] \qquad (2\text{-}40)$$

In the receiver, this signal is down-converted using a locally generated reference

$$h_R(t) = 2 \sum_{n=-\infty}^{\infty} p(t - \hat{T}_d - nT_c) \cos[\omega_n t + \phi_n - \omega_n \hat{T}_d] \qquad (2\text{-}41)$$

After bandpass filtering to recover the difference frequency component of the down-conversion mixer, the received signal $y(t)$ is, assuming that $\hat{T}_d = T_d$,

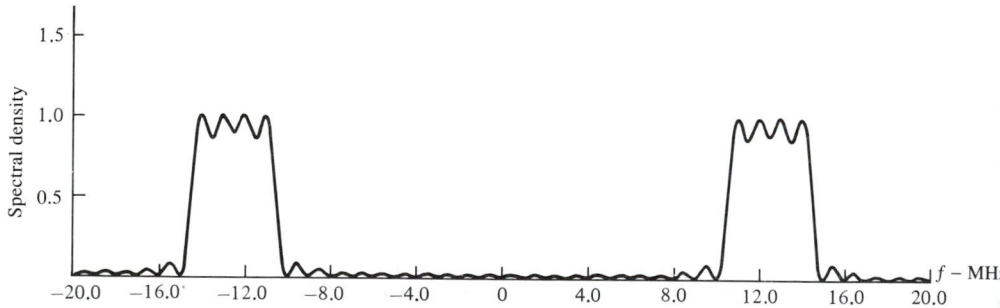

FIGURE 2-17. Transmitted power spectrum for a frequency-hop system with $T = 1.0\ \mu s$ and $f_1 = 11$, $f_2 = 12$, $f_3 = 13$, $f_4 = 14$ MHz.

$$y(t) = [s_t(t - T_d)h_R(t)]_{\text{LP}}$$

$$= \sqrt{2P} \sum_{n=-\infty}^{\infty} p(t - T_d - nT_c) \cos[\omega_0 t - \omega_0 T_d + \theta_d(t - T_d)] \quad (2\text{-}42)$$

$$= \sqrt{2P} \cos[\omega_0 t - \omega_0 T_d + \theta_d(t - T_d)]$$

and the data-modulated carrier has been recovered. If a tracking error exists, $\hat{T}_d \neq T_d$, the recovered carrier is phase modulated by terms having the form $\Sigma_n (T_d - \hat{T}_d)\omega_n$. This can be a serious problem in a fully coherent FH system unless a means is provided for coherent carrier tracking which is independent of the FH code tracking loop. One widely used method for coherent slow frequency hop is to estimate the received carrier phase for each frequency-hop dwell. Feedforward carrier phase estimation is typically used. A known synchronization word is prepended to the data for each frequency hop. The receiver estimates the received carrier phase using the synchronization word. For certain applications it may be necessary to change the synchronization word for each hop so that the synchronization word may not be exploited by a hostile jammer or interceptor.

2-4.2 Noncoherent Slow-Frequency-Hop Spread Spectrum

Because of the difficulty of building truly coherent frequency synthesizers as well as the code tracking requirements alluded to above, many frequency-hop spread-spectrum systems use either noncoherent or differentially coherent data modulation schemes. The modem block diagram is unchanged from that illustrated in Figure 2-16. In the receiver, however, no effort is made to precisely recover the phase of the data-modulated carrier since it is not required by the demodulator.

A common data modulation for FH systems is M-ary frequency shift keying. Suppose, for example, that the data modulator outputs one of 2^L tones each LT seconds, where T is the duration of one information bit. Usually, these tones are spaced far enough apart so that the transmitted signals are orthogonal. This implies that the data modulator frequency spacing is at least $1/LT$ and that the data modulator output spectral width is approximately $2^L/LT$. Each T_c seconds, the data modulator output is translated to a new frequency by the frequency-hop modulator. When $T_c \geq LT$ the FH system is called a *slow-frequency-hop* system. The output of this spread-spectrum modulator is illustrated in Figure 2-18. In this figure the "instantaneous" transmitted spectrum is shown as a function of time for a system with $L = 2$ and $k = 3$. Two data bits are collected each $2T = T_s$ seconds and one of four frequencies is generated by the data modulator. This frequency is translated to one of $2^k = 8$ frequency-hop bands by the FH modulator. In this example, a new frequency-hop band is selected after each group of 2 symbols or 4 bits is transmitted.

In the receiver, the transmitted signal is down-converted using a local oscillator which outputs the sequence of frequencies 0, $5W_d$, $6W_d$, $2W_d$, $7W_d$, ... and the output of the down-converter is a sequence of tones in the first (lowest) FH band representing the data. The down-converter output is illustrated in Figure 2-18b. This signal can be demodulated using the conventional methods for noncoherent MFSK (i.e., a bank of bandpass filters with energy detectors at their outputs).

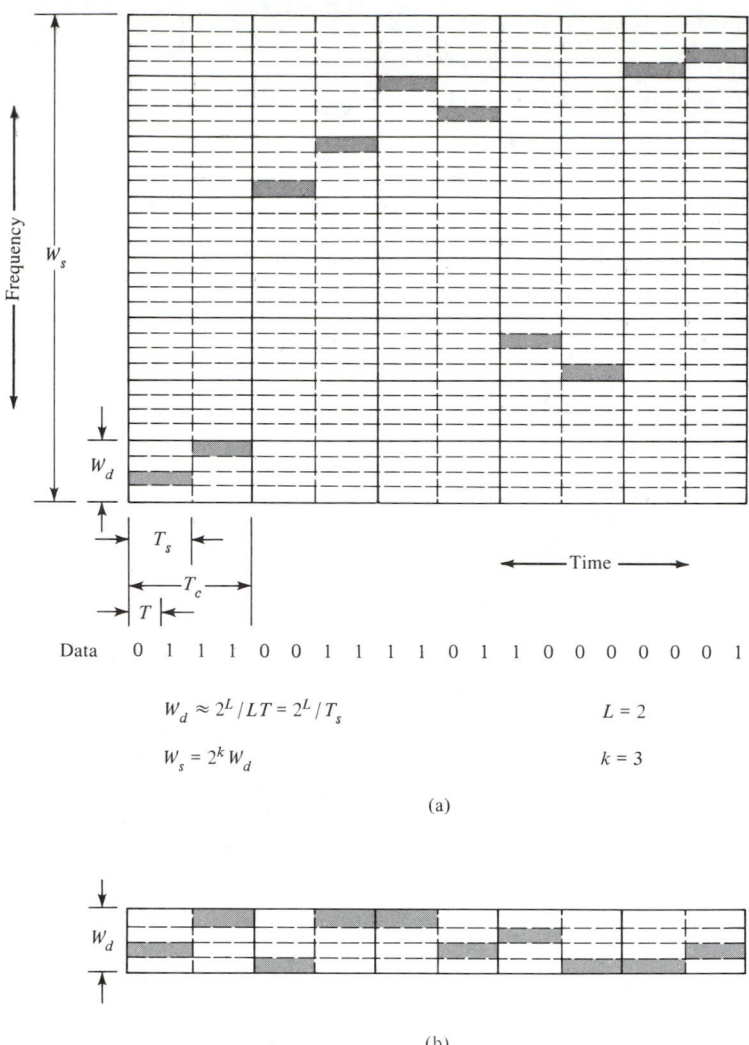

$$W_d \approx 2^L / LT = 2^L / T_s \qquad\qquad L = 2$$

$$W_s = 2^k W_d \qquad\qquad k = 3$$

(a)

(b)

FIGURE 2-18. Pictorial representation of (a) transmitted signal for an *M*-ary FSK slow-frequency-hop spread-spectrum system; (b) receiver down-converter output.

A very preliminary estimate of the processing gain of the FH system just de-scribed can be obtained by considering a noise jammer. In the absence of frequency hopping, the jammer chooses a bandwidth W_d centered on the proper carrier fre-quency and forces the receiver operating signal-to-noise ratio to $E_b/N_J = E_b W_d/J$, where J is the average jammer power. When frequency hopping is added, the jam-mer must place noise in all 2^k frequency-hop bands in order to cause the receiver to have the same performance as before. Thus the jammer requires a total power 2^k times as large as before and the processing gain is $2^k = W_s/W_d$.

2-4.3 Noncoherent Fast-Frequency-Hop Spread Spectrum

In contrast to the slow-FH system, where the hop-frequency band changes more slowly than symbols come out of the data modulator, the hop-frequency band can change many times per symbol in a *fast-frequency-hop* system. A significant benefit achieved when fast frequency hop is used is that frequency diversity gain is seen on each transmitted symbol. This is particularly beneficial in a partial-band jamming or when the transmission channel causes rapid signal fading as in microwave mobile telephony applications.

A representation of the transmitted signal for a fast-frequency-hop system is illustrated in Figure 2-19. The output of the MFSK modulator is one of 2^L tones as before, but now this tone is subdivided into K chips. After each chip, the MFSK modulator output is hopped to a different frequency. Since the chip duration T_c is shorter than the data modulator output symbol duration T_s, the minimum tone spacing for orthogonal signals is now $1/T_c = K/LT$. The receiver frequency-dehopping operation functions in exactly the same way as before. The output of the down-conversion operation is shown in Figure 2-19b.

The data demodulator can operate in several different modes in a fast-frequency-hop system. One mode is to make a hard decision on each frequency-hop chip as it is received and to make an estimate of the data modulator output based on all K chip decisions. The decision rule could be a simple majority vote. Another mode would be to calculate the likelihood of each data modulator output symbol as a function of the total signal received over K chips and to choose the largest. A receiver which calculates the likelihood that each symbol was transmitted is optimum in the sense that minimum error probability is achieved for a given E_b/N_0. Each of these possible operating modes performs differently and has different complexity. The spread-spectrum system designer must choose the mode of operation that best solves the particular problem. It will be shown later that fast frequency hop is a very useful technique in either a fading-signal environment or in a partial band-jamming environment, and its use with error correction coding is particularly convenient.

2-5 Hybrid Direct-Sequence/Frequency-Hop Spread Spectrum

A third method for spectrum spreading is to employ both direct-sequence and frequency-hop spreading techniques in a hybrid direct-sequence/frequency-hop system. One reason for using hybrid techniques is that some of the advantages of both types of systems are combined in a single system. Hybrid techniques are widely used in military spread-spectrum systems. Many methods of combining DS and FH spreading are possible. The method discussed here was selected because of its simplicity as an example of a hybrid system.

Figure 2-20 illustrates a hybrid DS/FH spread-spectrum modem which employs differential binary PSK data modulation [6]. Because noncoherent frequency hopping is used, the data modulation must be either noncoherent or differentially

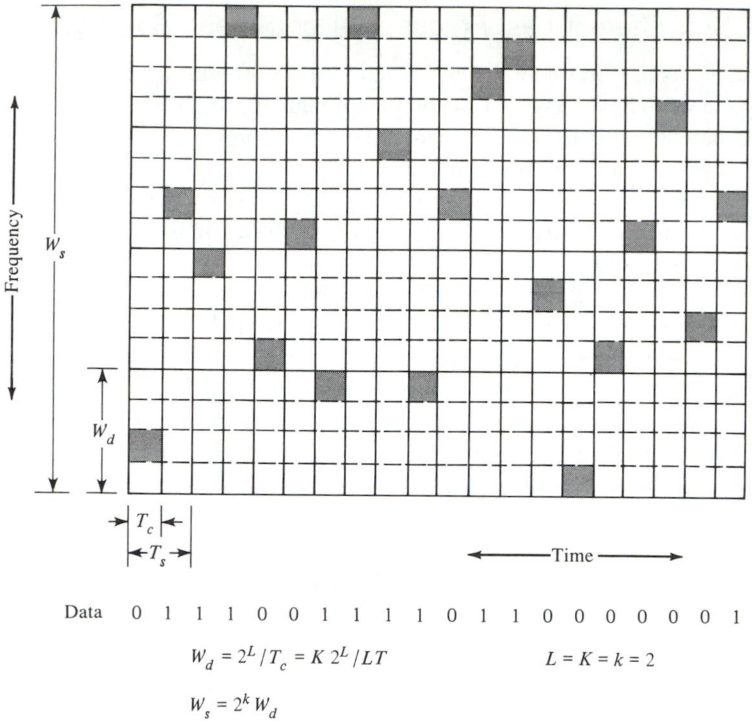

Data   0 1 1 1 0 0 1 1 1 1 0 1 1 0 0 0 0 0 0 1

$$W_d = 2^L/T_c = K\,2^L/LT \qquad\qquad L = K = k = 2$$

$$W_s = 2^k W_d$$

(a)

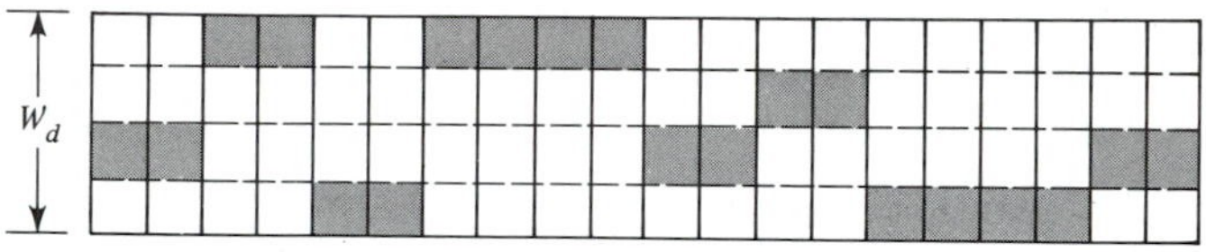

(b)

FIGURE 2-19. Pictorial representation of (a) transmitted signal for an *M*-ary FSK fast-frequency-hop spread-spectrum system and (b) receiver down-converter output.

coherent. DPSK modulation requires a differential data encoding prior to carrier modulation as shown. With this encoding the sampled output of the differential demodulator is the original data sequence. In Figure 2-20 the DPSK modulated carrier is first direct-sequence spread by multiplication with the DS spreading waveform $c(t)$, and then frequency hopped through up-conversion using the sequence of FH tones $h_T(t)$. The signal $h_T(t)$ is defined by (2-33). The power spectral density of the transmitted signal $s_t(t)$ is

$$S_t(f) = [S_{ds}(f) * S_h(f)]_{\substack{\text{sum freq.}\\ \text{terms}}} \qquad (2\text{-}43)$$

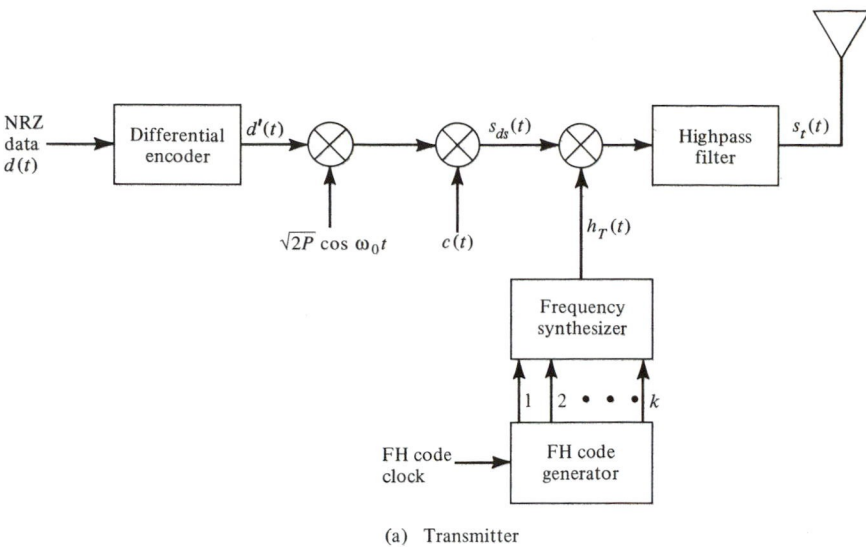

(a) Transmitter

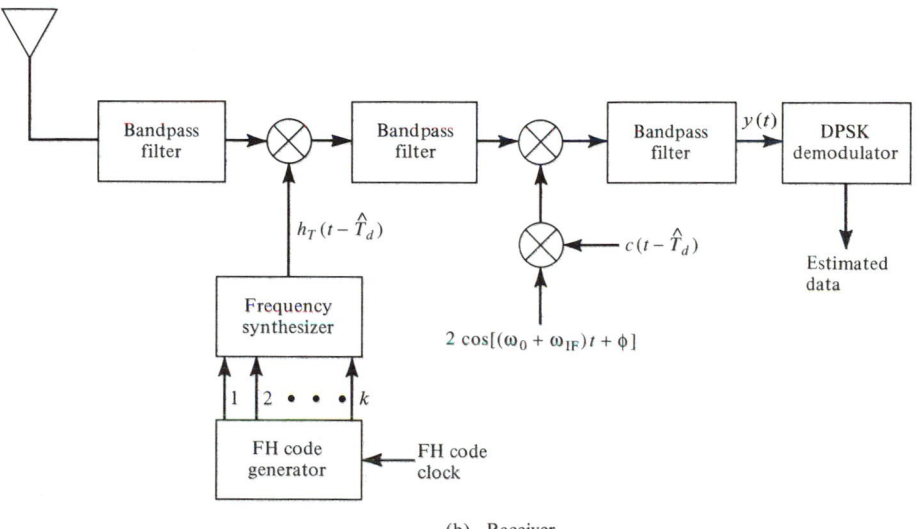

(b) Receiver

FIGURE 2-20. Hybrid direct-sequence/frequency-hop spread-spectrum modem.

where $S_h(f)$ is given by (⌐ 39) and $S_{ds}(f)$ is approximately (see Example 2-1)

$$S_{ds}(f) \approx \tfrac{1}{2} PT_c\{\text{sinc}^2[(f - f_0)T_c] + \text{sinc}^2[(f + f_0)T_c]\} \qquad (2\text{-}44)$$

where T_c is the DS spreading code chip duration. If the FH rate is slow relative to the DS bandwidth, the density $S_h(f)$ may be approximated by a sum of delta functions and (2-43) becomes

$$S_t(f) \cong \left[\int_{-\infty}^{\infty} \frac{1}{2} PT_c \{ \mathrm{sinc}^2[(f' - f_0)T_c] + \mathrm{sinc}^2[(f' + f_0)T_c] \} \right.$$

$$\left. \times \frac{1}{2^k} \sum_{m=1}^{2^k} \{ \delta(f - f' - f_m) + \delta(f - f' + f_m) \} \, df' \right]_{\substack{\text{sum freq.} \\ \text{terms}}} \qquad (2\text{-}45)$$

$$= \frac{PT_c}{2^{k+1}} \sum_{m=1}^{2^k} \{ \mathrm{sinc}^2[(f - f_m - f_0)T_c] + \mathrm{sinc}^2[(f + f_m + f_0)T_c] \}$$

The receiver recovers the DPSK modulated carrier by first frequency dehopping the received signal and then DS despreading the signal. Both of these operations have been explained in detail above, so that it can be easily demonstrated that the recovered data-modulated carrier is

$$y(t) = \sqrt{2P}\, d'(t - T_d) \cos[\omega_{\mathrm{IF}} t + \phi(t)] \qquad (2\text{-}46)$$

where $\phi(t)$ is a function that accounts for the random phase changes of the receiver FH synthesizer at each hop frequency. Of course, both the DS and the FH code sequences must be correctly phased for despreading to occur.

2-6 Complex-Envelope Representation of Spread-Spectrum Systems

All of the spread-spectrum systems discussed above can be conveniently represented mathematically using complex-envelope notation [17]. This common notation is useful not only as an aid in understanding the spreading/despreading process but also as an analytical and simulation tool. Any bandpass signal $v(t)$ whose Fourier spectrum is centered at frequency ω_0 can be expressed as

$$v(t) = \mathrm{Re}[\tilde{v}(t)e^{j\omega_0 t}] \qquad (2\text{-}47)$$

where $\tilde{v}(t)$ is the complex envelope of $v(t)$ and is a complex function of time, and $\mathrm{Re}\,(\cdot)$ is the real part of the argument. All the signal processing steps required in spread-spectrum systems (i.e., linear filtering and mixing) can be mathematically modeled as operations on the complex envelope of the signal of interest.

A generic complex envelope model of a spread-spectrum modem is illustrated in Figure 2-21. In this figure, all signals are complex functions of time, double lines present real and imaginary (in-phase and quadrature) paths, and the mixers perform complex multiplications. The actual (real) transmitted signal is $s(t) = \mathrm{Re}[\tilde{s}(t)e^{j\omega_0 t}]$, where power amplification has been ignored, and $\tilde{s}(t) = \tilde{d}(t)\tilde{c}(t)$. In writing this last expression, it has been assumed that the transmitter mixing operation is also part of an up-conversion process, so that $\omega_0 = \omega_1 + \omega_2$ where ω_1 and ω_2 are the actual center frequencies at the data modulator and spreading code generator outputs.

The receiver input $\tilde{r}(t)$ is the delayed transmitter output $\tilde{s}(t - T_d)$ plus interfer-

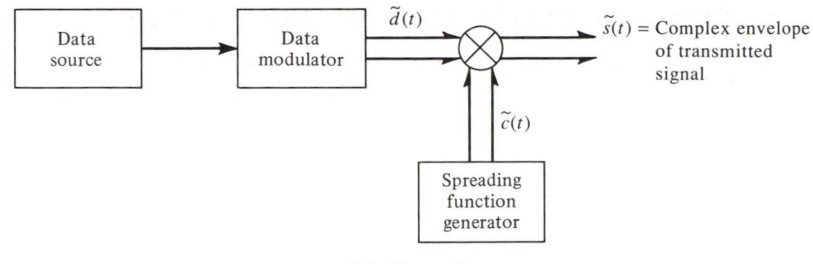

(a) Transmitter

(b) Receiver

FIGURE 2-21.   Generic complex-envelope model of spread-spectrum modem. (From Ref. 17.)

ence $\tilde{u}(t)$ plus thermal noise $\tilde{n}(t)$. In the receiver, the mixing operations are assumed to be part of the receiver down-conversion chain. Since the difference frequency components of the mixer outputs are required, the complex conjugate of the reference signal envelopes are used [17] as the mixer reference inputs. The first receiver mixing operation accounts for all frequency and phase differences between the received carrier (including Doppler effects) and the local reference carrier. If the system is coherent, a carrier tracking loop will force $\hat{\omega}_0 = \omega_0$ and $\hat{\phi} = \phi$ so that this complex-envelope multiplication has no effect. The second receiver mixing operation is the spread-spectrum despreading operation. In general, the input to the data demodulator is

$$\hat{\tilde{d}}(t - T_d) = \tilde{d}(t - T_d)\tilde{c}(t - T_d)\tilde{c}^*(t - \hat{T}_d)\exp\{-j[(\omega_0 - \hat{\omega}_0)t + \phi - \hat{\phi}]\}$$
$$+ \tilde{u}(t)\tilde{c}^*(t - \hat{T}_d)\exp\{-j[(\omega_0 - \hat{\omega}_0)t + \phi - \hat{\phi}]\} \qquad (2\text{-}48)$$
$$+ \tilde{n}(t)\tilde{c}^*(t - \hat{T}_d)\exp\{-j[(\omega_0 - \hat{\omega}_0)t + \phi - \hat{\phi}]\}$$

where the input bandpass filter has been assumed to have a bandwidth sufficiently wide to pass all signals without distortion.

This representation is completely general and, with proper selection of $\tilde{d}(t)$ and $\tilde{c}(t)$, can be used for any direct-sequence or frequency-hop spread-spectrum system with any type of data modulation. Table 2-1 gives the complex envelope for a

TABLE 2-1. Complex Envelope of Common Digital Modulation Types

Modulation Type	Complex Envelope — $\tilde{v}(t)$
Binary phase-shift keying (BPSK)	$\tilde{v}(t) = \sum_n a_n p_T(t - nT)$ $a_n \in \{+1, -1\}$
Quaternary phase-shift keying (QPSK)	$\tilde{v}(t) = \sum_n p_{2T}(t - 2nT) \exp(j\beta_n)$ $\beta_n \in \left\{0, \dfrac{\pi}{2}, \pi, \dfrac{3\pi}{2}\right\}$
Offset quaternary phase-shift keying (OQPSK)	$\tilde{v}(t) = \sum_n p_T(t - nT) \exp(j\beta_n)$ $\beta_n = \beta_{n-1} + a_n \dfrac{\pi}{2}$ $a_n \in \{+1, 0, -1\}$
M-ary phase-shift keying (MPSK)	$\tilde{v}(t) = \sum_n p_{mT}(t - nmT) \exp(j\beta_n)$ $m = \log_2 M$ $\beta_n \in \{\beta_1, \beta_2, \ldots, \beta_m\}$
Binary frequency-shift keying (FSK)	$\tilde{v}(t) = \sum_n p_T(t - nT) \exp[j(\omega_n t + \phi_n)]$ $\omega_n \in \{\omega_1, \omega_2\}$ $0 \le \phi_n < 2\pi$
Minimum-shift keying (MSK)	$\tilde{v}(t) = \sum_n p_T(t - nT) \exp\left[j\left(a_n \dfrac{\pi t}{2T} + x_n\right)\right]$ $a_n = \in \{+1, -1\}$ $x_n = x_{n-1} + (a_{n-1} - a_n)\dfrac{n\pi}{2}$
M-ary frequency-shift keying (MFSK)	$\tilde{v}(t) = \sum_n p_T(t - nT) \exp[j(\omega_n t - \phi_n)]$ $\omega_n \in \{\omega_1, \omega_2, \ldots, \omega_m\}$ $0 \le \phi_n < 2\pi$

number of the most common digital modulation types which can be used for either the data modulation or the spreading modulation. In this table, $p_T(t)$ and $p_{2T}(t)$ are unit pulses of duration T and $2T$ seconds, respectively. The variable T is the information bit duration when these envelopes are used for data modulation and the spreading code chip duration T_c when they are used for the spreading modulation.

EXAMPLE 2-3

Calculate the complex envelope of the data demodulator input for a fully coherent spread-spectrum modem that uses BPSK data modulation and MSK spreading modulation when the only interference is a single-tone jammer which is not at the system carrier frequency.

Solution: The complex envelope of the data modulation is

$$\tilde{d}(t) = \sum_n d_n p_T(t - nT)$$

and the complex envelope of the spreading modulation is

$$\tilde{c}(t) = \sum_m p_{T_c}(t - mT_c) \exp\left[j\left(c_m \frac{\pi t}{2T_c} + x_m \right) \right]$$

so that the transmitted signal is

$$\tilde{s}(t) = \sqrt{2P}\, \tilde{d}(t)\tilde{c}(t) = \sqrt{2P} \sum_n \sum_m d_n p_T(t - nT) p_{T_c}(t - mT_c)$$

$$\times \exp\left[j\left(c_m \frac{\pi t}{2T_c} + x_m \right) \right]$$

The transmitted signal power is P. The envelope of the jammer is

$$\tilde{u}(t) = \sqrt{2J}\, \exp(j\, \Delta\omega\, t)$$

where the offset frequency is $\Delta\omega$ and the jammer power is J.

Referring to Figure 2-21, and assuming perfect carrier tracking, the complex envelope at the input to the data demodulator is

$$\hat{\tilde{d}}(t - T_d) = \tilde{s}(t - T_d)\tilde{c}^*(t - \hat{T}_d)$$

$$+ \sqrt{2J}\, \tilde{c}^*(t - \hat{T}_d) \exp(j\, \Delta\omega t)$$

$$= \sqrt{2P} \sum_n d_n p_T(t - T_d - nT)$$

$$\times \sum_{n'} \sum_m p_{T_c}(t - T_d - n'T_c) p_{T_c}(t - \hat{T}_d - mT_c)$$

$$\times \exp\left\{ j\left[c_{n'} \frac{\pi}{2T_c}(t - T_d) + x_{n'} \right] - j\left[c_m \frac{\pi}{2T_c}(t - \hat{T}_d) + x_m \right] \right\}$$

$$+ \sqrt{2J} \sum_{m'} p_{T_c}(t - \hat{T}_d - m'T_c) \exp\left\{ -j\left[c_{m'} \frac{\pi}{2T_c}(t - \hat{T}_d) \right] + j\, \Delta\omega\, t \right\}$$

If perfect spreading code tracking is also assumed ($\hat{T}_d = T_d$), then

$$\hat{\tilde{d}}(t - T_d) = \sqrt{2P} \sum_n d_n p_T(t - T_d - nT)$$

$$+ \sqrt{2J} \sum_{m'} p_{T_c}(t - \hat{T}_d - m'T_c) \exp\left\{-j\left[c_{m'}\frac{\pi}{2T_c}(t - \hat{T}_d)\right] + j\,\Delta\omega\,t\right\}$$

and it is seen that the MSK spreading has been entirely removed from the desired signal and that the jamming tone has been MSK modulated via the despreading operation.

EXAMPLE 2-4

Calculate the complex envelope of the data demodulator input for a slow-frequency-hop spread-spectrum modem that uses differential binary PSK data modulation.

Solution: The envelope of the data modulation is the same as in Example 2-3 except that the differentially encoded data sequence is used. The envelope of the spreading modulation is

$$\tilde{c}(t) = \sum_m p_{T_c}(t - mT_c)\exp[j(\omega_m t + \phi_m)]$$

Assume that the receiver is able to perfectly track the frequency of the received signal using an AFC loop, but that no attempt is made to estimate the random phase changes associated with each frequency hop. Thus

$$\hat{\tilde{d}}(t - T_d) = \tilde{d}(t - T_d)\tilde{c}(t - T_d)\tilde{c}^*(t - \hat{T}_d)\exp[j(\phi - \hat{\phi})]$$

$$= \sqrt{2P} \sum_n d_n p_T(t - T_d - nT) \sum_m \sum_{m'} p_{T_c}(t - T_d - mT_c)$$

$$\times\, p_{T_c}(t - \hat{T}_d - m'T_c)\exp\{+j[\omega_m(t - T_d) + \phi_m]$$

$$-\, j[\omega_{m'}(t - \hat{T}_d) + \phi_{m'}]\}$$

If perfect code tracking is assumed ($\hat{T}_d = T_d$),

$$\hat{\tilde{d}}(t - T_d) = \sqrt{2P} \sum_n d_n p_T(t - T_d - nT) \sum_m \exp(j\theta_m)$$

where $\theta_m = \phi_m - \phi_{m'}$ is the difference between the transmitter and receiver frequency-hop phases.

2-7 Summary

This chapter introduced spread-spectrum system concepts. Spread-spectrum systems have many benefits as pointed out in the prologue. This chapter began by illustrating two of these—resistance to jamming and low probability of intercept. Both of these characteristics of spread-spectrum systems will be expanded upon in future chapters. The first concept was illustrated by Figure 2-1, which showed that a pulsed jammer could inflict considerable degradation on a spread-spectrum system. However, it will be shown in Chapter 7 that the proper application of coding will negate the detrimental effects of pulse jammers. The second concept was illustrated by (2-6) and (2-7). Cast in slightly different form, (2-6) becomes

$$P_d = \Phi \left[\frac{E_s}{N_0} \frac{1}{\sqrt{TW}} - \Phi^{-1}(1 - P_{fa}) \right]$$

where E_s is the total energy integrated. For fixed E_s/N_0, the probability of detection decreases with increasing TW because $\Phi(u)$ is an increasing function with u. Thus, the larger the bandwidth, for fixed observation time T, the lower the detectability of the signal, i.e., spread-spectrum signals are low probability of intercept relative to nonspread signals.

The different types of spread-spectrum methods introduced in this chapter were BPSK/DSSS (direct sequence spread spectrum), QPSK/DSSS, MSK/DSSS, coherent slow frequency-hop spread spectrum (FHSS), noncoherent slow-frequency-hop spread spectrum, noncoherent fast-frequency-hop spread spectrum, and hybrid DSSS/FHSS.

The chapter closed with a discussion of the representation of spread-spectrum systems in terms of complex envelopes.

References

[1] J. M. WOZENCRAFT and I. M. JACOBS, *Principles of Communication Engineering* (New York: Wiley, 1965).

[2] R. A. DILLARD, "Detectability of Spread-Spectrum Signals," *IEEE Trans. Aerosp. Electron. Syst.,* Vol. AES-15, pp. 526–537, July 1979.

[3] R. A. DILLARD and G. M. DILLARD, *Detectability of Spread-Spectrum Signals* (Norwood, Mass.: Artech House, 1989).

[4] M. K. SIMON, J. K. OMURA, R. A. SCHOLTZ, and B. K. LEVITT, *Spread Spectrum Communications Handbook* (New York: McGraw-Hill, 1994).

[5] D. L. NICHOLSON, *Spread Spectrum Signal Design: LPE and AJ Systems* (Rockville, Md.: Computer Science Press, 1988).

[6] R. E. ZIEMER and W. H. TRANTER, *Principles of Communications: Systems, Modulation, and Noise,* 4th ed. (Boston: Houghton Mifflin, 1995).

[7] J. K. HOLMES, *Coherent Spread Spectrum Systems* (New York: Wiley-Interscience, 1982).

[8] R. C. TITSWORTH and L. R. WELCH, *Power Spectra of Signals Modulated by Random and Pseudorandom Sequences,* Tech. Rep. 32–140, Jet Propulsion Laboratory, Pasadena, Calif., October 1961.

[9] W. C. LINDSEY and M. K. SIMON, *Telecommunication Systems Engineering* (Englewood Cliffs, N.J.: Prentice Hall, 1973).

[10] A. PAPOULIS, *Probability, Random Variables, and Stochastic Processes,* 3rd ed. (New York: McGraw-Hill, 1991).

[11] A. PAPOULIS, *The Fourier Integral and Its Applications* (New York: McGraw-Hill, 1962).

[12] B. K. LEVITT, "Effect of Modulation Format and Jamming Spectrum on Performance of Direct Sequence Spread Spectrum Systems," *Conf. Rec.,* IEEE Natl. Telecommun. Conf., 1980.

[13] F. AMOROSO and J. A. KIVETT, "Simplified MSK Signaling Technique," *IEEE Trans. Commun.,* Vol. COM-25, pp. 433–441, April 1977.

[14] S. PASUPATHY, "Minimum Shift Keying: A Spectrally Efficient Modulation," *IEEE Commun. Mag.,* Vol. 17, pp. 14–22, July 1979.

[15] R. E. ZIEMER and C. R. RYAN, "Minimum-Shift Keyed Modem Implementations for High Data Rates," *IEEE Commun. Mag.,* Vol. 21, pp. 28–37, October 1983.

[16] S. STEIN and J. J. JONES, *Modern Communication Principles* (New York: McGraw-Hill, 1967).

[17] R. A. SCHOLTZ, "The Spread Spectrum Concept," *IEEE Trans. Commun.,* Vol. COM-25, pp. 748–755, August 1977.

[18] R. L. HARRIS, "Introduction to Spread Spectrum Techniques," in *Spread Spectrum Communications,* NATO AGARD Lecture Ser. 58, July 6, 1973 (AD 766 914).

Problems

(2-1) Show that $z(t)$ of Figure 2-13b is the recovered data-modulated carrier at IF, that is,

$$z(t) = \sqrt{2P}\, d(t) \cos \omega_{IF} t$$

if perfect code and carrier tracking are assumed.

(2-2) Consider a non-spread-spectrum communication system employing differential binary PSK modulation. Suppose that this system is jammed by a narrowband pulse-noise jammer having total average power J and duty factor ρ.
(a) Find the optimum jammer duty factor ignoring thermal noise as a func-

tion of E_b/N_J and plot bit error probability versus E_b/N_J for the optimized jammer.

(b) Plot bit error probability versus E_b/N_J for nonoptimum jammer duty factors $\rho = 0.25$, 0.50, 0.75, and 1.0.

(2-3) Consider a BPSK direct-sequence spread-spectrum system using differential binary PSK data modulation. Suppose that this system is jammed by a narrowband pulse-noise jammer having a total average power J and duty factor ρ. Assume that the spreading code chip rate is 10 times the data bit rate.

(a) Find the optimum jammer duty factor, ignoring thermal noise as a function of E_b/N_J and plot bit error probability versus E_b/N_J for the optimized jammer.

(b) What is the processing gain of this spread-spectrum system?

(c) Compare the optimum duty factors for Problems 2-2 and 2-3.

(2-4) Consider a BPSK direct-sequence spread-spectrum system with arbitrary data modulation and continuous tone jamming. Assuming that the spreading code period is infinite, plot the processing gain of the system as a function of the jamming tone center frequency relative to the system carrier frequency. Assume that the spreading code rate is 100 times the data rate.

(2-5) Repeat Problem 2-4 assuming MSK direct-sequence modulation with a null-to-null bandwidth equal to the BPSK null-to-null bandwidth used in Problem 2-4. Compare the plots of Problems 2-4 and 2-3.

(2-6) Consider a direct-sequence BPSK spread-spectrum system which obtains its spreading code from the feedback shift register circuit shown below. Calculate the transmitter output power spectral density for arbitrary data modulation.

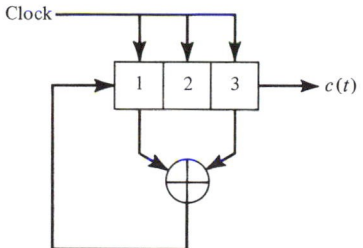

PROBLEM 2-6. Feedback shift register.

(2-7) An elementary random process comprises four sample functions each of which is assigned equal probability.

$$x_1(t) = 1$$

$$x_2(t) = -2$$

$$x_3(t) = \sin \pi t$$

$$x_4(t) = \cos \pi t$$

(a) Is the process stationary?

(b) Calculate $E[x(t)]$ and $E[x(t_1)x(t_2)]$.

(Wozencraft and Jacobs [1, Prob. 3.1])

(2-8) Let $x(t)$ and $y(t)$ be statistically independent, stationary random processes and define $z(t) = x(t)y(t)$. Is $z(t)$ stationary? Show that

$$S_z(f) = S_x(f) * S_y(f)$$

where, as usual, the symbol $*$ denotes covolution. (Wozencraft and Jacobs [1, Prob. 3.5])

(2-9) Consider the random process

$$s(t,\mathbf{a},T) = \sum_{n=-\infty}^{\infty} a_n p(t + T - nT_c)$$

where

$$p(t) = \sin\left(\frac{2\pi}{T_c} t\right) \qquad 0 \le t \le T_c$$

T is a random variable uniformly distributed on $(0,T_c)$, and $\mathbf{a} = (\ldots, a_{-1}, a_0, a_1, \ldots)$ is a doubly infinite sequence of equally likely binary random variables from the set $\{+1, -1\}$.

(a) Calculate the autocorrelation function of $s(t,\mathbf{a},T)$.

(b) Calculate the power spectral density of $s(t,\mathbf{a},T)$.

(2-10) Consider a frequency-hop spread-spectrum system that uses binary FSK data modulation and 64 orthogonal frequency-hop bands. Suppose that the data rate is 1.0 Mbps and that the frequency hop rate is 10^4 hops/s. Assume that the FH synthesizer is noncoherent from hop to hop and assume any convenient carrier frequency.

(a) Calculate the transmitter output power spectral density.

(b) Calculate the optimum number of frequency-hop bands for a partial-band noise jammer to jam assuming that the jammer has constant total power.

(2-11) Consider a frequency-hop spread-spectrum system that uses binary FSK data modulation and 64 orthogonal frequency hop bands. Suppose that the data rate is 10 kbps and the frequency hop rate is 1.0×10^6 hops/s. Assume that the FH synthesizer is noncoherent from hop to hop and assume any convenient carrier frequency. Calculate the transmitter output power spectral density.

(2-12) Consider a direct-sequence BPSK spread-spectrum receiver using arbitrary data modulation having the configuration illustrated below. The received thermal noise two-sided power spectral density is $N_0/2$ W/Hz and the noise bandwidths of the input bandpass filter is B hertz. Assuming that the spreading code chip rate is very large relative to the data rate, calculate the thermal noise spectral density at the data demodulator input for $B = 2/T_c$, $3/T_c$, and $4/T_c$. T_c is the spreading code chip duration.

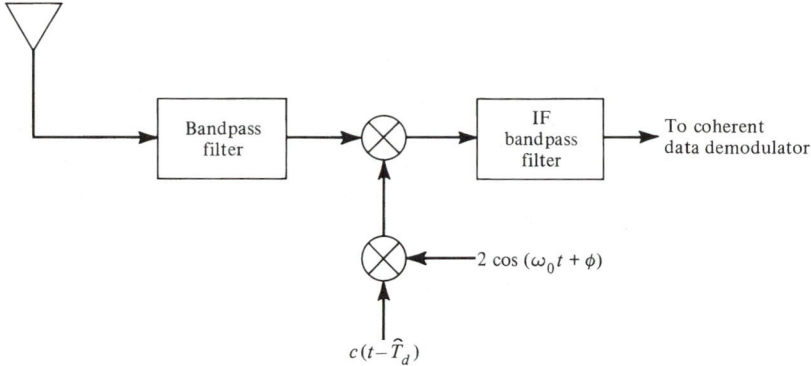

PROBLEM 2-12. Direct-sequence spread-spectrum receiver.

(2-13) Consider the direct-sequence BPSK spread-spectrum transmitter illustrated for this problem. Assume that the two code generators operate synchronously and that each generates an independent sequence of totally random binary (±1) symbols. Calculate the transmitted power spectral density as a function of τ, and plot results for $\tau = 0$, $T_c/4$, and $T_c/2$, where T_c is the code chip duration. (Harris [18])

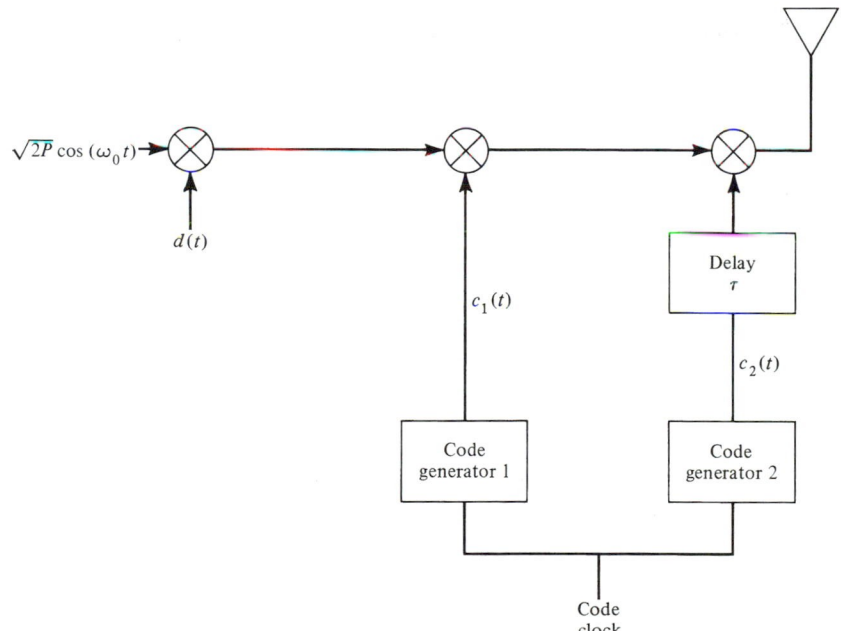

PROBLEM 2-13. Offset code direct-sequence spread-spectrum transmitter.

(2-14) It is known that a filter "matched" to the signaling waveform is optimum for data detection is an AWGN environment [1]. Show that in a BPSK direct-sequence spread-spectrum system, the receiver despreading function is part of that optimum matched filtering operation.

(2-15) Consider a direct-sequence BPSK spread-spectrum receiver that employs coherent BPSK data modulation as shown in the accompanying figure. Assume that a tone jammer is used against this receiver and that the jammer knows the correct carrier phase. Assume an infinite spreading code period, a jammer power J, and a spreading code chip rate that is N times the data rate. Determine the probability density function of the integrator output at the sampling instants. Write an expression for the bit error probability as a function of the jamming and signal noise powers, and N.

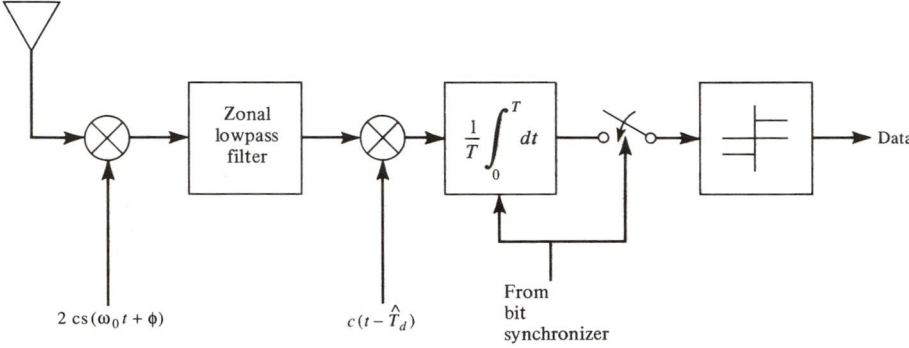

PROBLEM 2-15. Coherent BPSK spread-spectrum receiver.

CHAPTER 3

Binary Shift-Register Sequences for Spread-Spectrum Systems

3-1 Introduction

The waveform $c(t)$ used in the systems described in Chapter 2 to spread and despread the data-modulated carrier is usually generated using a shift register whose contents during each time interval is some linear or nonlinear combination of the contents of the register during the preceding time interval. In this chapter various techniques for generating $c(t)$ are described and analyzed.

For the spread-spectrum system to operate efficiently, the waveform $c(t)$ is selected to have certain desirable properties. For example, the phase of the received spreading code $c(t - T_d)$ must be initially determined and then tracked by the receiver. These functions are facilitated by choosing $c(t)$ to have a two-valued auto-correlation function as exhibited by the maximal-length sequences to be considered in detail later. It is often desirable to employ a $c(t)$ having a very wide bandwidth. This implies that electronically simple spreading code generators which can operate at very high speeds should be considered. When the spread-spectrum system is used for multiple access, sets of waveforms $c_1(t)$, $c_2(t)$, . . . , $c_m(t)$ must be found which have good cross-correlation properties. When jamming resistance is a concern, waveforms are used that have extremely long periods and are difficult for the jammer to generate. The spreading code generators discussed in this chapter have one or more of these properties. The codes discussed are the most commonly used codes in current spread-spectrum systems. Many other codes are possible and are discussed in the references.

In the following, the term *spreading code* is used to refer to the output of the binary shift-register generator and the term *spreading waveform* is reserved for the function $c(t)$, which takes on values ± 1 and is used as the actual input to the spreading or despreading modulator. The ideal spreading code would be an infinite sequence of equally likely random binary digits. Unfortunately, the use of an infinite random sequence implies infinite storage in both the transmitter and receiver. This is clearly not possible, so that the periodic *pseudorandom codes* (*PN codes*) as described in this chapter are always employed. In this book the term "PN code" is used for any periodic spreading code with noise-like properties. Specific PN codes include the maximal-length codes and Gold codes, among others.

The chapter begins with definitions and a review of the mathematics that will be needed to analyze the spreading codes. This is followed by a comprehensive discussion of maximal-length codes, which are by far the most widely used spreading codes. The chapter concludes with a discussion of Gold codes, which are combinations of maximal-length codes used in multiple-access systems.

3-2 Definitions, Mathematical Background, and Sequence Generator Fundamentals

3-2.1 Definitions

All of the spreading codes to be discussed are periodic sequences of ones and zeros with period N. It is convenient to represent a sequence of binary digits $\ldots, b_{-2}, b_{-1}, b_0, b_1, b_2, \ldots$ from the alphabet $\{0,1\}$ by a polynomial $b(D) = \cdots + b_{-2}D^{-2} + b_{-1}D^{-1} + b_0 + b_1D + b_2D^2 + \cdots$. The delay operator D implies simply that the binary symbol which multiplies D^j occurs during the jth time interval of the sequence. Because the code is periodic, $b_n = b_{N+n}$ for any n. The spreading waveform $c(t)$ derived from this spreading code is also periodic with period $T = NT_c$ and is specified by

$$c(t) = \sum_{n=-\infty}^{\infty} a_n p(t - nT_c) \tag{3-1}$$

where $a_n = (-1)^{b_n}$, and $p(t)$ is a unit pulse beginning at 0 and ending at T_c. The waveform $c(t)$ is deterministic, so that its *autocorrelation function* is defined by [1]

$$R_c(\tau) = \frac{1}{T} \int_0^T c(t)c(t + \tau) \, dt \tag{3-2}$$

Since $c(t)$ is periodic with period T, it follows that $R_c(\tau)$ is also periodic with period T. Consider two different spreading waveforms $c(t)$ and $c'(t)$. The *cross-correlation function* of these two deterministic waveforms is

$$R_{cc'}(\tau) = \frac{1}{T} \int_0^T c'(t)c(t + \tau) \, dt \tag{3-3}$$

where it has been assumed that both waveforms have the same period T. The cross-correlation function is also periodic with period T.

The variable τ in (3-2) and (3-3) can assume any value. That is, τ is not constrained to be an integral multiple of T_c. Substituting (3-1) into (3-3) yields

$$R_{cc'}(\tau) = \frac{1}{T} \sum_m \sum_n a_m a_n' \int_0^T p(t - mT_c)p(t + \tau - nT_c) \, dt \tag{3-4}$$

The integral in (3-4) is nonzero only when $p(t - mT_c)$ and $p(t + \tau - nT_c)$ overlap. The delay τ can be expressed as $\tau = kT_c + \tau_\epsilon$, where $0 \le \tau_\epsilon < T_c$. Using this substitution, the pulses overlap only for $n = k + m$ and $n = k + m + 1$, so that (3-4) becomes

$$R_{cc'}(\tau) = R_{cc'}(k, \tau_\epsilon)$$

$$= \frac{1}{N} \sum_{m=0}^{N-1} a_m a'_{k+m} \frac{1}{T_c} \int_0^{T_c - \tau_\epsilon} p(\lambda) p(\lambda + \tau_\epsilon) \, d\lambda \qquad (3-5)$$

$$+ \frac{1}{N} \sum_{m=0}^{N-1} a_m a'_{k+m+1} \frac{1}{T_c} \int_{T_c - \tau_\epsilon}^{T_c} p(\lambda) p(\lambda - T_c + \tau_\epsilon) \, d\lambda$$

where the substitution $\lambda = t - mT_c$ has also been employed. The *discrete periodic cross-correlation function* of two codes $b(D)$ and $b'(D)$ is defined by [2]

$$\theta_{bb'}(k) = \frac{1}{N} \sum_{n=0}^{N-1} a_n a'_{n+k} \qquad (3-6)$$

where $a_n = (-1)^{b_n}$. Using this definition, the cross-correlation function $R_{cc'}(\tau)$ becomes

$$R_{cc'}(k, \tau_\epsilon) = \left(1 - \frac{\tau_\epsilon}{T_c}\right) \theta_{bb'}(k) + \frac{\tau_\epsilon}{T_c} \theta_{bb'}(k + 1) \qquad (3-7)$$

This expression is often convenient since the theory used to analyze code sequences yields results exclusively in terms of unit delays; that is, $\theta_{bb'}(k)$ is calculated rather than $R_{cc'}(\tau)$. The discrete periodic cross-correlation function can be calculated by representing the sequences $b(D)$ and $b'(D)$ as binary vectors $\mathbf{b}$ and $\mathbf{b}'$ of length N. A delay of k time units of the original sequence is represented as a cyclic shift of k time units of the vector representation. The kth cyclic shift of $\mathbf{b}$ is represented by $\mathbf{b}(k)$. Using this notation, the function $\theta_{bb'}(k) = (N_A - N_D)/N$, where N_A is the number of places in which $\mathbf{b}(0)$ agrees and N_D the number of places in which $\mathbf{b}(0)$ disagrees with $\mathbf{b}'(k)$. Equivalently, N_A is the number of zeros and N_D the number of ones in the modulo-2 sum of $\mathbf{b}(0)$ and $\mathbf{b}(k)$. The *discrete periodic autocorrelation function* is denoted by $\theta_b(k)$ and is defined by (3-6) with $a'_n = a_n$. When the periodic autocorrelation function is used in place of the periodic cross-correlation function in (3-5), the result is the autocorrelation function

$$R_c(k, \tau_\epsilon) = \left(1 - \frac{\tau_\epsilon}{T_c}\right) \theta_b(k) + \frac{\tau_\epsilon}{T_c} \theta_b(k + 1) \qquad (3-8)$$

3-2.2 Finite-Field Arithmetic

Some of the manipulations that will be performed on the code sequences introduced later require an understanding of the mechanics of finite-field arithmetic. In particular, determining the initial load of a shift register generator which will produce a particular known delay of the code requires a knowledge of the various ways of representing the elements of a finite field. This same knowledge is required to determine what phases of a maximal-length sequence to add together to obtain a specific known third phase. This discussion is intended to provide adequate information to perform these basic manipulations. A complete introductory discussion of this subject can be found in Lin and Costello [3] and a comprehensive treatment can be found in Birkhoff and MacLane [4].

Consider a set $S = \{e_0, e_1, e_2, \ldots, e_{M-1}\}$ having M elements. A finite field is constructed by defining two binary operations on the set called *addition* and *multiplication* such that certain conditions are satisfied. Addition and multiplication of two elements e_j and e_k are denoted $e_j + e_k$ and $e_j \cdot e_k$, respectively. The conditions that must be satisfied for S and the two operations to be a finite field are:

1. The addition or multiplication of any two elements of S must yield an element of S. That is, the set is *closed* under both addition and multiplication.
2. Both addition and multiplication must be *commutative*.
3. There must exist an *additive identity element* which will always be denoted by 0. The result of the addition of any element of the field and the additive identity is the element [i.e., $e_j +$ (additive identity) $= e_j$].
4. The set S must contain an *additive inverse element* $-e_j$ for every element e_j. The addition of any element e_j and its additive inverse $-e_j$ is the additive identity element [i.e., $e_j + (-e_j) = 0$].
5. The set must contain a *multiplicative identity element* which will always be denoted by 1. The result of the multiplication of any element of the field and the multiplicative identity is the element [i.e., $e_j \cdot 1 = e_j$].
6. Excluding the element 0, the set must contain a *multiplicative inverse element* e_j^{-1} for every element e_j. The multiplication of any element e_j and its multiplicative inverse is the multiplicative identity element [i.e., $e_j \cdot e_j^{-1} = 1$].
7. Multiplication must be *distributive* over addition.
8. Both addition and multiplication must be *associative*.

EXAMPLE 3-1

Consider the set $S = \{0,1,2\}$ with addition and multiplication defined in Tables 3-1 and 3-2. It can easily be verified that this set, with the operations defined in Tables 3-1 and 3-2, satisfies all the conditions above and is therefore a field. By inspection it is seen that the set is closed under both operations. The symmetry of the tables shows that both operations are commutative. The additive inverse elements are $-0 = 0$, $-1 = 2$, and $-2 = 1$. The multiplicative inverse elements are $1^{-1} = 1$ and $2^{-1} = 2$. All other conditions can be verified similarly.

TABLE 3-1. Modulo-3 Addition

+	0	1	2
0	0	1	2
1	1	2	0
2	2	0	1

TABLE 3-2. Modulo-3 Multiplication

·	0	1	2
0	0	0	0
1	0	1	2
2	0	2	1

TABLE 3-3. Modulo-2 Addition

+	0	1
0	0	1
1	1	0

TABLE 3-4. Modulo-2 Multiplication

·	0	1
0	0	0
1	0	1

EXAMPLE 3-2

Consider the set $S = \{0,1\}$ with addition and multiplication defined in Tables 3-3 and 3-4. It can easily be verified that this set is a field of two elements. This is the binary number field that will be used extensively in what follows. Observe that addition can be accomplished electronically using an Exclusive-OR gate and multiplication can be accomplished using an AND gate.

It can be shown that the set of integers $\{0, 1, 2, \ldots, M - 1\}$, where M is prime and addition and multiplication are carried out modulo-M, is a field [3]. These fields are called *prime fields*. The operations of subtraction and division are also easily defined for any field using the addition and multiplication tables, just as is done with the real-number field. Subtraction is defined as the addition of the additive inverse and division is defined as multiplication by the multiplicative inverse. For example, 1-2 in the ternary field of Example 3-1 is defined by $1 + (-2) = 1 + 1 = 2$. Similarly, $1 \div 2 = 1 \cdot (2^{-1}) = 1 \cdot 2 = 2$. Note that nonprime fields do not necessarily employ modulo-M arithmetic.

Fields can be constructed having any prime number of elements p or any integral power of a prime number p^m of elements. A field having p^m elements is called an *extension field* of the field having p elements. Finite fields are often referred to as *Galois fields*, using the notation GF(M) for the field having M elements. The remainder of this discussion will be concerned exclusively with the binary number field GF(2) and its extensions GF(2^m). The reason for this is that the electronics used to implement the code generators is binary, and some of the shift register generators will be shown to generate the elements of GF(2^m).

Consider next the arithmetic of polynomials in the indeterminate D whose coefficients are elements of GF(2). A polynomial of degree m over GF(2) has the form $f(D) = f_0 + f_1 D + f_2 D^2 + \cdots + f_m D^m$, where f_j is an element of GF(2), that is, $f_j = 0$ or 1. The operations of addition, subtraction, multiplication, and division are defined for these polynomials in exactly the same way as for polynomials with real coefficients except that binary arithmetic is used. For example, the addition of $f(D)$ and $g(D) = g_0 + g_1 D + g_2 D^2 + \cdots + g_m D^m$ yields $h(D) = h_0 + h_1 D + h_2 D^2 + \cdots + h_m D^m$ where $h_0 = f_0 + g_0$, $h_1 = f_1 + g_1$, $h_2 = f_2 + g_2, \ldots, h_m = f_m + g_m$ and modulo-2 addition of coefficients is used. The multiplication of these same polynomials yields a product $h(D)$ with the following coefficients:

$$h_0 = f_0 g_0$$

$$h_1 = f_0 g_1 + f_1 g_0$$

$$h_2 = f_0 g_2 + f_1 g_1 + f_2 g_0$$
$$\vdots$$
$$h_m = f_0 g_m + f_1 g_{m-1} + \cdots + f_m g_0$$
$$h_{m+1} = f_1 g_m + f_2 g_{m-1} + \cdots + f_m g_1$$
$$\vdots$$
$$h_{2m} = f_m g_m$$

$$(3\text{-}9)$$

where all additions and multiplications are modulo-2. Since modulo-2 arithmetic is commutative, associative, and distributive, it is easy to show that polynomial arithmetic, as just defined, is also commutative, associative, and distributive.

The division of one polynomial over GF(2) by another yields a quotient $q(D)$ and a remainder $r(D)$ just as with ordinary long division of two polynomials. For example, suppose that $f(D) = 1 + D^5$ and $g(D) = 1 + D + D^3 + D^4$. The long division of $f(D)$ by $g(D)$ yields

$$(3\text{-}10)$$

$$
\begin{array}{r}
D + 1 \\
D^4 + D^3 + D + 1\overline{\big)D^5 \qquad\qquad\qquad\qquad + 1} \\
\underline{D^5 + D^4 \qquad\quad + D^2 + D} \\
D^4 \qquad\quad + D^2 + D + 1 \\
\underline{D^4 + D^3 \qquad\quad + D + 1} \\
D^3 + D^2
\end{array}
$$

so that $q(D) = 1 + D$ and $r(D) = D^2 + D^3$. This result can be verified by calculating $f(D) = q(D)g(D) + r(D)$. When $r(D) = 0$, $f(D)$ is said to be *divisible* by $g(D)$.

EXAMPLE 3-3

Let $f(D) = 1 + D + D^{10} + D^{19}$ and $g(D) = D^2 + D^{10}$. Then $h(D) = f(D) + g(D) = (1 + 0) + (1 + 0)D + (0 + 1)D^2 + (1 + 1)D^{10} + (1 + 0)D^{19} = 1 + D + D^2 + D^{19}$. Now subtract $f(D)$ from $h(D)$. Subtraction is defined as addition of the additive inverse. In the binary field, each element is its own additive inverse since $0 + 0 = 0$ and $1 + 1 = 0$, so that addition and subtraction are identical. Therefore, $h(D) - f(D) = h(D) + f(D) = (1 + 1) + (1 + 1)D + (1 + 0)D^2 + (0 + 1)D^{10} + (1 + 1)D^{19} = D^2 + D^{10} = g(D)$, as expected.

EXAMPLE 3-4

Divide $f(D) = 1 + D^6$ by $g(D) = 1 + D + D^3 + D^4$.

Solution:

$$(3\text{-}11)$$

$$
\begin{array}{r}
D^2 + D + 1 \\
D^4 + D^3 + D + 1\overline{\big)D^6 \qquad\qquad\qquad\qquad\qquad + 1} \\
\underline{D^6 + D^5 \qquad\quad + D^3 + D^2} \\
D^5 \qquad\quad + D^3 + D^2 \qquad + 1 \\
\underline{D^5 + D^4 \qquad\quad + D^2 + D} \\
D^4 + D^3 \qquad\quad + D + 1 \\
\underline{D^4 + D^3 \qquad\quad + D + 1} \\
0
\end{array}
$$

Thus $g(D)$ divides $f(D)$. This result is verified by multiplying the quotient $q(D) = 1 + D + D^2$ by $g(D)$.

$$(1 + D + D^2)(1 + D + D^3 + D^4)$$

$$
\begin{aligned}
= 1 + D &\qquad\quad + D^3 + D^4 \\
+ D &+ D^2 \qquad\quad + D^4 + D^5 \\
&+ D^2 + D^3 \qquad\quad + D^5 + D^6 \\
\hline
1 &\qquad\qquad\qquad\qquad + D^6 = f(D)
\end{aligned}
$$

(3-12)

Polynomials over GF(2) have roots and may be factorable. Substituting $D = 1$ into $g(D)$ of Example 3-4 will demonstrate that 1 is a root, so that $D - 1 = D + 1$ should be a factor of $g(D)$. To verify this, divide $g(D)$ by $D + 1$:

$$
\begin{array}{r}
D^3 \qquad\qquad + 1 \\
D + 1\,\overline{\smash)\,D^4 + D^3 \qquad + D + 1} \\
\underline{D^4 + D^3} \qquad\qquad\quad \\
D + 1 \\
\underline{D + 1} \\
0
\end{array}
$$

(3-13)

showing that $g(D) = (D + 1)(D^3 + 1)$. The roots of a polynomial over GF(2) may not be elements of GF(2), just as a real polynomial may not have real roots. For example, $D^4 + D^3 + 1$ does not have 0 or 1 as a root.

At this point, all the tools are in place to construct the extension field $GF(2^m)$ of the binary field GF(2) for $m > 1$. The extension field has 2^m elements. Consider all the polynomials of degree $m - 1$ over GF(2). There are 2^m such polynomials, so each polynomial can be used to represent a single element of the extension field $GF(2^m)$. For example, if $m = 2$, the extension field contains $2^2 = 4$ elements, and there are four polynomials of degree $m - 1 = 1$, which are 0, 1, D, and $1 + D$. The operations of addition and multiplication must now be defined such that all the properties of a field are satisfied. Suppose that addition of two elements of the field is defined as the normal modulo-2 polynomial addition of the two polynomials representing the field elements. It can be easily verified that, using this addition rule, the addition of any two elements of the field yields another element of the field so that the field is closed under addition. The additive identity element is 0 and the additive inverse of any element is the element itself. It has already been noted that addition of polynomials over GF(2) is commutative.

Multiplication of two elements of $GF(2^m)$ is defined using special polynomials of degree m which are called *primitive polynomials*. A polynomial $h(D)$ of degree m is said to be primitive if the smallest integer n for which $h(D)$ divides $D^n + 1$ is $n = 2^m - 1$. Primitive polynomials are said to be *irreducible* since they are not the product of any two polynomials of lower degree (i.e., they cannot be factored). Primitive polynomials of any degree m are known to exist and have been tabulated in Refs. 5 and 6. The product of two elements of $GF(2^m)$ is defined as the remainder when the normal polynomial product is divided by the primitive polynomial chosen to define multiplication. This multiplication is referred to as modulo-$h(D)$ multiplication. Observe that more than one primitive polynomial exists for most $m > 1$ and different multiplication rules will be obtained depending on which primitive polynomial is chosen. The polynomial product of two polynomials of degree $m - 1$ or less has a degree of at most $2m - 2$, and the remainder when dividing by a polynomial of degree m has a degree of at most $m - 1$. Since the remainder has degree at most $m - 1$, it is another element of GF(2^m), so that the field is closed under multiplication. Polynomial multiplication is associative and commutative, so that multiplication of elements of GF(2^m) is also associative and commutative.

If it can be demonstrated that a multiplicative inverse element exists for each nonzero element of GF(2^m), all the conditions defining a field will have been satisfied. To determine the multiplicative inverse elements, consider the sequence of nonzero elements of GF(2^m) beginning with 1 and such that each element is the modulo-$h(D)$ product of D and the preceding element [3]:

$$1$$
$$D$$
$$D \cdot D = D^2 \tag{3-14}$$
$$D^2 \cdot D = D^3$$
$$\vdots$$

For all products where the normal polynomial product has degree less than m (the degree of the primitive polynomial), the remainder obtained when dividing by $h(D)$ is the normal polynomial product. That is, the normal polynomial product equals the modulo-$h(D)$ product. At some point in this sequence the normal polynomial product D^m will appear. At this point the remainder or modulo-$h(D)$ product is $r(D) = 1 + h_1 D + h_2 D^2 + \cdots + h_{m-1} D^{m-1}$, as can be seen by long division

$$D^m + h_{m-1}D^{m-1} + \cdots$$

$$
\begin{array}{r}
1 \\
+ h_2D^2 + h_1D + 1\,\overline{\smash{\big)}\,D^m} \\
\underline{D^m + h_{m-1}D^{m-1} + \cdots + h_2D^2 + h_1D + 1} \\
h_{m-1}D^{m-1} + \cdots + h_2D^2 + h_1D + 1
\end{array}
\tag{3-15}
$$

Thus $D^{m-1} \cdot D = D^m = h_{m-1}D^{m-1} + \cdots + h_2D^2 + h_1D + 1$ and the sequence of powers of D can be written as polynomials of degree less than or equal to $m - 1$.

EXAMPLE 3-5 [3]

Let $m = 4$ and consider the multiplication law defined by $h(D) = D^4 + D + 1$. Performing the long division of D^4 by $h(D)$ yields a remainder of $r(D) = 1 + D$. Thus $D^4 = 1 + D$ and the sequence of nonzero elements of GF(2^4) written in the order described above becomes

D^0: 1

D^1: $1 \cdot D = D$

D^2: $D \cdot D = D^2$

D^3: $D^2 \cdot D = D^3$

D^4: $D^3 \cdot D = D^4 = 1 + D$

D^5: $(1 + D) \cdot D = D + D^2$

D^6: $(D + D^2) \cdot D = D^2 + D^3$

D^7: $(D^2 + D^3) \cdot D = D^3 + D^4 = 1 + D + D^3$

D^8: $(1 + D + D^3) \cdot D = D + D^2 + D^4 = 1 + D^2$

D^9: $(1 + D^2) \cdot D = D + D^3$

D^{10}: $(D + D^3)D = D^2 + D^4 = 1 + D + D^2$

D^{11}: $(1 + D + D^2)D = D + D^2 + D^3$

D^{12}: $(D + D^2 + D^3)D = D^2 + D^3 + D^4 = 1 + D + D^2 + D^3$

D^{13}: $(1 + D + D^2 + D^3)D = D + D^2 + D^3 + D^4 = 1 + D^2 + D^3$

D^{14}: $(1 + D^2 + D^3)D = D + D^3 + D^4 = 1 + D^3$

D^{15}: $(1 + D^3)D = D + D^4 = 1$

D^{16}: $1 \cdot D = D$

D^{17}: $D \cdot D = D^2$

$\vdots$

Observe that the sequence is periodic and there are 15 distinct elements.

A primitive polynomial of degree m divides $D^{2^m-1} + 1$ so that $D^{2^m-1} + 1 = q(D)h(D)$, which implies that $D^{2^m-1} = q(D)h(D) + 1$, and the remainder when dividing D^{2^m-1} by $h(D)$ is 1. Thus the sequence of elements of GF(2^m) written such that each element is D times the preceding element and using the proposed multiplication rule repeats after at most $2^m - 1$ elements. Since the smallest n for which a primitive polynomial divides $D^n + 1$ is $n = 2^m - 1$, the sequence of elements cannot repeat sooner than after $2^m - 1$ elements, so that the sequence always contains exactly $2^m - 1$ distinct elements. Thus there are two useful ways of representing the elements of the extension field GF(2^m). The first representation is as a polynomial of degree $m - 1$ over GF(2). The second representation is as power of D. The two representations are related through the primitive polynomial used to define the multiplication law. Using the second representation, the fact that $D^{2^m-1} = 1$ leads immediately to the multiplicative inverse of any element of GF(2^m). The multiplicative inverse of any element D^j is $(D^j)^{-1} = D^{2^m-1-j}$ since $D^j \cdot D^{2^m-1-j} = D^{2^m-1} = 1$.

EXAMPLE 3-6 ──

Derive a table of the multiplicative inverse elements in polynomial form for the nonzero elements of $GF(2^4)$ using the primitive polynomial $h(D) = D^4 + D + 1$.

Solution: The nonzero elements of this field are the first 15 terms of the sequence derived in Example 3-5. These elements, along with their inverses, are as follows:

Element		Inverse Element	
1	1	1	1
D	D	D^{14}	$1 + D^3$
D^2	D^2	D^{13}	$1 + D^2 + D^3$
D^3	D^3	D^{12}	$1 + D + D^2 + D^3$
D^4	$1 + D$	D^{11}	$D + D^2 + D^3$
D^5	$D + D^2$	D^{10}	$1 + D + D^2$
D^6	$D^2 + D^3$	D^9	$D + D^3$
D^7	$1 + D + D^3$	D^8	$1 + D^2$
D^8	$1 + D^2$	D^7	$1 + D + D^3$
D^9	$D + D^3$	D^6	$D^2 + D^3$
D^{10}	$1 + D + D^2$	D^5	$D + D^2$
D^{11}	$D + D^2 + D^3$	D^4	$1 + D$
D^{12}	$1 + D + D^2 + D^3$	D^3	D^3
D^{13}	$1 + D^2 + D^3$	D^2	D^2
D^{14}	$1 + D^3$	D	D

$$(3\text{-}16)$$

Several products of an element and its inverse will verify that the inverses are correct. Using the multiplication law defined by $h(D)$, $(D^7)(D^7)^{-1} = (D^7)(D^8) = (1 + D + D^3)(1 + D^2)$. The normal polynomial multiplication yields

$$
\begin{array}{l}
1 + D \phantom{+{}} + D^3 \\
 + D^2 + D^3 + D^5 \\
\hline
1 + D + D^2 + D^5
\end{array}
\qquad (3\text{-}17)
$$

and long division by $h(D)$ yields

$$
\begin{array}{r}
D \\
D^4 + D + 1\,\overline{\smash{\big)}\,D^5 + D^2 + D + 1} \\
\underline{D^5 + D^2 + D} \\
1
\end{array}
\qquad (3\text{-}18)
$$

so that $(D^7)(D^8) = 1$, as expected.

──

The preceding paragraphs have presented a brief introduction to finite-field arithmetic. The purpose of the discussion was to give the student sufficient proficiency with finite fields to enable certain manipulations with maximal-length codes to be defined later. In summary, a finite field or Galois field is a set of elements upon which two operations (addition and multiplication) are defined. The set and these operations must satisfy certain very precise rules. Integer fields having a prime number of elements and using modulo-M multiplication and addition exist. The most commonly used integer field used in this book is the binary number field with elements 0 and 1 and using modulo-2 addition and multiplication. Extension fields of the binary field having 2^m elements also exist. Two representations for the elements of these extension fields are convenient. One representation associates one element of GF(2^m) with each possible binary polynomial of degree less than or equal to $m - 1$. Addition is defined as normal polynomial addition. Multiplication is defined as the modulo-$h(D)$ product of the polynomials, where $h(D)$ is primitive. A second representation of the elements of GF(2^m) is as powers of D. It was shown that $D^{2^m-1} = 1$, which resulted in easy identification of the multiplicative inverse for any element of GF(2^m). Facility with finite-field arithmetic will be useful in the study of maximal-length codes. Error correction coding also makes extensive use of the concepts discussed here.

3-2.3 Sequence Generator Fundamentals

In this section the actual shift registers that are used to generate PN codes are examined. It will be shown that shift registers with feedback and/or feedforward connections can be used to multiply and divide polynomials over GF(2). Since these shift registers execute the mathematical operations described in the preceding section, that same mathematics becomes a powerful tool in the design and analysis of the generators. In what follows, modulo-2 polynomial arithmetic will be used to calculate the output of a shift register generator given a particular input and feedback/feedforward connections, to determine the maximum possible period of a linear feedback shift register, and to determine several different feedback schemes which provide identical outputs but which are different electronically. It is convenient now to limit the binary sequences under consideration to semi-infinite sequences rather than the doubly infinite sequences discussed previously. Thus all sequences are assumed to begin at time zero, and a code sequence $a(D)$ contains only positive powers of the delay operation D. That is,

$$a(D) = \sum_{j=0}^{\infty} a_j D^j \tag{3-19}$$

Consider the logic circuit illustrated in Figure 3-1. In this figure, the boxes represent unit delays or shift registers, circles containing subscripted letter coefficients represent a connection if the coefficient is a 1 or no connection if the coefficient is a 0, and circles containing a "+" represent modulo-2 adders or exclusive-OR gates. The circles containing subscripted coefficients may also be viewed as a modulo-2 multiplication of the input by the coefficient.

The output of the jth modulo-2 adder (number the adders from left to right

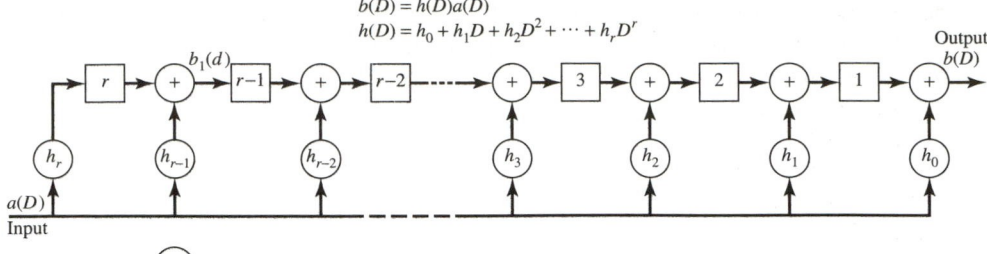

FIGURE 3-1. Circuit for multiplying polynomials. (From Ref. 5.)

starting with 1), denoted $b_j(D)$, is

$$b_j(D) = b_{j-1}(D)D + a(D)h_{r-j}$$

for $j = 2, \ldots, r$, and for $j = 1$,

$$b_1(D) = a(D)Dh_r + a(D)h_{r-1}$$

Thus, by iteration

$$
\begin{aligned}
b(D) &= b_r(D) \\
&= b_{r-1}(D)D + a(D)h_0 \\
&= b_{r-2}(D)D^2 + a(D)Dh_1 + a(D)h_0 \\
&\;\;\vdots \\
&= b_1(D)D^{r-1} + a(D)D^{r-2}h_{r-2} + \cdots + a(D)h_0 \\
&= a(D)D^r h_r + a(D)D^{r-1}h_{r-1} + \cdots + a(D)h_0 \\
&= \sum_{k=0}^{r} [a(D)D^k]h_k \\
&= h(D)a(D)
\end{aligned}
$$

(3-20)

and it is seen that the circuit performs the normal polynomial multiplication of the input sequence $a(D)$ and the transfer polynomial $h(D)$.

The circuit of Figure 3-1 can also be configured as a two-input multiplier as illustrated in Figure 3-2. The output of this circuit is

$$b(D) = a_1(D)h(D) + a_2(D)k(D)$$ (3-21)

where $k(D)$ is the transfer function for the second input $a_2(D)$. Suppose now that $k_0 = 0$ in the second transfer function $k(D)$, and suppose further that $a_2(D)$ is taken from the output of Figure 3-2 that is, $a_2(D) = b(D)$. Since $k_0 = 0$, define

$$k(D) = g(D) + 1$$ (3-22)

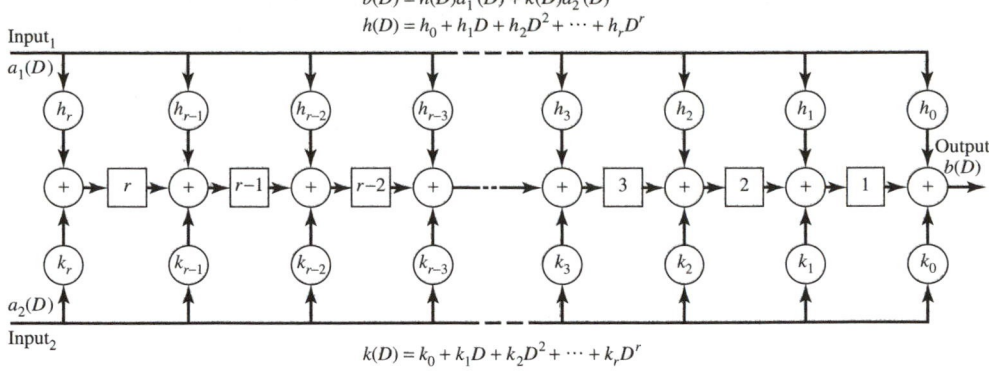

FIGURE 3-2. Two-input modulo-2 polynomial multiplier. (From Ref. 5.)

where $g(D)$ is a transfer function with $g_0 = 1$. The modified circuit is illustrated in Figure 3-3. The input–output relationship of this circuit is

$$b(D) = a_1(D)h(D) + b(D)[g(D) + 1] \qquad (3\text{-}23)$$

Adding $b(D)g(D) + b(D)$ to both sides of this equation yields

$$b(D)g(D) = a_1(D)h(D) \qquad (3\text{-}24)$$

which can be solved for $b(D)$ if a polynomial $c(D)$ can be found which satisfies the relationship $g(D)c(D) = 1$. With $c(D)$ so defined, (3-24) can be written

$$b(D) = a_1(D)h(D)c(D) \qquad (3\text{-}25)$$

The coefficients of $c(D)$ must satisfy the following relationships:

$$g_0 c_0 = 1 \qquad (3\text{-}26a)$$

$$\sum_{l=0}^{\min\{j,r\}} g_l c_{j-l} = 0 \qquad j = 1, 2, \ldots \qquad (3\text{-}26b)$$

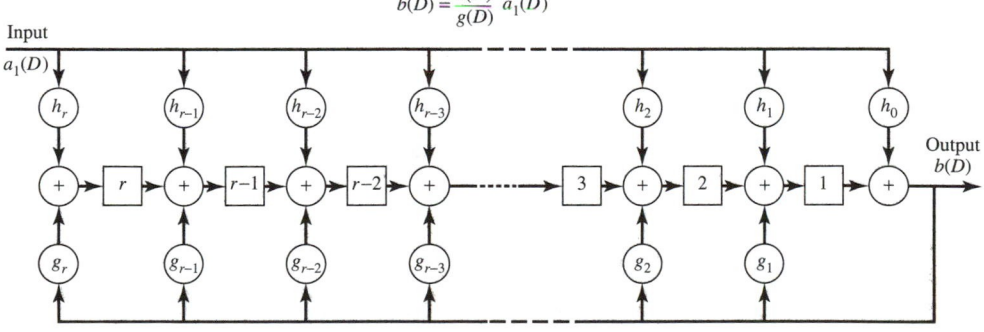

FIGURE 3-3. Circuit that simultaneously multiplies by $h(D)$ and divides by $g(D)$.

in order for the product $g(D)c(D)$ to equal unity. Note that the polynomial multiplication being considered here is *not* modulo a primitive polynomial but is normal multiplication using modulo-2 arithmetic to combine coefficients. Using (3-26), all of the coefficients of $c(D)$ can be found and then used in (3-25) to determine $b(D)$.

Consider the polynomial long division of 1 by $g(D)$, where $g(D)$ is written with low-order terms on the left:

$$
\begin{array}{r}
1 + g_1 D + (g_2 + g_1^2)D^2 + \cdots \\
\hline
1 + g_1 D + \cdots + g_r D^r \,\big|\, 1 \\
\end{array}
$$

$$
\begin{aligned}
&1 + g_1 D + \qquad g_2 D^2 + \qquad g_3 D^3 + \cdots + \qquad g_r D^r \\[-2pt]
\hline
&g_1 D + \qquad g_2 D^2 + \qquad g_3 D^3 + \cdots + \qquad g_r D^r \\
&g_1 D + \qquad g_1^2 D^2 + \qquad g_1 g_2 D^3 + \cdots + g_1 g_{r-1} D^r + g_1 D^{r+1} \\[-2pt]
\hline
&(g_2 + g_1^2)D^2 + (g_3 + g_1 g_2)D^3 + \cdots \\
&(g_2 + g_1^2)D^2 + g_1(g_2 + g_1^2)D^3 + \cdots \\[-2pt]
\hline
&\qquad \cdots
\end{aligned}
$$

$$(3\text{-}27)$$

Solving for the first few coefficients of $c(D)$ from (3-26) yields

$$
\begin{aligned}
c_0 &= g_0 = 1 \\
c_1 &= g_1 c_0 = g_1 \\
c_2 &= g_1 c_1 + g_2 c_0 = g_1^2 + g_2 \\
&\ \ \vdots
\end{aligned}
$$

$$(3\text{-}28)$$

Comparing (3-27) and (3-28) it can be seen that $c(D) = 1/g(D)$, where the long division is carried out in the manner of (3-27). Thus the circuit of Figure 3-3 has been shown to multiply an arbitrary input polynomial by $h(D)$ and divide it by $g(D)$ simultaneously, that is,

$$
b(D) = a_1(D)\,\frac{h(D)}{g(D)}
$$

$$(3\text{-}29)$$

EXAMPLE 3-7 _____

Suppose that $h(D) = D^6$ and $g(D) = 1 + D + D^2 + D^3 + D^6$. The multiplier/divider shift register for these polynomials is illustrated in Figure 3-4a. What is the output of this circuit given that the input is $a(D) = 1$ (i.e., the input is a 1 at time zero followed by an infinite string of zeros)?

Solution: From the previous discussion, the output is

$$
b(D) = a(D)\frac{h(D)}{g(D)}
$$

$$
= \frac{D^6}{1 + D + D^2 + D^3 + D^6}
$$

Performing the long division with high-order coefficients of the divisor on the

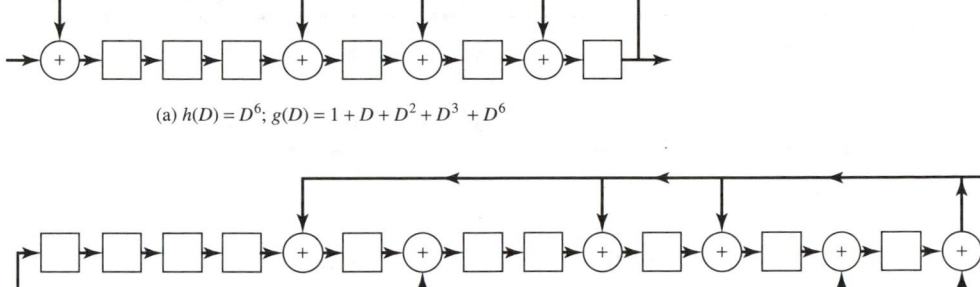

(a) $h(D) = D^6$; $g(D) = 1 + D + D^2 + D^3 + D^6$

(b) $h(D) = 1 + D + D^5 + D^{10}$; $g(D) = 1 + D^2 + D^3 + D^6$

FIGURE 3-4. Multiplier/divider circuits.

right yields

$$
\begin{array}{r}
D^6 + D^7 \qquad\qquad\quad + D^{10} + D^{11} + D^{12} + \cdots \\
1 + D + D^2 + D^3 + D^6\,\overline{)\,D^6} \\
D^6 + D^7 + D^8 + D^9 + \qquad\qquad + D^{12} \\
\hline
D^7 + D^8 + D^9 \qquad\qquad\quad + D^{12} \\
D^7 + D^8 + D^9 + D^{10} \qquad\qquad + D^{13} \\
\hline
D^{10} \qquad\quad + D^{12} + D^{13} \\
D^{10} + D^{11} + D^{12} + D^{13} \qquad\qquad + D^{16} \\
\hline
D^{11} \qquad\qquad\qquad + D^{16} \\
D^{11} + D^{12} + D^{13} + D^{14} \qquad\qquad\qquad D^{17} \\
\hline
D^{12} + D^{13} + D^{14} \qquad\qquad + D^{16} + D^{17} \\
\vdots
\end{array}
$$

This output sequence can be verified by manually calculating the contents of the shift register for the first few shifts.

EXAMPLE 3-8

The circuit for simultaneously multiplying by $h(D) = 1 + D + D^5 + D^{10}$ and dividing by $g(D) = 1 + D^2 + D^3 + D^6$ is illustrated in Figure 3-4b.

The input to the circuit of Figure 3-3 can be any binary sequence, including a sequence that ends at some finite time. After the input sequence ends, the circuit is equivalent to the feedback shift register shown in Figure 3-5. Suppose that the input sequence ends at time j. The highest power of D in $a_1(D)$ is D^j and the highest power of D in the product $a_1(D)h(D)$ in (3-24) is D^{j+r}, since $h(D)$ had degree r. Therefore, the coefficient of any power of D greater than $j + r$ on the left side of (3-24) must be zero, and the coefficients of $b(D)$ and $g(D)$ must satisfy

$$
\sum_{m=0}^{r} g_m b_{i-m} = 0 \tag{3-30}
$$

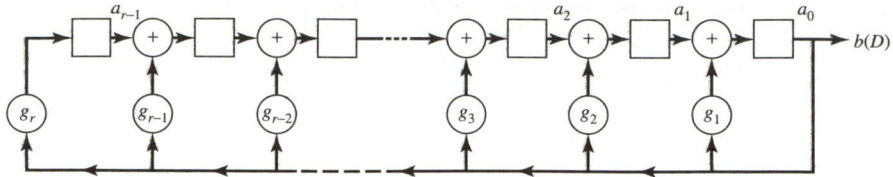

FIGURE 3-5. High-speed linear feedback shift-register generator.

for $i > j + r$. Since $g_0 = 1$, this relationship can be written

$$b_i = \sum_{m=1}^{r} g_m b_{i-m} \qquad (3\text{-}31)$$

This recurrence relation must be satisfied at all times subsequent to the end of the input sequence.

A second circuit configuration that satisfies the recurrence relationship of (3-31) and therefore generates an identical output sequence is illustrated in Figure 3-6. This can be verified by considering the output of the generator to be the input to the leftmost shift-register stage as shown. By inspection the output is

$$b(D) = g_1 D b(D) + g_2 D^2 b(D) + \cdots + g_r D^r b(D) \qquad (3\text{-}32)$$

Equating coefficients of D^i on both sides of this equation will show that (3-31) is satisfied. This feedback configuration is, in fact, a commonly used configuration. The configuration chosen for a particular application depends on such things as the speed at which the hardware must operate and whether delayed outputs are also required. Delayed outputs for all delays up to r are available from the configuration of Figure 3-6 but not from the configuration of Figure 3-5. The configuration of Figure 3-5, however, can function at higher speeds than that of Figure 3-6 since there is less propagation delay in the feedback path.

It is of interest to be able to determine the output $b(D)$ of either the circuit of Figure 3-5 or 3-6 given the initial contents of the shift register. This is easily accomplished for the circuit of Figure 3-5. Its output $b(D)$ is identical to the output $b'(D)$ beginning at time r of Figure 3-3 with $h(D) = D^r$ and with an input

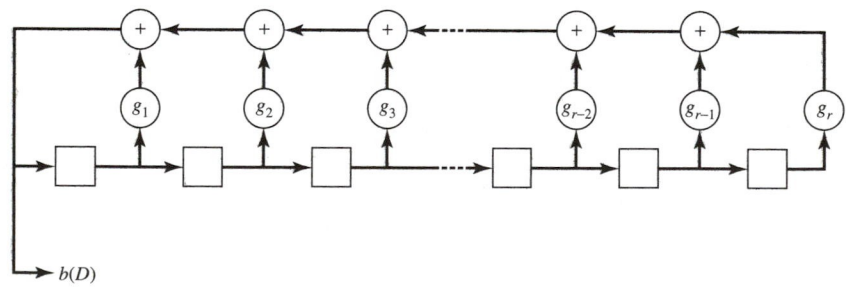

FIGURE 3-6. Linear feedback shift register whose output satisfies the same recurrence relationship as the generator of Figure 3-5.

$$a_1(D) = a_0 + a_1 D + a_2 D^2 + \cdots + a_{r-1} D^{r-1} \qquad (3\text{-}33)$$

This input is nonzero just long enough to load the shift register. The output, from (3-29) is,

$$b'(D) = \frac{D^r a_1(D)}{g(D)} \qquad (3\text{-}34)$$

The problem being considered is to find the output beginning at the time that the shift register is completely loaded. The loading process just described consumes r time units, so the desired result is $b'(D)$ of (3-34) beginning at time r. Observe that the output $b'(D)$ is zero for the first r time units while the shift register loads so that the output beginning at time r is simply $b'(D)$ shifted by r time units or

$$b(D) = \frac{a_1(D)}{g(D)} \qquad (3\text{-}35)$$

EXAMPLE 3-9

Find the output of the circuit of Figure 3-5 with $g(D) = 1 + D + D^3 + D^4$ and an initial shift-register load of 0001. The circuit and the initial load are illustrated in Figure 3-7.

Solution: The initial load is described by

$$a_1(D) = 1$$

and the output is

$$b(D) = \frac{1}{1 + D + D^3 + D^4}$$

Performing the polynomial long division yields

$$
\begin{array}{l}
\phantom{1 + D + D^3 + D^4 \overline{)}} 1 + D + D^2 \qquad\qquad\qquad + D^6 + D^7 + D^8 \qquad\qquad + \cdots \\
1 + D + D^3 + D^4 \,\overline{)\, 1} \\
 1 + D \qquad + D^3 + D^4 \\
 \overline{ D \qquad + D^3 + D^4} \\
 D + D^2 \qquad + D^4 + D^5 \\
 \overline{ D^2 + D^3 \qquad + D^5} \\
 D^2 + D^3 \qquad + D^5 + D^6 \\
 \overline{ D^6} \\
 D^6 + D^7 \qquad + D^9 + D^{10} \\
 \overline{ D^7 \qquad + D^9 + D^{10}} \\
 D^7 + D^8 \qquad + D^{10} + D^{11} \\
 \overline{ D^8 + D^9 \qquad + D^{11}} \\
 D^8 + D^9 \qquad + D^{11} + D^{12} \\
 \overline{\phantom{D^8 + D^9 + D^{11}} D^{12}} \\
 \qquad\qquad\qquad\qquad \vdots
\end{array}
$$

Observe that the output sequence is periodic with a period of six. This can be verified with a manual calculation of the contents of the shift register as a function of time.

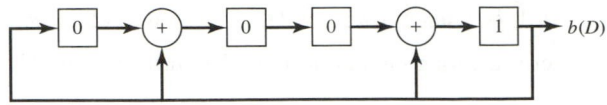

FIGURE 3-7. Circuit configuration for Example 3-9.

The procedure just described applies *only* to the configuration of Figure 3-5. Although the two shift-register configurations generate identical output sequences, the sequence of shift register states each goes through is different. A means of finding the output sequence for the shift register of Figure 3-6 is to find the equivalent initial state of the circuit of Figure 3-5 and then to use the results just described. Suppose that the initial state for the circuit of Figure 3-6 is

$$a(D) = a_0 + a_1 D + a_2 D^2 + \cdots + a_{r-1} D^{r-1} \qquad (3\text{-}36)$$

This means that a_0 is in the rightmost shift-register stage, a_1 in the second from the right, and so on. Define $c(D)$ to be the output of the rightmost shift register of Figure 3-6, that is, $c(D) = D^r b(D)$. The first r elements of $c(D)$ are $a_0, a_1, a_2, \ldots, a_{r-1}$. Since the circuits of Figures 3-5 and 3-6 are equivalent, the initial load of the circuit of Figure 3-5 can be chosen such that its output $b'(D) = c(D)$. Let the initial load that accomplishes this be

$$a'(D) = a'_0 + a'_1 D + a'_2 D^2 + \cdots + a'_{r-1} D^{r-1} \qquad (3\text{-}37)$$

Since the two output sequences are equal, (3-35) becomes

$$\begin{aligned} a_0 + a_1 D + \cdots + a_{r-1} D^{r-1} + b'_r D^r + b'_{r+1} D^{r+1} + \cdots \\ = \frac{a'_0 + a'_1 D + a'_2 D^2 + \cdots + a'_{r-1} D^{r-1}}{g_0 + g_1 D + g_2 D^2 + \cdots + g_r D^r} \end{aligned} \qquad (3\text{-}38)$$

Thus the initial state of the configuration of Figure 3-5, which produces the same output sequence as the configuration of Figure 3-6 with the initial load of $a(D)$, is found by equating the first r coefficients of

$$\{a(D) + b'_r D^r + b'_{r+1} D^{r+1} + \cdots\}g(D) = a'(D) \qquad (3\text{-}39)$$

Observe that none of the coefficients b'_j affect the calculation of a'_j, so that the desired shift-register load is simply the first r coefficients of the normal polynomial product $a(D)g(D)$. The entire output sequence can be found using (3-35) and the fact that $c(D) = b'(D) = D^r b(D)$.

EXAMPLE 3-10 _____

Find the output of the circuit of Figure 3-6 with $g(D) = 1 + D + D^3 + D^4$ and an initial shift-register load of 0001. The circuit and the initial load are illustrated in Figure 3-8.

Solution: The initial load in polynomial form is $a(D) = 1$ and the product on the left side of (3-39) is

$$g(D) + (b'_r D^r + b'_{r+1} D^{r+1} + \cdots)g(D) = a'_0 + a'_1 D + a'_2 D^2 + a'_3 D^3$$

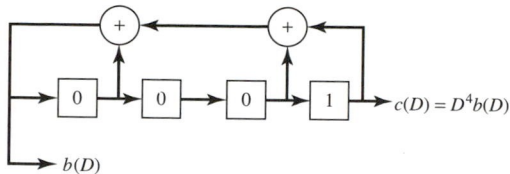

FIGURE 3-8. Circuit configuration for Example 3-10.

so that

$$a_0' = g_0 = 1$$

$$a_1' = g_1 = 1$$

$$a_2' = g_2 = 0$$

$$a_3' = g_3 = 1$$

Thus $a'(D) = 1 + D + D^3$ and the complete output sequence is $b'(D) = a'(D)/g(D)$ or

$$
\begin{array}{r}
1 \qquad\qquad + D^4 + D^5 + D^6 \qquad\qquad\qquad + D^{10} + D^{11} + D^{12} + \cdots \\
\hline
1 + D + D^3 + D^4\,\overline{)\,1 + D \qquad + D^3 } \\
1 + D \qquad + D^3 + D^4 \\
\hline
D^4 \\
D^4 + D^5 \qquad + D^7 + D^8 \\
\hline
D^5 \qquad + D^7 + D^8 \\
D^5 + D^6 \qquad + D^8 + D^9 \\
\hline
D^6 + D^7 \qquad + D^9 \\
D^6 + D^7 \qquad + D^9 + D^{10} \\
\hline
D^{10} \\
\vdots
\end{array}
$$

Multiplying by $D^{-r} = D^{-4}$ yields the output $b(D) = D^{-4} + 1 + D + D^2 + D^6 + D^7 + D^8 + \cdots$. Observe that the output is identical to the output of Example 3-9 except for beginning with the D^{-4} term. This means simply that the output sequence beginning at time -4 is known. These additional known output symbols are the load of the shift register.

Several observations about the circuits of Figures 3-5 and 3-6 are now made. First, given nonzero initial conditions, neither of the registers will ever reach an all-zeros state. This can be seen from the circuit of Figure 3-5, which would reach the state where all registers except the rightmost contain zeros just prior to reaching the all-zeros state. The single 1 in the rightmost register would be fed back to some other register. Since all g's cannot be zero, the all-zeros state is never reached. Second, since the register contains r stages and an r-stage shift register has at most $2^r - 1$ nonzero states, the output must be periodic with a period of *at most* $2^r - 1$. The period can be significantly less than $2^r - 1$.

Consider the determination of the maximum period of either of the shift-register circuits. Note that the same circuit may generate many different output sequences;

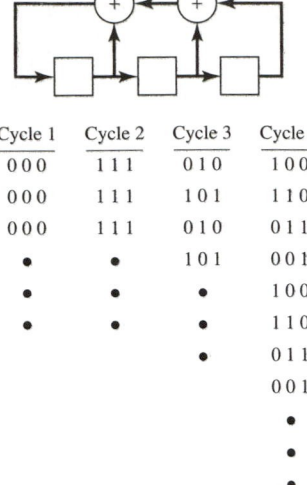

Cycle 1	Cycle 2	Cycle 3	Cycle 4
0 0 0	1 1 1	0 1 0	1 0 0
0 0 0	1 1 1	1 0 1	1 1 0
0 0 0	1 1 1	0 1 0	0 1 1
•	•	1 0 1	0 0 1
•	•	•	1 0 0
•	•	•	1 1 0
		•	0 1 1
			0 0 1
			•
			•
			•

FIGURE 3-9. Linear feedback shift-register cycles for four different initial conditions. (From Ref. 7.)

the particular output sequence generated depends on the initial state of the register. For example, consider the linear feedback shift-register generator illustrated in Figure 3-9. Four different sets of shift-register states are possible depending on the initial state. These four possible cycles have periods of 1, 1, 2, and 4 as shown. Since all possible shift-register states are included in one of the four cycles, there are no other cycles. The maximum possible period for an arbitrary feedback shift-register connection defined by $g(D)$ can be found [5] by defining the reciprocal polynomial of $g(D)$ by

$$g_r(D) = D^r g\left(\frac{1}{D}\right) \tag{3-40}$$

It can be shown [5] that the maximum possible period of the shift-register generator is the smallest possible integer N for which $D^N + 1$ is divisible by $g_r(D)$. That is, the maximum period is the smallest N for which a polynomial $h_r(D)$ exists such that

$$g_r(D)h_r(D) = D^N + 1 \tag{3-41}$$

EXAMPLE 3-11 _____

Suppose that $g(D) = 1 + D + D^3 + D^4$. Then

$$g_r(D) = D^4 g\left(\frac{1}{D}\right)$$

$$= D^4(1 + D^{-1} + D^{-3} + D^{-4})$$

$$= D^4 + D^3 + D + 1$$

By performing all the necessary long divisions, it can be shown that the smallest integer N for which $D^N + 1$ is evenly divisible by $g_r(D)$ is $N = 6$.

$$
\begin{array}{r}
D^2 + D \quad + 1 \\
D^4 + D^3 + D + 1 \overline{\smash{\big)}\ D^6 \hspace{5.5cm} + 1} \\
\underline{D^6 + D^5 \hspace{1.5cm} + D^3 + D^2} \\
D^5 \hspace{1.5cm} + D^3 + D^2 \hspace{1cm} + 1 \\
\underline{D^5 + D^4 \hspace{1.5cm} + D^2 + D} \\
D^4 + D^3 \hspace{1.5cm} + D + 1 \\
\underline{D^4 + D^3 \hspace{1.5cm} + D + 1} \\
0
\end{array}
$$

Thus the maximum possible period is 6. Observe that this period is less than the maximum possible period for a four-stage shift register, which is $2^4 - 1 = 15$.

Finally, note that the linear feedback shift-register configuration illustrated in Figure 3-5 is often referred to as the *Galois configuration* in the literature [22], while the configuration illustrated in Figure 3-6 is referred to as the *Fibonacci configuration*. Both Galois and Fibonacci were mathematicians whose research is the foundation for the operation of these code generators.

3-2.4 State-Machine Representation of Shift-Register Generators

Consider the shift-register generators illustrated in Figure 3-10 and 3-11, which are identical to Figures 3-5 and 3-6 with the addition of new labels for the contents of the shift registers. These generators may be viewed as a state machine whose state at time n is the contents of the shift register represented by a column vector. Define the state of the shift-register generator at time n by

$$
\mathbf{S}_n = \begin{bmatrix} s_{0,n} \\ s_{1,n} \\ s_{2,n} \\ \vdots \\ s_{r-2,n} \\ s_{r-1,n} \end{bmatrix}
$$

Consider Figure 3-10. Given the contents of the shift register at time n, the contents of the shift registers at time $n + 1$ are

$$
s_{0,n+1} = s_{1,n} + g_1 \cdot s_{0,n}
$$

$$
s_{1,n+1} = s_{2,n} + g_2 \cdot s_{0,n}
$$

$$
\vdots \hspace{5cm} \text{(3-42)}
$$

$$
s_{r-2,n+1} = s_{r-1,n} + g_{r-1} \cdot s_{0,n}
$$

$$
s_{r-1,n+1} = g_r \cdot s_{0,n}
$$

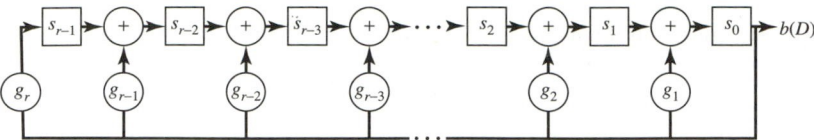

FIGURE 3-10. Galois feedback generator.

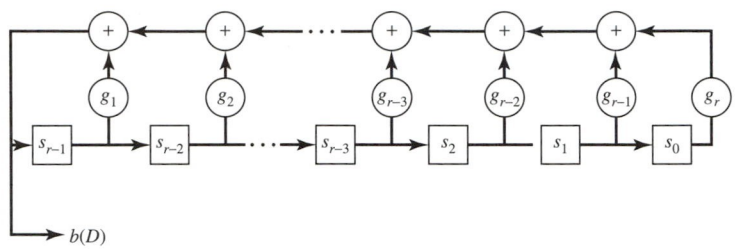

$b(D)$

FIGURE 3-11. Fibonacci feedback generator.

where all additions and multiplication are modulo-2. These equations can be conveniently written using matrix arithmetic. Define the Galois state transition matrix **G** as the $r \times r$ square matrix that defines (3-42). Specifically,

$$\mathbf{S}_{n+1} = \begin{bmatrix} s_{0,n+1} \\ s_{1,n+1} \\ s_{2,n+1} \\ \vdots \\ s_{r-2,n+1} \\ s_{r-1,n+1} \end{bmatrix} = \begin{bmatrix} g_1 & 1 & 0 & 0 & \cdots & 0 \\ g_2 & 0 & 1 & 0 & \cdots & 0 \\ g_3 & 0 & 0 & 1 & \cdots & 0 \\ \vdots & \vdots & & & \cdots & \vdots \\ g_{r-1} & 0 & 0 & 0 & \cdots & 1 \\ g_r & 0 & 0 & 0 & \cdots & 0 \end{bmatrix} \times \begin{bmatrix} s_{0,n} \\ s_{1,n} \\ s_{2,n} \\ \vdots \\ s_{r-2,n} \\ s_{r-1,n} \end{bmatrix} = \mathbf{G} \times \mathbf{S}_n \qquad (3\text{-}43)$$

$$(3\text{-}44)$$

The state at any time n can be found by applying (3-43) recursively beginning at time $n = 0$. The result is

$$\mathbf{S}_1 = \mathbf{G} \times \mathbf{S}_0$$
$$\mathbf{S}_2 = \mathbf{G} \times \mathbf{S}_1 = \mathbf{G} \times \mathbf{G} \times \mathbf{S}_0 \qquad (3\text{-}44)$$
$$\vdots$$
$$\mathbf{S}_n = \mathbf{G}^n \times \mathbf{S}_0$$

By premultiplying (3-43) by $\mathbf{G}^{-1}$, the inverse of $\mathbf{G}$, a matrix equation for finding the state of the shift register at time n from the state at time $n + 1$ is found.

$$\mathbf{G}^{-1} \times \mathbf{S}_{n+1} = \mathbf{G}^{-1} \times \mathbf{G} \times \mathbf{S}_n = \mathbf{S}_n \qquad (3\text{-}45)$$

This equation can also be applied recursively enabling the calculation of the state any number of time units in the past. The result is

$$\mathbf{S}_{n-k} = [\mathbf{G}^{-1}]^k \mathbf{S}_n \qquad (3\text{-}46)$$

The output of the shift-register generator at time n is the contents of the rightmost shift register. Using matrix arithmetic, the output b_n of the shift-register generator is

$$b_n = [1 \quad 0 \quad 0 \quad 0 \quad \cdots \quad 0] \times \begin{bmatrix} s_{0,n} \\ s_{1,n} \\ s_{2,n} \\ \vdots \\ s_{r-2,n} \\ s_{r-1,n} \end{bmatrix} \tag{3-47}$$

$$= [1 \quad 0 \quad 0 \quad 0 \quad \cdots \quad 0] \times \mathbf{G}^n \times \mathbf{S}_0$$

Similar results may be found for the Fibonacci shift register of Figure 3-11. Define the Fibonacci state-transition matrix in a manner similar to (3-43). The result is

$$\mathbf{S}_{n+1} = \begin{matrix} s_{0,n+1} \\ s_{1,n+1} \\ s_{2,n+1} \\ \vdots \\ s_{r-2,n+1} \\ s_{r-1,n+1} \end{matrix} = \begin{bmatrix} 0 & 1 & 0 & 0 & \cdots & 0 \\ 0 & 0 & 1 & 0 & \cdots & 0 \\ 0 & 0 & 0 & 1 & \cdots & 0 \\ \vdots & \vdots & \vdots & & \cdots & \vdots \\ 0 & 0 & 0 & 0 & \cdots & 1 \\ g_r & g_{r-1} & g_{r-2} & g_{r-3} & \cdots & g_1 \end{bmatrix} \times \begin{matrix} s_{0,n} \\ s_{1,n} \\ s_{2,n} \\ \vdots \\ s_{r-2,n} \\ s_{r-1,n} \end{matrix} \tag{3-48}$$

$$= \mathbf{F} \times \mathbf{S}_n$$

and the output b_n at time n is

$$b_n = [g_r \quad g_{r-1} \quad g_{r-2} \quad g_{r-3} \quad \cdots \quad g_1] \times \begin{bmatrix} s_{0,n} \\ s_{1,n} \\ s_{2,n} \\ \vdots \\ s_{r-2,n} \\ s_{r-1,n} \end{bmatrix} \tag{3-49}$$

$$= [g_r \quad g_{r-1} \quad g_{r-2} \quad g_{r-3} \quad \cdots \quad g_1] \times \mathbf{F}^n \times \mathbf{S}_0$$

EXAMPLE 3-12 _____

Calculate the first four outputs of the Galois shift-register generator illustrated in Figure 3-7.

Solution: The Galois state-transition matrix is

$$\mathbf{G} = \begin{bmatrix} 1 & 1 & 0 & 0 \\ 0 & 0 & 1 & 0 \\ 1 & 0 & 0 & 1 \\ 1 & 0 & 0 & 0 \end{bmatrix}$$

and the state at time $n = 0$ is

$$\mathbf{S}_0 = \begin{bmatrix} 1 \\ 0 \\ 0 \\ 0 \end{bmatrix}$$

At any time n, the output is

$$b_n = [1 \quad 0 \quad 0 \quad 0] \times \begin{bmatrix} 1 & 1 & 0 & 0 \\ 0 & 0 & 1 & 0 \\ 1 & 0 & 0 & 1 \\ 1 & 0 & 0 & 0 \end{bmatrix}^n \times \begin{bmatrix} 1 \\ 0 \\ 0 \\ 0 \end{bmatrix}$$

For $n = 0$,

$$b_0 = [1 \quad 0 \quad 0 \quad 0] \times \begin{bmatrix} 1 & 1 & 0 & 0 \\ 0 & 0 & 1 & 0 \\ 1 & 0 & 0 & 1 \\ 1 & 0 & 0 & 0 \end{bmatrix}^0 \times \begin{bmatrix} 1 \\ 0 \\ 0 \\ 0 \end{bmatrix} = [1 \quad 0 \quad 0 \quad 0] \times \begin{bmatrix} 1 \\ 0 \\ 0 \\ 0 \end{bmatrix} = 1$$

For $n = 1$,

$$b_1 = [1 \quad 0 \quad 0 \quad 0] \times \begin{bmatrix} 1 & 1 & 0 & 0 \\ 0 & 0 & 1 & 0 \\ 1 & 0 & 0 & 1 \\ 1 & 0 & 0 & 0 \end{bmatrix}^1 \times \begin{bmatrix} 1 \\ 0 \\ 0 \\ 0 \end{bmatrix}$$

$$= [1 \quad 1 \quad 0 \quad 0] \times \begin{bmatrix} 1 \\ 0 \\ 0 \\ 0 \end{bmatrix} = 1$$

For $n = 2$,

$$b_2 = [1 \quad 0 \quad 0 \quad 0] \times \begin{bmatrix} 1 & 1 & 0 & 0 \\ 0 & 0 & 1 & 0 \\ 1 & 0 & 0 & 1 \\ 1 & 0 & 0 & 0 \end{bmatrix} \times \begin{bmatrix} 1 & 1 & 0 & 0 \\ 0 & 0 & 1 & 0 \\ 1 & 0 & 0 & 1 \\ 1 & 0 & 0 & 0 \end{bmatrix} \times \begin{bmatrix} 1 \\ 0 \\ 0 \\ 0 \end{bmatrix}$$

$$= [1 \quad 0 \quad 0 \quad 0] \times \begin{bmatrix} 1 & 1 & 1 & 0 \\ 1 & 0 & 0 & 1 \\ 0 & 1 & 0 & 0 \\ 1 & 1 & 0 & 0 \end{bmatrix} \times \begin{bmatrix} 1 \\ 0 \\ 0 \\ 0 \end{bmatrix}$$

$$= [1 \quad 1 \quad 1 \quad 0] \times \begin{bmatrix} 1 \\ 0 \\ 0 \\ 0 \end{bmatrix} = 1$$

For $n = 3$,

$$b_3 = [1 \quad 0 \quad 0 \quad 0] \times \begin{bmatrix} 1 & 1 & 0 & 0 \\ 0 & 0 & 1 & 0 \\ 1 & 0 & 0 & 1 \\ 1 & 0 & 0 & 0 \end{bmatrix}^3 \times \begin{bmatrix} 1 \\ 0 \\ 0 \\ 0 \end{bmatrix}$$

$$= [1 \quad 0 \quad 0 \quad 0] \times \begin{bmatrix} 1 & 1 & 1 & 0 \\ 1 & 0 & 0 & 1 \\ 0 & 1 & 0 & 0 \\ 1 & 1 & 0 & 0 \end{bmatrix} \times \begin{bmatrix} 1 & 1 & 0 & 0 \\ 0 & 0 & 1 & 0 \\ 1 & 0 & 0 & 1 \\ 1 & 0 & 0 & 0 \end{bmatrix} \times \begin{bmatrix} 1 \\ 0 \\ 0 \\ 0 \end{bmatrix}$$

$$= [1 \quad 0 \quad 0 \quad 0] \times \begin{bmatrix} 0 & 1 & 1 & 1 \\ 0 & 1 & 0 & 0 \\ 0 & 0 & 1 & 0 \\ 1 & 1 & 1 & 0 \end{bmatrix} \times \begin{bmatrix} 1 \\ 0 \\ 0 \\ 0 \end{bmatrix}$$

$$= [0 \quad 1 \quad 1 \quad 1] \times \begin{bmatrix} 1 \\ 0 \\ 0 \\ 0 \end{bmatrix} = 0$$

These results agree with the results of Example 3-9.

3-3 Maximal-Length Sequences

All of the discussion in Section 3-1 was general in that, except for requiring that $g_0 = 1$, no restrictions were placed on the generator functions. In this section discussion is limited to linear feedback shift-register generators having the form of Figure 3-5 or 3-6 with $g(D)$ a primitive polynomial. Recall that the maximum possible period of a shift-register generator is the smallest N for which the reciprocal $g_r(D)$ of the generator polynomial $g(D)$ divides $D^N + 1$. It can be demonstrated [5] that the reciprocal of a primitive polynomial is also primitive. Thus the smallest N for which a primitive polynomial $g(D)$ of degree r divides $D^N + 1$ is $N = 2^r - 1$. This means that a shift-register initial condition exists which results in a cycle with period $N = 2^r - 1$. Since an r-stage shift register has a total of $2^r - 1$ nonzero states, all states are passed through in this cycle having period $N = 2^r - 1$, and there is only one possible cycle. Shift-register sequences having the maximum possible period for an r-stage shift register are called *maximal-length sequences* or *m-sequences*. Since the shift register passes through all possible states, each different initial condition results in a different phase of the same m-sequence.

3-3.1 Properties of *m*-Sequences

Maximal-length sequences have a number of properties which are useful in their application to spread-spectrum systems. Some of these properties are given here.

PROPERTY I. *A maximal-length sequence contains one more one than zero. The number of ones in the sequence is $\frac{1}{2}(N + 1)$.*

Proof: *Consider the generator of Figure 3-5, where the rightmost symbol of the shift-register state is the output symbol. The shift register passes through all possible nonzero states. Of these states, $2^{r-1} = \frac{1}{2}(N + 1)$ have a one in the right-*

most position, and $2^{r-1} - 1$ have a zero in the rightmost position. Thus there is one more one than zero in the output sequence.

PROPERTY II.   *The modulo-2 sum of an m-sequence and any phase shift of the same sequence is another phase of the same m-sequence* (shift-and-add property).

Proof: *Consider the shift-register generator of Figure 3-5. The output is given by (3-35) for any initial condition. Since any different initial condition results in a different phase of the same m-sequence, two phases $b(D)$ and $b'(D)$ of the same sequence can be written $b(D) = a(D)/g(D)$ and $b'(D) = a'(D)/g(D)$, where $a(D)$ and $a'(D)$ are distinct initial conditions. The modulo-2 sum $b(D) + b'(D) = [a(D) + a'(D)]/g(D) = a''(D)/g(D)$. Since the modulo-2 sum of any two distinct initial conditions is a third distinct initial condition, $a''(D)/g(D) = b''(D)$ is a third distinct phase of the original sequence $b(D)$.*

PROPERTY III.   *If a window of width r is slid along the sequence for N shifts, each r-tuple except the all zero r-tuple will appear exactly once.*

Proof: *Consider the shift-register generator of Figure 3-6. The sequence $b(D)$ passes through the shift register of this generator so that the window of width r is simply the state of the shift register. Since the shift register passes through all nonzero states exactly once, all possible r-tuples appear in the window exactly once.*

PROPERTY IV.   *The periodic autocorrelation function $\theta_b(k)$ is two-valued and is given by*

$$\theta_b(k) = \begin{cases} 1.0 & k = lN \\ -\dfrac{1}{N} & k \neq lN \end{cases} \tag{3-50}$$

where l is any integer and N is the sequence period.

Proof: *The value of the periodic autocorrelation function $\theta_b(k)$ is $(N_A - N_D)/N$, where N_A is the number of zeros and N_D is the number of ones in the modulo-2 sum of the sequence $\mathbf{b}$ and the kth cyclic shift of $\mathbf{b}$. For $k = lN$, the kth cyclic shift of $\mathbf{b}$ is identical to $\mathbf{b}$, since the sequence period is N, so that the modulo-2 sum contains all zeros and $N_A = N$, $N_D = 0$, and $\theta_b(lN) = N/N = 1.0$. For $k \neq lN$, the modulo-2 sum is some phase of the original sequence by Property II. Then, by Property I, there is one more one than zero in the modulo-2 sum, so that $N_A - N_D = -1$ and $\theta_b(k) = -1/N$.*

PROPERTY V.   *Define a run as a subsequence of identical symbols within the m-sequence. The length of this subsequence is the length of the run. Then, for any m-sequence, there is*

1. *1 run of ones of length r.*
2. *1 run of zeros of length r − 1.*
3. *1 run of ones and 1 run of zeros of length r − 2.*
4. *2 runs of ones and 2 runs of zeros of length r − 3.*
5. *4 runs of ones and 4 runs of zeros of length r − 4.*

$$\vdots$$

r. 2^{r-3} *runs of ones and* 2^{r-3} *runs of zeros of length* 1.

Proof [8]: *Consider the shift register of Figure 3-6. There can be no run of ones having length $l \geq r$ since this would require that the all-ones shift register state be followed by another all-ones state. This cannot occur since each shift register state occurs once and only once during N cycles. Thus there is a single run of r consecutive ones, and this run is preceded by a zero and followed by a zero.*

A run of $r − 1$ ones must be preceded by and followed by a zero. This requires that the shift register state which is $r − 1$ ones followed by a 0 be followed immediately by the state which is a 0 followed by $r − 1$ ones. These two states are also passed through in the generation of the run of r ones, where they are separated by the all-ones state. Since each state occurs only once, there can be no run of $r − 1$ ones. A run of $r − 1$ zeros must be preceded by and followed by 1's. Thus the shift register must pass through the state which is a 1 followed by $r − 1$ zeros. This state occurs only once, so there is a single run of $r − 1$ zeros.

Now consider a run of k ones where $1 \leq k < r − 1$. Each run of k ones must be preceded by and followed by a 0. Thus the shift register must pass through the state which is a 0 followed by k ones followed by a 0, with the $r − k − 2$ remaining positions taking on arbitrary values. There are 2^{r-k-2} possible ways to complete these remaining positions in the shift register, so there are 2^{r-k-2} runs of k ones. Similarly, there are 2^{r-k-2} runs of k zeros.

3-3.2 Power Spectrum of *m*-Sequences

The power spectrum of the spreading waveform $c(t)$ is frequently used in the analysis of the performance of spread-spectrum systems. This power spectrum is easily calculated using the Wiener–Khintchine theorem and Property IV above for maximal-length spreading codes. The power spectrum of $c(t)$ is the Fourier transform of the autocorrelation function $R_c(\tau)$, which is given by (3-8). For $0 \leq \tau \leq T_c$, $k = 0$, $\tau_\epsilon = \tau$, and (3-8) becomes

$$R_c(\tau) = \left(1 - \frac{\tau}{T_c}\right) - \frac{1}{N}\left(\frac{\tau}{T_c}\right) \qquad 0 \leq \tau \leq T_c$$

$$= 1 - \frac{\tau}{T_c}\left(1 + \frac{1}{N}\right)$$

(3-51a)

where (3-50) has been used to evaluate $\theta_b(k)$. For $T_c < \tau < (N − 1)T_c$, $k \neq lN$ and

$(k + 1) \neq lN$ for any integer l, so that

$$R_c(\tau) = \left(1 - \frac{\tau_\epsilon}{T_c}\right)\left(-\frac{1}{N}\right) + \left(\frac{\tau_\epsilon}{T_c}\right)\left(-\frac{1}{N}\right) \qquad T_c < \tau < (N-1)T_c$$

$$= -\frac{1}{N}$$

$(3\text{-}51b)$

For $(N-1)T_c \leq \tau < NT_c$, $k = N-1$ and $k + 1 = N$, so that (3-8) becomes

$$R_c(\tau) = \left(1 - \frac{\tau_\epsilon}{T_c}\right)\left(-\frac{1}{N}\right) + \frac{\tau_\epsilon}{T_c} \qquad (N-1)T_c \leq \tau < NT_c$$

$$= \frac{\tau_\epsilon}{T_c}\left(1 + \frac{1}{N}\right) - \frac{1}{N}$$

$(3\text{-}51c)$

where $\tau_\epsilon = \tau - (N-1)T_c$, so that $0 < \tau_\epsilon \leq T_c$. Since $\theta_b(k)$ of (3-50) is periodic, $R_c(\tau)$ is also periodic and has a period $T = NT_c$. Thus (3-51) defines one complete cycle of $R_c(\tau)$. The autocorrelation function $R_c(\tau)$ is illustrated in Figure 3-12.

The power spectrum is found by taking the Fourier transform of (3-51). The result is

$$S_c(f) = \sum_{m=-\infty}^{\infty} P_m \, \delta(f - mf_0) \qquad (3\text{-}52)$$

where $P_0 = 1/N^2$, $P_m = [(N+1)/N^2] \, \text{sinc}^2(m/N)$, and $f_0 = 1/NT_c$. The power spectrum as a function of f is illustrated in Figure 3-13. This power spectrum consists of discrete spectral lines at all harmonics of $1/NT_c$. The envelope of the amplitude of these lines is given by $[(N+1)/N^2] \, \text{sinc}^2(fT_c)$ except for the dc term, which has an amplitude $1/N^2$. Note that the ordinate in Figure 3-13 is absolute and is not decibels.

Suppose that the m-sequence $c(t)$ is used to biphase modulate a sinusoidal carrier having power P and frequency f_0. The modulated carrier is

$$s(t) = \sqrt{2P} \, c(t) \cos 2\pi f_0 t \qquad (3\text{-}53)$$

The power spectrum of this modulated carrier is the convolution of the power

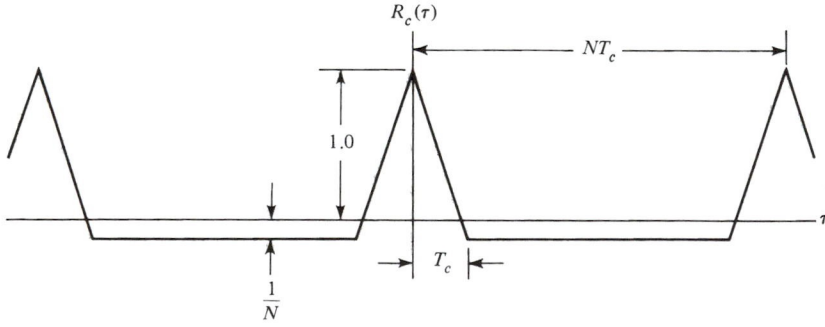

FIGURE 3-12. Autocorrelation function for a maximal-length sequence with chip duration T_c and period NT_c.

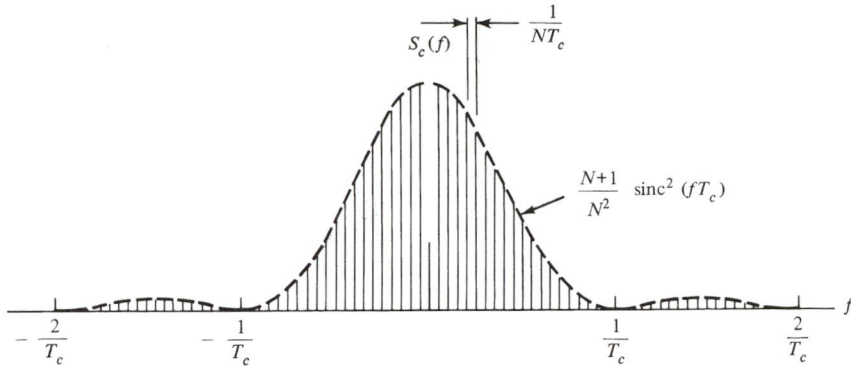

FIGURE 3-13. Power spectrum of a maximal-length sequence with chip duration T_c and period NT_c.

spectrum of the carrier and the power spectrum of the spreading code. Thus

$$S_s(f) = S_c(f) * \frac{P}{2} \delta(f - f_0) + S_c(f) * \frac{P}{2} \delta(f + f_0) \tag{3-54}$$

and the resultant power spectrum is a translation of the discrete spectrum $S_c(f)$ upward and downward by a frequency f_0. In most spread-spectrum systems the carrier is randomly modulated by data as well as the spreading code. In these cases the transmitted spectrum is continuous and not discrete.

3-3.3 Tables of Polynomials Yielding *m*-Sequences

It is often necessary to design circuits that generate *m*-sequences having a particular number of stages. Since finding the primitive polynomials used to generate these sequences is difficult, a number of authors have generated tables of primitive polynomials for quick reference. In particular, Peterson and Weldon [5] have an extensive table of polynomials in their Appendix C. Part of the table from Ref. 5 is reproduced here as Table 3-5. In the table, all polynomials are specified by an octal number that defines the coefficients of $g(D)$. The octal number gives the coefficients of $g(D)$ beginning with g_0 on the right and proceeding to g_r in the last nonzero position on the left.

TABLE 3-5. Primitive Polynomials Having Degree $r \le 34$

Degree	Octal Representation of Generator Polynomial (g_0 on right to g_r on left)
2	[7]*
3	[13]*
4	[23]*
5	[45]*, [75], [67]

TABLE 3-5. Primitive Polynomials Having Degree $r \leq 34$ (continued)

Degree	Octal Representation of Generator Polynomial (g_0 on right to g_r on left)
6	[103]*, [147], [155]
7	[211]*, [217], [235], [367], [277], [325], [203]*, [313], [345]
8	[435], [551], [747], [453], [545], [537], [703], [543]
9	[1021]*, [1131], [1461], [1423], [1055], [1167], [1541], [1333], [1605], [1751], [1743], [1617], [1553], [1157]
10	[2011]*, [2415], [3771], [2157], [3515], [2773], [2033], [2443], [2461], [3023], [3543], [2745], [2431], [3177]
11	[4005]*, [4445], [4215], [4055], [6015], [7413], [4143], [4563], [4053], [5023], [5623], [4577], [6233], [6673]
12	[10123], [15647], [16533], [16047], [11015], [14127], [17673], [13565], [15341], [15053], [15621], [15321], [11417], [13505]
13	[20033], [23261], [24623], [23517], [30741], [21643], [30171], [21277], [27777], [35051], [34723], [34047], [32535], [31425]
14	[42103], [43333], [51761], [40503], [77141], [62677], [44103], [45145], [76303], [64457], [57231], [64167], [60153], [55753]
15	[100003]*, [102043], [110013], [102067], [104307], [100317], [177775], [103451], [110075], [102061], [114725], [103251], [100021]*, [100201]*
16	[210013], [234313], [233303], [307107], [307527], [306357], [201735], [272201], [242413], [270155], [302157], [210205], [305667], [236107]
17	[400011]*, [400017], [400431], [525251], [410117], [400731], [411335], [444257], [600013], [403555], [525327], [411077], [400041]*, [400101]*
18	[1000201]*, [1000247], [1002241], [1002441], [1100045], [1000407], [1003011], [1020121], [1101005], [1000077], [1001361], [1001567], [1001727], [1002777]
19	[2000047], [2000641], [2001441], [2000107], [2000077], [2000157], [2000175], [2000257], [2000677], [2000737], [2001557], [2001637], [2005775], [2006677]
20	[4000011]*, [4001051], [4004515], [6000031], [4442235]
21	[10000005]*, [10040205], [10020045], [10040315], [10000635], [10103075], [10050335], [10002135], [17000075]
22	[20000003]*, [20001043], [22222223], [25200127], [20401207], [20430607], [20070217]

TABLE 3-5. Primitive Polynomials Having Degree $r \leq 34$ (continued)

Degree	Octal Representation of Generator Polynomial (g_0 on right to g_r on left)
23	[40000041]*, [40404041], [40000063], [40010061], [50000241], [40220151], [40006341], [40405463], [41103271], [41224445], [4043561]
24	[100000207], [125245661], [113763063]
25	[200000011]*, [200000017], [204000051], [200010031], [200402017], [252001251], [201014171], [204204057], [200005535], [200014731]
26	[400000107], [430216473], [402365755], [426225667], [510664323], [473167545], [411335571]
27	[1000000047], [1001007071], [1020024171], [1102210617], [1250025757], [1257242631], [1020560103], [1112225171], [1035530241]
28	[2000000011]*, [2104210431], [2000025051], [2020006031], [2002502115], [2001601071]
29	[4000000005]*, [4004004005], [4000010205], [4010000045], [4400000045], [4002200115], [4001040115], [4004204435], [4100060435], [4040003075], [4004064275]
30	[10,040,000,007], [10,104,264,207], [10,115,131,333], [11,362,212,703], [10,343,244,533]
31	[20,000,000,011]*, [20,000,000,017], [20,000,020,411], [21,042, 104, 211] [20,010,010,017], [20,005,000,251], [20,004,100,071], [20,202,040,217] [20,000,200,435], [20,060,140,231], [21,042,107,357]
32	[40,020,000,007], [40,460,216,667], [40,035,532,523], [42,003,247,143], [41,760,427,607]
33	[100,000,020,001]*, [100,020,024,001], [104,000,420,001], [100,020,224,401], [111,100,021,111], [100,000,031,463], [104,020,466,001], [100,502,430,041], [100,601,431,001]
34	[201,000,000,007], [201,472,024,107], [377,000,007,527], [225,213,433,257], [227,712,240,037], [251,132,516,577], [211,636,220,473], [200,000,140,003]
35	[400,000,000,005]*
36	[1,000,000,004,001]*
37	[2,000,000,012,005]
38	[4,000,000,000,143]
39	[10,000,000,000,021]*
40	[20,000,012,000,005]
61	[200,000,000,000,000,000,047]
89	[400,000,000,000,000,000,000,000,000,151]

Source: Refs. 5, 6, and 9.

EXAMPLE 3-13 ———————————————————————————————

The table contains the entry [367]. Expanding the octal entry 367 into binary form yields

	3			6			7		octal

$$0 \quad 1 \quad 1 \quad 1 \quad 1 \quad 0 \quad 1 \quad 1 \quad 1 \qquad \text{binary}$$

$$g_7 \quad g_6 \quad g_5 \quad g_4 \quad g_3 \quad g_2 \quad g_1 \quad g_0 \qquad \text{coefficient}$$

so that

$$g(D) = 1 + D + D^2 + D^4 + D^5 + D^6 + D^7$$

The shift register generator can be either the form of Figure 3-5 or 3-6.

Table 3-5 is a list of primitive polynomials of all degrees up to 40 and degrees 61 and 89. Each entry in brackets represents one primitive polynomial as a series of octal numbers, exactly as explained above. The entries followed by an asterisk correspond to circuit implementation with only two feedback connections. Two feedback connection implementations are very useful for high-speed applications. No reciprocal polynomials are listed in Table 3-5. Since the reciprocal polynomial of a primitive polynomial is also primitive, each entry of Table 3-5 can be used to generate two distinct m-sequences. It can be demonstrated that the sequences generated by the reciprocal polynomial $g_r(D)$ is equivalent to the reverse of the sequence generated by $g(D)$.

EXAMPLE 3-14 ———————————————————————————————

Consider the sequence generated by the polynomial corresponding to the entry [13] of Table 3-5. The primitive polynomial is $g(D) = 1 + D + D^3$ and its reciprocal is $D^3 g(1/D) = D^3 + D^2 + 1$. Using the configuration of Figure 3-5 with an initial load of $a(D) = 1$, the output sequence for $g(D)$ is, by the polynomial long division, $1 + D + D^2 + D^4 + D^7 + \cdots$. The output corresponding to the same initial state for the reciprocal polynomial is, again by long division, $1 + D^2 + D^3 + D^4 + D^7 + \cdots$.

The two output sequences are

$$g(D) \rightarrow 1110100,1110100,\ldots$$

$$g_r(D) \rightarrow 1011100,1011100,\ldots$$

which, except for a phase shift, are simply the reverse of one another.

The list of primitive polynomials in Table 3-5 is not complete. Additional polynomials may be found in the references [5,6]. The number N_p of primitive polyno-

TABLE 3-6. Number of Primitive Polynomials N_p of Degree r

r	N_p	r	N_p
2	1	16	2,048
3	2	17	7,710
4	2	18	8,064
5	6	19	27,594
6	6	20	24,000
7	18	21	84,672
8	16	22	120,032
9	48	23	356,960
10	60	24	276,480
11	176	25	1,296,000
12	144	26	1,719,900
13	630	27	4,202,496
14	756	28	4,741,632
15	1,800	29	18,407,808

Source: Ref. 22.

mials of degree r that exist may be found from [22]

$$N_p = \frac{2^r - 1}{r} \prod_{i=1}^{J} \frac{p_i - 1}{p_i} \tag{3-55}$$

where p_i are the prime factors of $2^r - 1$. That is,

$$2^r - 1 = \prod_{i=1}^{J} p_i^{e_i}$$

where e_i are positive integers and p_i are prime numbers. Consider, for example, $r = 9$. The prime factors of $2^9 - 1$ are 7 and 73. Thus $p_1 = 7$ and $p_2 = 73$, and

$$N_p = \frac{2^9 - 1}{9} \left(\frac{7 - 1}{7} \right) \left(\frac{73 - 1}{73} \right) = 48$$

Table 3-6 gives the value of N_p for r between 2 and 29.

3-3.4 Partial Autocorrelation Properties of m-Sequences

The autocorrelation properties of maximal-length sequences are defined over a complete cycle of the sequence. That is, the two-valued autocorrelation of Property IV can be guaranteed only when the integration of (3-2) is over a full period of the waveform $c(t)$. In Chapter 5, where the code synchronization problem of spread-spectrum communications is addressed, it will be shown that rapid synchronization of long codes often requires an estimate of the correlation between the received code and the receiver despreading code be made in less than a full code period. Thus the correlation estimate is based on a correlation over a partial period and is related to the partial autocorrelation properties of the code.

Since the partial autocorrelation is associated with an integration over a fraction of the code period, the partial autocorrelation function is dependent on the size of this fraction and the starting time of the integration. The *partial autocorrelation function* of the spreading waveform $c(t)$ is defined by

$$R_c(\tau,t,T_w) = \frac{1}{T_w} \int_t^{t+T_w} c(\lambda)c(\lambda + \tau) \, d\lambda \tag{3-56}$$

where T_w is the duration of the correlation and t is the starting time of the correlation. Using (3-1) for $c(t)$ and the substitution $\gamma = \lambda - t$ yields

$$R_c(\tau,t,T_w)$$

$$= \frac{1}{T_w} \sum_{n=-\infty}^{\infty} \sum_{m=-\infty}^{\infty} a_m a_n \int_0^{T_w} p(\gamma + t - mT_c)p(\gamma + t + \tau - nT_c) \, d\gamma \tag{3-57}$$

Now let $\tau = kT_c + \tau_\epsilon$, $T_w = WT_c$, and assume that $t = k'T_c$. Then the integral is nonzero only for $n = m + k$ or $n = m + k + 1$ and the autocorrelation can be written

$$R_c(\tau_\epsilon,k,k',W)$$

$$= \frac{1}{WT_c} \sum_{m=-\infty}^{\infty} a_m a_{m+k} \int_0^{WT_c} p[\gamma - (m - k')T_c]p[\gamma - (m - k')T_c + \tau_\epsilon] \, d\gamma \tag{3-58}$$

$$+ \frac{1}{WT_c} \sum_{m=-\infty}^{\infty} a_m a_{m+k+1} \int_0^{WT_c} p[\gamma - (m - k')T_c]p[\gamma - (m - k' + 1)T_c + \tau_\epsilon] \, d\gamma$$

The integrand of (3-58) is nonzero within the limits of integration only when $0 \leq (m - k')T_c \leq (W - 1)T_c$, which implies that the limits on the summations can be reduced to $k' \leq m \leq w + k' - 1$. For any fixed value of m, the integrand of the first integral is nonzero only for $(m - k')T_c \leq \gamma \leq (m + 1 - k')T_c - \tau_\epsilon$, and the integrand of the second integral is nonzero only for $(m + 1 - k')T_c - \tau_\epsilon \leq \gamma \leq (m + 1 - k')T_c$, so that (3-58) can be written

$$R_c(\tau_\epsilon,k,k',W) = \frac{1}{WT_c} \sum_{m=k'}^{W+k'-1} a_m a_{m+k} \int_{(m-k')T_c}^{(m+1-k')T_c - \tau_\epsilon} p[\gamma - (m - k')T_c]$$

$$\times \; p[\gamma - (m - k')T_c + \tau_\epsilon] \, d\gamma$$

$$+ \frac{1}{WT_c} \sum_{m=k'}^{W+k'-1} a_m a_{m+k+1} \int_{(m+1-k')T_c - \tau_\epsilon}^{(m+1-k')T_c} p[\gamma - (m - k')T_c]$$

$$\times \; p[\gamma - (m + 1 - k')T_c + \tau_\epsilon] \, d\gamma \tag{3-59}$$

$$= \frac{1}{W} \sum_{m=k'}^{W+k'-1} a_m a_{m+k} \left(1 - \frac{\tau_\epsilon}{T_c} \right)$$

$$+ \frac{1}{W} \sum_{m=k'}^{W+k'-1} a_m a_{m+k+1} \frac{\tau_\epsilon}{T_c}$$

for $|\tau_\epsilon| \leq T_c$.

The *discrete partial autocorrelation function* of a sequence $b(D)$ is defined by [2]

$$\theta_b(k,k',W) = \frac{1}{W} \sum_{m=k'}^{k'+W-1} a_m a_{m+k} \qquad (3\text{-}60)$$

so that the partial autocorrelation function can be written

$$R_c(\tau_\epsilon,k,k',W) = \left(1 - \frac{\tau_\epsilon}{T_c}\right)\theta_b(k,k',W) + \frac{\tau_\epsilon}{T_c} \theta_b(k+1, k', W) \qquad (3\text{-}61)$$

for $|\tau_\epsilon| \leq T_c$. Thus $R_c(\tau,t,T_w)$ can easily be calculated from knowledge of $\theta_b(k,k',W)$. The discrete partial autocorrelation function can be calculated in the same manner as the discrete periodic autocorrelation function. The value of $\theta_b(k,k',W)$ is the number of agreements N_A minus the number of disagreements between $\mathbf{b}(0)$ and $\mathbf{b}(k)$ over the window beginning at k' and ending at $k' + W$ divided by W. This is equivalent to the difference between the number of zeros and the number of ones over the same window of the modulo-2 sum of $\mathbf{b}(0)$ and $\mathbf{b}(k)$.

EXAMPLE 3-15

Evaluate $\theta_b(6,k',7)$ for the 15-bit m-sequence generated using the generator of Figure 3-5 with an initial condition $a(D) = 1$ and the primitive polynomial $g(D) = 1 + D + D^4$.

Solution: One cycle of $\mathbf{b}(0)$ and $\mathbf{b}(6)$ and their modulo-2 sum are

$$\mathbf{b}(0) \qquad = 1\ 1\ 1\ 1\ 0\ 1\ 0\ 1\ 1\ 0\ 0\ 1\ 0\ 0\ 0$$

$$\mathbf{b}(6) \qquad = 0\ 1\ 1\ 0\ 0\ 1\ 0\ 0\ 0\ 1\ 1\ 1\ 1\ 0\ 1$$

$$\mathbf{b}(0) + \mathbf{b}(6) = 1\ 0\ 0\ 1\ 0\ 0\ 0\ 1\ 1\ 1\ 1\ 0\ 1\ 0\ 1$$

Then $\theta_b(6,k',7)$ is the number of zeros minus the number of 1's in $\mathbf{b}(0) + \mathbf{b}(6)$ in a seven-unit window beginning at k'. The value of $\theta_b(6,k',7)$ is plotted in Figure 3-14 as a function of k'.

Observe that the partial autocorrelation function is not well behaved as was the full-period autocorrelation function. The partial-period autocorrelation is not two-valued and its variation as a function of window size and window placement can cause serious difficulties if not taken into account in the system design. The mean and variance over k' of $\theta_b(k,k',W)$ are useful quantities for the spread-spectrum system designer. Because of Property II of maximal-length sequences, the modulo-2 sum of $\mathbf{b}(0)$ and $\mathbf{b}(k)$ is another phase $\mathbf{b}(q)$ of the same m-sequence. Then

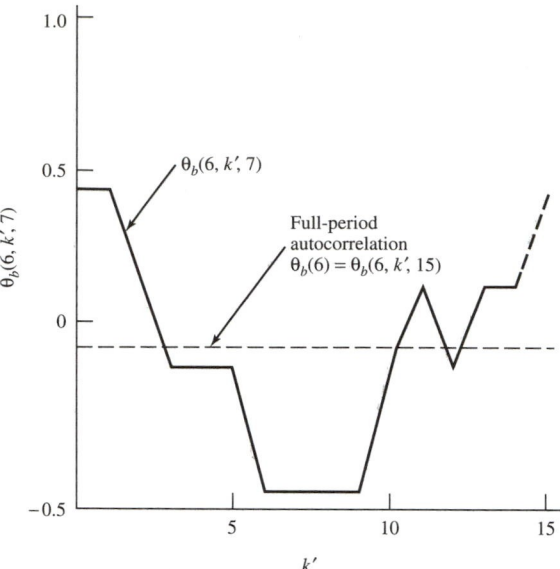

FIGURE 3-14. Discrete partial autocorrelation function $\theta_b(6,k',7)$ for 15-symbol m-sequence generated by $g(D) = 1 + D + D^4$.

$$\theta_b(k,k',W) = \frac{1}{W} \sum_{i=0}^{W-1} a_{i+q+k'}$$

where $a_i = (-1)^{b_i}$; and the average of $\theta_b(k,k',W)$ over all k' becomes

$$\overline{\theta_b(k,k',W)} = \frac{1}{N} \sum_{k'=0}^{N-1} \frac{1}{W} \sum_{i=0}^{W-1} a_{i+q+k'}$$

$$= \frac{1}{WN} \sum_{i=0}^{W-1} \sum_{k'=0}^{N-1} a_{i+q+k'} \tag{3-62}$$

Because of Property I of the m-sequence, the inner summation in the last line of (3-62) is equal to -1 for all k' and q so that [10]

$$\overline{\theta_b(k,k',W)} = -\frac{1}{N} \tag{3-63}$$

The second moment of the discrete partial autocorrelation function is [10]

$$\overline{\theta_b^2(k,k',W)} = \frac{1}{N} \sum_{k'=0}^{N-1} \theta_b^2(k,k',W)$$

$$= \frac{1}{W}\left(1 - \frac{W-1}{N}\right) \tag{3-64}$$

The variance over k' of $\theta_b(k,k',W)$ is then

$$\text{var}\,[\theta_b(k,k',W)] = \overline{\theta_b^2(k,k',W)} - [\overline{\theta_b(k,k',W)}]^2$$

$$= \frac{1}{W}\left(1 - \frac{W-1}{N}\right) - \frac{1}{N^2} \tag{3-65}$$

Observe that for $W = N$ the variance equals zero as expected.

EXAMPLE 3-16 ———————————————————————

Plot the mean and variance as a function of window size W for the family of 31-bit maximal-length sequences.

Solution: For any window size,

$$\overline{\theta_b(k,k',W)} = -\frac{1}{31}$$

Equation (3-65) for the variance is

$$\text{var}[\theta_b(k,k',W)] = \frac{1}{W}\left(1 - \frac{W-1}{31}\right) - \left(\frac{1}{31}\right)^2$$

These relationships are plotted in Figure 3-15.

The results just derived can be used to determine approximate thresholds when correlations are being performed in order to determine whether or not two code phases agree. Higher-order moments of $\theta_b(k,k',W)$ are calculated in Ref. 10. These higher-order moments are especially useful when a lowpass filtered m-sequence is being used as a noise source.

3-3.5 Power Spectrum of $c(t)c(t + \epsilon)$

In Chapter 2 the despreading operation in all the receivers was accomplished by correlating the received signal with a replica of the spreading waveform $c(t)$. When the receiver-generated code replica is at exactly the correct phase, despreading occurs and the data modulation can then be extracted using a conventional data demodulator. In Chapter 4 it will be shown that one method of maintaining or tracking the correct receiver code phase will involve correlating the received signal with a replica of the code waveform which is offset in phase by some fraction of a code period. The power spectrum of the output $b(t,\epsilon) = c(t)c(t + \epsilon)$ of the despreading correlator is calculated in Appendix D. The result calculated in Appendix D is valid only for $|\epsilon| \le T_c$. The result is

$$S_b(f,\epsilon) = \left[1 - \left(1 + \frac{1}{N}\right)\frac{|\epsilon|}{T_c}\right]^2 \delta(f)$$

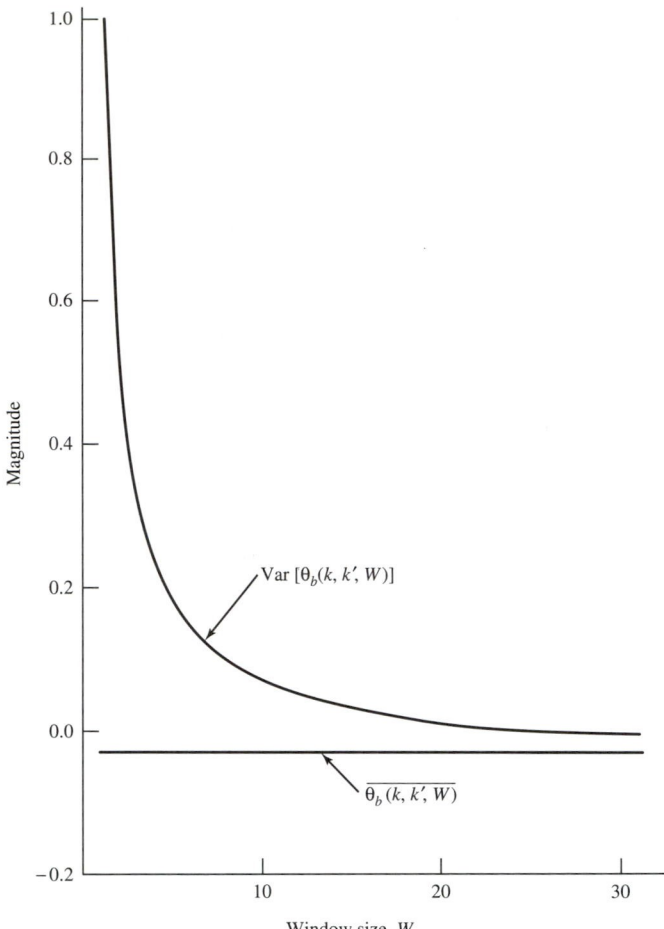

FIGURE 3-15. Mean and variance of the discrete partial autocorrelation function as a function of window size for 31-bit *m*-sequences.

$$+ \left(1 + \frac{1}{N}\right)\left(\frac{|\epsilon|}{T_c}\right)^2 \sum_{\substack{n=-\infty \\ n \neq 0}}^{\infty} \text{sinc}^2(nf_c|\epsilon|)\delta(f - nf_c) \qquad (3\text{-}66)$$

$$+ \frac{N+1}{N^2}\left(\frac{|\epsilon|}{T_c}\right)^2 \sum_{\substack{m=-\infty \\ m \neq 0}}^{\infty} \text{sinc}^2\left(\frac{mf_c}{N}|\epsilon|\right)\delta\left(f - \frac{mf_c}{N}\right)$$

where N is the *m*-sequence period and $T_c = 1/f_c$ is the code clock period. The power spectrum of $b(t,\epsilon)$ has been calculated for $|\epsilon| \geq T_c$ in Ref. 11.

The power spectrum $S_b(f,\epsilon)$ of $b(t,\epsilon)$ is illustrated in Figure 3-16 for $\epsilon = 0, 0.1T_c$, $0.5T_c$, and T_c. For $\epsilon = 0$ observe that all the spectral lines collapse into a single spectral line at zero frequency. This corresponds to complete despreading of the spread-spectrum signal. For $\epsilon = T_c$ the function $b(t,\epsilon)$ is simply a phase-shifted

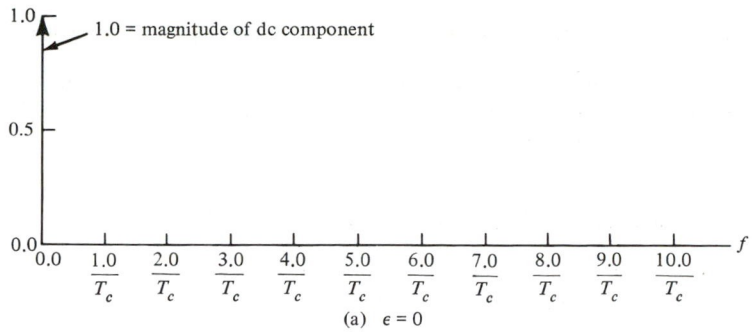

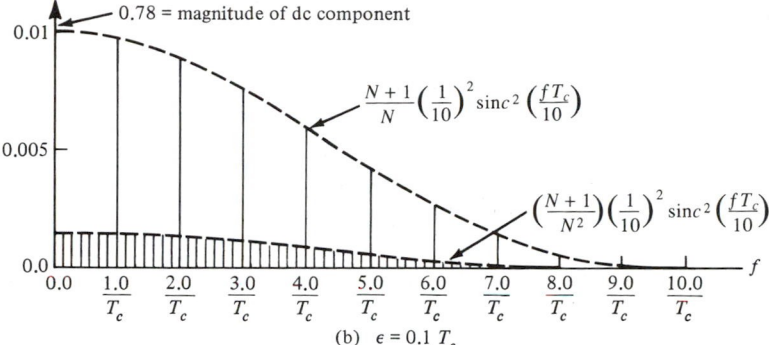

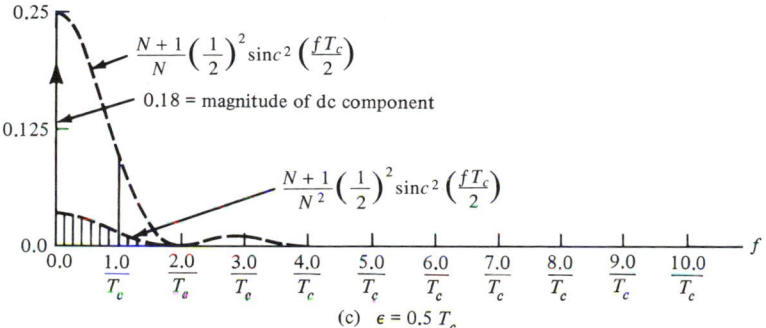

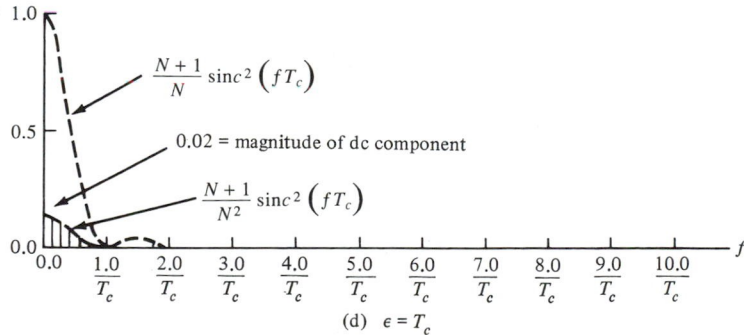

FIGURE 3-16. Power spectrum of $c(t)\,c(t+\epsilon)$ for various ϵ and $N = 7$.

replica of $c(t)$ by the shift-and-add property, so $S_b(f,\epsilon) = S_c(f)$. Finally, observe that the power spectrum for any $\epsilon \neq 0$ or T_c is significantly wider than the spectrum of the spreading waveform $c(t)$.

3-3.6 Generation of Specific Delays of an m-Sequence

It is sometimes useful to be able to generate two different phases of an m-sequence for phase differences that would be impractical to generate using a shift register or a delay line. For example, different phase shifts of a very long code may be used to distinguish different users in a direct-sequence spread-spectrum multiple-access communications system. The spread-spectrum receiver selects the desired received signal by changing the phase of the reference despreading code. The desired phase changes may be equivalent to thousands of code symbols thus making the use of straightforward delay-line phase shifters impractical. Rapid initial synchronization of a direct-sequence receiver may require the generation of multiple phases of a very long spreading code. In this case, multiple acquisition correlators (see Chapter 5) using different phases of the spreading code are employed. The initial phases of the reference spreading codes may be separated by thousands of code symbols again making shift-register phase shifters impractical. The methods discussed in this section are practical means of generating any desired phase shift of a maximal-length sequence.

Two techniques are discussed in this section for generating specific delays of an m-sequence. One of these methods is simply to calculate the shift-register initial conditions required to generate a sequence delayed by k chips from the sequence generated from another specific initial condition. The other method makes use of the shift-and-add property of m-sequence generators. Both methods require knowledge of the mechanics of finite-field arithmetic already discussed.

Determining the Initial Condition Yielding a Specific Delay. Consider the sequence generator of Figure 3-5. Given an initial condition $a(D)$, the output of this generator is $b(D) = a(D)/g(D)$. Another initial condition $a'(D)$ will produce another output sequence $b'(D)$, which is equal to $D^k b(D)$; that is, $b'(D)$ is the sequence $b(D)$ delayed by k chips. The task being addressed here is determining $a'(D)$ for a specific k.

This problem is most easily solved if the states $a(D)$ of the m-sequence generator are associated with elements of the extension field $\mathrm{GF}(2^m)$ defined by the primitive polynomial $g(D)$, where m is the number of elements in the shift register and $m - 1$ is the degree of $a(D)$. The degree of $g(D)$ is m. The state $a(D)$ will sequence through all elements of $\mathrm{GF}(2^m)$. Recall from an earlier discussion that each element in the sequence of elements of $\mathrm{GF}(2^m)$ in polynomial form is generated by multiplying the preceding element by D and taking the remainder when dividing this product by $g(D)$. Let $q(D)$ represent an element of $\mathrm{GF}(2^m)$. Then the element l units later, denoted by $q'(D)$, satisfies

$$D^l q(D) = p(D)g(D) + q'(D) \tag{3-67}$$

The contents of the shift register at time zero are specified by $a(D)$. From Figure

3-5 it is seen that the contents of the shift register at time 1, denoted by $a'(D)$, satisfies

$$a'(D) = D^{-1}a(D) + D^{-1}a_0 g(D) \tag{3-68}$$

or, equivalently,

$$D^{-1}a(D) = D^{-1}a_0 g(D) + a'(D) \tag{3-69}$$

Since the contents of the shift register represent elements of $GF(2^m)$, it is appropriate to consider D^{-1} as the multiplicative inverse element of D in $GF(2^m)$ so that $D^{-1} = D^{2^m-2}$ and (3-69) becomes

$$D^{2^m-2}a(D) = a_0 D^{2^m-2}g(D) + a'(D) \tag{3-70}$$

This equation has the same form as (3-67) with $l = 2^m - 2$, so that it is concluded that the contents of the shift register advance $2^m - 2$ steps through the sequence of elements of $GF(2^m)$ on each cycle. Since there are only $2^m - 1$ nonzero elements of $GF(2^m)$, the register may also be considered to cycle through the elements of $GF(2^m)$ in reverse order.

EXAMPLE 3-17

The elements of $GF(2^4)$ are given in Example 3-6 for the primitive polynomial $g(D) = 1 + D + D^4$. The shift-register generator corresponding to $g(D)$ is illustrated in Figure 3-17 together with the contents of the register $a(D)$ beginning with state $a(D) = 1$. The element of $GF(2^m)$ corresponding to $a(D)$ is found by comparing $a(D)$ with the polynomials of Example 3-6.

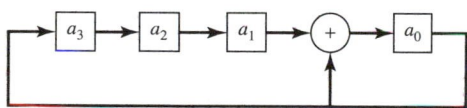

Cycle	Register state	$a(D)$	Element of $GF(2^4)$
0	0 0 0 1	1	D^0
1	1 0 0 1	$1 \quad\quad\quad + D^3$	D^{14}
2	1 1 0 1	$1 \quad\quad + D^2 + D^3$	D^{13}
3	1 1 1 1	$1 + D + D^2 + D^3$	D^{12}
4	1 1 1 0	$D + D^2 + D^3$	D^{11}
5	0 1 1 1	$1 + D + D^2$	D^{10}
6	1 0 1 0	$D \quad\quad + D^3$	D^9
7	0 1 0 1	$1 \quad + D^2$	D^8
8	1 0 1 1	$1 + D \quad\quad + D^3$	D^7
9	1 1 0 0	$D^2 + D^3$	D^6
10	0 1 1 0	$D + D^2$	D^5
11	0 0 1 1	$1 + D$	D^4
12	1 0 0 0	D^3	D^3
13	0 1 0 0	D^2	D^2
14	0 0 1 0	D	D^1

FIGURE 3-17. Comparison of shift-register states with elements of $GF(2^4)$ for a typical m-sequence.

At this point the problem of finding the shift-register initial conditions corresponding to a particular advance or delay reduces to a problem of manipulating elements of $GF(2^m)$ using the techniques developed earlier. With $a(D)$ defining one initial condition and $a'(D)$ defining the initial condition corresponding to an *advance* of k units, the discussion above implies that $a'(D)$ is the remainder found when dividing $D^{k(2^m-2)}a(D)$ by $g(D)$. The product $D^{k(2^m-2)}$ can be reduced using the fact that $D^{2^m-1} = 1$. The load corresponding to a *delay* of k units is the remainder found when dividing $D^k a(D)$ by $g(D)$.

EXAMPLE 3-18

Consider the generator of Example 3-17 with an initial condition of $a(D) = 1$. What initial condition $a'(D)$ will produce an advance of 20 units? What $a'(D)$ will produce a delay of 20 units?

Solution: The period of the m-sequence is 15 units, so that an advance of 20 units is equivalent to an advance of 5 units. $D^{k(2^m-2)} = D^{5\cdot14} = D^{70} = D^{10}$. Thus $a(D)D^{k(2^m-2)} = D^{10} \cdot 1 = D^{10}$. From Example 3-6, $D^{10} = 1 + D + D^2$, and since the degree of $1 + D + D^2$ is less than the degree of $g(D) = 1 + D + D^4$, the remainder desired is just $1 + D + D^2$. Thus $a'(D) = 1 + D + D^2$. The same remainder is found if D^{10} itself is divided by $g(D)$. The result can be verified by comparison with the cycle 5 shift register state of Figure 3-17.

A delay of 20 units is equivalent to a delay of 5 units, so that $a'(D)$ is the remainder when dividing $D^5 a(D)$ by $g(D)$. This remainder is $D^2 + D = a'(D)$. This result can also be verified by comparison with the table within Figure 3-17.

The technique just described works only for the shift-register configuration of Figure 3-5. The shift-register configuration of Figure 3-6 unfortunately does not cycle through the elements of $GF(2^m)$. The most convenient technique for finding the desired initial condition $a'(D)$ for the configuration of Figure 3-6 is to find the corresponding initial conditions for the configuration of Figure 3-5. Recall that this same technique was employed when the output sequence of this shift register configuration was calculated earlier. The procedure for generating the initial condition for one configuration from the other is described in (3-36) through (3-39).

EXAMPLE 3-19

Consider the shift register $g(D) = 1 + D + D^4$ and initial condition $a_1(D) = 1$. The generator is illustrated in Figure 3-18 together with all register states. Denote the corresponding initial condition for the configuration of Figure 3-5 by $e(D)$. Equation (3-39) becomes

$$e(D) = g(D)(a(D) + b_r D^r + \cdots)$$

and equating coefficients of D yields

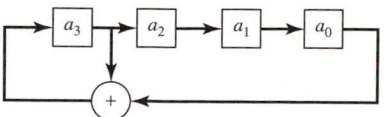

Cycle	Register state	$a(D)$
0	0 0 0 1	1
1	1 0 0 0	D^3
2	1 1 0 0	$D^2 + D^3$
3	1 1 1 0	$D + D^2 + D^3$
4	1 1 1 1	$1 + D + D^2 + D^3$
5	0 1 1 1	$1 + D + D^2$
6	1 0 1 1	$1 + D \qquad + D^3$
7	0 1 0 1	$1 \qquad + D^2$
8	1 0 1 0	$D \qquad + D^3$
9	1 1 0 1	$1 + \qquad + D^2 + D^3$
10	0 1 1 0	$D + D^2$
11	0 0 1 1	$1 + D$
12	1 0 0 1	$1 + \qquad D^3$
13	0 1 0 0	D^2
14	0 0 1 0	D

FIGURE 3-18. Shift-register states for Example 3-19.

$$e_0 = a_0 \qquad = 1$$

$$e_1 = a_1 + a_0 = 1$$

$$e_2 = a_2 + a_1 = 0$$

$$e_3 = a_3 + a_2 = 0$$

Thus $e(D) = 1 + D$. Suppose that $k = 20$ as in Example 3-18. Then $D^{k(2^m-2)} = D^{10}$ and $D^{10}e(D) = D^{10} + D^{11} = 1 + D + D^2 + D + D^2 + D^3 = 1 + D^3 = e'(D)$. Now (3-39) can be used to return to the original shift-register configuration. Thus

$$e'(D) = g(D)(a'(D) + b_r D^r + \cdots)$$

or

$$1 + D^3 = (1 + D + D^4)(a_0 + a_1 D + a_2 D^2 + a_3 D^3 + \cdots)$$

which implies

$$a_0 = 1$$

$$a_0 + a_1 = 0$$

$$a_2 + a_1 = 0$$

$$a_3 + a_2 = 1$$

or $a(D) = 1 + D + D^2$. This result can be verified by comparison with cycle 5 of Figure 3-18.

Determining the Phases of an *M*-sequence That Add to Produce a Particular Third Phase. Consider the sequence generator configuration with output shift register as illustrated in Figure 3-19, where there are r stages in the generator. The output $b(D)$ can be written $b(D) = a(D)/g(D)$. The present task is to determine the correct set of delayed outputs to add which will yield $b'(D)$ such that $b'(D) = D^k b(D)$. If the output shift register were k units in length, the problem would be trivial; however, the shift register is only $r - 1$ units long. The output $b'(D)$ is defined by

$$b'(D) = s_0 b(D) + s_1 Db(D) + s_2 D^2 b(D) + \cdots + s_{r-1} D^{r-1} b(D)$$

$$= s(D)b(D) \tag{3-71}$$

The polynomial $s(D)$ is a connection polynomial with binary coefficients representing whether or not a particular delay of $b(D)$ is included in the modulo-2 sum used to obtain $b'(D)$. The polynomial $s(D)$ must be found such that $b'(D) = D^k b(D)$. Making use of the results of the preceding section, both $b(D)$ and $b'(D)$ can be written as a function of the generator polynomial and a known initial condition. That is,

$$b'(D) = \frac{a'(D)}{g(D)} = s(D)b(D) = s(D)\frac{a(D)}{g(D)} \tag{3-72}$$

so that the initial conditions are related by

$$a'(D) = s(D)a(D) \tag{3-73}$$

Since the problem being addressed is concerned with the relative phase of two sequences and not the absolute phase of either, $a(D)$ can be arbitrarily chosen to be $a(D) = 1$. Then

$$s(D) = a'(D) \tag{3-74}$$

and the problem has been solved provided that $a'(D)$ can be determined. The initial conditions $a'(D)$ are found using the procedure described in the preceding section. The final output of this calculation is a connection polynomial $s(D)$ which can be used with either shift-register configuration (Figure 3-5 or 3-6).

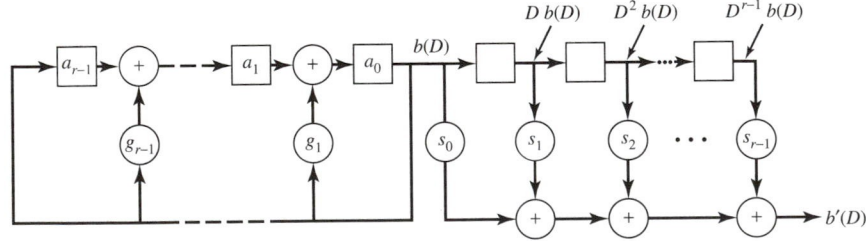

FIGURE 3-19. Generation of alternate phases of an *m*-sequence from short delays of $b(D)$.

EXAMPLE 3-20

Determine the proper phases of the output of the shift-register generator of Figure 3-18 which may be added to produce another sequence which is delayed from the original by 12 symbols.

Solution: Using the results of the preceding section, $a'(D)$ is the remainder after dividing $D^{12}a(D)$ by $g(D)$. Thus, with $g(D) = 1 + D + D^4$ and $a(D) = 1$,

$$a'(D) = 1 + D + D^2 + D^3 = s(D)$$

and all four delays of $b(D)$ must be added. Observe that in Figure 3-19 it was presumed that the output $b(D)$ is taken from the rightmost shift register of the generator. For the alternative shift-register configuration, the output is conveniently taken to be the input to the leftmost shift register stage, as shown in Figure 3-6. With this convention, the delays of the sequence $b(D)$ required for the sum to produce $b'(D)$ are available within the generator itself and no external delays are needed. Figure 3-20 shows the final configuration and associated outputs for a delay of 12 units. A single cycle of each sequence has been illustrated.

Security of Maximal-Length Sequences. Spread-spectrum systems are often used to protect digital transmissions from being jammed or to preclude unintended reception of the signal. Both of these objectives can only be met if the jammer or unintended receiver does not have knowledge of the spreading waveform $c(t)$. Unfortunately, when the jammer or intercepter can receive a relatively noise-free copy of the transmitted signal, the spreading code feedback connections and initial phase can be determined in a straightforward manner. For this reason, maximal-length sequences are a poor choice for the spreading code when a high level of security is required.

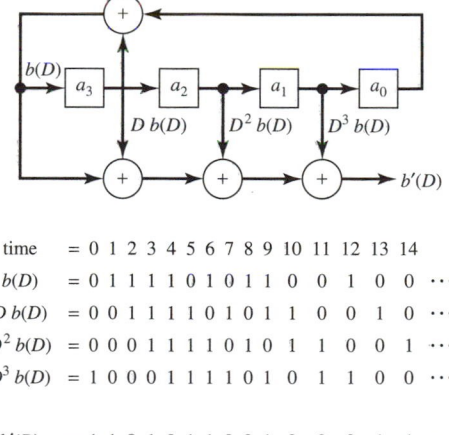

time	= 0 1 2 3 4 5 6 7 8 9 10 11 12 13 14
$b(D)$	= 0 1 1 1 1 0 1 0 1 1 0 0 1 0 0 ⋯
$D\,b(D)$	= 0 0 1 1 1 1 0 1 0 1 1 0 0 1 0 ⋯
$D^2\,b(D)$	= 0 0 0 1 1 1 1 0 1 0 1 1 0 0 1 ⋯
$D^3\,b(D)$	= 1 0 0 0 1 1 1 1 0 1 0 1 1 0 0 ⋯
$b'(D)$	= 1 1 0 1 0 1 1 0 0 1 0 0 0 1 1 ⋯

FIGURE 3-20. Shift-register configuration yielding a 12-symbol delay.

Suppose that the unintended party has access to an uncorrupted version of the transmitted spreading code. Thus the unintended party knows the sequence b_0, b_1, b_2, b_3, . . ., and would like to determine the shift-register feedback connections used to generate this sequence. The party knows that an m-sequence is being transmitted and can easily determine the period of the sequence by measuring the received power spectrum accurately. Each symbol of the m-sequence must satisfy the recursion relationship of (3-31), so that the unintended party can write the following series of equations:

$$b_i = b_{i-1}g_1 + b_{i-2}g_2 + \cdots + b_{i-m}g_m$$

$$b_{i+1} = b_i g_1 + b_{i-1}g_2 + \cdots + b_{i-m+1}g_m$$

$$b_{i+2} = b_{i+1}g_1 + b_i g_2 + \cdots + b_{i-m+2}g_m \qquad (3\text{-}75)$$
$$\vdots$$

After m such equations have been written, the unintended party will have m equations in the m unknowns g_1 through g_m which can be solved. Massey [12] has provided an efficient technique for solving this system of equations. The purpose of the present discussion is merely to make the student aware that algorithms exist for determining the shift-register generator feedback connections so that the details of Massey's algorithm are left as a reference. The system of (3-75) can also be solved by brute force, as demonstrated in the following example.

EXAMPLE 3-21

Suppose that the sequence 0 1 1 0 0 1 0 0 is received and that the known period of the m-sequence is 15. The first symbol received is the rightmost symbol in this sequence. Thus $m = 4$ and the set of equations to solve is

$$(1) \quad 0 = 0 \cdot g_1 + 1 \cdot g_2 + 1 \cdot g_3 + 0 \cdot g_4$$

$$(2) \quad 1 = 0 \cdot g_1 + 0 \cdot g_2 + 1 \cdot g_3 + 1 \cdot g_4$$

$$(3) \quad 0 = 1 \cdot g_1 + 0 \cdot g_2 + 0 \cdot g_3 + 1 \cdot g_4$$

$$(4) \quad 0 = 0 \cdot g_1 + 1 \cdot g_2 + 0 \cdot g_3 + 0 \cdot g_4$$

Adding (1) and (4) yields $0 = g_3$. Substituting $g_3 = 0$ into (1) yields $g_2 = 0$. Substituting $g_2 = g_3 = 0$ into (2) yields $g_4 = 1$; then substituting $g_2 = g_3 = 0$ and $g_4 = 1$ into (3) yields $g_1 = 1$. Therefore,

$$g(D) = 1 + D + D^4$$

This is the generator used in Example (3-20) and the received sequence is a subsequence of $b(D)$ in Figure 3-20.

Finally, note that the number of symbols which must be received is $2m$, where m is the degree of $g(D)$ and $2m$ is much shorter than the period $N = 2^m - 1$ of the m-sequence. The assumption that the $2m$ symbols be received without error can only be made in special circumstances so that the security of m-sequences may in fact be slightly better than is implied in this discussion.

3-4 Gold Codes

One of the applications of spread-spectrum systems is to provide a means other than frequency-division multiple access or time-division multiple access of sharing the scarce channel resources. When channel resources are shared using spread-spectrum techniques, all users are permitted to transmit simultaneously using the same band of frequencies. Users are each assigned a different spreading code so that they can be separated in the receiver despreading process. A goal of the spread-spectrum system designer for a multiple-access system is to find a set of spreading codes or waveforms such that as many users as possible can use a band of frequencies with as little mutual interference as possible.

Recall that the receiver despreading operation is a correlation operation with the spreading code of the desired transmitter. Ideally, a received signal that has been spread using a different spreading code will not be despread and will cause minimal interference in the desired signal. The specific amount of interference from a user employing a different spreading code is related to the cross-correlation between the two spreading codes. The Gold codes introduced in this section were invented in 1967 at the Magnavox Corporation specifically for multiple-access applications of spread spectrum. Relatively large sets of Gold codes exist which have well controlled cross-correlation properties. The treatment here is intended only to familiarize the student with the fact that this code family exists and with some of its most fundamental properties. A considerably more thorough discussion of these codes can be found in Sarwate and Pursley [2], from which the following information has been extracted. The reference also provides a large bibliography for further study.

The full-period cross-correlation between two spreading codes was defined in (3-6), which is repeated here:

$$\theta_{bb'}(k) = \frac{1}{N} \sum_{n=0}^{N-1} a_n a'_{n+k} \tag{3-6}$$

Although the detailed correlation could be evaluated for multiple-access code sets, in many cases adequate information for system analysis can be obtained from the cross-correlation spectrum. The *cross-correlation spectrum* is a list of all possible values of $\theta_{bb'}(k)$ and the number of values of k which yield that particular cross-correlation. When $b = b'$ the cross-correlation spectrum becomes the *autocorrelation spectrum*. The autocorrelation spectrum for an m-sequence is

$$1.0 \quad \text{occurs} \quad 1 \text{ time}$$

$$-\frac{1}{N} \quad \text{occurs} \quad N-1 \text{ times}$$

The Gold code sets to be defined shortly have a cross-correlation spectrum which is three-valued.

Consider an m-sequence that is represented by a binary vector $\mathbf{b}$ of length N, and a second sequence $\mathbf{b}'$ obtained by sampling every qth symbol of $\mathbf{b}$. The second sequence is said to be a *decimation* of the first, and the notation $\mathbf{b}' = \mathbf{b}[q]$ is used to indicate that $\mathbf{b}'$ is obtained by sampling every qth symbol of $\mathbf{b}$. The decimation of an m-sequence may or may not yield another m-sequence. When the decimation does yield an m-sequence, the decimation is said to be a *proper decimation.* It has been proven (Sarwate and Pursley [2]) that $\mathbf{b}' = \mathbf{b}[q]$ has period N if and only if gcd $(N, q) = 1$, where ''gcd'' denotes the greatest common divisor: Sarwate and Pursley [2] have also shown that proper decimation by odd integers q will give *all* of the m-sequences of period N. Thus any pair of m-sequences having the same period N can be related by $\mathbf{b}' = \mathbf{b}[q]$ for some q.

The cross-correlation spectrum of pairs of m-sequences can be three-valued, four-valued, or possibly many-valued. Certain special pairs of m-sequences whose cross-correlation spectrum is three-valued, where those three values are

$$-\frac{1}{N} t(n)$$

$$-\frac{1}{N} \tag{3-76}$$

$$\frac{1}{N} [t(n) - 2]$$

where

$$t(n) = \begin{cases} 1 + 2^{0.5(n+1)} & \text{for } n \text{ odd} \\ 1 + 2^{0.5(n+2)} & \text{for } n \text{ even} \end{cases}$$

where the code period $N = 2^n - 1$, are called *preferred pairs* of m-sequences. Finding preferred pairs of m-sequences is necessary in defining sets of Gold codes. The following conditions are sufficient to define a preferred pair $\mathbf{b}$ and $\mathbf{b}'$ of m-sequences:

1. $n \neq 0 \bmod 4$; that is, n is odd or $n = 2 \bmod 4$

2. $\mathbf{b}' = \mathbf{b}[q]$ where q is odd and either

$$q = 2^k + 1$$

or

$$q = 2^{2k} - 2^k + 1$$

3. gcd $(n, k) = \begin{cases} 1 & \text{for } n \text{ odd} \\ 2 & \text{for } n = 2 \bmod 4 \end{cases}$

EXAMPLE 3-22 _____

Find a preferred pair of m-sequences having a period of 31 units and evaluate their cross-correlation spectrum.

Solution: Since $N = 31$, $n = 5$. Table 3-5 contains the entry [45] which may be used to generate an m-sequence of length 31. The decimation $\mathbf{b}' = \mathbf{b}[3]$ is proper, so that the pair $(\mathbf{b}, \mathbf{b}[3])$ is a candidate pair. The first condition is satisfied since $n = 1$ mod 4. The second condition is satisfied also since q is odd and $q = 2^k + 1$ for $k = 1$. Finally, gcd $(5, 1) = 1$, so that all three conditions are satisfied and a preferred pair has been found. The m-sequence **b** and **b**[3] are

```
       0 1 2 3 4 5 6 7 8 9 10 11 12 13 14 15 16 17 18 19 20 21 22 23 24 25 26 27 28 29 30
b  = 1 0 1 0 1 1 1 0 1 1  0  0  0  1  1  1  1  1  0  0  1  1  0  1  0  0  1  0  0  0  0

b' = 1 0 1 1 0 1 0 1 0 0  0  1  1  1  0  1  1  1  1  1  0  0  1  0  0  1  1  0  0  0  0
```

A straightforward but tedious manual calculation of the cross-correlation function will show that for any phase shift the cross-correlation takes on one of the three values $-9/31$, $-1/31$, or $+7/31$.

Let $b(D)$ and $b'(D)$ represent a preferred pair of m-sequences having period $N = 2^n - 1$. The family of codes defined by $\{b(D), b'(D), b(D) + b'(D), b(D) + Db'(D), b(D) + D^2b(D), \ldots, b(D) + D^{N-1}b'(D)\}$ is called the set of *Gold codes* for this preferred pair of m-sequences. In this definition, the notation $D^jb'(D)$ represents a phase shift of the m-sequence $b'(D)$ by j units. Gold codes sets have the property that any pair of codes in the set, say **y** and **z**, have a three-valued cross-correlation spectrum which takes on the values defined in (3-76). A typical shift register configuration used to generate a family of Gold codes is illustrated in Figure 3-21. The particular preferred pair of m-sequences used in this figure are those found in Example 3-22. The complete family of Gold codes for this generator is obtained using different initial loads of either of the shift registers. The code $b(D)$ is obtained by choosing some nonzero $a(D)$ for the upper generator and setting $a'(D) = 0$ for the lower generator. Similarly, the code $b'(D)$ is obtained with $a(D) = 0$ and $a'(D)$ an arbitrary nonzero initial condition. Thirty-one other codes of the family are obtained using the same $a(D)$ used for $b(D)$ with all possible nonzero $a'(D)$. There are a total of $N + 2$ codes in any family of Gold codes.

In summary, Gold codes are families of codes with well-behaved cross-correlation properties which are constructed by a modulo-2 addition of specific relative phases of a preferred pair of m-sequences. The period of any code in the family is

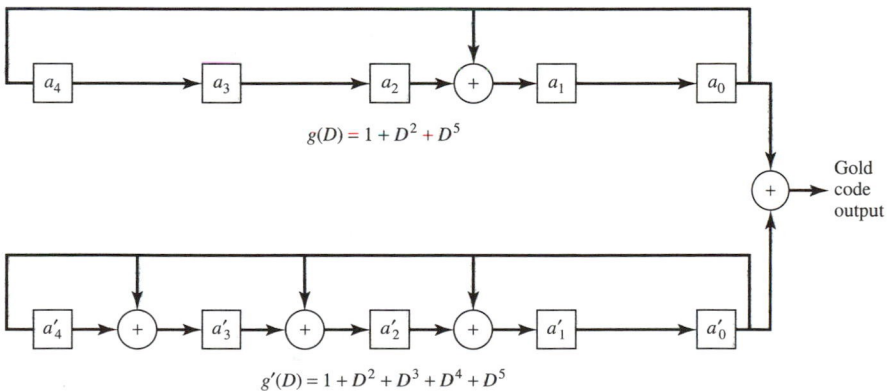

FIGURE 3-21. Typical Gold code generator.

N, which is the same as the period of the m-sequences. These codes are important and they have been selected by NASA for use on Tracking and Data Relay Satllite System (TDRSS). The particular set of Gold codes employed in TDRSS are defined in STDN 108 by the Goddard Space Flight Center. This brief description of Gold codes has not been rigorous and the student is referred to Sarwate and Pursley [2], Holmes [9], or Spilker [13] for further information on this interesting subject.

3-5 Nonlinear Code Generators

In spread-spectrum applications where security is important, it is necessary that an eavesdropper and/or jammer not be able to obtain complete knowledge of the spreading code being employed. If the spreading code is known, the eavesdropper can demodulate the transmitted signal just as the desired receiver does, and the jammer can transmit the spreading code as a highly effective jamming signal. Thus the difficulty of determining the specific code generator configuration from knowledge of the transmitted signal is an important issue for the spread-spectrum system designer. Recall that the feedback connections for an n-stage maximal-length code can be easily determined from knowledge of $2n$ successive code symbols. For this reason, m-sequences are never used when a high degree of security is required. The overall subject of security is beyond the scope of this text. Tutorial articles on communications privacy can be found in a special issue of the IEEE *Communications Society Magazine* (November 1978) devoted to this subject and a discussion of encryption techniques can be found in Ref. 14. The discussion of nonlinear spreading codes that follows indicates one way in which increased security through increased complexity can be achieved. One possible approach for the system designer is to develop codes that cannot be described by the simple linear recurrence relationship of (3-31) or codes for which r in (3-31) is so large that the solution of the system of (3-75) is computationally impossible. Fortunately, other methods of increasing complexity exist (Groth [15]) which are significantly more practical than either of the methods above. These methods make use of modulo-2 multiplications in addition to modulo-2 addition in order to increase effective complexity.

It is interesting to note that *any periodic sequence* of binary digits of period $2^r - 1$ can be generated by a *linear* feedback shift register. In the worst case, this generator is simply a recirculating shift register of length $2^r - 1$. In most cases, combinations of several short shift-register generators can be used. For example, consider the periodic sequence [15]

$$1\ 1\ 1\ 1\ 1\ 0\ 1\ 0\ 1\ 0\ 0\ 1\ 1\ 0\ 0\ 0\ 1\ 0\ 0\ 0\ 0\ldots$$

having a period of 21. This sequence can be generated using a 21-stage recirculating shift register. The output sequence of this recirculating shift register can be represented in polynomial form by

$$b(D) = \frac{1 + D + D^2 + D^3 + D^4 + D^6 + D^8 + D^{11} + D^{12} + D^{16}}{1 + D^{21}} \tag{3-77}$$

This ratio has the same form as (3-35) with $g(D) = 1 + D^{21}$, so that the shift-register feedback connections are simply one connection from the last stage of a 21-stage shift register to the input to the first stage. The numerator above is the initial load of the shift register $a(D)$. The denominator of (3-77) factors into two terms, one of which is identical to the numerator and can be canceled to obtain

$$b(D) = \frac{1}{1 + D + D^5}$$

Thus this 21-bit sequence can be more simply generated by a five-stage shift register with appropriate feedback connections. The denominator of the preceding expression can be further factored into

$$b(D) = \frac{1}{(1 + D + D^2)(1 + D^2 + D^3)} \tag{3-78}$$

indicating that the sequence can also be generated by two separate sequence generators whose outputs are modulo-2 summed. Not all sequences yield polynomials that can be factored as conveniently as those above. The important point of this discussion is that *any periodic sequence* can be generated by a *linear* feedback shift register. The number of stages in the generator, however, may be excessive and the sequence may be much more conveniently generated if the use of nonlinear elements is allowed.

The major contribution of Groth [15] is the demonstration of a technique that enables the construction of a linear equivalent of many nonlinear generators. This enables evaluation of the ''complexity'' of the nonlinear circuit. Consider only sequences of length $2^r - 1$ which are generated by linear and nonlinear operations on an r-stage shift register. The *complexity of a sequence* is defined as the number of stages in the linear equivalent feedback shift register. Maximal-length sequences have minimum possible complexity $(= r)$. In secure applications, the use of sequences which have very high complexity is desired so that the number of equations in (3-75) is large. Thus much computational power is required to solve (3-75).

One type of circuit used to generate high-complexity sequences is illustrated in Figure 3-22. In this figure, nonlinear feedforward logic has been added to a conventional linear feedback shift-register generator. For analytical simplicity, all binary multipliers have only two inputs and the $e_{i,j}$ coefficients indicate which shift register stages are connected to the multiplier. As before, $e_{i,j} = 0$ indicates no connections or that stage i and stage j are not multiplied. The outputs of all multipliers are added to form the final output sequence. A specific example of a high-complexity sequence generator is illustrated in Figure 3-23a. In Figure 3-23a all $e_{i,j} = 0$ except $e_{1,2}$, $e_{3,4}$, and $e_{5,6}$, which equal 1. Note that the period of the sequence of Figure 3-22 is $2^r - 1$; that is, the period has not been increased by using nonlinear feedforward logic.

Groth [15] gives specific procedures for determining the linear equivalent of a circuit such as that shown in Figure 3-22. This procedure is long and complicated but not conceptually difficult. As a specific example, the linear circuit of Figure 3-23b produces exactly the same output as the circuit of Figure 3-23a. Notice that Figure 3-23b contains 21 shift-register stages, so that 42 equations must be solved to

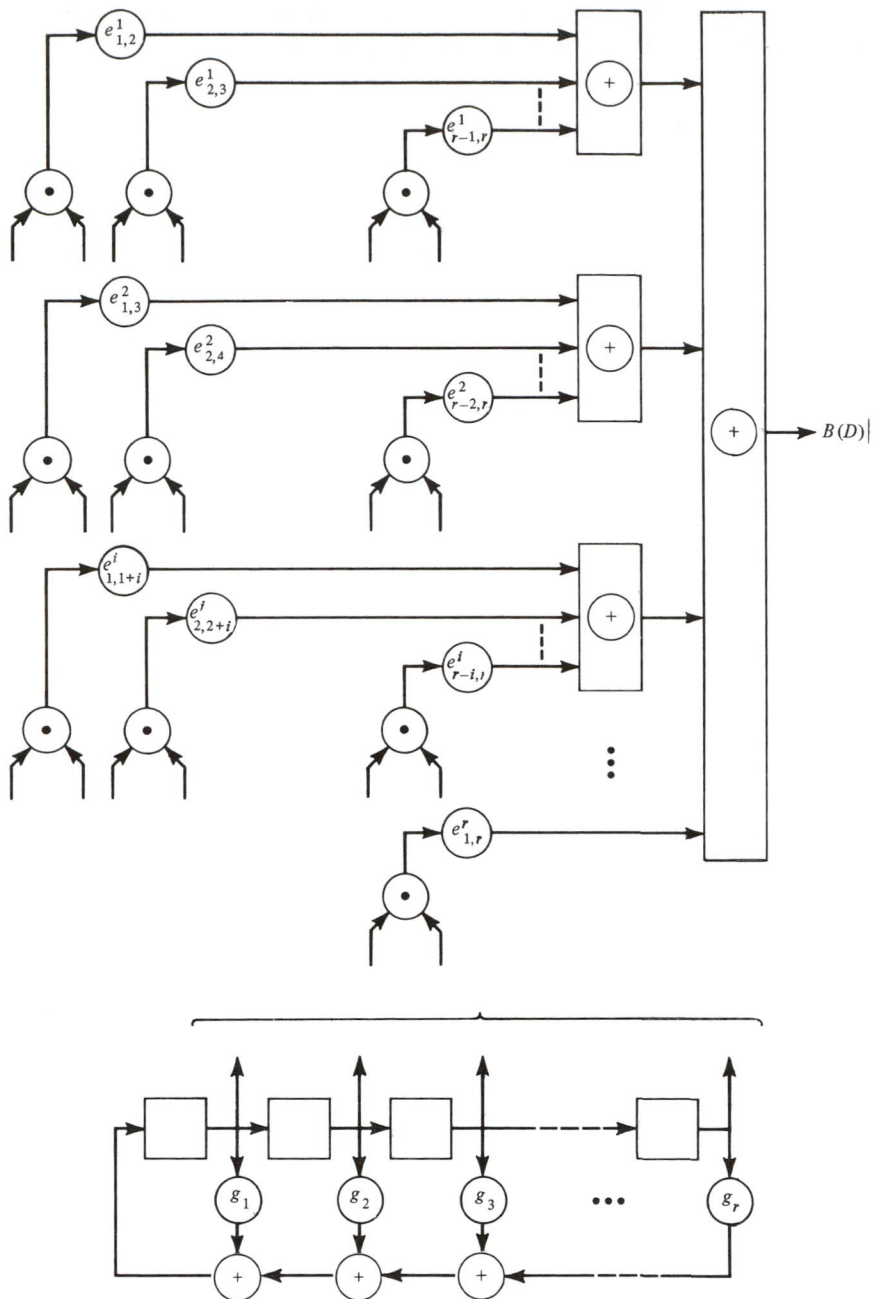

FIGURE 3-22.   Generalized feedforward nonlinear logic attached to linear generator. (From Ref. 15.)

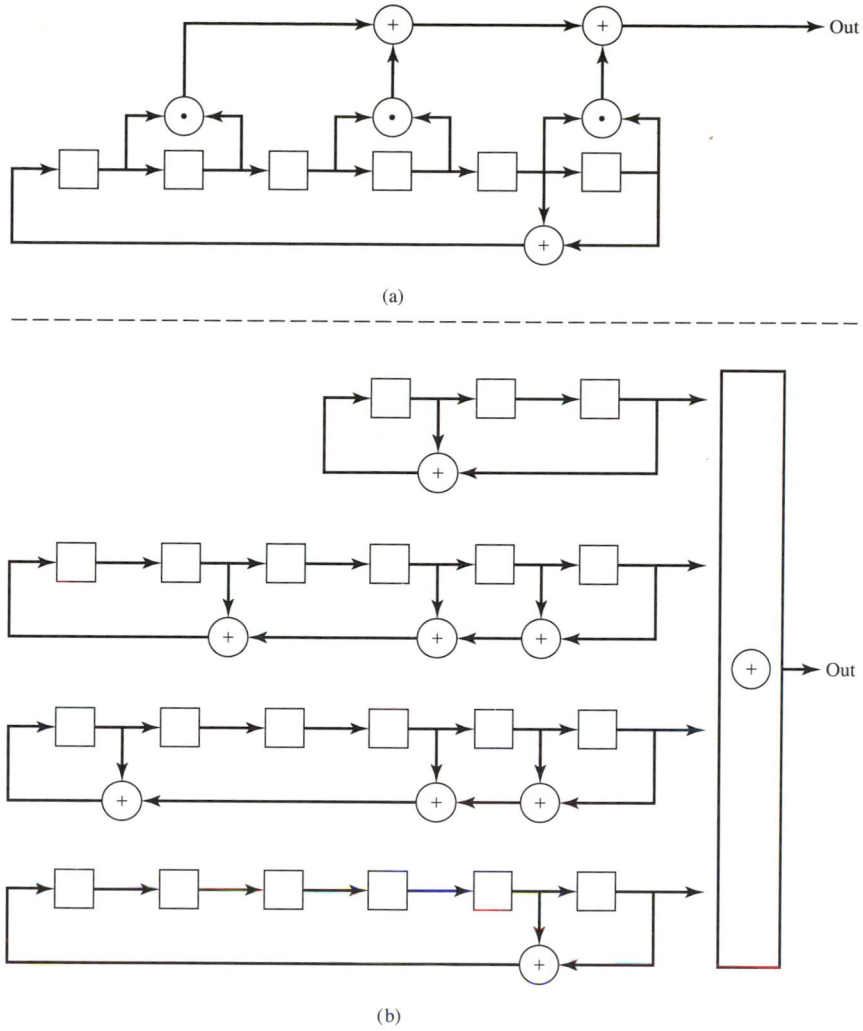

(a)

(b)

FIGURE 3-23. (a) Nonlinear generator; (b) linear equivalent. (From Ref. 15.)

determine the linear feedback connections. The original shift register contains only six shift registers and, if connected in a linear manner, would require 12 equations to specify.

This brief discussion of nonlinear sequence generators for secure applications is in no way comprehensive. Other nonlinear connections are possible, but significant analytical results are not available in the open literature. The basic concept, however, is always the same: Generate sequences that are very difficult to regenerate. Not considered above are schemes where the nonlinearities are placed in the feedback path rather than just in the feedforward path. Nonlinear generators of many types are discussed by Golomb [8].

Further Reading. Much has been written on the subject of codes for spread-spectrum systems. Comprehensive treatment of the mathematical background can be found in Birkhoff and MacLane [4] or Jacobson [16], while concise summaries of this material can be found in Peterson and Weldon [5] or Lin and Costello [3]. The text by Peterson and Weldon also contain a considerable amount of material on basic shift-register structures and their use in error correction coding. A comprehensive treatment of both linear and nonlinear codes can be found in the book by Golomb [8], which is the only book available devoted entirely to shift-register sequences. Concise summaries of the properties and mathematical structure of maximal-length sequences can be found in MacWilliams and Sloane [17] or Sarwate and Pursley [2]; the first reference also discusses pseudorandom arrays, while the second includes a great deal of information on the correlation properties of other code families as well as m-sequences. The partial correlation properties of m-sequences are analyzed by Lindholm [10]. The power spectrum of m-sequences and products of m-sequences is discussed in Gill [18] and Gill and Spilker [19]. Most of the references contain rather extensive reference lists which will lead the student to a rich variety of research material on code sequences.

The use of special codes for ranging and acquisition is discussed in Titsworth [20]. The nonlinear theory is discussed in Golomb [8] as well as in two more recent papers (Groth [15] and Key [21]). The latter two papers depend heavily on finite mathematics. Unfortunately, many of the current research results in the nonlinear theory remain in the classified literature. Finally, two recent texts on spread-spectrum communications (Holmes [9] and Simon [22]) contain chapters on spreading codes as well as further reference lists.

3-6 Summary

The subject of generating spreading codes for spread-spectrum communications systems was taken up in this chapter. After some mathematical preliminaries involving finite-field arithmetic, the concept of binary sequence generation with feedback shift-register circuits was discussed. Following this, the idea of a maximal-length sequence, or m-sequence, was defined as a feedback shift-register generated sequence where the feedback combination is linear and the feedback connection polynomial, $g(D)$, is obtained from a primitive polynomial. Their properties were discussed, where it was found that an m-sequence has properties similar to a random coin-toss sequence where heads is 1 and tails is 0. The autocorrelation and power spectrum of a periodically repeated m-sequence were presented, where it was seen that the periodic autocorrelation function is ideal in that the ratio of peak correlation to minimum magnitude correlation is N, where N is the length of the sequence. An abbreviated table of primitive polynomials for determining the feedback connections for m-sequence generators is given in Table 3-5. Although the periodic autocorrelation function of an m-sequence is nearly ideal, the correlation properties over part of an m-sequence are far from ideal, as shown in Figure 3-15 for a 31-bit m-sequence. Choice of initial conditions to generate a specific phase of an m-sequence was discussed as well as the addition of two m-sequences to produce a desired phase of a third m-sequence. Although m-sequences have many nice properties, they are not very secure in that an interceptor need only observe a number of bits equal to twice the shift register length in order to determine the feedback connections for the generator.

Gold codes are families of pseudo-random codes having well-behaved cross-correlation properties between member codes of the family of codes. A Gold-code family can be generated by modulo-2 addition of specific relative phases of a pair of m-sequences known as preferred pair sequences.

The final topic in this chapter was that of nonlinear code generation. These are codes which are formed by nonlinear feedback combinations. They are much more secure than m-sequences, but considerably less can be said about their properties, nor is a general technique available for finding suitable nonlinear codes.

References

[1] R. E. ZIEMER and W. H. TRANTER, *Principles of Communications: Systems, Modulation, and Noise* 4th ed. (Boston: Houghton Mifflin, 1995).

[2] D. V. SARWATE and M. B. PURSLEY, "Crosscorrelation Properties of Pseudo-random and Related Sequences," *Proc. IEEE,* Vol. 68, pp. 593–619, May 1980.

[3] S. LIN and D. J. COSTELLO, *Error Control Coding: Fundamentals and Applications* (Englewood Cliffs, N.J.: Prentice Hall, 1983).

[4] G. BIRKHOFF and S. MACLANE, *A Survey of Modern Algebra* (New York: Macmillan, 1965).

[5] W. W. PETERSON and E. J. WELDON, *Error-Correcting Codes* (Cambridge, Mass.: MIT Press, 1972).

[6] K. METZGER and R. J. BOUWENS, *An Ordered Table of Primitive Polynomials over GF(2) of Degrees 2 Through 19 for Use with Linear Maximal Sequence Generators,* TM107, Cooley Electronics Laboratory, University of Michigan, Ann Arbor, July 1972 [AD 746876].

[7] D. L. SCHILLING, B. H. BATSON, and R. PICKHOLTZ, *Spread Spectrum Communications,* Short Course Notes, Natl. Telecommun. Conf., 1980.

[8] S. W. GOLOMB, *Shift Register Sequences* (Laguna Hills, Calif.: Aegean Park Press, 1982).

[9] J. K. HOLMES, *Coherent Spread Spectrum Systems* (New York: Wiley-Interscience, 1982).

[10] J. H. LINDHOLM, "An Analysis of the Pseudo-randomness Properties of Subsequences of Long m-Sequences," *IEEE Trans. Inf. Theory,* Vol. IT-15, pp. 122–127, July 1968.

[11] S. V. GLISIC, "Power Density Spectrum of the Product of Two Time-Displaced Versions of a Maximum Length Binary Pseudonoise Signal," *IEEE Trans. Commun.,* Vol. COM-31, pp. 281–286, February 1983.

[12] J. L. MASSEY, ''Shift-Register Synthesis and BCH Decoding,'' *IEEE Trans. Inf. Theory,* Vol. IT-15, pp. 122–127, January 1969.

[13] J. J. SPILKER, JR., *Digital Communications by Satellite* (Englewood Cliffs, N.J.: Prentice Hall, 1977).

[14] W. DIFFIE and M. E. HELLMAN, ''New Directions in Cryptography,'' *IEEE Trans. Inf. Theory,* Vol. IT-22, pp. 644–654, November 1976.

[15] E. J. GROTH, ''Generation of Binary Sequences with Controllable Complexity,'' *IEEE Trans. Inf. Theory,* Vol. IT-17, pp. 288–296, May 1971.

[16] N. JACOBSON, *Basic Algebra 1* (San Francisco: W.H. Freeman, 1974).

[17] F. J. MACWILLIAMS and J. A. SLOANE, ''Pseudo-random Sequences and Arrays,'' *Proc. IEEE,* Vol. 64, pp. 1715–1729, December 1976.

[18] W. J. GILL, ''Effect of Synchronization Error in Pseudo-random Carrier Communications,'' *Conf. Rec.,* First Annual IEEE Commun. Conf., pp. 187–191, June 1965.

[19] W. J. GILL and J. J. SPILKER, ''An Interesting Decomposition Property for the Self-Products of Random or Pseudorandom Binary Sequences,'' *IEEE Trans. Commun. Syst.,* Vol. CS-11, pp. 246–247, June 1963.

[20] R. C. TITSWORTH, ''Optimal Ranging Codes,'' *IEEE Trans. Space Electron. Telem.,* Vol. SET-10, pp. 19–30, March 1964.

[21] E. L. KEY, ''An Analysis of the Structure and Complexity of Nonlinear Binary Sequence Generators,'' *IEEE Trans. Inf. Theory,* Vol. IT-22, pp. 732–736, November 1976.

[22] M. K. SIMON et al., *Spread-Spectrum Communications Handbook* (New York: McGraw-Hill, 1994).

Problems

(3-1) Demonstrate that the set of integers $\{0,1,2,3\}$ with addition and multiplication defined modulo-4 is not a field. Define the addition and multiplication tables which, together with this set, define a field of four elements. The polynomial $1 + D + D^2$ is primitive.

(3-2) Construct a Galois field having 32 elements and list all elements in polynomial format together with its multiplicative inverse element. The polynomial $1 + D + D^5$ is primitive.

(3-3) Prove that the polynomial $g(D) = 1 + D + D^3$ is primitive.

(3-4) Prove that the additive and multiplicative identity elements of a finite field are unique.

(3-5) Factor the polynomial $g(D) = 1 + D^2 + D^3 + D^4$, and evaluate $[g(D)]^4$.

(3-6) Let D^{10} and D^{12} represent elements of the Galois field having 16 elements and defined using the primitive polynomial $h(D) = 1 + D + D^4$.
   (a) Evaluate the product $D^{10} \cdot D^{12}$ using both the polynomial representation of these elements and the power-of-D representation.
   (b) Evaluate $D^{10} + D^{12}$.
   (c) Evaluate D^{10}/D^{12}.
   (d) Evaluate $D^{10} - D^{12}$.

(3-7) Let $g(D) = 1 + D + D^4$, $h(D) = 1 + D + D^2$, and $a(D) = 1 + D + D^2 + D^3$. Develop a shift-register configuration which generates the function

$$b(D) = \frac{h(D)}{g(D)} a(D)$$

and determine the circuits output as a function of time. Can this circuit be used to automate the calculation of Problem 3-6?

(3-8) Consider the feedback-shift-register configuration shown below. Determine the output of this circuit with the initial condition $a(D) = 1 + D + D^2 + D^3$.

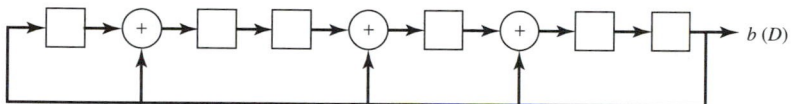

PROBLEM 3-8.   Feedback shift register.

(3-9) Consider the feedback-shift-register configuration shown below. Determine the output of this circuit with initial condition $a(D) = 1 + D + D^2 + D^3$ and compare with the result of Problem 3-8.

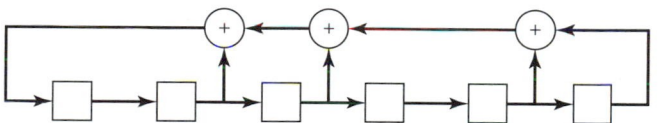

PROBLEM 3-9.   Feedback shift register.

(3-10) What is the maximum possible period of the binary sequence generated by the linear feedback-shift-register configuration below? Find all possible output cycles of this circuit.

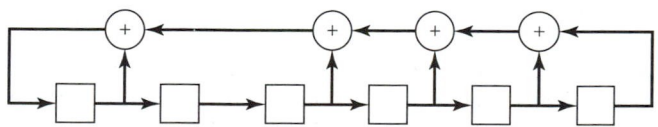

PROBLEM 3-10.   Non-maximal-length linear feedback shift register.

(3-11) Consider the maximal-length sequence generated using the primitive polynomial $g(D) = 1 + D + D^2 + D^4 + D^5$. Demonstrate that Properties I through V for m-sequences are satisfied for this particular m-sequence.

(3-12) Plot the autocorrelation function and the detailed power spectrum for the maximal-length sequence specified by $g(D) = 1 + D + D^4$ using a shift register clock rate of 1.0 kHz.

(3-13) The code of Problem 3-12 is used to BPSK modulate a 1.0-MHz carrier. Plot the power spectrum of the modulated carrier.

(3-14) The code of Problem 3-12 is used to MSK modulate a 1.0-MHz carrier. Plot the power spectrum of the modulated carrier. (*Hint:* Consider serial MSK implementations.)

(3-15) Two maximal-length shift-register generators having periods N_1 and N_2 are run off the same clock. A third sequence is generated by modulo-2 adding the outputs of these generators. What is the period of the third output sequence?

(3-16) Consider the m-sequence generator specified by the primitive polynomial 155_8. Define a shift-register configuration that will generate this sequence in the forward direction and a shift-register configuration that will generate this sequence in the reverse direction.

(3-17) The correlation-filter arrangement illustrated below is one part of a code tracking loop in a direct-sequence spread-spectrum modem. Suppose that the spreading code is an m-sequence having period N and that the bandpass filter is ideal and is specified by

$$H(\omega) = \begin{cases} 1.0 & |\omega \pm \omega_0| \le 0.2\pi/T_c \\ 0.0 & \text{elsewhere} \end{cases}$$

Plot the ratio of the filter output power at the carrier frequency to all other power at the filter output as a function of the code period N. All of the filter output power except that at the carrier frequency is called *code self-noise*.

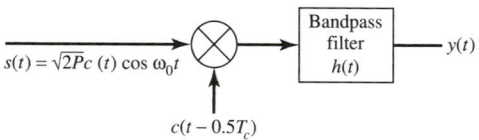

PROBLEM 3-17. Correlator illustrating spreading code self-noise.

(3-18) A direct-sequence spread-spectrum modem is used to communicate between a base station and a mobile station separated by a distance which yields a propagation delay of 0.15 μs. The spreading code is a maximal-length code specified by the generator $g(D) = 1 + D^3 + D^{10}$ and the code generator clock rate is 100 MHz. Determine a set of initial conditions for the transmitter and receiver code generators such that, if the generators are started simultaneously, the received signal will be despread.

(3-19) The receiver despreading code of Problem 3-18 is generated by modulo-2 addition of some number of delays of a code generator which is phase synchronous with the transmitter code generator. What delays should be added to achieve despreading in the receiver?

(3-20) The following sequence of symbols from a spread-spectrum transmitter are received:

$$1\ 0\ 1\ 1\ 1\ 1\ 1\ 0\ 1\ 1\ 0\ 0\ 1$$

Spectral analysis of the received signal indicates that the power spectrum consists of discrete lines which are spaced at 322.6 kHz and that the spreading code rate is 10 MHz. What code generator is being used in the transmitter?

(3-21) Solve the following set of simultaneous equations over GF(2):

$$1 = \quad x \quad + z$$
$$1 = w \quad + y$$
$$0 = w + x \quad + z$$
$$0 = \quad x + y$$

(3-22) Calculate the discrete periodic cross-correlation function for the pair of m-sequences defined by $g_1(D) = 1 + D^2 + D^5$ and $g_2(D) = 1 + D^2 + D^3 + D^4 + D^5$.

(3-23) Calculate the discrete partial autocorrelation function for an m-sequence defined by $g(D) = 1 + D^2 + D^5$ using a window size of five chips and a phase difference of two chips.

(3-24) Consider the Galois shift-register generator illustrated in Figure 3-17 with an initial load of $a(D) = 1$. Using the state-machine representation of the shift register, find the load of the shift register five cycles before and five cycles after the initial load.

(3-25) Prove that it is not possible to generate a maximal-length shift-register sequence using a shift register with an odd number of taps [9, p. 309].

(3-26) Consider a shift-register generator defined by $g(D) = 1 + D + D^3 + D^4$.
  (a) Does this shift register generate a maximal-length code?
  (b) Find all possible state sequences for the Fibonacci shift register for this code.
  (c) Does the shift-and-add property apply to the sequences generated by this generator?

(3-27) Consider the shift-register generator defined by $g(D) = 1 + D + D^3$.

    (a) Find the Fibonacci generator maxtrix $\mathbf{F}$ for this code generator.

    (b) Find the Galois generator matrix $\mathbf{G}$ for this code generator.

    (c) Prove that $\mathbf{F}^7 = \mathbf{1}$, which $\mathbf{1}$ is the identity matrix.

    (d) Prove that $\mathbf{G}^7 = \mathbf{1}$.

(3-28) Define the recurrence relationship for the code generator implied by (3-77).

CHAPTER 4

Code Tracking Loops

4-1 Introduction

Spread-spectrum communication requires that the transmitter and receiver spreading waveforms be synchronized. If the two waveforms are out of synchronization by as little as one chip, insufficient signal energy will reach the receiver data demodulator for reliable data detection. The task of achieving and maintaining code synchronization is always delegated to the receiver. There are two components of the synchronization problem. The first component is the determination of the initial code phase. This part of the problem is called *code acquisition* and is addressed in Chapter 5. The second component is the problem of maintaining code synchronization after initial acquisition. This problem, called *code tracking,* is addressed in this chapter.

Code tracking is accomplished using phase-locked techniques very similar to those used for generation of coherent carrier references. The principal difference between the phase-locked loops used for carrier tracking and those discussed here is in the implementation of the phase discriminator. For carrier tracking, the discriminator is often as simple as a multiplier, whereas for modern code tracking loops, several multipliers and usually pairs of filters and envelope detectors will be employed in the phase discriminator. Since code tracking loops are phase-locked loops (PLLs), the goal of the analysis to follow is to develop models of the various code tracking loops which are identical to the conventional PLL model and then to draw on the vast store of PLL results. Appendix A summarizes PLL results.

Code tracking loops for spread-spectrum systems can be categorized in several ways. First, there are coherent and noncoherent loops. Coherent loops make use of received carrier phase information whereas noncoherent loops do not. All but one of the phase discriminators to be discussed make use of correlation operations between the received signal and two different phases (early and late) of the receiver-generated spreading waveform. These two correlation operations can be accomplished using two independent channels or using a single channel that is time shared. A tracking loop that makes use of two independent correlators is called a *delay-lock tracking loop,* which will be referred to as a DLL, and a tracking loop that time shares a single correlator is called a *tau-dither tracking loop,* which will be referred to as a TDL. All the categories of tracking loops mentioned above are discussed in this chapter.

The tracking loops discussed are designed to achieve low root mean square (rms) tracking jitter in the presence of AWGN while tracking the dynamics of the received spreading waveform. The transmission delay, T_d, is actually a function of time, $T_d(t)$, when there is relative motion between the transmitter and the receiver. This function $T_d(t)$ must be tracked by the code tracking loop. The same issues arise in the design of a code tracking loop which arise in the design of a carrier tracking phase-locked loop. The loop bandwidth will be selected to be a compromise between a wide bandwidth, which facilitates tracking the dynamics of $T_d(t)$, and a narrow bandwidth, which minimizes the tracking jitter due to interference. The problem of cycle skipping is more serious for code tracking than it is for carrier tracking. A cycle skip in a code tracking loop may cause the system to reenter the acquisition mode.

The chapter begins with a discussion of the optimum tracking loop for direct-sequence spread-spectrum systems. The baseband delay-lock tracking loop is discussed next since many important concepts can be introduced with this simple code tracking loop. The baseband DLL operates directly on the spreading waveform itself rather than on a modulated carrier. Thus it is assumed that a demodulator preceding the tracking loop has accurately recovered the spreading waveform from the modulated carrier. Next, the noncoherent DLL tracking loop is discussed. This loop is similar to the baseband loop except for a more complex phase discriminator. The noncoherent TDL is discussed following the DLL because, even though it is electronically simpler than the DLL, it is analytically more complicated. Tracking loops for frequency-hop spread-spectrum systems are discussed last.

Some of the earliest work on code tracking discriminators was done by Spilker and Magill [1]. This early work was very general in nature and showed that the proper error signal for tracking could be derived from a correlation of the received signal with a receiver generated first derivative (with respect to time) of the spreading waveform. Later work [2–6] specialized Spilker and Magill's early work to the types of tracking loops currently employed. Code tracking loops are also discussed in Refs. 7 to 10. Simon et al. [10] present a highly detailed comprehensive analysis of the DLL and TDL which includes some considerations not covered in this book and which is highly recommended for readers having applications with stringent specifications.

4-2 Optimum Tracking of Wideband Signals

The transmitted waveform $s(t)$ in a spread-spectrum communication system is a wideband signal. It has been stated [1] that the optimum tracking discriminator for an arbitrary wideband signal received with additive white Gaussian noise (AWGN) is a multiplier that forms the product of the received signal plus noise and the first derivative with respect to time of the receiver generated replica of the transmitted signal. This discriminator is optimum in that its output is a maximum likelihood estimate of the phase difference between the two wideband signals in an AWGN environment. This means that the output phase error estimate is the most probable phase error, given the available received information. A tracking loop making use of this optimum discriminator is illustrated in Figure 4-1. The received signal $r(t) =$

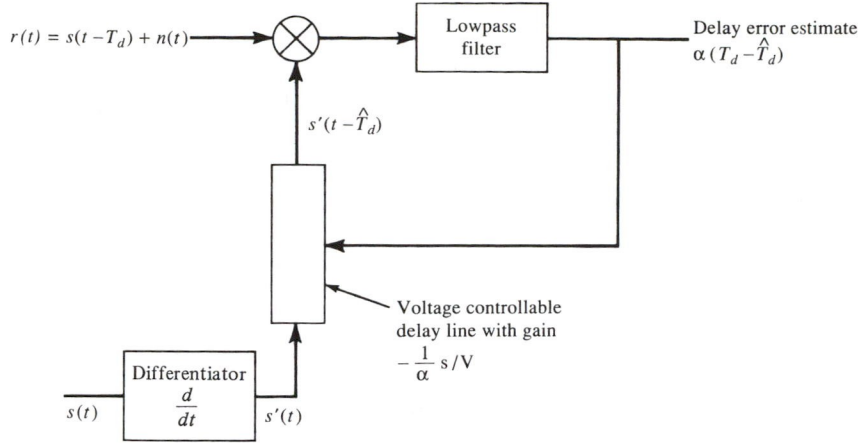

FIGURE 4-1. Tracking loop for arbitrary wideband signal. (From Ref. 1.)

$s(t - T_d) + n(t)$ is multiplied by a differentiated and delayed receiver-generated replica of $s(t)$. The multiplier output contains a dc component related to the delay error $(T_d - \hat{T}_d)$, where $\hat{T}_d$ is the receiver estimate of the transmission delay. This dc component is extracted by the lowpass filter and used to correct the delay of the voltage controllable delay line.

This tracking loop configuration will not be analyzed in detail here since it is not commonly used in modern spread-spectrum systems. However, because it is optimum and because modern tracking loop configurations are usually discrete approximations of this loop, the operation of the loop will be described briefly for a single special case. Suppose that the received signal is the baseband spreading waveform $c(t - T_d)$ itself† and suppose that thermal noise is ignored. Let the spreading waveform be derived from an m-sequence which has a period N of seven symbols of length T_c. The received signal is shown in Figure 4-2a. The function of the tracking loop is to produce the signal $c(t - \hat{T}_d)$ with $\delta = (T_d - \hat{T}_d)/T_c$ as small as possible. The signal $c(t - \hat{T}_d)$ is shown in Figure 4-2b and its first derivative with respect to time is shown in Figure 4-2c. The derivative is a series of impulse functions. The tracking loop multiplier output is shown in Figure 4-2d. Since, in this example, $\hat{T}_d > T_d$ and $|T_d - \hat{T}_d| < T_c$, all the impulse functions at the multiplier output are positive. The dc component of the multiplier output is the time average of

$$c(t - T_d) \frac{d}{dt} [c(t - \hat{T}_d)]$$

which is $(N + 1)/NT_c$. When $\hat{T}_d < T_d$ and $|T_d - \hat{T}_d| < T_c$, all of the impulses at the multiplier output are negative and the dc component is $-(N + 1)/NT_c$. When $|T_d - \hat{T}_d| \geq T_c$, there is an equal number of positive and negative multiplier output impulses, and the dc level is zero. The dc output of the multiplier is shown in Figure

---

† The technique described here will not work if the received signal is data modulated since the discriminator slope will change sign with the data modulation.

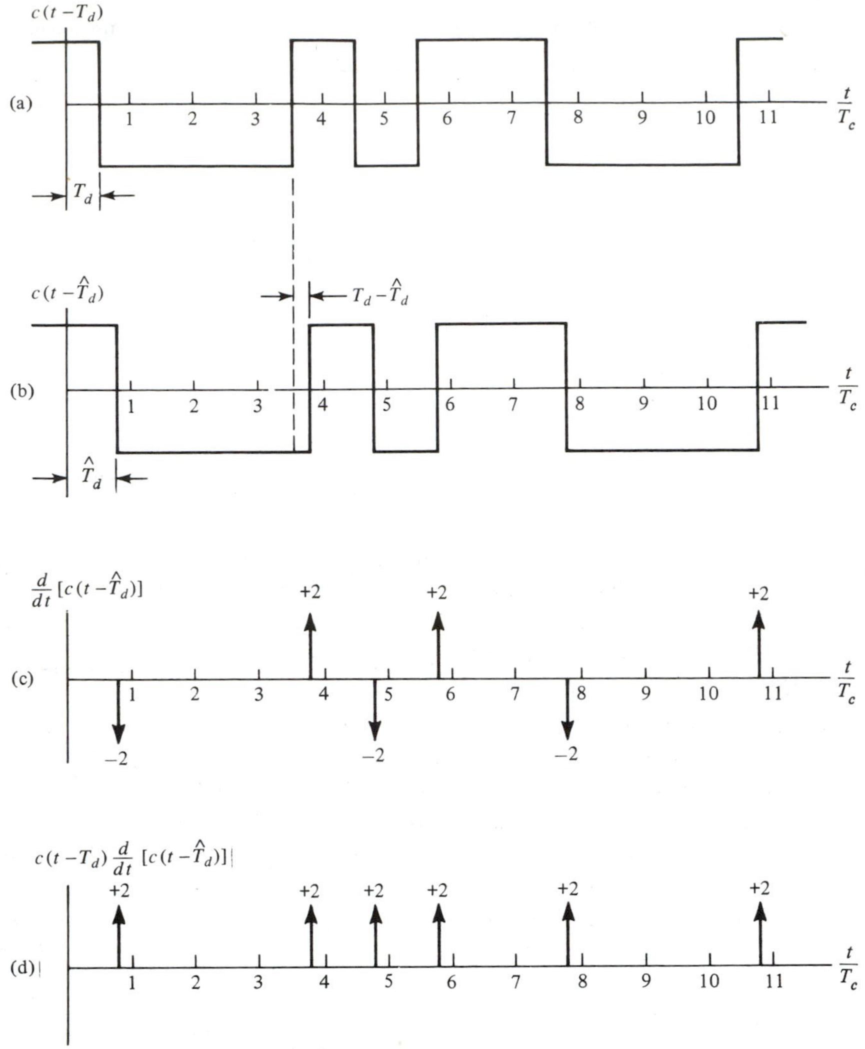

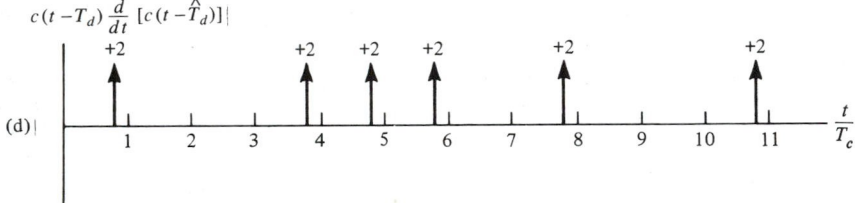

FIGURE 4-2. Waveforms of optimum code tracking loop: (a) received waveform; (b) receiver-generated replica of spreading waveform; (c) derivative of b; (d) multiplier output.

4-3 as a function of the normalized delay difference $\delta = (T_d - \hat{T}_d)/T_c$. Observe that when $\delta < 0$, the delay line input will be positive and that this will decrease $\hat{T}_d$ as required to drive δ to zero. Whenever $|T_d - \hat{T}_d| < T_c$, a voltage exists which pushes the delay $\hat{T}_d$ in the correct direction.

The optimum delay discriminator operates similarly for any wideband signal. The reader is referred to Spilker [1,7] for a complete analysis of this discriminator. Finally, observe that the conventional phase-locked loop employs exactly the discriminator just described. In a conventional PLL, the input signal $\sin(\omega t + \phi)$ is

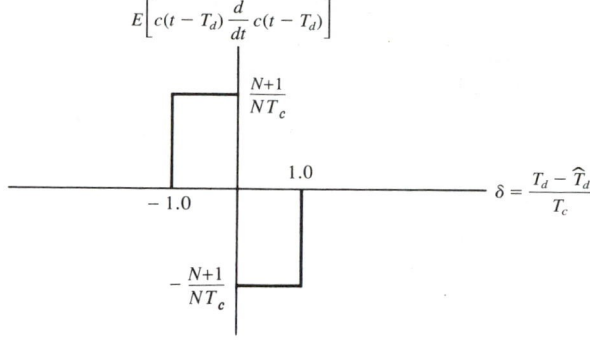

FIGURE 4-3. Optimum delay discriminator output dc component for *m*-sequence baseband tracking loop.

correlated with a signal cos $(\omega t + \hat{\phi})$, which is, within a constant, the derivative of the received signal.

4-3 Baseband Delay-Lock Tracking Loop

The function of a baseband DLL is to track the time-varying phase of the received spreading waveform $c(t - T_d)$. The function $\hat{T}_d(t)$ will denote the receiver estimate of $T_d(t)$, and T_d and $\hat{T}_d$ are always functions of time, whether or not this dependence is written explicitly. The received signal consists of the spreading waveform $\sqrt{P}\, c(t - T_d)$ with power P and additive white Gaussian noise $n(t)$ with two-sided power spectral density $N_0/2$ W/Hz. That is,

$$s_r(t) = \sqrt{P}\, c(t - T_d) + n(t) \tag{4-1}$$

Figure 4-4 is a conceptual block diagram of the tracking loop. It consists of a phase discriminator, a loop filter, a voltage-controlled oscillator, and a spreading-waveform generator.

The received signal is input to the delay-lock discriminator where, after power division, it is correlated with an early spreading waveform $c(t - \hat{T}_d + (\Delta/2)T_c)$, and a late spreading waveform, $c(t - \hat{T}_d - (\Delta/2)T_c)$. The parameter Δ is the total normalized time difference between the early and late discriminator channels.

Consider the operation of the delay-lock discriminator as a static phase-measuring device in a noiseless environment. That is, let T_d and $\hat{T}_d$ be fixed and determine the output of the discriminator. This output will contain a component which is a function of $\delta = (T_d - \hat{T}_d)/T_c$ and is suitable for driving the VCO just as the phase-locked-loop multiplier output contained a component $\sin(\theta_i - \theta_0)$. In the static case, it is convenient to write $y_1(t)$, $y_2(t)$, and $\epsilon(t,\delta)$ as explicit functions of T_d, $\hat{T}_d$, and t. Thus the early-correlator output is

$$y_1(t,T_d,\hat{T}_d) = K_1 \sqrt{\frac{P}{2}}\, c(t - T_d) c\left(t - \hat{T}_d + \frac{\Delta}{2}T_c\right) \tag{4-2a}$$

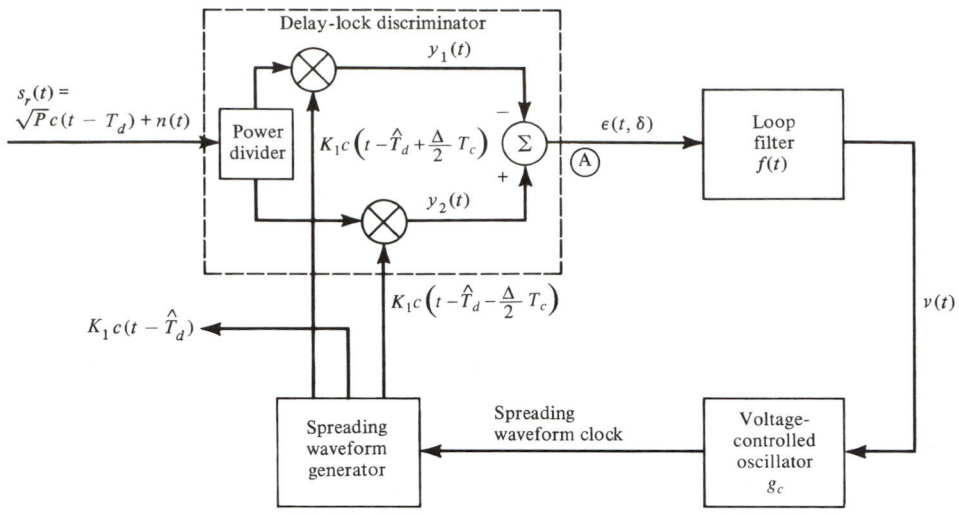

FIGURE 4-4. Conceptual block diagram: baseband delay-lock tracking loop.

and the late-correlator output is

$$y_2(t,T_d,\widehat{T}_d) = K_1 \sqrt{\frac{P}{2}} c(t - T_d) c\left(t - \widehat{T}_d - \frac{\Delta}{2} T_c\right) \tag{4-2b}$$

In these expressions, K_1 is the multiplier gain and is dependent on the particular multiplier hardware implementation. The input signal has been divided by $\sqrt{2}$ to account for the power division, and noise has been ignored. The delay-lock discriminator output is the difference of $y_2(t)$ and $y_1(t)$ and is

$$\epsilon(t,T_d,\widehat{T}_d) = y_2(t,T_d,\widehat{T}_d) - y_1(t,T_d,\widehat{T}_d)$$

$$= K_1 \sqrt{\frac{P}{2}} c(t - T_d)\left[c\left(t - \widehat{T}_d - \frac{\Delta}{2} T_c\right) - c\left(t - \widehat{T}_d + \frac{\Delta}{2} T_c\right)\right] \tag{4-3}$$

The dc component of this signal is used for code tracking. The time-varying component, which is also a function of δ, is called *code self-noise*.

The dc component of $\epsilon(t,T_d,\widehat{T}_d)$ is denoted $K_1 \sqrt{P/2}\, D_\Delta(T_d,\widehat{T}_d)$ and is the time average of $\epsilon(t,T_d,\widehat{T}_d)$. Thus

$$K_1 \sqrt{\frac{P}{2}} D_\Delta(T_d,\widehat{T}_d) = \frac{1}{NT_c} \int_{-NT_c/2}^{NT_c/2} K_1 \sqrt{\frac{P}{2}} c(t - T_d)$$

$$\times \left[c\left(t - \widehat{T}_d - \frac{\Delta}{2} T_c\right) - c\left(t - \widehat{T}_d + \frac{\Delta}{2} T_c\right)\right] dt \tag{4-4}$$

where NT_c is the period of $c(t)$. Recalling the definition of the autocorrelation function of $c(t)$ from (3-2),

$$D_\Delta(T_d, \hat{T}_d) = R_c\left(T_d - \hat{T}_d - \frac{\Delta}{2}T_c\right) - R_c\left(T_d - \hat{T}_d + \frac{\Delta}{2}T_c\right)$$

$$= R_c\left[\left(\delta - \frac{\Delta}{2}\right)T_c\right] - R_c\left[\left(\delta + \frac{\Delta}{2}\right)T_c\right] \qquad (4\text{-}5)$$

$$\triangleq D_\Delta(\delta)$$

This function is plotted in Figure 4-5 for four values of Δ, where $c(t)$ is the waveform derived from a maximal-length sequence and $R_c(\tau)$ is defined in (3-51). Observe from Figure 4-5 that there is a range of δ near zero for which $D_\Delta(\delta)$ is linearly related to δ. This region is always selected as the normal operating region for the tracking loop. The slope of the discriminator S-curve near $\delta = 0$ is $2(1 + 1/N)$ for all $0 < \Delta < 2.0$. The range of δ for which the discriminator characteristic has the slope $2(1 + 1/N)$ is $|\delta| < \Delta/2$ for $\Delta \le 1.0$ and $|\delta| < 1 - \Delta/2$ for $1 \le \Delta < 2$. This range decreases to zero at $\Delta = 2.0$, and is maximum for $\Delta = 1.0$.

In most analyses of code tracking loops the time-varying component of $\epsilon(t, T_d, \hat{T}_d)$, which is denoted $K_1\sqrt{P/2}\,N_\Delta(t, T_d, \hat{T}_d)$, can be safely ignored since most of this self-noise power is at frequencies which are well outside the bandwidth of the tracking loop. A detailed proof of this statement requires calculation of the power spectrum of $\epsilon(t, T_d, \hat{T}_d)$. This calculation is tedious and will not be given here. Throughout the remainder of this chapter code self-noise will be ignored.

Consider now the operation of the phase discriminator in an AWGN environment. With noise included,

$$y_1(t, T_d, \hat{T}_d) = K_1\sqrt{\frac{P}{2}}\,c(t - T_d)c\left(t - \hat{T}_d + \frac{\Delta}{2}T_c\right)$$

$$+ \frac{K_1}{\sqrt{2}}c\left(t - \hat{T}_d + \frac{\Delta}{2}T_c\right)n(t) \qquad (4\text{-}6a)$$

$$y_2(t, T_d, \hat{T}_d) = K_1\sqrt{\frac{P}{2}}\,c(t - T_d)c\left(t - \hat{T}_d - \frac{\Delta}{2}T_c\right)$$

$$+ \frac{K_1}{\sqrt{2}}c\left(t - \hat{T}_d - \frac{\Delta}{2}T_c\right)n(t) \qquad (4\text{-}6b)$$

The discriminator output is

$$\epsilon(t, T_d, \hat{T}_d) = K_1\sqrt{\frac{P}{2}}[D_\Delta(\delta) + N_\Delta(t, T_d, \hat{T}_d)]$$

$$+ \frac{K_1}{\sqrt{2}}n(t)\left[c\left(t - \hat{T}_d - \frac{\Delta}{2}T_c\right) - c\left(t - \hat{T}_d + \frac{\Delta}{2}T_c\right)\right] \qquad (4\text{-}7)$$

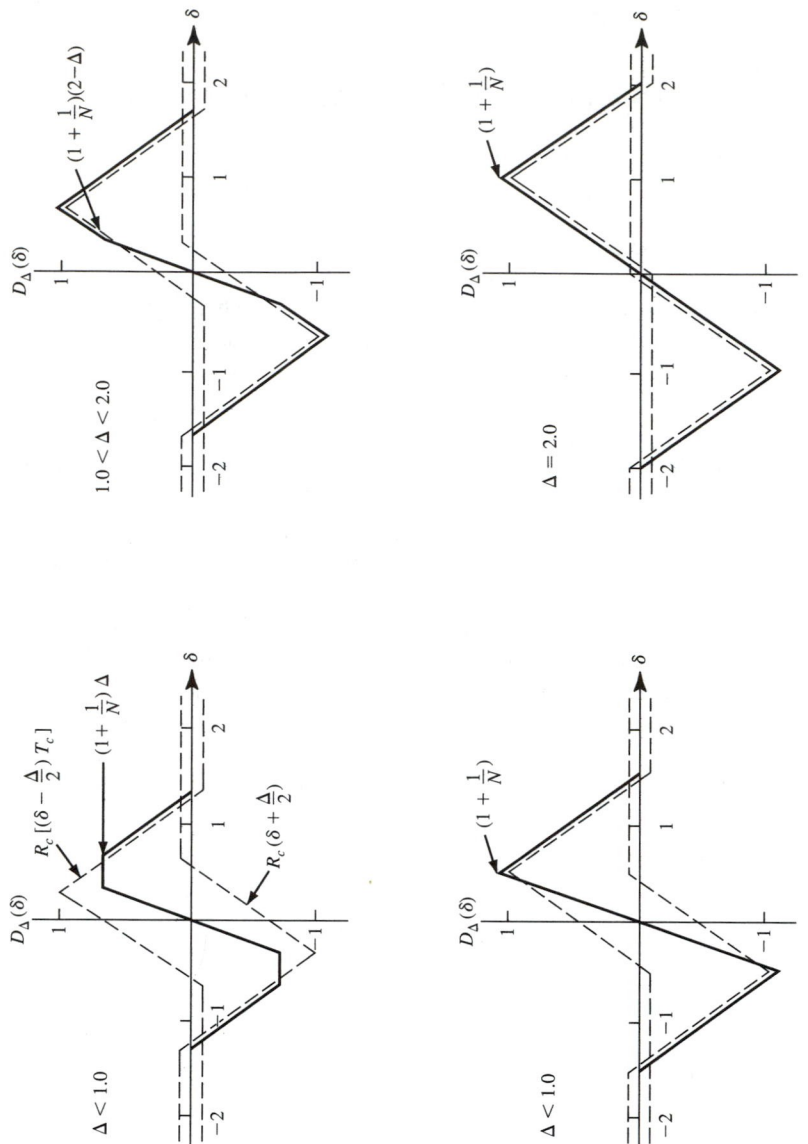

FIGURE 4-5. Delay-lock discriminator dc outputs for maximal-length sequence spreading codes and various values of Δ.

Assuming that code self-noise can be ignored, this can be written

$$\epsilon(t,T_d,\hat{T}_d) = K_1 \sqrt{\frac{P}{2}} \left[D_\Delta(\delta) + \frac{1}{\sqrt{P}} n'(t) \right] \tag{4-8}$$

where

$$n'(t) = n(t) \left[c\left(t - \hat{T}_d - \frac{\Delta}{2}T_c\right) - c\left(t - \hat{T}_d + \frac{\Delta}{2}T_c\right) \right] \tag{4-9}$$

To evaluate the noise performance of the tracking loop, the power spectrum of $n'(t)$ must be calculated. This calculation is accomplished by first evaluating the autocorrelation function of $n'(t)$. Care must be exercised in defining the autocorrelation function since, with $\hat{T}_d$ and Δ fixed, the random process $n'(t)$ defined above is not wide-sense stationary and therefore does not possess a power spectrum in the normal sense. With $\hat{T}_d$ interpreted as a random variable, stationarity is achieved and

$$R_{n'}(\tau) = E\left\{ n(t)n(t + \tau) \left[c\left(t - \hat{T}_d - \frac{\Delta}{2}T_c\right) - c\left(t - \hat{T}_d + \frac{\Delta}{2}T_c\right) \right] \right.$$
$$\left. \times \left[c\left(t + \tau - \hat{T}_d - \frac{\Delta}{2}T_c\right) - c\left(t + \tau - \hat{T}_d + \frac{\Delta}{2}T_c\right) \right] \right\} \tag{4-10}$$

The white noise $n(t)$ is independent of the spreading code $c(t)$, so that the expected value can be factored to obtain

$$R_{n'}(\tau) = E[n(t)n(t + \tau)]E\left\{ \left[c\left(t - \hat{T}_d - \frac{\Delta}{2}T_c\right) - c\left(t - \hat{T}_d + \frac{\Delta}{2}T_c\right) \right] \right. \tag{4-11}$$
$$\left. \times \left[c\left(t + \tau - \hat{T}_d - \frac{\Delta}{2}T_c\right) - c\left(t + \tau - \hat{T}_d + \frac{\Delta}{2}T_c\right) \right] \right\}$$

The autocorrelation function of the noise is a delta function, that is,

$$E[n(t)n(t + \tau)] = \frac{N_0}{2}\delta(\tau) \tag{4-12}$$

Substituting (4-12) into (4-11) and setting $\tau = 0$, since $\delta(\tau)$ is zero for all $\tau \neq 0$, results in

$$R_{n'}(\tau) = \frac{N_0}{2}\delta(\tau)E\left\{ \left[c\left(t - \hat{T}_d - \frac{\Delta}{2}T_c\right) - c\left(t - \hat{T}_d + \frac{\Delta}{2}T_c\right) \right]^2 \right\}$$
$$= \frac{N_0}{2}\delta(\tau)\left\{ E\left[c^2\left(t - \hat{T}_d - \frac{\Delta}{2}T_c\right) \right] \right.$$
$$\left. - 2E\left[c\left(t - \hat{T}_d - \frac{\Delta}{2}\right)c\left(t - \hat{T}_d + \frac{\Delta}{2}T_c\right) \right] \right. \tag{4-13}$$
$$\left. + E\left[c^2\left(t - \hat{T}_d + \frac{\Delta}{2}T_c\right) \right] \right\}$$

The spreading waveforms take on only values of ± 1, so that $c^2(t) = 1.0$. Assume that the spreading waveforms are derived from maximal-length sequences. Because of the shift-and-add property of m-sequences, the function $c(t - \hat{T}_d - (\Delta/2)T_c)$ $c(t - \hat{T}_d + (\Delta/2)T_c)$ may be viewed as a signal that switches between two phases of the same m-sequence for $\Delta \geq 1.0$ or between the all-ones sequence and some phase of the m-sequence for $\Delta < 1.0$. For $\Delta > 1.0$, the expected value of the product signal is $-1/N$. For $\Delta < 1.0$, the all-ones sequence is output for $(1 - \Delta)T_c$ seconds and some phase of the m-sequence is output for ΔT_c seconds of every chip. The expected value of this signal is $(1 - \Delta) - \Delta(1/N)$. Thus

$$R_{n'}(\tau) = \frac{N_0}{2} \delta(\tau) \left\{ 2 - 2E \left[c\left(t - \hat{T}_d - \frac{\Delta}{2} T_c \right) c\left(t - \hat{T}_d + \frac{\Delta}{2} T_c \right) \right] \right\}$$

$$= \begin{cases} N_0 \, \delta(\tau) \left(1 + \dfrac{1}{N} \right) & \text{for } \Delta \geq 1.0 \\[3mm] N_0 \, \delta(\tau) \, \Delta \left(1 + \dfrac{1}{N} \right) & \text{for } \Delta \leq 1.0 \end{cases} \qquad (4\text{-}14)$$

The two-sided power spectral density of $n'(t)$ is the Fourier transform of $R_{n'}(\tau)$ and is

$$S_{n'}(f) = \begin{cases} N_0 \left(1 + \dfrac{1}{N} \right) & \text{for } \Delta \geq 1.0 \\[3mm] N_0 \Delta \left(1 + \dfrac{1}{N} \right) & \text{for } \Delta < 1.0 \end{cases} \qquad (4\text{-}15)$$

Thus the random process $n'(t)$ is also white. The process is not, however, Gaussian since the second term in the product of (4-9) is periodic and takes on only values of ± 2 and 0.

At this point the delay-lock discriminator of Figure 4-4 has been adequately characterized to enable the development of a linear model of the entire tracking loop which will be valid for small tracking errors δ. Consider the voltage-controlled oscillator. The output frequency of this oscillator is $f_0 + g_c v(t)$, where f_0 is the rest or quiescent frequency and g_c is the VCO gain in Hz/V. The instantaneous output phase of this oscillator is the integral of frequency and is

$$2\pi f_0 t + 2\pi \frac{\hat{T}_d(t)}{T_c} = 2\pi f_0 t + 2\pi \int_0^t g_c v(\lambda) \, d\lambda + \theta_0 \qquad (4\text{-}16)$$

where θ_0 is the phase at time zero. The initial phase is set equal to zero in all that follows. The left side of the equation is the oscillatory output phase written directly as a function of $\hat{T}_d(t)$. Subtracting $2\pi f_0 t$ from both sides of (4-16) and dividing by 2π yields

$$\frac{\hat{T}_d(t)}{T_c} = g_c \int_0^t v(\lambda) \, d\lambda \qquad (4\text{-}17)$$

and taking the Laplace transform yields

$$\frac{\widehat{T}_d(s)}{T_c} = \frac{g_c V(s)}{s}$$ (4-18)

where $\widehat{T}(s)$ represents the Laplace transform of $\widehat{T}(t)$.

The tracking loop filter is assumed to be passive, linear, and time invariant, so that its input–output relationship can be described by a differential equation having the form

$$a_m \frac{d^m \epsilon}{dt^m} + a_{m-1} \frac{d^{m-1} \epsilon}{dt^{m-1}} + \cdots + a_0 \epsilon = b_n \frac{d^n v}{dt^n} + b_{n-1} \frac{d^{n-1} v}{dt^{n-1}} + \cdots + b_0 v$$ (4-19)

where $m \leq n$ and ϵ represents $\epsilon(t, \delta)$. This linear difference equation can be solved using classical techniques. Using Laplace transform techniques, the ratio of the Laplace transform of the output, $V(s)$, to the Laplace transform of the input $E(s, \delta)$ is found to be

$$\frac{V(s)}{E(s, \delta)} = \frac{a_m s^m + a_{m-1} s^{m-1} + \cdots + a_0}{b_n s^n + b_{n-1} s^{n-1} + \cdots + b_0}$$ (4-20)

$$\triangleq F(s)$$

where $F(s)$ is the transfer function of the filter. This filter can also be described using its impulse response function $f(t)$. The filter output is the convolution of the input signal and the impulse response. That is,

$$v(t) = \int_{-\infty}^{t} \epsilon(\lambda, \delta) f(t - \lambda)\, d\lambda$$ (4-21)

The impulse response $f(t)$ is the inverse Laplace transform of the transfer function $F(s)$.

Using (4-8) to represent the delay-lock discriminator output, and (4-17) and (4-21) to represent the VCO and loop filter characteristics, the nonlinear integral equation representing the operation of the tracking loop of Figure 4-4 is

$$\frac{\widehat{T}_d(t)}{T_c} = g_c \int_0^t v(\lambda)\, d\lambda$$

$$= g_c \int_0^t \int_{-\infty}^{\lambda} \epsilon(\alpha, \delta) f(\lambda - \alpha)\, d\alpha\, d\lambda$$ (4-22)

$$= g_c \int_0^t \int_{-\infty}^{\lambda} \left\{ K_1 \sqrt{\frac{P}{2}} D_\Delta[\delta(\alpha)] + \frac{K_1}{\sqrt{2}} n'(\alpha) \right\} f(\lambda - \alpha)\, d\alpha\, d\lambda$$

where $n'(t)$ is defined in (4-9). In this equation, the dependence of the normalized phase error $\delta(t) = [T_d(t) - \widehat{T}_d(t)]/T_c$ on time has been shown explicitly. Equation (4-22) suggests the equivalent circuit illustrated in Figure 4-6. The equation that describes this circuit can be written down by inspection, and will be identical to

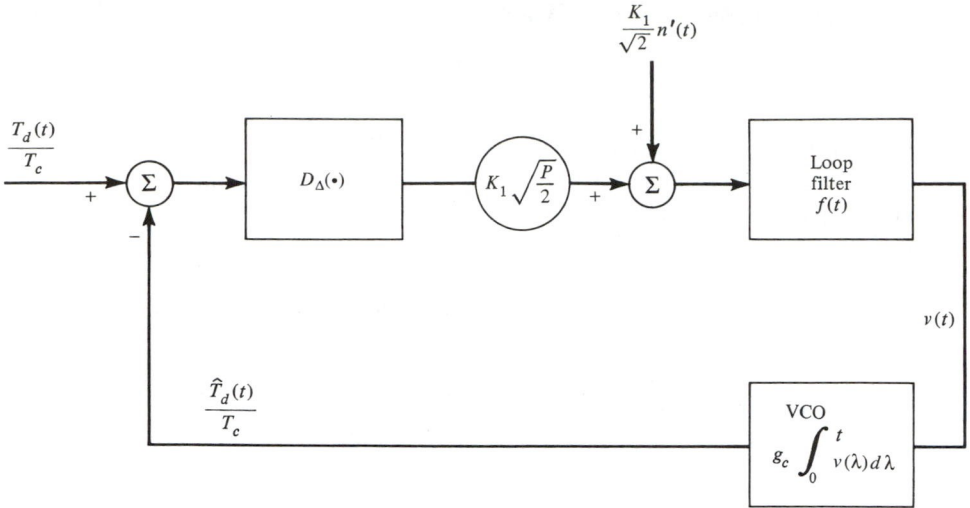

FIGURE 4-6.   Nonlinear equivalent circuit for the baseband DLL.

(4-22). This circuit is the nonlinear [because of the function $D_\Delta(\delta)$] equivalent circuit for the baseband DLL.

For small tracking error, the phase discriminator output is a linear function of the tracking error and (4-22) can be written

$$\frac{\widehat{T}_d(t)}{T_c} = g_c \int_0^t \int_{-\infty}^\lambda \left[K_1 \sqrt{\frac{P}{2}} 2\left(1 + \frac{1}{N}\right) \frac{T_d(\alpha) - \widehat{T}_d(\alpha)}{T_c} + \frac{K_1}{\sqrt{2}} n'(\alpha) \right]$$
$$\times f(\lambda - \alpha)\, d\alpha\, d\lambda \qquad (4\text{-}23)$$

Consider the system illustrated in Figure 4-7. The equation that describes the operation of this system can be written down by inspection, and is identical to (4-23), where K_d is defined in Figure 4-7. Therefore the system of Figure 4-7 describes the operation of the baseband DLL for small tracking errors. The noise function at the input to the circuit of Figure 4-7 will produce noise at point A, which is identical to the noise at point A of Figure 4-4 when the signal-to-noise ratio is adequately large that the linear approximation to the delay-lock discriminator is valid. The power spectral density of the white noise function $n'(t)$ was given in (4-15). The power spectral density of $K_1 n'(t)/\sqrt{2}K_d$ is $(K_1/\sqrt{2}K_d)^2$ times the result of (4-15).

The systems of Figures 4-4 and 4-7 are equivalent, so that either can be analyzed to find the tracking loop's noise and dynamic tracking performance. To illustrate the equivalence of the two systems, the loop filter was characterized by its impulse response and the VCO was characterized by an integral of its control voltage. These functions can also be characterized by the Laplace transform relationships of (4-18) and (4-20). The Laplace transform is a linear operation, so that the input difference circuit can be characterized by the difference of the Laplace transforms of $T_d(t)/T_c$ and $\widehat{T}_d(t)/T_c$. Using transform notation, the system of Figure 4-7 can be represented by the system of Figure 4-8.

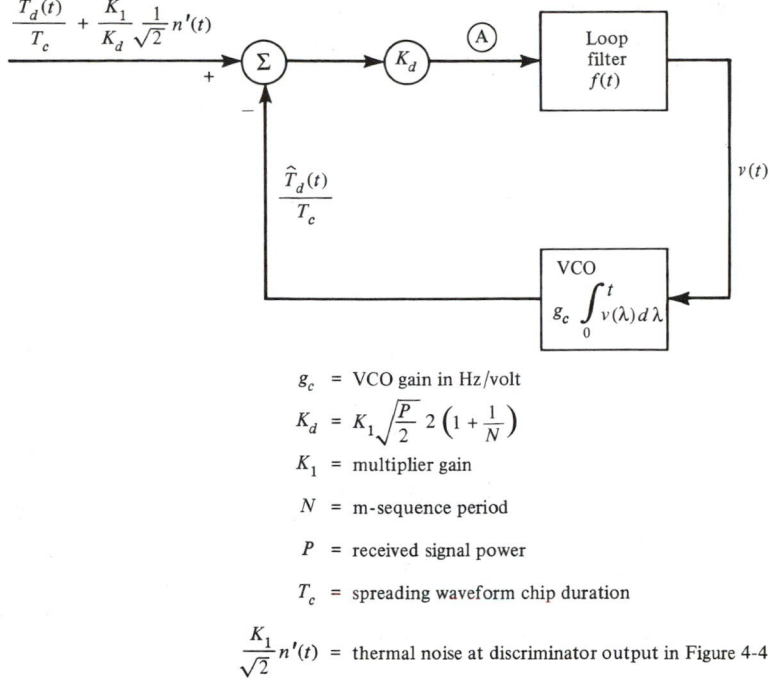

$$g_c = \text{VCO gain in Hz/volt}$$

$$K_d = K_1 \sqrt{\frac{P}{2}} \, 2 \left(1 + \frac{1}{N}\right)$$

$$K_1 = \text{multiplier gain}$$

$$N = \text{m-sequence period}$$

$$P = \text{received signal power}$$

$$T_c = \text{spreading waveform chip duration}$$

$$\frac{K_1}{\sqrt{2}} n'(t) = \text{thermal noise at discriminator output in Figure 4-4}$$

FIGURE 4-7. Linear equivalent circuit for the baseband DLL.

Figure 4-8 is identical to the Laplace-transform model of a classical phase-locked loop which may be found in many references (e.g., [11–14]). In these references, the input noise process is always assumed to be Gaussian. In the analysis of this chapter, $n'(t)$ is not Gaussian. The pdf of $n'(t)$ contains an impulse function at zero amplitude. Since noise of zero amplitude has no effect on tracking, and since the remaining pdf is Gaussian, accurate results will be obtained if the Gaussian analysis is used.

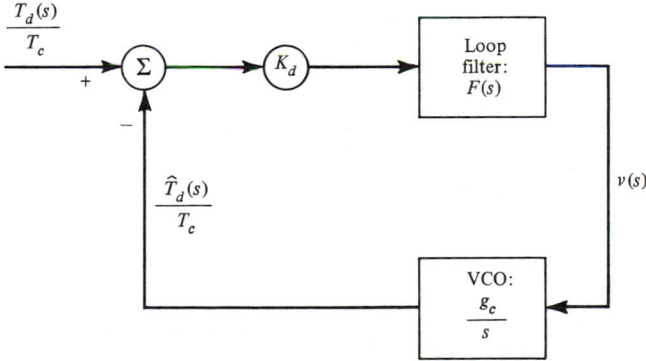

FIGURE 4-8. Laplace-transform model of linear equivalent circuit for baseband DLL.

The tracking loop output $\widehat{T}_d(s)$ can be related directly to its input $T_d(s)$ using Figure 4-8. This relationship is written by inspection and is

$$\frac{\widehat{T}_d(s)}{T_c} = \frac{T_d(s) - \widehat{T}_d(s)}{T_c}\left[K_d g_c \frac{F(s)}{s}\right] \tag{4-24}$$

Solving this equation for $\widehat{T}_d(s)/T_d(s)$ yields

$$\frac{\widehat{T}_d(s)}{T_d(s)} = \frac{K_d g_c F(s)}{s + K_d g_c F(s)} \triangleq H(s) \tag{4-25}$$

and solving for $[\widehat{T}_d(s) - T_d(s)]/T_c$, which is the code tracking error, yields

$$\frac{T_d(s) - \widehat{T}_d(s)}{T_c} = \frac{T_d(s)}{T_c}\left[\frac{s}{s + K_d g_c F(s)}\right] \tag{4-26}$$

These equations are the classical closed-loop servomechanism equations and from this point on the analysis of this tracking loop is identical to the analysis of any other servo loop. In particular, the response of the tracking loop to steps, ramps, and parabolas of input phase can be determined for various loop filter configurations. In addition, the model of Figure 4-8 is the same, except for the definition of some of the constants, as the PLL model, so that much of the PLL analysis also applies to the code tracking loop.

A result that is particularly important to the spread-spectrum system designer is the relationship between the mean-square tracking error or tracking jitter and the received signal-to-noise ratio in the loop bandwidth. The power spectrum of the tracking jitter is given by [12]

$$S_\delta(f) = |H(j2\pi f)|^2 S_{n''}(f) \tag{4-27}$$

where $H(s)$ is the closed-loop transfer function defined by (4-25) and $S_{n''}(f)$ is the two-sided power spectrum of the Gaussian noise process at the input to the loop model of Figure 4-7. The power spectrum of $n''(t)$ is $K_1^2/2K_d^2$ times the result of (4-15) and is

$$S_{n''}(f) = \begin{cases} \dfrac{1}{2}\left(\dfrac{K_1}{K_d}\right)^2 N_0\left(1 + \dfrac{1}{N}\right) & \text{for } \Delta \geq 1.0 \\[3ex] \dfrac{1}{2}\left(\dfrac{K_1}{K_d}\right)^2 N_0\,\Delta\left(1 + \dfrac{1}{N}\right) & \text{for } \Delta < 1.0 \end{cases} \tag{4-28}$$

Then the variance of δ which is denoted σ_δ^2 is

$$\sigma_\delta^2 = \int_{-\infty}^{\infty} S_{n''}(f)|H(j2\pi f)|^2 \, df \tag{4-29}$$

Since $S_{n''}(f)$ is constant over all f [i.e., $n''(t)$ is a white noise process],

$$\sigma_\delta^2 = S_{n''}(0) \int_{-\infty}^{\infty} |H(j2\pi f)|^2 \, df \tag{4-30}$$

The integral on the right side of this equation is defined as the two-sided noise bandwidth

$$W_L = \int_{-\infty}^{\infty} |H(j2\pi f)|^2 \, df \tag{4-31}$$

The units of W_L are hertz. The final result for tracking jitter is then

$$\sigma_\delta^2 = \begin{cases} \dfrac{1}{2}\left(\dfrac{K_1}{K_d}\right)^2 N_0\left(1 + \dfrac{1}{N}\right)W_L & \text{for } \Delta \geq 1.0 \\[3ex] \dfrac{1}{2}\left(\dfrac{K_1}{K_d}\right)^2 N_0\,\Delta\left(1 + \dfrac{1}{N}\right)W_L & \text{for } \Delta < 1.0 \end{cases} \tag{4-32}$$

Observe that the tracking error variance is reduced as Δ decreases for $\Delta \leq 1.0$. This can be explained by noticing that as Δ decreases, the slope of the discriminator characteristic remains constant while some of the input noise cancels in (4-9) when the early and late codes overlap. Evaluating K_d as a function of K_1, P, and N in (4-32), and defining $\rho_L = 2P/N_0 W_L$ yields

$$\sigma_\delta^2 = \begin{cases} \dfrac{1}{2}\left(\dfrac{K_1}{K_d}\right)^2 N_0\left(1 + \dfrac{1}{N}\right)W_L & \text{for } \Delta \geq 1.0 \\[3ex] \dfrac{1}{2}\left(\dfrac{K_1}{K_d}\right)^2 N_0\,\Delta\left(1 + \dfrac{1}{N}\right)W_L & \text{for } \Delta < 1.0 \end{cases} \tag{4-32}$$

where ρ_L is the signal-to-noise ratio in the loop bandwidth.

EXAMPLE 4-1

A commonly used tracking loop filter is the simple lead-lag filter whose transfer function is

$$F(s) = \frac{1 + \tau_1 s}{\tau_1 s} \tag{4-34}$$

With this loop filter, the closed-loop transfer function of (4-25) becomes

$$H(s) = \frac{[K_d g_c(\tau_2/\tau_1)]s + K_d g_c/\tau_1}{s^2 + [K_d g_c(\tau_2/\tau_1)]s + K_d g_c/\tau_1}$$

$$= \frac{2\zeta\omega_n s + \omega_n^2}{s^2 + 2\zeta\omega_n s + \omega_n^2} \tag{4-35}$$

where the usual loop natural frequency and damping factor have been defined by

$$\omega_n = \sqrt{\frac{K_d g_c}{\tau_1}} \tag{4-36}$$

$$\zeta = \frac{\tau_2}{2}\omega_n \tag{4-37}$$

Since K_d is a function of the input signal level, both K_d and ζ are functions of P. This implies that automatic gain control is an important issue in tracking loop design. The two-sided noise bandwidth for a second-order loop with this loop filter is given by [12]

$$W_L = \omega_n \left(\zeta + \frac{1}{4\zeta} \right) \tag{4-38}$$

This result is derived using contour integration to evaluate (4-31). The units of W_L are hertz.

This completes the analysis of the baseband DLL. Fundamentally the same techniques will be used in the following sections to analyze other types of tracking loops. The two major components of this analysis were the derivation of the phase discriminator characteristic or S-curve, and the characterization of the thermal noise at the discriminator output. These two analyses were used to develop a linear model of the tracking loop which then enables the application of the vast store of control loop theory to the code tracking problem.

4-4 Noncoherent Delay-Lock Tracking Loop

Two difficulties arise when the baseband loop of Section 4-3 is applied to actual spread-spectrum communication systems. First, since the tracking loop input is the spreading waveform $c(t)$, this spreading waveform must be recovered from the carrier prior to code tracking. That is, the received signal must be demodulated prior to code tracking. Since spread-spectrum systems typically operate at very low signal-to-noise ratios in the transmission bandwidth, this demodulation will be difficult. In addition, the modulation is coherent and therefore a coherent carrier reference must be generated prior to demodulation. The generation of this coherent reference at extremely low signal-to-noise ratios is also difficult. The second difficulty stems from the fact that any communication system must convey information from the transmitter to the receiver. This implies that the carrier is in some way modulated with this information. The baseband signals analyzed in Section 4-3 conveniently ignored any data modulation principally because the baseband loop would not function properly had the received signal been $d(t - T_d)c(t - T_d)$ rather than $c(t - T_d)$.

Neither of these difficulties are present for the noncoherent delay-lock tracking loop discussed in this section. The noncoherent loop employs a phase discriminator which is significantly different from that used in the baseband loop. The discriminator contains two energy detectors, which are, of course, not sensitive to data modulation or carrier phase, and thus enable the discriminator to ignore data modulation and carrier phase. Figure 4-9 is a conceptual block diagram of the tracking loop for the special case where the spreading modulation is binary phase-shift keying. With

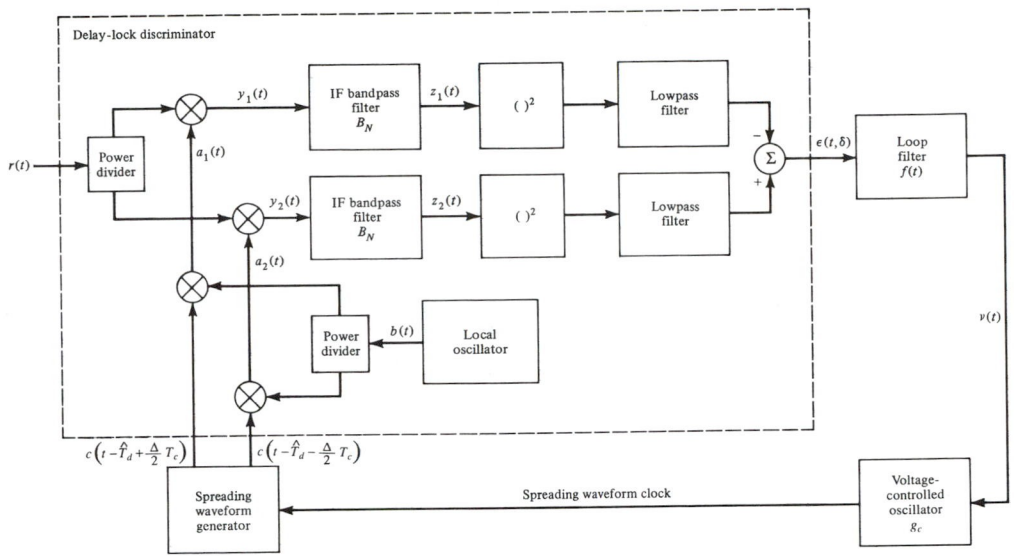

FIGURE 4-9. Noncoherent delay-lock code tracking loop.

minor modifications, which will be discussed later, this tracking loop can be used for any direct-sequence spreading modulation. BPSK is used to introduce the student to this tracking loop because the analysis is simplified.

The received signal in this case is a data and spreading code-modulated carrier in bandlimited AWGN. This signal is represented by

$$r(t) = \sqrt{2P}\,c(t - T_d)\cos[\omega_0 t + \theta_d(t - T_d) + \phi] + n(t) \qquad (4\text{-}39)$$

where P is the received signal power, $\theta_d(t - T_d)$ is the arbitrary data phase modulation, T_d is the transmission delay, ϕ is the random received carrier phase, ω_0 is the carrier radian frequency, and

$$n(t) = \sqrt{2}\,n_I(t)\cos\omega_0 t - \sqrt{2}\,n_Q(t)\sin\omega_0 t \qquad (4\text{-}40)$$

represents the bandlimited zero-mean white Gaussian noise process. The received noise is assumed to have a two-sided power spectral density of $N_0/2$ W/Hz. The functions $n_I(t)$ and $n_Q(t)$ are independent zero-mean lowpass white Gaussian noise processes each having a two-sided power spectral density of $N_0/2$. Other authors choose to omit the $\sqrt{2}$ factors in (4-40). When these factors are omitted the baseband processes have a two-sided power spectral density of N_0 W/Hz.

The received signal is power divided and then correlated with early and late spreading waveform modulated local oscillator signals. The reference local oscillator output is

$$b(t) = 2\sqrt{2K_1}\cos[(\omega_0 - \omega_{1F})t + \phi'] \qquad (4\text{-}41)$$

so that

$$a_1(t) = 2\sqrt{K_1}c\left(t - \hat{T}_d + \frac{\Delta}{2}T_c\right)\cos[(\omega_0 - \omega_{IF})t + \phi'] \tag{4-42a}$$

$$a_2(t) = 2\sqrt{K_1}c\left(t - \hat{T}_d - \frac{\Delta}{2}T_c\right)\cos[(\omega_0 - \omega_{IF})t + \phi'] \tag{4-42b}$$

where ω_{IF} is the intermediate radian frequency, and ϕ' is the random local oscillator phase. The amplitude of these signals has been chosen so that the difference frequency outputs of the early and late correlator mixers due to signal will have power $PK_1/2$. Thus K_1 may be interpreted as the RF-to-IF conversion loss of the mixers. Any gain in an actual system between the input and the input to the squaring circuit may also be included in K_1. The IF bandpass filters are assumed to have center frequency ω_{IF} rad/s and a one-sided noise bandwidth of B_N hertz, so that the sum frequency mixer output components are rejected by these filters. Therefore, only the difference frequency mixer outputs need to be calculated. These components are

$$y_1(t) = \sqrt{K_1 P}\, c(t - T_d)c\left(t - \hat{T}_d + \frac{\Delta}{2}T_c\right)\cos[\omega_{IF}t + \phi - \phi' + \theta_d(t - T_d)]$$

$$+ \sqrt{K_1}\, n_I(t)c\left(t - \hat{T}_d + \frac{\Delta}{2}T_c\right)\cos(\omega_{IF}t - \phi') \tag{4-43a}$$

$$- \sqrt{K_1}\, n_Q(t)c\left(t - \hat{T}_d + \frac{\Delta}{2}T_c\right)\sin(\omega_{IF}t - \phi')$$

$$y_2(t) = \sqrt{K_1 P}\, c(t - T_d)c\left(t - \hat{T}_d - \frac{\Delta}{2}T_c\right)\cos[\omega_{IF}t + \phi - \phi' + \theta_d(t - T_d)]$$

$$+ \sqrt{K_1}\, n_I(t)c\left(t - \hat{T}_d - \frac{\Delta}{2}T_c\right)\cos(\omega_{IF}t - \phi') \tag{4-43b}$$

$$- \sqrt{K_1}\, n_Q(t)c\left(t - \hat{T}_d - \frac{\Delta}{2}T_c\right)\sin(\omega_{IF}t - \phi')$$

These correlator outputs are the sum of a component due to signal and a component due to noise. Define the noise components by

$$n_{1,in}(t) = \sqrt{\frac{K_1}{2}}c\left(t - \hat{T}_d + \frac{\Delta}{2}T_c\right)n'(t) \tag{4-44a}$$

$$n_{2,in}(t) = \sqrt{\frac{K_1}{2}}c\left(t - \hat{T}_d - \frac{\Delta}{2}T_c\right)n'(t) \tag{4-44b}$$

where

$$n'(t) = \sqrt{2}n_I(t)\cos(\omega_{IF}t - \phi') - \sqrt{2}n_Q(t)\sin(\omega_{IF}t - \phi') \tag{4-45}$$

The signal component is again the sum of a desired signal, which will be used for code tracking, and a code self-noise term. which is negligible when the processing gain is sufficiently high. Thus there are three components of interest at the correlator output.

To determine the signal and self-noise components of the discriminator output, consider the noiseless case. Assume that the spreading code is independent of the data modulation and local oscillator carrier phase so that the power spectrum of $y_1(t)$ is the convolution of the power spectrum of $c(t - T_d)c(t - \hat{T}_d + (\Delta/2)T_c)$ and the power spectrum $S_d(\omega)$ of $\sqrt{K_1 P} \cos[\omega_{IF} t + \phi - \phi' + \theta_d(t - T_d)]$. The power spectrum of the product of two maximal-length codes is derived in Appendix D and the result is given in (3-66), which is repeated here for convenience using radian frequency and defining $\alpha = T_d - \hat{T}_d + (\Delta/2)T_c$:

$$
S_{cc'}(\omega, \alpha) = \left[1 - \left(1 + \frac{1}{N} \right) \frac{|\alpha|}{T_c} \right]^2 2\pi \delta(\omega)
$$

$$
+ \left(1 + \frac{1}{N} \right) \left(\frac{|\alpha|}{T_c} \right)^2 \sum_{\substack{n=-\infty \\ n \neq 0}}^{\infty} \mathrm{sinc}^2 \left(\frac{n\omega_c}{2\pi} |\alpha| \right) 2\pi \delta(\omega - n\omega_c) \quad (3\text{-}66)
$$

$$
+ \frac{N+1}{N^2} \left(\frac{|\alpha|}{T_c} \right)^2 \sum_{\substack{m=-\infty \\ m \neq 0}}^{\infty} \mathrm{sinc}^2 \left(\frac{m\omega_c}{2\pi N} |\alpha| \right) 2\pi \delta \left(\omega - \frac{m\omega_c}{N} \right)
$$

In this expression cc' denotes the maximal-length sequence product, and α is the phase difference in seconds between c and c'. The power spectrum, $S_{y_1}(\omega)$, of the data-modulated IF carrier is the convolution of $S_d(\omega)$ and $S_{cc'}(\omega)$:

$$
S_{y_1}(\omega) = \frac{1}{2\pi} \int_{-\infty}^{\infty} S_d(\omega') S_{cc'}(\omega - \omega') \, d\omega'
$$

$$
= \left[1 - \left(1 + \frac{1}{N} \right) \frac{|\alpha|}{T_c} \right]^2 S_d(\omega)
$$

$$
+ \left(1 + \frac{1}{N} \right) \left(\frac{|\alpha|}{T_c} \right)^2 \sum_{\substack{n=-\infty \\ n \neq 0}}^{\infty} \mathrm{sinc}^2 \left(\frac{n\omega_c}{2\pi} |\alpha| \right) S_d(\omega - n\omega_c) \tag{4-46}
$$

$$
+ \frac{N+1}{N^2} \left(\frac{|\alpha|}{T_c} \right)^2 \sum_{\substack{m=-\infty \\ m \neq 0}}^{\infty} \mathrm{sinc}^2 \left(\frac{m\omega_c}{2\pi N} |\alpha| \right) S_d \left(\omega - \frac{m\omega_c}{N} \right)
$$

Observe that this is a sum of frequency translations of $S_d(\omega)$ by all frequency components of cc'. The bandwidth of the IF filter has been selected to be just wide enough to pass the data-modulated IF carrier. Simon [10] has shown that code tracking performance may be improved by selecting the IF filter bandwidth such that squaring loss is minimized. This alternative bandwidth selection strategy will result in distortion of the data-modulated carrier by the IF filter. Simon's results will be discussed later in this chapter. To simplify this analysis, distortion of the data-modulated IF carrier by the IF filter is ignored.

In all spread-spectrum systems, the spreading waveform chip frequency ω_c is much greater than the maximum significant data modulation frequency. Therefore, all translations of $S_d(\omega)$ by harmonics of ω_c [i.e., the second line of (4-46)], will lie outside the IF filter passband and will be rejected. The components of the third line of (4-46) are translated, not by harmonics of ω_c but by harmonics of ω_c/N. The code period N is usually large, so that translations of $S_d(\omega)$ by ω_c/N may result in components within the IF passband. Fortunately, these components have a magnitude proportional to $(N + 1)/N^2 \approx 1/N$, so that their effect on the discriminator may be minimal.

To gain a preliminary idea of the magnitude relative to the desired term of the terms on the third line of (4-46), assume that $S_d(\omega)$ has a uniform density over the frequency range $0 \le |\omega \pm \omega_{IF}| < \pi B_N$, and calculate the power spectrum of the third term at $\omega = \pm\omega_{IF}$ for the limiting case where N approaches infinity. Denote this limiting value at $\omega = +\omega_{IF}$ by β, where

$$\beta = \lim_{N\to\infty} \frac{N+1}{N^2}\left(\frac{|\alpha|}{T_c}\right)^2 \sum_{\substack{m=-\infty \\ m\neq 0}}^{\infty} \text{sinc}^2\left(\frac{m\omega_c}{2\pi N}|\alpha|\right)S_d\left(\omega_{IF} - \frac{m\omega_c}{N}\right) \quad (4\text{-}47)$$

It can be shown that β is approximately given by

$$\beta = \left(\frac{\Delta}{2}\right)^2 \frac{B_N}{f_c} S_d(\omega_{IF}) \quad (4\text{-}48)$$

An identical expression can be derived for the value of the power spectral density at $\omega = -\omega_{IF}$. The most important characteristic of (4-48) is that β decreases inversely with increasing f_c, so that, with sufficiently high code rate, β can be safely ignored. Unless otherwise stated, the code rate of the spread-spectrum system will be assumed adequately high that both the second and third lines of (4-46) can be ignored.

The preceding arguments have shown that for high processing gains, the only component of interest is

$$y_1(t) = \sqrt{K_1 P}c(t - T_d)c\left(t - \hat{T}_d + \frac{\Delta}{2}T_c\right)\cos[\omega_{IF}t + \phi - \phi' + \theta_d(t - T_d)]$$

is the component resulting from the product of the dc component of $c(t - T_d)c(t - \hat{T}_d + (\Delta/2)T_c)$ with the cosine. By definition, the dc component of the spreading waveform product is the autocorrelation function $R_c(\tau)$ of the spreading waveform evaluated at $\tau = T_d - \hat{T}_d + (\Delta/2)T_c$. Thus, for high processing gains,

$$y_1(t) \cong \sqrt{K_1 P}R_c\left[\left(\delta + \frac{\Delta}{2}\right)T_c\right]\cos[\omega_{IF}t + \phi - \phi' + \theta_d(t - T_d)]$$

$$\triangleq x_1(t) \quad (4\text{-}49a)$$

where $\delta = (T_d - \hat{T}_d)/T_c$. An identical development for the late channel will yield

$$y_2(t) \cong \sqrt{K_1 P}R_c\left[\left(\delta - \frac{\Delta}{2}\right)T_c\right]\cos[\omega_{IF}t + \phi - \phi' + \theta_d(t - T_d)]$$

$$\triangleq x_2(t) \quad (4\text{-}49b)$$

If IF bandpass filters have been designed to pass these signals with negligible distortion so that, in the noiseless case being considered, the square-law device inputs are exactly these signals.

The input to the square-law devices of Figure 4-9 is a narrowband signal centered at $\omega = \omega_{IF}$. The output of the squaring circuit has a component at baseband and a component centered at $\omega = 2\omega_{IF}$. The lowpass filters reject all components near $2\omega_{IF}$. Therefore, the signal component of the delay-lock discriminator output is

$$\epsilon(t,\delta) = [x_2^2(t) - x_1^2(t)]_{\text{lowpass}}$$

$$= \frac{1}{2} K_1 P \left\{ R_c^2 \left[\left(\delta - \frac{\Delta}{2} \right) T_c \right] - R_c^2 \left[\left(\delta + \frac{\Delta}{2} \right) T_c \right] \right\} \qquad (4\text{-}50)$$

$$\triangleq \frac{1}{2} K_1 P D_\Delta(\delta)$$

where

$$D_\Delta(\delta) \triangleq R_c^2 \left[\left(\delta - \frac{\Delta}{2} \right) T_c \right] - R_c^2 \left[\left(\delta + \frac{\Delta}{2} \right) T_c \right] \qquad (4\text{-}51)$$

For maximal-length sequence spreading waveforms, the autocorrelation function is given by (3-51) and, after straightforward algebraic simplification.

$$D_\Delta(\delta) = \begin{cases} 0 & \text{for } -N + 1 + \frac{\Delta}{2} < \delta \leq -\left(1 + \frac{\Delta}{2}\right) \\[2ex] \dfrac{1}{N^2} - \left[1 + \left(1 + \dfrac{1}{N}\right)\left(\delta + \dfrac{\Delta}{2}\right)\right]^2 & \text{for } -\left(1 + \dfrac{\Delta}{2}\right) < \delta \leq -\dfrac{\Delta}{2} \\[2ex] \dfrac{1}{N^2} - \left[1 - \left(1 + \dfrac{1}{N}\right)\left(\delta + \dfrac{\Delta}{2}\right)\right]^2 & \text{for } -\dfrac{\Delta}{2} < \delta \leq -\left(1 - \dfrac{\Delta}{2}\right) \\[2ex] 2\left(1 + \dfrac{1}{N}\right)\left[2 - \left(1 + \dfrac{1}{N}\right)\Delta\right]\delta & \text{for } -\left(1 - \dfrac{\Delta}{2}\right) < \delta \leq \left(1 - \dfrac{\Delta}{2}\right) \\[2ex] \left[1 + \left(1 + \dfrac{1}{N}\right)\left(\delta - \dfrac{\Delta}{2}\right)\right]^2 - \dfrac{1}{N^2} & \text{for } \left(1 - \dfrac{\Delta}{2}\right) < \delta \leq \dfrac{\Delta}{2} \\[2ex] \left[1 - \left(1 + \dfrac{1}{N}\right)\left(\delta - \dfrac{\Delta}{2}\right)\right]^2 - \dfrac{1}{N^2} & \text{for } \dfrac{\Delta}{2} < \delta \leq 1 + \dfrac{\Delta}{2} \end{cases} \qquad (4\text{-}52a)$$

for $\Delta \geq 1.0$ and

$$D_\Delta(\delta) = \begin{cases} 0 & \text{for } -N + 1 + \dfrac{\Delta}{2} < \delta < -\left(1 + \dfrac{\Delta}{2}\right) \\[2ex] \dfrac{1}{N^2} - \left[1 + \left(\delta + \dfrac{\Delta}{2}\right)\left(1 + \dfrac{1}{N}\right)\right]^2 & \text{for } -\left(1 + \dfrac{\Delta}{2}\right) < \delta < \left(\dfrac{\Delta}{2} - 1\right) \\[2ex] -2\left(1 + \dfrac{1}{N}\right)\Delta\left[1 + \left(1 + \dfrac{1}{N}\right)\delta\right] & \text{for } \left(\dfrac{\Delta}{2} - 1\right) < \delta < -\dfrac{\Delta}{2} \\[2ex] 2\left(1 + \dfrac{1}{N}\right)\left[2 - \left(1 + \dfrac{1}{N}\right)\Delta\right]\delta & \text{for } -\dfrac{\Delta}{2} < \delta < +\dfrac{\Delta}{2} \\[2ex] 2\left(1 + \dfrac{1}{N}\right)\Delta\left[1 - \left(1 + \dfrac{1}{N}\right)\delta\right] & \text{for } \dfrac{\Delta}{2} < \delta < \left(1 - \dfrac{\Delta}{2}\right) \\[2ex] \left[1 - \left(1 + \dfrac{1}{N}\right)\left(\delta - \dfrac{\Delta}{2}\right)\right]^2 - \dfrac{1}{N^2} & \text{for } \left(1 - \dfrac{\Delta}{2}\right) < \delta < \left(1 + \dfrac{\Delta}{2}\right) \end{cases} \tag{4-52b}$$

for $\Delta \leq 1.0$. This function is periodic with period N. Equation (4-52) defines a single cycle of $D_\Delta(\delta)$. Observe that in the area near $\delta = 0$, $D_\Delta(\delta)$ is a linear function of δ despite the fact that nonlinear operations have been used in the discriminator. The slope of $D_\Delta(\delta)$ near $\delta = 0$ is a function of Δ and equals zero when $\Delta = 2.0$.

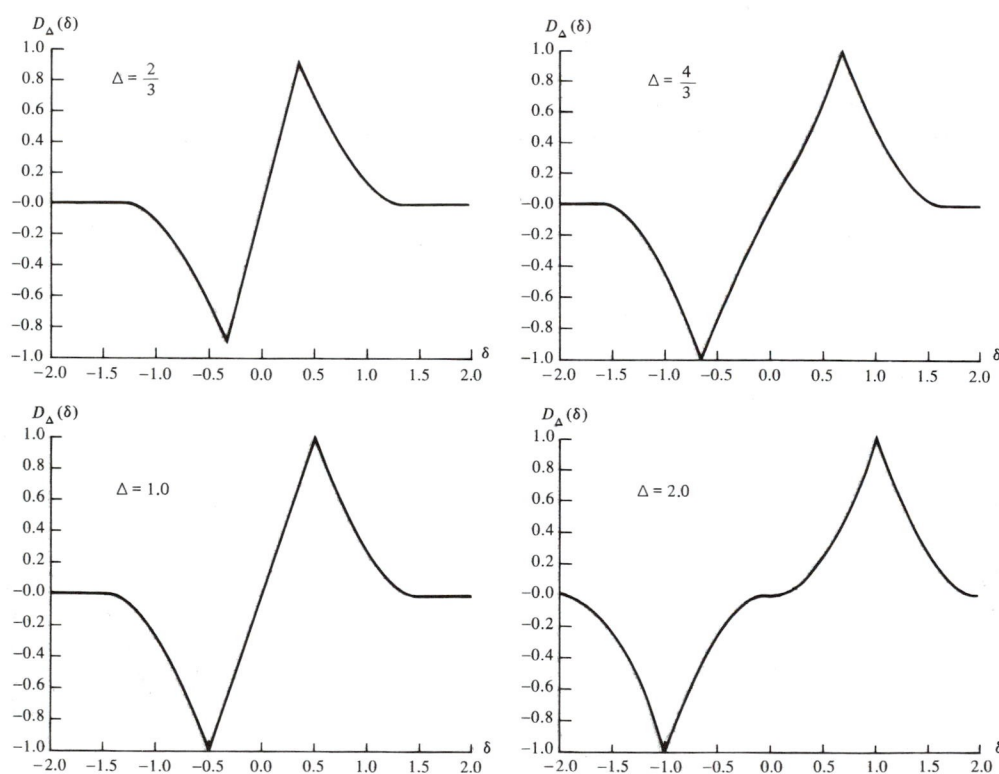

FIGURE 4-10. Noncoherent delay-lock discriminator S-curves for various Δ.

Because of this, the noncoherent delay-lock discriminator is never used with an early-to-late delay difference of two code chip times. Figure 4-10 illustrates $D_\Delta(\delta)$ for the same four values of Δ which were used in Figure 4-5 for the baseband tracking loop. This completes the analysis of the signal component of the discriminator output $\epsilon(t,\delta)$.

Consider next the noise component $\epsilon(t,\delta)$. The noise process at the correlator outputs have already been defined in (4-44) and (4-45). These noise processes, $n_{1,\text{in}}(t)$ and $n_{2,\text{in}}(t)$, are the products of a bandlimited white Gaussian noise process $n'(t)$ centered at $\omega = \omega_{\text{IF}}$ and the spreading waveforms $\sqrt{K_1/2}\,c(t - \hat{T}_d \pm (\Delta/2)T_c)$. Since the noise and the spreading functions are independent, the power spectrum of $n_{1,\text{in}}(t)$ or $n_{2,\text{in}}(t)$ can be calculated by convolving the spectrum of $n'(t)$ and the spectrum of the spreading waveform $\sqrt{K_1/2}\,c(t)$. That is,

$$S_{n_{j,\text{in}}}(\omega) = \frac{K_1}{4\pi} S_{n'}(\omega) * S_c(\omega) \qquad j = 1,2 \qquad (4\text{-}53)$$

The effect of this convolution is to spread the noise power over a bandwidth wider than its original bandwidth and thereby reduce the power spectral density in the frequency range passed by the IF filter. The IF filter is narrowband relative to the result of (4-53), so that the convolution needs to be evaluated only for frequencies near ω_{IF}. In the case of interest, the power spectrum of $n'(t)$ has a much wider bandwidth than the power spectrum of $c(t)$, so that the convolution has little effect and $S_{n_{j,\text{in}}}(\omega_{\text{IF}})$ may be approximated by

$$S_{n_{j,\text{in}}}(\omega_{\text{IF}}) = \begin{cases} \dfrac{K_1}{2}\left(\dfrac{N_0}{2}\right) & 0 \le |\omega \pm \omega_{\text{IF}}| < \pi B \\[2mm] 0 & \text{elsewhere} \end{cases} \qquad (4\text{-}54)$$

where B is the one-sided bandwidth of the received noise process. The $K_1/2$ factor accounts for the conversion gain of the mixer and the fact that noise power has been divided equally between the early and late correlator channels.

The noise processes $n_{1,\text{in}}(t)$ and $n_{2,\text{in}}(t)$ are Gaussian since $n'(t)$ is zero-mean Gaussian and the multiplying function takes on values of ± 1 with equal probability. The two processes are not, however, independent since one can always be derived from the other through multiplication by $+1$ or -1. Consider the noise output of the IF bandpass filters, which will be denoted by $n_1(t)$ and $n_2(t)$. The bandpass filters will be characterized by their impulse response $h(t)$, so that

$$n_j(t) = \int_{-\infty}^{\infty} n_{j,\text{in}}(\lambda) h(t - \lambda)\, d\lambda \qquad j = 1,2 \qquad (4\text{-}55)$$

Therefore,

$$n_j(t) = \int_{-\infty}^{\infty} \sqrt{\frac{K_1}{2}}\, c\left(\lambda - \hat{T}_d \pm \frac{\Delta}{2} T_c\right) n'(\lambda) h(t - \lambda)\, d\lambda \qquad (4\text{-}56)$$

The processes $n_1(t)$ and $n_2(t)$ are Gaussian since the filter inputs are Gaussian. If they are also uncorrelated, they are independent. The cross-correlation between $n_1(t)$ and $n_2(t)$ is

$$R_{n_1 n_2}(\tau) \triangleq E[n_1(t)n_2(t + \tau)] \tag{4-57}$$

This function is difficult to calculate for the general case where the spreading waveform is assumed to be a deterministic function. The calculation is considerably simplified if $c(t)$ is assumed to be random. In that case

$$R_{n_1 n_2}(\tau) = E\left[\frac{K_1}{2}\int_{-\infty}^{\infty}\int_{-\infty}^{\infty} c\left(\gamma - \hat{T}_d + \frac{\Delta}{2}T_c\right)n'(\gamma)h(t - \gamma)\right.$$

$$\left.\times\, c\left(\lambda - \hat{T}_d - \frac{\Delta}{2}T_c\right)n'(\lambda)h(t + \tau - \lambda)\, d\gamma\, d\lambda\right]$$

$$= \frac{K_1}{2}\int_{-\infty}^{\infty}\int_{-\infty}^{\infty} h(t - \gamma)h(t + \tau - \lambda) \tag{4-58}$$

$$\times\, E\left[c\left(\gamma - \hat{T}_d + \frac{\Delta}{2}T_c\right)c\left(\lambda - \hat{T}_d - \frac{\Delta}{2}T_c\right)\right]E[n'(\gamma)n'(\lambda)]\, d\gamma\, d\lambda$$

$$= \frac{K_1}{2}\int_{-\infty}^{\infty}\int_{-\infty}^{\infty} h(t - \gamma)h(t + \tau - \lambda)R_c(\gamma - \lambda + \Delta T_c)R_{n'}(\gamma - \lambda)\, d\gamma\, d\lambda$$

The evaluation of this autocorrelation function is further simplified by assuming that the bandwidth of $n'(t)$ is significantly wider than the bandwidth of $c(t)$. With this assumption $R_c(\gamma - \lambda + \Delta T_c)$ is nearly constant over the range of significant values of $R_{n'}(\gamma - \lambda)$ and the desired cross-correlation can be approximated by

$$R_{n_1 n_2}(\tau) = \frac{K_1}{2}R_c(\Delta T_c)\int_{-\infty}^{\infty}\int_{-\infty}^{\infty} h(t - \gamma)h(t + \tau - \lambda)R_{n'}(\gamma - \lambda)\, d\gamma\, d\lambda$$

$$= \frac{K_1}{2}R_c(\Delta T_c)\int_{-\infty}^{\infty}\int_{-\infty}^{\infty} h(\gamma')h(\lambda')R_{n'}(\lambda' - \gamma' - \tau)\, d\gamma'\, d\lambda'$$

$$= \frac{K_1}{2}R_c(\Delta T_c)\int_{-\infty}^{\infty}\int_{-\infty}^{\infty}\int_{-\infty}^{\infty} S_{n'}(f)e^{j2\pi f(\lambda' - \gamma' - \tau)}h(\gamma')h(\lambda')\, df\, d\lambda'\, d\gamma'$$

$$= \frac{K_1}{2}R_c(\Delta T_c)\int_{-\infty}^{\infty} S_{n'}(f)e^{-j2\pi f\tau}\int_{-\infty}^{\infty} h(\gamma')e^{-j2\pi f\gamma'} \tag{4-59}$$

$$\times\int_{-\infty}^{\infty} h(\lambda')e^{+j2\pi f\lambda'}\, d\lambda'\, d\gamma'\, df$$

$$= \frac{K_1}{2}R_c(\Delta T_c)\int_{-\infty}^{\infty} |H(j2\pi f)|^2 S_{n'}(f)e^{-j2\pi f\tau}\, df$$

Recall that the spreading code was assumed to be random for this calculation, so that $R_c(\Delta T_c)$ is defined by (C-6). In particular, $R_c(0) = 1.0$ and $R_c(T_c) = 0.0$. Thus the cross-correlation function of $n_1(t)$ and $n_2(t)$ is equal to zero when $\Delta = 1.0$.

The point of the preceding analysis is that $R_{n_1 n_2}(\tau)$ can be made small with the proper choice of the delay difference between the early and late correlator channels. A similar result is conjectured for maximal-length spreading codes, with the slight difference that the minimum cross-correlation would not be zero but would instead be related to $1/N$. The IF filter output-noise processes may or may not be strictly independent. In either case, they are narrowband white Gaussian noise processes and can be represented by

$$n_j(t) = \sqrt{2}n_{jI}(t) \cos \omega_{\text{IF}}t - \sqrt{2}n_{jQ}(t) \sin \omega_{\text{IF}}t \qquad (4\text{-}60)$$

with $j = 1, 2$. The power transfer function of either of the IF filters is $|H(f)|^2$, where $H(f)$ is the Fourier transform of $h(t)$. Therefore, the power spectrum of $n_j(t)$ is

$$S_{n_j}(f) = S_{n_{j,\text{in}}}(f)|H(f)|^2 \qquad (4\text{-}61)$$

and $S_{n_{j,\text{in}}}(f)$ is defined by (4-53) and (4-54). Suppose, for example, that $H(f)$ is an ideal bandpass filter with bandwidth B_N. Then $S_{n_j}(f)$ at the output of the IF filter has magnitude $\frac{1}{4}K_1N_0$ and bandwidth B_N centered on $\pm f_{\text{IF}}$, and the power spectra of $n_{1I}(t)$, $n_{1Q}(t)$, $n_{2I}(t)$, or $n_{2Q}(t)$ have the same magnitude and bandwidth as $S_{n_j}(f)$ but centered on zero frequency.

The output of the delay-lock discriminator $\epsilon(t,\delta)$ is given by

$$\epsilon(t,\delta) = [x_2(t) + n_2(t)]_{\text{LP}}^2 - [x_1(t) + n_1(t)]_{\text{LP}}^2 \qquad (4\text{-}62)$$

where $x_1(t)$ and $x_2(t)$ are defined in (4-49), and the subscript LP denotes the lowpass components of the squared function. Expanding this equation and eliminating all double-frequency terms yields

$$\epsilon(t,\delta) = \frac{1}{2}K_1P\left\{ R_c^2\left[\left(\delta - \frac{\Delta}{2}\right)T_c\right] - R_c^2\left[\left(\delta + \frac{\Delta}{2}\right)T_c\right]\right\}$$

$$+ \sqrt{2K_1P}\left\{ R_c\left[\left(\delta - \frac{\Delta}{2}\right)T_c\right]n_{2I}(t) - R_c\left[\left(\delta + \frac{\Delta}{2}\right)T_c\right]n_{1I}(t)\right\}$$

$$\times \cos[\phi - \phi' + \theta_d(t - T_d)] \qquad (4\text{-}63)$$

$$+ \sqrt{2K_1P}\left\{ R_c\left[\left(\delta - \frac{\Delta}{2}\right)T_c\right]n_{2Q}(t) - R_c\left[\left(\delta + \frac{\Delta}{2}\right)T_c\right]n_{1Q}(t)\right\}$$

$$\times \sin[\phi - \phi' + \theta_d(t - T_d)]$$

$$+ [n_{2I}(t)]^2 + [n_{2Q}(t)]^2 - [n_{1I}(t)]^2 - [n_{1Q}(t)]^2$$

The first term of this equation is the desired signal defined in (4-50). The second and third terms are (signal $\times$ noise) terms, and the last four terms are (noise)^2 terms. The power spectrum of $\epsilon(t,\delta)$ is denoted by $S_\epsilon(f)$ and is the Fourier transform of

$$R_\epsilon(\tau) = E[\epsilon(t,\delta)\epsilon(t + \tau,\delta)] \qquad (4\text{-}64)$$

where the expectation is over all sample functions of the Gaussian noise random process and over all received carrier phase angles ϕ and reference phase angles ϕ'. Equation (4-64) is evaluated in detail in Appendix E; the result is

$$R_\epsilon(\tau) = \frac{1}{4}K_1^2P^2\left\{ R_c^2\left[\left(\delta - \frac{\Delta}{2}\right)T_c\right] - R_c^2\left[\left(\delta + \frac{\Delta}{2}\right)T_c\right]\right\}^2$$

$$+ 2K_1P\left\{ R_c^2\left[\left(\delta - \frac{\Delta}{2}\right)T_c\right] + R_c^2\left[\left(\delta + \frac{\Delta}{2}\right)T_c\right]\right\}E[n_b(t)n_b(t + \tau)]$$

$$\times E\{\cos[\theta_d(t) - \theta_d(t + \tau)]\} \qquad (4\text{-}65)$$

$$+ 4E\{[n_b(t)n_b(t + \tau)]^2\} - 4E^2\{[n_b(t)]^2\}$$

In this expression, $n_b(t)$ has been used to represent any one of the four baseband processes n_{1I}, $n_{1Q}(t)$, $n_{2I}(t)$, or $n_{2Q}(t)$. Since all four have identical statistics, the statistics of any one can be used for calculations of the autocorrelation, power, and so on. In deriving (4-65) it has been assumed that $n_{1I}(t)$, $n_{1Q}(t)$, $n_{2I}(t)$, and $n_{2Q}(t)$ are all independent. This assumption results in accurate results when the input noise bandwidth is large relative to the signal bandwidth and simultaneously $\Delta \geq 1.0$.

The purpose of calculating $R_\epsilon(\tau)$ is to enable the calculation of the delay-lock discriminator output power spectrum with signal and noise at its input. The desired power spectrum is the Fourier transform of $R_\epsilon(\tau)$. This Fourier transform is also calculated in Appendix E. The result is

$$
S_\epsilon(f) = \frac{1}{4} K_1^2 P^2 \left\{ R_c^2 \left[\left(\delta - \frac{\Delta}{2} \right) T_c \right] - R_c^2 \left[\left(\delta + \frac{\Delta}{2} \right) T_c \right] \right\}^2 \delta(f)
$$

$$
+ 2K_1 P \left\{ R_c^2 \left[\left(\delta - \frac{\Delta}{2} \right) T_c \right] + R_c^2 \left[\left(\delta + \frac{\Delta}{2} \right) T_c \right] \right\} S_{n_b}(f) * S_{\theta_d}(f) \quad (4\text{-}66)
$$

$$
+ 8 S_{n_b}(f) * S_{n_b}(f)
$$

where $S_{\theta_d}(f)$ is the Fourier transform of twice the real part of the complex autocorrelation of the complex envelope of a carrier having random phase β and phase modulation $\theta_d(t)$. That is,

$$
S_{\theta_d}(f) = \int_{-\infty}^{\infty} 2 \operatorname{Re}[R_A(\tau)] e^{-j2\pi f\tau} \, d\tau \quad (4\text{-}67)
$$

where

$$
R_A(\tau) = \tfrac{1}{2} E[A^*(t)A(t + \tau)] \quad (4\text{-}68)
$$

$$
A(t) = \exp[j\theta_d(t) + j\beta] \quad (4\text{-}69)
$$

Equation (4-66) can be evaluated for any data modulation to obtain the desired signal and noise power spectra at the phase discriminator output. The result has been derived assuming BPSK direct-sequence spreading modulation and is therefore not valid for arbitrary spreading modulation.

To derive a simple model for the tracking loop, consider the special case where the received carrier is not modulated with data. This special case, although somewhat unrealistic for a communication system, will result in the maximum possible noise component in the power spectral density, $S_\epsilon(f)$, within the tracking loop bandwidth. To verify this, observe that with no data modulation $S_{\theta_d}(f)$ is a delta function at $f = 0$ and the convolution does no spectral widening of the (signal $\times$ noise) term. Specifically, with no data modulation

$$
A(t) = \exp(j\beta) \quad (4\text{-}70)
$$

$$
R_A(\tau) = \tfrac{1}{2} E[A^*(t)A(t + \tau)] = \tfrac{1}{2} \quad (4\text{-}71)
$$

From (4-67), then,

$$
S_{\theta_d}(f) = 2 \int_{-\infty}^{\infty} \tfrac{1}{2} e^{-j2\pi f\tau} \, d\tau = \delta(f) \quad (4\text{-}72)
$$

Assume also that the IF filters are perfect brick wall filters. Then $S_{n_b}(f)$ has magnitude $\frac{1}{4}K_1N_0$, two-sided bandwidth B_N, and the convolution of (4-66) can be performed by inspection to obtain the final result for $S_\epsilon(f)$ illustrated in Figure 4-11.

The calculation of the discriminator output power spectrum gives no information about the probability density function of this output. Although the input noise is Gaussian, the nonlinear devices used in the detector assure that the output statistics are not Gaussian. Fortunately, the loop filter following the discriminator has a bandwidth which is much smaller than the bandwidth of the discriminator output random process, so that the central limit theorem (see Ref. 15) may be applied to show that the loop filter output is nearly Gaussian. Therefore, little error is made by assuming that the input is Gaussian. Also because the loop filter bandwidth is small, the discriminator output power spectral density may be assumed to be white with a value equal to the zero-frequency noise component of Figure 4-11. When evaluating the second line of (4-66) assume that the tracking loop is closed and is tracking the received code phase accurately so that δ is small. Thus setting $\delta = 0.0$ in the evaluation of the magnitude of the second term of (4-66) results in negligible error.

The remainder of the analysis of the noncoherent tracking loop is identical to the analysis of the baseband tracking loop performed earlier. Specifically, the VCO is characterized in the time domain by (4-17) and in the frequency domain by (4-18). The loop filter is again assumed to be a lumped linear time-variant passive filter which is characterized by a differential equation having the form of (4-19). This filter can also be described by its Laplace transform transfer function $F(s)$ as in (4-20) or by its impulse response $f(t)$ in the time domain. Using (4-17), with $v(t)$ defined in Figure 4-9, the VCO can be described by

$$\frac{\hat{T}_d(t)}{T_c} = g_c \int_0^t v(\lambda)\,d\lambda \tag{4-73}$$

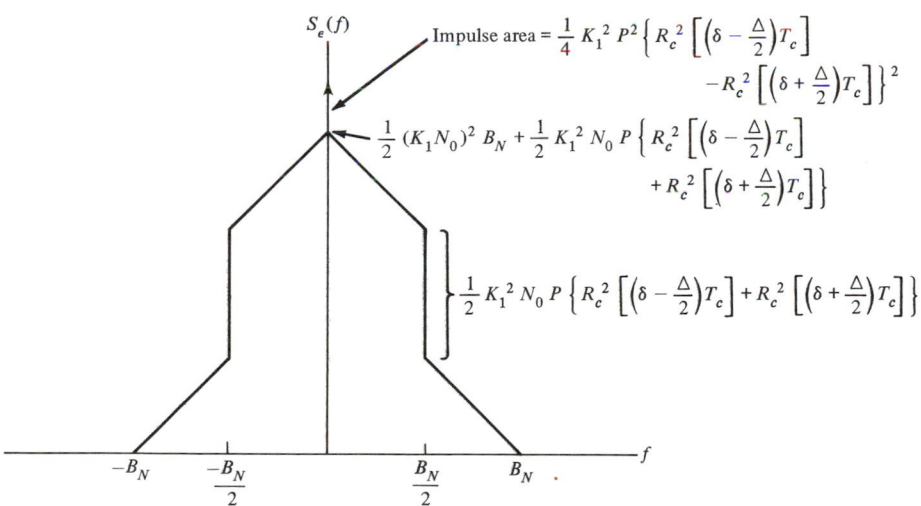

FIGURE 4-11. Complete DLL discriminator output power spectrum.

where the units of g_c are Hz/V. The VCO input is the convolution of the loop filter input and the filter impulse response, so that

$$\frac{\hat{T}_d(t)}{T_c} = g_c \int_0^t \int_{-\infty}^{\lambda} \epsilon(\lambda,\delta)f(\lambda - \alpha) \, d\alpha \, d\lambda \qquad (4\text{-}74)$$

In this expression, $\epsilon(t,\delta)$ is the complete signal plus noise discriminator output waveform. The noise component of this signal is assumed to be a lowpass white Gaussian noise process having a two-sided power spectral density $\eta/2$ given by the last two terms of (4-66) evaluated at $f = 0$. In the special (worst) case where there is no data modulation, the noise power spectrum is illustrated in Figure 4-11 and

$$\frac{\eta}{2} = \frac{1}{2}(K_1 N_0)^2 B_N + \frac{1}{2} K_1^2 N_0 P \left\{ R_c^2 \left[\left(\delta - \frac{\Delta}{2} \right) T_c \right] + R_c^2 \left[\left(\delta + \frac{\Delta}{2} \right) T_c \right] \right\} \quad (4\text{-}75)$$

The noise component of $\epsilon(t,\delta)$ will be denoted by $n_\epsilon(t)$. The signal component of $\epsilon(t,\delta)$ was given by (4-50). Combining these results yields

$$\epsilon(t,\delta) = \tfrac{1}{2} K_1 P D_\Delta(\delta) + n_\epsilon(t) \qquad (4\text{-}76)$$

and (4-74) becomes

$$\frac{\hat{T}_d(t)}{T_c} = g_c \int_0^t \int_{-\infty}^{\lambda} [\tfrac{1}{2} K_1 P D_\Delta(\delta) + n_\epsilon(\alpha)]f(\lambda - \alpha) \, d\alpha \, d\lambda \qquad (4\text{-}77)$$

where $\delta = [T_d(\alpha) - \hat{T}_d(\alpha)]/T_c$. The model of Figure 4-12 is also described by (4-77), so that it is a correct model for the noncoherent code tracking loop.

For small δ the discriminator S-curve $D_\Delta(\delta)$ can be approximated by a linear

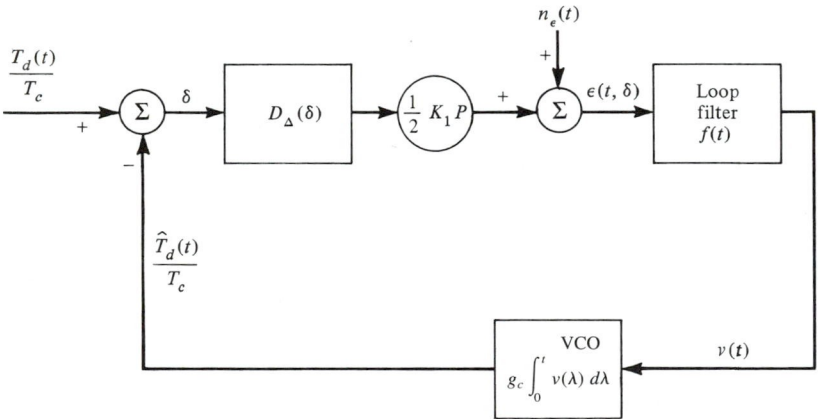

Notes: 1. $n_\epsilon(t)$ is lowpass white Gaussian noise process
with two-sided psd $\eta/2$ given by Equation (4-75).
2. $D_\Delta(\delta)$ is given by Equation (4-51).

FIGURE 4-12. Nonlinear equivalent circuit for the noncoherent delay-lock tracking loop.

function of δ. From (4-52), and for small δ,

$$D_\Delta(\delta) = 4\left(1 + \frac{1}{N}\right)\left[1 - \left(1 + \frac{1}{N}\right)\frac{\Delta}{2}\right]\delta \qquad (4\text{-}78)$$

This linear model for the phase discriminator permits moving the noise process $n_\epsilon(t)$ to the loop input with appropriate gain adjustment. Letting

$$K_d = 4\left(1 + \frac{1}{N}\right)\left[1 - \left(1 + \frac{1}{N}\right)\frac{\Delta}{2}\right]\left(\frac{1}{2}K_1P\right) \qquad (4\text{-}79)$$

the linearized equivalent circuit for the tracking loop can be represented as illustrated in Figure 4-13. The Laplace-transform model for this tracking loop is identical to the model shown in Figure 4-8 for the baseband tracking loop with appropriate changes in the definition of K_d.

The derivation of the linear model for this tracking has been long and many assumptions have been made. The two fundamental calculations that led to the linear model were the calculation of the noiseless discriminator output characteristic

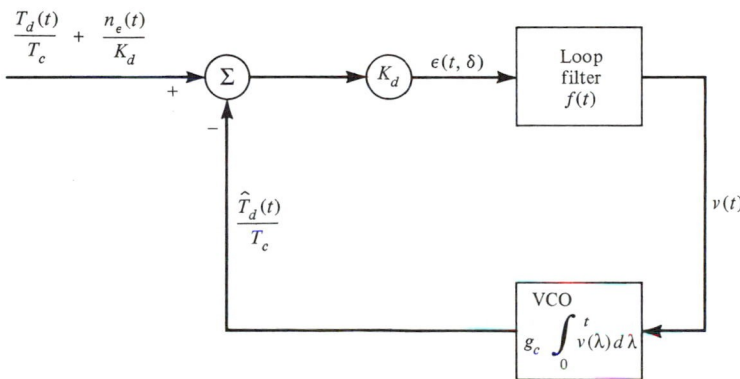

g_c = VCO gain in Hz / volt

$K_d = \dfrac{1}{2}K_1P\left[\dfrac{d}{d\delta}D_\Delta(\delta)\right]_{\delta=0}$

$K_d = \left(1 + \dfrac{1}{N}\right)\left[1 - \left(1 + \dfrac{1}{N}\right)\dfrac{\Delta}{2}\right](2K_1P)$ for BPSK only

K_1 = mixer conversion gain/loss

Δ = normalized difference between early and late correlator channels

N = m−sequence spreading code period

P = received signal power

T_c = spreading waveform chip duration

$n_\epsilon(t)$ = thermal noise component at discriminator output (see note on Figure 4-12)

FIGURE 4-13. Linear equivalent circuit for the noncoherent delay-lock tracking loop.

and the modeling of the discriminator output noise process. The significant assumptions made in this development are:

1. The code self-noise and clock components of (3-66) have been ignored. As a preliminary step in including the effects of these terms, their spectra can be modeled as continuous spectra and added to the thermal noise spectrum.
2. The thermal noise processes at the two IF bandpass filter outputs have been assumed to be independent. This assumption is quite accurate for $\Delta \geq 1.0$. For $\Delta < 1.0$, the assumption is not accurate; however, it is conjectured that the correlated components will cancel one another in the difference circuit at the discriminator output.
3. The discriminator output noise has been assumed to be Gaussian. Because of the small bandwidth of the loop filter relative to the discriminator output bandwidth, this assumption is quite good.
4. The effect of the data modulation in the (signal × noise) term at the discriminator output has been ignored. The effect of data modulation can be included by using (4-66) to calculate the power spectrum of $n_\epsilon(t)$. Ignoring data modulation results in pessimistic performance estimates.

These assumptions are consistent with those made in the existing literature on this subject.

Using the linear model, the closed-loop transfer function $H(s)$ is given by (4-25), which is repeated here for convenience.

$$H(s) = \frac{K_d g_c F(s)}{s + K_d g_c F(s)} = \frac{\hat{T}_d(s)}{T_d(s)} \tag{4-25}$$

and the Laplace transform of the tracking error is given by (4-26):

$$\frac{T_d(s) - \hat{T}_d(s)}{T_c} = \frac{T_d(s)}{T_c}\left[\frac{s}{s + K_d g_c F(s)}\right] \tag{4-26}$$

The two-sided loop noise bandwidth W_L is defined by (4-31) and the rms tracking jitter is given by

$$\sigma_\delta^2 = \int_{-\infty}^{\infty} S_{n''}(f)|H(j2\pi f)|^2 \, df \tag{4-29}$$

where $S_{n''}(f)$ is the noise power spectrum at the input to the model of Figure 4-13. This power spectrum is approximately flat over the loop bandwidth. Its magnitude is given by $\eta/2K_d^2$, where $\eta/2$ is defined in (4-75). Thus

$$\sigma_\delta^2 = \frac{\eta}{2K_d^2} W_L \tag{4-80}$$

EXAMPLE 4-2 _____

Find an expression for the tracking jitter for the noncoherent tracking loop for the special case where $\Delta = 1.0$ and $N \gg 1$.

Solution: When $\Delta = 1.0$ and $\delta \approx 0$, (4-75) simplifies to

$$\frac{\eta}{2} = \frac{1}{2}(K_1 N_0)^2 B_N + \frac{1}{2}K_1^2 N_0 P\left(\frac{1}{4} + \frac{1}{4}\right) = \frac{1}{2}K_1^2 N_0\left(N_0 B_N + \frac{P}{2}\right) \quad (4\text{-}81)$$

and for $N \gg 1$ and $\Delta = 1.0$.

$$K_d \cong K_1 P \quad (4\text{-}82)$$

Therefore,

$$\sigma_\delta^2 = \frac{\frac{1}{2}K_1^2 N_0(N_0 B_N + P/2)}{(K_1 P)^2} W_L = \frac{1}{2\rho_L}\left(1 + \frac{2}{\rho_{IF}}\right) \quad (4\text{-}83)$$

where

$$\rho_L = \frac{2P}{N_0 W_L} \quad (4\text{-}84)$$

$$\rho_{IF} = \frac{P}{N_0 B_N} \quad (4\text{-}85)$$

The variables ρ_L and ρ_{IF} are the signal-to-noise ratios in the loop bandwidth and in the IF filter bandwidth, respectively. Observe that (4-83) is similar to (4-33). The difference in the two expressions is the second term within the parentheses of (4-83). This term is the result of the squaring operation used in the noncoherent loop.

This completes the analysis of the noncoherent code tracking loop. This tracking loop is often used in modern spread-spectrum systems. Simon [6,10] has analyzed the noncoherent DLL taking into account the details of the bandpass filter. The choice of the bandpass filter is a design issue that must be addressed by the system designer. Making the bandwidth of this filter too large permits excessive noise to reach the squaring devices and degrades loop tracking. Making the bandwidth of this filter too small limits the amount of desired signal energy passed by the filter causing reduced discriminator gain and thus degrading loop tracking performance. Simon's analysis shows that for a particular received signal-to-noise ratio, there is an optimum IF filter bandwidth. Using a filter with this optimum filter bandwidth will minimize the code tracking jitter. In Ref. 10, Simon correctly notes that the simplified analysis presented above is approximately 0.9 dB optimistic with respect to tracking loop noise performance.

4-5 Tau-Dither Noncoherent Tracking Loop

The delay-lock loop discussed in Section 4-4 performs the task of code tracking efficiently and is widely used. It has two problems, however, which led spread-spectrum researchers to invent the tau-dither code tracking loop discussed in this section. The first problem with the DLL is that the early and late IF channels must be precisely amplitude balanced. When they are not properly balanced, the discriminator characteristic is offset and does not produce zero volts out when the tracking error is zero. The other problem is that the DLL uses costly components somewhat freely. Both of these problems are solved in the tau-dither tracking loop (TDL) by time sharing a single correlation channel for both early and late channel use. The price paid to solve these problems is slightly worse noise performance and considerably more difficult analysis. The application of time-shared tracking discriminators to noncoherent spread-spectrum code tracking was first proposed by Hartman [5], whose analysis is followed closely in this section. The tau-dither tracking loop is also analyzed by Simon [6,10].

A functional block diagram of the tau-dither tracking loop is illustrated in Figure 4-14. Except for the discriminator, the loop is the same as the DLL of Figure 4-9. The discriminator has a single channel which is switched between use as an early correlator and a late correlator by a switching signal $q(t)$. The signal $q(t)$ is a square wave of frequency f_q which takes on values of ± 1. When $q(t) = -1$, the switch at the spreading waveform generator is in the position shown and the correlator functions as a late correlator. When $q(t) = +1$, the correlator functions as an early correlator. The signal $q(t)$ is also used to multiply the squaring circuit output. This multiplication provides the sign inversion necessary to generate the discriminator S-curve from the early and late autocorrelation functions. The analysis of the dithering loop is similar to the tracking loop analyses performed in Sections 4-3 and 4-4. The discriminator output signal component will be modeled as a linear function of

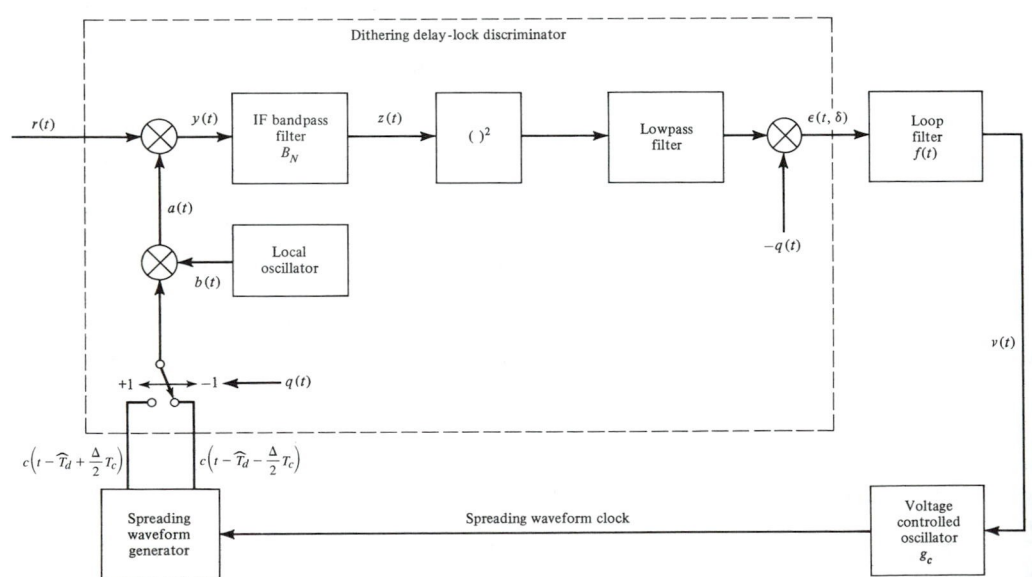

FIGURE 4-14. Tau-dither noncoherent code tracking loop.

the normalized tracking error δ, and the discriminator output noise power spectral density will be calculated. These calculations are facilitated by developing a discriminator model which is very similar to the DLL discriminator, so that some of the previous analysis can be used directly.

The input to the tau-dither loop $r(t)$ is a data- and direct-sequence-modulated carrier plus bandlimited AWGN with two-sided power spectral density $N_0/2$. That is,

$$r(t) = \sqrt{2P}\,c(t - T_d)\cos[\omega_0 t + \theta_d(t - T_d) + \phi] + n(t) \qquad (4\text{-}86)$$

where

$$n(t) = \sqrt{2}\,n_I(t)\cos\omega_0 t - \sqrt{2}\,n_Q(t)\sin\omega_0 t \qquad (4\text{-}87)$$

and the baseband in-phase and quadrature noise functions are independent lowpass white Gaussian noise processes with two-sided power spectral densities $N_0/2$. The random received carrier phase is denoted by ϕ. Consideration has been limited to BPSK direct-sequence spreading modulation, but the data modulation $\theta_d(t)$ is an arbitrary phase modulation. The reference local oscillator output is

$$b(t) = 2\sqrt{K_1}\cos[(\omega_0 - \omega_{\mathrm{IF}})t + \phi'] \qquad (4\text{-}88)$$

The signal $a(t)$ depends on $q(t)$ and is

$$a(t) = 2\sqrt{K_1}\,c\!\left(t - \hat{T}_d + q(t)\frac{\Delta}{2}T_c\right)\cos[(\omega_0 - \omega_{\mathrm{IF}})t + \phi'] \qquad (4\text{-}89)$$

Assume now that the frequency of $q(t)$ is sufficiently low relative to the IF filter bandwidth that filter transients may be ignored. In this case, an equivalent discriminator is a two-channel discriminator as shown in Figure 4-15, whose output is switched between the early and late channels by functions $q_1(t)$ and $q_2(t)$. The functions $q_1(t)$ and $q_2(t)$ are defined by

$$q_1(t) = \tfrac{1}{2}[1 + q(t)] \qquad (4\text{-}90\text{a})$$

$$q_2(t) = \tfrac{1}{2}[1 - q(t)] \qquad (4\text{-}90\text{b})$$

and are plotted in Figure 4-16. In Figure 4-15 the functions $a_1(t)$ and $a_2(t)$ are

$$a_1(t) = 2\sqrt{K_1}\,c\!\left(t - \hat{T}_d + \frac{\Delta}{2}T_c\right)\cos[(\omega_0 - \omega_{\mathrm{IF}})t + \phi'] \qquad (4\text{-}91\text{a})$$

$$a_2(t) = 2\sqrt{K_1}\,c\!\left(t - \hat{T}_d - \frac{\Delta}{2}T_c\right)\cos[(\omega_0 - \omega_{\mathrm{IF}})t + \phi'] \qquad (4\text{-}91\text{b})$$

Observe that the power dividers that were shown in the DLL discriminator of Figure 4-9 are not shown in Figure 4-15 since this figure is used only as an analytical model. In this model, the received signal is always correlated with the early code in the upper channel, and always correlated with the late code in the lower channel. Because $q_1(t)$ and $q_2(t)$ take on values of +1 or 0, the discriminator output is alternately switched between the early and late channels. Thus, except for the transient which occurs in the IF filter of Figure 4-14 when the early/late transition occurs, the discriminator output is identical in Figures 4-14 and 4-15. Except for the

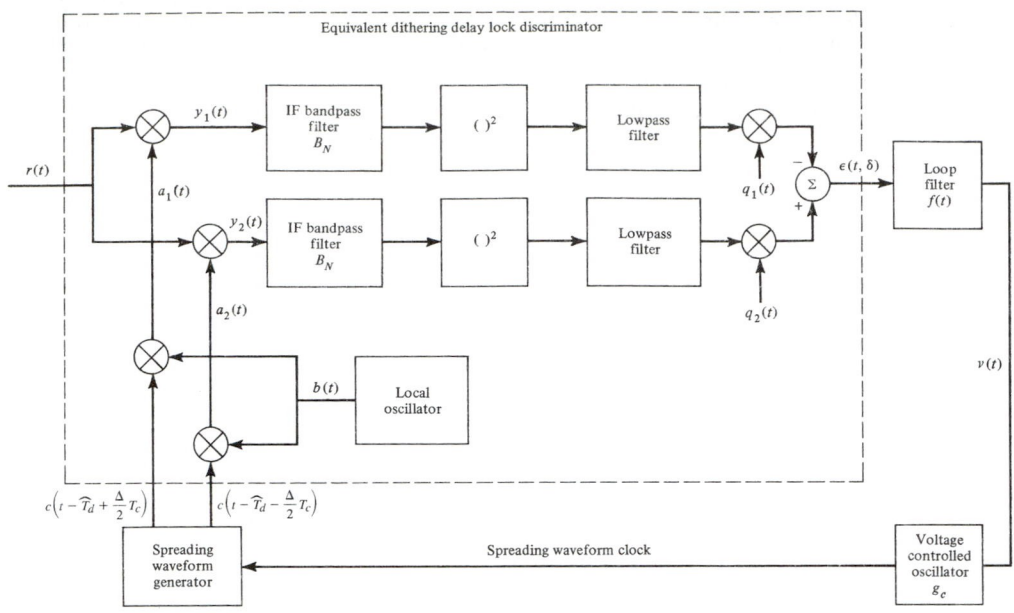

FIGURE 4-15.   Equivalent tau-dither noncoherent code tracking loop.

multipliers for $q_1(t)$ and $q_2(t)$ and a factor of 2 in input power, Figure 4-15 is analytically identical to the DLL of Figure 4-9.

Calculation of the discriminator output in the noiseless case follows the same steps described earlier for the DLL. In particular, code clock and self-noise components can be ignored under the same conditions as those described earlier. The mixer output signals are given by (4-49) except for a factor of 2 in power; thus

$$y_1(t) = \sqrt{2K_1 P} R_c\left[\left(\delta + \frac{\Delta}{2}\right)T_c\right] \cos[\omega_{\mathrm{IF}}t + \phi - \phi' + \theta_d(t - T_d)]$$
$$\triangleq x_1(t) \tag{4-92a}$$

$$y_2(t) = \sqrt{2K_1 P} R_c\left[\left(\delta + \frac{\Delta}{2}\right)T_c\right] \cos[\omega_{\mathrm{IF}}t + \phi - \phi' + \theta_d(t - T_d)]$$
$$\triangleq x_2(t) \tag{4-92b}$$

Again, the IF bandpass filters are assumed to pass the data-modulated IF carrier without distortion, so that the discriminator output signal component is

$$\epsilon(t,\delta) = [x_2^2(t)q_2(t) - x_1^2(t)q_1(t)]_{\mathrm{lowpass}} \tag{4-93}$$

Writing $q_1(t)$ and $q_2(t)$ as functions of $q(t)$ and simplifying (4-93) yields

$$\epsilon(t,\delta) = \{\tfrac{1}{2}[x_2^2(t) - x_1^2(t)] - \tfrac{1}{2}q(t)[x_2^2(t) + x_1^2(t)]\}_{\mathrm{lowpass}} \tag{4-94}$$

Substituting for $x_1(t)$ and $x_2(t)$ from (4-92) and simplifying by eliminating all double-frequency terms results in

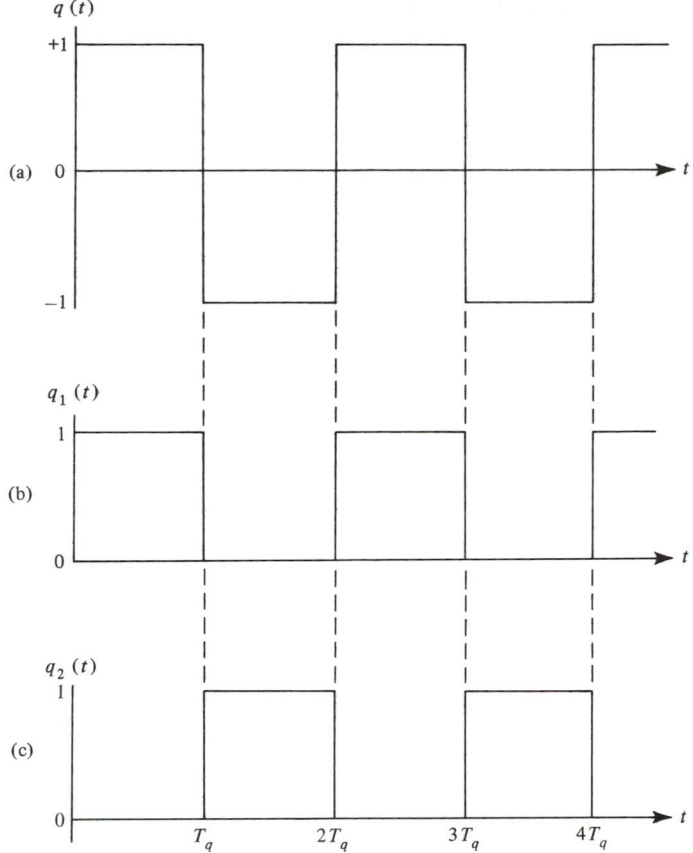

FIGURE 4-16. Dithering loop switching functions: (a) $q(t)$; (b) $q_1(t)$; and (c) $q_2(t)$.

$$\epsilon(t,\delta) = \frac{1}{2}K_1 P\left\{R_c^2\left[\left(\delta - \frac{\Delta}{2}\right)T_c\right] - R_c^2\left[\left(\delta + \frac{\Delta}{2}\right)T_c\right]\right\}$$
$$- \frac{1}{2}q(t)K_1 P\left\{R_c^2\left[\left(\delta + \frac{\Delta}{2}\right)T_c\right] + R_c^2\left[\left(\delta - \frac{\Delta}{2}\right)T_c\right]\right\} \tag{4-95}$$

The first term of this equation is identical to (4-50) and is the desired tracking error. The second term consists of harmonics of the dithering frequency. At this point the assumption is made that the dithering frequency is significantly higher than the bandwidth of the loop filter. When this is true, the second term of (4-95) is rejected by this filter and

$$\epsilon(t,\delta) \cong \frac{1}{2}K_1 P\left\{R_c^2\left[\left(\delta - \frac{\Delta}{2}\right)T_c\right] - R_c^2\left[\left(\delta + \frac{\Delta}{2}\right)T_c\right]\right\}$$
$$\triangleq \frac{1}{2}K_1 PD_\Delta(\delta) \tag{4-96}$$

Thus, in the noiseless case, the discriminator output is the same as the discriminator output of the DLL. A factor of 2 decrease in output does not occur since the input power divider used in the DLL is unnecessary in the dithering loop. The discriminator S-curve is defined by (4-52) and is illustrated in Figure 4-10.

The next step in the analysis is to calculate the power spectral density of the noise component at the discriminator output. The noise analysis up to the output of the IF bandpass filter is identical to the noise analysis for the DLL. The noise at the output of these filters is specified in (4-60).

Analysis in Section 4-4 showed that the noise processes of (4-60) are not independent in general but are very nearly independent when $\Delta \geq 1.0$ and the system input noise bandwidth is wide relative to the received signal bandwidth. In this section it will be assumed that these processes are independent. To simplify the following analysis, the IF bandpass filter is assumed to be an ideal brick wall filter. Then the noise power spectrum is given by (4-54) and (4-61):

$$S_{n_1}(f) = S_{n_2}(f) = \begin{cases} 2\dfrac{K_1}{2}\left(\dfrac{N_0}{2}\right) = \dfrac{K_1 N_0}{2} & 0 \leq |f \pm f_{\mathrm{IF}}| < \dfrac{B_N}{2} \\ 0 & \text{elsewhere} \end{cases} \tag{4-97}$$

where the density of (4-54) has been increased by a factor of 2 to account for the absence of a power divider in the tau-dither loop. In summary, the noise components at the IF bandpass filter outputs are independent, bandlimited, white Gaussian noise processes with two-sided power spectral densities given by (4-97).

With signal components defined in (4-92) and noise components defined in (4-60), the discriminator output signal is

$$\epsilon(t,\delta) = [x_2(t) + n_2(t)]_{\mathrm{LP}}^2 q_2(t) - [x_1(t) + n_1(t)]_{\mathrm{LP}}^2 q_1(t) \tag{4-98}$$

where the subscript LP denotes that the double-frequency terms will be ignored. Expanding (4-98) and then simplifying yields

$$\epsilon(t,\delta) = K_1 P\left\{ R_c^2\left[\left(\delta - \frac{\Delta}{2}\right)T_c\right]q_2(t) - R_c^2\left[\left(\delta + \frac{\Delta}{2}\right)T_c\right]q_1(t)\right\} \tag{4-99}$$

$$+ 2\sqrt{K_1 P}\,\cos[\phi - \phi' + \theta_d(t - T_d)]\left\{n_{2I}(t)R_c\left[\left(\delta - \frac{\Delta}{2}\right)T_c\right]q_2(t)\right.$$

$$\left. - n_{1I}(t)R_c\left[\left(\delta + \frac{\Delta}{2}\right)T_c\right]q_1(t)\right\}$$

$$+ 2\sqrt{K_1 P}\,\sin[\phi - \phi' + \theta_d(t - T_d)]\left\{n_{2Q}(t)R_c\left[\left(\delta - \frac{\Delta}{2}\right)T_c\right]q_2(t)\right.$$

$$\left. - n_{1Q}(t)R_c\left[\left(\delta + \frac{\Delta}{2}\right)T_c\right]q_1(t)\right\}$$

$$+ [n_{2I}(t)]^2 q_2(t) + [n_{2Q}(t)]^2 q_2(t) - [n_{1I}(t)]^2 q_1(t) - [n_{1Q}(t)]^2 q_1(t)$$

To calculate the discriminator output power spectrum, the autocorrelation function

$$R_\epsilon(\tau) = E[\epsilon(t,\delta)\epsilon(t + \tau,\delta)] \qquad (4\text{-}100)$$

must be evaluated. This is done in Appendix E, where it is shown that

$$
\begin{aligned}
R_\epsilon(\tau) = K_1^2 P^2 &\left\{ R_c^2\left[\left(\delta - \frac{\Delta}{2}\right)T_c\right] + R_c^2\left[\left(\delta + \frac{\Delta}{2}\right)T_c\right] \right\}^2 R_{q1}(\tau) \\
&- K_1^2 P^2 R_c^2\left[\left(\delta - \frac{\Delta}{2}\right)T_c\right] R_c^2\left[\left(\delta + \frac{\Delta}{2}\right)T_c\right] \\
&+ 8K_1 P \sigma_n^2 \left\{ R_c^2\left[\left(\delta - \frac{\Delta}{2}\right)T_c\right] + R_c^2\left[\left(\delta + \frac{\Delta}{2}\right)T_c\right] \right\} R_{q1}(\tau) \\
&- 2K_1 P \sigma_n^2 \left\{ R_c^2\left[\left(\delta - \frac{\Delta}{2}\right)T_c\right] + R_c^2\left[\left(\delta + \frac{\Delta}{2}\right)T_c\right] \right\} \\
&+ 4K_1 P \left\{ R_c^2\left[\left(\delta - \frac{\Delta}{2}\right)T_c\right] + R_c^2\left[\left(\delta + \frac{\Delta}{2}\right)T_c\right] \right\} R_{n_b}(\tau) R_{q1}(\tau) \\
&\times E\{\cos[\theta_d(t - T_d) - \theta_d(t + \tau - T_d)]\} \\
&+ 4E\{[n_b(t)n_b(t + \tau)]^2\} R_{q1}(\tau) \\
&+ 12\sigma_n^4 R_{q1}(\tau) - 4\sigma_n^4
\end{aligned}
\qquad (4\text{-}101)
$$

In this equation, $n_b(t)$ represents any one of the four signals $n_{1I}(t)$, $n_{1Q}(t)$, $n_{2I}(t)$, $n_{2Q}(t)$, and σ_n^2 is the mean square value of $n_b(t)$.

The Fourier transform of $R_\epsilon(\tau)$ is the power spectrum needed to complete the analysis of the tau-dither tracking loop. The Fourier transform is also calculated in Appendix E; the result is†

$$
\begin{aligned}
S_\epsilon(f) = \frac{1}{4} K_1^2 P^2 &\left\{ R_c^2\left[\left(\delta - \frac{\Delta}{2}\right)T_c\right] - R_c^2\left[\left(\delta + \frac{\Delta}{2}\right)T_c\right] \right\}^2 \delta(f) \\
&+ \left[K_1 P \left\{ R_c^2\left[\left(\delta - \frac{\Delta}{2}\right)T_c\right] + R_c^2\left[\left(\delta + \frac{\Delta}{2}\right)T_c\right] \right\} + 4\sigma_n^2 \right]^2 \\
&\times \left[\frac{1}{\pi^2}\delta(f - f_q) + \frac{1}{\pi^2}\delta(f + f_q) + \frac{1}{9\pi^2}\delta(f + 3f_q) + \frac{1}{9\pi^2}\delta(f - 3f_q) \right] \\
&+ \left[K_1 P \left\{ R_c^2\left[\left(\delta - \frac{\Delta}{2}\right)T_c\right] + R_c^2\left[\left(\delta + \frac{\Delta}{2}\right)T_c\right] \right\} \right] \\
&\times \left[S_{n_b}(f) + \frac{4}{\pi^2}S_{n_b}(f - f_q) + \frac{4}{\pi^2}S_{n_b}(f + f_q) + \frac{4}{9\pi^2}S_{n_b}(f - 3f_q) \right. \\
&\left. + \frac{4}{9\pi^2}S_{n_b}(f + 3f_q) \right]
\end{aligned}
\qquad (4\text{-}102)
$$

† It has been assumed in deriving this result that only the first three harmonics of the dither frequency cause significant noise components within the tracking loop bandwidth.

$$+ 2S_T(f) + \frac{8}{\pi^2}S_T(f - f_q) + \frac{8}{\pi^2}S_T(f + f_q)$$

$$+ \frac{8}{9\pi^2}S_T(f - 3f_q) + \frac{8}{9\pi^2}S_T(f + 3f_q)$$

where $S_T(f) = S_{n_b}(f) * S_{n_b}(f)$. The first term of this equation is the square of (4-96). The square is expected since this is a power spectrum. This term is the desired error correction signal for code tracking. The remainder of the equation is undesired dither frequency clock components plus (signal × noise) and (noise × noise) components. The dither clock frequency must be selected to be large enough that the clock components of (4-102) are outside the tracking loop bandwidth. The power spectrum of $n_b(f)$ is

$$S_{n_b}(f) = \begin{cases} \dfrac{K_1 N_0}{2} & |f| < \tfrac{1}{2}B_N \\ 0 & \text{elsewhere} \end{cases} \tag{4-103}$$

The power spectra $S_{n_b}(f)$ and $S_T(f)$ are illustrated in Figure 4-17, and the complete power spectrum $S_\epsilon(f)$ is illustrated in Figure 4-18 for the particular case where $f_q = \tfrac{1}{4}B_N$.

Denote the total noise at the output of the tau-dither delay discriminator by $n_\epsilon(t)$. The signal component at the discriminator output was given in (4-96), and the total

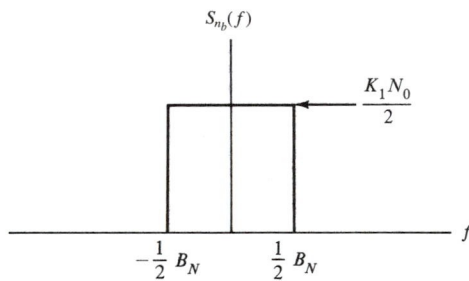

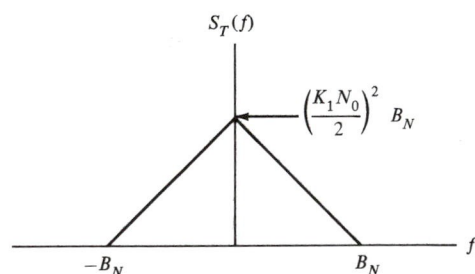

FIGURE 4-17. Power spectra $S_{n_b}(f)$ and $S_T(f) = S_{n_0}(f) * S_{n_0}(f)$.

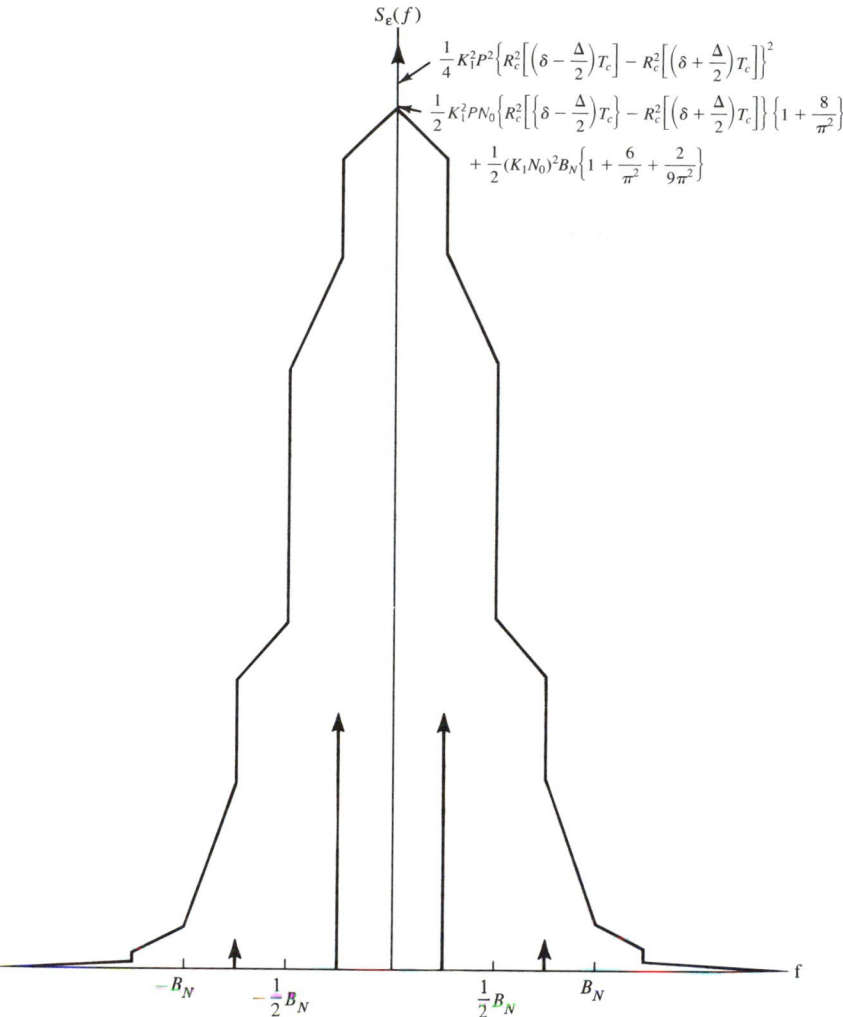

FIGURE 4-18.   Typical power spectrum of noise and clock components at output of tau-dither discriminator.

discriminator output is

$$\epsilon(t,\delta) = \tfrac{1}{2}K_1 P D_\Delta(\delta) + n_\epsilon(t) \tag{4-104}$$

Although $n_\epsilon(t)$ is considerably different, (4-73), (4-74), and (4-77), which were derived for the DLL, also apply directly to the tau-dither loop. Therefore, the nonlinear model of Figure 4-12 is also an accurate model for the tau-dither loop with $D_\Delta(\delta)$ defined in (4-96). Since $D_\Delta(\delta)$ for the DLL and the TDL are identical, the linearization of the model for $\delta \ll 1$ is the same and Figure 4-13 is also a valid model for the tau-dither loop. It follows that (4-25) and (4-26) also apply to the tau-dither loop.

The code tracking jitter for the tau-dither loop is calculated using (4-29):

$$\sigma_\delta^2 = \int_{-\infty}^{\infty} S_{n''}(f)|H(j2\pi f)|^2 \, df \tag{4-29}$$

where $n''(t)$ is the noise at the input to the linear model of Figure 4-13. That is, $n''(t) = n_\epsilon(t)/K_d$. In all cases of interest, the tracking loop noise bandwidth is very much smaller than the IF filter noise bandwidth or the dither frequency, so that $n''(t)$ may be assumed to be a lowpass white Gaussian noise process having a two-sided power spectral density $\eta/2$ given by the (signal $\times$ noise) and (noise $\times$ noise) components of (4-102) evaluated at $f = 0$ and divided by K_d^2. For $B_N/6 < f_q < B_N/3$ evaluation of these components of (4-102) yields

$$\frac{\eta}{2} = \frac{K_1^2 N_0 P}{2K_d^2} \left\{ R_c^2\left[\left(\delta - \frac{\Delta}{2}\right)T_c \right] + R_c^2\left[\left(\delta + \frac{\Delta}{2}\right)T_c \right] \right\} \left(1 + \frac{8}{\pi^2} \right)$$
$$+ \frac{1}{2}(K_1 N_0)^2 \frac{B_N}{K_d^2}\left[1 + \frac{8}{\pi^2}\left(1 - \frac{f_q}{B_N} \right) + \frac{8}{9\pi^2}\left(1 - \frac{3f_q}{B_N} \right) \right] \tag{4-105}$$

The phase detector gain K_d is given by (4-79). For $N \gg 1$.

$$K_d \cong \left(1 - \frac{\Delta}{2} \right) 2K_1 P$$

so that

$$\sigma_\delta^2 = \frac{\eta}{2} W_L$$

$$\cong \frac{N_0 W_L}{8P(1 - \Delta/2)^2}\left\{ R_c^2\left[\left(\delta - \frac{\Delta}{2}\right)T_c \right] + R_c^2\left[\left(\delta + \frac{\Delta}{2}\right)T_c \right] \right\} \left(1 + \frac{8}{\pi^2} \right) \tag{4-106}$$
$$+ \frac{N_0^2 B_N W_L}{8P^2(1 - \Delta/2)^2}\left[1 + \frac{8}{\pi^2}\left(1 - \frac{f_q}{B_N} \right) + \frac{8}{9\pi^2}\left(1 - \frac{3f_q}{B_N} \right) \right]$$

EXAMPLE 4-3 _____

Consider a tau-dither tracking loop for which $\Delta = 1.0, N \gg 1$, and $f_q = B_N/4$. In this case (4-106) becomes

$$\sigma_\delta^2 = \frac{N_0 W_L}{2P}\left(\frac{1}{4} + \frac{1}{4} \right)\left(1 + \frac{8}{\pi^2} \right)$$
$$+ \frac{N_0^2 B_N W_L}{2P^2}\left[1 + \frac{8}{\pi^2}\left(\frac{3}{4} \right) + \frac{8}{9\pi^2}\left(\frac{1}{4} \right) \right] \tag{4-107}$$
$$= \frac{1}{2\rho_L}\left(1.811 + \frac{3.261}{\rho_{\mathrm{IF}}} \right)$$

where

$$\rho_L = \frac{2P}{N_0 W_L} \tag{4-108}$$

$$\rho_{\text{IF}} = \frac{P}{N_0 B_N} \tag{4-109}$$

are the signal-to-noise ratios in the loop and IF bandwidth, respectively. Comparing (4-107) to (4-83) for the DLL shows that noise performance has indeed been sacrificed to attain the benefits of using the tau-dither tracking loop.

This completes the rather lengthy discussion of the tau-dither code tracking loop. The assumptions that have been made in this analysis are the same as those made in the analysis of the delay-lock tracking loop. Spreading code clock and self-noise components have been ignored in the discriminator and the noise processes at the output of the two IF filters in the model of Figure 4-15 have been assumed to be independent. The latter assumption is very good for the tau-dither loop since the two outputs affect the error signal at different times.

4-6 Double-Dither Noncoherent Tracking Loop

In certain applications, the noise performance degradation of the TDL loop relative to the DLL is unacceptable. The double-dither noncoherent tracking loop proposed in Hopkins [16] can be used in these cases to solve the gain-imbalance problem of the DLL. The double-dither loop-noise performance is the same as the DLL noise performance. The price paid for simultaneously achieving good noise performance and solving the amplitude balance problem is increased hardware complexity.

The hardware configuration of the double-dither loop is illustrated in Figure 4-19. Observe the similarity between the loop of Figure 4-9 and the double-dither loop. The only difference is that the use of the two channels in the double-dither loop alternates between early and late channel correlation. When $q(t) = +1$ in Figure 4-19, the configuration is identical to Figure 4-9. When $q(t) = -1$, the switches are reversed, as is the sign of the differencing circuit output, so that the output $\epsilon(t,\delta)$ is unchanged. If the transient that occurs when $q(t)$ changes states can be ignored, and if the two channels are identical, the double-dither loop performance is identical to the DLL performance.

To understand the mechanism used by the double-dither loop to solve the gain-imbalance problem, consider the operation of the discriminator in the noiseless case. Code self-noise and code clock components at the IF bandpass filter inputs can be ignored under the same conditions that they are ignored in the DLL and TDL discriminators. Attributing all the gain imbalance between the two channels to the

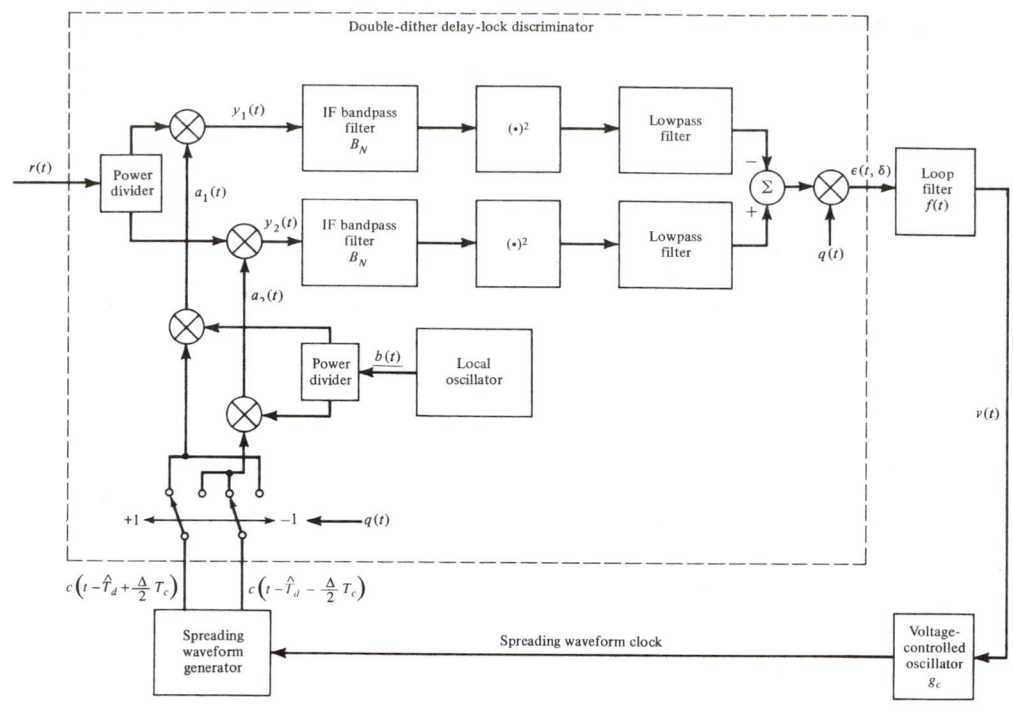

FIGURE 4-19. Double-dither noncoherent code tracking loop. (From Ref. 16.)

mixer conversion gain, analysis identical to that which resulted in (4-49) can be used to show that

$$y_1(t) = \sqrt{K_1 P} R_c \left[\left(\delta + q(t) \frac{\Delta}{2} \right) T_c \right] \cos[\omega_{IF} t + \phi - \phi' + \theta_d(t - T_d)] \quad (4\text{-}110a)$$

$$y_2(t) = \sqrt{K_2 P} R_c \left[\left(\delta - q(t) \frac{\Delta}{2} \right) T_c \right] \cos[\omega_{IF} t + \phi - \phi' + \theta_d(t - T_d)] \quad (4\text{-}110b)$$

where K_1 is the upper channel mixer conversion gain and K_2 is the lower channel mixer conversion gain. It immediately follows that the noiseless discriminator output is

$$\epsilon(t, \delta) = \frac{P}{2} \left\{ K_2 R_c^2 \left[\left(\delta - q(t) \frac{\Delta}{2} \right) T_c \right] - K_1 R_c^2 \left[\left(\delta + q(t) \frac{\Delta}{2} \right) T_c \right] \right\} q(t) \quad (4\text{-}111)$$

This expression can also be written

$$\epsilon(t, \delta) = \frac{P}{2} \left\{ [K_2 q_1(t) + K_1 q_2(t)] R_c^2 \left[\left(\delta - \frac{\Delta}{2} \right) T_c \right] \right.$$

$$\left. - [K_1 q_1(t) + K_2 q_2(t)] R_c^2 \left[\left(\delta + \frac{\Delta}{2} \right) T_c \right] \right\} \quad (4\text{-}112)$$

where $q_1(t)$ and $q_2(t)$ are defined by (4-90). If the two channels are identical (i.e., $K_1 = K_2$), then (4-112) reduces to (4-50) showing that the double-dither loop is identical to the DLL. If $K_1 \neq K_2$ and there is no dither [i.e., $q(t) = 1$ for all t], (4-111) becomes

$$\epsilon(t,\delta) = \frac{P}{2}\left\{K_2 R_c^2\left[\left(\delta - \frac{\Delta}{2}\right)T_c\right] - K_1 R_c^2\left[\left(\delta + \frac{\Delta}{2}\right)T_c\right]\right\} \qquad (4\text{-}113)$$

Discriminator S-curves for several values of (K_1,K_2) and $\Delta = 1.0$ are illustrated in Figure 4-20. Observe that the curves do not cross the origin. This is the problem that

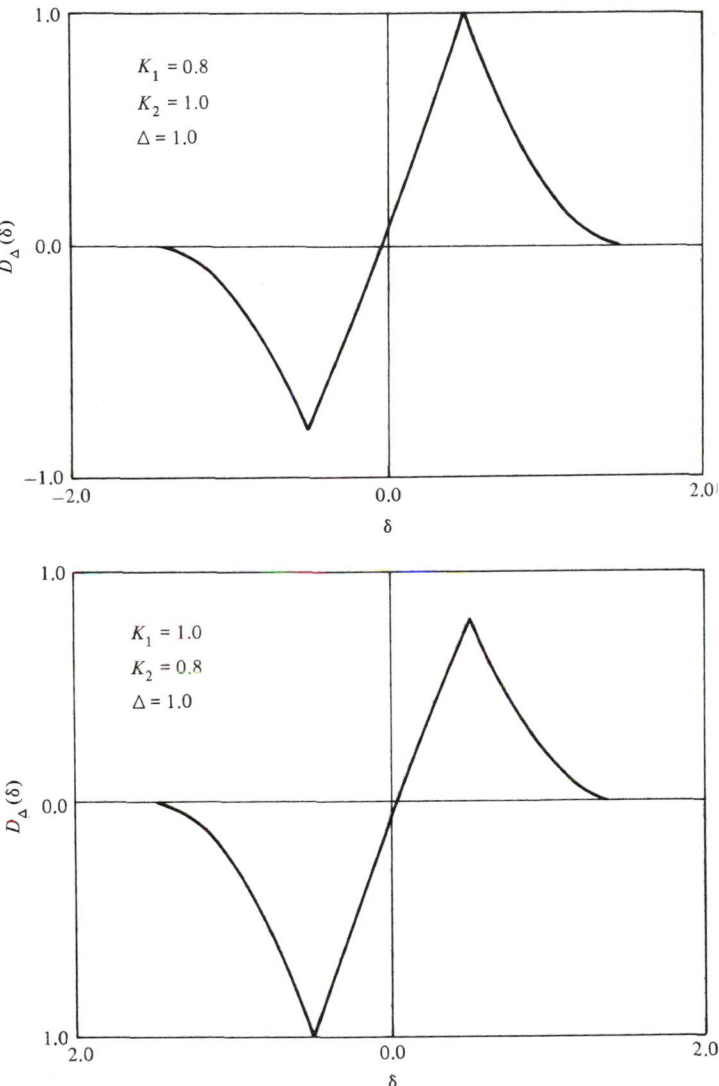

FIGURE 4-20. DLL discriminator S-curve with gain imbalance.

is solved by the double-dither discriminator. When $q(t)$ is a square wave whose frequency is well above the bandwidth of the loop filter, the time average of (4-112) is the desired tracking error.

$$\langle \epsilon(t,\delta) \rangle = \frac{P}{2}\left(\frac{K_1 + K_2}{2}\right)\left\{ R_c^2\left[\left(\delta - \frac{\Delta}{2}\right)T_c\right] - R_c^2\left[\left(\delta + \frac{\Delta}{2}\right)T_c\right]\right\} \quad (4\text{-}114)$$

In this case, the discriminator S-curve crosses the origin as in Figure 4-10; however, the magnitude of the S-curve is the average magnitude over the two gains K_1 and K_2.

Space does not permit derivation of the noise performance of this loop here. The student, at this point, has all the tools necessary to do this derivation. The principal task of this derivation is to generate the complete power spectrum at the double-dither discriminator output. Finally, note that the double-dither technique can also be applied to other dual-channel phase detectors. The student is referred to Hopkins [16] for a complete analysis of this code tracking loop.

4-7 Noncoherent Delay-Lock Tracking Loop with Arbitrary Data and Spreading Modulation

The discussion of Section 4-4 was limited to systems that employ BPSK direct-sequence spreading modulation. In this section it will be shown that the same results apply to a system having an arbitrary constant-envelope phase modulation when the BPSK autocorrelation function is replaced by the autocorrelation function for the modulation being considered. The code tracking loop configuration for arbitrary constant envelope direct-sequence modulation is illustrated in Figure 4-21. This configuration is identical to that of Figure 4-9 except that the two mixers which were used as BPSK modulators have been replaced by more general phase modulators.

The received signal $r(t)$ is the sum of the data-modulated and direct-sequence spread carrier and bandlimited white Gaussian noise

$$r(t) = \sqrt{2P}\cos[\omega_0 t + \theta_{ss}(t - T_d) + \theta_d(t - T_d) + \phi] + n(t) \quad (4\text{-}115)$$

where $n(t)$ is the AWGN, $\theta_{ss}(t - T_d)$ is the direct-sequence spreading modulation, $\theta_d(t - T_d)$ is the information modulation, and ϕ is the random received carrier phase. The received signal has power P. The early and late reference signals are

$$a_1(t) = 2\sqrt{K_1}\cos\left[(\omega_0 - \omega_{IF})t + \theta_{ss}\left(t - \hat{T}_d + \frac{\Delta}{2}T_c\right) + \phi'\right] \quad (4\text{-}116a)$$

and

$$a_2(t) = 2\sqrt{K_1}\cos\left[(\omega_0 - \omega_{IF})t + \theta_{ss}\left(t - \hat{T}_d - \frac{\Delta}{2}T_c\right) + \phi'\right] \quad (4\text{-}116b)$$

where K_1 is the small-signal conversion loss of the mixers, ϕ' is the random reference oscillator phase, and Δ is the normalized delay difference between the early and late correlator channels.

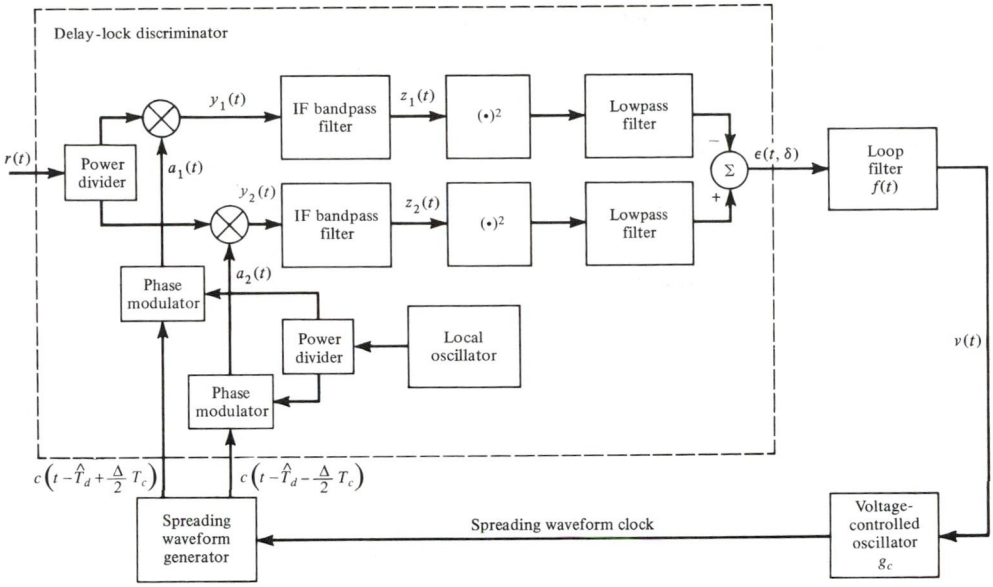

FIGURE 4-21. Noncoherent delay-lock tracking loop for arbitrary direct-sequence spreading modulation.

The despreading mixer output signals are the products of (4-116) and (4-115) with a gain adjustment to account for the power divider. Because of the bandpass filter, only the difference frequency component of this product is of interest. The desired outputs are

$$y_1(t) = \sqrt{PK_1} \cos\left[\omega_{\mathrm{IF}}t + \theta_{\mathrm{ss}}(t - T_d) - \theta_{\mathrm{ss}}\left(t - \hat{T}_d + \frac{\Delta}{2} T_c \right) \right.$$
$$\left. + \theta_d(t - T_d) + \phi - \phi' \right]$$
$$+ \sqrt{K_1} n_I(t) \cos\left[\omega_{\mathrm{IF}}t - \theta_{\mathrm{ss}}\left(t - \hat{T}_d + \frac{\Delta}{2} T_c \right) - \phi' \right] \quad (4\text{-}117a)$$
$$- \sqrt{K_1} n_Q(t) \sin\left[\omega_{\mathrm{IF}}t - \theta_{\mathrm{ss}}\left(t - \hat{T}_d + \frac{\Delta}{2} T_c \right) - \phi' \right]$$

$$y_2(t) = \sqrt{PK_1} \cos\left[\omega_{\mathrm{IF}}t + \theta_{\mathrm{ss}}(t - T_d) - \theta_{\mathrm{ss}}\left(t - \hat{T}_d - \frac{\Delta}{2} T_c \right) \right.$$
$$\left. + \theta_d(t - T_d) + \phi - \phi' \right]$$
$$+ \sqrt{K_1} n_I(t) \cos\left[\omega_{\mathrm{IF}}t - \theta_{\mathrm{ss}}\left(t - \hat{T}_d - \frac{\Delta}{2} T_c \right) - \phi' \right] \quad (4\text{-}117b)$$
$$- \sqrt{K_1} n_Q(t) \sin\left[\omega_{\mathrm{IF}}t - \theta_{\mathrm{ss}}\left(t - \hat{T}_d - \frac{\Delta}{2} T_c \right) - \phi' \right]$$

It is convenient at this point to express $y_1(t)$ and $y_2(t)$ using complex-envelope notation. Thus

$$
\begin{aligned}
\tilde{y}_1(t) = \sqrt{PK_1}\, &\exp\left[j\theta_{ss}(t - T_d) - j\theta_{ss}\left(t - \hat{T}_d + \frac{\Delta}{2}T_c\right)\right] \\
&\times \exp[j\theta_d(t - T_d)]\exp[j(\phi - \phi')] \\
+ \sqrt{K_1}\, &\exp\left[-j\theta_{ss}\left(t - \hat{T}_d + \frac{\Delta}{2}T_c\right)\right]\exp[-j\phi'][n_I(t) + jn_Q(t)]
\end{aligned}
\tag{4-118a}
$$

$$
\begin{aligned}
\tilde{y}_2(t) = \sqrt{PK_1}\, &\exp\left[j\theta_{ss}(t - T_d) - j\theta_{ss}\left(t - \hat{T}_d - \frac{\Delta}{2}T_c\right)\right] \\
&\times \exp[j\theta_d(t - T_d)]\exp[j(\phi - \phi')] \\
+ \sqrt{K_1}\, &\exp\left[-j\theta_{ss}\left(t - \hat{T}_d - \frac{\Delta}{2}T_c\right)\right]\exp[-j\phi'][n_I(t) + jn_Q(t)]
\end{aligned}
\tag{4-118b}
$$

Adding and subtracting appropriate quantities from (4-118a), $\tilde{y}_1(t)$ can be expressed as

$$
\begin{aligned}
\tilde{y}_1(t) = \sqrt{PK_1}\,E&\left\{\exp\left[j\theta_{ss}(t - T_d) - j\theta_{ss}\left(t - \hat{T}_d + \frac{\Delta}{2}T_c\right)\right]\right\} \\
&\times \exp[j\theta_d(t - T_d)]\exp[j(\phi - \phi')] \\
+ \sqrt{PK_1}&\left[\exp\left[j\theta_{ss}(t - T_d) - j\theta_{ss}\left(t - \hat{T}_d + \frac{\Delta}{2}T_c\right)\right]\right. \\
&\left. - E\left\{\exp\left[j\theta_{ss}(t - T_d) - j\theta_{ss}\left(t - \hat{T}_d + \frac{\Delta}{2}T_c\right)\right]\right\}\right] \\
&\times \exp[j\theta_d(t - T_d)]\exp[j(\phi - \phi')] \\
+ \sqrt{K_1}&\exp\left[-j\theta_{ss}\left(t - \hat{T}_d + \frac{\Delta}{2}T_c\right)\right]\exp(-j\phi')[n_I(t) + jn_Q(t)]
\end{aligned}
\tag{4-119}
$$

where the expected value is over all possible transmission delays T_d with fixed $\delta = (T_d - \hat{T}_d)/T_c$. For any useful spreading modulation, the power in the second term of this equation is approximately uniformly spread over a bandwidth which is one or two times the bandwidth of the transmitted spread-spectrum signal. When the system processing gain is large, the data modulation bandwidth is very much smaller than the spreading bandwidth. The IF filter bandwidth is approximately equal to the data modulation bandwidth, so that the majority of the power in the second term of (4-119) will be rejected by the IF bandpass filter. This term contains the code clock and self-noise components which were shown to be negligible for

BPSK spreading in Section 4.4. In the following, these terms are ignored. Thus

$$\tilde{y}_1(t) \cong \sqrt{K_1 P} E\left\{ \exp\left[j\theta_{ss}(t - T_d) - j\theta_{ss}\left(t - \hat{T}_d + \frac{\Delta}{2}T_c\right)\right]\right\}$$

$$\times \exp[j\theta_d(t - T_d)] \exp[j(\phi - \phi')] \tag{4-120}$$

$$+ \sqrt{K_1} \exp\left[-j\theta_{ss}\left(t - \hat{T}_d + \frac{\Delta}{2}T_c\right)\right] \exp(-j\phi')[n_I(t) + jn_Q(t)]$$

The complex autocorrelation function of the spreading modulation envelope is [17]

$$R_{\theta_{ss}}(\tau) = \tfrac{1}{2}E\{\exp[j\theta_{ss}(j) - j\theta_{ss}(t + \tau)]\} \tag{4-121}$$

so that

$$\tilde{y}_1(t) \cong \sqrt{K_1 P}\, 2R_{\theta_{ss}}\left[\left(\delta + \frac{\Delta}{2}\right)T_c\right] \exp[j\theta_d(t - T_d)] \exp[j(\phi - \phi')]$$

$$+ \sqrt{K_1} \exp\left[-j\theta_{ss}\left(t - \hat{T}_d + \frac{\Delta}{2}T_c\right)\right] \exp(-j\phi')[n_I(t) + jn_Q(t)] \tag{4-122a}$$

where $\delta = (T_d - \hat{T}_d)/T_c$. Similarly,

$$\tilde{y}_2(t) \cong \sqrt{K_1 P}\, 2R_{\theta_{ss}}\left[\left(\delta - \frac{\Delta}{2}\right)T_c\right] \exp[j\theta_d(t - T_d)] \exp[j(\phi - \phi')]$$

$$\tag{4-122b}$$

$$+ \sqrt{K_1} \exp\left[-j\theta_{ss}\left(t - \hat{T}_d - \frac{\Delta}{2}T_c\right)\right] \exp(-j\phi')[n_I(t) + jn_Q(t)]$$

Denote the complex noise components of $\tilde{y}_1(t)$ and $\tilde{y}_2(t)$ by $\tilde{n}_1(t)$ and $\tilde{n}_2(t)$, respectively. The power spectra of $\tilde{n}_1(t)$ and $\tilde{n}_2(t)$, denoted $S_{n1}(f)$ and $S_{n2}(f)$, are calculated from their complex autocorrelation functions. Consider the early channel. The complex noise autocorrelation function is

$$R_{n1}(\tau) = \tfrac{1}{2}E[\tilde{n}_1(t)\tilde{n}_1^*(t + \tau)]$$

$$= \tfrac{1}{2}K_1 E\left\{ \exp\left[-j\theta_{ss}\left(t - \hat{T}_d + \frac{\Delta}{2}T_c\right) + j\theta_{ss}\left(t + \tau - \hat{T}_d + \frac{\Delta}{2}T_c\right)\right]\right\}$$

$$\times E\{[n_I(t) + jn_Q(t)][n_I(t + \tau) - jn_Q(t + \tau)]\} \tag{4-123}$$

where the expectation factors because the spreading modulation and the noise are independent. In (4-123) the expected value is over all $\hat{T}_d$ and all sample functions of the noise processes. Using the fact that $n_I(t)$ and $n_Q(t)$ are independent and identically distributed, and using (4-121), equation (4-123) becomes

$$R_{n1}(\tau) = 2K_1 R_{\theta_{ss}}(\tau)R_{nI}(\tau) \tag{4-124}$$

The power spectrum of $\tilde{n}_1(t)$ is the Fourier transform of this autocorrelation. Using the Fourier transform convolution theorem,

$$S_{n1}(f) = 2K_1 S_{\theta_{ss}}(f) * S_{nI}(f) \tag{4-125}$$

When the bandwidth of the noise is much larger than the spreading modulation bandwidth, $R_{\theta_{ss}}(\tau)$ may be approximated by its value at $\tau = 0$,

$$R_{\theta_{ss}}(\tau) \approx \tfrac{1}{2}E[\exp(j0)] = \tfrac{1}{2} \tag{4-126}$$

and

$$S_{n1}(f) \cong 2K_1 \tfrac{1}{2} S_{n_I}(f) = \frac{K_1 N_0}{2} \qquad |f| < B_{\text{in}}$$
$$= 0 \tag{4-127}$$

where $N_0/2$ is the input noise two-sided power spectral density, and an input bandwidth of $2B_{\text{in}}$ has been assumed. An identical expression can be derived for $S_{n2}(f)$.

Equation (4-122) takes the IF bandpass filter partially into account since code self-noise has been ignored. The remaining terms are also processed by the IF filters. As in Section 4-4 it is assumed that the data modulation passes through the IF filter without distortion. The noise components, on the other hand, are wideband relative to the IF filter bandwidth, so that the filters have a significant effect on these signals. The IF bandpass filters are characterized by their impulse response $h(t)$ or the lowpass equivalent complex envelope impulse response $h_l(t)$, where

$$h(t) = 2\,\mathrm{Re}[h_l(t)\exp(j\omega_{\text{IF}}t)] \tag{4-128}$$

Expressions for the complex envelope filter output noise which are analogous to (4-56) for BPSK spreading are then

$$\tilde{n}_1^0(t) = \sqrt{K_1}\exp(-j\phi')\int_{-\infty}^{+\infty}\exp\left[-j\theta_{ss}\left(\lambda - \hat{T}_d + \frac{\Delta}{2}T_c\right)\right]$$
$$\times\,[n_I(\lambda) + jn_Q(\lambda)]h_l(t-\lambda)\,d\lambda \tag{4-129a}$$

$$\tilde{n}_2^0(t) = \sqrt{K_1}\exp(-j\phi')\int_{-\infty}^{\infty}\exp\left[-j\theta_{ss}\left(\lambda - \hat{T}_d - \frac{\Delta}{2}T_c\right)\right]$$
$$\times\,[n_I(\lambda) + jn_Q(\lambda)]h_l(t-\lambda)\,d\lambda \tag{4-129b}$$

These filter outputs are Gaussian since the filter inputs are Gaussian. Thus, if $\tilde{n}_1^0(t)$ and $\tilde{n}_2^0(t)$ are uncorrelated, they are also independent. The complex cross-correlation of the two noise components is

$$R_{n,\text{out}}(\tau) \triangleq \frac{1}{2}E\{[\tilde{n}_1^0(t)][\tilde{n}_2^0(t+\tau)]^*\}$$

$$= \frac{1}{2}K_1 E\left\{\int_{-\infty}^{\infty}\int_{-\infty}^{\infty}\exp\left[-j\theta_{ss}\left(\lambda - \hat{T}_d + \frac{\Delta}{2}T_c\right) + j\theta_{ss}\left(\gamma - \hat{T}_d - \frac{\Delta}{2}T_c\right)\right]\right.$$
$$\left.\times\,[n_I(\lambda) + jn_Q(\lambda)][n_I(\gamma) - jn_Q(\gamma)]h_l(t-\lambda)h_l^*(t+\tau-\gamma)\,d\lambda\,d\gamma\right\} \tag{4-130}$$

$$= \frac{1}{2}K_1 \int_{-\infty}^{\infty} \int_{-\infty}^{\infty} E\left\{\exp\left[-j\theta_{ss}\left(\lambda - \hat{T}_d + \frac{\Delta}{2}T_c\right) + j\theta_{ss}\left(\gamma - \hat{T}_d - \frac{\Delta}{2}T_c\right)\right]\right\}$$

$$\times \{E[n_I(\lambda)n_I(\gamma)] + E[n_Q(\lambda)n_Q(\gamma)]\}h_I(t - \lambda)h_I^*(t + \tau - \gamma)\, d\lambda\, d\gamma$$

where the expected value is over all estimated propagation delays $\hat{T}_d$ and all sample functions of the random processes $n_I(t)$ and $n_Q(t)$. Using (4-121) and remembering that $n_I(t)$ and $n_Q(t)$ are identically distributed, (4-130) becomes

$$R_{n,\text{out}}(\tau) = 2K_1 \int_{-\infty}^{\infty} \int_{-\infty}^{\infty} R_{\theta_{ss}}(\lambda - \gamma + \Delta T_c)R_{n_I}(\lambda - \gamma)$$

$$\times h_I(t + \tau - \gamma)h_I^*(t - \gamma)\, d\lambda\, d\gamma \tag{4-131}$$

This equation is comparable to (4-58) for BPSK spreading. Using similar steps to those following (4-58), it may be shown that with the assumption that $R_{\theta_{ss}}(\tau)$ is slowly varying relative to $R_{n_I}(\tau)$, $R_{n,\text{out}}(\tau)$ is directly proportional to $R_{\theta_{ss}}(\Delta T_c)$. For $\Delta = 0$, it is clear that the two filter output noise processes are identical and therefore not independent. The two processes may be assumed independent only when Δ is sufficiently large. Although the analysis is more complicated when the filter output noise processes are dependent, it is conjectured that loop noise performance is unaffected since the correlated components will cancel at the discriminator output. In the following it will be assumed that the noise processes at the IF filter outputs are independent.

The complex envelopes of the filter output noise processes will be represented by

$$\tilde{n}_k^0(t) = \sqrt{2}\,n_{kI}^0(t) + j\sqrt{2}\,n_{kQ}^0(t) \tag{4-132}$$

where $n_{kI}^0(t)$ and $n_{kQ}^0(t)$, $k = 1, 2$, are assumed independent lowpass Gaussian noise processes. The power spectrum of $\tilde{n}_k^0(t)$ is the product of the power transfer function of the filter and the power spectrum of $\tilde{n}_k(t)$ given in (4-127). Assuming that the IF filter bandwidth is B_N,

$$S_{n,\text{out}}(f) = \begin{cases} \frac{1}{2}K_1 N_0 & |f| < \frac{1}{2}B_N \\ 0 & \text{elsewhere} \end{cases} \tag{4-133}$$

This expression is valid for both $\tilde{n}_1^0$ and $\tilde{n}_2^0$. The power spectrum of any one of the in-phase or quadrature noise components of (4-132) is one-half the magnitude of (4-133). Combining all of the above, the filter output complex envelopes are given by

$$\tilde{z}_1(t) = 2\sqrt{K_1 P}\,R_{\theta_{ss}}\left[\left(\delta + \frac{\Delta}{2}\right)T_c\right]\exp[j\theta_d(t - T_d)]\exp[j(\phi - \phi')]$$

$$+ \sqrt{2}\,n_{1I}^0(t) + j\sqrt{2}\,n_{1Q}^0(t) \tag{4-134a}$$

$$\tilde{z}_2(t) = 2\sqrt{K_1 P}\,R_{\theta_{ss}}\left[\left(\delta - \frac{\Delta}{2}\right)T_c\right]\exp[j\theta_d(t - T_d)]\exp[j(\phi - \phi')]$$

$$+ \sqrt{2}\,n_{2I}^0(t) + j\sqrt{2}\,n_{2Q}^0(t) \tag{4-134b}$$

These signals are input to the square-law detectors. The desired output of the square-law detector is the difference frequency component whose complex envelope is $\frac{1}{2}\tilde{z}_1(t)\tilde{z}_1^*(t)$ or $\frac{1}{2}\tilde{z}_2(t)\tilde{z}_2^*(t)$. The delay-lock discriminator output is then

$$\epsilon(t,\delta) = \frac{1}{2}\{\tilde{z}_2(t)\tilde{z}_2^*(t) - \tilde{z}_1(t)\tilde{z}_1^*(t)\}$$

$$= 2K_1P\left\{\left|R_{\theta_{ss}}\left[\left(\delta - \frac{\Delta}{2}\right)T_c\right]\right|^2 - \left|R_{\theta_{ss}}\left[\left(\delta + \frac{\Delta}{2}\right)T_c\right]\right|^2\right\}$$

$$+ 2\sqrt{2K_1P}\left\{\text{Re}\left[R_{\theta_{ss}}\left[\left(\delta - \frac{\Delta}{2}\right)T_c\right]\right]n_{2I}^0(t) + \text{Im}\left[R_{\theta_{ss}}\left[\left(\delta - \frac{\Delta}{2}\right)T_c\right]\right]n_{2Q}^0(t)\right.$$

$$\left. - \text{Re}\left[R_{\theta_{ss}}\left[\left(\delta + \frac{\Delta}{2}\right)T_c\right]\right]n_{1I}^0(t) - \text{Im}\left[R_{\theta_{ss}}\left[\left(\delta + \frac{\Delta}{2}\right)T_c\right]\right]n_{1Q}^0(t)\right\}$$

$$\times \cos[\theta_d(t - T_d) + \phi - \phi']$$

$$\hspace{6cm}(4\text{-}135)$$

$$+ 2\sqrt{2K_1P}\left\{\text{Re}\left[R_{\theta_{ss}}\left[\left(\delta - \frac{\Delta}{2}\right)T_c\right]\right]n_{2Q}^0(t) + \text{Im}\left[R_{\theta_{ss}}\left[\left(\delta - \frac{\Delta}{2}\right)T_c\right]\right]n_{2I}^0(t)\right.$$

$$\left. - \text{Re}\left[R_{\theta_{ss}}\left[\left(\delta + \frac{\Delta}{2}\right)T_c\right]\right]n_{1Q}^0(t) - \text{Im}\left[R_{\theta_{ss}}\left[\left(\delta + \frac{\Delta}{2}\right)T_c\right]\right]n_{1I}^0(t)\right\}$$

$$\times \sin[\theta_d(t - T_d) + \phi - \phi']$$

$$+ [n_{2I}^0(t)]^2 + [n_{2Q}^0(t)]^2 - [n_{1I}^0(t)]^2 - [n_{1Q}^0(t)]^2$$

The first term of $\epsilon(t,\delta)$ is the desired phase correcting component, the second and third terms are the $s \times n$ components, and the remaining terms are the $n \times n$ components. The discriminator S-curve is given by the first term. The S-curve is a function of the specific spreading modulation employed.

For all constant-envelope phase modulations of interest, the complex autocorrelation is real. In this special case, (4-135) is identical to (4-63) with the exception that $R_c[(\delta \pm \Delta/2)T_c]$ has been replaced by $2\,\text{Re}[R_{\theta_{ss}}[(\delta \pm \Delta/2)T_c]]$, and $R_c^2[(\delta \pm \Delta/2)T_c]$ has been replaced by $4|R_{\theta_{ss}}[(\delta \pm \Delta/2)T_c]|^2$. This difference is expected since the complex envelope autocorrelation function for BPSK is

$$R_{\theta_{ss}}(\tau) \triangleq \frac{1}{2}E[c(t)c^*(t + \tau)] = \frac{1}{2}R_c(\tau) \hspace{2cm}(4\text{-}136)$$

The similarity of (4-63) and (4-135) enables direct application of the analysis for the BPSK system to the more general case. In particular, the autocorrelation function of the delay-lock discriminator output is derived from (4-65) and is

$$R_\epsilon(\tau) = \frac{1}{4}K_1^2P^2\left\{4\left|R_{\theta_{ss}}\left[\left(\delta - \frac{\Delta}{2}\right)T_c\right]\right|^2 - 4\left|R_{\theta_{ss}}\left[\left(\delta + \frac{\Delta}{2}\right)T_c\right]\right|^2\right\}^2$$

$$+ 2K_1P\left\{4\,\text{Re}\left[R_{\theta_{ss}}\left[\left(\delta - \frac{\Delta}{2}\right)T_c\right]\right]^2 \hspace{2cm}(4\text{-}137)\right.$$

$$\left. + 4\,\text{Re}\left[R_{\theta_{ss}}\left[\left(\delta + \frac{\Delta}{2}\right)T_c\right]\right]^2\right\}E[n^0(t)n^0(t + \tau)]$$

$$\times E\{\cos[\theta_d(t) - \theta_d(t + \tau)]\}$$
$$+ 4E\{[n^0(t)n^0(t + \tau)]^2\} - 4E^2\{[n^0(t)]^2\}$$

where $n^0(t)$ has been used to represent any one of the four IF filter output processes $n^0_{1I}(t)$, $n^0_{1Q}(t)$, $n^0_{2I}(t)$, or $n^0_{2Q}(t)$. The Fourier transform of (4-137) is the discriminator output power spectrum. This spectrum is given by a simple modification of (4-66); the result is

$$S_\epsilon(f) = \frac{1}{4}K_1^2P^2\left\{4\left|R_{\theta_{ss}}\left[\left(\delta - \frac{\Delta}{2}\right)T_c\right]\right|^2 - 4\left|R_{\theta_{ss}}\left[\left(\delta + \frac{\Delta}{2}\right)T_c\right]\right|^2\right\}^2\delta(f)$$

$$+ 2K_1P\left\{4\,\text{Re}\left[R_{\theta_{ss}}\left[\left(\delta - \frac{\Delta}{2}\right)T_c\right]\right]^2 \right. \qquad (4\text{-}138)$$

$$+ 4\,\text{Re}\left[R_{\theta_{ss}}\left[\left(\delta + \frac{\Delta}{2}\right)T_c\right]\right]^2\right\}S_{n^0}(f) * S_{\theta_d}(f)$$

$$+ 8S_{n^0}(f) * S_{n^0}(f)$$

where $S_{\theta_d}(f)$ is the Fourier transform of the complex-envelope autocorrelation function $R_A(\tau)$ defined in (4-68). Under the assumption that the data modulation is narrowband relative to the noise, the discriminator output power spectrum is illustrated in Figure 4-11 with the proper substitution for the spreading modulation autocorrelation.

Define $D_\Delta(\delta)$ and $n_\epsilon(t)$ as the desired and undesired components of the discriminator output as in (4-76). That is,

$$D_\Delta(\delta) = 4\left|R_{\theta_{ss}}\left[\left(\delta - \frac{\Delta}{2}\right)T_c\right]\right|^2 - 4\left|R_{\theta_{ss}}\left[\left(\delta + \frac{\Delta}{2}\right)T_c\right]\right|^2 \qquad (4\text{-}139)$$

$$\epsilon(t,\delta) = \tfrac{1}{2}K_1PD_\Delta(\delta) + n_\epsilon(t) \qquad (4\text{-}140)$$

The discriminator S-curve $D_\Delta(\delta)$ is dependent only on the spreading phase modulation characteristics. With these definitions, the nonlinear loop model is given in Figure 4-12 and the linear model is given in Figure 4-13. Specific results for tracking jitter are dependent on the particular spreading modulation employed.

EXAMPLE 4-4

Consider the DLL with MSK spreading modulation and a normalized early-late difference of $\Delta = 1.0$. Assume that the spreading code is a totally random binary sequence. The complex autocorrelation function for MSK phase modulation is

$$R_{\theta_{ss}}(\tau) = \frac{1}{2\pi}\left[\pi\left(1 - \frac{|\tau|}{2T_c}\right)\cos\frac{\pi|\tau|}{2T_c} + \sin\frac{\pi|\tau|}{2T_c}\right] \qquad (4\text{-}141)$$

The discriminator S-curve is calculated from (4-141) and (4-139) and the result is illustrated in Figure 4-22. Using straightforward algebraic manipulations, the

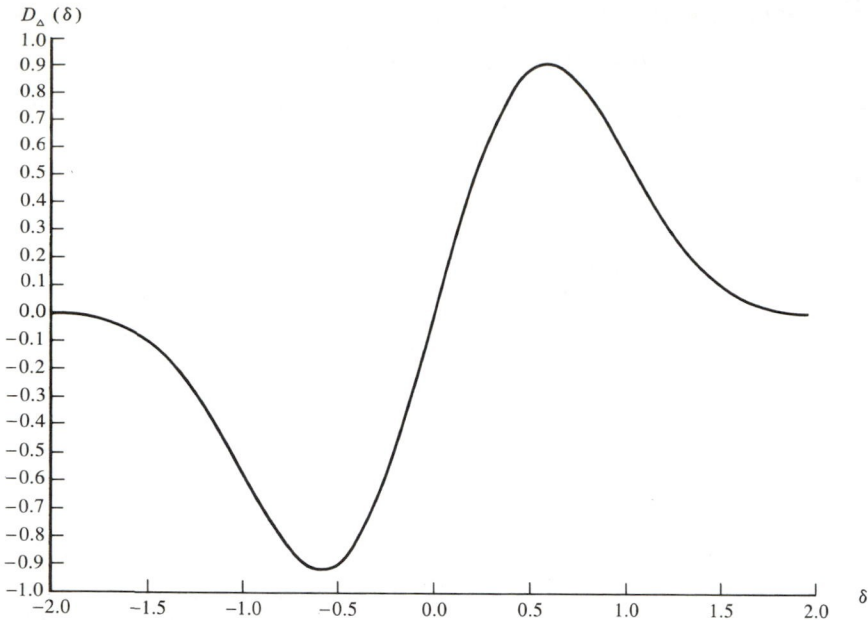

FIGURE 4-22. Noncoherent delay-lock discriminator S-curve using MSK spreading modulation.

slope of $D_\Delta(\delta)$ at $\delta = 0$ can be shown to be

$$\left[\frac{d}{d\delta}D_\Delta(\delta)\right]_{\substack{\delta=0\\\Delta=1.0}} = \frac{3}{2}\left(\frac{3\pi}{4}\cos\frac{\pi}{4} + \sin\frac{\pi}{4}\right)\sin\frac{\pi}{4} \tag{4-142}$$

$$= 2.517$$

so that

$$K_d = 1.259 K_1 P \tag{4-143}$$

The comparable result for BPSK spreading is

$$\left[\frac{d}{d\delta}D_\Delta(\delta)\right]_{\substack{\delta=0\\\Delta=1.0}} = 2.0 \tag{4-144}$$

$$K_d = K_1 P \tag{4-145}$$

The magnitude of the delay-lock discriminator output noise power spectrum is given in (4-75) for the case where data modulation has been ignored. Thus

$$S_{n_\epsilon}(f) = \frac{1}{2}K_1^2 N_0^2 B_n$$

$$+ \frac{1}{2}K_1^2 N_0 P\left\{4R_{\theta_{ss}}^2\left[\left(\delta - \frac{\Delta}{2}\right)T_c\right] + 4R_{\theta_{ss}}^2\left[\left(\delta + \frac{\Delta}{2}\right)T_c\right]\right\}$$

$$= \frac{1}{2} K_1^2 N_0^2 B_N + \frac{1}{2} K_1^2 N_0 P \frac{2}{\pi^2} \left(\frac{3\pi}{4} \cos \frac{\pi}{4} + \sin \frac{\pi}{4} \right)^2 \qquad (4\text{-}146)$$

$$= \frac{1}{2} K_1^2 N_0 (N_0 B_N + 1.41 P)$$

and the equivalent white noise power spectral density at the input to the linear equivalent circuit is

$$\frac{S_{n_\epsilon}(t)}{K_d^2} = \frac{1}{2} \left(\frac{N_0}{2P} \right) \left[1.44 + 1.26 \left(\frac{N_0 B_N}{P} \right) \right] \qquad (4\text{-}147)$$

The tracking jitter variance is equal to the product of this noise density and the two-sided loop noise bandwidth W_L, so that

$$\sigma_\delta^2 = \frac{1}{2} \left(\frac{N_0 W_L}{2P} \right) \left[1.44 + 1.26 \left(\frac{N_0 B_N}{P} \right) \right]$$

$$= \frac{1}{2\rho_L} \left(1.44 + \frac{1.26}{\rho_{IF}} \right) \qquad (4\text{-}148)$$

Comparing the last expression to (4-83) for BPSK spreading modulation, it is observed that, for this special case where $\Delta = 1.0$, the optimum choice of modulation for minimum tracking jitter depends on whether or not the squaring loss component is dominant. Note, however, that the choice may be more clear for other delay differences.

4-8 Code Tracking Loops for Frequency-Hop Systems

The noncoherent DLL can also be used in a frequency-hopping spread-spectrum system. In particular, the conceptual block diagram of Figure 4-21 can be used if the phase modulators are replaced with frequency synthesizers. The tracking loop configuration for a frequency-hopping system is illustrated in Figure 4-23. Conceptually, the early and late reference signals at the output of the frequency synthesizer may be generated from a single synthesizer and an appropriate delay line. In practice, the delay line approach may not be possible due to the difficulty of building delay lines having adequate delay.

Observe that the ''spreading waveform generator'' of Figure 4-21 has been replaced by a ''spreading code generator,'' indicating that its output is not a waveform $c(t) = \pm 1$ but a digital signal which controls the frequency of the synthesizer. The code generator outputs k binary digits at a time, implying that the synthesizer generates 2^k frequencies. As seen in all the other tracking loops analyzed in this chapter, the delay-lock discriminator output consists of a desired phase correction signal and noise. The most difficult part of the loop analysis is the characterization of these components. Although the analysis of Section 4-7 was general, an assumption was

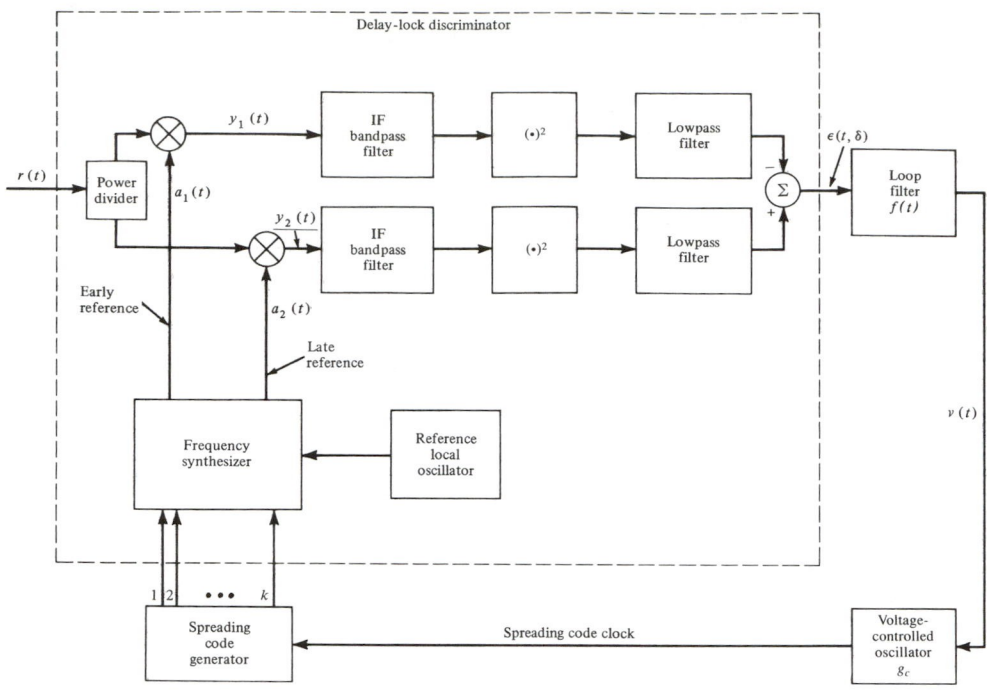

FIGURE 4-23. Noncoherent tracking loop for frequency-hopping spread-spectrum system.

made that the transmitter and receiver spreading code phase modulators were identical. In the present case, this assumption implies that the transmitter and receiver frequency synthesizers are precisely phase matched. Because of the hardware difficulties that this implies, that assumption is not made in the analysis that follows. Discussion is limited to tracking loops having an early-late channel difference of one chip, and systems that employ a slow hop relative to the data modulation.

The received signal is the sum of the frequency-hopped and data-modulated carrier having power P and bandlimited additive white Gaussian noise having two-sided power spectral density $N_0/2$. That is,

$$r(t) = \sqrt{2P} \cos \left[\omega_0 t + \sum_{n=-\infty}^{\infty} (\omega_n t + \phi_n) p_{T_c}(t - T_d - nT_c) + \theta_d(t - T_d) \right]$$
$$+ \sqrt{2} n_I(t) \cos \omega_0 t - \sqrt{2} n_Q(t) \sin \omega_0 t \qquad (4\text{-}149)$$

In this equation, $(\omega_0 + \omega_n)$ is the transmission frequency during time interval n, ϕ_n is the frequency synthesizer random phase during this same interval, $p_{T_c}(t)$ is a unit pulse beginning at $t = 0$ and ending at $t = T_c$, and $\theta_d(t)$ is the arbitrary data phase modulation. The reference signals $a_1(t)$ and $a_2(t)$ are frequency hopped using the same hop pattern as used in the transmitter, but using frequencies that are offset from the transmitter frequencies by the IF. The hopping patterns of the reference signals are offset in phase from the receiver estimate of the transmission delay $\hat{T}_d$ by $\pm\frac{1}{2}$ chip, where a chip is the frequency-hop dwell time T_c. Thus

$a_1(t) =$

$$2\sqrt{K_1}\cos\left[(\omega_0 + \omega_{\text{IF}})t + \sum_{n=-\infty}^{\infty}(\omega_n t + \phi_n')p_{T_c}\left(t - \hat{T}_d + \frac{\Delta}{2}T_c - nT_c\right)\right] \quad (4\text{-}150\text{a})$$

$a_2(t) =$

$$2\sqrt{K_1}\cos\left[(\omega_0 + \omega_{\text{IF}})t + \sum_{n=-\infty}^{\infty}(\omega_n t + \phi_n')p_{T_c}\left(t - \hat{T}_d - \frac{\Delta}{2}T_c - nT_c\right)\right] \quad (4\text{-}150\text{b})$$

where K_1 is the mixer conversion loss, ϕ_n' is the receiver synthesizer random phase during the nth time interval, and all other terms have been defined earlier.

Consider the early channel mixer output signal $y_1(t)$. The sum frequency terms of this mixing operation will be rejected by the IF bandpass filter, so that the signal of interest is

$$y_1(t) = \sqrt{K_1 P}\cos\left[\omega_{\text{IF}}t - \sum_{n=-\infty}^{\infty}(\omega_n t + \phi_n)p_{T_c}(t - T_d - nT_c) - \theta_d(t - T_d)\right.$$

$$\left. + \sum_{m=-\infty}^{\infty}(\omega_m t + \phi_m')p_{T_c}\left(t - \hat{T}_d + \frac{\Delta}{2}T_c - mT_c\right)\right] \quad (4\text{-}151)$$

$$+ \sqrt{K_1}\,n_I(t)\cos\left[\omega_{\text{IF}}t + \sum_{n=-\infty}^{\infty}(\omega_n t + \phi_n')p_{T_c}\left(t - \hat{T}_d + \frac{\Delta}{2}T_c - nT_c\right)\right]$$

$$- \sqrt{K_1}\,n_Q(t)\sin\left[\omega_{\text{IF}}t + \sum_{n=-\infty}^{\infty}(\omega_n t + \phi_n')p_{T_c}\left(t - \hat{T}_d + \frac{\Delta}{2}T_c - nT_c\right)\right]$$

The bandwidth of the received noise covers the entire received signal spectrum. During each hop dwell, a different portion of the received noise spectrum is translated to near the IF and will pass through the IF filter. It will be assumed that the frequency hop rate is slow enough that hop transients may be neglected. Thus the mixer output noise may be approximated by a narrowband white Gaussian noise process centered at the IF (i.e., not hopped), whose two-sided power spectral density is $K_1 N_0/4$.

The signal component of (4-151) is centered on the IF only when the two summations within the argument of the cosine are generating the same frequency. This occurs whenever $p_{T_c}(t - T_d - nT_c)$ and $p_{T_c}(t - \hat{T}_d + (\Delta/2)T_c - mT_c)$ overlap. If any overlap of these pulses occurs, it will occur on every hop interval. During that portion of time where no overlap occurs, the signal component is translated to a frequency that will not pass through the IF filter and the signal may be ignored during this time. Thus the signal component may be replaced by an equivalent (as far as the tracking loop is concerned) pulsed signal and (4-151) may be approximated by

$$y_1'(t) = \sqrt{K_1 P} \sum_{n=-\infty}^{\infty} p_{T_c}(t - T_d - nT_c) p_{T_c}\left(t - \hat{T}_d + \frac{\Delta}{2} T_c - nT_c\right)$$

$$\times \cos[\omega_{IF} t - (\phi_n - \phi_n') - \theta_d(t - T_d)] \tag{4-152a}$$

$$+ \sqrt{K_1}\, n_{1I}(t) \cos \omega_{IF} t - \sqrt{K_1}\, n_{1Q}(t) \sin \omega_{IF} t$$

Figure 4-24 illustrates typical early channel mixer input and output spectra. The portions of the early mixer output spectrum marked with the letter "A" will not pass through the IF filter and are ignored in (4-152). Similarly, the late channel mixer equivalent output is

$$y_2'(t) = \sqrt{K_1 P} \sum_{n=-\infty}^{\infty} p_{T_c}(t - T_d - nT_c) p_{T_c}\left(t - \hat{T}_d - \frac{\Delta}{2} T_c - nT_c\right)$$

$$\times \cos[\omega_{IF} t - (\phi_n - \phi_n') - \theta_d(t - T_d)] \tag{4-152b}$$

$$+ \sqrt{K_1}\, n_{2I}(t) \cos \omega_{IF} t - \sqrt{K_1}\, n_{2Q}(t) \sin \omega_{IF} t$$

The discussion in this section is limited to systems with $\Delta = 1.0$. Thus the early and late reference signals are never simultaneously at the same frequency and, at any instant of time, the early and late channel noise processes come from different portions of the received signal band and are therefore independent.

The frequency-hop rate is assumed to be very slow relative to the IF filter bandwidth so that a quasi-static analysis similar to that used in the dithering loop analysis of Section 4-5 may be used. With this assumption, the pulse modulation of (4-152) passes through the filter without distortion. The filter output noise processes are Gaussian and independent since the inputs are Gaussian and independent. Thus the filter outputs are approximately

$$z_1(t) = \sqrt{K_1 P} \sum_{n=-\infty}^{\infty} p_{T_c}(t - T_d - nT_c) p_{T_c}\left(t - \hat{T}_d + \frac{\Delta}{2} T_c - nT_c\right)$$

$$\times \cos[\omega_{IF} t - (\phi_n - \phi_n') - \theta_d(t - T_d)] \tag{4-153a}$$

$$+ \sqrt{2}\, n_{1I}^0(t) \cos \omega_{IF} t - \sqrt{2}\, n_{1Q}^0(t) \sin \omega_{IF} t$$

$$z_2(t) = \sqrt{K_1 P} \sum_{n=-\infty}^{\infty} p_{T_c}(t - T_d - nT_c) p_{T_c}\left(t - \hat{T}_d - \frac{\Delta}{2} T_c - nT_c\right)$$

$$\times \cos[\omega_{IF} t - (\phi_n - \phi_n') - \theta_d(t - T_d)] \tag{4-153b}$$

$$+ \sqrt{2}\, n_{2I}^0(t) \cos \omega_{IF} t - \sqrt{2}\, n_{2Q}^0(t) \sin \omega_{IF} t$$

The noise signals $n_{1I}^0(t)$, $n_{1Q}^0(t)$, $n_{2I}^0(t)$, and $n_{2Q}^0(t)$ are all independent, lowpass, white Gaussian noise processes with two-sided power spectral densities of $K_1 N_0 / 4$. Their two-sided bandwidth equals the IF filter one-sided bandwidth.

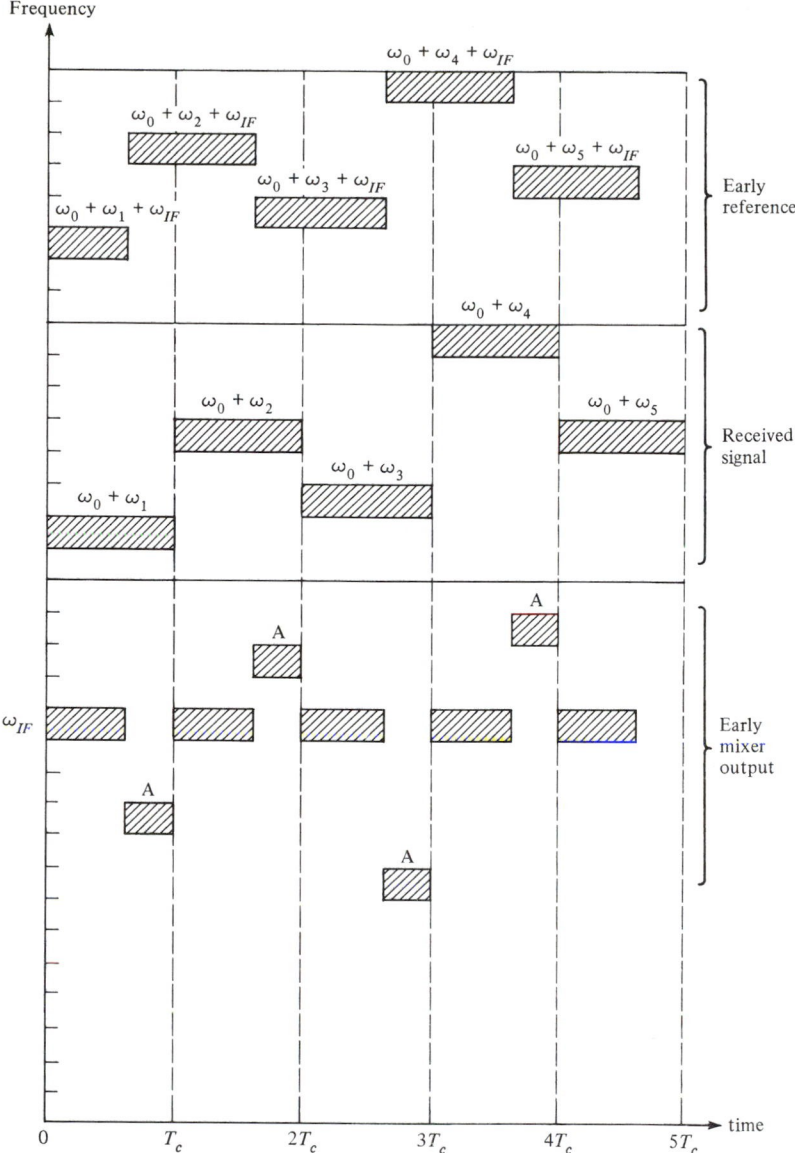

FIGURE 4-24. Illustration of typical early channel mixer input and output signal spectra.

The slow hop assumption also enables simplification of the calculation of the output of the lowpass filters which follow the squaring devices. The outputs are calculated for the signal present (i.e., pulses overlap) and signal absent cases independently and combined at the output. As usual, the sum frequency terms at the squaring device outputs are rejected by the lowpass filters. The (signal × signal) and (signal $\times$ noise) output terms are present only when the pulse signals overlap; however, the (noise $\times$ noise) term is always present. The early channel lowpass filter output is

$$[z_1^2(t)]_{LP} = \{\tfrac{1}{2}K_1P + n_{1I}^0(t)\sqrt{2K_1P}\cos[\phi_n - \phi_n' - \theta_d(t - T_d)]$$
$$- n_{1Q}^0(t)\sqrt{2K_1P}\sin[\phi_n - \phi_n' - \theta_d(t - T_d)]\}$$
$$\times \left\{ \sum_{n=-\infty}^{\infty} p_{T_c}(t - T_d - nT_c)p_{T_c}\left(t - \hat{T}_d + \frac{\Delta}{2}T_c - nT_c\right) \right\} \qquad (4\text{-}154)$$
$$+ [n_{1I}^0(t)]^2 + [n_{1Q}^0(t)]^2$$

A similar expression is found for $|z_2^2(t)|_{LP}$.

Straightforward algebraic and trigonometric manipulation on $[z_2^2(t)]_{LP} - [z_1^2(t)]_{LP}$ yields the following result for the discriminator output $\epsilon(t,\delta)$:

$$\epsilon(t,\delta) = \frac{1}{2}K_1P\left[\sum_{n=-\infty}^{\infty} p_{T_c}(t - T_d - nT_c)p_{T_c}\left(t - \hat{T}_d - \frac{\Delta}{2}T_c - nT_c\right) \right.$$
$$\left. - \sum_{m=-\infty}^{\infty} p_{T_c}(t - T_d - mT_c)p_{T_c}\left(t - \hat{T}_d + \frac{\Delta}{2}T_c - mT_c\right) \right]$$
$$+ \sqrt{2K_1P}\cos[\theta_d(t - T_d)][n_{2I}^0(t)\beta_1(t) - n_{1I}^0(t)\beta_2(t)]$$
$$+ \sqrt{2K_1P}\sin[\theta_d(t - T_d)][n_{2I}^0(t)\beta_3(t) - n_{1I}^0(t)\beta_4(t)] \qquad (4\text{-}155)$$
$$+ \sqrt{2K_1P}\sin[\theta_d(t - T_d)][n_{2Q}^0(t)\beta_1(t) - n_{1Q}^0(t)\beta_2(t)]$$
$$- \sqrt{2K_1P}\cos[\theta_d(t - T_d)][n_{2Q}^0(t)\beta_3(t) - n_{1Q}^0(t)\beta_4(t)]$$
$$+ [n_{2I}^0(t)]^2 + [n_{2Q}^0(t)]^2 - [n_{1I}^0(t)]^2 + [n_{1Q}^0(t)]^2$$

where
$$\beta_1(t) = \sum_{n=-\infty}^{\infty} \cos(\phi_n - \phi_n')p_{T_c}(t - T_d - nT_c)p_{T_c}\left(t - \hat{T}_d - \frac{\Delta}{2}T_c - nT_c\right)$$

$$\beta_2(t) = \sum_{n=-\infty}^{\infty} \cos(\phi_n - \phi_n')p_{T_c}(t - T_d - nT_c)p_{T_c}\left(t - \hat{T}_d + \frac{\Delta}{2}T_c - nT_c\right)$$

$$\beta_3(t) = \sum_{n=-\infty}^{\infty} \sin(\phi_n - \phi_n')p_{T_c}(t - T_d - nT_c)p_{T_c}\left(t - \hat{T}_d - \frac{\Delta}{2}T_c - nT_c\right)$$

$$\beta_4(t) = \sum_{n=-\infty}^{\infty} \sin(\phi_n - \phi_n')p_{T_c}(t - T_d - nT_c)p_{T_c}\left(t - \hat{T}_d + \frac{\Delta}{2}T_c - nT_c\right)$$

Consider the term of this equation defined by

$$e(t) = \sqrt{2K_1P}\cos[\theta_d(t - T_d)]$$
$$\times \left[n_{2I}^0(t) \sum_{n=-\infty}^{\infty} \cos(\phi_n - \phi_n')p_{T_c}(t - T_d - nT_c) \right.$$

$$\times p_{T_c}\!\left(t - \hat{T}_d - \frac{\Delta}{2}T_c - nT_c\right) \tag{4-156}$$

$$- n_{11}^0(t) \sum_{m=-\infty}^{\infty} \cos(\phi_m - \phi_m') p_{T_c}(t - T_d - mT_c) p_{T_c}\!\left(t - \hat{T}_d + \frac{\Delta}{2}T_c - mT_c\right)\Bigg]$$

Discussion is limited to cases where $\Delta = 1.0$. Thus $p_{T_c}(t - T_d - nT_c) p_{T_c}(t - \hat{T}_d - (\Delta/2)T_c - nT_c)$ is nonzero only where $p_{T_c}(t - T_d - mT_c) p_{T_c}(t - \hat{T}_d + (\Delta/2)T_c - mT_c)$ is zero, and vice versa, and the factor within parentheses switches between its first term and its second term twice each T_c seconds. Assume that the frequency-hop rate is slow enough that the transient due to the switching may be neglected. Then, since $n_{21}^0(t)$ and $n_{11}^0(t)$ have identical statistics, and since the arguments of the cosines are identical, the function $e(t)$ may be approximated by

$$e(t) \cong \sqrt{2K_1P}\,\cos[\theta_d(t - T_d)]n_{11}^0(t)\cos[\phi(t)] \tag{4-157}$$

where $\phi(t)$ represents the phase modulation due to the noncoherence of the frequency synthesizers. That is,

$$\phi(t) = \sum_{n=-\infty}^{\infty} (\phi_n - \phi_n') p_{T_c}(t - T_d - nT_c) \tag{4-158}$$

Identical arguments may be used to simplify three other terms of $\epsilon(t,\delta)$, resulting in

$$\epsilon(t,\delta) \cong \tfrac{1}{2}K_1P\Bigg[\sum_{n=-\infty}^{\infty} p_{T_c}(t - T_d - nT_c)p_{T_c}(t - \hat{T}_d - \tfrac{1}{2}T_c - nT_c)$$

$$- \sum_{m=-\infty}^{\infty} p_{T_c}(t - T_d - mT_c)p_{T_c}(t - \hat{T}_d + \tfrac{1}{2}T_c - mT_c)\Bigg]$$

$$+ \sqrt{2K_1P}\,\cos[\theta_d(t - T_d)]n_{11}^0(t)\cos\phi(t)$$
$$+ \sqrt{2K_1P}\,\sin[\theta_d(t - T_c)]n_{11}^0(t)\sin\phi(t) \tag{4-159}$$
$$+ \sqrt{2K_1P}\,\sin[\theta_d(t - T_d)]n_{1Q}^0(t)\cos\phi(t)$$
$$- \sqrt{2K_1P}\,\cos[\theta_d(t - T_d)]n_{1Q}^0(t)\sin\phi(t)$$
$$+ [n_{21}^0(t)]^2 + [n_{2Q}^0(t)]^2 - [n_{11}^0(t)]^2 - [n_{1Q}^0(t)]^2$$

The first term of this equation contains the desired phase correction information in its dc component. Calculating the dc component and denoting the remaining signal by $n_s(t)$ yields

$$\epsilon(t,\delta) = \tfrac{1}{2}K_1P\{R[(\delta - \tfrac{1}{2})T_c] - R[(\delta + \tfrac{1}{2})T_c]\} + n_s(t)$$
$$+ \sqrt{2K_1P}\,\cos[\theta_d(t - T_d) - \phi(t)]n_{11}^0(t)$$
$$+ \sqrt{2K_1P}\,\sin[\theta_d(t - T_d) - \phi(t)]n_{1Q}^0(t) \tag{4-160}$$
$$+ [n_{21}^0(t)]^2 + [n_{2Q}^0(t)]^2 - [n_{11}^0(t)]^2 - [n_{1Q}^0(t)]^2$$

where

$$R(\tau) = \begin{cases} 0 & \text{for } \tau < -T_c \\[2mm] \dfrac{\tau}{T_c} + 1.0 & \text{for } -T_c \le \tau < 0 \\[2mm] -\dfrac{\tau}{T_c} + 1.0 & \text{for } 0 \le \tau < T_c \\[2mm] 0 & \text{for } T_c \le \tau \end{cases} \tag{4-161}$$

The immediate goal of this analysis is the calculation of the power spectrum of $\epsilon(t,\delta)$. This power spectrum is calculated in the usual manner through calculation of the autocorrelation function. In this calculation, $n_s(t)$ will be ignored since its power spectrum consists of impulses at harmonics of the hop frequency, which is assumed to be much higher than the bandwidth of the loop filter. The calculation of the autocorrelation function is simplified by taking advantage of the fact that $\theta_d(t)$, $\phi(t)$, and all noise components are independent. Thus the expectations factor and many terms are immediately seen to equal zero. The calculation is very similar to the calculation detailed in Appendix E. The result, after some manipulation, is

$$\begin{aligned} R_\epsilon(\tau) = \tfrac{1}{4}K_1^2 P^2 \{ &R[(\delta - \tfrac{1}{2})T_c] - R[(\delta + \tfrac{1}{2})T_c] \}^2 \\ &+ 2K_1 P R_{n^0}(\tau) E\{\cos[\phi(t) - \phi(t+\tau)]\} E\{\cos[\theta_d(t) - \theta_d(t+\tau)]\} \\ &- 2K_1 P R_{n^0}(\tau) E\{\sin[\phi(t) - \phi(t+\tau)]\} E\{\sin[\theta_d(t) - \theta_d(t+\tau)]\} \\ &+ 4E\{[n^0(t)]^2[n^0(t+\tau)]^2\} - 4E^2\{[n^0(t)]^2\} \end{aligned} \tag{4-162}$$

In this equation, $n^0(t)$ represents any one of the four independent processes $n_{1I}^0(t)$, $n_{1Q}^0(t)$, $n_{2I}^0(t)$, or $n_{2Q}^0(t)$.

Consider the expectations $E\{\cos[\phi(t) - \phi(t+\tau)]\}$ and $E\{\sin[\phi(t) - \phi(t+\tau)]\}$. The expected value is over all phase angles ϕ_n and ϕ_n' of (4-158) as well as all transmission delays T_d. From (4-158),

$$\phi(t) - \phi(t+\tau) = \sum_{n=-\infty}^{\infty} (\phi_n - \phi_n') p_{T_c}(t - T_d - nT_c) \tag{4-163}$$

$$- \sum_{m=-\infty}^{\infty} (\phi_m - \phi_m') p_{T_c}(t + \tau - T_d - mT_c)$$

For $|\tau| < T_c$, the unit pulses overlap for a fraction $(T_c - |\tau|)/T_c$ of all T_d. For these T_d, $\phi(t) - \phi(t+\tau) = 0$, and for all other T_d, $\phi(t) - \phi(t+\tau)$ is uniformly distributed over $(0,2\pi)$. Thus

$$E\{\cos[\phi(t) - \phi(t+\tau)]\} = \frac{T_c - |\tau|}{T_c} \qquad \text{for } |\tau| < T_c \tag{4-164}$$

$$E\{\sin[\phi(t) - \phi(t+\tau)]\} = 0 \qquad \text{for } |\tau| < T_c \tag{4-165}$$

For $T_c < |\tau|$, the unit pulses do not overlap, $\phi(t) - \phi(t + \tau)$ is uniformly distributed over $(0, 2\pi)$ for all T_d, and both expectations are zero. Thus

$$R_\epsilon(\tau) = \tfrac{1}{4}K_1^2 P^2\{R[(\delta - \tfrac{1}{2})T_c] - R[(\delta + \tfrac{1}{2})T_c]\}^2$$
$$+ 2K_1 P R_{n^0}(\tau) R_h(\tau) 2 \operatorname{Re}[R_A(\tau)] \tag{4-166}$$
$$+ 4E\{[n^0(t)]^2[n^0(t + \tau)]^2\} - 4E^2\{[n^0(t)]^2\}$$

where $R_A(\tau)$ is the data modulation autocorrelation, and

$$R_h(\tau) = \begin{cases} 1 - \dfrac{|\tau|}{T_c} & |\tau| \leq T_c \\[2mm] 0 & \text{elsewhere} \end{cases} \tag{4-167}$$

The Fourier transform of (4-166) yields the desired delay-lock discriminator output power spectral density. The first term is a constant and its Fourier transform is a delta function at zero frequency. The second term is the product of three auto-correlations and its Fourier transform is the convolution of the Fourier transforms of the individual terms. The transform of $R_{n^0}(\tau)$ is the power spectrum of any one of the lowpass white noise processes given by

$$S_{n^0}(f) = \begin{cases} \tfrac{1}{4}K_1 N_0 & |f| < \tfrac{1}{2}B_N \\ 0 & \text{elsewhere} \end{cases} \tag{4-168}$$

The transform of $R_h(\tau)$ is

$$S_h(f) = T_c \operatorname{sinc}^2 f T_c \tag{4-169}$$

and the transform of $2 \operatorname{Re}\{R_A(\tau)\}$ is denoted by $S_{\theta_d}(f)$ and is given by (4-67). The transform of the last two terms of (4-166) are evaluated in Appendix E; the result is $8S_{n^0}(f) * S_{n^0}(f)$. Combining these results yields

$$S_\epsilon(f) = \tfrac{1}{4}K_1^2 P^2\{R[(\delta - \tfrac{1}{2})T_c] - R[(\delta + \tfrac{1}{2})T_c]\}^2 \delta(f)$$
$$+ 2K_1 P S_{n^0}(f) * S_h(f) * S_{\theta_d}(f) + 8S_{n^0}(f) * S_{n^0}(f) \tag{4-170}$$

In order to generate specific code tracking jitter results for comparison with the results of the other analyses of this chapter, consider the case where there is no data modulation. In this case, $S_{\theta_d}(f) = \delta(f)$. It was assumed earlier than the frequency-hop rate is slow relative to the IF filter bandwidth. Therefore, $S_{n^0}(f)$ is nearly constant over the range of significant values of $S_h(f)$ and

$$S_{n^0}(f) * S_h(f) = \int_{-\infty}^{\infty} S_{n^0}(\lambda) S_h(f - \lambda)\, d\lambda \tag{4-171}$$
$$\approx S_{n^0}(f) \int_{-\infty}^{\infty} S_h(f - \lambda)\, d\lambda = S_{n^0}(f)$$

Thus

$$S_\epsilon(f) \cong \tfrac{1}{4}K_1^2 P^2\{R[(\delta - \tfrac{1}{2})T_c] - R[(\delta + \tfrac{1}{2})T_c]\}^2 \delta(f)$$
$$+ 2K_1 P S_{n^0}(f) + 8S_{n^0}(f) * S_{n^0}(f) \tag{4-172}$$

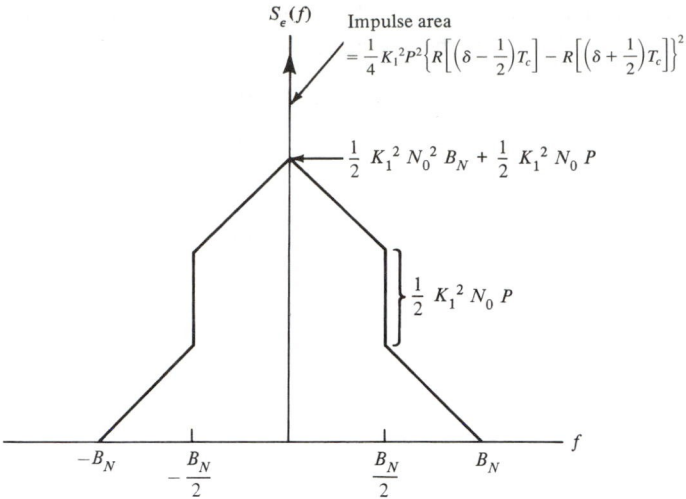

FIGURE 4-25.   Frequency-hop code tracking loop discriminator output power spectrum.

which is illustrated in Figure 4-25. This figure should be compared to Figure 4-11 for the DLL for direct-sequence code tracking.

The discriminator S-curve is defined by the discriminator dc output component. The normalized S-curve, $D_\Delta(\delta)$, is plotted in Figure 4-26. It is defined by

$$D_\Delta(\delta) = R[(\delta - \tfrac{1}{2})T_c] - R[(\delta + \tfrac{1}{2})T_c]$$

with $R(\tau)$ defined by (4-161). The total discriminator output is then

$$\epsilon(t,\delta) = \tfrac{1}{2}K_1 P D_\Delta(\delta) + n_\epsilon(t) \tag{4-173}$$

where $n_\epsilon(t)$ represents all output noise. From this point on the loop analysis is identical to that of Section 4-4. The linearized model of the loop is given in Figure 4-13 with

$$K_d = \tfrac{1}{2}K_1 P \left[\frac{d}{d\delta} D_\Delta(\Delta) \right]_{\delta=0} = K_1 P \tag{4-174}$$

and where the input noise power spectrum has a magnitude given by the zero-frequency value of the continuous component of Figure 4-25 multiplied by $1/K_d^2$. Finally, the variance of the code tracking jitter is calculated using (4-80); the result is

$$\sigma_\delta^2 = \frac{\tfrac{1}{2}K_1^2 N_0^2 B_N + \tfrac{1}{2}K_1^2 N_0 P}{(K_1 P)^2} W_L$$

$$= \frac{1}{\rho_L}\left(1 + \frac{1}{\rho_{IF}}\right) \tag{4-175}$$

This result should be compared to (4-83). The increased tracking jitter is the result

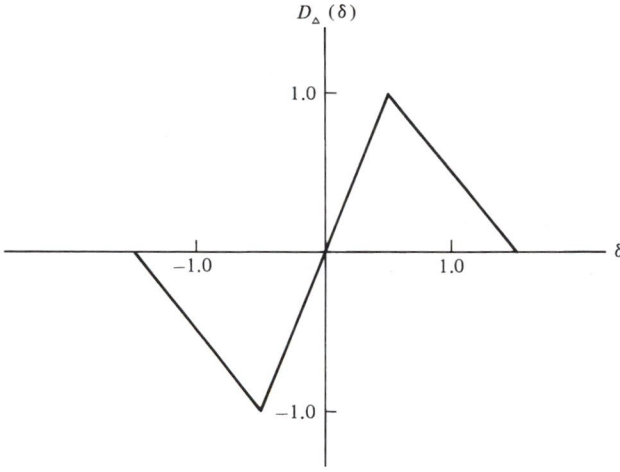

FIGURE 4-26. Normalized S-curve for the noncoherent code tracking loop discriminator.

of an increased (signal $\times$ noise) term in the power spectrum of Figure 4-25, which is, in turn, the result of the fact that slow frequency hop is used.

This completes the analysis of the delay-lock frequency-hop code tracking loop. Other tracking loop configurations are possible. A configuration that is similar to the direct-sequence dithered tracking loop is illustrated in Figure 4-27. This configura-

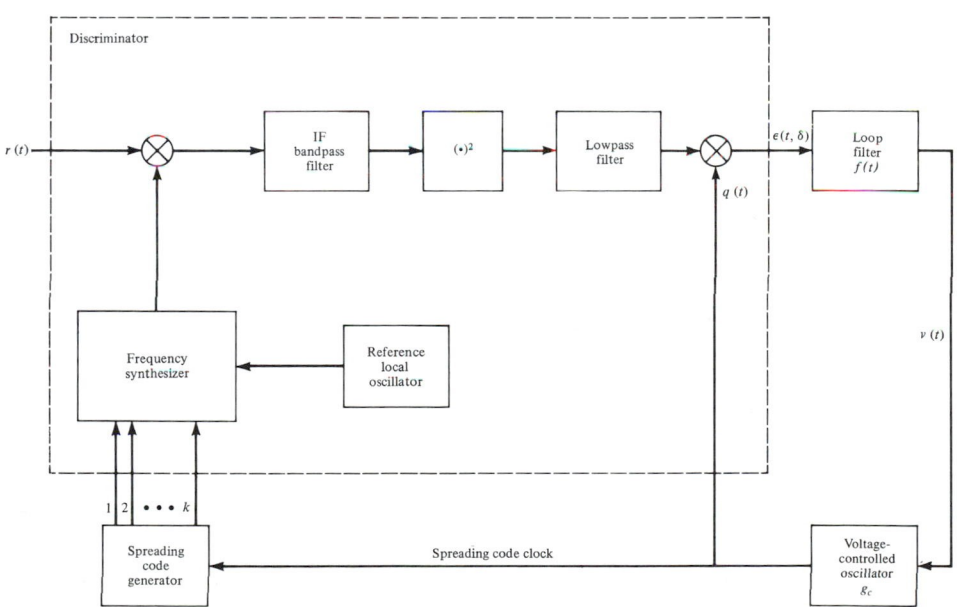

FIGURE 4-27. Noncoherent code tracking loop configuration for slow frequency hop. (From Ref. 18.)

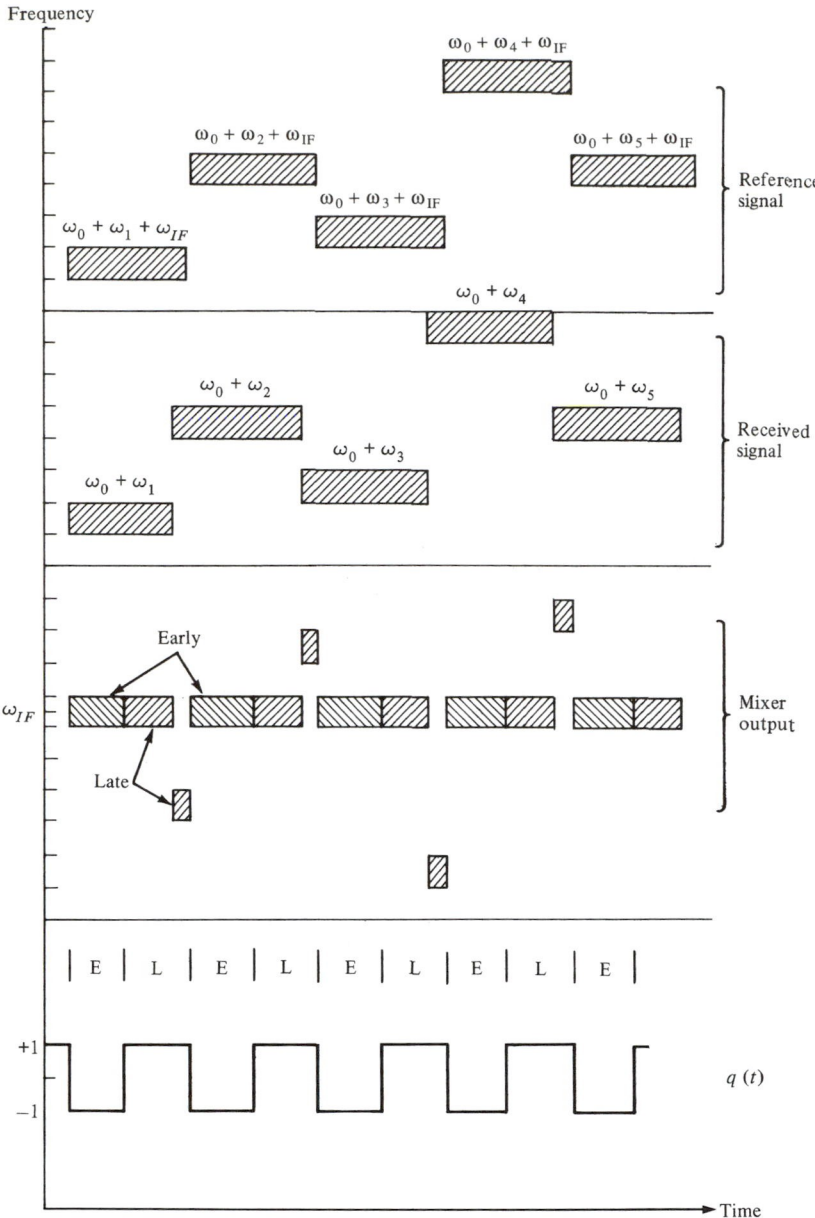

FIGURE 4-28. Typical dehopping mixer inputs and output for FH tracking loop configuration of Figure 4-27.

tion can be derived from the loop of Figure 4-23 by observing that, when δ is small, one-half of the early or late channel mixer output power is out of band and is rejected by the IF bandpass filter. In addition, the two channels never translate the received signal to IF at the same time so that time sharing a single filter and square-law detector appears possible. Figure 4-28 illustrates typical dehopping mixer input and output spectra. The reference signal may be thought of as an early reference for one-half time and a late reference for the other one-half time. This is illustrated by dividing the output spectrum into two regions denoted by different crosshatching in Figure 4-28. When the reference is late as illustrated, the duration of the late half of the mixer output is reduced and a dc correction is generated at the output of the second mixer. Similarly, when the reference is early, the early portion of the mixer output is reduced. Comparison of Figures 4-24 and 4-28 will show that the slope of the discriminator characteristics is larger by a factor of 2 for the DLL. This is seen by observing that whereas the early or late channel output is reduced by a tracking error in the time-shared loop, in the DLL, the output power lost in one channel is added to the other. It is conjectured that the performance of the time-shared loop is slightly inferior to that of the DLL.

4-9 Summary

The primary goal of this chapter has been to present a method of analysis that can be applied to all types of code tracking loops. In all cases, the analysis was aimed at developing a linear model of the tracking loop identical to that of Figure 4-7. This model was developed through characterization of the signal and noise components of the discriminator output. Accomplishing this task required numerous assumptions to be made about the independence of certain noise processes and the relative bandwidths of certain signals. The analysis is sufficiently detailed that the point of departure for extensions should be clear. A factor that has been consistently ignored is the effect of distortions due to the IF bandpass filter for the noncoherent loops. The student is referred to Simon [6,10] for an analysis that does not make this assumption.

Table 4-1 summarizes a few of the most significant results of this chapter. For noncoherent systems, the delay-lock loops using either BPSK or MSK spreading clearly provide the best noise performance. MSK spreading appears to be a good choice for high data rates where the IF bandwidth is large, resulting in low IF signal-to-noise ratios. The tau-dither tracking loop provides inferior noise performance but is significantly simpler in hardware and also solves the gain-balance problem of the delay-lock loops. In any system design, many parameters must be optimized which are not explicitly shown in Table 4-1. For example, early-late delay differences other than $\Delta = 1.0$ may be desirable in some cases. For the dithering loop, other dithering frequencies may provide superior performance under some conditions.

Additional tracking loop configurations have been discussed in the literature. In all cases these alternative configurations are designed to solve one or more of the problems of the loops discussed in this chapter. For example, Yost and Boyd [19,20] have proposed a modified PN code tracking loop which has the hardware simplicity of the tau-dither loop and tracking performance better than the delay-lock loop. The motivation for the design of the phase detector of the modified loop was the mitigation of the gain-balance problem of the DLL and the simplification of tracking loop

TABLE 4-1. Summary of Direct-Sequence Code Tracking Loop Noise Performance Results for $\Delta = 1.0$

Type of Loop	Discriminator Gain, K_d	Equivalent Input Noise PSD	Code Tracking Jitter Variance
Baseband coherent delay-lock loop	$K_1\sqrt{2P}$	$\dfrac{N_0}{4P}$	$\sigma_\delta^2 = \dfrac{1}{2\rho_L}$
Noncoherent delay-lock loop with BPSK spreading	$K_1 P$	$\dfrac{N_0}{4P} + \dfrac{B_N N_0^2}{2P^2}$	$\sigma_\delta^2 = \dfrac{1}{2\rho_L}\left(1 + \dfrac{2}{\rho_{IF}}\right)$
Tau-dither noncoherent loop with BPSK spreading and $f_q = B_N/4$	$K_1 P$	$\dfrac{N_0}{4P}(1.81) + \dfrac{B_N N_0^2}{2P^2}(1.63)$	$\sigma_\delta^2 = \dfrac{1}{2\rho_L}\left(1.811 + \dfrac{3.261}{\rho_{IF}}\right)$
Noncoherent delay-lock loop with MSK spreading	$1.26 K_1 P$	$\dfrac{N_0}{4P}(1.44) + \dfrac{B_N N_0^2}{2P^2}(0.63)$	$\sigma_\delta^2 = \dfrac{1}{2\rho_L}\left(1.44 + \dfrac{1.26}{\rho_{IF}}\right)$

hardware. The analysis of Ref. 19 accounts for all bandpass filtering operations in detail. Another effort to solve the DLL gain-balance problem [21] resulted in the invention of a tracking loop which generates the phase error signal using a product operation rather than the DLL sum operation. The noise performance of this loop is equivalent to that of the DLL and the gain-balance problem has been eliminated. More recently, Gaudenzi and Luise [22] have proposed a coherent delay-lock loop configuration which is claimed to solve the gain balance problem and to have, by virtue of its coherent processing, performance better than the Yost and Boyd modified loop.

This concludes the discussion of code tracking loops for spread-spectrum systems. The references provide a large selection of material for further reading.

References

[1] J. J. SPILKER and D. T. MAGILL, "The Delay-Lock Discriminator: An Optimum Tracking Device," *Proc. IRE,* Vol. 49, pp. 1403–1416, September 1961.

[2] J. J. SPILKER, "Delay-Lock Tracking of Binary Signals," *IEEE Trans. Space Electron. Telem.,* Vol. Set-9, pp. 1–8, March 1963.

[3] W. J. GILL, "A Comparison of Binary Delay-Lock Tracking Loop Implementations," *IEEE Trans. Aerosp. Electron. Syst.,* Vol. AES-2, pp. 415–424, July 1966.

[4] R. B. WARD, "Digital Communications on a Pseudonoise Tracking Link Using Sequence Inversion Modulation." *IEEE Trans. Commun. Technol.,* Vol. CT-15, pp. 69–78, February 1967.

[5] H. P. HARTMAN, "Analysis of a Dithering Loop for PN Code Tracking," *IEEE Trans. Aerosp. Electron. Syst.,* Vol. AES-10, pp. 2–9, January 1974.

[6] M. K. SIMON, "Noncoherent Pseudonoise Code Tracking Performance of Spread Spectrum Receivers," *IEEE Trans. Commun.,* Vol. COM-25, pp. 327–345, March 1977.

[7] J. J. SPILKER, *Digital Communications by Satellite* (Englewood Cliffs, N.J.: Prentice Hall, 1977).

[8] J. K. HOLMES, *Coherent Spread Spectrum Systems* (New York: Wiley-Interscience. 1982).

[9] R. C. DIXON, *Spread Spectrum Systems* (New York: Wiley-Interscience, 1976).

[10] M. K. SIMON, J. K. OMURA, R. A. SCHOLTZ, and B. K. LEVITT, *Spread Spectrum Communications Handbook* (New York: McGraw-Hill, 1994).

[11] R. E. ZIEMER and R. L. PETERSON, *Digital Communications and Spread Spectrum Systems* (New York: Macmillan, 1985).

[12] F. M. GARDNER, *Phaselock Techniques,* 2nd ed. (New York: Wiley, 1979).

[13] A. J. VITERBI, *Principles of Coherent Communications* (New York: McGraw-Hill, 1966).

[14] W. C. LINDSEY and M. K. SIMON, *Telecommunication Systems Engineering,* (Englewood Cliffs, N.J.: Prentice Hall, 1973).

[15] A. PAPOULIS, *Probability Random Variables and Stochastic Processes* 3rd ed. (New York: McGraw-Hill, 1991).

[16] P. M. HOPKINS, "Double Dither Loop for Pseudonoise Code Tracking," *IEEE Trans. Aerosp. Electron. Syst.,* Vol. AES-13, pp. 644–650, November 1977.

[17] S. STEIN and J. JONES, *Modern Communications Principles* (New York: McGraw-Hill, 1967).

[18] R. L. PICKHOLTZ, D. L. SCHILLING, and L. B. MILSTEIN, "Theory of Spread Spectrum Communications: A Tutorial," *IEEE Trans. Commun.,* Vol. COM-30, pp. 855–844, May 1982.

[19] R. A. YOST and R. W. BOYD, "A Modified PN Code Tracking Loop: Its Performance and Implementation Sensitivities," *Proc. Natl. Telecommun. Conf.,* December 1980.

[20] R. A. YOST and R. W. BOYD, "A Modified PN Code Tracking Loop: Its Performance Analysis and Comparative Evaluation," *IEEE Trans. Commun.,* Vol. COM-30, pp. 1027–1036, May 1982.

[21] DAVID T. LaFLAME, "A Delay-Lock Loop Implementation Which Is Insensitive to Arm Gain Imbalance," *IEEE Trans. Commun.,* Vol. COM-27, pp. 1632–1633, October 1979.

[22] R. DE GAUDENZI and M. LUISE, "Decision-Directed Coherent Delay-Lock Tracking Loop for DS-Spread-Spectrum Signals," *IEEE Trans. Commun.,* Vol. COM-39, pp. 758–765, May 1991.

Problems

(4-1) Calculate the variance of the code tracking error for a noncoherent delay-lock code tracking loop in a system that utilizes offset-QPSK spreading modulation. Express the result as a function of the signal-to-noise ratios in the loop bandwidth and in the IF filter bandwidth.

(4-2) Consider a spread-spectrum communications system which is used between two platforms that move relative to one another. Suppose that BPSK spreading modulation is used and that the tracking loop is tracking the correct code phase. At time $t = 0$ the platforms begin accelerating at a rate of $1.5g$ relative to one another. This acceleration continues until the relative velocity is 3 mach, at which time the acceleration ceases. Derive a complete expression for the normalized tracking error as a function of time with tracking loop bandwidth as a parameter. Assume that the carrier frequency is 10 GHz, the spreading code rate is 100 MHz, and that the tracking loop is second order and critically damped.

(4-3) Derive an expression for the variance of the tracking jitter for a system using BPSK spreading modulation and a tau-dither noncoherent delay-lock tracking loop using a dither frequency $f_q \ll B_N$. Assume that $\Delta = 1.0$. Compare the result with the result derived in the text for $f_q = B_N/4$.

(4-4) Consider a spread-spectrum communication system using a noncoherent delay-lock code tracking loop having a noise bandwidth of 1 kHz and an IF filter noise bandwidth of 100 kHz. Plot the variance of the tracking jitter as a function of loop signal-to-noise ratio for $\Delta = 1.0$, 0.5, and 0.1.

(4-5) A tone jammer is being used to attempt to disrupt communication over a link which employs a tau-dither tracking loop with $f_q = B_N/4$ and $\Delta = 1.0$. Assume that the spreading code rate is 100 MHz, the IF filter noise bandwidth is 100 kHz, and that the tracking loop two-sided noise bandwidth is 1 kHz. Derive an expression for the variance of the tracking error as a function of the ratio of received signal power to received jammer power.

(4-6) Calculate the sensitivity of the loop bandwidth and damping to changes in received power level for a second-order noncoherent delay-lock tracking loop and a second-order tau-dither tracking loop. Assume that the design point damping is $\xi = 1/\sqrt{2}$.

(4-7) A system designer has a choice between BPSK and MSK spreading modulation. The available transmission bandwidth and the data modulation bandwidth are fixed. What is the best modulation for a system that must achieve minimum absolute tracking error variance?

(4-8) Plot the normalized discriminator S-curve for a noncoherent delay-lock tracking loop using MSK spreading modulation for $\Delta = 0.5$, 1.0, 1.5, and 2.0.

(4-9) Consider a received signal

$$r(t) = \sqrt{2P} \cos[\omega_0 t + \theta(t)] + \sqrt{2} n_I(t) \cos \omega_0 t - \sqrt{2} n_Q(t) \sin \omega_0 t$$

where $n_I(t)$ and $n_Q(t)$ are independent lowpass white Gaussian noise processes with two-sided bandwidth B_N and two-sided noise power spectral density $N_0/2$. Derive an alternative representation for the noise using $\cos[\omega_0 t + \theta(t)]$ and $\sin[\omega_0 t + \theta(t)]$ as quadrature carriers and determine what conditions are necessary for $n_I'(t)$ and $n_Q'(t)$ of this representation to be independent.

(4-10) Define the sum of two random processes $y_1(t)$ and $y_2(t)$ by

$$z(t) = y_1(t) + y_2(t)$$

Derive an expression for the power spectrum of $z(t)$ as a function of the individual and cross power spectra of $y_1(t)$ and $y_2(t)$.

(4-11) Consider a linear second-order tracking loop as shown in Figure 4-8 with a loop filter defined by

$$F(s) = \frac{1 + \tau_2 s}{1 + \tau_1 s}$$

Derive expressions for the dynamic tracking error when the input is defined by **(a)** $T_d(t)/T_c = AU(t)$; **(b)** $T_d(t)/T_c = BtU(t)$; **(c)** $T_d(t)/T_c = Ct^2U(t)$. In these expressions, $U(t)$ is the unit step function.

(4-12) Repeat Problem 4-11 for

$$F(s) = \frac{1 + \tau_2 s}{\tau_1 s}$$

(4-13) A BPSK transmitter is moving at a constant velocity V_r relative to its intended receiver. The transmitted carrier frequency is f_t and the transmitted bit rate is R_t. The receiver carrier synchronization and bit synchronization loops operate perfectly. What is the receiver output bit rate R_r? Prove your result.

(4-14) The figure below illustrates three different mechanizations for a second-order loop filter.
(a) Derive the filter transfer function $F(s)$ for each filter shown.
(b) Derive the closed-loop transfer function for each filter shown.
Assume that the operational amplifier has infinite input impedance, zero output impedance, but finite gain.

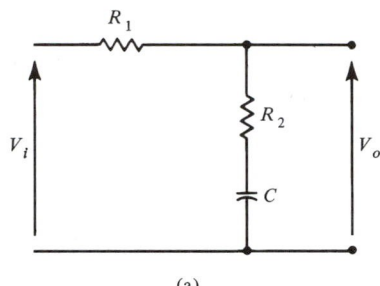

(a)

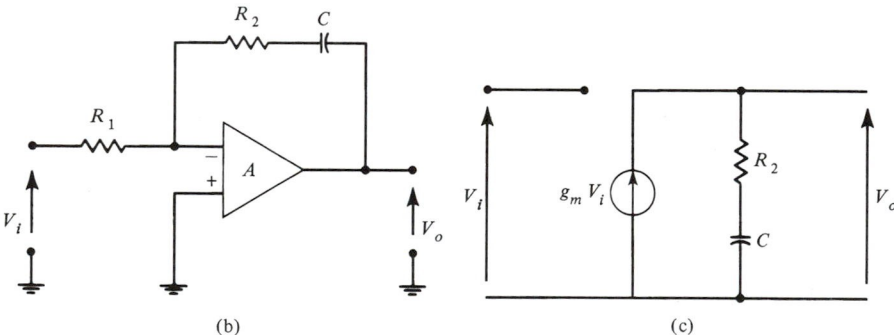

(b)   (c)

PROBLEM 4-14.   Loop filters for second-order tracking loops.

(4-15) Two bandpass signals $x_1(t)$ and $x_2(t)$ have complex envelopes $\tilde{x}_1(t)$ and $\tilde{x}_2(t)$ and different carrier frequencies f_1 and f_2. Let $y(t) = x_1(t)x_2(t)$. Prove that the complex envelope of the difference frequency component is

$$\tilde{y}_d(t) = \tfrac{1}{2}\tilde{x}_1(t)\tilde{x}_2^*(t)$$

and that the complex envelope of the sum frequency component is

$$\tilde{y}_s(t) = \tfrac{1}{2}\tilde{x}_1(t)\tilde{x}_2(t)$$

(4-16) Consider the *RLC* bandpass filter illustrated below. Derive the impulse response of the equivalent lowpass filter under the assumption of high-Q components.

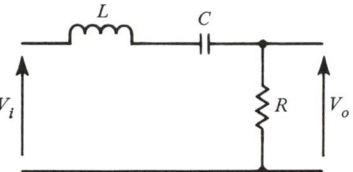

PROBLEM 4-16. *RLC* bandpass filter.

(4-17) Reconsider the bandpass filter of Problem 4-16. The input to this circuit is a rectangular pulse of a carrier at frequency f, whose complex envelope is

$$p_T(t) = \begin{cases} A & 0 < t < T \\ 0 & \text{elsewhere} \end{cases}$$

Derive and plot an expression for the filter output.

(4-18) The receiver front end illustrated in the figure below is used in a spread-spectrum communication system. At point 1, the received signal power is -116.3 dBm and the received noise single-sided noise power spectral density is -174 dBm per hertz.
  (a) Using the component gains and noise figures shown, calculate the value of E_b/N_0 at the input to the delay-lock discriminator. Assume a data rate of 10 kbps.
  (b) Assume a filter noise bandwidth of 10 MHz and calculate the signal-to-noise ratio P_s/P_N at the input to the delay-lock discriminator.

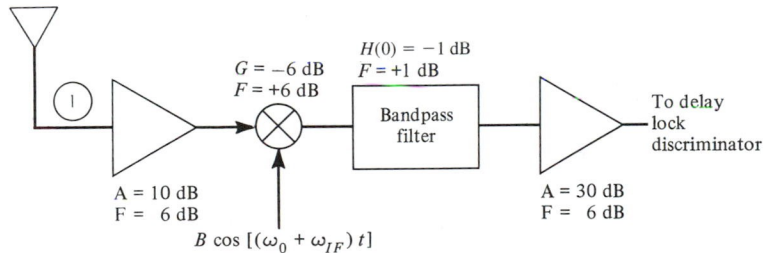

PROBLEM 4-18. Typical receiver front end.

(4-19) Consider a noncoherent delay-lock code tracking loop designed so that the ratio of the signal power to thermal noise power at the IF bandpass filter output is $+10$ dB. When spreading code self-noise is taken into account, calculate the additional signal power required to achieve $+10$ dB SNR as a function of the ratio of spreading code chip rate to IF filter bandwidth. Assume that $\Delta = 1.0$ and that the data modulation spectrum is uniform over the IF filter bandwidth.

(4-20) In actual spread-spectrum systems, the squaring device in the energy detectors of the phase discriminators are usually crystal detectors characterized by their sensitivity in units of millivolts of output per milliwatt of input. These devices can be modeled by an ideal squaring device either preceded by or followed by an ideal amplifier. Calculate the gain of the amplifier in both cases as a function of the detector sensitivity K.

(4-21) A BPSK spread-spectrum system is to be used for determining the range between a satellite whose position is accurately known and a vehicle on the surface of the earth. This range is measured by measuring the propagation time of the spread spectrum signal using the spreading code epoch as a time reference. Select a spreading code, spreading code rate, tracking loop configuration, and estimate the required signal-to-noise ratio in the tracking loop bandwidth to achieve a variance of 3 m for the range measurement. The system should be designed to have a minimum range ambiguity of 10,000 km.

5

Initial Synchronization of the Receiver Spreading Code

5-1 Introduction

The analysis of the code tracking loops in Chapter 4 presumed that the received spreading waveform and receiver-generated replica of the spreading waveform are initially synchronized in both phase and frequency. This initial synchronization of the spreading waveforms is a significant problem in spread-spectrum system design and is the subject of the present chapter. Phase/frequency synchronization is difficult because typical spreading waveform periods are long and bandwidths are large. Thus uncertainty in the estimated propagation delay $\hat{T}_d$ translates into a large number of symbols of code phase uncertainty. Oscillator instabilities and doppler frequency shifts result in frequency uncertainties which must also be resolved. The correct code phase/frequency must be found quickly using the minimum amount of hardware possible. In many cases this process must be accomplished at very low signal-to-noise ratios or in the presence of jamming.

A widely used technique for initial synchronization is to search serially through all potential code phases and frequencies until the correct phase and frequency are identified. Each reference phase/frequency is evaluated by attempting to despread the received signal. If the estimated code phase and frequency are correct, despreading will occur and will be sensed. If the estimated code phase or frequency is incorrect, the received signal will not be despread, and the reference waveform will be stepped to a new phase/frequency for evaluation. This technique is called *serial search*. Because of the widespread use of serial-search techniques, a large fraction of this chapter is devoted to the analysis of this technique. Specifically, strategies for rapidly evaluating the correctness of the trial phase/frequency are discussed in detail. In addition, a method for calculating mean synchronization time is given. This calculation requires knowledge of the probability that the detector indicates that the trial phase/frequency is correct when, in fact, it is not (i.e., the probability of false alarm P_{fa}) and the probability that the detector indicates that the trial phase/frequency is correct when it is indeed correct (i.e., the probability of correct detection P_d) for various methods of detecting the presence of a signal in noise. Both the probability of false alarm and the probability of correct detection are a function of evaluation (integration) time and signal-to-noise ratio. Methods of calculating P_d and P_{fa} are described. The serial-search signal detectors discussed in this book are all noncoherent detectors and are therefore not sensitive to received carrier phase.

Coherent detectors are, of course, possible but are of limited application since coherent carrier recovery is not typically accomplished until after code synchronization. Serial-search acquisition schemes have been referred to in the literature [1–3] as *low-decision-rate* detectors since a large number of spreading code symbols must typically be received to make a correct/incorrect decision on the hypothesized reference spreading code phase/frequency.

A highly efficient method of initial synchronization is to matched filter detect the received signal. The matched filter is designed to output a pulse when a particular sequence of code symbols is received. When this pulse is sensed, the receiver code generator is started using an initial condition corresponding to the received code phase and synchronization is complete. This technique requires matched filters with extremely large time–bandwidth products. In the usual case where the spreading code period is long, the matched filter must also be easily programmable. For example, the spreading code period may be longer than the entire mission being carried out. Thus a matched filter designed for a single sequence of code symbols may never receive these symbols. A programmable matched filter can be programmmed for a particular code phase which is dependent on an estimate of the propagation delay. The requirement for programmability has led researchers to investigate high-speed digital correlators as well as programmable convolvers for this application.

Consider a direct-sequence spread-spectrum system employing BPSK spreading. When the spreading code is an *m*-sequence that is generated using the generator of Figure 3-6 a convenient method of synchronization is to demodulate the received symbol stream directly and to load the shift register generator with this symbol stream. At high signal-to-noise ratios is it likely that the demodulated symbols are correct so that the shift register initial condition is also correct. At reduced signal-to-noise ratios, the shift-register load may not be correct and additional loads will have to be attempted. Each load is evaluated by attempting to despread the received signal just as was described for serial search techniques. This technique is called *rapid acquisition by sequential estimation* (RASE) and was first described by Ward [4]. A slightly modified version of this technique performs a preliminary evaluation of the shift-register load by comparing its output with several received symbols prior to attempting the full correlation despreading evaluation. This modified technique is called *recursion-aided RASE* (RARASE) and is described in Ward and Yiu [5]. Both of these techniques are described in this chapter.

The number of code phases and frequencies that must be evaluated to obtain initial synchronization is proportional to propagation delay uncertainty expressed in spreading code chips and the relative dynamics of the transmitter and receiver. Since the code chip duration is inversely proportional to the chip rate, synchronization time is also directly proportional to the clock rate used for the spreading code generators. Since a frequency-hop spread-spectrum system may employ a clock rate which is much lower than the transmission bandwidth, synchronization time for frequency-hop systems is typically much lower than for direct-sequence systems having the same transmission bandwidth. This factor is a principal reason why frequency hopping has been selected for some current spread-spectrum systems. Initial synchronization techniques for frequency-hop spread-spectrum systems are the same as those used for direct-sequence systems except that RASE techniques are not applicable.

The techniques described in this chapter are generally capable of determining the received spreading code phase to within an accuracy of $\pm\frac{1}{2}$ to $\pm\frac{1}{4}$ of a chip. When the code tracking loop is closed, there may therefore be a phase error of $\pm\frac{1}{2}$ chip which the tracking loop is expected to eliminate. This transition from the completion of the initial synchronization function to fine code tracking is called *tracking loop pull-in* and is important because it affects the required loop bandwidth selection. The study of the tracking loop pull-in characteristic is a nonlinear analysis problem. A method for evaluating the loop pull-in trajectories is discussed.

5-2 Problem Definition and the Optimum Synchronizer

In general, both the phase and frequency of the received spread-spectrum signal will be unknown to the receiver. In this chapter it will be assumed that a priori information is available to the receiver which bounds the phase uncertainty to a range of ΔT seconds and the frequency uncertainty to a range $\Delta\Omega$ rad/s. The method almost universally employed to evaluate the receiver reference phase and frequency is to attempt to despread the received signal using that phase and frequency. An energy detector at the despreader output measures the signal plus noise energy in a narrow bandwidth at a known frequency. If the phase and frequency of the receiver-generated replica of the spreading waveform are correct, the received signal will be collapsed in bandwidth, translated to the center frequency of the bandpass filter, and the energy detector will detect the presence of signal. Figure 5-1 is a simplified conceptual block diagram of the functions needed to evaluate a single reference phase and frequency. The control logic shown in dashed lines may be added to select values of $\hat{T}_d$ and $\hat{\omega}$ for evaluation.

The system of Figure 5-1 will yield a "correct phase/frequency" or "hit" decision over a range of phase/frequency values near the correct values. The size of this

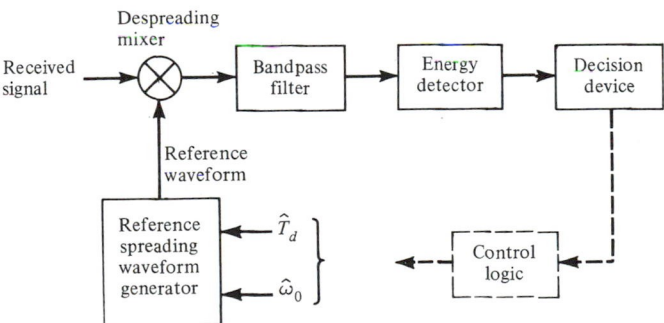

FIGURE 5-1. System used to evaluate a single spreading waveform phase and frequency.

range is calculated from the spreading waveform autocorrelation function and the bandwidth of the IF filter. For example, suppose that the received signal $r(t)$ is a carrier which is BPSK modulated by an m-sequence waveform $c(t - T_d)$ plus AWGN. Thus

$$r(t) = c(t - T_d) \cos(\omega_0 t + \phi) + n(t) \tag{5-1}$$

where ω_0 is the carrier frequency and ϕ is the random received carrier phase. The reference waveform is

$$a(t) = c(t - \hat{T}_d) \cos[(\hat{\omega}_0 + \omega_{IF})t] \tag{5-2}$$

and the despreader output difference frequency term is

$$x(t) = c(t - T_d)c(t - \hat{T}_d) \cos[\omega_{IF}t + (\hat{\omega}_0 - \omega_0)t - \phi] + n'(t) \tag{5-3}$$

The power spectrum of $c(t - T_d)c(t - \hat{T}_d)$ is given in (3-66) and plotted in Figure 3-16 for m-sequence spreading codes. This power spectrum contains an impulse at zero frequency which corresponds to a single spectral line at $\omega = \omega_{IF} + \hat{\omega}_0 - \omega_0$ at the output of the despreader. The power in this component is $R_c^2(\tau)$, where $R_c(\tau)$ is the autocorrelation of the spreading waveform and $\tau = \hat{T}_d - T_d$. The power in this component will be sensed by the energy detector if τ is sufficiently small so that $R_c(\tau)$ is near unity and if $\hat{\omega}_0 - \omega_0$ is not so large that the desired component will lie outside the passband of the bandpass filter. The phase/frequency uncertainty region may be graphically depicted as a rectangle with dimensions $\Delta\Omega \times \Delta T$ as in Figure 5-2. This rectangle can be subdivided into smaller rectangles whose dimensions Δt and $\Delta\omega$ are the range around the correct phase/frequency over which the system of Figure 5-1 will yield a "hit." Thus a single test in each cell of Figure 5-2 will be sufficient to determine the correct received phase and frequency.

An initial synchronization system which is optimum in the sense that it achieves synchronization with a given probability in the minimum possible time is one that evaluates all cells of Figure 5-2 simultaneously. This system requires a subsystem similar to that shown in Figure 5-1 for every phase/frequency cell of Figure 5-2 and is thus not optimum in a minimum-hardware sense. This minimum-acquisition-time

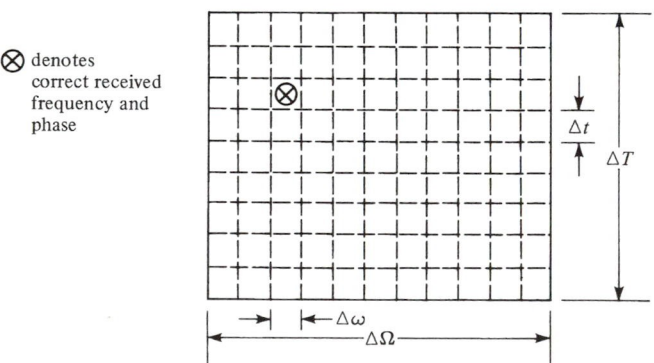

FIGURE 5-2. Phase/frequency uncertainty region.

system is rarely implemented because of its hardware complexity. A direct-sequence code acquisition system which uses digital processing technology to search all cells of the uncertainty region simultaneously has been reported in Ref. 6. This system correlates the received signal with all phases and Doppler shifts of a reference signal simultaneously. The correlator outputs are evaluated and the received code phase/frequency is estimated to be the reference code phase/frequency which generated the maximum correlation value. This acquisition strategy therefore produces a maximum-likelihood estimate of the received code phase/frequency. All the systems discussed in this chapter are designed to achieve a compromise between acquisition time and reasonable complexity without compromising jamming resistance or any other important system characteristic. Synchronization time is proportional to the total number of cells that must be evaluated and it makes no difference whether these cells were all generated from frequency uncertainty or all from phase uncertainty or a combination. Thus, in some of the following analysis, frequency uncertainty will be ignored.

5-3 Serial-Search Synchronization Techniques

Serial-search techniques are by far the most commonly used spread-spectrum synchronization techniques. Any synchronization system that evaluates the phase/frequency cells of Figure 5-2 serially (i.e., one after another) until the correct cell is found is said to use serial search. In this section the mean and variance of the synchronization time are calculated for a very general serial-search system. The result is simple and the system designer will be able to use it to make design trade-off studies to minimize mean synchronization time. Inputs to the mean synchronization time result are the probability of detection when the correct cell is being evaluated. P_d, and the probability of false alarm when an incorrect cell is being evaluated, P_{fa}, as a function of the integration time T_i and SNR. The calculation of P_d and P_{fa} is also discussed in detail.

5-3.1 Calculation of the Mean and Variance of Synchronization Time

For simplicity assume that no frequency uncertainty exists and that the correct phase is uniformly distributed over the region ΔT. Based on the spreading waveform autocorrelation function, a phase step size of Δt seconds has been chosen. For convenience assume that $\Delta T/\Delta t$ is an integer C. Because it is equally probable that the correct phase is in any cell, the serial search can begin anywhere. Let the search begin at one boundary of the uncertainty region. The search will advance through one cell at a time until C cells have been evaluated. If synchronization has not been achieved at that time, a retrace will start the search over again.

The mean synchronization time is calculated in a straightforward manner by considering all possible sequences of events leading to a correct synchronization. An event in the probability space being considered is defined by a particular location n for the correct phase cell, a particular number of missed detections j of the correct phase cell, and a particular number of false alarms k in all incorrect phase cells evaluated. The total synchronization time for a particular event defined by (n, j, k) is

$$T(n, j, k) = nT_i + jCT_i + kT_{\text{fa}} \tag{5-4}$$

where T_i is the (fixed) integration time for evaluation of each cell and T_{fa} is the time required to reject an incorrect cell when a false alram occurs. The false-alarm penalty T_{fa} may be many times larger than T_i, so that false alarms are undesirable events. The total number of cells evaluated for this event is $(n + jC)$, the total number of correct cells is $(j + 1)$, and the total number of incorrect cells is $(n + jC - j - 1)$. The probability of the correct cell being the nth cell is $1/C$ and the probability of j missed detections followed by a correct detection is $P_d(1 - P_d)^j$. The k false alarms can occur in any order within the $(n + jC - j - 1) \triangleq K$ incorrect cells. The probability of a particular ordering is $P_{\text{fa}}^k(1 - P_{\text{fa}})^{K-k}$, and there are $\binom{K}{k}$ orderings of k false alarms in K cells. Combining all of the above, the probability of the event (n, j, k) is

$$\Pr (n, j, k) = \frac{1}{C} P_d (1 - P_d)^j \binom{K}{k} P_{\text{fa}}^k (1 - P_{\text{fa}})^{K-k} \tag{5-5}$$

The mean synchronization time is

$$\overline{T}_s = \sum_{n, j, k} T(n, j, k) \Pr (n, j, k) \tag{5-6}$$

The correct cell number can range over $(1, C)$, the number of missed detections can range from zero to infinity, and the number of false alarms ranges over $(0, K)$. Thus the mean synchronization time is

$$\overline{T}_s = \frac{1}{C} \sum_{n=1}^{C} \sum_{j=0}^{\infty} \sum_{k=0}^{K} [(n + jC)T_i + kT_{\text{fa}}] \binom{K}{k} P_{\text{fa}}^k (1 - P_{\text{fa}})^{K-k} P_d (1 - P_d)^j$$

$$= \frac{1}{C} \sum_{n=1}^{C} \sum_{j=0}^{\infty} (n + jC)T_i \left[\sum_{k=0}^{K} \binom{K}{k} P_{\text{fa}}^k (1 - P_{\text{fa}})^{K-k} \right] P_d (1 - P_d)^j \tag{5-7}$$

$$+ \frac{1}{C} \sum_{n=1}^{C} \sum_{j=0}^{\infty} T_{\text{fa}} \left[\sum_{k=0}^{K} \binom{K}{k} k P_{\text{fa}}^k (1 - P_{\text{fa}})^{K-k} \right] P_d (1 - P_d)^j$$

The first summation over k is evaluated using the identity

$$\sum_{k=0}^{K} \binom{K}{k} a^k b^{K-k} = \binom{K}{0} b^K + \binom{K}{1} ab^{K-1} + \cdots + \binom{K}{K} a^K$$

$$= (b + a)^K$$

With $a = P_{\text{fa}}$ and $b = 1 - P_{\text{fa}}$, $b + a = 1.0$ and the first summation over k equals unity. The second sum over k is the mean of a discrete random variable which has a binomial distribution. The mean value is KP_{fa} [7]. Thus (5-7) simplifies to

$$\overline{T}_s = \frac{1}{C} \sum_{n=1}^{C} \sum_{j=0}^{\infty} [(n + jC)T_i + KT_{\text{fa}}P_{\text{fa}}]P_d(1 - P_d)^j \tag{5-8}$$

Expanding this expression using $K = n + jC - j - 1$ and simplifying yields

$$\overline{T}_s = \frac{1}{C} \sum_{n=1}^{C} \sum_{j=0}^{\infty} [(n - 1)T_{\text{da}} + (j + 1)T_i + j(C - 1)T_{\text{da}}]P_d(1 - P_d)^j \tag{5-9}$$

where

$$T_{\text{da}} = T_i + T_{\text{fa}}P_{\text{fa}} \tag{5-10}$$

is the average dwell time at an incorrect phase cell. Equation (5-9) can be evaluated completely using the identities [8]

$$\sum_{i=1}^{L} i = L\left(\frac{L + 1}{2}\right) \tag{5-11a}$$

$$\sum_{i=0}^{\infty} m^i = \frac{1}{1 - m} \tag{5-11b}$$

$$\sum_{i=0}^{\infty} im^i = \frac{m}{(1 - m)^2} \tag{5-11c}$$

After some straightforward algebraic manipulation, the result is

$$\overline{T}_s = (C - 1)T_{\text{da}}\left(\frac{2 - P_d}{2P_d}\right) + \frac{T_i}{P_d} \tag{5-12}$$

This expression for mean synchronization time is a function of P_{fa} through the definition of T_{da}. This result can also be derived using flow graph techniques [1,9–11]. Although generating this particular result is considerably more difficult using flow graph techniques, the flow graph technique is more powerful in that, among other things, it may be generalized to include the case of a nonuniform probability distribution for the received code phase n. The result of (5-12) is identical to Eq. 1.3 in Ref. 1 with the exception that in Ref. 1 the false-alarm penalty time has been limited to be K times the initial integration time.

Synchronization time for any serial search strategy is a random variable. The mean of this random variable has been calculated above. The second moment or variance will now be calculated. Recall that the variance of a random variable x with probability distribution $p_x(\alpha)$ is given by

$$\sigma_x^2 = E[(x - \overline{x})^2] = E[x^2] - E^2[x] \tag{5-13}$$

Since the mean of the synchronization time has already been calculated, the variance will be known if the mean-square value can be calculated. The synchronization time for a received phase which lies in the nth cell and which is detected after j missed detections and k false alarms is given by (5-4). The mean-square value is

$$\overline{T_s^2} = \sum_{n,j,k} T^2(n,j,k) \Pr\,[n,j,k] \tag{5-14}$$

This expression simplifies by recognizing that the sums over k are the moments of a binomially distributed random variable. The first sum is [7]

$$\overline{T_s^2} = \frac{1}{C} \sum_{n=1}^{C} \sum_{j=0}^{\infty} \sum_{k=0}^{K} [(n+jC)T_i + kT_{\text{fa}}]^2 \binom{K}{k} P_{\text{fa}}^k (1-P_{\text{fa}})^{K-k} P_d (1-P_d)^j \tag{5-15}$$

Expanding the square and grouping like powers of k results in

$$\overline{T_s^2} = \frac{P_d}{C} \sum_{n=1}^{C} \sum_{j=0}^{\infty} \sum_{k=0}^{K} (A_2 k^2 + A_1 k + A_0) \binom{K}{k} P_{\text{fa}}^k (1-P_{\text{fa}})^{K-k} (1-P_d)^j \tag{5-16}$$

where $A_2 = T_{\text{fa}}^2$
$\quad\quad A_1 = 2(n+jC)T_i T_{\text{fa}}$
$\quad\quad A_0 = (n+jC)^2 T_i^2$

This expression simplifies by recognizing that the sums over k are the moments of a binomially distributed random variable. The first sum is [7]

$$\sum_{k=0}^{K} A_2 k^2 \binom{K}{k} P_{\text{fa}}^k (1-P_{\text{fa}})^{K-k} = A_2 [K^2 P_{\text{fa}}^2 + K P_{\text{fa}} (1-P_{\text{fa}})] \tag{5-17}$$

and the other two terms were evaluated above. Thus (5-16) simplifies to

$$\overline{T_s^2} = \frac{P_d}{C} \sum_{n=1}^{C} \sum_{j=0}^{\infty} \{A_2 [K^2 P_{\text{fa}}^2 + K P_{\text{fa}} (1-P_{\text{fa}})] + A_1 [K P_{\text{fa}}] + A_0\}(1-P_d)^j \tag{5-18}$$

This expression is evaluated by grouping like powers of j after substituting $K = n + jC - j - 1$, which results in

$$\overline{T_s^2} = \frac{P_d}{C} \sum_{n=1}^{C} \sum_{j=0}^{\infty} (B_2 j^2 + B_1 j + B_0)(1-P_d)^j \tag{5-19}$$

where $B_2 = C^2 T_i^2 + 2C(C-1)T_i T_{\text{fa}} P_{\text{fa}} + T_{\text{fa}}^2 P_{\text{fa}}^2 (C-1)^2$
$\quad\quad B_1 = 2nC T_i^2 + 2(2Cn - n - C)T_i T_{\text{fa}} P_{\text{fa}}$
$\quad\quad\quad\quad + 2T_{\text{fa}}^2 P_{\text{fa}}^2 (n-1)(C-1) + T_{\text{fa}}^2 (C-1) P_{\text{fa}} (1-P_{\text{fa}})$
$\quad\quad B_0 = n^2 T_i^2 + 2n(n-1)T_i T_{\text{fa}} P_{\text{fa}} + T_{\text{fa}}^2 (n-1)^2 P_{\text{fa}}^2$
$\quad\quad\quad\quad + (n-1)P_{\text{fa}}(1-P_{\text{fa}})T_{\text{fa}}^2$

The summations over j can be evaluated using (5-11) together with [8]

$$\sum_{i=0}^{\infty} i^2 m^i = \frac{m(1+m)}{(1-m)^3} \tag{5-11d}$$

The result is

$$\overline{T_s^2} = \frac{1}{C} \sum_{n=1}^{C} \left[B_2 \frac{(1 - P_d)(2 - P_d)}{P_d^2} + B_1 \frac{1 - P_d}{P_d} + B_0 \right] \tag{5-20}$$

The summation over n is evaluated by grouping like powers of n, which yields

$$\overline{T_s^2} = \frac{1}{C} \sum_{n=1}^{C} (D_2 n^2 + D_1 n + D_0) \tag{5-21}$$

where $D_2 = (T_i + T_{fa} P_{fa})^2$

$$D_1 = -2 T_i T_{fa} P_{fa} - 2 T_{fa}^2 P_{fa}^2 + T_{fa}^2 P_{fa}(1 - P_{fa})$$
$$+ [2 C T_i^2 + 2(2C - 1) T_i T_{fa} P_{fa} + 2(C - 1) T_{fa}^2 P_{fa}^2] \left(\frac{1 - P_d}{P_d} \right)$$

$$D_0 = T_{fa}^2 P_{fa}^2 - T_{fa}^2 P_{fa}(1 - P_{fa})$$
$$+ [-2 C T_i T_{fa} P_{fa} - 2(C - 1) T_{fa}^2 P_{fa}^2$$
$$+ (C - 1) T_{fa}^2 P_{fa}^2 (1 - P_{fa})] \left(\frac{1 - P_d}{P_d} \right)$$
$$+ [C^2 T_i^2 2C(C - 1) T_i T_{fa} P_{fa} + (C - 1)^2 T_{fa}^2 P_{fa}^2] \frac{(1 - P_d)(2 - P_d)}{P_d^2}$$

The summations over n are evaluated using [8]

$$\sum_{n=1}^{C} n = \frac{C(C + 1)}{2} \tag{5-22a}$$

$$\sum_{n=1}^{C} n^2 = \frac{C(C + 1)(2C + 1)}{6} \tag{5-22b}$$

The remainder of the calculation is simply a bookkeeping exercise. Using (5-12) to obtain the square of the mean value of T_s, the final result is

$$\sigma_{T_s}^2 = \overline{T_s^2} - (\overline{T_s})^2$$
$$= \left[\frac{C^2 - 1}{12} - \frac{(C - 1)^2}{P_d} + \frac{(C - 1)^2}{P_d^2} \right] T_{da}^2$$
$$+ (2C - 1) \frac{1 - P_d}{P_d^2} T_i^2 + 2(C - 1) \frac{1 - P_d}{P_d^2} T_i T_{fa} P_{fa} \tag{5-23}$$
$$- (C - 1) \frac{2 - P_d}{2 P_d} T_{fa}^2 P_{fa}^2 + (C - 1) \frac{2 - P_d}{2 P_d} T_{fa}^2 P_{fa}$$

This result is similar to the result derived in Holmes and Chen [9] using signal flow graph techniques. For $C \gg 1$, $1 - P_d \ll 1$, and $P_{fa} \ll 1.0$, the variance is approximated by

$$\sigma_{T_s}^2 \approx T_{\text{da}}^2 C^2 \left(\frac{1}{12} - \frac{1}{P_d} + \frac{1}{P_d^2} \right) \qquad (5\text{-}24)$$

which agrees exactly with Holmes and Chen [9] and Simon et al. [1, Eq. 1.7]. For $P_d = 1.0$, $P_{\text{fa}} = 0.0$, and $C \gg 1$, the variance is

$$\sigma_{T_s}^2 = T_{\text{da}}^2 C^2 (1/12) \qquad (5\text{-}25)$$

which is equal to the variance of a random variable which is uniformly distributed over the range $(0, CT_{\text{da}})$. The variance increases as P_d falls below unity.

One of the goals of the spread-spectrum system designer will be to design a synchronization system that minimizes mean synchronization time. Equation (5-12) indicates that mean synchronization is a function of P_d, P_{fa}, T_i, T_{fa}, and C. The designer has some degree of control over all these variables, including C. Even though the phase uncertainty region in seconds is fixed by system requirements, the region can be subdivided into any number of cells C by the system designer. The remaining four variables are not independent of one another and therein lies the difficulty in the design. It will be seen quantitatively later that high P_d together with low P_{fa} implies large T_i. Thus there will be an optimum set of P_d, P_{fa}, T_i, T_{fa} which minimizes mean synchronization time. It is not correct to assume that minimum average synchronization time will always be achieved with $1 - P_d \ll 1.0$, so that the correct phase cell is detected on the first sweep. In some cases the selection of a moderate P_d will result in a much lower T_i than a high P_d, and will thus result in reduced average T_s, even though several sweeps of the uncertainty region may be required. In some instances the system designer is interested not only in the mean and variance of acquisition time but also the details of the probability density or the cumulative probability density of the acquisition time. The cumulative probability density of the acquisition time has been calculated for the fixed integration time serial search strategy by DiCarlo [12]. Another analysis of the probability density of acquisition time is presented in Ref. 13. The analysis technique used in Ref. 12 is different from that presented in Ref. 13, so that both works may be of interest to the reader. The results of these references will enable the system designer to evaluate the probability that the system has completed the acquisition process within a specified time. The results of Ref. 13 are more general than those of Ref. 12 since Ref. 13 considers an arbitrary probability distribution for the received code phase. In contrast, Ref. 12 considers only a uniform distribution for the received code phase.

5-3.2 Modified Sweep Strategies

All the results up to this point have been derived assuming that the received spreading waveform phase is uniformly distributed over a particular uncertainty region. Suppose now that the received phase distribution is defined by $p(n)$, the probability that the received phase is within the nth phase cell. When $p(n)$ is known, the sweep strategy should be modified to search the most likely phase cells first and then the less likely cells. For example, if the received phase distribution were Gaussian, a reasonable search strategy would be (Braun [14]) to search the cells within one standard deviation of the most likely cell first and then expand the search to cells within two standard deviations, and so on. Note that $p(n)$ is a discrete distribution and therefore cannot be Gaussian. The distribution of the received phase is, however, continuous; the function $p(n)$ is easily derived from the continuous density and the cell boundaries. Figure 5-3 illustrates the Gaussian distribution and the search strategy proposed by Braun [14]. Observe that the received phase density has been truncated at three standard deviations.

The average synchronization time for a system based on the assumptions illustrated in Figure 5-3 is calculated using the same technique used above for the uniform distribution. Equation (5-6) applies directly. The proper limits for all summations as well as $T(n,j,k)$ and $\Pr[n,j,k]$ must be determined. The limits for the first summation are the same as before except that the phase cells will be numbered differently. The limits are $-C/2 \leq n \leq +C/2$ where cell number zero is at the center of the uncertainty region. The discrete probability of the nth phase cell being

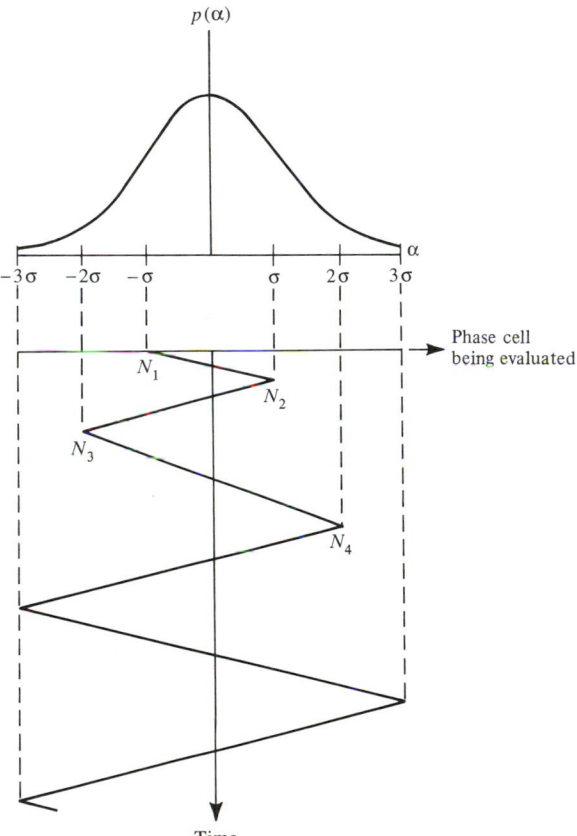

FIGURE 5-3. Received spreading waveform phase probability density and possible sweep strategy. (From Ref. 14.)

correct is

$$p(n) = A \exp\left(-\frac{n^2}{2T^2}\right) \qquad -\tfrac{1}{2}C \le n \le +\tfrac{1}{2}C \qquad (5\text{-}26)$$

where the constant A is calculated from

$$\sum_{n=-C/2}^{C/2} p(n) = 1.0$$

$$T = \tfrac{1}{6}C$$

Since all phase cells are not evaluated on the first sweep, the probability of finding the correct cell on the first sweep is zero for some n. Each pass (partial or complete) through the uncertainty region will be numbered using the variable b. Let N_b, and N_{b+1} denote the starting and ending cell number for the bth pass. For the strategy of Figure 5-3, $N_1 = -C/6, N_2 = +C/6, N_3 = -C/3, N_4 = +C/3$, and so on, where all fractions are rounded to the nearest integer. Let $f(n)$ denote the number of the pass that evaluates cell n for the first time. For example, if $C/6 < n < C/3$, the cell will not be evaluated until the third sweep. Recall that j denotes the number of missed detections prior to a correct detection of the correct cell. The limits of the sum over j are unchanged from above (i.e., $0 \le j \le \infty$). The variable k is the total number of false alarms in all incorrect phase cells evaluated prior to synchronization. The total of all incorrect cells is denoted $K(n, j)$, and is given by

$$K(n,j) = \sum_{b=1}^{f(n)+j-1} |N_{b+1} - N_b| + |n - N_{f(n)+j}| - j \qquad (5\text{-}27)$$

As before, the false alarms can occur in any order within the $K(n,j)$ incorrect phase cells and the number of false alarms is binomially distributed. Combining all of the above yields

$$\Pr[n,j,k] = A \exp\left(-\frac{n^2}{2T^2}\right) P_d(1 - P_d)^j \binom{K(n,j)}{k} P_{\text{fa}}^k (1 - P_{\text{fa}})^{K(n,j)-k} \qquad (5\text{-}28)$$

$$\overline{T}_s = \sum_{n=-C/2}^{C/2} \sum_{j=0}^{\infty} \sum_{k=0}^{K(n,j)} \{[K(n,j) + j]T_i + kT_{\text{fa}}\}A \exp\left(-\frac{n^2}{2T^2}\right)$$

$$\times P_d(1 - P_d)^j \binom{K(n,j)}{k} P_{\text{fa}}^k (1 - P_{\text{fa}})^{K(n,j)-k} \qquad (5\text{-}29)$$

This expression has been evaluated in Braun [14] using the characteristic function. For reasonably high P_d, the expression can be evaluated on a digital computer to obtain specific numerical results.

In addition to the mean and variance of the acquisition time, the system designer may also need to know the probability density of the acquisition time for modified search strategies. The probability density must be known when there is a system requirement for the completion of code synchronization within a specified time.

Research results by Meyr and Polzer [15], Jovanovic [16,17], and Weinberg [18] give, under slightly different analytical assumptions and approximations, techniques for calculating the probability density of interest under a wide variety of conditions. Using these results it is possible to calculate the pdf for any search strategy (e.g., linear search as in Section 5-3.1 or expanding window search as in Section 5-3.2 or others [2,14]) using any available prior information about the received code phase.

Modified sweep strategies of the type described here have been used for many years in radar. Their use results in reduced average synchronization time when the distribution of the received phase is nonuniform. The magnitude of synchronization time savings is a function of the variance of the received phase distribution.

5-3.3 Continuous Linear Sweep of Uncertainty Region

Most current spread-spectrum systems employ a sweep system which moves from one uncertainty cell to the next in discrete steps. An obvious variation of this strategy is to offset the clock frequency of the reference waveform generator slightly so that the phase of the waveform slips linearly past the receiver waveform phase. Recall that phase is the integral of frequency. The output of the despreading mixer is then a wideband signal except when the received and reference waveform phases have slipped sufficiently close to one another that despreading occurs. When despreading occurs, the despread energy can be detected and the sweep terminated. This technique was analyzed in Sage [19]. Although more sophisticated search strategies have been developed since this early paper, Sage's results are a reasonable baseline to which other strategies may be compared.

Consider a system that employs BPSK direct-spreading modulation. The received waveform is

$$r(t) = \sqrt{2P}\, c(t - T_d) \cos \omega_0 t + n(t) \tag{5-30}$$

where $n(t)$ is the usual bandlimited white Gaussian noise. The receiver configuration for synchronization is illustrated in Figure 5-4. For simplicity, the center frequency of the bandpass filter is approximately equal to the received carrier frequency. The

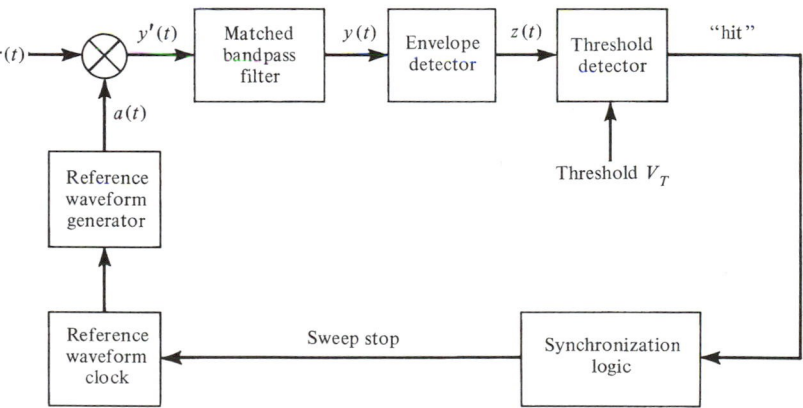

FIGURE 5-4. Synchronization system for linear sweep of uncertainty region.

reference waveform is then simply the spreading waveform, that is, $a(t) = c(t - \hat{T}_d)$. The despreading mixer output is

$$y'(t) = \sqrt{2P}\, c(t - T_d)c(t - \hat{T}_d) \cos(\omega_0 t + \theta) + n(t)c(t - \hat{T}_d) \quad (5\text{-}31)$$

where θ is a random phase that is assumed unknown. The reference waveform clock is offset in frequency so that the propagation delay estimate is varying linearly with time; that is, $\hat{T}_d = \hat{T}_d(t) = \hat{T}_{do} + Kt$, where $\hat{T}_{do}$ is an arbitrary fixed initial condition, $K = T_c/T_{ss}$, and T_{ss} is the time required to search one chip.

Recalling previous analysis of waveforms of the type of (5-31), the signal component at the mixer output is wideband when $T_d - \hat{T}_d(t) > T_c$ and has a sinusoidal component at frequency ω_0 whose magnitude is $\sqrt{2P}\, R_c(\tau)$ when $T_d - \hat{T}_d(t) = \tau < T_c$. $R_c(\tau)$ is the autocorrelation function of $c(t)$ and may be approximated by the function of Figure 2-9 for the present discussion. Since T_d is varying linearly with t, so is τ. Therefore, the sinusoidal component at frequency ω_0 at the mixer output is amplitude modulated by the autocorrelation function and is

$$x'(t) = \sqrt{2P}\, R_c(T_d - \hat{T}_{do} - Kt) \cos(\omega_0 t + \theta) \quad (5\text{-}32)$$

which is illustrated in Figure 5-5.

The purpose of the bandpass filter, envelope detector, and threshold detector is to detect this signal approximately when the peak amplitude is reached. The filter is selected to have an impulse response which is matched to the signal $x'(t)$. The matched impulse response $h(t)$ is a time-reversed and delayed replica of the input waveform [20]. The delay is included so that the filter will be causal. In this case,

$$h(t) = bR_c(Kt - T_c) \cos \hat{\omega}_0 t \quad (5\text{-}33)$$

Since the frequency of the received signal is unknown during synchronization, $\hat{\omega}_0$ has been used in (5-33).

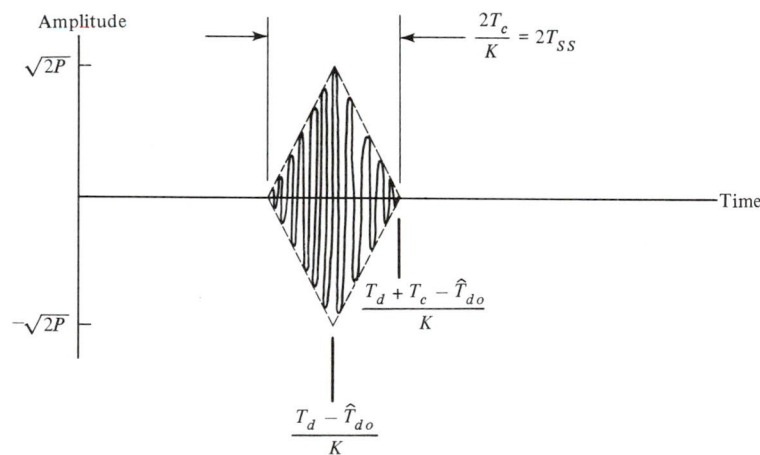

FIGURE 5-5. Signal component at despreading mixer output.

The matched filter output due to signal is the convolution of $h(t)$ and $x'(t)$. This output is

$$x(t) = \int_{-\infty}^{\infty} h(\alpha)x'(t - \alpha)\, d\alpha$$

$$= \sqrt{\frac{P}{2}}\, b \int_{-\infty}^{\infty} R_c(K\alpha - T_c)R_c(T_d - \hat{T}_{\mathrm{do}} - Kt + K\alpha)$$

$$\times \cos(\Delta\omega_0\alpha + \omega_0 t + \theta)\, d\alpha \tag{5-34}$$

$$+ \sqrt{\frac{P}{2}}\, b \int_{-\infty}^{\infty} R_c(K\alpha - T_c)R_c(T_d - \hat{T}_{\mathrm{do}} - Kt + K\alpha)$$

$$\times \cos[(\omega_0 + \hat{\omega}_0)\alpha - \omega_0 t - \theta]\, d\alpha$$

where $\Delta\omega_0 = \hat{\omega}_0 - \omega_0$. The carrier frequency is assumed to be large relative to the maximum rate of change of the autocorrelation function product above, so that the second integral in the last line of (5-34) may be ignored. Expanding the cosine within the first integral yields

$$x(t) \cong \sqrt{\frac{P}{2}}\, b \cos(\omega_0 t + \theta) \int_{-\infty}^{\infty} R_c(K\alpha - T_c)R_c(T_d - \hat{T}_{\mathrm{do}} - Kt + K\alpha)$$

$$\times \cos(\Delta\omega_0\alpha)\, d\alpha$$

$$- \sqrt{\frac{P}{2}}\, b \sin(\omega_0 t + \theta) \int_{-\infty}^{\infty} R_c(K\alpha - T_c)R_c(T_d - \hat{T}_{\mathrm{do}} - Kt + K\alpha) \tag{5-35}$$

$$\times \sin(\Delta\omega_0\alpha)\, d\alpha$$

At this point the assumption is made that the frequency error $\Delta f_0 = 2\pi\,\Delta\omega_0$ is small enough that the sine and cosine within the integrals of (5-35) are approximately constant over the range of nonzero values of the autocorrelation products. Using $\Delta\omega_0\alpha = \Delta\omega_0 T_c/K$ for the arguments of the sine and cosine, (5-35) becomes

$$x(t) \cong \sqrt{\frac{P}{2}}\, b \cos\left(\omega_0 t + \theta + \Delta\omega_0 \frac{T_c}{K}\right)$$

$$\times \int_{-\infty}^{\infty} R_c(K\alpha - T_c)R_c(T_d - \hat{T}_{\mathrm{do}} - Kt + K\alpha)\, d\alpha \tag{5-36}$$

This signal is a cosine with frequency ω_0 and arbitrary phase and with an amplitude given by $\sqrt{P/2}\,b$ times the integral of the autocorrelation function product. The integral is the output of a filter matched to the envelope of the signal of Figure 5-5. Although the absolute time of the maximum output of this filter is not known, it is known that the maximum occurs at the end of the triangular pulse which occurs at

$t = (T_d + T_c - \hat{T}_{do})/K$. Thus the maximum filter output is

$$x_{max}(t) = \sqrt{\frac{P}{2}}\, b \cos\left(\omega_0 t + \theta + \Delta\omega_0 \frac{T_c}{K}\right) \int_{-\infty}^{\infty} R_c^2(K\alpha - T_c)\, d\alpha$$

$$= \frac{\sqrt{2P}}{3}\, bT_{ss} \cos(\omega_0 t + \theta + \Delta\omega_0 T_{ss}) \tag{5-37}$$

for the autocorrelation function of Figure 2-9. The search rate for this system is $1/T_{ss}$ chips per second.

The noise power at the output of the matched filter is calculated using the baseband equivalent filter transfer function. The impulse response $h_t(t)$ of the baseband equivalent filter is calculated from (5-33) and the relationship

$$h(t) = 2\,\mathrm{Re}[\tilde{h}(t)e^{j\tilde{\omega}_0 t}] \tag{5-38}$$

The result is

$$\tilde{h}(t) = \frac{b}{2} R_c(Kt - T_c) \tag{5-39}$$

The Fourier transform of $\tilde{h}(t)$ is the filter transfer function $\tilde{H}(f)$. This Fourier transform was calculated earlier and the result is, ignoring the delay term,

$$\tilde{H}(f) = \frac{b}{2} T_{ss}\, \mathrm{sinc}^2 f T_{ss} \tag{5-40}$$

The narrowband noise at the input to the matched filter is

$$n_i(t) = n(t)c(t - \hat{T}_d)$$

$$= \sqrt{2}\, n_I(t)c(t - \hat{T}_d) \cos \hat{\omega}_0 t - \sqrt{2}\, n_Q(t)c(t - \hat{T}_d) \sin \hat{\omega}_0 t \tag{5-41}$$

and the complex envelope of this noise process is

$$\tilde{n}(t) = c(t - \hat{T}_d)[\sqrt{2}\, n_I(t) + j\sqrt{2}n_Q(t)] \tag{5-42}$$

The power spectrum of $\tilde{n}(t)$ is the Fourier transform of its autocorrelation function. Since $n_I(t)$ and $n_Q(t)$ are independent of one another and both are independent of $c(t)$,

$$R_{\tilde{n}}(\tau) = \tfrac{1}{2}E[\tilde{n}(t)\tilde{n}^*(t + \tau)]$$

$$= 2R_c(\tau)S_{n_I}(\tau) \tag{5-43}$$

The Fourier transform of (5-43) is

$$S_{\tilde{n}}(f) = 2S_c(f)*S_{n_I}(f) \tag{5-44}$$

Assume now that the input noise spectrum is wideband relative to the spectrum of $c(t)$ so that

$$S_{\tilde{n}}(f) \cong 2S_{n_I}(f) = \begin{cases} N_0 & |f| < B \\ 0 & \text{elsewhere} \end{cases} \tag{5-45}$$

where $1/T_c \ll B \ll f_0$. The noise power at the filter output is

$$N = \int_{-\infty}^{\infty} |\tilde{H}(f)|^2 S_{\tilde{n}}(f) \, df$$

$$= \tfrac{1}{4} b^2 T_{ss}^2 N_0 \int_{-B}^{+B} \text{sinc}^4 f T_s \, df \tag{5-46}$$

$$\cong \frac{b^2}{6} T_{ss} N_0$$

The last equality is from Gradshteyn and Ryzhik [8]. Finally, the maximum signal-to-noise power ratio at the matched filter output is calculated from (5-37) and (5-46) and is

$$\text{SNR}_{\max} = \frac{\tfrac{1}{2}[(\sqrt{2P}/3)bT_{ss}]^2}{\tfrac{1}{6}b^2 T_{ss} N_0} = \frac{2}{3}\left(\frac{PT_{ss}}{N_0}\right) \tag{5-47}$$

Observe that signal-to-noise ratio increases with increasing T_{ss} or, equivalently, with decreasing sweep rate.

The matched filter output calculated above is the sum of a sinusoid and Gaussian noise. The peak signal-to-noise ratio occurs at the end of the pulse of Figure 5-5 and has a magnitude given by (5-47). The function of the envelope detector and threshold comparator in Figure 5-4 is to detect the presence of the sinusoid. Since the phase of the sinusoid is unknown, the envelope detector is the optimum detection device. Both the probability of detecting the signal P_d and the probability of falsely declaring signal present P_{fa} can be calculated if the probability density function of the envelope detector output is known. This density function is the well-known Ricean pdf (Rice [21]) of the envelope of a sinusoid in Gaussian noise. Denote the envelope detector output by $z(t)$; then

$$p_z(\alpha) = \begin{cases} \dfrac{\alpha}{N} \exp\left(-\dfrac{\alpha^2 + A^2}{2N}\right) I_0\left(\dfrac{\alpha A}{N}\right) & \text{for } \alpha \geq 0 \\ \\ 0 & \text{elsewhere} \end{cases} \tag{5-48}$$

where N is the noise power or variance, A is the amplitude of the sinusoid, and $I_0(\cdot)$ is the zeroth-order modified Bessel function. The probability of detection is the integral from V_T to infinity of this pdf evaluated for the maximum signal-to-noise ratio. Probability of false alarm is the same integral evaluated at the minimum signal-to-noise ratio (i.e., for $A = 0$). These calculations are depicted graphically in Figure 5-6, which shows the pdf for SNRs of zero and eight. The area which is crosshatched is P_d and the area which is double crosshatched is P_{fa}. Both P_d and P_{fa} are functions of SNR and V_T and cannot be independently selected. For a specific SNR, the selection of P_{fa} implies a particular threshold V_T which then sets P_d. Figure 5-7 illustrates these relationships.

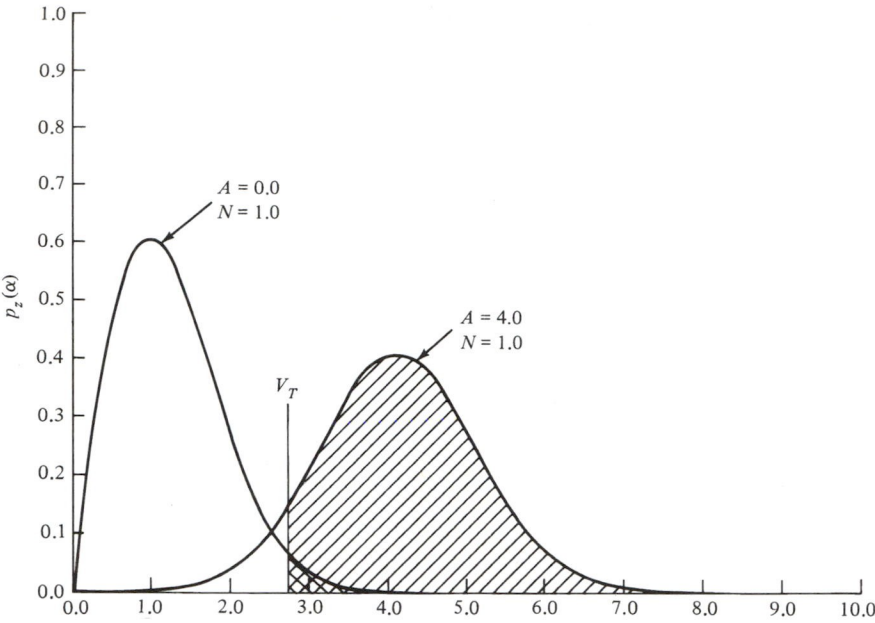

FIGURE 5-6. Example of the calculation of P_d and P_{fa}.

EXAMPLE 5-1

Consider a spread-spectrum system using a spreading code clock frequency $f_c = 3$ MHz. Suppose that the received carrier power-to-noise power spectral density ratio is 46.25 dB-Hz $= 10 \log(P/N_0)$ and that the propagation delay uncertainty is ± 1.2 ms. What sweep rate is required to yield a single sweep (P_d, P_{fa}) pair equal to $(0.9, 10^{-6})$, $(0.8, 10^{-6})$, $(0.9, 10^{-3})$, and $(0.8, 10^{-3})$? What is the average synchronization time for each case assuming a false-alarm penalty of $100T_i$?

Solution: From Figure 5-7, the required signal-to-noise ratios for these four (P_d, P_{fa}) pairs are 13.4, 12.8, 11.0, and 10.3 dB, respectively. From (5-47)

$$10 \log(\text{SNR}_{\max}) = 10 \log \frac{2}{3} + 10 \log \frac{P}{N_0} + 10 \log T_{ss}$$

Solving for T_{ss} for each of the four cases yields

$$T_{ss} = \begin{cases} 778 \ \mu s & \text{for } (0.9, 10^{-6}) \\ 678 \ \mu s & \text{for } (0.8, 10^{-6}) \\ 448 \ \mu s & \text{for } (0.9, 10^{-3}) \\ 381 \ \mu s & \text{for } (0.8, 10^{-3}) \end{cases}$$

Equation (5-12) for mean synchronization time applies to the discrete phase step analysis. At low P_{fa}, it also applies approximately to the present case. In (5-12)

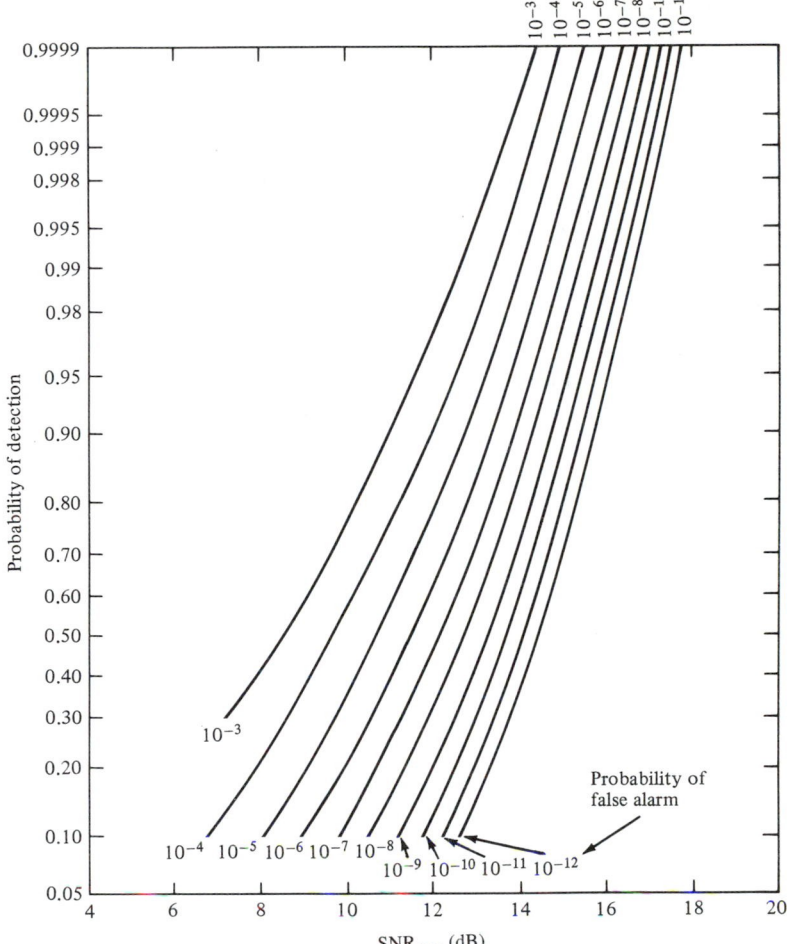

FIGURE 5-7. Probability of detection for a sinusoid in Gaussian noise as a function of signal-to-noise ratio and the probability of false alarm. (From Ref. 22.)

use

$$T_i = T_{ss}$$

$$T_{da} = T_{ss} + P_{fa}(100T_{ss}) = T_{ss}(1 + 100P_{fa})$$

$$C = 2(1.2 \text{ ms})(3 \times 10^6 \text{ chips/s}) = 7.2 \times 10^3$$

With these substitutions, (5-12) becomes

$$\overline{T}_s = (7199)T_{ss}(1 + 100P_{fa})\left(\frac{2 - P_d}{2P_d}\right) + \frac{T_{ss}}{P_d}$$

and the final result is

$$\overline{T}_s = \begin{cases} 3.42 \text{ s} & \text{for } (0.9, 10^{-6}) \\ 3.66 \text{ s} & \text{for } (0.8, 10^{-6}) \\ 2.17 \text{ s} & \text{for } (0.9, 10^{-3}) \\ 2.26 \text{ s} & \text{for } (0.8, 10^{-3}) \end{cases}$$

Observe that permitting P_{fa} to increase by three orders of magnitude can decrease synchronization time significantly.

Finally, a method of calculation of additional points on the curves of Figure 5-7 is desired. These points are calculated by integrating the Ricean density function over the limits of V_T to infinity using the desired signal-to-noise ratio. This integral is called the Marcum Q-function defined by

$$Q(a,b) \triangleq \int_b^\infty \alpha \exp[-\tfrac{1}{2}(\alpha^2 + a^2)]I_0(a\alpha)\, d\alpha \tag{5-49}$$

An efficient algorithm for calculating $Q(a,b)$ has been given in Shnidman [23]. Equation (5-49) is related to the desired integral through straightforward normalization (see also Appendix F).

5-3.4 Detection of a Signal in Additive White Gaussian Noise

All synchronization systems that employ a discrete step serial search evaluate a particular phase/frequency cell by estimating whether or not signal energy is present at the output of a filter following the despreading mixer. Calculation of the mean and variance of the synchronization time requires knowledge of P_d and P_{fa} for this estimate as a function of evaluation time T_i and received signal-to-noise ratio. Three somewhat different methods of detecting signal energy are discussed in this section. Each method results in a different relationship between P_d, P_{fa}, T_i, and SNR. For each method, this relationship will also depend on other factors, such as detection thresholds. In the first case, a single integration of the output of a square-law envelope detector for T_i seconds will be used to detect signal energy. In the second case, multiple integrations together with some logic will be used to detect the signal. This will result in an evaluation time that is a random variable. The final method considered will integrate the square-law envelope detector output for a variable length of time until a selected P_d and P_{fa} are achieved. All these methods are used in current spread-spectrum systems.

Fixed Integration Time Detection. The simplest method of detecting the presence of signal energy at the output of a narrowband filter is illustrated in Figure 5-8. The input to the bandpass filter is taken from the output of the despreading mixer. When the reference waveform phase is correct, the received signal will be despread and $s(t)$ will appear at the filter output. The AWGN process $n(t)$ will always be present at the filter output. The filter output is squared and then lowpass filtered to

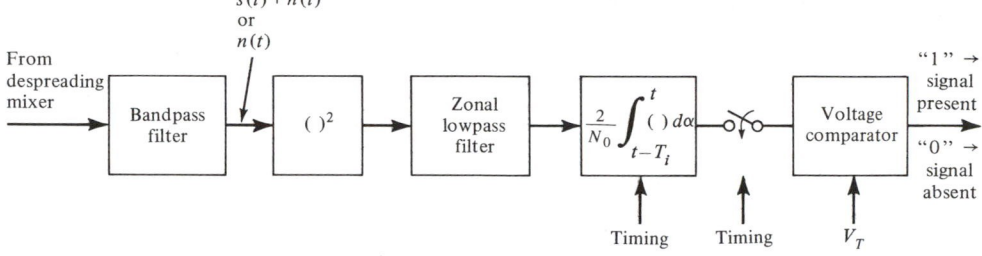

FIGURE 5-8.   Fixed-integration-time energy detector.

eliminate the double-frequency terms that result from the squaring operation. The lowpass squared output is then integrated for T_i seconds, and the output of this operation is compared to a fixed threshold V_T. The factor $2/N_0$ is associated with the integrator, so that the statistics of the integrator output may be stated in terms of standard functions. If the integrator output is above the threshold, the signal is declared present. Otherwise, the signal is declared absent. This system has been analyzed in detail in Urkowitz [24].

To evaluate the performance of this energy detector, the pdf of the integrator output must be calculated. The exact calculation of this pdf is difficult and systems analysts always settle for approximate results. Several different approximations will be discussed in the following. Consider first the output due to noise alone. The filter output noise process is represented by

$$n(t) = \sqrt{2}\, n_I(t) \cos \omega_{\text{IF}} t - \sqrt{2}\, n_Q(t) \sin \omega_{\text{IF}} t \qquad (5\text{-}50)$$

Assume that the IF filter bandwidth is B so that both $n_I(t)$ and $n_Q(t)$ are baseband white Gaussian noise processes with two-sided power spectral densities $N_0/2$ over the frequency range $|f| < B/2$. Squaring (5-50) and retaining only the baseband components of the result yields

$$[n^2(t)]_{\text{LP}} = n_I^2(t) + n_Q^2(t) \qquad (5\text{-}51)$$

and the integrator output at the sampling instant is

$$V = \frac{2}{N_0} \int_0^{T_i} [n^2(t)]_{\text{LP}}\, dt = \frac{2}{N_0} \int_0^{T_i} n_I^2(t)\, dt + \frac{2}{N_0} \int_0^{T_i} n_Q^2(t)\, dt \qquad (5\text{-}52)$$

Consider the baseband noise processes $n_I(t)$ and $n_Q(t)$. These signals are band-limited to the frequency range $-\frac{1}{2}B < f < +\frac{1}{2}B$. The sampling theorem [25] states that these bandlimited signals can be exactly represented by a sum of orthonormal sampling functions appropriately weighted and time translated. In particular,

$$n_I(t) = \sum_{k=-\infty}^{\infty} \frac{1}{\sqrt{B}}\, n_I\!\left(\frac{k}{B}\right) \sqrt{B}\, \text{sinc}(Bt - k) \qquad (5\text{-}53)$$

A similar expression is obtained for $n_Q(t)$. A finite number of terms of this infinite sum can be used as approximation to the baseband noise process over the range

$0 \le t \le T_i$. For example, let $n_I'(t)$ be defined by

$$n_I'(t) = \begin{cases} n_I(t) & 0 \le t \le T_i \\ 0 & \text{elsewhere} \end{cases} \tag{5-54}$$

as illustrated in Figure 5-9. Formally applying the sampling theorem to the signal $n_I'(t)$ yields

$$n_I'(t) = \sum_{k=1}^{BT_i} \frac{1}{\sqrt{B}} n_I'\left(\frac{k}{B}\right) \sqrt{B} \operatorname{sinc}(Bt - k) \tag{5-55}$$

since samples of $n_I'(t)$ outside of the indicated range of k are zero. Unfortunately, the signal defined by (5-54) is not bandlimited, since it is time limited, and the sampling theorem does not apply. For large BT_i it can be argued that the range of significant frequency components of $n_I'(t)$ is not much greater than $-\frac{1}{2}B \le f \le +\frac{1}{2}B$, so that the result of (5-55) is approximately correct. With this understanding, the first integral of the right side of (5-52) can be approximately represented by

$$\frac{2}{N_0} \int_0^{T_i} n_I^2(t)\, dt = \frac{2}{N_0} \int_0^{T_i} n_I'^2(t)\, dt = \frac{2}{N_0} \int_{-\infty}^{\infty} n_I'^2(t)\, dt$$

$$\cong \frac{2}{N_0} \int_{-\infty}^{\infty} \left[\sum_{k=1}^{BT_i} \frac{1}{\sqrt{B}} n_I'\left(\frac{k}{B}\right) \sqrt{B} \operatorname{sinc}(Bt - k) \right]$$

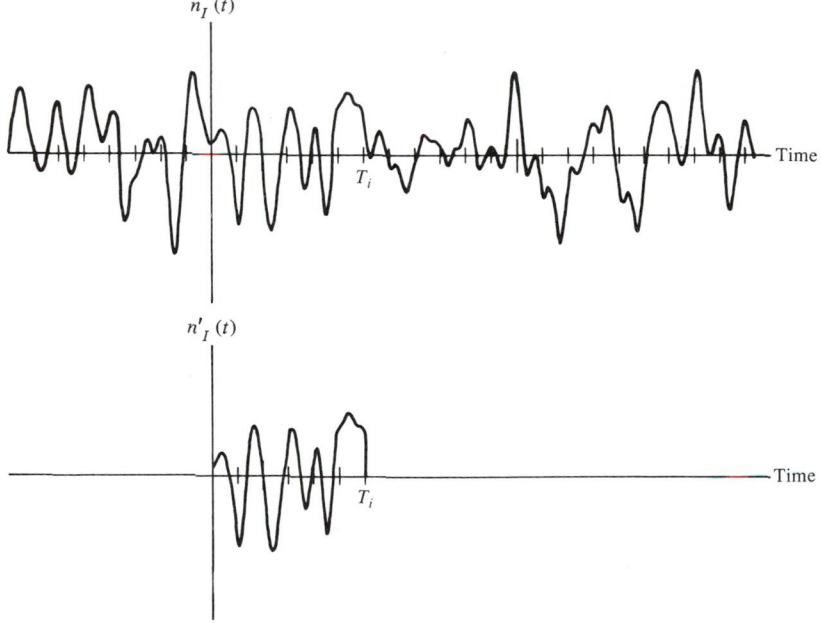

FIGURE 5-9. Baseband noise processes $n_I(t)$ and $n_I'(t)$.

$$\times \left[\sum_{k'=1}^{BT_i} \frac{1}{\sqrt{B}} n_I'\left(\frac{k'}{B}\right) \sqrt{B} \operatorname{sinc}(Bt - k') \right] dt \quad (5\text{-}56)$$

The sampling functions $\sqrt{B} \operatorname{sinc}(Bt - k)$ are orthonormal over the limits $-\infty < t < +\infty$, that is,

$$\int_{-\infty}^{\infty} \sqrt{B} \operatorname{sinc}(Bt - k) \sqrt{B} \operatorname{sinc}(Bt - k') \, dt = \begin{cases} 1.0 & k = k' \\ 0 & k \neq k' \end{cases} \quad (5\text{-}57)$$

Using (5-57), equation (5-56) simplifies to

$$\frac{2}{N_0} \int_0^{T_i} n_I^2(t) \, dt \approx \frac{2}{N_0 B} \sum_{k=1}^{BT_i} n_I'^2\left(\frac{k}{B}\right) \quad (5\text{-}58)$$

A similar expression results for the second integral on the right side of (5-52) and V can be approximately represented by

$$V = \frac{2}{N_0} \int_0^{T_i} [n^2(t)]_{\text{LP}} \, dt \cong \frac{2}{N_0 B} \sum_{k=1}^{BT_i} \left[n_I^2\left(\frac{k}{B}\right) + n_Q^2\left(\frac{k}{B}\right) \right] \quad (5\text{-}59)$$

which is a constant times the sum of the squares of $2BT_i$ independent Gaussian random variables. Taking into account the normalization factor $2/N_0 B$, the variance of each of the Gaussian random variables in the sum is unity. The fact that the samples of the noise processes are independent is deduced directly from the autocorrelation function of $n_I(t)$ or $n_Q(t)$, which is

$$R_n(\tau) = \tfrac{1}{2} N_0 B \operatorname{sinc}(B\tau) \quad (5\text{-}60)$$

and indicates that time samples spaced $1/B$ apart are uncorrelated and therefore independent.

The sum of the squares of $2BT_i$ Gaussian random variables with unit variance has a chi-squared probability density function with $n = 2BT_i$ degrees of freedom [7,26]. The chi-squared probability density is

$$p_c(\alpha) = \begin{cases} \dfrac{1}{2^{n/2} \Gamma\left(\dfrac{n}{2}\right)} \alpha^{(n/2)-1} \exp\left(-\dfrac{\alpha}{2}\right) & \alpha \geq 0 \\[4mm] 0 & \alpha < 0 \end{cases} \quad (5\text{-}61)$$

The exact result for the pdf of V has been calculated in Grenander et al. [27]. This calculation is well beyond the scope of this book; however, it is instructive to compare the exact result with the chi-squared approximation. Figure 5-10 compares the exact result to the chi-squared result for time–bandwidth products of $8/\pi$ and $16/\pi$. Observe that even for these small time–bandwidth products, the approximation is quite good. The approximation improves as time–bandwidth product increases.

When the input to the squaring device of Figure 5-8 is signal plus noise, indicating that the correct spreading waveform phase is being evaluated, the analysis is

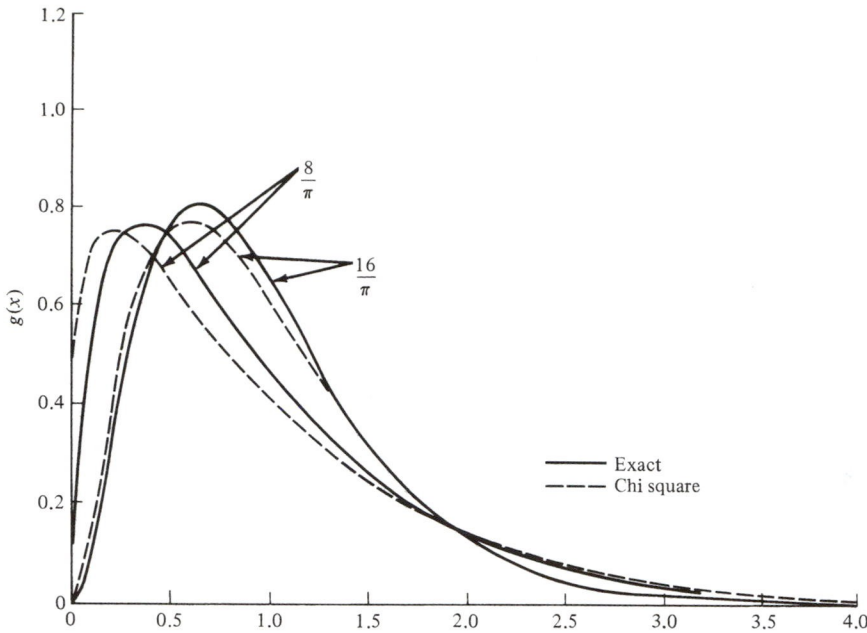

FIGURE 5-10. Comparison of exact and approximate (chi-squared) integrator output probability density functions. (From Ref. 27.)

somewhat more complicated. Assume for simplicity that the signal is a sinusoid with amplitude $\sqrt{2P}$ and arbitrary phase θ which can be represented by

$$s(t) = \sqrt{2P} \cos \theta \cos \omega_{IF}t - \sqrt{2P} \sin \theta \sin \omega_{IF}t \qquad (5\text{-}62)$$

The integrator output is

$$
\begin{aligned}
V &= \frac{2}{N_0} \int_0^{T_i} [s(t) + n(t)]_{LP}^2 \, dt \\
&= \frac{2}{N_0} \int_0^{T_i} \{[\sqrt{2P} \cos \theta + \sqrt{2}\, n_I(t)] \cos \omega_{IF}t \\
&\qquad\qquad - [\sqrt{2P} \sin \theta + \sqrt{2}\, n_Q(t)] \sin \omega_{IF}t\}_{LP}^2 \, dt \\
&= \frac{2}{N_0} \int_0^{T_i} [\sqrt{P} \cos \theta + n_I(t)]^2 \, dt + \frac{2}{N_0} \int_0^{T_i} [\sqrt{P} \sin \theta + n_Q(t)]^2 \, dt
\end{aligned}
\qquad (5\text{-}63)
$$

Using identical reasoning to that used above to develop an approximation to the integrator output in the noise-alone case, the two integrals in the last line of (5-63) are approximately given by

$$\frac{2}{N_0} \int_0^{T_i} [\sqrt{P} \cos \theta + n_I(t)]^2 \, dt \cong \frac{2}{N_0 B} \sum_{k=1}^{BT_i} \left[\sqrt{P} \cos \theta + n_I\left(\frac{k}{B}\right) \right]^2 \qquad (5\text{-}64a)$$

$$\frac{2}{N_0} \int_0^{T_i} [\sqrt{P} \sin \theta + n_Q(t)]^2 \, dt \cong \frac{2}{N_0 B} \sum_{k=1}^{BT_i} \left[\sqrt{P} \sin \theta + n_Q\left(\frac{k}{B}\right) \right]^2 \qquad (5\text{-}64b)$$

Thus the integrator output V may be approximated by the sum of $2BT$, independent Gaussian random variables all with unit variance. Half of the random variables have mean $\sqrt{2P/N_0 B} \cos \theta$ and half have a mean $\sqrt{2P/N_0 B} \sin \theta$. The phase θ is also a random variable and is uniformly distributed over the range $(0, 2\pi)$. For a particular integration, however, this phase may be considered constant. The probability density function of this sum has a noncentral chi-squared density [26,28] with $n = 2T_i B$ degrees of freedom and noncentrality parameter

$$\lambda = \frac{2}{N_0 B} \sum_{k=1}^{BT_i} [(\sqrt{P} \cos \theta)^2 + (\sqrt{P} \sin \theta)^2] = \frac{2BT_i P}{N_0 B} = n \frac{P}{N_0 B} \qquad (5\text{-}65)$$

which is n times the signal-to-noise power ratio at the output of the bandpass filter in Figure 5-8.

The noncentral chi-squared probability density function is [26]

$$p_{nc}(\alpha) = \frac{1}{2} \left(\frac{\alpha}{\lambda}\right)^{(n-2)/4} \exp\left(-\frac{\lambda}{2} - \frac{\alpha}{2}\right) I_{(n/2)-1} \left(\sqrt{\alpha\lambda}\right) \qquad (5\text{-}66)$$

where λ is the noncentrality parameter and $I_N(\cdot)$ is the modified Bessel function of the first kind of order N.

Knowledge of the probability density function of the integrator output theoretically permits evaluation of P_d and P_{fa} for a particular fixed T_i, B, P, N_0, and V_T. This calculation is difficult because of the complexity of (5-61) and (5-66). The desired P_d is given by

$$P_d = \Pr(V > V_T | \text{signal present})$$

$$= \int_{V_T}^{\infty} p_{nc}(\alpha) \, d\alpha \qquad (5\text{-}67a)$$

and the desired P_{fa} is given by

$$P_{fa} = \Pr(V > V_T | \text{no signal present})$$

$$= \int_{V_T}^{\infty} p_c(\alpha) \, d\alpha \qquad (5\text{-}67b)$$

Tables of both the central [28] and noncentral [29] chi-squared probability integrals are available. In addition, selected receiver operating characteristics (ROC) are plotted in Urkowitz [24], which give sets of (P_d, P_{fa}, λ) for time–bandwidth products of 2, 10, 20, 30, and 50. These ROCs are similar to the curves of Figure 5-7. A third alternative for calculating the integrals for P_d and P_{fa} is through the use of a nomogram presented in Urkowitz [24], which was derived from a nomogram in Smirnov and Potapov [30]. This nomogram is reproduced here as Figure 5-11. In this nomogram, the integrator output voltage is V and the detection threshold (see Figure 5-8) is denoted V_T. The probability of false alarm is calculated from Figure

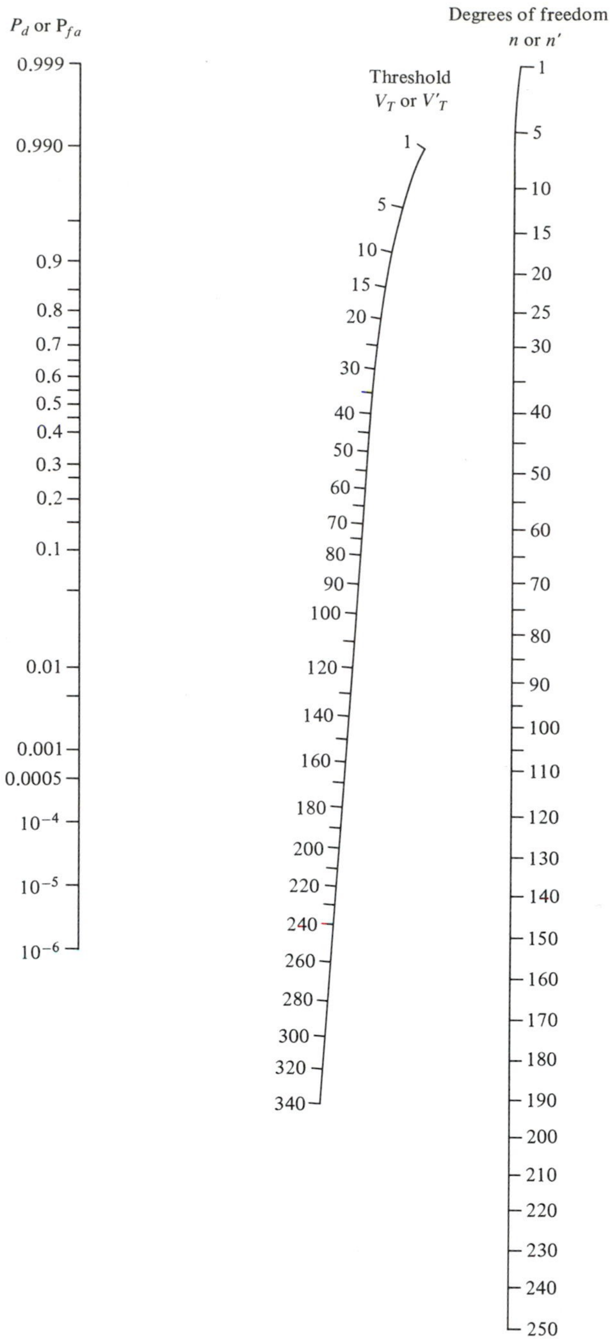

FIGURE 5-11. Nomogram for evaluating P_d and P_{fa} for the detector of Figure 5-8. (From Ref. 24. Copyright © 1967, IEEE. Reprinted with permission.)

5-11 using the given time–bandwidth product $2T_iB = n$ and the selected normalized threshold V_T. The probability of detection is calculated using a modified number of degrees of freedom n' given by

$$n' = \frac{(n + \lambda)^2}{n + 2\lambda} \tag{5-68}$$

and a modified threshold V_T' given by

$$V_T' = V_T\frac{n + \lambda}{n + 2\lambda} \tag{5-69}$$

EXAMPLE 5-2 [24]

Suppose that $n = 2T_iB = 40$, $\lambda = 30$, and $P_{\text{fa}} = 10^{-4}$ are given. The normalized threshold is $V_T = 81$, which is read directly from the nomogram. For the given noncentrality parameter λ, the modified number of degrees of freedom is calculated from (5-68) and the modified threshold from (5-69). Thus

$$n' = \frac{(40 + 30)^2}{40 + 60} = 49$$

$$V_T' = 81\left(\frac{40 + 30}{40 + 60}\right) = 56.7$$

Using n' and V_T', $P_d \cong 0.23$ is read directly from the nomogram.

Although the nomogram of Figure 5-11 is useful for making a few calculations, the system designer must find alternative means of evaluating P_d and P_{fa} as a function of integration time, received signal-to-noise ratio, and threshold when a large number of calculations must be made. Fortunately, efficient numerical techniques exist for evaluation of the integral of the chi-squared and noncentral chi-squared probability density. Ross [31] has developed convenient computer programs for evaluating the integrals of the chi-squared and noncentral chi-squared densities. The source code for these programs is available on the Internet and the programs are described briefly in Appendix F. These programs are based on and are an extension of published results (see Appendix F) and produce highly accurate results.

Yet another method for determining the detection and false-alarm probabilities for the detector of Figure 5-8 is to approximate the probability density at the output of the integrator by a Gaussian probability density with the same mean and variance and to use the Q-function [26] to evaluate the integral of the Gaussian pdf. This approximation is quite accurate when the integrator time–bandwidth product T_iB is large. The Q-function is easy to calculate using polynomial approximations. Denote the mean of the integrator output by E_V and the variance of the integrator output by σ_V^2. Then the approximate probability density for the voltage V at the integrator output is

$$p_G(\alpha) = \frac{1}{\sqrt{2\pi\sigma_V^2}} \exp\left[-\frac{(\alpha - E_V)^2}{2\sigma_V^2}\right] \tag{5-70}$$

The integral of this density from V_T to infinity yields the desired probabilities P_d and P_{fa}. Specifically, both P_d and P_{fa} are calculated from

$$\frac{1}{\sqrt{2\pi\sigma_V^2}} \int_{V_T}^{\infty} \exp\left[-\frac{(\alpha - E_V)^2}{2\sigma_V^2}\right] d\alpha = \frac{1}{\sqrt{2\pi}} \int_{(V_T - E_V)/\sigma_V}^{\infty} \exp\left(-\frac{1}{2}\alpha^2\right) d\alpha$$

$$= Q\left(\frac{V_T - E_V}{\sigma_V}\right) \tag{5-71}$$

Appendix B gives a convenient and quite accurate rational approximation to $Q(\cdot)$. In order to use (5-71), the mean and variance of the integrator output must be known. The mean of the integrator output is

$$E_V = E\left\{\frac{2}{N_0} \int_0^{T_i} [s(t) + n(t)]_{LP}^2 \, dt\right\} \tag{5-72}$$

where the expected value is over all sample functions of $n(t)$ and all received signal phases. Replacing $s(t)$ and $n(t)$ with their narrowband representations of (5-50) and (5-62) and simplifying yields

$$
\begin{aligned}
E_V &= \frac{2}{N_0} \int_0^{T_i} E\{[\sqrt{P}\cos\theta + n_I(t)]^2\} \, dt + \frac{2}{N_0} \int_0^{T_i} E\{[\sqrt{P}\sin\theta + n_Q(t)]^2\} \, dt \\
&= \frac{2T_i}{N_0} E\{[\sqrt{P}\cos\theta + n_I(t)]^2\} + \frac{2T_i}{N_0} E\{[\sqrt{P}\sin\theta + n_Q(t)]^2\} \\
&= \frac{2T_i}{N_0}\{P + E[n_I^2(t)] + E[n_Q^2(t)]\} \tag{5-73} \\
&= \frac{2T_i}{N_0}\left\{\frac{PB}{B} + \frac{N_0 B}{2} + \frac{N_0 B}{2}\right\} \\
&= n\left(\frac{P}{N_0 B} + 1\right)
\end{aligned}
$$

Calculations of the variance of the integrator output is given in Papoulis [7]. The result is

$$\sigma_V^2 = \frac{4}{N_0} \int_0^{T_i} \left(1 - \frac{\lambda}{T_i}\right)\{R_w(\lambda) - E^2[w(t)]\} \, d\lambda \tag{5-74}$$

where $w(t)$ is the integrator (baseband) input and $R_w(\tau)$ is the autocorrelation function of $w(t)$. In the present case,

$$
\begin{aligned}
w(t) &= [s(t) + n(t)]_{LP}^2 \\
&= P + 2\sqrt{P}\cos\theta \, n_I(t) + 2\sqrt{P}\sin\theta \, n_Q(t) + n_I^2(t) + n_Q^2(t)
\end{aligned}
\tag{5-75}
$$

The mean of $w(t)$ is

$$E[w(t)] = P + E[n_I^2(t)] + E[n_Q^2(t)] = P + N_0 B \tag{5-76}$$

The autocorrelation function is

$$\begin{aligned}
R_w(\tau) &= E[w(t)w(t + \tau)] \\
&= E\{[P + 2\sqrt{P}\cos\theta\, n_I(t) + 2\sqrt{P}\sin\theta\, n_Q(t) + n_I^2(t) + n_Q^2(t)] \\
&\quad \times [P + 2\sqrt{P}\cos\theta\, n_I(t + \tau) + 2\sqrt{P}\sin\theta\, n_Q(t + \tau) \\
&\quad + n_I^2(t + \tau) + n_Q^2(t + \tau)]\} \\
&= P^2 + 4PE[n_I^2(t)] + 4PE[\cos^2\theta]E[n_I(t + \tau)n_I(t)] \\
&\quad + 4PE[\sin^2\theta]E[n_I(t)n_I(t + \tau)] \\
&\quad + 2E[n_I^2(t)n_I^2(t + \tau)] + 2E^2[n_I^2(t)]
\end{aligned} \tag{5-77}$$

The last step follows from the independence of $n_I(t)$, $n_Q(t)$ and θ and the fact that $E[n_I(t)] = E[n_Q(t)] = E[\cos\theta] = E[\sin\theta] = 0$. It has also been recognized that statistics of $n_I(t)$ and $n_Q(t)$ are identical and both processes are stationary, so that terms such as $E[n_Q^2(t)]$ can be equated to $E[n_I^2(t)]$. The expected values over θ are

$$E(\cos^2\theta) = E(\tfrac{1}{2} + \tfrac{1}{2}\cos 2\theta) = \tfrac{1}{2} \tag{5-78a}$$

and

$$E(\sin^2\theta) = E(\tfrac{1}{2} - \tfrac{1}{2}\cos 2\theta) = \tfrac{1}{2} \tag{5-78b}$$

so that (5-77) simplifies to

$$R_w(\tau) = P^2 + 4P\sigma_n^2 + 4PR_n(\tau) + 2R_{n^2}(\tau) + 2\sigma_n^4 \tag{5-79}$$

where σ_n^2 denotes the variance of either $n_I(t)$ or $n_Q(t)$.

The autocorrelation function $R_{n^2}(\tau)$ is [32]

$$R_{n^2}(\tau) = 2R_n^2(\tau) + \sigma_n^4$$

so that

$$R_w(\tau) = (P + 2\sigma_n^2)^2 + 4PR_n(\tau) + 4R_n^2(\tau) \tag{5-80}$$

Substituting (5-76) and (5-80) into (5-74) yields

$$\sigma_V^2 = \frac{8T_i}{N_0^2} \int_0^{T_i} \left(1 - \frac{\lambda}{T_i}\right)[4PR_n(\lambda) + 4R_n^2(\lambda)]\, d\lambda \tag{5-81}$$

The autocorrelation function of $n_I(t)$ or $n_Q(t)$ is

$$R_n(\tau) = \tfrac{1}{2}N_0 B \operatorname{sinc} B\tau \tag{5-82}$$

so that

$$\sigma_V^2 = \frac{16BT_i}{N_0} \int_0^{T_i} \left(1 - \frac{\lambda}{T_i}\right)(P \operatorname{sinc} B\lambda + \tfrac{1}{2}N_0 B \operatorname{sinc}^2 B\lambda)\, d\lambda \tag{5-83}$$

For small BT_i products, the variance may be evaluated numerically. For large BT_i, (5-83) can be approximated by

$$\sigma_V^2 \approx \frac{16BT_i}{N_0} \int_0^\infty (P \text{ sinc } B\lambda + \tfrac{1}{2}N_0 B \text{ sinc}^2 B\lambda)\, d\lambda \qquad (5\text{-}84)$$

The integrals are evaluated in Gradshteyn and Ryzhik [8]. The final approximate result is

$$\sigma_V^2 = \frac{8T_i}{N_0}(P + \tfrac{1}{2}N_0 B)$$

$$= \frac{4(2T_i B)}{N_0 B}(P + \tfrac{1}{2}N_0 B) \qquad (5\text{-}85)$$

$$= 2n\left[2\left(\frac{P}{N_0 B}\right) + 1\right]$$

For time–bandwidth products larger than about 3, the approximate result is within 10% of the exact result. With both E_V and σ_V^2 determined, (5-71) can be easily evaluated.

In summary, the performance of the energy detector of Figure 5-8 is difficult to calculate exactly. A number of approximations to the probability density function of the integrator output exist with varying degrees of accuracy and computational difficulty. The central and noncentral chi-squared approximation is the most accurate but evaluation of the integral of the pdf must be via tables, nomograms, or numerical evaluation. The most convenient and accurate means of evaluating the integral of the chi-squared and noncentral chi-squared densities for system design is to use the computer programs discussed in Appendix F. For large time–bandwidth products, the integrator output pdf is nearly Gaussian. In this case, polynomial approximations for the probability integral exist and are convenient for systems analysis. The choice of which approximation to use requires a trade between required accuracy and computational effort.

EXAMPLE 5-3 _____

Reconsider the spread-spectrum system of Example 5-1, which uses a spreading code rate of $f_c = 3$ MHz. The design point SNR $= 46.25$ dB-Hz and the propagation delay uncertainty is ± 1.2 ms, which corresponds to ± 3600 chips. Suppose that the serial-search step size is $\tfrac{1}{2}$ chip, so that the total number of phase cells which can be searched is $C = 14,400$. Evaluate the average synchronization time and comparator threshold for $P_d = 0.9$ and 0.7 and $BT_i = 10$ and 50. Assume that the IF bandpass filter bandwidth is 24 kHz and that the false-alarm penalty is $100T_i$.

Solution: The integration times for the BT_i products of 10 and 50 are $T_i = 10/(24 \times 10^3) = 417\ \mu s$ and $T_i = 50/(24 \times 10^3) = 2.08$ ms. The normalized noncentrality parameter is calculated from (5-65). Substituting the design point

$P/N_0 = 4.22 \times 10^4$ into this equation yields $\lambda = 35.1$ for $BT_i = 10$ and $\lambda = 176$ for $BT_i = 50$. The modified number of degrees of freedom is calculated from (5-68). For $BT_i = 10$, $n' = 33.7$ and for $BT_i = 50$, $n' = 169$. Using n' with the nomogram of Figure 5-11 yields the modified thresholds V'_T given in the following table:

P_d	BT_i	V'_T	V_T	P_{fa}
0.9	10	24	39.3	0.007
0.9	50	144	236	$<10^{-6}$
0.7	10	29	47.5	0.0006
0.7	50	159	260	$<10^{-6}$

The thresholds V_T are calculated using (5-69) and the results are also given in the table. Finally, P_{fa} is found using the nomogram with the threshold V_T and unmodified number of degrees of freedom $n = 20$ and 100. The nomogram stops at $P_{\text{fa}} = 10^{-6}$, so these values are used in the synchronization-time calculations.

Substituting C, P_d, P_{fa}, T_i, and $T_{\text{fa}} = 100T_i$ into (5-10) and (5-12) yields the mean synchronization times given in the following table:

P_d	BT_i	$\overline{T}_s$
0.9	10	6.24 s
0.9	50	18.3 s
0.7	10	5.91 s
0.7	50	27.8 s

Comparing the results of Examples 5-1 and 5-3 points to a conclusion that a linear sweep with matched-filter detection is superior to the scheme of this example. This conclusion is incorrect, however, because no attempt has been made to optimize the selection of P_d, P_{fa}, and BT_i for the stepped search. In addition, the matched filter approach ignores the fact that the despread signal is modulated by a data signal. This fact will certainly degrade performance of the matched filter approach. The IF filter bandwidth of the energy detector for the stepped phase serial search is selected to be wide enough to pass the data-modulated despread carrier. It should be clear from the results of this example that mean synchronization time can vary over a wide range as P_d, P_{fa}, and BT_i are varied, so that some optimization procedure is necessary.

In modern spread-spectrum systems, the fixed-integration-time detector of Figure 5-8 would probably be implemented digitally using the conceptual block diagram illustrated in Figure 5-12. The bandpass filter of Figure 5-12 has a bandwidth B and the signal $s(t)$ and noise $n(t)$ at the output of the bandpass filter are represented by

$$s(t) = \sqrt{2P} \cos(\omega_{\text{IF}}T + \phi) \tag{5-86a}$$

$$n(t) = \sqrt{2}n_I(t) \cos(\omega_{\text{IF}}t + \phi) - \sqrt{2}n_Q(t) \sin(\omega_{\text{IF}}t + \phi) \tag{5-86b}$$

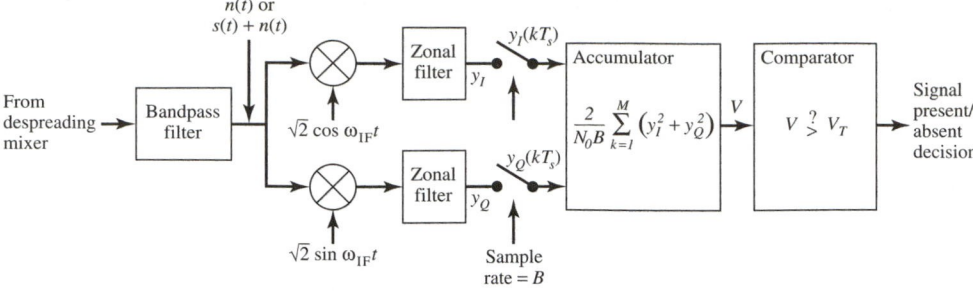

FIGURE 5-12. Digitally implemented fixed-integration-time detector.

The zonal filter outputs are calculated by multiplying (5-86a) or the sum of (5-86a) and (5-86b) by the down-conversion reference signals and simplifying the result by eliminating the double frequency terms. The results when signal plus noise are present at the bandpass filter output are

$$y_I(t) = [\sqrt{P} + n_I(t)] \cos \phi - [n_Q(t)] \sin \phi \tag{5-87a}$$

$$y_Q(t) = -[\sqrt{P} + n_I(t)] \sin \phi - [n_Q(t)] \cos \phi \tag{5-87b}$$

The lowpass bandwidth of either $y_I(t)$ or $y_Q(t)$ is $B/2$. These signals are sampled at the minimum Nyquist rate of B samples per second and the sampled signals are input to the digital processor with appropriate analog-to-digital conversion (not illustrated). The samples are multiplied by a normalization factor and the sum of the squares is accumulated. The output of the accumulator is

$$
\begin{aligned}
V &= \frac{2}{N_0 B} \sum_{k=1}^{M} \left[y_I^2\left(\frac{k}{B}\right) + y_Q^2\left(\frac{k}{B}\right) \right] \\
&= \frac{2}{N_0 B} \sum_{k=1}^{M} \left[\left\{ \sqrt{P} + n_I\left(\frac{k}{B}\right) \right\}^2 + n_Q^2\left(\frac{k}{B}\right) \right]
\end{aligned}
\tag{5-88}
$$

The comparable result for noise alone at the bandpass filter output is found by settting $P = 0$ in (5-88). The variance of either n_I or n_Q is $N_0 B/2$. Thus, taking the normalization factor into account, (5-88) is seen to be the sum of the squares of $n = 2M$ unit variance random variables. When no signal is present, the probability density of V is therefore chi-squared with n degrees of freedom. When signal is present, the density of V is noncentral chi-squared with n degrees of freedom and the noncentrality parameter is

$$\lambda = \frac{2}{N_0 B} \sum_{k=1}^{M} P = 2M\left(\frac{P}{N_0 B}\right) = n\left(\frac{P}{N_0 B}\right) \tag{5-89}$$

which is n times the signal-to-noise power ratio at the output of the bandpass filter. This result is identical to (5-65).

The analysis presented above for the fixed integration time detector has been idealized in that many degradations have been ignored. For example, it was presumed that the spreading code chip rate was known exactly. If, in fact, the code chip rate is known only approximately, the correlation between the received sequence and the receiver reference sequence may be changing during the time that the same correlation is being evaluated. Thus integration time may be limited by imprecise knowledge of the spreading code chip rate. Further, the average search rate and thus the average synchronization time is affected by imprecise knowledge of the chip rate. Suppose, for example, that the time required to evaluate a reference spreading code phase is $T_i = 0.001$ s and that the serial search strategy steps through code phases in $\frac{1}{2}$-chip increments. Then the code search rate is $\frac{1}{2}T_i = 500$ chips per second. Now suppose that receiver reference chip rate and the actual received chip rate differ by 100 chips/s. This chip rate error causes the relative phase of the received and reference codes to shift by $100 \times 0.001 = 0.1$ chip during the time required to evaluate a reference phase. Thus, instead of evaluating 0.5 chip in T_i seconds, the system evaluates either 0.6 chip or 0.4 chip depending on the sign of the code clock error, and the effective code search rate is either 600 or 400 chips/s. Various analyses on the subject of spread-spectrum acquisition with imprecise chip rate knowledge have been published [1,6,9,33]; some of these references also present strategies for mitigating the negative effects of knowing the chip rate only approximately.

The analyses presented above also ignored the effect of the post-despreading bandpass filter on the despread data-modulated IF carrier. In many systems, the bandwidth of the IF filter is selected such that most, but not all, of the power of the data-modulated carrier passes through the filter. The selection of the IF bandwidth is part of the overall system design. When the bandwidth is too large, excessive noise will reach the energy detector and performance will be degraded. When the bandwidth is too small, the data-modulated IF carrier is partially rejected by the filter, thus reducing signal-to-noise ratio at the filter output and degrading performance. The degradation due to data modulation and the power transfer function of the IF filter is addressed in Ref. 1.

Finally, recall from Chapter 4 that the direct-sequence despreading operation must be considered in detail to determine the noise power spectral density N_0 at the input to the bandpass filter of Figure 5-8. Specifically, the noise power spectrum at the output of the despreader is the convolution of the input noise power spectrum and the power spectrum of the despreading reference. Since the thermal noise at the input to the despreading mixer does not have infinite bandwidth, the noise power spectrum at the output of the despreading mixer has a wider bandwidth and lower peak value than at the mixer input.

Multiple-Dwell Detection. Fixed-integration-time detection is limited in that only two parameters, T_i and V_T, can be varied to reduce synchronization time. With signal-to-noise ratio and the number of uncertainty cells fixed, the selection of T_i and V_T completely determine P_d, P_{fa}, T_{da} and therefore average synchronization time. There is only one correct cell within the uncertainty region. Thus most of the cells evaluated by the energy detector are noise alone. An energy detection scheme that is capable of rejecting these incorrect phase cells rapidly while not letting P_{fa} become so large that false-alarm penalty time dominates synchronization time is desirable. The multiple-dwell detection scheme, discussed in this section, accomplishes this using multiple evaluations of the same phase cell. The first evaluation is very short and results in immediate rejection of many incorrect cells. The short integration time of this first evaluation also results in a high false-alarm probability. When a false alarm occurs on the first evaluation, a second evaluation of the same cell begins, using a longer integration time. This second evaluation is capable of rejecting most of the false alarms of the first evaluation. The second evaluation may be followed by a third, fourth, or as many as desired to achieve a particular performance level. Since a particular phase cell may be rejected after one or more integrations, the time required to evaluate a cell is a random variable. In this section a very general technique for evaluating the mean of this random variable is described. All integration times and thresholds as well as the logic followed by the multiple-dwell detection scheme are chosen to yield minimum average synchronization time. Fixed-integration-time detection is a special case (single-dwell) of the multiple-dwell systems described here.

Figure 5-13 is a simplified conceptual block diagram of a multiple-dwell energy detector. The detector is similar to the detector in Figure 5-8 except for the addition of a detection logic function which selects integration times T_j and thresholds V_j for each integration, and which outputs signal present/absent signals.

Figure 5-14 is a typical flow diagram for a multiple-dwell system. The system begins by selecting a code phase for evaluation using whatever a priori information available. Received signal despreading is attempted with this code phase resulting in noise alone at the bandpass filter output if the phase is incorrect or signal plus noise at the filter output if the phase is correct. The filter output is squared and lowpass filtered, and the result is integrated for T_1 seconds. At the end of the integration, the integrator output V is compared with a threshold V_1. If $V < V_1$, a "miss" (incorrect

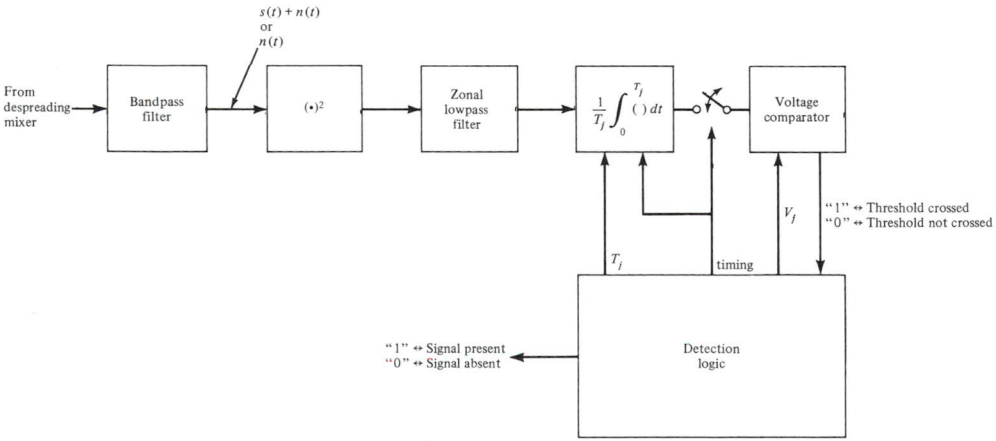

FIGURE 5-13. Simplified conceptual block diagram of multiple-dwell energy detector.

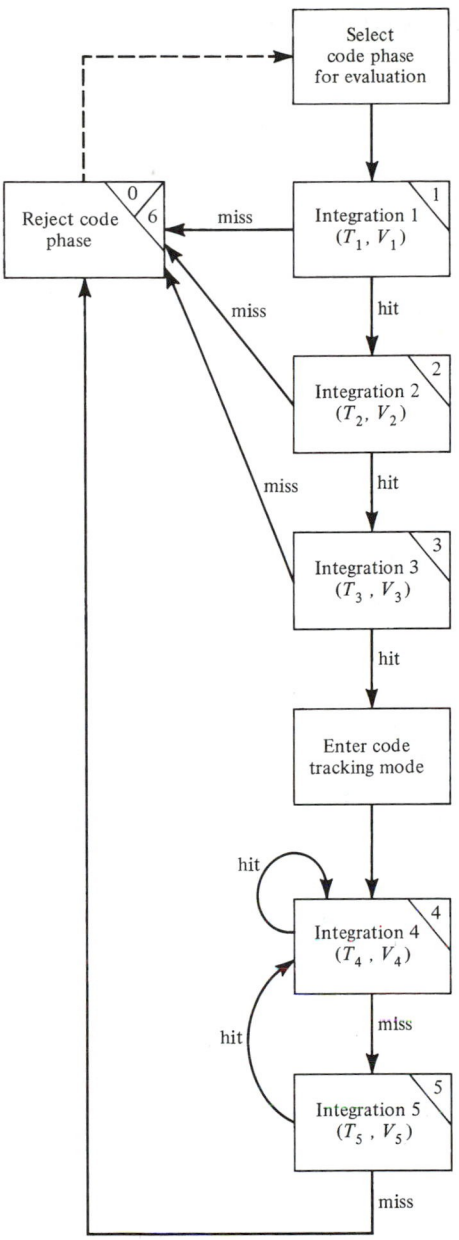

FIGURE 5-14. Typical logic flow diagram for a multiple-dwell detector.

phase) is declared and the phase cell is rejected. If $V \geq V_1$ a ''hit'' (potential correct phase) is declared and a second integration begins. This second integration is T_2 seconds long. At the end of the second integration, the integrator output V is compared with a threshold V_2. If $V < V_2$, the phase is rejected and, if $V \geq V_2$, the phase remains a potentially correct phase. When $V \geq V_2$, a third integration begins, after which the phase is either rejected or declared correct. The "hit" output from integration 3 generates the "signal present" output from the detection logic of Figure 5-13. The "signal absent" output is generated when a "miss" occurs on any of the first three integrations.

The logic of Figure 5-14 also includes a provision for continuing to evaluate the spreading code phase during code tracking. Integrations 4 and 5 provide a means for detecting when the code tracking loop has lost lock. During normal code tracking, the logic continues to cycle around the loop connecting integration 4 with itself. When a false alarm occurs, integrations 4 and 5 must occur before the code phase can be rejected. These two integration times may be quite large since an incorrect dismissal of the correct code phase is very serious. The average time required to go between block 4 and block 6, after a false alarm has occurred, is the false-alarm penalty referred to earlier. The following analysis will include the false-alarm event explicitly. The flow diagram of Figure 5-14 is one example of a multiple-dwell system. Many other examples are possible. Other systems may differ in the total number of integrations and also in the interconnections between states. For example, a "miss" on integration 2 may return the system to integration 1 rather than to the "reject code phase" condition.

The goal of the analysis that follows is to calculate the average dwell time T_{da} at an incorrect phase cell, the probability of detecting the correct phase cell P_d when it is evaluated, and the average time required to evaluate the correct phase cell T_i. Knowledge of these variables will permit calculations of the mean synchronization time using (5-12).

Consider first the calculation of T_{da}. This average dwell time is the average time required by the system to cycle through the logic diagram of Figure 5-14 from the "integration 1" function to the "reject code phase" function with noise alone present. For convenience, some of the blocks of Figure 5-14 have been numbered and will denote the system states. The blocks that are not numbered are those blocks which are included to clarify the logic flow but which do not enter into the analysis to follow. Thus T_{da} is the average time required for the system to progress from state 1 to state 0/6. The reject code phase block has been numbered as both state 0 and state 6 because, in what follows, it will be convenient to distinguish between routes going between state 1 and 0/6 which do or do not pass through states 4 and 5. The flow diagram of Figure 5-14 can be represented by a state transition diagram as shown in Figure 5-15. In this figure, the functions that do not affect the analysis to follow have been deleted and the 0/6 state has been split into two states. Each transition of the diagram has been labeled with the probability of occurrence of the transition and z raised to a power equal to the integration time associated with the "from" state. At this point it may be recognized that the multiple-dwell system is exactly represented by a finite-state Markov chain with absorbing boundaries [34]. The transition probabilities p_{jk} are the probabilities of passing from state j to state k and are calculated from the integration time T_j, the threshold V_j, and the SNR using the techniques described previously for the fixed-integration-time detector.

There are an infinite number of paths which can be followed through the state transition diagram between state 1 and state 0 or 6. The mean time required to reject

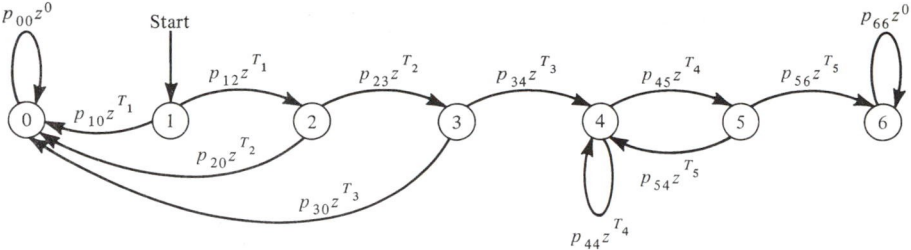

FIGURE 5-15. State-transition diagram of a typical multiple-dwell detector.

an incorrect phase is given by

$$T_{da} = \sum_{i \in L} \Pr(l) T_l \tag{5-90}$$

where L denotes the set of all paths beginning at state 1 and ending at either state 0 or 6, and l denotes a particular path in L. $\Pr(l)$ is the probability of following path l, and T_l is the time required to transverse path l. Consider any specific path through the state diagram of Figure 5-15, say path 1–2–3–0, which will be referred to as path l_0. The probability of following this path is the product of the state transition probabilities encountered while traversing this path, that is,

$$\Pr(l_0) = p_{12} p_{23} p_{30} \tag{5-91}$$

The time required to traverse this path is

$$T_{l_0} = T_1 + T_2 + T_3 \tag{5-92}$$

and the term of (5-90) corresponding to path l_0 is

$$\Pr(l_0) T_{l_0} = p_{12} p_{23} p_{30}(T_1 + T_2 + T_3) \tag{5-93}$$

Let $B(l,z)$ denote the product of the path labels encountered as path l is traversed. Thus for path l_0,

$$\begin{aligned} B(l_0,z) &= p_{12} z^{T_1} p_{23} z^{T_2} p_{30} z^{T_3} \\ &= p_{12} p_{23} p_{30} z^{T_1 + T_2 + T_3} \end{aligned} \tag{5-94}$$

The sum of integration times in the exponent can be transformed into a product term by taking the derivative with respect to z. The result is

$$\frac{d}{dz} B(l_0,z) = p_{12} p_{23} p_{30}(T_1 + T_2 + T_3) z^{T_1 + T_2 + T_3 - 1} \tag{5-95}$$

This equation is identical to (5-93) if $z = 1.0$. That is,

$$\Pr(l_0) T_{l_0} = \left[\frac{d}{dz} B(l_0,z) \right]_{z=1} \tag{5-96}$$

This result is true in general for any path l, so that

$$T_{\text{da}} = \sum_{i \in L} \Pr(l)T_l = \sum_{i \in L} \left[\frac{d}{dz} B(l,z) \right]_{z=1} \tag{5-97}$$

The state-transition diagram can also be described by a transition matrix whose rows correspond to starting states, whose columns correspond to ending states, and whose elements are the path labels on the transition from the row state to the column state. The transition matrix Q' corresponding to the system described in Figure 5-15 is

$$Q' = \begin{matrix} & \begin{matrix} 0 & 6 & & 1 & 2 & 3 & 4 & 5 \end{matrix} \leftarrow \text{``to''} \\ & & & & & & & & \text{state} \\ \begin{matrix} 0 \\ 6 \\ 1 \\ 2 \\ 3 \\ 4 \\ 5 \end{matrix} & \left[\begin{matrix} p_{00}z^0 & 0 & \vert & 0 & 0 & 0 & 0 & 0 \\ 0 & p_{66}z^0 & \vert & 0 & 0 & 0 & 0 & 0 \\ p_{10}z^{T_1} & 0 & \vert & 0 & p_{12}z^{T_1} & 0 & 0 & 0 \\ p_{20}z^{T_2} & 0 & \vert & 0 & 0 & p_{23}z^{T_2} & 0 & 0 \\ p_{30}z^{T_3} & 0 & \vert & 0 & 0 & 0 & p_{34}z^{T_3} & 0 \\ 0 & 0 & \vert & 0 & 0 & 0 & p_{44}z^{T_4} & p_{45}z^{T_4} \\ 0 & p_{56}z^{T_5} & \vert & 0 & 0 & 0 & p_{54}z^{T_5} & 0 \end{matrix} \right] \end{matrix}$$

$$\underset{\substack{\uparrow \\ \text{``from''} \\ \text{state}}}{}$$

$$\triangleq \left[\begin{array}{c|c} \mathbf{U} & \mathbf{0} \\ \hline \mathbf{R} & \mathbf{Q} \end{array} \right] \tag{5-98}$$

Following Hopkins [35], the end states 0 and 6 have been assigned to the first two rows and columns, and the matrix has been partitioned into four submatrices $\mathbf{U}$, $\mathbf{R}$, $\mathbf{Q}$, and $\mathbf{0}$. The submatrix $\mathbf{Q}$ contains information about all the internal states of the system. These definitions are similar to those in Hopkins [35] except for the use of the delay operator z which permits the selection of different integration times for each decision. Define a new matrix $\mathbf{X} = \mathbf{Q}^n \mathbf{R}$. Since $\mathbf{Q}$ is a square matrix, this new matrix is proper. The elements of $\mathbf{X}$ are denoted x_{jk}. The rows of $\mathbf{X}$ correspond to the same states as the rows of $\mathbf{Q}$, that is, the inner states, and the columns of $\mathbf{X}$ correspond to the same states as the columns of $\mathbf{R}$, that is, the end states. Each term of each element x_{jk} will correspond to a path of length $n + 1$ through the state-transition diagram from state j to state k. If there are no paths of length $n + 1$ from state j to state k, then $x_{jk} = 0$. If there is one path, then x_{jk} is equal to the product of path labels encountered when traversing this path. If there are two paths, x_{jk} will have two terms each equal to the product of path labels encountered on the associated path.

EXAMPLE 5-4

Consider the state transition diagram of Figure 5-16. The matrix Q' is

$$
Q' = \begin{array}{c} \\ 0 \\ 4 \\ 1 \\ 2 \\ 3 \end{array}
\begin{array}{c}
\begin{array}{ccccc} 0 & 4 & 1 & 2 & 3 \end{array} \\
\left[\begin{array}{ccc|cc}
1 & 0 & 0 & 0 & 0 \\
0 & 1 & 0 & 0 & 0 \\
\hline
p_{10}z^{T_1} & 0 & 0 & p_{12}z^{T_1} & 0 \\
0 & 0 & p_{21}z^{T_2} & 0 & p_{23}z^{T_2} \\
0 & p_{34}z^{T_3} & 0 & 0 & p_{33}z^{T_3}
\end{array}\right]
\end{array}
$$

Therefore,

$$
R = \begin{bmatrix}
p_{10}z^{T_1} & 0 \\
0 & 0 \\
0 & p_{34}z^{T_3}
\end{bmatrix}
$$

and

$$
Q = \begin{bmatrix}
0 & p_{12}z^{T_1} & 0 \\
p_{21}z^{T_2} & 0 & p_{23}z^{T_2} \\
0 & 0 & p_{33}z^{T_3}
\end{bmatrix}
$$

The product Q^3R is calculated to be

$$
X = Q^3R =
\begin{array}{c} 1 \\ 2 \\ 3 \end{array}
\begin{array}{c}
\begin{array}{cc} \quad 0 & \qquad\qquad\qquad 4 \end{array} \\
\left[\begin{array}{cc}
0 & p_{12}p_{23}p_{33}p_{34}z^{T_1+T_2+2T_3} \\
p_{21}p_{12}p_{21}p_{10}z^{2T_1+2T_2} & p_{21}p_{12}p_{23}p_{34}z^{T_1+2T_2+T_3} + p_{23}p_{33}^2p_{34}z^{T_2+3T_3} \\
0 & p_{33}^3p_{34}z^{4T_3}
\end{array}\right]
\end{array}
$$

From this matrix it is seen that there are no paths of length exactly 4 between state 1 and state 0. This can be verified with a study of Figure 5-16. There is, however, one path of length 4 between state 2 and state 0; the path followed is 2–1–2–1–0. There are two paths of length 4 between state 2 and state 4; the paths followed are 2–1–2–3–4 and 2–3–3–3–4.

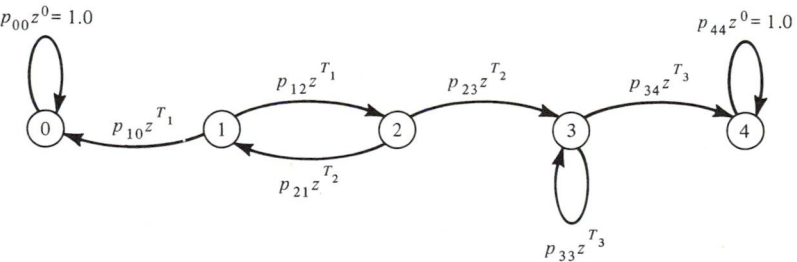

FIGURE 5-16. State-transition diagram for Example 5-4.

Example 5-4 leads to the conclusion that the products of the path labels on all possible paths from inner states to end states through the state transition diagram are enumerated using matrix products of the form $\mathbf{Q}^n\mathbf{R}$. In addition, the infinite matrix sum

$$\mathbf{Y} = \mathbf{R} + \mathbf{QR} + \mathbf{Q}^2\mathbf{R} + \mathbf{Q}^3\mathbf{R} + \cdots \tag{5-99}$$

enumerates all paths of all lengths between inner states and end states. Specifically,

$$y_{jk} = \sum_{l \in L(j,k)} B(l,z) \tag{5-100}$$

where $L(j,k)$ is the set of all paths beginning at state j and ending at state k.

Return now to the system defined by Figures 5-14 and 5-15. The average time to reject an incorrect cell is given by (5-97). In this equation, L denotes all paths between state 1 and state 0 or 6; thus

$$L = L(1,0) + L(1,6) \tag{5-101}$$

and (5-97) becomes

$$
\begin{aligned}
T_{\text{da}} &= \sum_{l \in L(1,0)} \left[\frac{d}{dz} B(l,z) \right]_{z=1} + \sum_{l \in L(1,6)} \left[\frac{d}{dz} B(l,z) \right]_{z=1} \\
&= \left\{ \frac{d}{dz} \left[\sum_{l \in L(1,0)} B(l,z) \right] + \frac{d}{dz} \left[\sum_{l \in L(1,6)} B(l,z) \right] \right\}_{z=1} \\
&= \left[\frac{d}{dz} (y_{10}) + \frac{d}{dz} (y_{16}) \right]_{z=1}
\end{aligned}
\tag{5-102}
$$

The derivative of a matrix is the matrix of the derivatives of the elements of the original matrix. Thus, each of the terms on the last line of (5-102) can be calculated from $\dfrac{d}{dz}\{\mathbf{Y}\}$. Equation (5-99) can be rewritten as

$$
\begin{aligned}
\mathbf{Y} &= (\mathbf{1} + \mathbf{Q} + \mathbf{Q}^2 + \cdots)\mathbf{R} \\
&= (\mathbf{1} - \mathbf{Q})^{-1}\mathbf{R}
\end{aligned}
\tag{5-103}
$$

The last equality is derived by defining

$$\mathbf{A} = \mathbf{1} + \mathbf{Q} + \mathbf{Q}^2 + \cdots \tag{5-104}$$

Then

$$\mathbf{AQ} = \mathbf{Q} + \mathbf{Q}^2 + \mathbf{Q}^3 + \cdots \tag{5-105}$$

and

$$\mathbf{A} - \mathbf{AQ} = \mathbf{A}(\mathbf{1} - \mathbf{Q}) = \mathbf{1} \tag{5-106}$$

so that

$$\mathbf{A} = (\mathbf{I} - \mathbf{Q})^{-1} \tag{5-107}$$

The derivative with respect to z of (5-103) is

$$\frac{d}{dz}(\mathbf{Y}) = (\mathbf{I} - \mathbf{Q})^{-1}\left(\frac{d}{dz}\mathbf{R}\right) + \left[\frac{d}{dz}(\mathbf{I} - \mathbf{Q})^{-1}\right]\mathbf{R}$$

$$= (\mathbf{I} - \mathbf{Q})^{-1}\left(\frac{d}{dz}\mathbf{R}\right) - (\mathbf{I} - \mathbf{Q})^{-1}\left[\frac{d}{dz}(\mathbf{I} - \mathbf{Q})\right](\mathbf{I} - \mathbf{Q})^{-1}\mathbf{R} \tag{5-108}$$

where the last equality follows from the matrix identity

$$\frac{d}{dz}\mathbf{A}^{-1} = -\mathbf{A}^{-1}\left(\frac{d}{dz}\mathbf{A}\right)\mathbf{A}^{-1} \tag{5-109}$$

At this point it is convenient to define a diagonal $n \times n$ matrix of integration times by

$$\mathbf{T} = \begin{bmatrix} T_1 & & & \\ & T_2 & & 0 \\ & & \ddots & \\ & 0 & & T_n \end{bmatrix} \tag{5-110}$$

where n is the number of internal states in the state transition diagram. With this definition it is seen that

$$\left(\frac{d}{dz}\mathbf{R}\right)_{z=1} = (\mathbf{TR})_{z=1} \tag{5-111}$$

and that

$$\left[\frac{d}{dz}(\mathbf{I} - \mathbf{Q})\right]_{z=1} = (-\mathbf{TQ})_{z=1} \tag{5-112}$$

Using these equations in (5-108) with $z = 1$,

$$\left(\frac{d}{dz}\mathbf{Y}\right)_{z=1} = \{(\mathbf{I} - \mathbf{Q})^{-1}\mathbf{TR} + (\mathbf{I} - \mathbf{Q})^{-1}\mathbf{TQ}(\mathbf{I} - \mathbf{Q})^{-1}\mathbf{R}\}_{z=1}$$

$$= \{(\mathbf{I} - \mathbf{Q})^{-1}\mathbf{T}[\mathbf{R} + \mathbf{Q}(\mathbf{I} - \mathbf{Q})^{-1}\mathbf{R}]\}_{z=1}$$

$$= \{(\mathbf{I} - \mathbf{Q})^{-1}\mathbf{T}[\mathbf{I} + \mathbf{Q}(\mathbf{I} - \mathbf{Q})^{-1}]\mathbf{R}\}_{z=1} \tag{5-113}$$

$$= \{(\mathbf{I} - \mathbf{Q})^{-1}\mathbf{T}[\mathbf{I} + \mathbf{Q}(\mathbf{I} + \mathbf{Q} + \mathbf{Q}^2 + \cdots)]\mathbf{R}\}_{z=1}$$

$$= [(\mathbf{I} - \mathbf{Q})^{-1}\mathbf{T}(\mathbf{I} - \mathbf{Q})^{-1}\mathbf{R}]_{z=1}$$

Finally, combining (5-102) and (5-113), the average time required to reject an incorrect phase cell is the sum of the elements on the first row of the last line above. Knowledge of all transition probabilities and integration times is required to evaluate this matrix equation. Since incorrect phase cells are being considered, the transition probabilities are calculated using the techniques described previously for the fixed-integration-time detector with noise alone as the detector input. It is assumed that the thresholds and integration times for each state are given.

The probability of detecting the correct phase cell when it is tested. P_d, is evaluated using the same techniques used to calculate T_{da}. In the example system of Figure 5-14, the probability of detection is the probability of passing from state 1 to the "enter code tracking mode" function. In any case except the noiseless case, arrival in the code tracking mode also implies that at some future time the system will pass through states 4 and 5 and will arrive at state 6. The probability of this event is unity since there is no other path to an end state and the system is guaranteed to eventually reach an end state [34]. Therefore, P_d is the probability of passing from state 1 to state 6. This probability is

$$P_d = \sum_{l \in L(1,6)} \Pr(l) = \sum_{l \in L(1,6)} B(l,z)_{z=1} \tag{5-114}$$

Using (5-100),

$$P_d = (y_{16})_{z=1}$$

and from (5-103), P_d is the element of the first row and second column of

$$(\mathbf{Y})_{z=1} = [(\mathbf{I} - \mathbf{Q})^{-1}\mathbf{R}]_{z=1} \tag{5-115}$$

When P_d is evaluated, the transition probabilities must be calculated for signal plus noise input to the energy detector. Thus two sets of transition probabilities must be calculated in order to evaluate average synchronization time. One set is calculated for the noise alone case and is used to evaluate T_{da}. The other set is calculated for the signal plus noise case and is used to calculate P_d.

To complete the calculation of average synchronization time, the average time used in evaluating the correct phase cell, T_i, must be determined. This calculation is identical to the calculation of T_{da} except that the transition probabilities for signal plus noise input to the detector are used, and the integration times in all states following the "enter tracking mode" block are set to zero. These integration times are set to zero so that drop-lock time will not be included in T_i. The *drop-lock time* is the average time that the system will remain cycling through the states following the tracking mode block when the signal is indeed present. Equation (5-113) yields the average times required to pass from state 1 to state 0 or 6. The path from state 1 to state 6 includes the drop-lock time unless the integration times mentioned are set to zero.

All the tools are now in place to calculate mean synchronization time for a spread-spectrum system using stepped serial search with multiple-dwell detection. First, the transition probabilities are calculated for the noise alone and signal plus noise cases. Integration times and thresholds are assumed known. Next, T_{da} is evaluated using (5-113) and (5-102). P_d is evaluated using (5-115), and T_i is evaluated using (5-113). Finally, T_{da}, P_d, and T_i are substituted into (5-12) to calculate $\overline{T}_s$. Note that average values for T_{da} and T_i can be used in (5-12) since both terms appear as linear factors. The student is cautioned that the same substitution cannot be used to calculate the variance of the synchronization time.

EXAMPLE 5-5

Reconsider the system of Examples 5-1 and 5-3. The spreading code chip rate is $f_c = 3$ MHz, the design point SNR $= 46.25$ dB-Hz, and the propagation delay uncertainty is ± 1.2 ms. Assume again that the serial-search step size is $\frac{1}{2}$ chip, so that $C = 14{,}400$. Assume that a double-dwell detection system is used which has a state-transition diagram as shown in Figure 5-16. The IF filter bandwidth is 24 kHz. Arbitrarily choose time–bandwidth products for integrations 1, 2, and 3 of 4, 10, and 50, respectively. Also, arbitrarily choose signal plus noise transition probabilities of $p_{12} = p_{23} = 0.9$ and $p_{33} = 0.99$. Calculate the mean synchronization time for this system.

Solution: The integration times for the stated time–bandwidth products are $T_1 = 167\ \mu s$, $T_2 = 417\ \mu s$, and $T_3 = 2.08$ ms. The signal-to-noise ratio is $P/N_0 = 4.22 \times 10^4$. The normalized noncentrality parameters are calculated from (5-65); the results are $\lambda_1 = 14.07$, $\lambda_2 = 35.1$, and $\lambda_3 = 176$. The modified number of degrees of freedom, calculated from (5-68) are $n'_1 = 13.5$, $n'_2 = 33.7$, and $n'_3 = 169$. Using these values along with the specified P_d's in the nomogram of Figure 5-11 yields normalized modified thresholds $V'_{T1} = 7.7$, $V'_{T2} = 24$, and $V'_{T3} = 129$. Using (5-69) the thresholds are found to be $V_{T1} = 12.6$, $V_{T2} = 39.3$, and $V_{T3} = 211.5$. The transition probabilities with noise alone at the detector input will be distinguished with a prime. The transition probabilities p'_{12}, p'_{23}, and p'_{33} are read from the nomogram using the given time–bandwidth products and thresholds V_T. The results are $p'_{12} = 0.12$, $p'_{23} = 0.007$, and $p'_{33} < 10^{-6}$. Using all of these values, the state-transition diagrams for the noise alone input and the signal plus noise input are shown in Figure 5-17.

Equation (5-113) is evaluated with $z = 1.0$ to find T_{da}, the result is

$$\left(\frac{d}{dz}\mathbf{Y}\right)_{z=1} = \begin{bmatrix} 1.000 & -0.120 & 0 \\ -0.993 & 1.000 & -0.007 \\ 0 & 0 & 0.999999 \end{bmatrix}^{-1} \begin{bmatrix} 167\ \mu s & 0 & 0 \\ 0 & 417\ \mu s & 0 \\ 0 & 0 & 2.08\ \text{ms} \end{bmatrix}$$

$$\times \begin{bmatrix} 1.000 & -0.120 & 0 \\ -0.993 & 1.000 & -0.007 \\ 0 & 0 & 0.999999 \end{bmatrix}^{-1} \begin{bmatrix} 0.880 & 0 \\ 0 & 0 \\ 0 & 0.999999 \end{bmatrix}$$

$$= \begin{bmatrix} 246\ \mu s & 2.62\ \mu s \\ 489\ \mu s & 20.3\ \mu s \\ 0 & 2.08\ \text{ms} \end{bmatrix}$$

The average dwell time at an incorrect phase cell is $246\ \mu s + 2.62\ \mu s \approx 249\ \mu s$. The probability of detection of the correct phase cell is calculated using

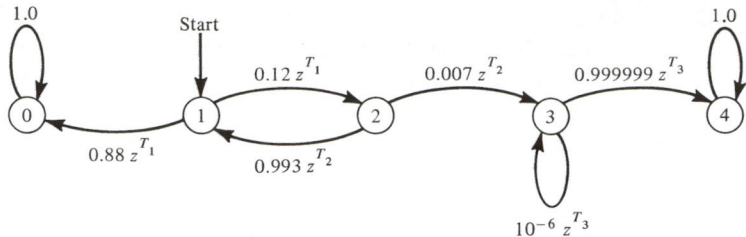

(a) Noise alone input

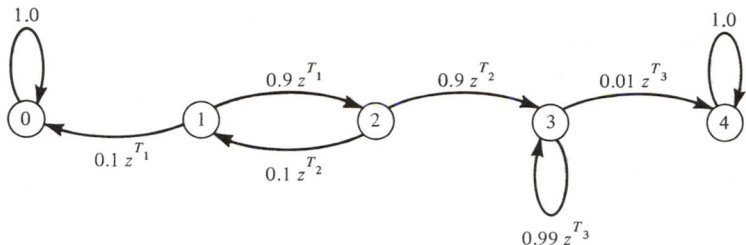

(b) Signal plus noise input

FIGURE 5-17. State-transition diagram for Example 5-5.

(5-115) with $z = 1$. The result is

$$(\mathbf{Y})_{z=1} = \begin{bmatrix} 1.0 & -0.90 & 0 \\ -0.10 & 1.0 & -0.90 \\ 0 & 0 & 0.01 \end{bmatrix}^{-1} \begin{bmatrix} 0.1 & 0 \\ 0 & 0 \\ 0 & 0.01 \end{bmatrix}$$

$$= \begin{bmatrix} 0.11 & 0.89 \\ 0.0011 & 0.99 \\ 0 & 1.0 \end{bmatrix}$$

In this calculation the matrix $(\mathbf{I} - \mathbf{Q})^{-1}\mathbf{R}$ has been evaluated using the transition probabilities of Figure 5-17b. The detection probability is the element in the first row and last column of this matrix; thus $P_d = 0.89$.

Next, T_i is evaluated using (5-113). The transition probabilities of Figure 5-17b are used, and T_3 is set to zero. The result is

$$\left(\frac{d}{dz}\mathbf{Y}\right)_{z=1} = \begin{bmatrix} 1.0 & -0.90 & 0 \\ -0.10 & 1.0 & -0.90 \\ 0 & 0 & 0.01 \end{bmatrix}^{-1} \begin{bmatrix} 167\,\mu s & 0 & 0 \\ 0 & 417\,\mu s & 0 \\ 0 & 0 & 0 \end{bmatrix}$$

$$\times \begin{bmatrix} 1.0 & -0.90 & 0 \\ -0.10 & 1.0 & -0.90 \\ 0 & 0 & 0.01 \end{bmatrix}^{-1} \begin{bmatrix} 0.10 & 0 \\ 0 & 0 \\ 0 & 0.01 \end{bmatrix}$$

$$= \begin{bmatrix} 24.7\,\mu s & 572\,\mu s \\ 7.1\,\mu s & 470\,\mu s \\ 0 & 0 \end{bmatrix}$$

The time T_i is the sum of terms in the first row of the final result; thus $T_i = 572\ \mu s + 24.7\ \mu s = 597\ \mu s$.

The final result for mean synchronization time from (5-12) is

$$\overline{T}_s = (14.399)(249\ \mu s)\left[\frac{2 - 0.89}{2(0.89)}\right] + \frac{597\ \mu s}{0.89}$$

$$= 2.24\ s + 671\ \mu s$$

$$\approx 2.24\ s$$

The relative magnitude of the two terms of (5-12) seen above is typical when a large number of cells are being evaluated. For this reason, the last term is usually not calculated.

Comparing the final results of Examples 5-3 and 5-5, it is concluded that a significant saving in average synchronization time is possible using multiple-dwell detection. Note, however, that neither example used optimum integration times and that the false-alarm penalty in Example 5-3 is larger than the equivalent penalty in Example 5-5. The conclusion that multiple-dwell systems do result in reduced synchronization times with respect to a fixed-integration-time system is valid.

Finally, the state transition diagrams for multiple-dwell detectors discussed above are typical, but many others are possible. Several types of transition diagrams are illustrated in Figure 5-18. Still other types are possible. It appears from some of these diagrams that time could be saved by beginning several integrations at the same time or, equivalently, using a single integrator which is sampled at appropriate times. The analysis of acquisition time presented above, however, is no longer applicable because it was presumed that the integrations for all dwells of the multiple-dwell detector are independent. Hall [36] and Hall and Weber [37] have presented the results of analysis of the multiple-dwell detector for which the integrator of Figure 5-13 is not reset for each dwell. Their result for the two-dwell case indicates that the performance gained by not resetting the integrator after the first dwell corresponds to less than 1 dB in input signal-to-noise ratio.

Other analytical techniques are available for predicting the performance of serial search systems that employ multiple-dwell detectors. DiCarlo [12] and DiCarlo and Weber [38] use flow graph techniques to calculate the mean and standard deviation of the synchronization time for a serial-search system that uses a multiple-dwell detector. Note that their result is the mean and variance of all synchronization time T_s (see Section 5-3.1) directly and not the average dwell time of the multiple-dwell detector itself. DiCarlo and Weber [38] develop the following formula for average synchronization time $\overline{T}_s$ for a serial-search system using an N-dwell detector.

$$\overline{T}_s = \frac{1}{2P_D}\sum_{j=1}^{N}\left[(C - 1)(2 - P_D)\left(kP_F\delta_{iN} + \prod_{i=1}^{j-1}p'_{i-1,i}\right) + 2\prod_{i=1}^{j-1}p_{i-1,i}\right]T_j \quad (5\text{-}116)$$

where $p'_{i-1,i}$ denoted a state-transition probability (see Figures 5-14 and 5-15) with

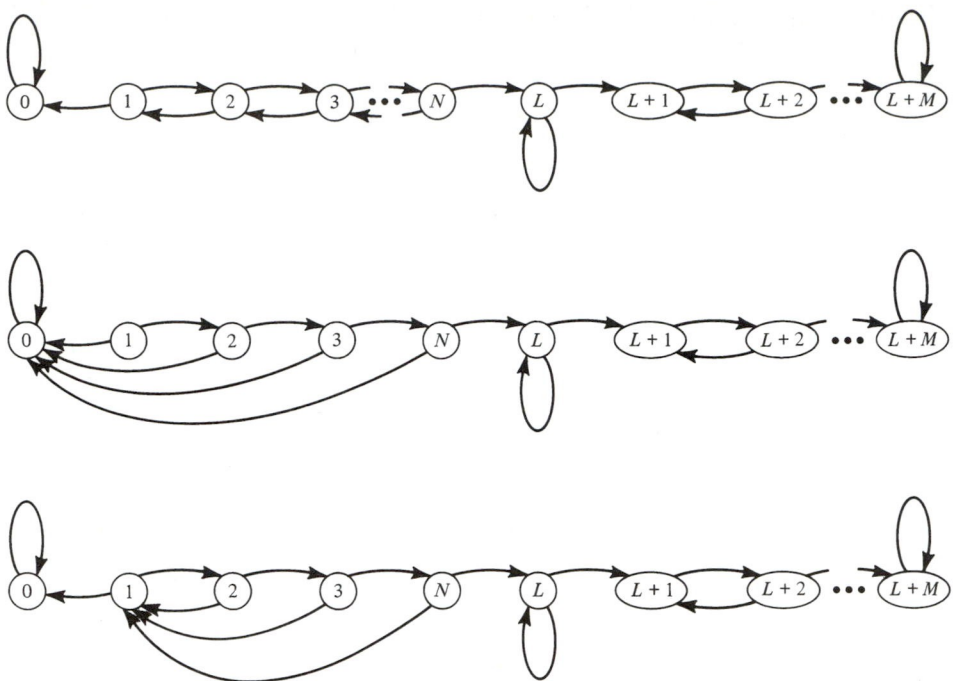

FIGURE 5-18. Typical state-transition diagram types for multiple-dwell detection.

noise alone at the output of the post-despreading IF filter (i.e., the false-alarm probabilities), $p_{i-1,i}$ denotes a state-transition transition probability with signal and noise at the IF filter output (i.e., the detection probabilities), δ_{ij} is the Kronecker delta function, T_j is the integration time for the jth integration, C is the number of phase cells, kT_N is the false-alarm penalty time (previously denoted T_{fa}),

$$P_D = \prod_{i=1}^{N} p_{i-1,i}$$

$$P_F = \prod_{i=1}^{N} p'_{i-1,i}$$

and the following definitions apply:

$$\prod_{i=1}^{0} p_{i-1,i} \equiv 1$$

$$\prod_{i=1}^{0} p'_{i-1,i} \equiv 1$$

This result applies to a multiple-dwell detector similar to that defined in Figure 5-14 except that N, rather than three, dwells are permitted. This result is based on the assumption that all integrations are independent and that a failure to detect a signal at any dwell will reject the trial code phase as shown in Figure 5-14.

Numerical results [12] obtained using (5-116) show that under certain conditions, synchronization-time improvements (relative to the synchronization time for a single-dwell detector) of more than a factor of 2 are possible using a three-dwell detector. Simon et al. [1, Sec. 1.3.2] present graphs illustrating the performance improvement possible using two- and three-dwell detectors; the improvement factor is plotted as a function of false-alarm probability with detection probability as a parameter. DiCarlo and Weber [38] also develop a formula for the standard deviation of the synchronization time. Other analyses of multiple-dwell serial-search code synchronization may be found in [16,17,39], where yet another analysis technique is used, which permits the calculation of the cumulative probability distribution of the synchronization time.

Sequential Detection. Consider the system illustrated in Figure 5-19 for detecting whether the phase of the receiver generated spreading waveform is correct. Once again, if the phase is correct, the received waveform will be despread, and signal power will appear at the output of the bandpass filter. The purpose of the envelope detector and the sequential detection processor is to detect this signal power reliably but without generating excessive false alarms when no signal is present. For any $K = 1, 2, \ldots$, let $\mathbf{x}_K = (x_1, x_2, x_3, \ldots, x_K)$ denote a sequence of samples of the envelope detector output, and assume that the joint pdf of these samples is known. This pdf is denoted $p_s(\mathbf{x}_K)$ when signal is present and $p_n(\mathbf{x}_K)$ when noise alone is present.

During the evaluation of a single spreading waveform phase, samples of the detector output are taken one at a time and input to the sequential detection proces-

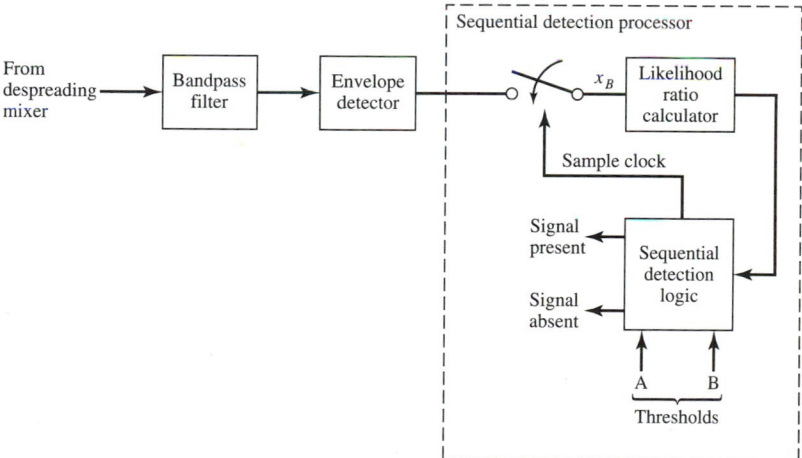

FIGURE 5-19. Sequential detector.

sor. After each sample (including the first), the likelihood ratio calculator computes

$$\Lambda(x_1, x_2, \ldots, x_K) = \frac{p_s(\mathbf{x}_K)}{p_n(\mathbf{x}_K)} \tag{5-117}$$

This likelihood ratio is input to the sequential detection logic. The logic function compares $\Lambda(\mathbf{x}_K)$ with an upper threshold A and a lower threshold B where $A > B$. If $\Lambda(\mathbf{x}_K) > A$, the signal is declared present and the test ends. If $\Lambda(\mathbf{x}_K) < B$, the signal is declared absent and the test ends. However, if $B < \Lambda(\mathbf{x}_K) < A$, no decision is made about the presence or absence of the signal and test continues by taking another sample, calculating another likelihood ratio, and so on.

The likelihood ratio indicates whether the sequence of samples is more likely to have resulted from signal and noise at the filter output, $\Lambda(\mathbf{x}_K) > 1.0$, or from noise alone, $\Lambda(\mathbf{x}_K) < 1.0$. When $\Lambda(\mathbf{x}_K) \gg 1.0$, the sequence of samples is much more likely to have come from signal and noise and a decision that the spreading waveform phase is correct is very reliable. Similarly, when $\Lambda(\mathbf{x}_K) \ll 1.0$ it is much more likely that the sequence of samples are the result of noise alone. The thresholds A and B are selected so that a specific reliability is achieved on the signal present/ absent decision. Figure 5-20 illustrates $p_s(x_1)$ and $p_n(x_1)$ for three possible first sample values. For sample S_1, $\Lambda(x_1) \ll 1.0$; for sample S_2, $\Lambda(x_1) = 1.0$; for sample S_3, $\Lambda(x_1) \gg 1.0$. The probability density function for signal plus noise shown in this figure is calculated for a signal-to-noise ratio of about $+3$ dB.

Assume that the samples of the envelope detector output are spaced sufficiently

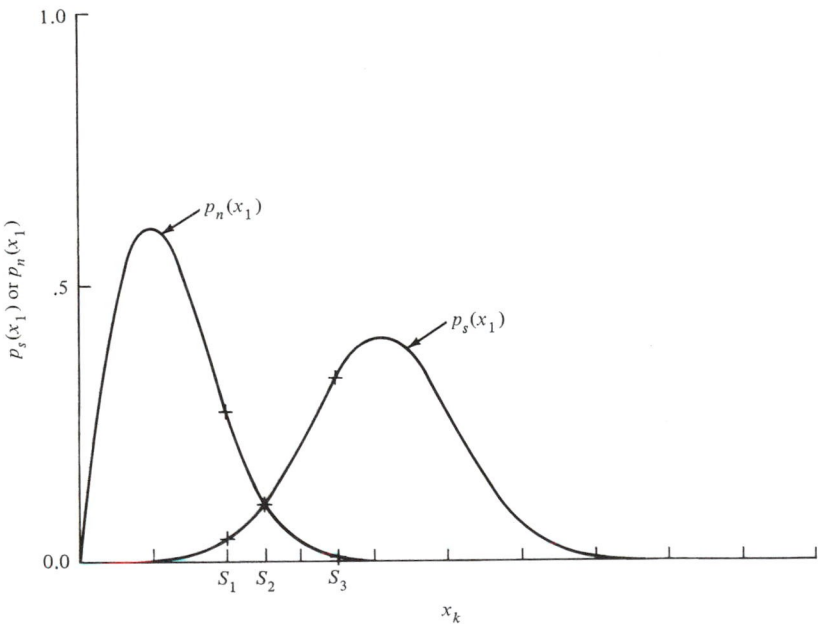

FIGURE 5-20. Envelope detector output probability density function.

in time that they may be considered independent. Then the joint pdf's are products of single sample pdf's, that is,

$$p_s(\mathbf{x}_K) = \prod_{k=1}^{K} p_s(x_k) \tag{5-118a}$$

$$p_n(\mathbf{x}_K) = \prod_{k=1}^{K} p_n(x_k) \tag{5-118b}$$

and

$$\Lambda(\mathbf{x}_K) = \prod_{k=1}^{K} \lambda(x_k) \tag{5-119}$$

where

$$\lambda(x_k) \triangleq \frac{p_s(x_k)}{p_n(x_k)} \tag{5-120}$$

When signal is present at the design point SNR, envelope detector samples usually result in $\lambda(x_k) > 1.0$ and the product $\Lambda(\mathbf{x}_K)$ grows with increasing K. When no signal is present, the envelope detector samples usually result in $\lambda(x_k) < 1.0$ and the product $\Lambda(\mathbf{x}_K)$ approaches zero as K increases. In the limit as $K \to \infty$, $\Lambda(\mathbf{x}_K)$ will always cross the upper threshold when signal is present at the design point SNR or the lower threshold when no signal is present. For finite K, errors can be made. A missed detection occurs whenever $\Lambda(\mathbf{x}_K)$ crosses the lower threshold before crossing the upper threshold when signal is indeed present. The probability of this event is $1.0 - P_d$. A false alarm occurs whenever $\Lambda(\mathbf{x}_K)$ crosses the upper threshold prior to crossing the lower threshold when no signal is present. The probability of this event is P_{fa}. Typical sequences of values of $\Lambda(\mathbf{x}_K)$ for signal plus noise and noise alone are illustrated in Figure 5-21. No errors occur in the cases shown. Observe that the sample values move toward one of the thresholds with a step size that is a random variable. The statistics of this random variable and the absolute value of the thresholds will determine the average number of samples that must be processed to cross one of the thresholds. The average sample number (ASN) is directly related to the quantities T_{da} and T_i, which must be evaluated to calculate the average synchronization time. The ASN is a function of the received signal-to-noise ratio, the detector characteristic, and the thresholds A and B.

Detectors having the general form shown in Figure 5-19 are called sequential detectors and were first discussed in Wald [40]. A great deal of literature is available on sequential detection. Two particularly good discussions are found in Helstrom [41] and Hancock and Wintz [42]. Sequential detection is sometimes referred to as the *sequential probability ratio test* (SPRT). It has been proven that sequential detection is optimum in the sense that it yields the minimum average detection time for a specified P_d and P_{fa} [43].

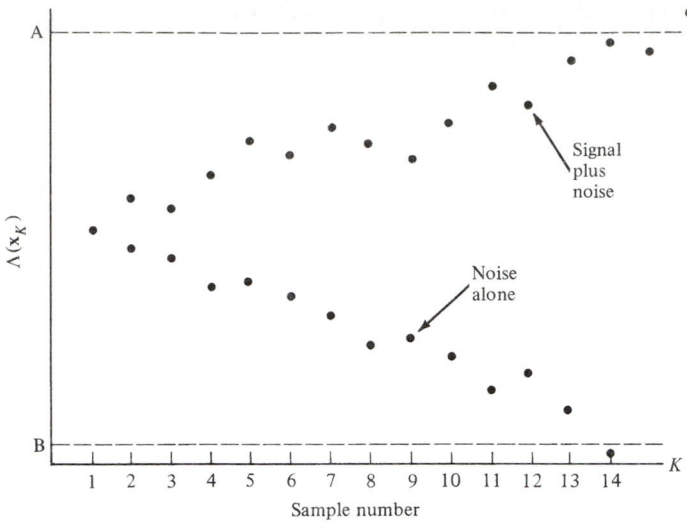

FIGURE 5-21. Typical sequences of values of $\Lambda(\mathbf{x}_K)$ for sequential detection.

Relationship Between Thresholds and Error Probabilities

To design systems that use sequential detection, it is necessary to be able to relate the thresholds and the probabilities P_d and P_{fa} for any signal-to-noise ratio. Consider the set of all real vectors $\mathbf{x}_K$ for all $K = 1, 2, \ldots$, and assume that the functions $p_s(\mathbf{x}_K)$ and $p_n(\mathbf{x}_K)$ are known for the particular SNR. Let Γ_1 denote the set of all vectors that result in a decision that signal is present at the output of the IF bandpass filter. Thus, for every $\mathbf{x}_K \in \Gamma_1$,

$$A < \frac{p_s(\mathbf{x}_K)}{p_n(\mathbf{x}_K)} \tag{5-121}$$

Note that all vectors in Γ_1 do not have the same dimensionality since different sequences $\mathbf{x}_K$ may result in a threshold crossing after different numbers of samples are taken. Assume now that the bandpass filter output is noise alone. A false alarm occurs for any $\mathbf{x}_K \in \Gamma_1$ and the probability of this event is

$$P_{\text{fa}} = \int_{\mathbf{x}_K \in \Gamma_1} p_n(\mathbf{x}_K) \, d\mathbf{x}_K \tag{5-122a}$$

Similarly, when the filter output is signal plus noise, a correct detection occurs whenever $\mathbf{x}_K \in \Gamma_1$, and the probability of this event is

$$P_d = \int_{\mathbf{x}_K \in \Gamma_1} p_s(\mathbf{x}_K) \, d\mathbf{x}_K \tag{5-122b}$$

Multiplying both sides of (5-121) by $p_n(\mathbf{x}_K)$, integrating over Γ_1, and using the results of (5-122) yields

$$P_d > AP_{\text{fa}} \tag{5-123}$$

Let Γ_2 denote the set of all vectors which result in a decision that noise alone is present at the output of the filter. For every $\mathbf{x}_K \in \Gamma_2$,

$$\frac{p_s(\mathbf{x}_K)}{p_n(\mathbf{x}_K)} < B \tag{5-124}$$

Assume that the bandpass filter output is noise alone. A correct dismissal occurs for all $\mathbf{x}_K \in \Gamma_2$, and this event occurs with probability

$$1 - P_{\text{fa}} = \int_{\mathbf{x}_K \in \Gamma_2} p_n(\mathbf{x}_K) \, d\mathbf{x}_K \tag{5-125a}$$

When the filter output is signal plus noise, a missed detection occurs when $\mathbf{x}_K \in \Gamma_2$. This event occurs with probability

$$1 - P_d = \int_{\mathbf{x}_K \in \Gamma_2} p_s(\mathbf{x}_K) \, d\mathbf{x}_K \tag{5-125b}$$

Combining (5-124) and (5-125) results in

$$1 - P_d < B(1 - P_{\text{fa}}) \tag{5-126}$$

If (5-123) and (5-126) were equalities, the thresholds could be solved for as functions of P_d and P_{fa}. For very low signal-to-noise ratios, these equalities can be assumed with little error. To see this, reconsider the set Γ_1 and (5-121). Suppose that Γ_1' is defined as the set of all $\mathbf{x}_K$ which lead to a decision that signal is present and for which

$$A < \frac{p_s(\mathbf{x}_K)}{p_n(\mathbf{x}_K)} < A + \Delta A \tag{5-127}$$

The set Γ_1' is a subset of Γ_1. For small signal-to-noise ratios, there is a correspondingly small difference in the pdf's $p_n(x_1)$ and $p_s(x_1)$, so that samples of the likelihood ratio are typically near unity, although large or small values do occur from time to time. Thus, from one sample to the next, small changes in $\Lambda(\mathbf{x}_K)$ are typical. Since the changes in $\Lambda(\mathbf{x}_K)$ are small, it is likely that the value of $\Lambda(\mathbf{x}_K)$ after the sample, which results in the threshold crossing is very close to the threshold. This in turn implies that the most likely $\mathbf{x}_K$ which result in an upper threshold crossing are in Γ_1'. Let $\overline{\Gamma}_1'$ denote the set of all $\mathbf{x}_K$ in Γ_1 but not in Γ_1', that is, $\Gamma_1' + \overline{\Gamma}_1' = \Gamma_1$. Equation (5-122) can be rewritten as

$$P_{\text{fa}} = \int_{\mathbf{x}_K \in \Gamma_1'} p_n(\mathbf{x}_K) \, d\mathbf{x}_K + \int_{\mathbf{x}_K \in \overline{\Gamma}_1'} p_n(\mathbf{x}_K) \, d\mathbf{x}_K \tag{5-128a}$$

$$P_d = \int_{\mathbf{x}_K \in \Gamma_1'} p_s(\mathbf{x}_K) \, d\mathbf{x}_K + \int_{\mathbf{x}_K \in \overline{\Gamma}_1'} p_s(\mathbf{x}_K) \, d\mathbf{x}_K \tag{5-128b}$$

Based on the discussion above, it is reasonable to assume that the rightmost term in

(5-125) vanishes as SNR approaches zero. Multiplying both sides of (5-127) by $p_n(\mathbf{x}_K)$ and integrating over Γ'_1 yields

$$AP_{\text{fa}} < P_d < (A + \Delta A)P_{\text{fa}} \qquad (5\text{-}129)$$

As SNR approaches zero, this equation remains valid for vanishingly small ΔA and

$$AP_{\text{fa}} = P_d \qquad (5\text{-}130)$$

is valid in the limit. A similar argument leads to

$$1 - P_d = B(1 - P_{\text{fa}}) \qquad (5\text{-}131)$$

These equations can be solved for P_d and P_{fa}, the result is

$$P_{\text{fa}} = \frac{1 - B}{A - B} \qquad (5\text{-}132\text{a})$$

$$P_d = A\left(\frac{1 - B}{A - B}\right) \qquad (5\text{-}132\text{b})$$

It cannot be overemphasized that the (5-130) and (5-131) are valid only for SNR approaching zero. Inaccuracies occur when the value of $\Lambda(\mathbf{x}_K)$ that first crosses the threshold is significantly different from the threshold. This problem has been discussed in detail in Wald [40], Helstrom [41], and Hancock and Wintz [42] and is called the *excess over boundary problem*.

When the SNR is not small and the excess over boundary problem cannot be ignored, the designer normally resorts to simulations to characterize the system. Simulations of these systems are straightforward and provide the only reasonable method of performance characterization when SNR is medium to high. The computer simulation also enables the designer to include the effects of different detector types as well as the effects of dependent samples. Additional discussion of simulators for sequential detection can be found in Cobb and Darby [44].

Operating Characteristic Function

In the discussion above it was assumed that the SNR was fixed and known. It is usually necessary to be able to calculate the performance of the system at SNRs other than the design point. Examination of (5-132) indicates that the SNR does not enter in the relationship between the thresholds and P_d and P_{fa}. The test itself, however, is a function of SNR since $p_n(\mathbf{x}_K)$ and $p_s(\mathbf{x}_K)$ are functions of SNR. Thus, the relationships of (5-132) are valid only when the likelihood ratio calculation and the received SNR are matched. Suppose that the system is designed for a particular design point SNR and that the actual SNR is different from the design point. When noise-alone tests are being performed, the system will operate as designed; that is, P_{fa} is not affected by the change in SNR. When signal is present, however, modified performance will be observed and P_d will not necessarily equal the design point P_d. The actual P_d as a function of SNR for a sequential test designed for SNR_0 is related to the *operating characteristic function* (OCF), which is usually denoted $L(\text{SNR})$.

The OCF has been derived in Wald [40] and is also discussed in Helstrom [41] and Hancock and Wintz [42]. The OCF is defined as the probability that the sequential test will result in a decision that noise alone is present when actually signal is present at a specified SNR. Suppose that the SPRT is designed using $p_n(x_1)$ and $p_{s_0}(x_1)$ but that the actual detector output pdf with signal present is $p_s(x_1)$. Consider a new pdf $p(x_1)$, which is defined by

$$p(x_1) = \left[\frac{p_{s_0}(x_1)}{p_n(x_1)} \right]^h p_s(x_1) \tag{5-133}$$

where the parameter h is selected so that

$$\int_{-\infty}^{\infty} p(x_1) \, dx_1 = 1.0 \tag{5-134}$$

With this choice of h, $p(x_1)$ is a valid probability density function. An SPRT is designed to distinguish between the density functions $p(x_1)$ and $p_s(x_1)$. Assuming independent samples, this SPRT employs the likelihood ratio

$$\Lambda(\mathbf{x}_K) = \prod_{k=1}^{K} \frac{p(x_k)}{p_s(x_k)} \tag{5-135}$$

and compares the result with two thresholds $A' = A^h$ and $B' = B^h$, where A and B are thresholds used in the original SPRT. The new SPRT estimates that $p_s(x_1)$ is the correct pdf if $\Lambda(\mathbf{x}_k) < B'$ and that $p(x_1)$ is the correct pdf if $\Lambda(\mathbf{x}_K) > A'$. The likelihood ratio for the original test is

$$\Lambda_0(\mathbf{x}_K) = \prod_{k=1}^{K} \frac{p_{s_0}(x_k)}{p_n(x_k)} \tag{5-136}$$

From (5-133), (5-134), and (5-135) it is found that $[\Lambda_0(\mathbf{x}_K)]^h = \Lambda(\mathbf{x}_K)$. Thus, whenever $\Lambda(\mathbf{x}_K)$ crosses the threshold A' or B', $\Lambda_0(\mathbf{x}_K)$ crosses the threshold A or B, respectively.

The value of the operating characteristic function is the probability that the original test estimates that $p_n(x_1)$ is the correct pdf when $p_s(x_1)$ is correct. This probability is equal to the probability that threshold B is crossed by the original test when $p_s(x_1)$ is true, which is equivalent to the probability that threshold B' is crossed by the new test when $p_s(x_1)$ is true. The latter probability is the probability that $p_s(x_1)$ is estimated to be the correct pdf by the new test when $p_s(x_1)$ is true. The derived test is matched to its input statistics so that (5-132a) can be used to determine the probability of its estimating that $p_s(x_1)$ is correct when $p_s(x_1)$ is indeed correct. This probability is

$$1 - P_{\text{fa}} = 1 - \frac{1 - B^h}{A^h - B^h} = \frac{A^h - 1}{A^h - B^h} = L(\text{SNR}) \tag{5-137}$$

The probability of detecting the signal at the received SNR is the probability of

$\Lambda_0(\mathbf{x}_K)$ crossing the upper threshold when $p_s(x_1)$ is true. This probability is

$$P_d(\text{SNR}) = 1 - L(\text{SNR}) = \frac{1 - B^h}{A^h - B^h} \tag{5-138}$$

Observe that when the received SNR is zero, $p_s(x_1) = p_n(x_1)$ and $h = +1$. From (5-138)

$$P_d(0) = \frac{1 - B}{A - B} \tag{5-139}$$

which is seen to equal P_{fa} for the original test as expected. Similarly, when the received SNR is equal to the design point SNR, $p_s(x_1) = p_{s_0(x_1)}$, and $h = -1$. From (5-138)

$$P_d(\text{SNR}_0) = \frac{1 - B^{-1}}{A^{-1} - B^{-1}} = A\left(\frac{1 - B}{A - B}\right) \tag{5-140}$$

which is equal to the design point P_d of (5-132b).

Unfortunately, calculation of the operating characteristic function and $P_d(\text{SNR})$ is often extremely difficult because of the difficulty of determining h from (5-133) and (5-134). When the detector output pdf's are Gaussian, however, a simple solution exists [42]. In most other cases simulations can be used to evaluate $P_d(\text{SNR})$ with little difficulty. An example operating characteristic function is shown in Figure 5-22. This example is the OCF for the SPRT of the mean of a Gaussian distribution. The design point mean is denoted μ and the actual mean is denoted θ. The test is designed to determine whether the detector output samples have a mean of zero

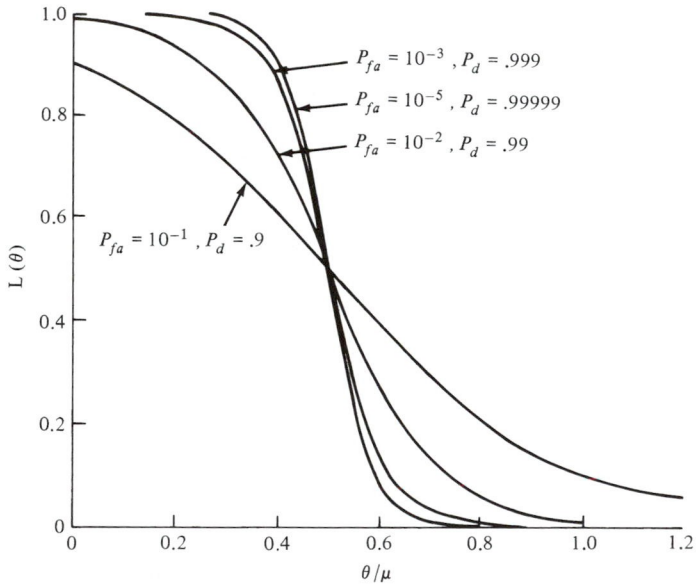

FIGURE 5-22. Operating characteristic function for the SPRT of the mean of a Gaussian distribution. (From Ref. 42.)

(noise alone) or a mean of θ (signal plus noise). Observe that at the design point $(\theta/\mu = 1.0)$, $L(\theta) = 1 - P_d$.

SPRT Using a Linear Envelope Detector and the Log-Likelihood Ratio

Although the sequential test can be analyzed in its most basic form where the actual likelihood ratio is calculated, the multiplications of the likelihood ratios as in (5-119) are inconvenient and not necessary. The performance of the SPRT is unchanged if, instead of using the likelihood ratio, any monotonic strictly increasing function of the likelihood ratio is used. Of course, the thresholds are also modified by this same monotonic function. Since the likelihood ratio is positive, the logarithm is a convenient function to use since it also converts the undesirable product functions to summations. At this point it is also convenient to limit attention to the specific detector and associated probability densities of most interest for application to spread-spectrum systems. The detector is a linear envelope detector and the synchronization hardware is trying to detect the presence/absence of a sinusoid at the bandpass filter output. Therefore, the pdf of a single sample of the envelope output is given by (5-48), which is due to Rice. The likelihood ratio, using (5-48) and (5-120), with slightly modified notation, is

$$\lambda(x_k) = \frac{p_s(x_k)}{p_n(x_k)}$$

$$= \frac{(x_k/N)\exp(-(x_k^2 + 2P)/2N)I_0(x_k\sqrt{2P}/N)}{(x_k/N)\exp(-x_k^2/2N)} \tag{5-141}$$

where $p_n(x_k)$ is found from (5-48) by setting the signal power $P = 0$. The output of the bandpass filter has been assumed to be AWGN with power N or to be AWGN with power N plus a sinusoid with power P and amplitude $\sqrt{2P}$. Simplifying (5-141) and taking the logarithm yields the log-likelihood ratio for a single sample

$$\ln[\lambda(x_k)] = -\frac{P}{N} + \ln\left[I_0\left(\frac{x_k\sqrt{2P}}{N}\right)\right] \tag{5-142}$$

The log-likelihood ratio for a sequence of samples is the logarithm of (5-119), which is

$$\ln[\Lambda(\mathbf{x}_K)] = \ln\left[\prod_{k=1}^{K}\lambda(x_k)\right]$$

$$= \sum_{k=1}^{K}\ln[\lambda(x_k)] \tag{5-143}$$

$$= \sum_{k=1}^{K}\left[\ln\left[I_0\left(\frac{x_k\sqrt{2P}}{N}\right)\right] - \frac{P}{N}\right]$$

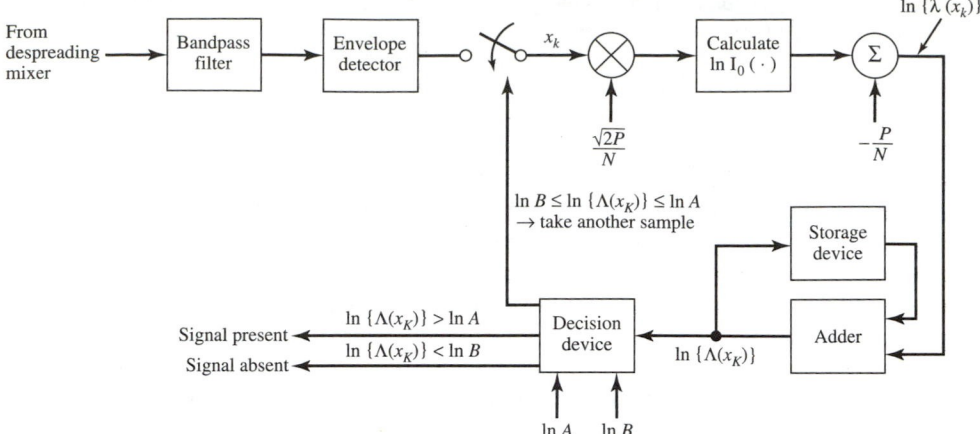

FIGURE 5-23. Sequential detector using linear envelope detector and log-likelihood ratio.

Figure 5-23 is a conceptual block diagram of a sequential detector that uses the simplifications discussed above. Samples of the linear envelope detector output are scaled by $\sqrt{2P}/N$ and input to the $\ln[I_0(\cdot)]$ calculator. The output of this function is input to the adder after the bias P/N is subtracted. The adder adds the new value $\ln[\lambda(x_k)]$ to the sum of all previous values to obtain $\ln[\Lambda(\mathbf{x}_K)]$. The final result is compared with an upper threshold $\ln A$ and a lower threshold $\ln B$. If the result is above the upper threshold, the signal is declared present. If the result is below the lower threshold, signal is declared absent. If the result is between the two thresholds, another sample is taken and the process repeats. This sequential detector is a special case of the sequential detector described earlier.

When the log-likelihood ratio is used in place of the likelihood ratio, the upper threshold, which was larger than unity is positive, but the lower threshold, which was less than unity, is negative. Figure 5-24 illustrates typical sequences of values of $\ln[\Lambda(\mathbf{x}_K)]$ leading to upper and lower threshold crossings for the log-likelihood SPRT. With this detector, sequences corresponding to signal plus noise typically have a positive slope and sequences corresponding to noise alone typically have a negative slope. It will be seen later that at SNRs lower than the design point SNR, the positive slope sequence of Figure 5-24 may have nearly a zero slope with the effect that many samples are required to reach either threshold.

Average Sample Number

The last performance measure that needs to be calculated in order to evaluate the mean synchronization time for the spread-spectrum system is the average sample number (ASN). The ASN is the average number of samples taken by the detector prior to crossing one of the thresholds. With noise alone at the filter output, ASN determines the average dwell time T_{da} at an incorrect spreading waveform phase.

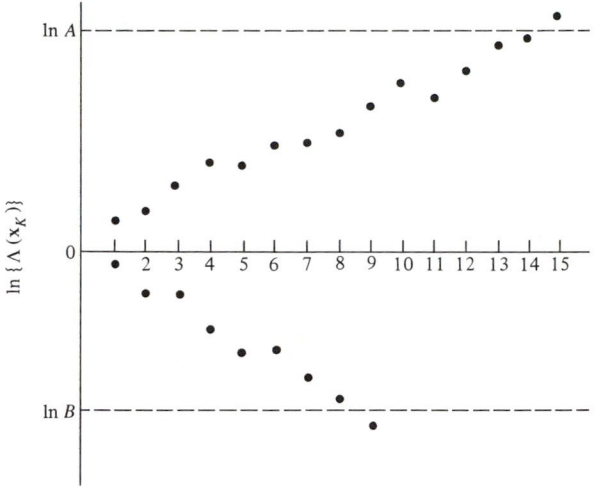

FIGURE 5-24. Typical sequences of values of $\ln[\Lambda(\mathbf{x}_K)]$ leading to upper and lower threshold crossing for SPRT using log-likelihood ratio.

Denote the ASN for noise alone by $\overline{K}_n$ and the sampling interval by T_s. Then

$$T_{da} = \overline{K}_n T_s + T_{fa} P_{fa} \tag{5-144}$$

where the false-alarm penalty time T_{fa} is assumed known. With signal and noise at the filter output, ASN determines the average time to evaluate the correct spreading waveform phase, T_i. Denote the ASN for signal plus noise by $\overline{K}_s$; then

$$T_i = \overline{K}_s T_s \tag{5-145}$$

In the calculation of either $\overline{K}_N$ or $\overline{K}_s$ the possibilities of the test ending with either an upper or a lower threshold crossing must be taken into account.

The derivation of the ASN begins with a short development from Kendall and Stuart [45], which is also presented in Helstrom [41]. Define a random variable y_k which is equal to 1 if the SPRT has not finished prior to the kth sample and thus observes x_k, and is equal to 0 otherwise. Let P_i denote the probability that a SPRT finishes at exactly the ith stage. The mean value of y_k is $E(y_k)$, where the expected value is over all possible runs of the sequential test. The value of $y_k = 1$ for all those tests which finish at or after the kth sample. The value of $y_k = 0$ for all other tests. Therefore,

$$E(y_k) = \sum_{i=1}^{k-1} 0 \cdot P_i + \sum_{i=k}^{\infty} 1 \cdot P_i$$

$$= \sum_{i=k}^{\infty} P_i \tag{5-146}$$

For a test that finishes on the Kth sample, the log-likelihood ratio at the end of the test is

$$\ln[\Lambda(\mathbf{x}_K)] = \sum_{k=1}^{K} \ln[\lambda(x_k)]$$

$$= \sum_{k=1}^{\infty} y_k \ln[\lambda(x_k)]$$

$$(5\text{-}147)$$

since $y_k = 1$ for $1 \le k \le K$ and $y_k = 0$ for $K < k$. The random variable y_k depends only on the values of the detector output x_1 through x_{k-1} since they determine whether or not a test finishes on or before the $(k - 1)$th stage and therefore whether a kth sample is required. Therefore, the random variables y_k and x_k are independent and the expected value of (5-147) over all tests can be written

$$E\{\ln[\Lambda(\mathbf{x}_K)]\} = E\left\{ \sum_{k=1}^{\infty} y_k \ln[\lambda(x_k)] \right\}$$

$$= \sum_{k=1}^{\infty} E(y_k)E\{\ln[\lambda(x_k)]\}$$

$$(5\text{-}148)$$

Since the detector output pdf is the same for every sample x_k, the right expected value above is not a function of k and may be factored, resulting in

$$E\{\ln[\Lambda(\mathbf{x}_K)]\} = E\{\ln[\lambda(x_k)]\} \sum_{k=1}^{\infty} E(y_k)$$

$$= E\{\ln[\lambda(x_k)]\} \sum_{k=1}^{\infty} \sum_{i=k}^{\infty} P_i$$

$$= E\{\ln[\lambda(x_k)]\} \sum_{k=1}^{\infty} kP_k$$

$$= E\{\ln[\lambda(x_k)]\}E(k)$$

$$(5\text{-}149)$$

This equation is Kendall's result stating that the mean value of the log-likelihood ratio at the end of the SPRT is equal to the product of the mean value of the log-likelihood ratio for each sample and the expected number of samples taken to complete the test. The expected value $E[k]$ is the average sample number being calculated.

Once again, it is necessary to make the assumption that excess over boundaries can be neglected. This implies tests using a large number of samples on the average are being considered which in turn implies low SNR and/or $1 - P_d \ll 1$ and $P_{\mathrm{fa}} \ll 1$. With this assumption, the value of the log-likelihood ratio at the end of the test is either $\ln A$ or $\ln B$. When the SPRT is evaluating incorrect phase cells (i.e., noise alone at the filter output), the value is $\ln A$ with probability P_{fa} and ln B with probability $1 - P_{fa}$. In this case,

$$E\{\ln[\Lambda(\mathbf{x}_k)]\} = P_{fa} \ln A + (1 - P_{fa}) \ln B \tag{5-150}$$

and the average sample number is

$$\overline{K}_n = \frac{P_{fa} \ln A + (1 - P_{fa}) \ln B}{E\{\ln[\lambda(x_k)]\}} \tag{5-151}$$

When the SPRT is evaluating the correct phase cell, the final log-likelihood ratio is ln A with probability P_d and ln B with probability $1 - P_d$. Recall that the probability of detection is given, in general, by the operating characteristic function; specifically $P_d = 1 - L(\text{SNR})$. Using the OCF,

$$E\{\ln[\Lambda(\mathbf{x}_k)]\} = L(\text{SNR}) \ln B + [1 - L(\text{SNR})] \ln A \tag{5-152}$$

and the ASN is

$$\overline{K}_s = \frac{L(\text{SNR}) \ln B + [1 - L(\text{SNR}) \ln A}{E\{\ln[\lambda(x_k)]\}} \tag{5-153}$$

At the design point SNR,

$$\overline{K}_s = \frac{(1 - P_d) \ln B + P_d \ln A}{E\{\ln[\lambda(x_k)]\}} \tag{5-154}$$

The equations above are all based on the assumption that excess over boundaries can be ignored.

The average sample numbers of (5-151), (5-153), and (5-154) can be evaluated if the average of the log-likelihood ratio for a single sample can be determined. The single sample log-likelihood ratio is given by (5-142). Since the ASN equations apply principally at low SNR, it is appropriate to attempt to evaluate the mean of (5-142) at low SNR. The zeroth-order modified Bessel function may be represented by the series [46]

$$I_0(z) = \sum_{m=0}^{\infty} \frac{z^{2m}}{2^{2m}(m!)^2}$$

$$= 1 + \frac{z^2}{4} + \frac{z^4}{64} + \frac{z^6}{2304} + \cdots \tag{5-155}$$

For small z, $I_0(z)$ can be approximated by the first three terms of the series. The logarithm of (5-155) can be simplified with the well-known series expansion

$$\ln(1 + \alpha) = \alpha - \tfrac{1}{2}\alpha^2 + \tfrac{1}{3}\alpha^3 - \cdots \tag{5-156}$$

Let $\alpha = \tfrac{1}{4}z^2 + \tfrac{1}{64}z^4$ and combine (5-155) and (5-156) to obtain

$$\ln[I_0(z)] \cong \left(\frac{z^2}{4} + \frac{z^4}{64}\right) - \frac{1}{2}\left(\frac{z^2}{4} + \frac{z^4}{64}\right)^2 + \frac{1}{3}\left(\frac{z^2}{4} + \frac{z^4}{64}\right)^3 - \cdots$$

$$\cong \frac{z^2}{4} - \frac{z^4}{64} \tag{5-157}$$

where all powers of z greater than 4 have been ignored. Substituting $z = x_k\sqrt{2P}/N$ yields

$$\ln\left[I_0\left(x_k\frac{\sqrt{2P}}{N}\right)\right] \approx \left(\frac{P}{2N^2}\right)x_k^2 - \left(\frac{P^2}{16N^4}\right)x_k^4 \tag{5-158}$$

The problem being addressed is the calculation of the mean of (5-142). Substituting (5-158) into (5-142) and taking the expected value results in

$$E\{\ln[\lambda(x_k)]\} \approx E\left[-\frac{P}{N} + \left(\frac{P}{2N^2}\right)x_k^2 - \left(\frac{P^2}{16N^4}\right)x_k^4\right] \tag{5-159}$$

$$= -\frac{P}{N} + \frac{P}{2N^2}E(x_k^2) - \left(\frac{P^2}{16N^4}\right)E(x_k^4)$$

The expected values in this equation are over all noise processes at the bandpass filter output. The random variables x_k are samples of the output of the envelope detector and have a pdf given by (5-48). This is the well-known Ricean pdf, and the expected values above are the second and fourth moments of this pdf. These moments have been calculated in Rice [21] and the results are

$$E(x_k^2) = 2N\Gamma(2)_1F_1\left(-1;\ 1;\ -\frac{P'}{N}\right) \tag{5-160a}$$

and

$$E(x_k^4) = (2N)^2\Gamma(3)_1F_1\left(-2;\ 1;\ -\frac{P'}{N}\right) \tag{5-160b}$$

where $_1F_1(a;\ b;\ z)$ is the confluent hypergeometric function, which has a series representation [21]

$$_1F_1(a,b,z) = 1 + \frac{az}{b} + \frac{a(a+1)}{b(b+1)}\frac{z^2}{2!} + \frac{a(a+1)(a+2)}{b(b+1)(b+2)}\frac{z^3}{3!} + \cdots \tag{5-161}$$

$\Gamma(n) = (n-1)!$ is the gamma function and P'/N is the SNR being considered. When the expected values are being evaluated for noise-alone phase cells, $P'/N = 0$, and when the expected values are being evaluated for the correct phase cells, $P'/N =$ (operating point SNR). The SNR used in (5-159) is the design point SNR, which may or may not equal the operating point SNR. Observe that the infinite series is truncated for $a = -1, -2$, so that

$$E(x_k^2) = 2N\left(1 + \frac{P'}{N}\right) = 2(N + P') \tag{5-162a}$$

$$E(x_k^4) = (2N)^2 2\left[1 + \frac{2P'}{N} + \frac{1}{2}\left(\frac{P'}{N}\right)^2\right] = 8(N^2 + 2P'N + \tfrac{1}{2}[P']^2) \tag{5-162b}$$

Substituting these relationships into (5-159) and simplifying yields

$$E\{\ln[\lambda(x_k)]\} \cong -\frac{1}{2}\left(\frac{P}{N}\right)^2 + \left(\frac{P}{N}\right)\left(\frac{P'}{N}\right) - \left(\frac{P}{N}\right)^2\left(\frac{P'}{N}\right) - \frac{1}{4}\left(\frac{P}{N}\right)^2\left(\frac{P'}{N}\right)^2$$

$$\cong -\frac{1}{2}\left(\frac{P}{N}\right)^2 + \left(\frac{P}{N}\right)\left(\frac{P'}{N}\right)$$

(5-163)

Observe that the average increment for incorrect phase cells is $-P^2/2N^2$ and that at the design point SNR (i.e., $P' = P$) the average increment is $+P^2/2N^2$. Combining this result with (5-151) and substituting into (5-144) gives the average dwell time at an incorrect phase cell for low SNR.

$$T_{\text{da}} = \frac{P_{\text{fa}}\ln A + (1 - P_{\text{fa}})\ln B}{-\frac{1}{2}(P/N)^2}T_s + T_{\text{fa}}P_{\text{fa}}$$

(5-164)

where P'/N is set equal to zero since noise alone phase cells are being considered. Similarly, the average time required to evaluate a correct phase cell is calculated from (5-163), (5-153), and (5-145)

$$T_i = \frac{L(\text{SNR})\ln B + [1 - L(\text{SNR})]\ln A}{-\frac{1}{2}(P/N)^2 + (P/N)(P'/N)}T_s$$

(5-165)

For $P'/N = P/2N$ this equation is indeterminate. The average evaluation time in this special case is calculated in Helstrom [41]. In this equation $L(\text{SNR})$ is evaluated at the operating point SNR.

At this point the student has all the tools necessary to calculate the mean synchronization time for a spread-spectrum system using stepped serial search with sequential detection. The sequential test is designed for a particular operating point SNR which is assumed to be low. P_d and P_{fa} are assumed to be given although their selection is part of a long optimization process. The thresholds A and B are calculated using (5-130) and (5-131). The sampling interval T_s is given but must be sufficiently long that independent samples result. The dwell times T_{da} and T_i are calculated from (5-164) and (5-165), where $L(\text{SNR}) = 1 - P_d$ at the design point. Average synchronization time is calculated by substituting P_d, T_{da}, and T_i into (5-12) together with the given C.

A typical plot of the normalized average sample number versus SNR when signal plus noise is being evaluated is illustrated in Figure 5-25. The particular curve shown is from Helstrom [41] and is calculated using a SPRT designed to distinguish two Gaussian probability density functions. The design point SNR is 1.0. Observe that ASN increases significantly in the midrange of SNR. At midrange SNRs the mean of the single sample log-likelihood ratio of (5-159) is near zero and the cumulative log-likelihood ratio wanders near zero. For SNRs in region 1, (5-159) is negative and mostly missed detections occur. For SNRs in region 2, (5-159) is positive and the signal is usually detected. The design point probability of detection is achieved only when the actual normalized signal-to-noise ratio is equal to or greater than 1.0.

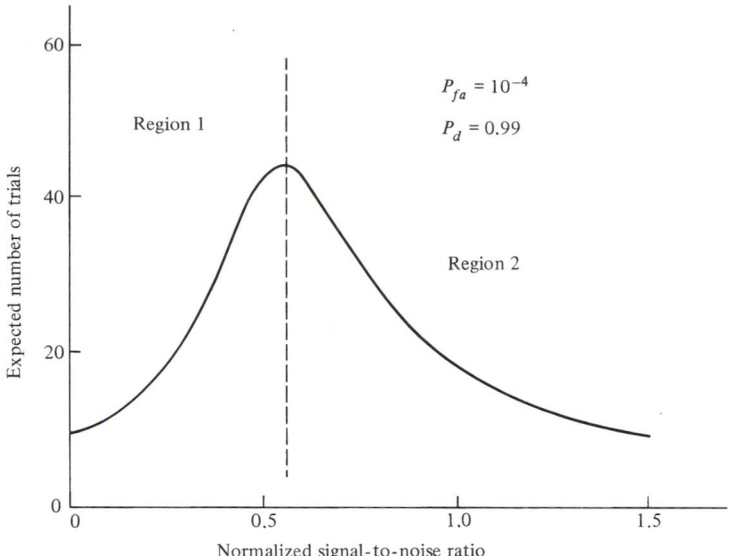

FIGURE 5-25. Typical plot of average sample number versus normalized signal-to-noise ratio. (From Ref. 41.)

Practical Considerations

Calculation of the log-likelihood ratio in a system using the SPRT is a difficult practical problem. The calculation can be made using a microprocessor that implements (5-155) and (5-156). Even with fast microprocessors, the time consumed by this calculation may dominate the overall synchronization time. A system using a microprocessor implementation of the SPRT is reported on in Carson [47]. An alternative digital implementation of the log-likelihood ratio calculation is a table lookup using a high-speed read-only memory (ROM). This technique permits higher processing speed and has been implemented.

For very low SNR, the same approximations that were used in the calculation of ASN can be used to simplify the calculation. In particular,

$$\ln[\lambda(x_k)] \cong -\frac{P}{N} + \left(\frac{P}{2N^2}\right)x_k^2 - \left(\frac{P^2}{16N^4}\right)x_k^4 \tag{5-166}$$

for small SNR. It was seen earlier that the x_k^4 term contributes to the mean of this function and therefore cannot be ignored. However, common practice is to include the mean of this term at the design point SNR with the bias term $(-P/N)$ resulting in

$$\ln[\lambda(x_k)] \cong \left[-\frac{P}{N} - \frac{1}{2}\left(\frac{P}{N}\right)^2 - \frac{1}{8}\left(\frac{P}{N}\right)^3 - \frac{1}{32}\left(\frac{P}{N}\right)^4\right] - \frac{P}{2N^2}x_k^2 \tag{5-167}$$

With this simplification, the linear envelope detector and log-likelihood calculator can be replaced by a *square-law* envelope detector and a bias circuit. The student is cautioned that terms up to at least the $(P/N)^2$ terms of the bias must be retained for proper operation [48]. The detailed performance of this simplified detector is discussed in Kendall [49].

Finally, some implementations of sequential detection replace the upper threshold comparator with a time-out mechanism on the total number of samples. After each sample is taken and processed, the log-likelihood ratio is compared with a lower threshold and the total number of samples is compared with a fixed limit. If the lower threshold is exceeded, the signal is declared absent. If the sample count is exceeded before the lower threshold is crossed, the signal is declared present. Discussion of this sequential detector can be found in Holmes [50]. Su and Weber [51] present algorithms for evaluating the performance of this modified sequential detection strategy.

Although performance prediction and design of the sequential detector is complex, its implementation complexity is reasonable and potential synchronization-time performance improvement is significant. The sequential detector has been used in many spread-spectrum systems, including the Tracking and Data Relay Satellite System, which provides multiple-access communications for low-earth-orbit satellites as well as the space shuttle. Simon et al. [1], Hall [36], and Lee [52] present graphs illustrating the synchronization-time improvements, relative to fixed-integration-time detectors, which are possible using sequential detectors. Under certain conditions these gains can be as large as a factor of 10 reduction in average synchronization time. Synchronization-time improvement is larger when probability of false alarm must be extremely low. The analysis in Ref. 52 includes the effects of partial correlation of the received and reference spreading codes for maximal-length sequences.

5-4 Generalized Analysis of Average Synchronization Time

In Section 5-3 mean synchronization time for single-dwell serial-search systems was calculated using a straightforward ensemble average over all possible sequences of code phase dismissals, false alarms, and missed detections which eventually resulted in code synchronization. Specifically,

$$\overline{T}_s = \sum_{n,j,k} T(n,j,k)\,\mathrm{Pr}(n,j,k) \tag{5-6}$$

where n denotes the location of the correct phase cell, j the number of times that the correct phase cell is not detected when that cell is evaluated, and k the number of false alarms in all incorrect phase cells. The summation in (5-6) is over all possible combinations (n,j,k) that lead to code synchronization. The probability of a particular (n,j,k) is $\mathrm{Pr}(n,j,k)$. The synchronization time associated with (n,j,k) is

$$T(n,j,k) = nT_i + jCT_i + kT_{\mathrm{fa}} \tag{5-4}$$

where T_i is the fixed integration time used to evaluate a phase cell, C the total number of uncertainty cells, and T_{fa} the time necessary to recover from a false detection at an incorrect phase cell. In general, (5-6) states that average synchronization time is the sum, over all possible sequences of events leading to synchronization, of the products of the probability and the duration of those events.

In the analysis of multiple-dwell serial synchronization the state-transition diagram and the state-transition matrix were used to enumerate all sequences of events leading from the start of the evaluation of an incorrect phase cell to the dismissal of that phase cell. Recall that the state-transition diagram branches (see Figure 5-15) were labeled with the product of the transition probability and z raised to a power equal to the duration of time associated with the transition. A function $B(l,z)$ was defined for any path l through the state-transition diagram; this function is the product of the labels on the branches of the state-transition diagram associated with the path l through the diagram. The average time to dismiss an incorrect phase cell T_{da} was

$$T_{da} = \sum_{l \in L} \left[\frac{d}{dz} B(l,z) \right]_{z=1} \tag{5-97}$$

The state-transition matrix was used to enumerate the set of all possible paths L leading from the start to either of the states corresponding to dismissal of the incorrect phase. The path enumeration was a list of path labels $B(l,z)$ for all $l \in L$.

The concepts defined by (5-6) and (5-97) can be extended to find a generalized solution to the average-synchronization-time problem. This generalized solution was first presented by DiCarlo [12] and DiCarlo and Weber [13] and subsequently analyzed by Polydoros [53] and Polydoros and Weber [2,3]. The notation used in Ref. 2 is used in this presentation to assist the reader who wishes to review a more comprehensive presentation of these concepts.

Consider the state-transition diagram of Figure 5-26a and the set of all possible paths L starting at state 1.0 and ending at state 2.0. Define a function $H(z)$ to represent the sum over all $l \in L$ of the branch label products $B(l,z)$. Thus

$$H(z) \equiv \sum_{l \in L} B(l,z) \tag{5-168}$$

The function $H(z)$ can be determined by using several basic principles of flow diagrams. Consider first the set of all paths from state 1.1 to state 2.0 which go through state 1.2. The product of the labels on any of these paths is $A_3(z)A_1^n(z)A_5(z)$, where n represents the number of times that the path traverses the loop labeled with $A_1(z)$. The sum of the labels on all of these paths is

$$\begin{aligned} A_7(z) &\equiv A_3(z)A_5(z) + A_3(z)A_1(z)A_5(z) + A_3(z)A_1^2(z)A_5(z) \\ &\quad + A_3(z)A_1^3(z)A_5(z) + \cdots + A_3(z)A_1^n(z)A_5(z) + \cdots \\ &= A_3(z)A_5(z)[1 + A_1(z) + A_1^2(z) + \cdots] \\ &= A_3(z)A_5(z)\frac{1}{1 - A_1(z)} \end{aligned} \tag{5-169}$$

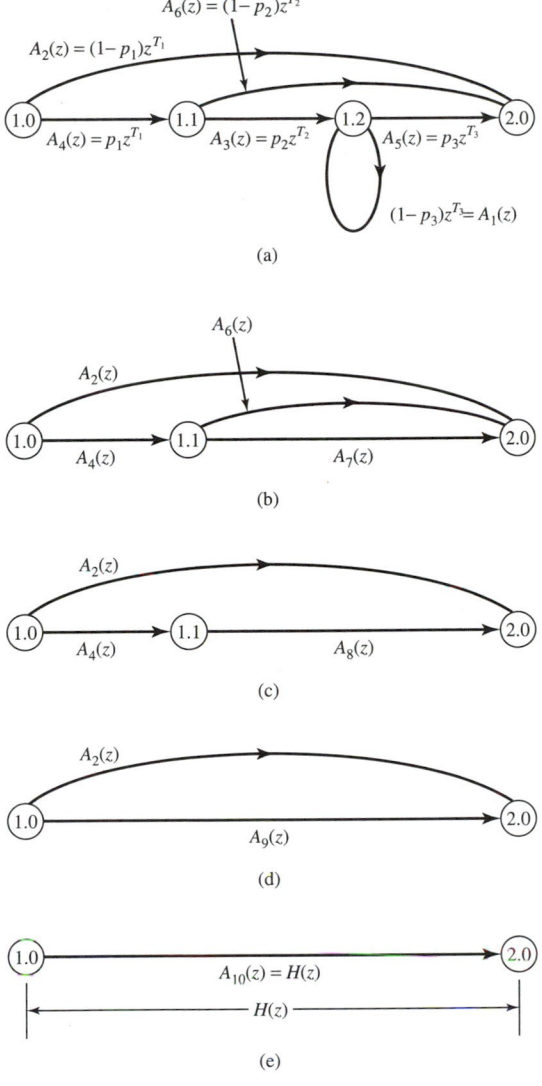

FIGURE 5-26. State-transition diagram reduction: (a) initial diagram; (b) first reduction; (c) second reduction; (d) third reduction; (e) final reduction to a single branch.

Examination of Figure 5-26a and (5-169) will demonstrate (1) that any branches through the state-transition diagram which are in series may be replaced by a single branch whose label is the product of the labels on the individual branches, and (2) a node that has a loop to itself and is labeled A may be replaced by a branch whose label is $1/(1 - A)$. These replacements do not affect the function $A_7(z)$. The reduced flow diagram of Figure 5-26b is the result of making these replacements. Next consider the parallel branches between states 1.1 and 2.0 in Figure 5-26b. By inspection it is evident that the sum of the labels on all paths between states 1.1 and 2.0 in Figure 5-26b is

$$A_8(z) = A_6(z) + A_7(z) \tag{5-170}$$

and it may be deduced that parallel branches in the state-transition diagram may be replaced by a single branch labeled with the sum of the labels on the individual branches without changing the function $A_8(z)$. Continuing in this manner, the state-transition diagram of Figure 5-26a is reduced to a single branch in Figure 5-26e. The branch label for Figure 5-26e is

$$H(z) = A_2(z) + A_4(z)\left[A_6(z) + \frac{A_3(z)A_5(z)}{1 - A_1(z)}\right] \tag{5-171}$$

The function $H(z)$ accounts for all possible paths between states 1.0 and 2.0. Expansion of (5-171) is precisely the sum of (5-168). The average time to proceed from state 1.0 to state 2.0 is calculated from (5-97):

$$T_{1,2} = \sum_{l \in L}\left[\frac{d}{dz}B(l,z)\right]_{z=1} = \frac{d}{dz}\sum_{l \in L}B(l,z)\bigg|_{z=1} = \frac{d}{dz}H(z)\bigg|_{z=1} \tag{5-172}$$

The concept of representing all possible paths through a state-transition diagram by a function $H(z)$ leads to the generalized solution for average synchronization time. The determination of $H(z)$ is facilitated by using the state-transition diagram reduction techniques demonstrated in Figure 5-26.

Figure 5-27 is a state-transition diagram that represents the entire direct-sequence code synchronization process. For a system having C cells in the code phase uncertainty region, the state-transition diagram of Figure 5-27 has $C + 2$ states. $C - 1$ of these states correspond to the $C - 1$ incorrect code phase states, one state corre-

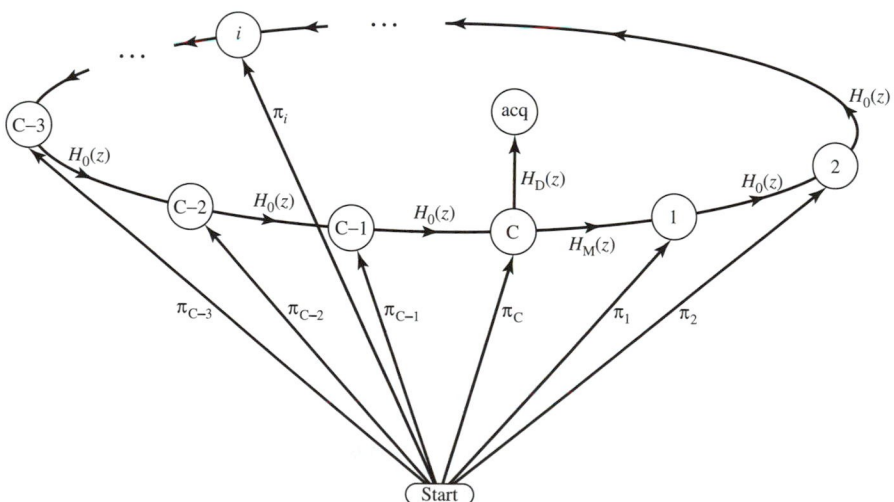

FIGURE 5-27. Serial-search circular state-transition diagram.

sponds to the correct code phase state (labeled state "C"), one state corresponds to the synchronization or acquisition state (labeled "acq"), and one state is the start state (labeled "start"). The complete synchronization process begins at "start" and ends at "acq." Examination of Figure 5-27 shows that there are many possible paths between these two states.

For this description it is assumed that the receiver has no prior knowledge about the correct code phase. Therefore, the receiver selects a code phase state randomly and begins evaluating that code phase. This random selection is represented by the branches labeled π_i leading from "start" to all other states except "acq." The labels on these branches represent the probability that the synchronization process follows that branch. In this case, all of these probabilities are identical and are $\pi_i = C^{-1}$. No synchronization time is used to make this random selection so that these branches do not include a factor z^T.

Having randomly selected a code phase, say phase i, the synchronization processor uses one of the detection techniques described in Section 5-3.4 to evaluate that phase. Assume that the selected code phase is an incorrect phase. The evaluation may consist of a single integration followed by a decision (fixed-integration-time detection) or a sequence of integrations with multiple decision points (multiple-dwell detection). In either case the code phase may be dismissed immediately or a false alarm may occur. In the case of an immediate dismissal, the synchronization process moves to state "i + 1" and begins evaluating that phase. In the case of a false alarm, the synchronization process also moves to state "i + 1"; however, this transition is delayed by the time required to detect the false alarm. Thus the branch between state "i" and state "i + 1" in Figure 5-27 actually represents multiple paths between the two states in the same manner that $H(z)$ of Figure 5-26e represented the multiple paths of Figure 5-26a.

All possible sequences of events leading from state "i" to state "i + 1" are represented by the function $H_0(z)$ in Figure 5-27. Figure 5-28 illustrates the state-

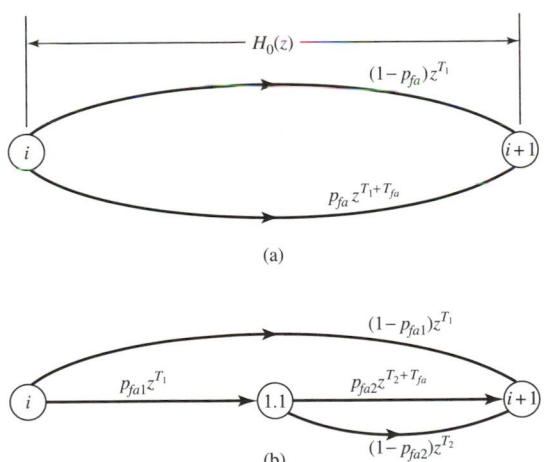

(a)

(b)

FIGURE 5-28. State-transition diagrams: (a) single-dwell system; (b) two-dwell system.

transition diagram for advancing from state "i" to state "i + 1" for a fixed-integration-time detector and for a two-dwell detector. For the fixed-integration-time detector of Figure 5-28a, there are only two paths between state "i" and "i + 1." The upper path corresponds to an integration time T_1, which results in a correct dismissal of the incorrect phase cell. The probability of following this path is $(1 - p_{fa})$ and the time required to follow this path is T_1. The lower path corresponds to an integration for time T_1 which results in a false alarm. The probability of following this path is p_{fa} and the time required to follow this path is $T_1 + T_{fa}$, where T_{fa} is the false-alarm penalty time. The function $H_0(z)$ representing Figure 5-28a is

$$H_0(z) = (1 - p_{fa})z^{T_1} + p_{fa}z^{T_1 + T_{fa}} \tag{5-173}$$

The two-dwell detector may be described similarly. The function $H_0(z)$ is determined from Figure 5-28 and the state-transition diagram reduction techniques discussed in conjunction with Figure 5-26. For the two-dwell detector,

$$H_0(z) = (1 - p_{fa1})z^{T_1} + p_{fa1}z^{T_1}[(1 - p_{fa2})z^{T_2} + p_{fa2}z^{T_2 + T_{fa}}] \tag{5-174}$$

The synchronization process advances from one incorrect state to the next following the arrows in Figure 5-27 until it reaches the correct code phase state "C." State "C" is the only state having a branch to the "acq" state. Although the detector used for the evaluation of the correct phase is identical to the detector used for incorrect phases, the function $H_0(z)$ is no longer applicable. $H_0(z)$ is not applicable since the transition probabilities have changed and since the branch to the acquisition state must be accommodated. In Figure 5-27 the function $H_D(z)$ represents all possible paths from state "C" to the acquisition state, while the function $H_M(z)$ represents all possible paths from state "C" to state "1." Figure 5-29 illustrates the state-transi-

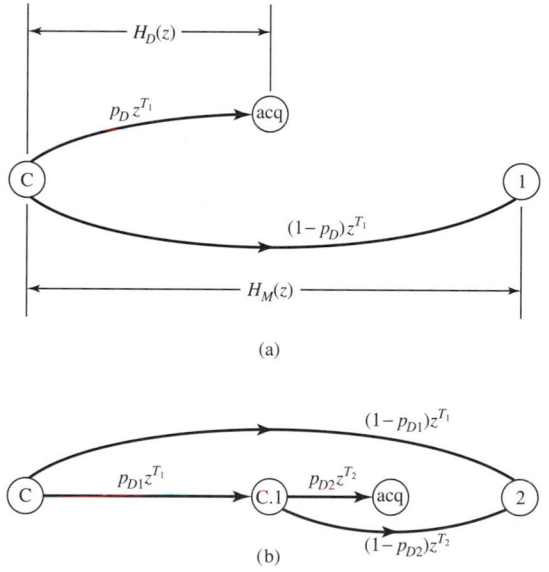

(a)

(b)

FIGURE 5-29. State-transition diagrams: (a) single-dwell system; (b) two-dwell system.

tion diagrams for the fixed-integration-time detector and the two-dwell detector for state "C." In Figure 5-29a, $H_D(z)$ and $H_M(z)$ are identified. The same functions for the two-dwell detector of Figure 5-29b are

$$H_D(z) = p_{D1}p_{D2}z^{T_1+T_2}$$

$$H_M(z) = p_{D1}(1 - p_{D2})z^{T_1+T_2} + (1 - p_{D1})z^{T_1} \qquad (5\text{-}175)$$

The synchronization process may or may not end with the evaluation of the correct code phase in state "C." If the detector fails to detect the correct phase, the acquisition process advances to state "1" and continues the search, eventually returning to the state "i" and then to state "C" for another evaluation and possible detection of the correct phase. Thus the state diagram is circular and the acquisition process may continue indefinitely.

To determine mean synchronization time using (5-172), the function $H(z)$ representing all possible paths L from "start" to "acq" through the state-transition diagram of Figure 5-27 must be calculated. Observe that all of paths in L begin with one of the branches labeled π_i from "start." Partition the set L into C subsets denoted $L_i = 1, 2, 3, \ldots, C$, such that all of the paths in L_i begin with the branch labeled π_i. This partitioning of L implies that the function $H(z)$ may be partitioned into C subfunctions, that is,

$$H(z) = \sum_{i=1}^{C} H_i(z) \qquad (5\text{-}176)$$

where each subfunction $H_i(z)$ represents all of the paths in L_i. The problem of calculating $H(z)$ has thus been partitioned into the problem of finding the C functions $H_i(z)$.

Figure 5-30 is the circular state-transition diagram representing all of the paths L_i

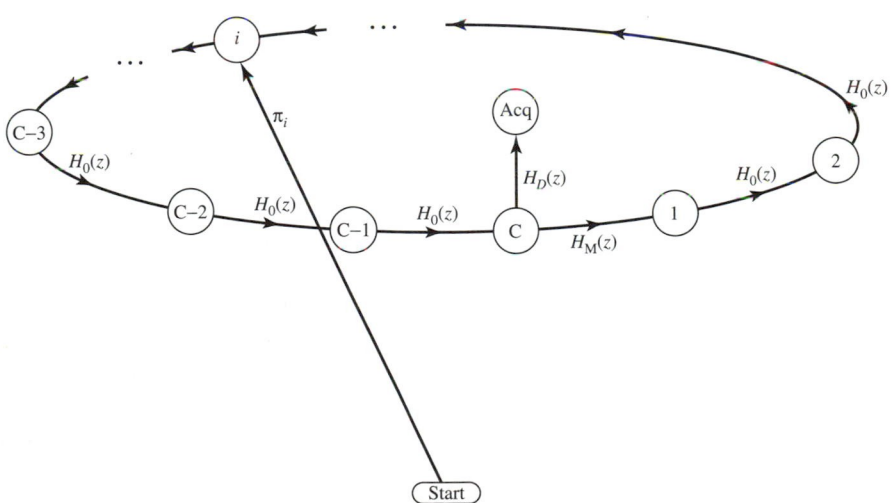

FIGURE 5-30. Serial-search circular state-transition diagram.

from "start" to state "i" and eventually to state "acq." All possible paths from "start" to "acq." in Figure 5-30 can be enumerated by inspection. All paths must begin with the branch labeled π_i, must include the C-i branches labeled $H_0(z)$ from state "i" to state "C" and must end with the branch labeled $H_D(z)$ from state "C" to "acq." The product of labels $h_{i0}(z)$ on the shortest path from "start" to "acq." is therefore

$$h_{i0}(z) = \pi_i[H_0(z)]^{C-i}H_D(z) \tag{5-177a}$$

The next-longer path occurs when there is a missed detection of the correct code phase on the first evaluation. The product of labels on this path is the product of (5-177a) and the product of labels on the branches traversed when the synchronization process advances completely around the circular state diagram from state "C" to the second occurrence of state "C." The product of labels on this path is

$$h_{i1}(z) = \pi_i[H_0(z)]^{C-i}H_D(z)\{H_M(z)[H_0(z)]^{C-1}\}^1 \tag{5-177b}$$

In general, the product of labels on the path corresponding to k missed detections of the correct code phase is

$$\begin{aligned} h_{ik}(z) &= \pi_i[H_0(z)]^{C-i}H_D(z)\{H_M(z)[H_0(z)]^{C-1}\}^k \\ &= \pi_i[H_0(z)]^{C(k+1)-i-k}[H_M(z)]^k H_D(z) \end{aligned} \tag{5-178}$$

and

$$\begin{aligned} H_i(z) &= \sum_{k=0}^{\infty} h_{ik}(z) \\ &= \pi_i H_D(z)[H_0(z)]^{C-i} \sum_{k=0}^{\infty} [H_0(z)]^{(C-1)k}[H_M(z)]^k \tag{5-179} \\ &= \pi_i H_D(z)[H_0(z)]^{C-i} \frac{1}{1 - H_M(z)H_0^{C-1}(z)} \end{aligned}$$

The desired function $H(z)$ is the sum of the functions $H_i(z)$. Specifically,

$$\begin{aligned} H(z) &= \sum_{i=1}^{C} H_i(z) \\ &= H_D(z) \frac{1}{1 - H_M(z)H_0^{C-1}(z)} \sum_{i=1}^{C} \pi_i H_0^{C-i}(z) \end{aligned} \tag{5-180}$$

Having found the function $H(z)$, the average acquisition time is calculated from

$$\overline{T}_s = \left[\frac{d}{dz}H(z)\right]_{z=1} \tag{5-181}$$

This result applies to all serial-search systems which can be represented by the state-transition diagram of Figure 5-27. The functions $H_D(z)$, $H_M(z)$, and $H_0(z)$ may represent a fixed-integration-time single-dwell detector or a multiple-dwell detector with any number of integration states. This result has been used extensively in the literature on direct-sequence spread-spectrum synchronization since it was developed by DiCarlo, Polydoros, and Weber. The result of (5-116) may be derived directly from (5-180).

5-5 Synchronization Using a Matched Filter

In all of the synchronization schemes discussed above, the received waveform plus interference was correlated with a receiver-generated replica of the spreading waveform to determine whether the phase of the replica was correct. The duration of this correlation was a function of the desired acquisition performance, the received signal-to-interference ratio, and the detection strategy (fixed integration time, multiple-dwell, or sequential detection); in all cases the correlation was over many spreading code chips. For simplicity, assume that the duration of the correlation is fixed and is $T_i = KT_c$, where T_c is the duration of a single spreading code chip and K is a positive integer. The value of K may be anywhere from 10 to several thousand. The hypothesized spreading code phase is declared correct or incorrect after the correlation. If the code phase is incorrect the synchronization processor steps the replica spreading code phase a fraction of a chip to another phase for evaluation. The average synchronization time, given a code phase uncertainty of M chips, is approximately†

$$\overline{T}_s = \frac{2M + 1}{2} T_i = \frac{2M + 1}{2} KT_c \approx MKT_c \tag{5-182}$$

Synchronization time can be reduced if the time to evaluate each phase cell, KT_c, can be reduced. The matched-filter synchronizers described in this section reduce the time required to evaluate a phase cell from KT_c to approximately T_c, thus reducing the average synchronization time by a factor approximately equal to K. This reduction in time to evaluate a phase cell is achieved using the well-known principles of matched filter detection [20,25,26,54,55].

Figure 5-31 is a top-level block diagram of a direct-sequence spread-spectrum receiver which uses a matched filter synchronization processor. The received signal, including the desired signal and interference, is input to the usual spreading code tracking loop and to the on-time despreader for data detection. In addition, the received signal is input to a bandpass filter, which is matched to a segment of the direct-sequence-spread waveform. When the waveform segment to which the filter is matched is received, the matched filter produces an output pulse. The synchronizer detects this pulse using an envelope detector followed by a threshold comparator. The spreading code generator is started at the correct phase when the matched-filter output pulse is sensed. The starting phase for the code generator is a function of the selected matched-filter waveform segment and the matched-filter and detec-

---

† In (5-12) assume that $P_d = 1.0$, $P_{fa} = 0.0$, and that the serial-search step size is $\frac{1}{2}$ chip, so that $C = 2M$.

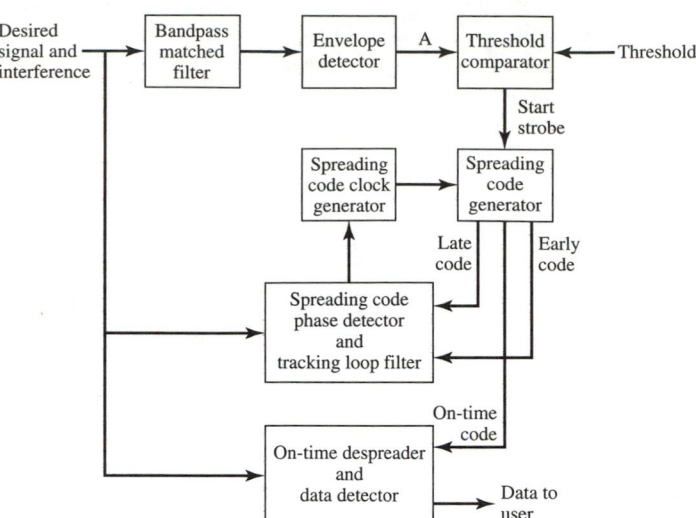

FIGURE 5-31. Direct-sequence spread-spectrum receiver using matched-filter code acquisition.

tion delays. The code generator starting phase is calculated prior to the start of the synchronization process and in some applications may be a constant set in the receiver design.

To gain a preliminary understanding of the performance of the matched-filter synchronizer, consider an ideal case. Suppose that the matched-filter synchronizer is able to detect the waveform segment the first time it is received and suppose that no detectable matched-filter output pulses are generated for other matched-filter input signals. That is, suppose that the probability of detection of the waveform segment is unity and the probability of false alarm on other waveform segments is zero. In this ideal case, average synchronization time is equal to the average time required for the selected waveform segment to arrive at the receiver. If the spreading code period is MT_c seconds and the received spreading code phase at the commencement of synchronization is random and uniformly distributed, the average time for the selected waveform segment to arrive is $MT_c/2$. In this ideal case, the average synchronization time is $\overline{T}_s = MT_c/2$. Comparing this result with (5-182), the average synchronization time has been reduced by a factor $2K$. This reduction in synchronization time has been achieved by freezing the phase of the receiver reference spreading code and waiting for a particular chip (identified by the matched-filter impulse response) of the received waveform to arrive. In contrast, the stepped serial search synchronization schemes advance the receiver reference spreading code at the same rate that the received waveform code is advancing, so that at any instant a single relative code phase is being evaluated. The active correlation and energy detection of the stepped serial search synchronizers is replaced by the passive correlation within the filter of the matched-filter synchronizer. Each received chip is used in the evaluation of a single code phase in stepped serial search. The passive correlator in the matched-filter synchronization scheme uses each received chip in the evaluation of many code phases.

To use the matched-filter synchronizer, the system designer must know the performance of the synchronizer as a function of the design parameters. As was the case for stepped serial-search synchronizers, the designer must know average synchronization time and may need to know the variance and perhaps the probability distribution of synchronization time. A detailed analysis of matched-filter synchronizers has been performed by Polydoros [3,53] and similar results are presented in Ref. 1. These results are reviewed later in this section following a simplified analysis of the matched-filter output waveform, which provides insight into some design issues. Also, the following simplified analysis will provide a basis for a digital implementation of the matched-filter synchronizer.

Suppose that BPSK direct-sequence spreading modulation is used and, for simplicity, ignore data modulation. The received spread-spectrum waveform in the noiseless case is

$$s(t) = \sqrt{2P}\, c(t - T_d)\cos(\omega_0' t + \phi') \qquad \text{(5-183a)}$$

where the received signal power is P, the carrier frequency and phase are ω_0' and ϕ', respectively, and the unknown delay to be found by the synchronizer is T_d. Consider an analog bandpass matched filter whose impulse response $h(t)$ is

$$h(t) = 2c_R(t)\cos \omega_0 t \qquad 0 \le t \le KT_c \qquad \text{(5-184a)}$$

where

$$c_R(t) \equiv c(KT_c - t) \qquad 0 \le t \le KT_c \qquad \text{(5-185)}$$

Thus $c_R(t)$ is a time-reversed and delayed version of K chips of the spreading waveform. The complex envelope representations for $s(t)$ and $h(t)$ using a reference carrier frequency ω_0 are

$$\tilde{s}(t) = \sqrt{2P}\, c(t - T_d)\exp[j(\Delta\omega t + \phi')] \qquad \text{(5-183b)}$$

and

$$\tilde{h}(t) = c(KT_c - t) \qquad 0 \le t \le KT_c \qquad \text{(5-184b)}$$

where $\omega_0' = \omega_0 + \Delta\omega$, so that $\Delta\omega$ represents the carrier frequency error of the received signal. The complex envelope of the output $y(t)$ of the matched filter is

$$\tilde{y}(t) = \tilde{s}(t) * \tilde{h}(t)$$

$$= \int_{-\infty}^{t} \tilde{s}(\lambda)\tilde{h}(t - \lambda)\, d\lambda \qquad \text{(5-186)}$$

$$= \int_{t-KT_c}^{t} \sqrt{2P}\, c(\lambda - T_d)\exp[j(\Delta\omega\lambda + \phi')]c(KT_c - t + \lambda)\, d\lambda$$

The digital implementation of the matched filter will employ a sampling rate that is a factor of N greater than the code chip rate. Denote the sampling period for the digital implementation by $T_s = T_c/N$ and assume for convenience that the unknown delay T_d is an integer multiple of the sampling period, $T_d = LT_s$. Using these relationships, the complex envelope at time $t = nT_s$ is

$$\tilde{y}(nT_s)$$
$$= \sqrt{2P} \int_{(n-KN)T_s}^{nT_s} c(\lambda - LT_s)c(\lambda + [KN - n]T_s) \exp[j(\Delta\omega\lambda + \phi')] \, d\lambda \quad (5\text{-}187)$$

This integral may be partitioned into a sum of integrals each over one sampling interval. The result is

$$\tilde{y}(nT_s) = \sqrt{2P} \sum_{m=0}^{KN-1} \int_{(n-KN+m)T_s}^{(n-KN+m+1)T_s} c(\lambda - LT_s)c(\lambda + [KN - n]T_s)$$

$$\times \exp[j(\Delta\omega\lambda + \phi')] \, d\lambda$$
$$\qquad\qquad (5\text{-}188)$$

$$= \sqrt{2P} \sum_{m=0}^{KN-1} c([n - KN + m - L]T_s)c(mT_s)$$

$$\times \int_{(n-KN+m)T_s}^{(n-KN+m+1)T_s} \exp[j(\Delta\omega\lambda + \phi') \, d\lambda$$

where the product of spreading waveforms has been removed from the integral since, because of the assumptions above relating T_c, T_s, and T_d, this product is constant over any sampling period. The last integral of (5-188) may be evaluated directly to obtain

$$\tilde{y}(nT_s)$$
$$= \sqrt{2P} \, T_s e^{j\theta} \frac{\sin(\frac{1}{2}\Delta\omega T_s)}{\frac{1}{2}\Delta\omega T_s} \sum_{m=0}^{KN-1}$$

$$\times c([n - KN + m - L]T_s)c(mT_s) \exp[j\,\Delta\omega mT_s] \quad (5\text{-}189)$$

where

$$\theta = \phi' + \Delta\omega(n + \tfrac{1}{2} - KN)T_s$$

Two special cases of (5-189) are of interest. The first special case is the particular time instant denoted $t_0 = n_0 T_s$, where the received signal waveform segment is synchronous with the filter impulse response and the filter produces its peak output. This time instant is defined by $n_0 = L + KN$, at which time the entire waveform segment corresponding to the matched-filter impulse response has been input to the filter. The matched-filter output at n_0 is

$$\tilde{y}(n_0 T_s) = \sqrt{2P} \, T_s e^{j\theta} \frac{\sin(\frac{1}{2}\Delta\omega T_s)}{\frac{1}{2}\Delta\omega T_s} \sum_{m=0}^{KN-1} c(mT_s)c(mT_s) \exp[j\,\Delta\omega mT_s]$$
$$\qquad\qquad (5\text{-}190)$$

$$= \sqrt{2P} \, T_s \, e^{j\theta} \frac{\sin(\frac{1}{2}\Delta\omega T_s)}{\frac{1}{2}\Delta\omega T_s} \sum_{m=0}^{KN-1} \exp[j\,\Delta\omega mT_s]$$

The summation in (5-190) may be simplified using [8]

$$\sum_{k=0}^{K-1} q^k = \frac{q^n - 1}{q - 1} \tag{5-191}$$

Using (5-191) to simplify (5-190) yields

$$\tilde{y}(n_0 T_s) = \sqrt{2P}\, T_s\, e^{j\theta} \frac{\sin(\frac{1}{2}\Delta\omega T_s)}{\frac{1}{2}\Delta\omega T_s} e^{j\theta'} \frac{\sin(\frac{1}{2}\Delta\omega KNT_s)}{\sin(\frac{1}{2}\Delta\omega T_s)}$$

$$= \sqrt{2P}\, KNT_s\, e^{j(\theta+\theta')} \frac{\sin(\frac{1}{2}\Delta\omega KNT_s)}{\frac{1}{2}\Delta\omega KNT_s} \tag{5-192}$$

where

$$\theta' = \tfrac{1}{2}\Delta\omega(KN - 1)T_s$$

so that

$$\theta + \theta' = \phi' + (L + \tfrac{1}{2}KN)\,\Delta\omega T_s$$

By definition the envelope detector of Figure 5-31 is insensitive to the phase of its input. Therefore, the phase in (5-192) may be ignored. When $\Delta\omega = 0$, the peak magnitude of the matched filter output envelope is $\sqrt{2P}\, KNT_s$. This peak magnitude is reduced when there is a carrier frequency error. This degradation is one of the primary problems in the application of matched-filter synchronizers. Observe that this problem becomes more difficult as system processing gain increases. In (5-192) the processing gain is K and the matched filter integration time is KNT_s. A 3-dB loss in envelope detector output occurs when the carrier frequency error in hertz is $\Delta f = 0.443/KNT_s$. The magnitude of (5-192) is important since it is a key component in determining the probability P_d that the matched filter detects the selected waveform segment when it actually arrives at the receiver.

The second special case for (5-189) occurs when the carrier frequency is known exactly, so that $\Delta\omega = 0$. In this case (5-189) becomes

$$\tilde{y}(nT_s) = \sqrt{2P}\, T_s\, e^{j\theta} \sum_{m=0}^{KN-1} c([n - KN + m - L]T_s)c(mT_s) \tag{5-193}$$

Now rewrite (5-193) as a double summation first over K chips of the waveform segment and then over the N sampling intervals of each chip. Specifically,

$$\tilde{y}(nT_s) = \sqrt{2P}\, T_s\, e^{j\theta} \sum_{k=0}^{K-1} \sum_{l=0}^{N-1} c([n + kN + l - KN - L]T_s)c([kN + l]T_s)$$

$$= \sqrt{2P}\, T_s\, e^{j\theta} \sum_{l=0}^{N-1} \sum_{k=0}^{K-1} c([n + kN + l - KN - L]T_s)c([kN + l]T_s) \tag{5-194}$$

It is convenient to define the phase difference $n - KN - L$ between the reference waveform $c([kN + l]T_s)$ and the received waveform $c([n + kN + l - KN - L]T_s)$ at time nT_s as the sum of an integer number of spreading code chips plus a chip phase offset. Define $n - KN - L \equiv K_0 N + l_0$ so that (5-194) becomes

$$\tilde{y}(nT_s) = \sqrt{2P}\, T_s\, e^{j\theta} \sum_{l=0}^{N-1} \sum_{k=0}^{K-1} c([kN + K_0 N + l + l_0]T_s)c([kN + l]T_s) \quad (5\text{-}195)$$

Because of the definition of l_0, the inner summation over index k is constant for all $l \in [0, N - l_0 - 1]$. The inner summation is also constant for $l \in [N - l_0, N - 1]$. Using these facts, the outer summation may be removed, resulting in

$$\tilde{y}(nT_s) = \sqrt{2P}\, T_s\, e^{j\theta}(N - l_0) \sum_{k=0}^{K-1} c([kN + K_0 N]T_s)c([kN]T_s)$$

$$(5\text{-}196)$$

$$+ \sqrt{2P}\, T_s\, e^{j\theta} l_0 \sum_{k=0}^{K-1} c([kN + K_0 N + 1]T_s)c([kN]T_s)$$

The two summations of (5-196) are partial autocorrelations of the spreading code. Denoting the value of the spreading waveform during chip interval k by c_k, the envelope of the matched-filter output is

$$\tilde{y}(nT_s) = \sqrt{2P}\, T_s\, e^{j\theta} \left[(N - l_0) \sum_{k=0}^{K-1} c_{k+K_0} c_k + l_0 \sum_{k=0}^{K-1} c_{k+K_0+1} c_k \right] \quad (5\text{-}197)$$

The fact that the matched-filter output envelope is the sum of partial autocorrelations of the spreading code must be accounted for in the analysis of its performance. The result of (5-197) is used in the analysis of probability of false alram, which is part of the analysis of average synchronization time. The partial autocorrelations of (5-197) are also present in all of the stepped serial-search systems analyzed earlier in this chapter. They were ignored, however, in the earlier analysis. The basis for this was that processing gain was sufficiently high that when the received and reference spreading waveforms are *not* synchronous, signal components at the output of the post-despreading bandpass filter were small relative to the noise components at the same point. If sufficient processing gain could be achieved for the matched-filter synchronizer, this same assumption would apply. Unfortunately, carrier frequency errors and the presence of data modulation limit the achievable processing gain for the matched-filter synchronizer, and (5-197) must usually be considered.

Average synchronization time for matched-filter synchronizers is calculated using the generalized analysis of Section 5-4. The circular state diagram illustrated in Figure 5-27 applies to the matched-filter detector. Average synchronization time is calculated using (5-181). Consider the synchronizer of Figure 5-31 and assume that the envelope of the matched-filter output is compared to a threshold after each sampling interval T_s. If the envelope exceeds the threshold, the receiver code generator is started and the code tracking loop is closed. If the threshold crossing was due to a correct detection of the synchronization waveform segment, synchronization is complete. If, however, the threshold crossing was due to a false alarm, the synchronization process is delayed by a false-alarm penalty time T_{fa}, after which the synchronization process continues. If the envelope does not exceed the threshold, the synchronizer waits T_s seconds and samples the envelope detector output again. Compare this process to the fixed-integration-time stepped-serial-search synchronizer. The matched-filter synchronizer evaluates one code phase each T_s seconds while the stepped-serial-search synchronizer evaluates one code phase each T_i seconds. During the time T_s between envelope detector output samples, the received spreading waveform advances $1/N$ chips. Thus the stepped-serial-search code-phase step size (usually $\frac{1}{2}$ chip) corresponds to the step size of $1/N$ chips in the matched-filter synchronizer. Thus the circular state diagram for the matched-filter synchronizer has $MN + 2$ states, where M is the code-phase initial uncertainty in chips. The branch labels $H_0(z)$, $H_D(z)$, and $H_M(z)$ of Figures 5-28a and 5-29a are directly applicable, with T_1 replaced by T_s. Using (5-180) and Figures 5-28a and 5-29a, $H(z)$ for the matched-filter synchronizer is

$$H(z) = p_D z^{T_s} \frac{\sum_{i=1}^{MN}(1/MN)[p_{fa}z^{T_s+T_{fa}} + (1 - p_{fa})z^{T_s}]^{MN-i}}{1 - (1 - p_D)z^{T_s}[p_{fa}z^{T_s+T_{fa}} + (1 - p_{fa})z^{T_s}]^{MN-1}} \quad (5\text{-}198)$$

Taking the derivative of (5-198) with respect to z and setting $z = 1$ yields the average synchronization time. After some straightforward manipulations, the result is

$$\overline{T}_s = \frac{T_s}{p_D} + \frac{2 - p_D}{2p_D}(MN - 1)(T_s + T_{fa}p_{fa}) \quad (5\text{-}199)$$

For $MN \gg 1$, $p_D \approx 1$, and $p_{fa} \approx 0$, this reduces to

$$\overline{T}_s \approx \frac{T_c}{N} + \frac{MT_c}{2}$$

where $T_c = NT_s$ has also been used. For other cases, evaluation of (5-199) requires knowledge of p_D and p_{fa}. Calculation of p_D and p_{fa} is beyond the scope of this book; the reader is directed to [1,3] for the results of this calculation. Both exact and approximate formulas for p_D and p_{fa} are presented in Refs. 1 and 3. These calculations take into account the partial autocorrelations of (5-197).

Finally, modern matched-filter synchronization systems are most often implemented digitally. The digital implementation of the synchronizer uses a lowpass implementation of the bandpass matched filter. Figure 5-32 illustrates both the bandpass and lowpass matched filters and the associated envelope detectors in analog form. These two implementations are equivalent and produce identical outputs $A(t)$ when the inputs $r(t)$ are identical. Consider an arbitrary narrowband (with respect to the carrier frequency) input signal $r(t)$, which may include both the signal of interest and interference. The complex envelope of this signal is denoted $\tilde{r}(t) \equiv r_x(t) + jr_y(t)$. The reference carrier for this complex envelope is ω_0, so that

$$r(t) = r_x(t) \cos \omega_0 t - r_y(t) \sin \omega_0 t \quad (5\text{-}200)$$

The bandpass matched filter is defined by either its bandpass impulse response

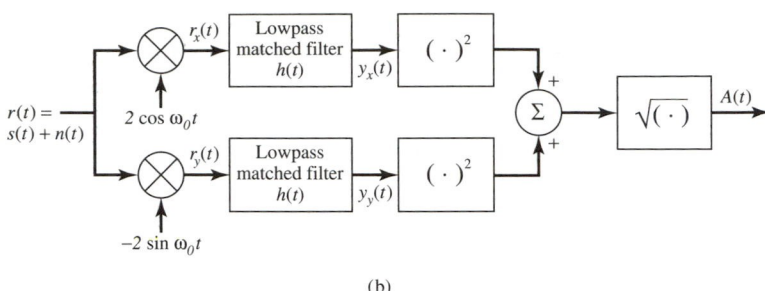

(a)

(b)

FIGURE 5-32.   Matched-filter implementations: (a) bandpass; (b) lowpass.

defined in (5-184a) or its complex envelope defined in (5-184b). Observe that the complex envelope is a real signal, implying that the bandpass filter is symmetric with respect to its center frequency ω_0. The complex envelope of the bandpass filter output is

$$\tilde{y}(t) = \int_{-\infty}^{t} \tilde{r}(\lambda)\tilde{h}(t - \lambda)\, d\lambda$$

$$\equiv y_x(t) + jy_y(t) \tag{5-201}$$

$$\equiv A(t)e^{j\beta(t)}$$

where

$$A(t) = \sqrt{y_x^2(t) + y_y^2(t)} \tag{5-202}$$

By definition, the envelope detector output is $A(t)$. Since the complex envelope $\tilde{h}(t)$ is real, (5-201) may also be written

$$\tilde{y}(t) = \int_{-\infty}^{t} [r_x(\lambda) + jr_y(\lambda)]\tilde{h}(t - \lambda)\, d\lambda$$

$$= \int_{-\infty}^{t} r_x(\lambda)\tilde{h}(t - \lambda)\, d\lambda + j\int_{-\infty}^{t} r_y(\lambda)\tilde{h}(t - \lambda)\, d\lambda \tag{5-203}$$

Now consider the lowpass circuit of Figure 5-32b. Ignoring the sum frequency terms, the outputs of the upper and lower mixers of Figure 5-32b are $r_x(t)$ and $r_y(t)$, respectively. The impulse response of either of the lowpass matched filters is $\tilde{h}(t)$. Thus the outputs of the lowpass filters are

$$\int_{-\infty}^{t} r_x(\lambda)\tilde{h}(t - \lambda)\, d\lambda = y_x(t)$$

$$\int_{-\infty}^{t} r_y(\lambda)\tilde{h}(t-\lambda)\, d\lambda = y_y(t) \tag{5-204}$$

so that the output $A(t)$ of Figure 5-32b is exactly the same as the output $A(t)$ of Figure 5-32a which was calculated in (5-202).

With the equivalence of the bandpass and lowpass implementations of the matched filter established, consider only the upper channel of the lowpass implementation. The output of this filter is the real part of (5-203).

$$\tilde{y}_x(t) = \int_{-\infty}^{t} r_x(\lambda)\tilde{h}(t-\lambda)\, d\lambda \tag{5-205}$$

Using (5-185) for the impulse response and considering the output only at time $t = nT_s$,

$$\tilde{y}_x(nT_s) = \int_{nT_s - KNT_s}^{nT_s} r_x(\lambda)c(KNT_s + \lambda - nT_s)\, d\lambda$$

$$= \sum_{m=0}^{KN-1} \int_{(n-KN+m)T_s}^{(n-KN+m+1)T_s} r_x(\lambda)c(KNT_s + \lambda - nT_s)\, d\lambda \tag{5-206}$$

$$= \sum_{m=0}^{KN-1} c(m) \int_{(n-KN+m)T_s}^{(n-KN+m+1)T_s} r_x(\lambda)\, d\lambda$$

This equation provides the basis for the digital implementation of the matched-filter envelope detector illustrated in Figure 5-33. In this figure, a sampling rate of twice the spreading code chip rate has been assumed.

This completes the discussion of matched-filter synchronizers. This discussion was not comprehensive and the reader is directed to Refs. 1, 3, and 53 for detailed analyses of matched-filter synchronizers and to Refs. 54 and 55 for a detailed analysis of matched filters and digitally implemented matched filters. The degradation due to carrier frequency error is the principal issue limiting the use of matched-filter synchronizers. Matched-filter coherent integration time is also limited by data transitions. Techniques are available that mitigate, but do not completely solve, these problems. A brute-force solution to carrier frequency error problem is, of course, to use a bank of matched-filter synchronizers. Another means of mitigating the problem is to partition the matched-filter coherent integration time and to noncoherently combine the results of the shorter coherent integrations. In the references cited, this problem is acknowledged by assuming that the matched-filter integration time will be shorter than that which would be possible with a noncoherent stepped-serial-search system. This shorter integration time results in detection and false-alarm probabilities which are not acceptable at the system level. To solve this problem a multiple-dwell system is proposed in which a detection by the matched-filter detector triggers a coincidence detector to validate the matched-filter hit. This multiple-dwell system is analyzed completely in the references. Ignoring the carrier frequency error issue, matched-filter synchronizers are capable of improving the synchronization time of direct-sequence spread-spectrum system by orders of magnitude and are therefore extremely important to the system designer.

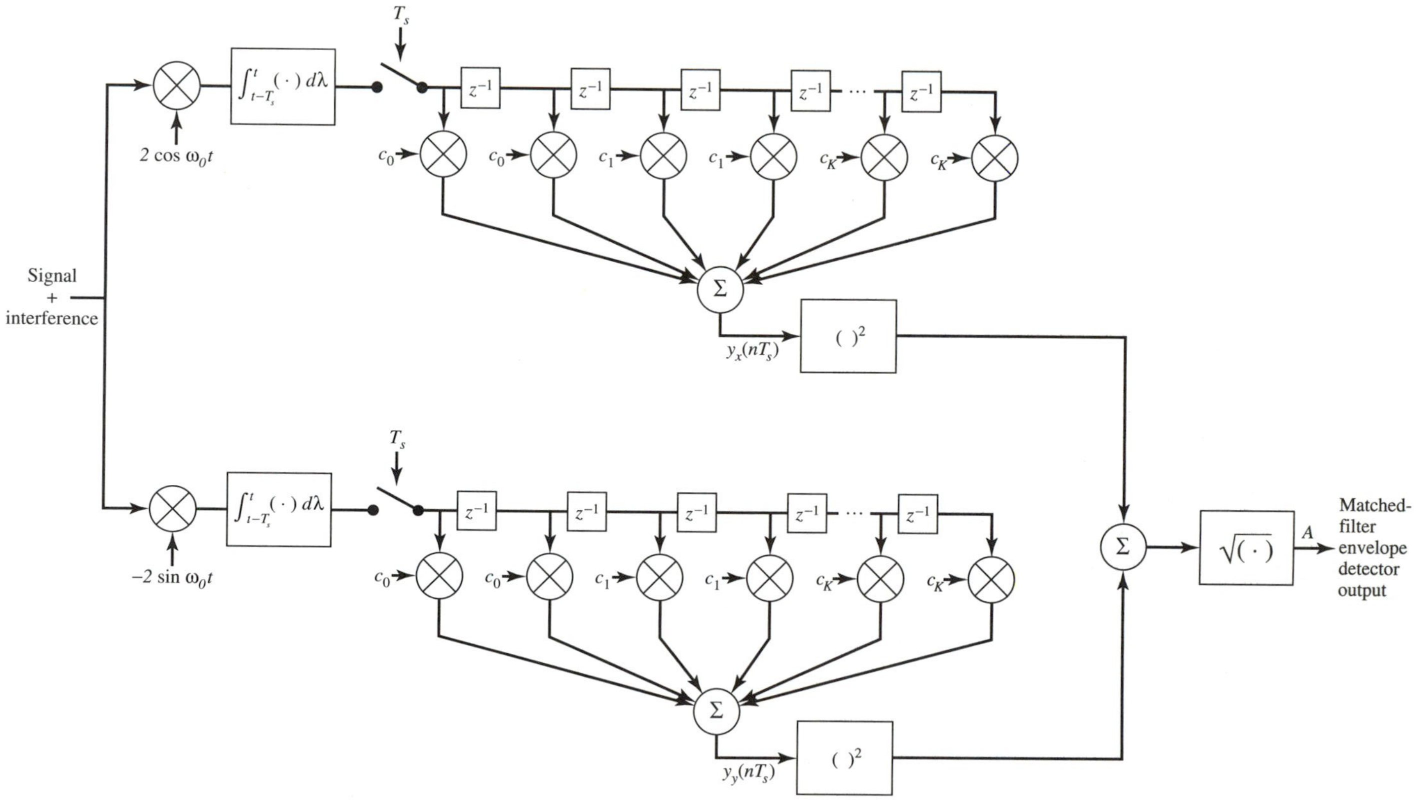

FIGURE 5-33. Digital implementation of noncoherent matched filter.

5-6 Synchronization by Estimating the Received Spreading Code

Recall from Chapter 3 that the shift register for one form (Figure 3-6) of a maximal-length sequence generator always contains r symbols of the code sequence. Thus, if r symbols of the spreading code can be estimated from the received waveform, these symbols can be loaded into the shift register generator to synchronize the system. At medium to high received SNR, these r symbols can be estimated with sufficient accuracy that this method of initial synchronization outperforms all the serial search techniques described earlier. This technique was first described in Ward [4] and was later expanded in Kilgus [56] and Ward and Yiu [5]. Although these references and the discussion to follow are limited to baseband spread-spectrum systems, the technique can in principle be used for any spreading modulation.

The system first described in Ward [4] is called recursion aided sequential estimation (RASE) and is described for a baseband signaling scheme. Figure 5-34

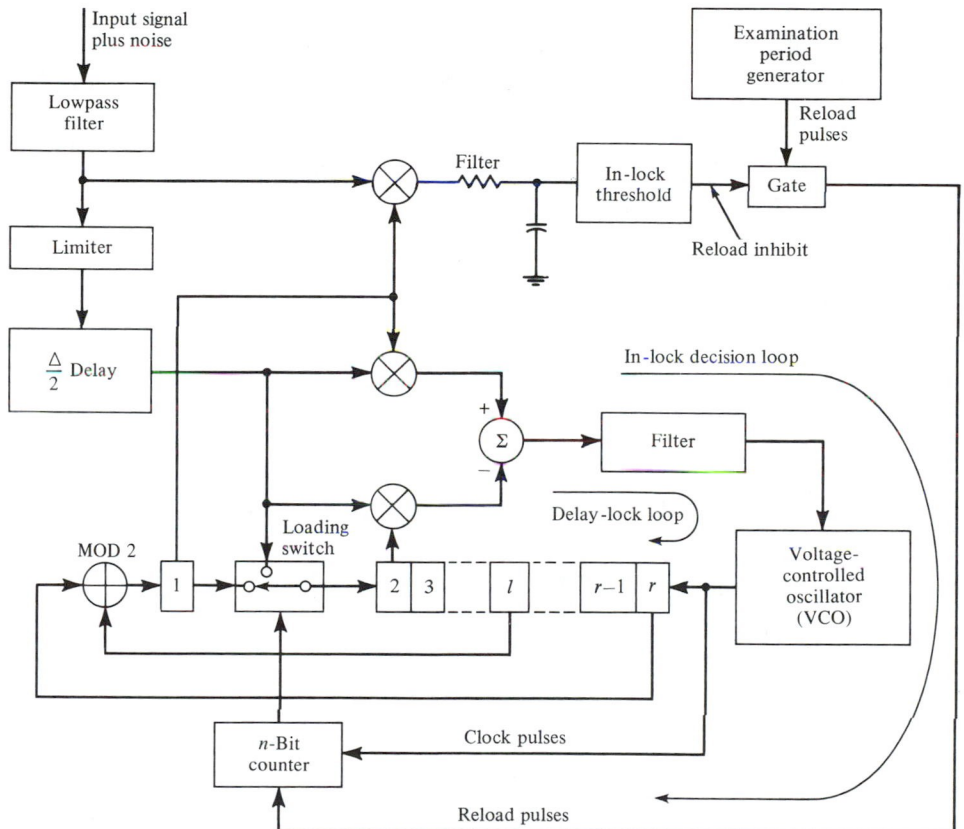

FIGURE 5-34. RASE synchronization system with DLL code tracking loop. (From Ref. 4.)

illustrates the RASE synchronization system along with a delay-lock tracking loop and an on-time lock detector correlator. The input signal is assumed to be a sequence of symbols from the alphabet ± 1 plus AWGN. The input is lowpass filtered to eliminate the majority of the noise and then limited. The limiter output is the estimate of the received sequence that is loaded into the second stage of the linear feedback shift register generator. After r symbol estimates are loaded into the shift register, the contents of all stages, including the first, are a function of the received signal. At this time the shift register loading switch is repositioned so that the code generator operates normally. If the load was correct, the reference code generator outputs will be properly phased to begin code tracking using stages 1 and 2 for the early and late reference signals. The $\frac{1}{2}$-chip delay in the received signal path is required to minimize the tracking loop pull-in transient which occurs when the tracking loop is closed. Correct code tracking is sensed by the on-time correlator, lowpass filter, and threshold comparator. The lowpass filter integrates over many code symbols and thus provides an accurate estimate of whether or not the tracking loop is functioning. If the shift register load was incorrect, the phase of the reference code will be incorrect and tracking loop pull-in will be impossible. The RASE system reloads the shift register each time an incorrect load is sensed. This process continues until a correct load and code tracking is achieved.

Suppose that the probability of correctly estimating a particular received symbol is denoted p. This probability is a function of received SNR. Then the probability of correctly loading the shift register of length r in a single trial is p^r, and the probability of an incorrect shift register load is $1 - p^r$. The probability of obtaining a correct load on exactly the kth trial is [4]

$$\Pr(k) = p^r(1 - p^r)^{k-1} \tag{5-207}$$

The average number of trials required to achieve a correct load is

$$\bar{k} = \sum_{k=1}^{\infty} k \Pr(k) \tag{5-208}$$

$$= \sum_{k=1}^{\infty} kp^r(1 - p^r)^{k-1}$$

Using the substitutions $q = 1 - p^r$, and $m = k - 1$ along with the identity [8]

$$\sum_{m=0}^{\infty} (a + m)q^m = \frac{a}{1 - q} + \frac{q}{(1 - q)^2} \tag{5-209}$$

(5-208) can be evaluated. The result is

$$\bar{k} = \frac{1}{p^r} \tag{5-210}$$

Each shift register load requires rT_c seconds plus T_c seconds to evaluate the correctness of the load. Ignoring the possibility that a correct load could be misidentified as an incorrect load, the mean synchronization time for this system is

$$\overline{T}_s = \overline{k}(rT_c + T_c) = \frac{rT_c + T_c}{p^r} \tag{5-211}$$

For the example system, the symbol error probability is calculated from the statistics of the signal at the output of the lowpass filter. Prior to synchronization no symbol synchronization signal is available so that an average error probability over all sample times must be calculated. The reference contains a plot of p versus received SNR. For very low and very high SNR the probability p is 0.5 and 1.0, respectively. Thus for very low SNR the mean synchronization time is

$$\overline{T}_s = 2^r(rT_c + T_e) \tag{5-212a}$$

and for very high SNR the mean synchronization time is

$$\overline{T}_s = rT_c + T_e \tag{5-212b}$$

A crude comparison can be made between a serial search synchronization system and RASE by assuming that both use identical evaluation times. Ignoring false-alarm penalty time and assuming that $1 - P_d \ll 1$, the average synchronization time for the serial search system is approximately

$$\overline{T}_s \approx 2\frac{2^r - 1}{2}T_e \approx 2^rT_e \tag{5-213}$$

No a priori information has been assumed about the received code phase, so that the phase uncertainty is the full period of the code. The serial search step size is $\frac{1}{2}$ chip. For very low SNR, comparison of (5-213) and (5-212a) show that average synchronization time is approximately the same for RASE and serial search. If a priori information about the received code phase is available, serial search performs better than RASE. At high SNR, however, the RASE system will achieve synchronization in much less time than serial search. Detailed numerical comparisons are given in Ward [4].

Extensions of the RASE system are possible. The most straightforward of these [5] compares the sequence generator output with received symbols for several chip times prior to initiating a comprehensive evaluation requiring T_c seconds. The goal of this check is to discard rapidly the majority of the incorrect loads, thus speeding up the search. This concept is exactly the same as that used in multiple-dwell synchronization. Another extension [56] employs forward error correction techniques to correct the shift register load.

5-7 Tracking Loop Pull-In

The synchronization schemes described above are designed to position the phase of the receiver-generated spreading waveform within a fraction of a chip of the received spreading waveform phase. It is assumed that this positioning is accurate enough that the tracking loop can take over the synchronization process at this point and pull the receiver spreading waveform to precisely the correct phase. This pull-in process is identical to the pull-in process of a conventional phase-locked loop, and it is analyzed using the same nonlinear analysis techniques [57]. Since the tracking loop error may be large during the pull-in transient, the linear analysis described in Chapter 4 is not applicable. The nonlinear analytical technique used in this case is to generate a family of curves relating the rate of change of the tracking error and the tracking error itself. Each member of this family is associated with a different set of initial conditions and/or different dynamics of the received waveform. Stable tracking will occur at points on these curves where the rate of change of the error signal is zero and where small perturbations away from the point will result in the proper direction of change of the error signal to return to the original point. The plots generated are called *phase-plane plots* and are described in detail in Viterbi [57] and Cunningham [58]. The general idea is to generate a family of phase plane trajectories related to a particular synchronization scenario and to see what trajectories lead to stable lock points. This will in turn lead to a set of initial conditions and input dynamics for which tracking loop pull-in is possible.

The analysis begins with the nonlinear equivalent circuit for the tracking loop illustrated in Figure 5-35. This model is very general in that it can be used to represent any of the tracking loops discussed in Chapter 4 with proper selection of the discriminator characteristic $D_\Delta(\delta)$ and loop gain K. This model was derived in Chapter 4. In this model, $\delta(t) = [T_d(t) - \hat{T}_d(t)]/T_c$ is the normalized tracking error which is of primary interest in this analysis. The discriminator characteristic $D_\Delta(\delta)$ and gain K relate the voltage at the input to the loop filter to the normalized tracking error. The loop filter is described by (4-19), repeated here for convenience:

$$a_m\frac{d^m\epsilon}{dt^m} + a_{m-1}\frac{d^{m-1}\epsilon}{dt^{m-1}} + \cdots + a_0\epsilon = b_n\frac{d^nv}{dt^n} + b_{n-1}\frac{d^{n-1}v}{dt^{n-1}} + \cdots + b_0v \quad (4\text{-}19)$$

where $\epsilon = \epsilon(t)$ is the filter input and $v = v(t)$ is the filter output. By inspection of Figure 5-35 it is seen that

$$g_c\int_0^t v(\lambda)\,d\lambda = \frac{\hat{T}_d(t)}{T_c} = \frac{T_d(t)}{T_c} - \delta(t) \quad (5\text{-}214)$$

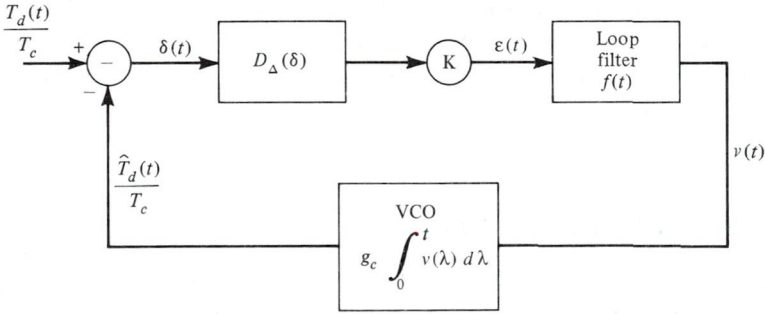

FIGURE 5-35. Noiseless nonlinear-equivalent circuit of spread-spectrum code tracking loop.

and taking one derivative with respect to time yields

$$g_c v(t) = \frac{d}{dt}\left[\frac{T_d(t)}{T_c}\right] - \frac{d}{dt}[\delta(t)] \tag{5-215}$$

The loop filter is $\epsilon(t) = KD_\Delta(\delta)$, so that the complete nonlinear differential equation describing the tracking loop is

$$\left(a_m \frac{d^m}{dt^m} + a_{m-1}\frac{d^{m-1}}{dt^{m-1}} + \cdots + a_0\right)KD_\Delta(\delta)$$

$$= \left(b_n \frac{d^n}{dt^n} + b_{n-1}\frac{d^{n-1}}{dt^{n-1}} + \cdots + b_0\right)\frac{1}{g_c}\left\{\frac{d}{dt}\left[\frac{T_d(t)}{T_c}\right] - \frac{d}{dt}[\delta(t)]\right\} \tag{5-216}$$

To simplify the following discussion, only second-order tracking loops will be considered which have a loop filter transfer function

$$F(s) = \frac{s\tau_2 + 1}{s(\tau_1 + \tau_2) + 1} = \frac{sa_1 + a_0}{sb_1 + b_0} \tag{5-217}$$

It is convenient to use normalized variables in the desired phase-plane plots. The customary [59–61] normalized variables are

$$\tau = \omega_n t \tag{5-218}$$

$$y(t) = \frac{T_d(t)}{T_c} \tag{5-219}$$

and

$$x(t) = \delta(t) \tag{5-220}$$

where ω_n is the natural radian frequency of the linearized loop. Using this normalization and the specific filter of (5-217), equation (5-216) becomes

$$\left(\tau_2\omega_n \frac{d}{d\tau} + 1\right)KD_\Delta[x(\tau)]$$

$$= \left[(\tau_2 + \tau_1)\omega_n \frac{d}{d\tau} + 1\right]\frac{1}{g_c}\left\{\omega_n \frac{d}{d\tau}[y(\tau)] - \omega_n \frac{d}{d\tau}[x(\tau)]\right\} \tag{5-221}$$

or

$$\tau_2\omega_n KD'_\Delta[x(\tau)]\dot{x}(\tau) + KD_\Delta[x(\tau)]$$

$$= (\tau_2 + \tau_1)\frac{\omega_n^2}{g_c}\ddot{y}(\tau) + \frac{\omega_n}{g_c}\dot{y}(\tau) - (\tau_2 + \tau_1)\frac{\omega_n^2}{g_c}\ddot{x}(\tau) - \frac{\omega_n}{g_c}\dot{x}(\tau) \tag{5-222}$$

where

$$\dot{x}(\tau) = \frac{dx(\tau)}{d\tau} \qquad \ddot{x}(\tau) = \frac{d^2x(\tau)}{d\tau^2}$$

$$\dot{y}(\tau) = \frac{dy(\tau)}{d\tau} \qquad \ddot{y}(\tau) = \frac{d^2 y(\tau)}{d\tau^2}$$

$$D'_\Delta[x(\tau)] = \frac{d}{dx} D_\Delta[x(\tau)]$$

Algebraic manipulation of (5-222) yields

$$\frac{\ddot{x}}{\dot{x}} = -\frac{Kg_c D_\Delta(x)}{(\tau_2 + \tau_1)\omega_n^2 \dot{x}} - \left[\frac{1}{(\tau_2 + \tau_1)\omega_n} + \left(\frac{\tau_2}{\tau_2 + \tau_1} \right) \frac{g_c}{\omega_n} KD'_\Delta(x) \right]$$

$$+ \frac{\ddot{y}}{\dot{x}} + \frac{1}{(\tau_2 + \tau_1)\omega_n} \frac{\dot{y}}{\dot{x}} \tag{5-223}$$

Simplification of (5-223) is accomplished by using the relationships between the second-order loop damping, ζ, natural frequency, ω_n, and the gains and time constants of the linearized loop. The linearized loop model is illustrated in Figure 5-36. In this figure, K_n is the slope of the discriminator characteristic $D_\Delta(x)$ in volts per chip at $x = \delta = 0$. This linear servo-loop model has been analyzed in many texts. With $F(s)$ given by (5-217) the loop damping and natural frequency are [61]

$$\zeta = \frac{1}{2} \left(\frac{K_n Kg_c}{\tau_2 + \tau_1} \right)^{1/2} \left(\tau_2 + \frac{1}{K_n Kg_c} \right) \tag{5-224a}$$

$$\omega_n = \left(\frac{K_n Kg_c}{\tau_2 + \tau_1} \right)^{1/2} \tag{5-224b}$$

For high loop gain, $\zeta = \frac{1}{2}\omega_n \tau_2$, (5-223) can be written

$$\frac{\ddot{x}}{\dot{x}} = -\frac{K_n Kg_c}{(\tau_2 + \tau_1)\omega_n^2} \left(\frac{D_\Delta(x)}{K_n} \right) \frac{1}{\dot{x}}$$

$$- \left[\frac{K_n Kg_c}{(\tau_2 + \tau_1)\omega_n} \left(\frac{1}{K_n Kg_c} \right) + \frac{K_n Kg_c}{(\tau_2 + \tau_1)\omega_n} \tau_2 \left(\frac{D'_\Delta(x)}{K_n} \right) \right] \tag{5-225}$$

$$+ \frac{\ddot{y}}{\dot{x}} + \frac{K_n Kg_c}{(\tau_2 + \tau_1)\omega_n} \frac{1}{K_n Kg_c} \frac{\dot{y}}{\dot{x}}$$

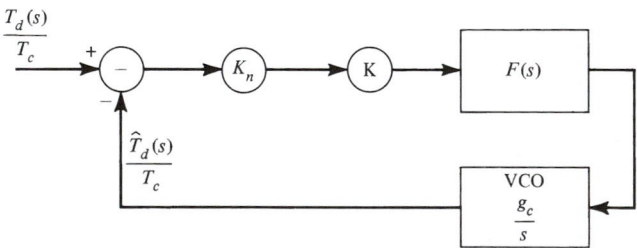

FIGURE 5-36. Noiseless linear-equivalent circuit of spread-spectrum code tracking loop.

Defining a normalized loop gain $g = K_n K g_c / \omega_n$ and using (5-224), this equation simplifies to

$$\frac{\ddot{x}}{\dot{x}} = -\frac{D_\Delta(x)}{K_n}\frac{1}{\dot{x}} - \left[\frac{1}{g} + 2\zeta\frac{D'_\Delta(x)}{K_n}\right] + \frac{\ddot{y}}{\dot{x}} + \frac{1}{g}\frac{\dot{y}}{\dot{x}}$$

$$(5\text{-}226)$$

$$= \frac{D_n(x) - [(1/g) + 2\zeta D'_n(x)]\dot{x} + \ddot{y} + (1/g)\dot{y}}{\dot{x}}$$

where $D_n(x) = D_\Delta(x)/K_n$ is the normalized discriminator characteristic which has a slope of 1.0 at $x = 0$. Noting that $\ddot{x}/\dot{x} = d\dot{x}/dx$, this equation is identical to the result derived in [59,61,62] and commented on in Nielson [60]. Similar relations can be derived for other loop filters.

Equation (5-226) is used to plot paths through the phase plane. This is done by choosing an arbitrary initial point $(x,\dot{x})$ and calculating the slope $d\dot{x}/dx$ of the trajectory passing through this point using (5-226). An appropriately small value of dx is chosen and the next point,

$$\left(x + dx, \dot{x} + \frac{d\dot{x}}{dx}dx\right)$$

is calculated. A new slope is calculated for the new point on the trajectory and the process repeats. At the end of many iterations, a single curve in the desired family will have been generated. Other curves are generated in an identical manner using other initial conditions and/or different normalized input signal dynamics $\dot{y}$ and $\ddot{y}$. Example phase-plane plots have been given in Ref. 60; these plots have been reproduced as Figures 5-37 through 5-39 for the convenience of the student. In all these plots, the loop is a critically damped second-order loop, and normalized input is $y(\tau) = T_d(\tau)/T_c = K_1\tau + K_0$, so that $\dot{y} = K_1$ and $\ddot{y} = 0$. Recall from Chapter 4 that for large phase errors the discriminator output is zero so that no correction signal exists in the tracking loop and the rate of change of the error signal $\dot{x}$ is equal to the rate of change of the input $\dot{y}$. Therefore, $\dot{y}$ determines the value of $\dot{x}$ for large x. Figures 5-37 and 5-38 illustrate the phase-plane trajectories for baseband loops having early-late delay differences of 1.0 chip and 2.0 chips, respectively. These curves illustrate that the loop is unable to pull-in at all if the input dynamics are too large. Those trajectories that do not converge to $\dot{x} = 0$ never achieve code tracking. Thus, even if the synchronization hardware placed the receiver code phase in exactly the correct position, large signal dynamics due to Doppler could cause the loop to never achieve lock. Figure 5-39 illustrates phase trajectories for the DLL using a delay difference of 1.0 chip. This same technique can be used to calculate the transient response to input dynamics after the loop is locked. In this case, the initial condition is $(x,\dot{x}) = (0,0)$.

Finally, the phase-plane plots give no indication of the time required for tracking loop pull-in. If pull-in time is important, plots of x versus τ can be easily generated from the same data used to calculate the phase trajectories. Example results are given in Nielsen [60].

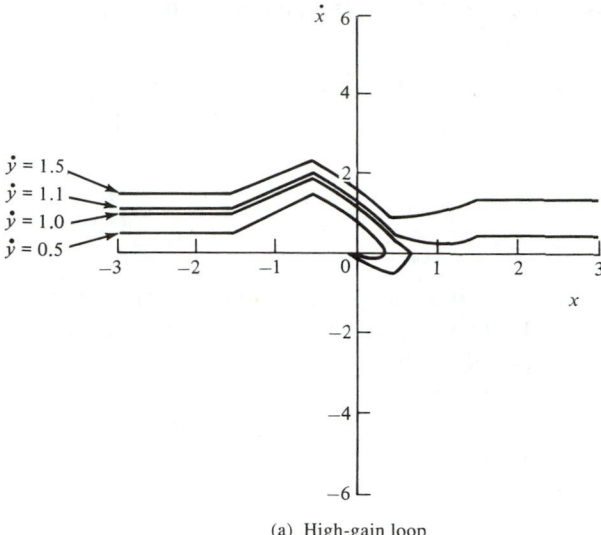

(a) High-gain loop

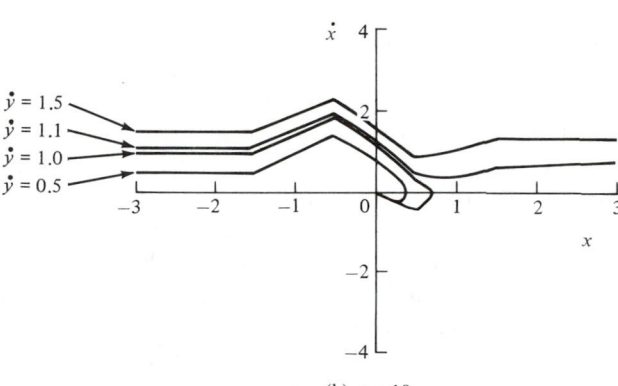

(b) $g = 10$

FIGURE 5-37. Phase-plane trajectories for baseband DLL using a delay difference of 1.0 chip. (From Ref. 60. Copyright © 1975, IEEE. Reprinted with permission.)

5-8 Summary

Initial spreading waveform synchronization is an extremely important problem in spread-spectrum system design. In fact, overall system performance is often limited by the performance of the synchronizer. The simplest synchronization scheme is to sweep the receiver spreading waveform phase until the proper phase is sensed. Stepped serial search over all potential waveform phases will usually achieve lower synchronization times than the swept search. After each phase step in a stepped serial search system, the correctness of the phase must be evaluated. This is accom-

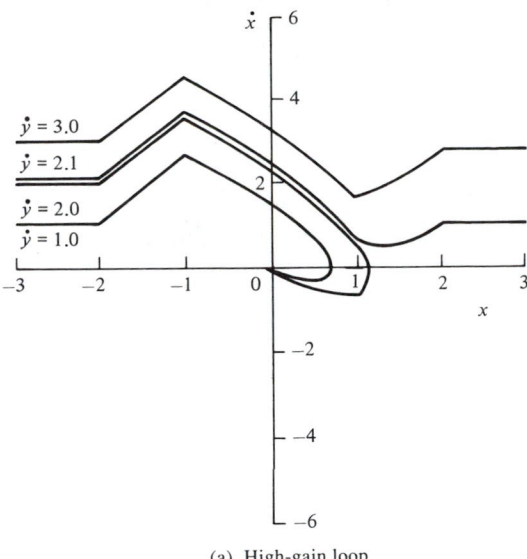

(a) High-gain loop

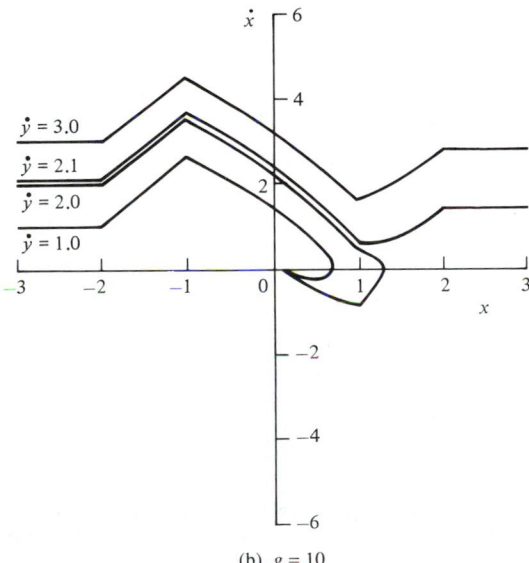

(b) $g = 10$

FIGURE 5-38. Phase-plane trajectories for baseband DLL loops using a delay difference of 2.0 chips. (From Ref. 60. Copyright © 1975, IEEE. Reprinted with permission.)

plished by attempting to despread the received waveform using the trial spreading waveform phase. When the phase is correct, the input signal spectrum is collapsed and energy appears at the output of a narrowband filter. This energy is sensed using one of a number of techniques. Fixed integration-time energy detection is the simplest technique. Improved performance is achieved using multiple-dwell techniques,

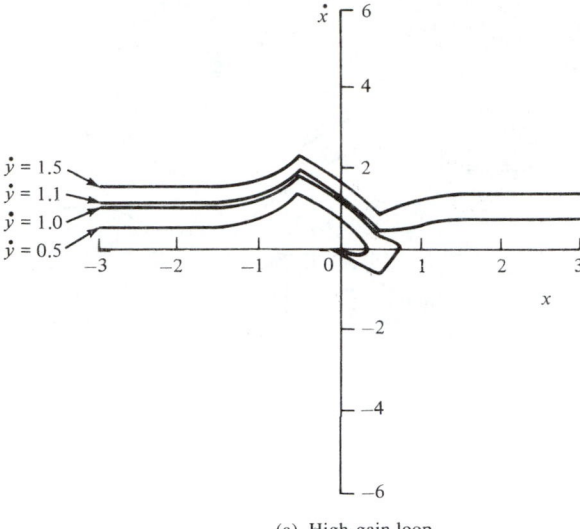

(a) High-gain loop

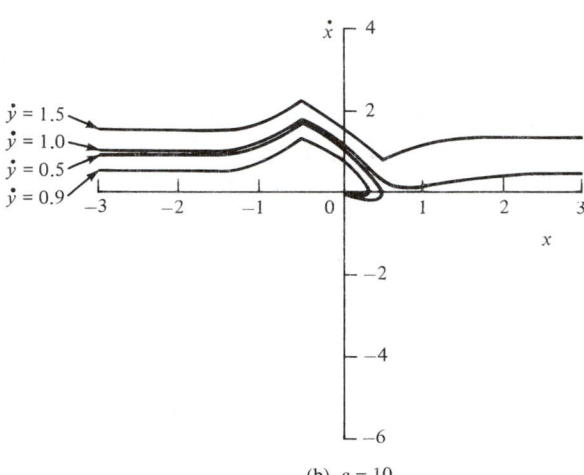

(b) $g = 10$

FIGURE 5-39. Phase-plane trajectories for noncoherent DLL using a delay difference of 1.0 chip. (From Ref. 60. Copyright © 1975, IEEE. Reprinted with permission.)

since incorrect phases can be quickly discarded without serious risk of paying a large false-alarm penalty. The most sophisticated energy detection technique is the Wald sequential probability ratio test. The performance of all these energy detection techniques can be calculated using the methods described in this chapter.

Matched filter synchronization promises to yield the lowest average synchronization time of all the schemes studied. At high SNR, sequential estimation of the code generator state from the wideband received signal is a viable synchronization technique. Tracking loop pull-in has been discussed since all the synchronization techniques finish with a phase error which is potentially a large fraction of a chip. Pull-in must occur before the linear tracking results of Chapter 4 apply.

References

[1] M. K. SIMON, J. K. OMURA, R. A. SCHOLTZ, and B. K. LEVITT, *Spread Spectrum Communications,* Vol. III (Rockville, Md.: Computer Science Press, 1985).

[2] A. POLYDOROS and C. WEBER, "A Unified Approach to Serial Search Spread-Spectrum Code Acquisition: Part I. General Theory," *IEEE Trans. Commun.,* Vol. COM-32, pp. 542–549, May 1984.

[3] A. POLYDOROS and C. WEBER, "A Unified Approach to Serial Search Spread-Spectrum Code Acquisition: Part II. A Matched Filter Receiver," *IEEE Trans. Commun.,* Vol. COM-32, pp. 550–560, May 1984.

[4] R. B. WARD, "Acquisition of Pseudonoise Signals by Sequential Estimation," *IEEE Trans. Commun. Technol.,* Vol. CT-13, pp. 475–483, December 1965.

[5] R. B. WARD and K. P. YIU, "Acquisition of Pseudonoise Signals by Recursion-Aided Sequential Estimation," *IEEE Trans. Commun.,* Vol. COM-25, pp. 784–794, August 1977.

[6] L. DAVISSON and P. FLIKKEMA, "Fast Single-Element PN Acquisition for the TDRSS MA System," *IEEE Trans. Commun.,* Vol. COM-36, pp. 1226–1235, November 1988.

[7] A. PAPOULIS, *Probability, Random Variables, and Stochastic Processes,* 2nd ed. (New York: McGraw-Hill, 1991).

[8] I. S. GRADSHTEYN and I. W. RYZHIK, *Tables of Integrals, Series, and Products* (New York: Academic Press, 1965).

[9] J. K. HOLMES and C. C. Chen, "Acquisition Time Performance of PN Spread Spectrum Systems," *IEEE Trans. Commun.,* Vol. COM-25, pp. 778–784, August 1977.

[10] W. K. ALEM, "Advanced Techniques for Direct-Sequence Spread-Spectrum Acquisition," Ph.D. dissertation, Department of Electrical Engineering, University of Southern California, Los Angeles, February 1977.

[11] W. K. ALEM and C. L. WEBER, "Acquisition Techniques for PN Sequences," *Natl. Telecommun. Conf. Rec.,* December 1977.

[12] D. DICARLO, "Multiple Dwell Serial Synchronization of Pseudonoise Signals," Ph.D. dissertation, Department of Electrical Engineering, University of Southern California, Los Angeles, May 1977.

[13] D. DICARLO and C. WEBER, "Statistical Performance of Single Dwell Serial Synchronization Systems," *IEEE Trans. Commun.,* Vol. COM-36, pp. 1382–1388, August 1988.

[14] W. R. Braun, "Performance Analysis for the Expanding Search PN Acquisition Algorithm," *IEEE Trans. Commun.,* Vol. COM-30, pp. 424–435, March 1982.

[15] H. Meyr and G. Polzer, "Performance Analysis for General PN Spread-Spectrum Acquisition Techniques," *IEEE Trans. Commun.,* Vol. COM-31, pp. 1317–1319, December 1983.

[16] V. Jovanović, "Analysis of Strategies for Serial-Search Spread-Spectrum Code Acquisition," *IEEE Trans. Commun.,* Vol. COM-36, pp. 1208–1220, August 1988.

[17] V. Jovanović, "On the Distribution Function of the Spread-Spectrum Code Acquisition Time," *IEEE J. Selected Areas Commun.,* Vol. 10, pp. 760–769, May 1992.

[18] A. Weinberg, "Generalized Analysis for the Evaluation of Search Strategy Effects on PN Acquisition Performance," *IEEE Trans. Commun.,* Vol. COM-31, pp. 37–49, January 1983.

[19] G. F. Sage, "Serial Synchronization of Pseudonoise Systems," *IEEE Trans. Commun. Technol.,* Vol. CT-12, pp. 123–127, December 1964.

[20] R. E. Ziemer and W. H. Tranter, *Principles of Communications,* 4th ed. (Boston: Houghton Mifflin, 1995).

[21] S. O. Rice, "Mathematical Analysis of Random Noise," *Bell Syst. Tech. J.,* Vol. 23, pp. 282–332, 1944, and Vol. 24, pp. 46–156, 1945.

[22] M. J. Skolnik, *Introduction to Radar Systems,* 2nd ed. (New York: McGraw-Hill, 1980).

[23] D. A. Shnidman, "Evaluation of the Q-Function," *IEEE Trans. Commun.,* Vol. COM-22, pp. 746–751, March 1974.

[24] H. Urkowitz, "Energy Detection of Unknown Deterministic Signals," *Proc. IEEE,* Vol. 55, pp. 523–531, April 1967.

[25] J. M. Wozencraft and I. M. Jacobs, *Principles of Communication Engineering* (New York: Wiley, 1965).

[26] A. Whalen, *Detection of Signals in Noise* (Orlando, Fla.: Academic Press, 1971).

[27] U. Grenander, H. Pollak, and D. Slepian, "The Distribution of Quadratic Forms in Normal Variates: A Small Sample Theory with Applications to Spectral Analysis," *J. Soc. Ind. Appl. Math.,* Vol. 7, pp. 374–401, December 1959.

[28] M. Zelen and N. C. Severo, "Probability Functions," in *Handbook of Mathematical Functions,* M. Abramowitz and I. Stegun, eds. (New York: Dover, 1972) (originally published in 1964 as NBS Applied Mathematics, Series 55).

[29] E. Fix, "Tables of Noncentral Chi-Square," *Publications in Statistics,* Vol. 1, No. 2 (Berkeley, Calif.: University of California Press, 1949).

[30] S. V. Smirnov and M. K. Potapov, "A Nomogram for the Chi-Square Probability Function," *J. SIAM,* Vol. 6, p. 124, January 1961.

[31] Arthur H. M. Ross, Qualcomm, Inc., San Diego, Calif., personal communications, December 1993.

[32] W. B. Davenport and W. L. Root, *An Introduction to the Theory of Random Signals and Noise* (New York: McGraw-Hill, 1958).

[33] U. Cheng, W. J. Hurd, and J. I. Stratman, "Spread-Spectrum Code Acquisition in the Presence of Doppler Shift and Data Modulation," *IEEE Trans. Commun.,* Vol. COM-38, pp. 241–250, February 1990.

[34] W. Feller, *An Introduction to Probability Theory and Its Applications* (New York: Wiley, 1950).

[35] P. M. Hopkins, "A Unified Analysis of Pseudonoise Synchronization by Envelope Correlation," *IEEE Trans. Commun.,* Vol. COM-25, pp. 770–778, August 1977.

[36] D. Hall, "Noncoherent Direct-Sequence Acquisition Techniques," Ph.D. dissertation, Department of Electrical Engineering, University of Southern California, Los Angeles, May 1985.

[37] D. Hall and C. L. Weber, "Noncoherent Sequential Acquisition of DS Waveforms," *Milcom '86 Conf. Rec.,* pp. 13.3.1-13.3.5, October 1986.

[38] D. M. DiCarlo and C. L. Weber, "Multiple Dwell Serial Search: Performance and Application to Direct Sequence Code Acquisition," *IEEE Trans. Commun.,* Vol. COM-31, pp. 650–659, May 1983.

[39] S.-M. Pan, D. E. Dodds, and S. Kumar, "Acquisition Time Distribution for Spread-Spectrum Receivers," *IEEE J. Selected Areas Commun.,* Vol. SAC-8, pp. 800–808, June 1990.

[40] A. Wald, *Sequential Analysis* (New York: Wiley, 1947) (also reprinted in 1973 by Dover, New York).

[41] C. W. Helstrom, "Sequential Detection," in *Communication Theory,* A. V. Balakrishnan, ed. (New York: McGraw-Hill, 1968).

[42] J. C. Hancock and P. A. Wintz, *Signal Detection Theory* (New York: McGraw-Hill, 1966).

[43] A. Wald and J. Wolfowitz, "Optimum Character of the Sequential Probability Ratio Test," *Ann. Math. Statist.,* Vol. 19, p. 326, 1948.

[44] R. F. Cobb and A. D. Darby, "Acquisition Performance of Simplified Implementations of the Sequential Detection Algorithm," *IEEE Natl. Telecommun. Conf. Rec.,* pp. 43.4.1–43.4.5, December 1978.

[45] M. G. KENDALL and A. STUART, *The Advanced Theory of Statistics,* Vol. 2 (New York: Hafner Press, 1961).

[46] F. W. J. OLIVER, "Bessel Functions of Integer Order," in *Handbook of Mathematical Functions,* M. Abramowitz and I. Stegun, eds. (New York: Dover, 1972) (originally published in 1964 as NBS Applied Mathematics, Series 55).

[47] L. CARSON, "A Microprocessor-Based Spread Spectrum Processor for Low Signal-to-Noise Ratios," *Phoenix Conf. Comput. Commun. Rec.,* pp. 230–234, March 1982.

[48] J. J. BUSSGANG and W. L. MUDGETT, "A Note of Caution on the Square-Law Approximation to an Optimum Detector," *IEEE Trans. Inf. Theory,* Vol. IT-6, pp. 504–505, September 1960.

[49] W. B. KENDALL, "Performance of the Biased Square-Law Sequential Detector in the Absence of Signal," *IEEE Trans. Inf. Theory,* Vol. IT-11, pp. 83–90, January 1965.

[50] J. K. HOLMES, *Coherent Spread Spectrum Systems* (New York: Wiley, 1982).

[51] Y. T. SU and C. L. WEBER, "A Class of Sequential Tests and Its Application," *IEEE Trans. Commun.,* Vol. COM-38, pp. 165–171, February 1990.

[52] Y.-H. LEE and S. TANTARATANA, "Sequential Acquisition of PN Sequences for DS/SS Communications: Design and Performance," *IEEE J. Selected Areas Commun.,* Vol. SAC-10, pp. 750–759, May 1992.

[53] A. POLYDOROS, "On the Synchronization Aspects of Direct-Sequence Spread Spectrum Systems," Ph.D. dissertation, Department of Electrical Engineering, University of Southern California, Los Angeles, August 1982.

[54] G. L. TURIN, "An Introduction to Matched Filters," *IRE Trans. Inf. Theory,* Vol. IT-6, pp. 311–329, June 1960.

[55] G. L. TURIN, "An Introduction to Digital Matched Filters," *Proc. IEEE,* Vol. 64, pp. 1092–1112, July 1976.

[56] C. C. KILGUS, "Pseudonoise Code Acquisition Using Majority Logic Decoding," *IEEE Trans. Commun.,* Vol. COM-21, pp. 772–774, June 1973.

[57] A. J. VITERBI, *Principles of Coherent Communication* (New York: McGraw-Hill, 1966).

[58] W. J. CUNNINGHAM, *Introduction to Nonlinear Analysis* (New York: McGraw-Hill, 1958).

[59] J. J. SPILKER, "Delay-Lock Tracking of Binary Signals," *IEEE Trans. Space Electron. Telem.,* Vol. SET-9, pp. 1–8, March 1963.

[60] P. T. NIELSEN, "On the Acquisition Behavior of Binary Delay-Lock Loops," *IEEE Trans. Aerosp. Electron. Syst.,* Vol. AES-11, pp. 415–418, May 1975.

[61] F. M. GARDNER, *Phaselock Techniques,* 2nd ed. (New York: Wiley, 1979).

[62] J. J. SPILKER, *Digital Communications by Satellite* (Englewood Cliffs, N.J.: Prentice Hall, 1977).

Problems

(5-1) Consider a BPSK spread-spectrum system using a spreading code clock $f_c = 100$ MHz. Suppose that the spreading code is a maximal-length sequence generated using a shift register of length 11, and that no a priori information is available about the transmitter to receiver propagation delay. The system data rate is 100 bps and the maximum uncertainty in the received carrier frequency is ± 5 kHz. Assume that the post despreading IF filter bandwidth is large enough to accommodate the frequency uncertainty. Calculate the average synchronization time for a single-dwell serial-search synchronizer which uses $\frac{1}{2}$-chip steps and which is designed for a single sweep probability of detection of $P_d = 0.8$, and for a probability of false alarm $P_{fa} = 10^{-3}$. Assume that the false-alarm penalty is $100T_i$ where T_i is the fixed integration time and that the received SNR in the data bandwidth is $+10$ dB. Plot the cumulative probability of being synchronized as a function of time.

(5-2) Repeat Problem 5-1 assuming that the received carrier frequency is precisely known.

(5-3) A synchronizer uses a stepped serial search with step size of $\frac{1}{2}$ chip and a fixed-integration-time energy detector. It is designed to achieve $P_d = 0.9$ with a time–bandwidth product of 25 when the reference waveform phase error is zero and the SNR is 0 dB in the post-despreading bandwidth. Assume BPSK spreading modulation and calculate the worst-case total probability of detection on a single sweep. Note that the synchronizer does not necessarily step the reference waveform phase to exactly zero phase error.

(5-4) Repeat Problem 5-3 assuming MSK spreading modulation.

(5-5) A BPSK direct-sequence spread-spectrum system is used to provide line-of-sight communications between two aircraft whose maximum relative velocity is ± 2-mach. The nominal carrier frequency is 950 MHz and the spreading rate is 5 M chips/s. The receiver synchronization system uses a correlator, bandpass filter, square-law envelope detector, and integrator. Calculate the output of the integrator as a function of integration time and transmitter-to-receiver relative velocity. Ignore noise. Explain your result and predict how it might affect system noise performance.

(5-6) Compare the times required to synchronize a pure frequency-hop spread-spectrum system and a pure BPSK direct-sequence spread-spectrum system. The data modulation is differential binary PSK for both systems and the data rate is 4800 bps. The frequency-hop rate is 100×10^3 hops/s and the direct chip rate is 100×10^6 chips/s. Assume that the initial system timing uncertainty is ± 0.5 ms and that both spreading code periods are long.

(5-7) Calculate the average number of sweeps through the phase uncertainty region for a serial-search synchronizer as a function of single sweep probability of detection P_d.

(5-8) Plot the probability density function of the output of a linear envelope detector with a sinusoid plus additive white Gaussian noise input for signal-to-noise ratios of -20, 0, 3, and 6 dB.

(5-9) A direct-sequence spread-spectrum system uses BPSK spreading modulation and a chip rate of 50×10^6 chips/s. The received carrier power/noise power spectral density ratio is $+35$ dB-Hz. The receiver synchronizer uses a swept serial search with a matched filter detector. What is the maximum sweep rate that may be used to obtain $P_d = 0.99$ and $P_{fa} = 10^{-6}$? Does this result apply if the spreading modulation was changed to MSK? If not, what modifications to the analysis of Section 5-2 must be made?

(5-10) The input to a fixed-integration-time energy detector identical to that illustrated in Figure 5-8 is taken from the RF receiver front end illustrated below. The energy detector filter is a third-order Butterworth filter with a -3-dB bandwidth of 5 kHz. The received signal power is -125 dBm and the receiver input stage is at room temperature. What is the signal-to-noise ratio at the energy detector filter output? What is the variance of the noise waveform at the filter output?

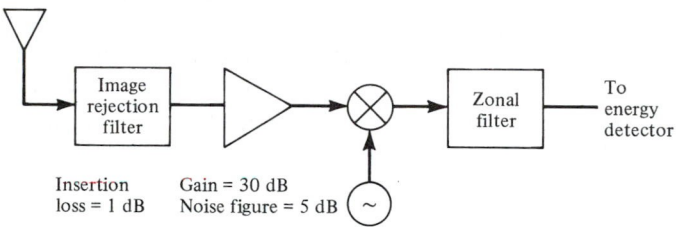

PROBLEM 5-10. Receiver front end.

(5-11) Reconsider the system of Problem 5-10.
  (a) Determine the absolute energy detector threshold required to obtain $P_{fa} = 10^{-3}$ assuming an integrator time–bandwidth product of 40. Do not assume that the integrator output pdf is Gaussian. What is the resultant P_d?
  (b) Repeat part (a) assuming that the integrator output pdf is Gaussian.

(5-12) Develop the state-transition diagram and matrix representation for the multiple-dwell energy detector defined by the logic diagram below.

(5-13) Evaluate the matrix expressions for T_{da} and P_d for the state-transition dia-

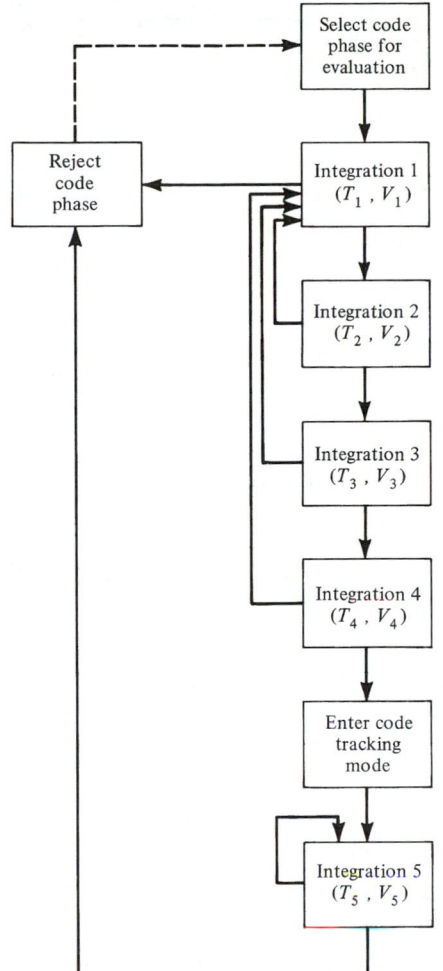

PROBLEM 5-12.   Multiple-dwell detector.

grams of the figure on the next page in general. That is, develop algebraic expressions for T_{da} and P_d in terms of the integration times and transition probabilities.

(5-14) Consider a direct-sequence spread-spectrum system that uses BPSK spreading modulation and a chip rate of 3×10^6 chips/s. Suppose that no timing information is available to the receiver and that the spreading code is an m-sequence generated using a 13-stage shift register. The synchronizer uses a double-dwell detector configured as in part (b) of the figure for Problem 5-13. The energy detector bandpass filter noise bandwidth is 5 kHz and the received $C/N_0 = 32$ dB-Hz. The detector thresholds are chosen to yield $p_{12}' = 0.1$ and $p_{23}' = 10^{-2}$ in the noise-alone case, and $p_{33} = 0.999$ with

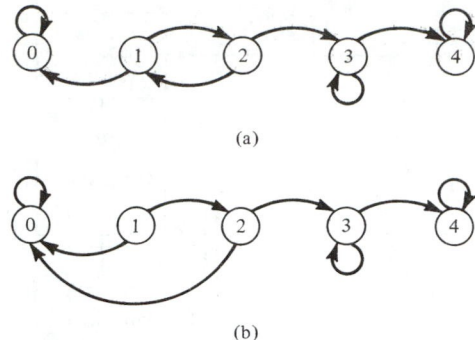

(a)

(b)

PROBLEM 5-13. State-transition diagrams.

signal present. Assume that the integrator time–bandwidth products for states 1, 2, and 3 and 100, 100, and 200, respectively, and calculate the average synchronization time.

(5-15) Repeat Problem 5-14 for a synchronizer using sequential detection designed for $P_d = 0.90$ and $P_{fa} = 10^{-3}$.

(5-16) The sequential detector discussed in Section 5-2 uses a linear envelope detector at the IF filter output. What modifications must be made to the log-likelihood ratio calculation when the linear detector is replaced by a square-law envelope detector?

(5-17) Determine the magnitude of the error made in calculating average synchronization time for a serial-search synchronization system using sequential detection when the fourth-order term of (5-157) is ignored.

(5-18) Design a sequential detector used to distinguish between two Gaussian probability density functions. Both pdf's have variance σ^2. One pdf has zero mean and the other has mean μ. Derive the operating characteristic function for this test.

(5-19) Develop a detailed procedure for calculating P_d and P_{fa} as a function of received SNR, frequency offset, and filter integration time for a matched-filter synchronizer. Assume BPSK direct-sequence spread-spectrum modulation.

(5-20) Consider a BPSK spread-spectrum system. Assume that the receiver has carrier synchronization and must obtain code synchronization using recursion aided sequential estimation. The spreading code generator is an 11-stage maximal-length feedback shift register. Each shift register load is evaluated using a fixed-integration-time detector employing an IF filter with a 10-kHz noise bandwidth. Integration time and thresholds are chosen to yield $P_d = 0.95$ and $P_{fa} = 10^{-3}$ at a received SNR of 0 dB in the IF bandwidth. The code clock rate is 10 MHz. Determine average synchronization time as a function of received SNR.

6

Performance of Spread-Spectrum Systems in a Jamming Environment

6-1 Introduction

The purpose of most spread-spectrum systems is to transfer information from one place to another. One figure of merit for these systems is the probability of correctly communicating a message in a particular noise and/or jamming environment. The techniques described in this chapter will enable the student to evaluate the transmission error probabilities for many common spectrum spreading and data modulation combinations in a number of different friendly and unfriendly signal environments. It will be shown in this chapter and the next that even the most sophisticated jammer can be almost completely countered by a combination of spectrum spreading, diversity, interleaving, and forward error correction (FEC).

In this chapter the error probabilities for systems without FEC will be derived. Spectrum spreading by itself will be shown to provide large performance improvements; however, the smart jammer is still able to produce significant performance degradations. In particular, the familiar exponential relationship between error probability and signal-to-noise ratio will be degraded to an inverse linear relationship between error probability and signal-to-jammer power ratio by the smart jammer. Further performance improvement requires FEC, diversity, and interleaving and is the subject of Chapter 7.

At the time of the original publication of this material [1], the predominant application of spread-spectrum technology was to military and space communications systems. Although the application to military and space communications remains extremely important, current research in spread-spectrum communications is focused on the application of spread-spectrum techniques to multiple-access communications. In particular, spread-spectrum technology is now being applied to cellular telephony, where it will provide significant increase, relative to the current frequency-division techniques, in the number of users that can be supported in a given bandwidth. The channel and interference models for the application of spread-spectrum technology to cellular telephony are very different from those discussed in this chapter. The gain of the channel assumed in this chapter is constant and time invariant across the entire transmission band, whereas the channel for cellular telephony is a time-varying fading channel. The interference assumed in this chapter is, for the most part, hostile, whereas the source of interference in cellular telephony is other users of the system. The application of spread spectrum to cellular telephony is the subject of Chapters 8 and 9.

6-2 Spread-Spectrum Communication System Model

Figure 6-1 illustrates the components of a spread-spectrum (SS) communication system. The discrete memoryless source (DMS) is assumed to output a sequence of independent equally likely symbols $\mathbf{u} = \ldots, u_{-1}, u_0, u_1, \ldots$ from the alphabet $\{0,1\}$. The channel coder receives a sequence of input symbols $\mathbf{u}$ from the DMS and outputs a sequence of channel symbols $\mathbf{x} = \ldots, x_{-1}, x_0, x_1, \ldots$ from the alphabet $\{0,1,\ldots,X-1\}$. The sequence of encoder output symbols has a very carefully controlled structure which enables the detection and correction of some transmission errors by the channel decoder. The channel coding may be either block coding or convolutional coding; both are discussed in Chapter 7.

In some interference environments, channel transmission errors will occur in bursts. Many FEC schemes have been designed for the stationary additive white Gaussian noise (AWGN) channel where error events are independent from one channel use to the next. When errors occur in bursts, the performance of FEC systems designed for the AWGN channel is degraded. The function of the interleaver/

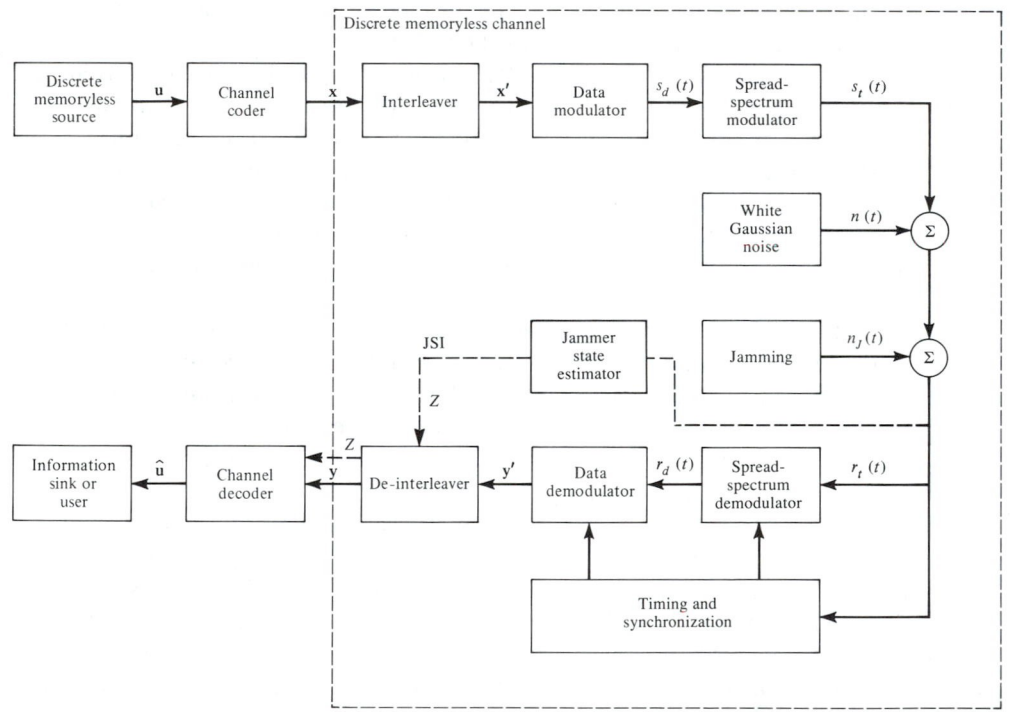

FIGURE 6-1. Spread-spectrum communication system model.

de-interleaver is to distribute channel errors randomly throughout the decoder input sequence $\mathbf{y}$, thereby enabling the use of FEC systems designed for AWGN to function well when the jammer is creating error bursts. The interleaver changes the order in which the coder output symbols are transmitted over the channel. With the transmission order changed, it is unlikely that a burst of contiguous channel errors will affect contiguous coder output symbols. Prior to decoding, the reordering performed by the interleaver is inverted by the de-interleaver. In the de-interleaving process, the burst of channel errors is distributed throughout the decoder input sequence. In all cases, the interleaver input and output alphabets are identical and are denoted $\mathbf{x}$ and $\mathbf{x}'$, respectively. The de-interleaver input and output alphabets are also identical but may be different from the interleaver alphabet. The de-interleaver input and output are denoted $\mathbf{y}'$ and $\mathbf{y}$, respectively.

The data modulator generates a signal waveform $s_d(t)$ in response to an input symbol sequence $\mathbf{x}'$. The modulation scheme may be any one of the digital modulation schemes discussed in Chapter 1. However, because of the large number of potential combinations of data modulation, spreading modulation, and jamming strategies, attention is limited in this chapter to the following data modulations: (1) coherent binary phase-shift keying (BPSK), (2) coherent quaternary phase-shift keying (QPSK), (3) differentially coherent binary phase-shift keying (DPSK), and (4) noncoherent M-ary frequency-shift keying (MFSK). The modulator output $s_d(t)$ is transmitted to the data demodulator input over a channel which includes the spread-spectrum spreading and despreading function. The demodulator observes the received waveform $r_d(t)$ and produces either an estimate of the data modulator input or a symbol that includes the data modulator input estimate and reliability information. To perform the demodulation function, timing information must be available. This information is simply the carrier phase and symbol timing for coherent modulations, and the symbol timing for differentially coherent and noncoherent modulations. This information is obtained from the timing and synchronization hardware shown. Note that, for spread-spectrum systems, symbol timing can usually be obtained from the spreading waveform phase, which must be known for the system to function.

The spread-spectrum modulator is one of the modulators discussed in Chapter 2. Attention in this chapter is limited to the following types of spreading modulations: (1) direct-sequence (DS) coherent binary phase-shift keying, (2) direct-sequence coherent quaternary phase-shift keying, (3) direct-sequence coherent minimum-shift keying, (4) noncoherent slow frequency hop (SFH), (5) noncoherent fast frequency hop, and (6) hybrid binary phase-shift keying direct sequence and non-coherent slow frequency hop. These spreading modulations combined with the data modulations listed above are typical of the modulation schemes used in modern spread-spectrum communication systems.

The spreading modulator output, denoted $s_t(t)$, is transmitted via the waveform channel which includes the transmitter up-conversion and power amplifier, the antennas, the receiver front end, and the receiver down-conversion. In the waveform channel the signal may be distorted, and Gaussian noise $n(t)$ and perhaps jamming $n_J(t)$ are added to the signal. The additive noise process is assumed to have a one-sided power spectral density of N_0 W/Hz.

A number of jamming signals have been postulated for examination in this chapter. Some of the postulated threats are clearly real, whereas others are considered only because of their optimal nature. The most benign jammer is the *barrage noise jammer*. This jammer transmits bandlimited white Gaussian noise with one-sided power spectral density (psd) of N_J W/Hz, as shown in Figure 6-2a. It is usually assumed that the barrage noise jammer power spectrum covers exactly the same frequency range as the SS signal. The effect of the barrage noise jammer on the system is simply to increase the Guassian noise level at the output of the receiver down-converter.

When the SS modulation has a frequency-hop component, jamming power can be more efficiently used by transmitting all the available power in a limited bandwidth which is smaller than the SS signal bandwidth. The jammer which uses this

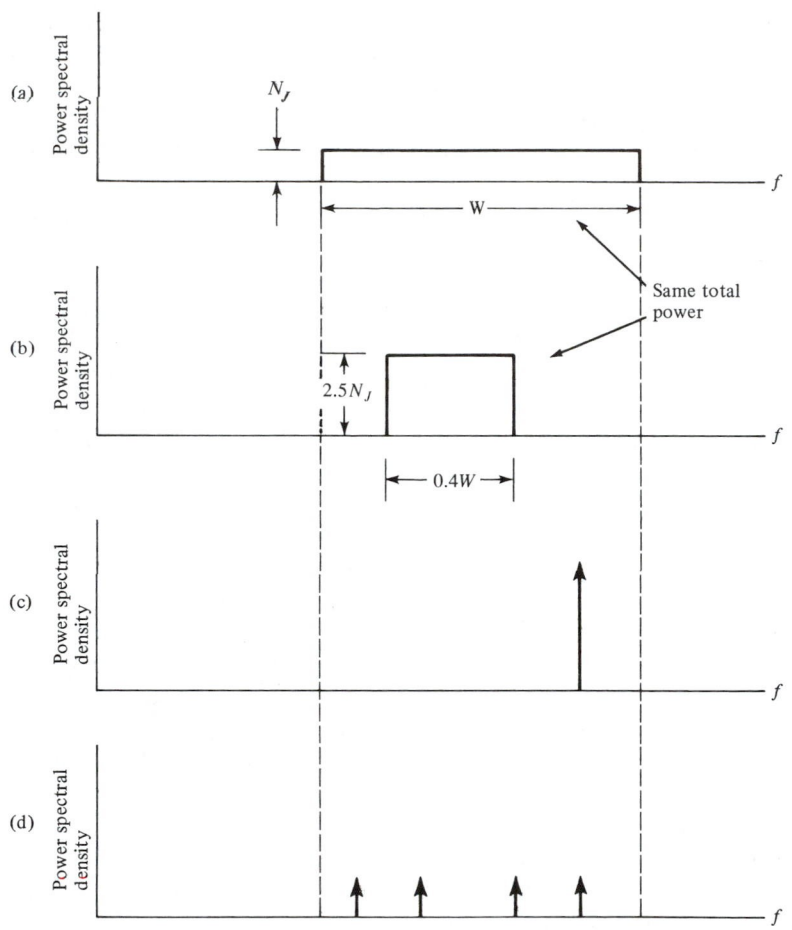

FIGURE 6-2. Typical jammer one-sided power spectral densities: (a) barrage noise jammer; (b) partial-band noise jammer; (c) tone jammer; (d) multiple-tone jammer.

strategy is called a *partial-band jammer* and the fraction of the SS signal bandwidth which is jammed is denoted by ρ. If the total jammer power is J and the SS signal bandwidth is W, the barrage noise jammer one-sided psd is $N_J = J/W$ over the entire band, while the partial-band jammer one-sided psd is $N'_J = N_J/\rho = J/\rho W$ over a bandwidth ρW as illustrated in Figure 6-2b for the special case $\rho = 0.4$. The partial-band jammer is particularly effective against a FH SS system because the signal will hop in and out of the jamming band and can be seriously degraded when in the jamming band. It will be shown later that there is an optimum (for the jammer) ρ which is a function of the ratio of signal power to total jammer power (P/J).

A third type of jammer is the single-tone jammer. The *single-tone jammer* transmits an unmodulated carrier with power J somewhere in the SS signal bandwidth. The one-sided power spectrum of this jamming signal is shown in Figure 6-2c. The single-tone jammer is important because the jamming signal is easy to generate and is rather effective against DS SS systems. The analysis of this jammer with coherent SS systems will show that the jammer should place the tone at the center of the SS signal bandwidth to achieve maximum effectiveness. The single-tone jammer is somewhat less effective against a FH signal since the FH instantaneous bandwidth is small and for large processing gains the probability of being jammed on any one hop is small. For FH systems, a better tone jamming strategy is to use several tones which share the power of the single-tone jammer; a jammer using this technique is called a *multiple-tone jammer* and the one-sided power spectrum of a typical jammer is shown in Figure 6-2d for a four-tone jammer. The jammer selects the number of tones so that the optimum degradation occurs when the SS signal hops to a tone frequency. There is little to be gained by annihilating the use of a single hop frequency when the same total power can be used to degrade a number of frequencies by a smaller but still significant amount. The optimum number of tones is a function of the received ratio of signal power to jammer power (P/J). The multiple-tone jammer is also effective against hybrid direct-sequence frequency-hop systems.

Another technique for concentrating the jamming power is to pulse the jammer "on" and "off" as discussed briefly in Chapter 2. The jamming philosophy is the same as for partial-band and multiple-tone jamming. Specifically, the jammer turns "on" with just sufficient power to degrade SS system performance significantly, but does not totally annihilate system performance when "on." The *pulsed noise jammer* transmits a pulsed bandlimited white Gaussian noise signal whose psd just covers the SS system bandwidth W. The duty factor for the jammer is the fraction of time during which the jammer is "on" and is denoted by ρ. A subscript will be used on ρ whenever necessary to distinguish between the pulse duty factor and the fractional bandwidth of a partial band jammer. When the jammer is "on," the received jammer power spectral density is $N'_J = J/\rho W$. Figure 6-3a and b compare the transmitted waveforms of the full-time barrage noise jammer and the pulsed noise jammer. The pulse duty factor used in Figure 6-3b is $\rho = 0.5$ and the rms amplitude of the signal is $\sqrt{2}$ larger than the rms amplitude of the full-time jammer. It is possible to use partial band techniques simultaneously with pulse techniques, although in this text they are considered independently. Note that pulsed jamming assumes a jammer final power amplifier which is average-power limited rather than peak-power limited. In the analysis to follow, peak-power limitations are ignored to simplify the

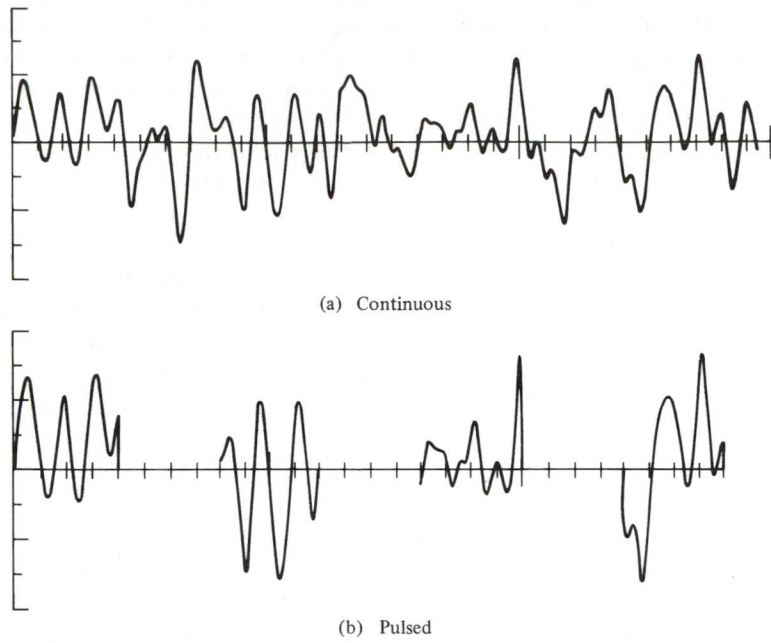

(a) Continuous

(b) Pulsed

FIGURE 6-3. Typical waveforms for continuous and pulsed noise jammer.

calculations. These limits must be considered by the spread-spectrum system designer. Pulse techniques can also be used for tone or multiple-tone jammers.

An extremely effective jamming technique for frequency-hop communications systems is to transmit frequency-hopped narrowband interference using the same frequency-hop sequence used by the communicator. A jammer using this strategy is called a follower or repeater jammer. The jammer attempts to determine the communicator's transmission frequency during each frequency-hop dwell using spectrum analysis techniques. To be effective, the jammer must be able to reliably estimate the communicator's transmission frequency, and the jammer's output must arrive at the FH communications receiver before the communicator's signal hops to a new carrier frequency. These constraints imply geometric bounds on the location of the jammer relative to the communicator's transmitter and receiver as well as bounds on the time to perform the necessary spectrum analysis. Torrieri [2] analyzes these and other fundamental limits on this type of jammer. Repeater jamming is not considered further in this book.

For the jammer to be most effective, the jamming signal must be tailored to the SS communication system and to the actual received signal power P. A jammer that has knowledge of the type of signaling being used, can accurately predict the received signal power, and can adapt to transmit the optimum jamming signal is called a *smart jammer*. A smart jammer is usually assumed in all worst-case system designs. The field of study that includes the design and analysis of jammers and jamming strategies is called *electronic counter measures* (ECM). Since one of the purposes of spread-spectrum communication systems is to counter specific jamming threats, spread-spectrum communications systems are sometimes called *electronic counter counter measures* (ECCM).

There are two other important types of interference to SS communication systems. The first of these is multipath interference. Multipath exists when there is more than one transmission path between the transmitter and receiver. If the secondary transmission paths can be characterized by the summation of several delayed and attenuated replicas of the desired signal, the multipath is called *specular multipath*. This type of multipath can occur when the receiving antenna sees reflections from such obstacles as buildings, the earth, the ionosphere, and so on. When the minimum delay between the desired and any delayed path is larger than the chip duration of a DS SS system, the receiver despreading operation will not despread the delayed signals and they can be rejected. A second type of nonhostile interference occurs when the spectrum spreading is used to provide a multiple-access capability. Multiple-access interference is discussed in detail in Chapters 8 and 9.

Returning now to the discussion of Figure 6-1, the waveform channel output is input to the spread-spectrum demodulator and the timing and synchronization hardware. These functions have been discussed in great detail in Chapters 2 through 5. The purpose of the spread-spectrum demodulator is to remodulate the received waveform with the SS spreading waveform to remove the spreading modulation from the received signal. The despreading operation is a key function of any SS system. Despreading can be accomplished only if accurate synchronization information is available. This information is derived in the timing and synchronization hardware. The operations of spreading waveform synchronization, carrier recovery, symbol synchronization, and frame synchronization are all included in the synchronization block of Figure 6-1.

Given the despread received signal $r_d(t)$ and symbol and carrier timing information, the data demodulator generates a discrete-time hard- or soft-decision output y'_j for each modulator input x'_j. The demodulator output alphabet may be continuous or discrete and may be the same or different from the modulator input alphabet $\{0,1,\ldots,X\}$. When the demodulator output alphabet and the modulator input alphabets are identical, the demodulator is a *hard-decision demodulator*. When the demodulator output alphabet is different from the modulator input alphabet, the demodulator is a *soft-decision demodulator*. In either case, denote the demodulator output alphabet by $\{0,1,\ldots,Y\}$. The demodulator output sequence is denoted $\mathbf{y}' = \ldots y'_{-1}, y'_0, y'_1, \ldots$. As described previously, this sequence is reordered by the deinterleaver, resulting in the sequence $\mathbf{y} = \ldots y_{-1}, y_0, y_1, \ldots$, which is input to the channel decoder.

The reason for choosing different alphabets for the modulator input and the demodulator output is to assist the channel decoder by making additional information available to it. This additional information tells the decoder the reliability of the demodulator's output estimate $\mathbf{y}$. A decoder that makes use of reliability information from the demodulator is called a *soft-decision decoder;* soft-decision decoders and their performance are discussed in Chapter 7. Additional reliability information can be obtained for the decoder by measuring the channel state as shown by the dashed lines of Figure 6-1. The jammer state estimator assists the decoder by indicating whether demodulator output symbols are or are not jammed. The output of the jammer state estimator will be denoted $\mathbf{z} = \ldots, z_{-1}, z_0, z_1, \ldots$ where $z_j = 1$ if channel use j is jammed and $z_j = 0$ otherwise. In its decision process the decoder then assigns less weight to those symbols that are known to be jammed, thereby improving the reliability of the decision process. The decoder therefore has two potential sources of reliability information. The first source is the demodulator itself and the second is from a separate block whose function is to monitor the channel. When jammer state information (JSI) is used by the decoder, the de-interleaver must be designed to associate the JSI with the proper decoder input symbols.

All of the functions within the dashed box of Figure 6-1 define a discrete memoryless channel (DMC) with input $\mathbf{x}$ and outputs $\mathbf{y}$ and $\mathbf{z}$. The channel is approximately memoryless, due to the interleaver. The DMC may be completely characterized by the probability that the channel output is $\mathbf{y}$ given that the channel input is $\mathbf{x}$ and given jammer state $\mathbf{z}$. That is, the channel is completely characterized by

$$p(\mathbf{y} \mid \mathbf{x}, \mathbf{z}) = \prod_{j=-\infty}^{\infty} p(y_j \mid x_j, z_j) \qquad (6\text{-}1)$$

where $p(y_j \mid x_j, z_j)$ is the probability that for channel use j, the demodulator output is y_j given that the input was x_j and the jammer state is z_j. The probabilities $p(y_j \mid x_j, z_j)$ are found by analysis of the data modulation/demodulation, the spreading modulation/demodulation, and the waveform channel, including jamming. When the demodulator output is continuous, the probabilities of (6-1) are replaced by probability density functions. Characterization of the channel using (6-1) enables the communications systems analyst to decouple the analysis of the waveform channel from the analysis of the forward error correction system. The discrete memoryless channel model is used extensively in Chapter 7.

The principal goal of the remainder of this chapter and the next is to describe the tools necessary to enable the student to predict the transmission error probability for most spread-spectrum communication systems in most interference environments. This will be accomplished using the system model described in this section. In most instances perfect synchronization will be presumed and the student is cautioned that some systems will be synchronization limited rather than transmission error probability limited. When system decisions are made, failure to consider the synchronization problem as well as transmission errors may lead to serious difficulty.

6-3 Performance of Spread-Spectrum Systems Without Coding

The results of this section predict the message error probability for many common spread-spectrum systems. These results are grouped by interference type rather than by system type since a number of system types respond in the same way to a particular interference.

6-3.1 Performance in AWGN or Barrage Noise Jamming

Because the barrage noise jammer transmits bandlimited white Gaussian noise at high power, the performance of any SS system is the same in either AWGN or barrage noise. When the noise is unintentional (i.e., thermal), the AWGN one-sided psd is denoted by N_0, and when the noise is intentional (i.e., jamming), the AWGN psd is denoted by N_J.

Coherent DS Systems. Consider an arbitrary coherent DS system transmitting a signal described by

$$s_t(t) = \sqrt{2P} \cos[2\pi f_0 + \theta_d(t) + \theta_{SS}(t)] \qquad (6\text{-}2)$$

where $\theta_d(t)$ is an arbitrary coherent data modulation and $\theta_{SS}(t)$ is an arbitrary coherent spreading modulation. The interference is bandlimited AWGN and the received signal is, assuming zero transmission delay,

$$r(t) = s_t(t) + \sqrt{2}\, n_I(t) \cos 2\pi f_0 t - \sqrt{2}\, n_Q(t) \sin 2\pi f_0 t \qquad (6\text{-}3)$$

The noise power spectrum is assumed to overlap the received signal spectrum completely so that $n_I(t)$ and $n_Q(t)$ are independent and each has a two-sided psd of $N_0/2$ or $N_J/2$ W/Hz. Both the direct-sequence spread-spectrum signal and interference are assumed to be bandlimited to the null-to-null main-lobe bandwidth of the spread-spectrum signal. This bandlimiting is accomplished by a receiver pre-despreading bandpass filter.

A simplified model of the receiver for this signal is illustrated in Figure 6-4. In this figure, the first mixer performs the despreading operation and its output near the IF is

$$x(t) = \sqrt{2P} \cos[2\pi f_{IF}t + \theta_d(t)]$$
$$+ [\sqrt{2}n_I(t) \cos 2\pi f_0 t - \sqrt{2}n_Q(t) \sin 2\pi f_0 t] \qquad (6\text{-}4)$$
$$\times \{2 \cos[2\pi(f_0 + f_{IF})t + \theta_{SS}(t)]\}$$

Perfect synchronization has been assumed and the distortion of the data-modulated carrier due to the pre-despreading bandpass filter has been ignored. The second term of this equation is the product of two independent noise-like waveforms and has a psd equal to the convolution of the psd's of each term. The noise psd $N(f)$ is $\frac{1}{2}N_0$ over frequency limits defined by $|f \pm f_0| \le \frac{1}{2}W$, where W is the null-to-null spread-

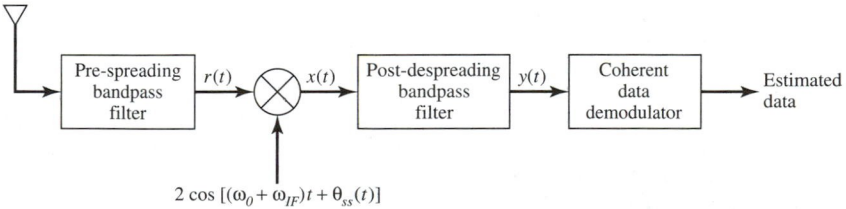

FIGURE 6-4. Simplified model of a coherent spread-spectrum receiver.

spectrum bandwidth. The psd of the despreading reference signal is a function of the specific type of spreading used. For BPSK or QPSK spreading the reference psd has the form

$$S_R(f) = \frac{2}{W} \operatorname{sinc}^2 \left[(f - f_0 - f_{\mathrm{IF}}) \frac{2}{W} \right] + \frac{2}{W} \operatorname{sinc}^2 \left[(f + f_0 + f_{\mathrm{IF}}) \frac{2}{W} \right] \quad (6\text{-}5)$$

The bandpass filter following the despreader eliminates all frequency components not close to f_{IF}, and these terms may be ignored when evaluating the convolution. The terms of interest of $N(f) * S_R(f)$ are

$$
\begin{aligned}
N(f) * S_R(f) &\cong \frac{N_0}{W} \int_{f_0 - (1/2)W}^{f_0 + (1/2)W} \operatorname{sinc}^2 \left[(f - \lambda + f_0 + f_{\mathrm{IF}}) \frac{2}{W} \right] d\lambda \\
&+ \frac{N_0}{W} \int_{-f_0 - (1/2)W}^{-f_0 + (1/2)W} \operatorname{sinc}^2 \left[(f - \lambda - f_0 - f_{\mathrm{IF}}) \frac{2}{W} \right] d\lambda
\end{aligned}
\quad (6\text{-}6)
$$

Making the change of variables $\gamma' = f - \lambda + f_0 + f_{\mathrm{IF}}$ in the first integral and $\gamma = f - \lambda - f_0 - f_{\mathrm{IF}}$ in the second integral results in

$$
\begin{aligned}
N(f) * S_R(f) &\cong \frac{N_0}{W} \int_{f + f_{\mathrm{IF}} - (1/2)W}^{f + f_{\mathrm{IF}} + (1/2)W} \operatorname{sinc}^2 \left(\gamma' \frac{2}{W} \right) d\gamma' \\
&+ \frac{N_0}{W} \int_{f - f_{\mathrm{IF}} - (1/2)W}^{f - f_{\mathrm{IF}} + (1/2)W} \operatorname{sinc}^2 \left(\gamma \frac{2}{W} \right) d\gamma
\end{aligned}
\quad (6\text{-}7)
$$

This convolution is an even function of f. When f_{IF} is large relative to W, the numerical value of the convolution is approximately determined by one of the integrals of (6-7). For positive frequencies near f_{IF},

$$N(f) * S_R(f) \cong \frac{N_0}{W} \int_{f - f_{\mathrm{IF}} - (1/2)W}^{f - f_{\mathrm{IF}} + (1/2)W} \operatorname{sinc}^2 \left(\gamma \frac{2}{W} \right) d\gamma \qquad \text{for } f > 0 \quad (6\text{-}8)$$

Assuming now that the data modulation bandwidth and hence the IF bandpass filter bandwidth is small relative to W, the value of (6-8) is approximately constant over the passband of the post-despreading bandpass filter. Therefore, the power spectral density of the Gaussian interference at the output of this filter may be approximated by its value at $f = f_{\mathrm{IF}}$. Denote this value of $\frac{1}{2}N_n$, which is given by

$$
\begin{aligned}
\tfrac{1}{2}N_n &= N(f_{\mathrm{IF}}) * S_R(f_{\mathrm{IF}}) \\
&= \frac{N_0}{W} \int_{-(1/2)W}^{+(1/2)W} \operatorname{sinc}^2 \left(\gamma \frac{2}{W} \right) d\gamma = K N_0
\end{aligned}
\quad (6\text{-}9)
$$

where $K(W/2) = 0.903 W/2$ is the area under the main lobe of the $\operatorname{sinc}^2(\cdot)$ function. The limits of integration in (6-9) are determined by the bandwidth of the pre-despreading bandpass filter of Figure 6-4, which was assumed to equal the main lobe of the direct-sequence spread-spectrum signal. If the bandwidth of this filter were, instead, much larger than the received signal bandwidth, the limits of integration would be increased and the value of K would approach unity. Thus the value of

$N_n/2$ would approach $N_0/2$ or $N_J/2$ when the interference is AWGN or barrage jamming, respectively. When both barrage jamming and AWGN are present and the pre-despreading bandpass filter bandwidth equals the main-lobe signal bandwidth,

$$\tfrac{1}{2}N_n = K(\tfrac{1}{2}N_0 + \tfrac{1}{2}N_J) \tag{6-10}$$

As stated above, for BPSK direct-sequence spreading, $K = 0.903$. A similar analysis for MSK direct-sequence spreading yields $K = 0.995$.

The noise process at the filter output is Gaussian. Thus the input to the coherent data demodulator is

$$y(t) = \sqrt{2P} \cos[2\pi f_{\mathrm{IF}}t + \theta_d(t)] + n_n(t) \tag{6-11}$$

where $n_n(t)$ is a bandlimited white Gaussian noise process with two-sided psd of $N_n/2$. From this point on, the analysis is identical to the analysis of any phase coherent digital modulation scheme. For the particular case where BPSK or QPSK data modulation is assumed, the bit error probability is

$$P_b = Q\left(\sqrt{\frac{2E_b}{N_n}}\right) = Q\left(\sqrt{\frac{2}{K[(N_0 R/P) + (J/P)(R/W)]}}\right) \tag{6-12}$$

where (6-10) is used with $N_J = J/W$, $E_b = P/R$, and R the bit rate.

In AWGN, the spectrum spreading has no effect except for a slight decrease, specified by the value of K, in the noise psd at the post-despreading bandpass output due to the pre-despreading bandpass filter. However, when the noise is due to a jammer, the total jammer power required to reduce system performance to a given level is greatly increased by spreading the spectrum. Suppose, for example, that the desired system performance is achieved with a demodulator input signal-to-noise ratio of $\mathrm{SNR}_a = E_b/N_{\mathrm{na}}$. For the normal non-spread-spectrum system the transmission bandwidth is W_d, which is equal to one or two times the data symbol rate, and the jammer power required to achieve SNR_a is $J_1 = N_{\mathrm{na}}W_d$. For a spread-spectrum system the transmission bandwidth is $W \gg W_d$ and the jammer power required to achieve SNR_a is approximately $J_2 = K N_{\mathrm{na}}W$ where $K \approx 1$ and is a function of the spreading modulation. The processing gain of the system is $J_2/J_1 \approx W/W_d$ and is due to the fact that the jammer must fill a much larger bandwidth with noise for the SS system than for the non-SS system. Figure 6-5 is a plot of (6-12) as a function of $(P/J)(W/R)$ for various signal-to-thermal noise ratios $P/N_0 R = E_b/N_0$ using the approximation $K = 1.0$. Observe that processing gain is included in this figure through the use of the factor W/R in the independent variable.

FH/MFSK. With slow FH spreading modulation, the transmitted signal is

$$s_t(t) = \sqrt{2P} \sum_{n=-\infty}^{\infty} p_{T_c}(t - nT_c) \cos[2\pi f_n t + \phi_n + \theta_d(t)] \tag{6-13}$$

where f_n is the hop frequency during the nth hop interval, $p_{T_c}(t)$ is a unit pulse of duration T_c, ϕ_n is a random phase during the nth hop interval, and $\theta_d(t)$ is the M-ary FSK data modulation. The received signal is given by (6-3) with f_0 equal to the

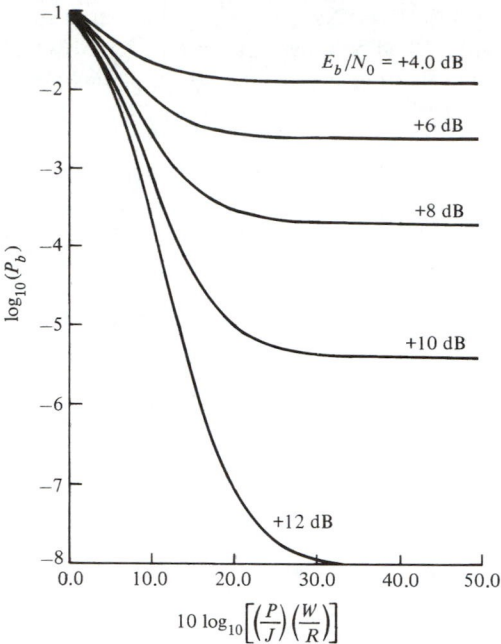

FIGURE 6-5. Performance of a coherent direct-sequence spread-spectrum system in barrage noise jamming.

center frequency of the transmission bandwidth and the despreading reference signal is

$$f_R(t) = 2 \sum_{n=-\infty}^{\infty} p_{T_c}(t - nT_c) \cos[2\pi(f_n + f_{IF})t] \qquad (6\text{-}14)$$

A simplified block diagram of the receiver is shown in Figure 6-6. The bandwidth of the pre-despreading bandpass filter is assumed to be equal to W, the bandwidth of received frequency-hop spread-spectrum signal.

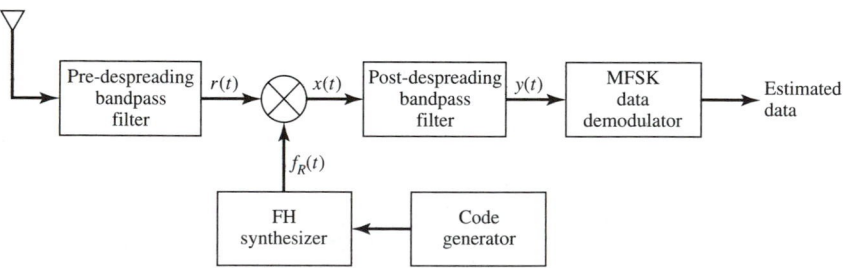

FIGURE 6-6. Simplified model of a FH/MFSK spread-spectrum receiver.

The components of the output of the despreading mixer near the IF are

$$x(t) = \sqrt{2P} \sum_{n=-\infty}^{\infty} p_{T_c}(t - nT_c) \cos[2\pi f_{IF} t + \phi_n + \theta_d(t)]$$

$$+ [\sqrt{2}\, n_I(t) \cos(2\pi f_0 t) - \sqrt{2} n_Q(t) \sin(2\pi f_0 t)] \qquad (6\text{-}15)$$

$$\times \left\{ 2 \sum_{n=-\infty}^{\infty} p_T(t - nT_c) \cos[2\pi(f_n + f_{IF})t] \right\}$$

The first term above is a M-ary FSK signal with random phase on each data symbol. The second term is a frequency-hopped noise signal with instantaneous one-sided bandwidth W.

The despreader output noise psd is the convolution of the pre-despreading band-pass filter output noise psd and the power spectrum of the synthesizer output. Since slow-frequency hopping has been assumed, a quasi-static analysis may be done with little loss in accuracy. Thus the desired despreader output psd may be calculated by considering one frequency-hop dwell at a time. For a single frequency-hop dwell, the output of the synthesizer is a single tone whose frequency is $f_n + f_{IF}$, and the despreader output psd is the pre-despreading bandpass filter output psd translated in frequency such that the components at f_n at the despreader input are at frequency f_{IF} at the despreader output. Since the pre-despreading bandpass filter output noise psd is constant and equal to $N_0/2 + N_J/2$ for all f_n, the despreader output noise psd at $f = f_{IF}$ is $N_n/2 = N_0/2 + N_J/2$ for all frequency-hop dwells. Thus the post-despreading bandpass filter output may be written

$$y(t) = \sqrt{2P} \cos[2\pi f_{IF} t + \theta_d(t)] + n_n(t) \qquad (6\text{-}16)$$

where the FH random phase has been included with the MFSK random phase and $n_n(t)$ is bandlimited AWGN with two-sided psd of $\tfrac{1}{2}N_n$. Thus the data demodulator input noise psd is the same as the received noise psd.

The symbol error probability for M-ary FSK in an AWGN environment is calculated in Arthurs and Dym [3]. The result is

$$P_s = \frac{1}{M} \exp\left(-\frac{E_s}{2N_n}\right) \sum_{q=2}^{M} \binom{M}{q} (-1)^q \exp\left[\frac{E_s(2 - q)}{2N_n q}\right] \qquad (6\text{-}17)$$

where E_s is the symbol energy and orthogonal signaling is assumed. Orthogonal signaling means that the symbol tone spacing for adjacent tones is at least $1/T_s$, where T_s is the symbol duration. The demodulator assumed in deriving this result is illustrated in Figure 6-7. Both implementations shown in Figure 6-7 perform identically and both are optimal noncoherent detectors. When orthogonal signaling is used, the distance in signal space between any signal and all others is the same. Thus, when an error is made, it is equally likely that the error symbol is any of the $M - 1$ other symbols.

The bit error probability is calculated by noting that there are $l = \log_2 M$ bits associated with each data symbol. Since all possible symbol errors are equally

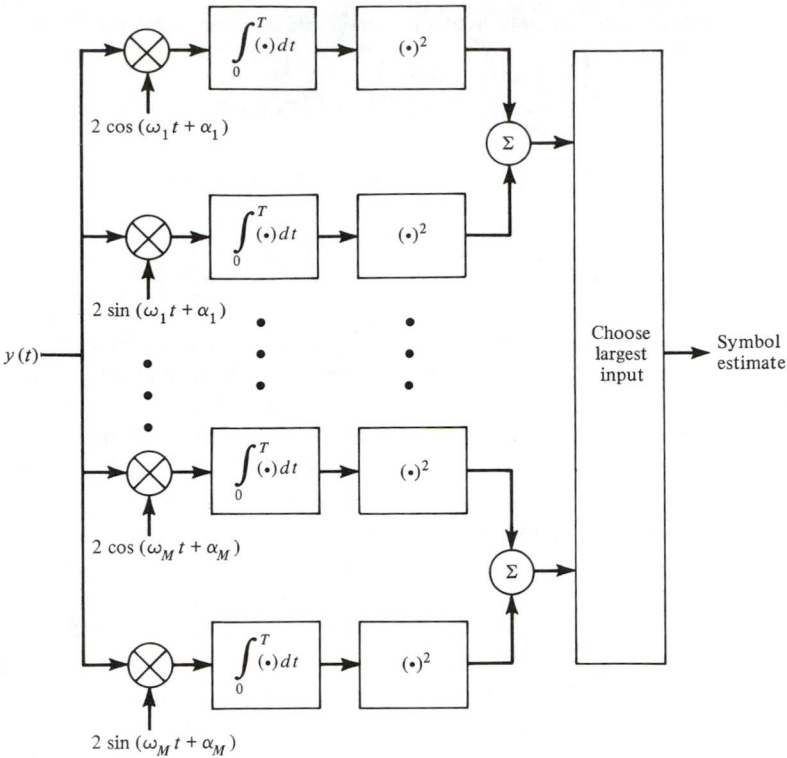

(a) Correlate and integrate implementation

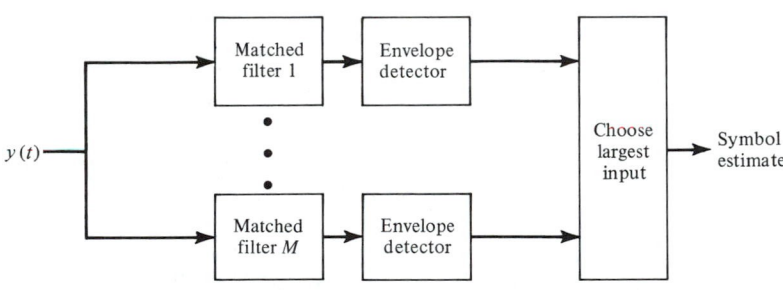

(b) Bandpass filter and envelope detect implementation

FIGURE 6-7. *M*-ary FSK noncoherent demodulators.

likely, the average number of bit errors made when a symbol error occurs is given by

$$\frac{1}{M-1} \sum_{i=1}^{l} \binom{l}{i} i = \frac{l 2^{l-1}}{M-1} = \frac{M}{2(M-1)} l \qquad (6\text{-}18)$$

Given that a symbol error has occurred, this expression is the sum over all possible symbol error events of the probability of that error event times the number of bit errors in that error event. Since l bits are demodulated with each symbol, the bit error probability, or average number of bit errors for each bit which is demodulated, is

$$P_b = \frac{1}{l}\left[\frac{M}{2(M-1)}l\right]P_s$$

$$= \frac{1}{2(M-1)}\exp\left(-\frac{lE_b}{2N_n}\right)\sum_{q=2}^{M}\binom{M}{q}(-1)^q\exp\left[\frac{lE_b(2-q)}{2N_nq}\right]$$

$$(6\text{-}19)$$

Note that $E_s = lE_b$. Using the relations $\frac{1}{2}N_n = \frac{1}{2}N_0 + \frac{1}{2}N_J$, $E_b = P/R$, and $N_J = J/W$, it can be shown that

$$\frac{E_b}{N_n} = \frac{1}{(N_0R/P) + (J/P)(R/W)}$$

$$(6\text{-}20)$$

Figure 6-8 is a plot of (6-19) using $(P/J)(W/R)$ as the independent variable for various signal-to-thermal noise ratios and $l = 1$.

Once again the spectrum spreading and despreading operation has not affected the form of the bit error probability expression. The system processing gain may be

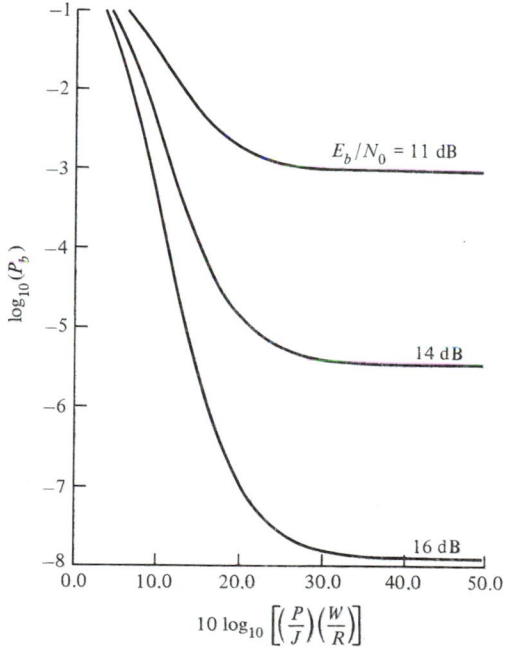

FIGURE 6-8. Performance of a FH/BFSK spread-spectrum system in barrage noise jamming.

calculated using the same steps used above for coherent systems. Specifically, suppose that $\mathrm{SNR}_a = E_b/N_{\mathrm{na}}$ is required to achieve the desired performance level. Without spread spectrum, the signal bandwidth is $W_d \cong MR/\log_2 M$ and the jammer power required is $J_1 = N_{\mathrm{na}}W_d$. With spread spectrum, the jammer must fill the entire transmission bandwidth with noise having the same psd, and the jammer power is $J_2 = N_{\mathrm{na}}W$. The FH/MSK processing gain is $J_2/J_1 = W/W_d$.

FH/DPSK. The receiver for a frequency-hop spread-spectrum system using differential PSK data modulation is identical to the receiver of Figure 6-6 with the MFSK demodulator replaced by a DPSK demodulator. The analysis of the effect of the despreading operation on AWGN or the barrage noise jamming is identical to the FH/MFSK analysis. The conclusion is that the noise psd at the data demodulator input is the same as the receiver input noise psd over the transmission bandwidth.

In this chapter, consideration is limited to binary DPSK modulation. The optimum demodulator for binary DPSK, illustrated in Figure 6-9a. This demodulator is analyzed in Arthurs and Dym [3], Park [4], Miller [5], and Lindsey and Simon [6], where the bit error probability is shown to be

$$P_b = \frac{1}{2}\exp\left(-\frac{E_b}{N_n}\right) = \frac{1}{2}\exp\left[\frac{-1}{(N_0 R/P) + (J/P)(R/W)}\right] \qquad (6\text{-}21)$$

Note that Arthurs and Dym and Lindsey and Simon also consider M-ary DPSK systems, and Miller provides some interesting practical results on degradations due to timing and frequency offsets.

An alternative demodulator for binary DPSK is shown in Figure 6-9b. This demodulator is discussed in many modern texts, where it is usually concluded that its noise performance is also given by (6-21). Park [4] has shown that (6-21) applies

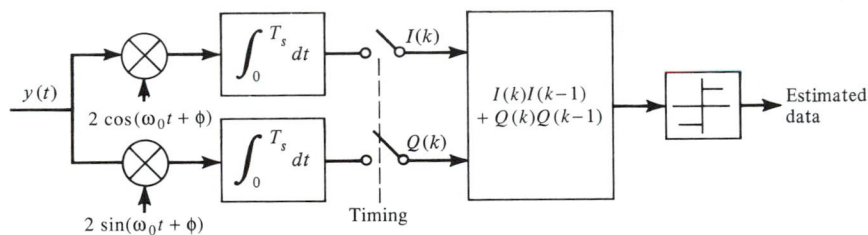

(a) Optimum demodulator

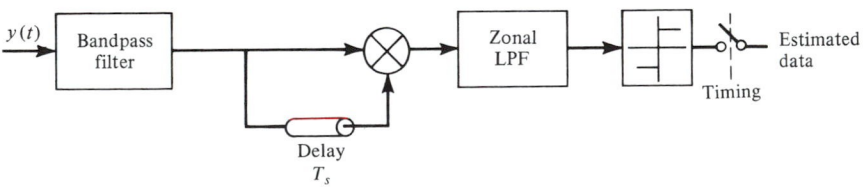

(b) Suboptimum demodulator

FIGURE 6-9. Binary differential phase-shift-keeping demodulators.

to Figure 6-9b only if intersymbol interference is ignored. When ISI is considered, the corrected expression is

$$P_b = \frac{1}{2} \exp\left(-k\frac{E_b}{N_n}\right) \tag{6-22}$$

where k depends on the specific characteristics of the input bandpass filter. In all cases, the analysis of the demodulator of Figure 6-9b presumes that the length of the delay line is simultaneously equal to a symbol time T_s and chosen so that $\omega_0 T_s = 2\pi l$, where l is any integer. For high carrier frequencies this implies extremely precise control of the length of the delay line. If this factor is a problem in system design, the equivalent baseband implementations can be used. Figure 6-10 is a plot of (6-21) using E_b/N_n from (6-20).

The effect of spectrum spreading on this system is identical to its effect on FH/MFSK. Frequency hopping forces the barrage jammer to fill the full spread transmission band W with noise rather than just the data bandwidth W_d. Thus the processing gain is approximately W/W_d. Finally, recall that slow-frequency-hop systems are being considered. The demodulators discussed each compare the phases of pairs of transmitted waveforms. If the transmission frequency changes between a transmitted symbol waveform and its reference waveform, phase comparison information is lost and a transmission error may result. Thus it is highly desirable to transmit a large number of data symbols on each hop frequency.

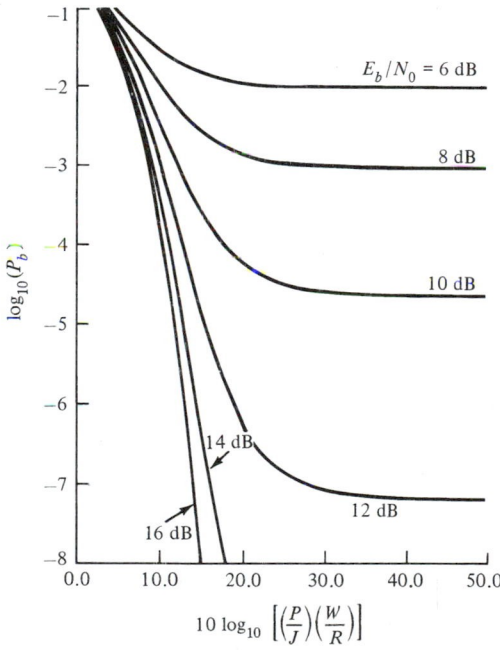

FIGURE 6-10. Performance of FH/DPSK spread-spectrum system in barrage noise jamming.

DS-FH/MFSK or DS-FH/DPSK. With the barrage noise jammer or in AWGN, there is little performance improvement which can be realized using hybrid DS-FH spreading modulation over FH spreading modulation. However, it may be more convenient for the hardware designer to obtain very wide transmission bandwidths using hybrid modulations. The presence of the frequency-hop component and its associated random phase shifts at each frequency hop makes it more convenient to use noncoherent or differentially coherent data modulation. The demodulator for DS-FH/MFSK is illustrated in Figure 6-11. The received signal is first mixed with a hopped reference signal to remove the frequency-hop component of the spreading modulation. The output of the first mixer is centered at the first IF and is still DS spread and has a phase modulation due to the noncoherence of the transmitter and receiver frequency synthesizers. The bandwidth of the first post-despreading band-pass filter must be large enough to pass the DS modulation without significant distortion. The second mixer removes the DS phase modulation and downconverts the signal to the second IF where filtering to the data bandwidth and data demodulation occurs.

The same arguments used for SFH to show that the noise psd at the first bandpass filter output is the same as the noise psd over the transmission band apply here. Thus the signal $y(t)$ may be written

$$y(t) = \sqrt{2P}\cos[2\pi f_{\text{IF}}t + \theta_{\text{SS}}(t) + \theta_d(t)] + n_n(t) \qquad (6\text{-}23)$$

where $\theta_{\text{SS}}(t)$ is the DS modulation, $\theta_d(t)$ is the data modulation including a random phase component due to the dehopping, and $n_n(t)$ is a bandlimited white Gaussian noise process with two-sided psd of $\frac{1}{2}N_n = \frac{1}{2}N_0 + \frac{1}{2}N_J$. The bandwidth of this noise process is obviously set by the bandpass filter. Since the filter bandwidth is roughly the same as the DS spreading bandwidth, the analysis of the effect of the DS despreading on the noise process is identical to the analysis for coherent systems given above. The result is that the despreader output noise psd at the second IF is given by a convolution of the input noise psd and the spreading waveform psd. The resultant density is dependent on the details of the spreading waveform and is given by $KN_n/2$, where K is a constant near unity.

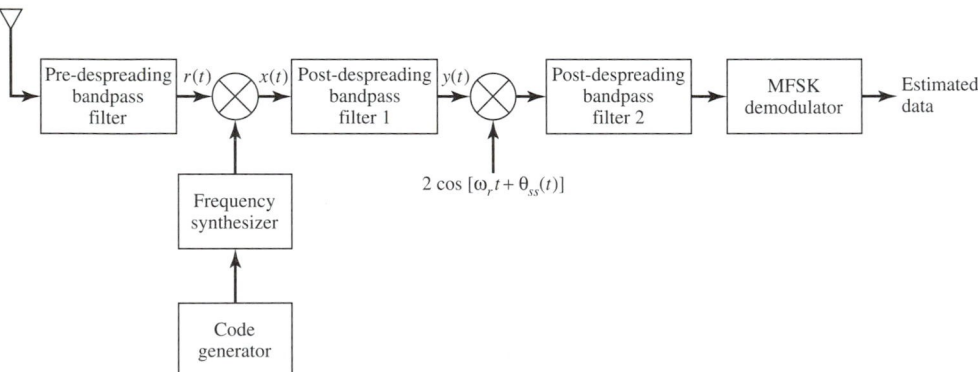

FIGURE 6-11. Simplified model of a DS-FH/MFSK spread-spectrum receiver.

The performance of the system is given by (6-19) with N_n replaced by KN_n. Once again the spectrum spreading has little effect in an AWGN environment, but does force the barrage jammer to use a very wide bandwidth. The processing gain of the system is approximately W/W_d. An identical result can be derived for DS-FH/DPSK except that the noise performance is given by (6-21) or (6-22), depending on which DPSK demodulator is used.

6-3.2 Performance in Partial-Band Jamming

Except for the transmission bandwidth, the partial-band jammer is the same as the wideband barrage noise jammer. Because of the smaller bandwidth, the partial-band jamming signal is easier to generate than the barrage jamming signal, and with some types of spectrum spreading, the partial-band jammer is considerably more effective than the barrage noise jammer. For these reasons, partial-band jamming is a commonly used ECM technique. In the analysis that follows, the communication system bandwidth is denoted by W, the nonspread data transmission bandwidth is denoted by W_d, and the jammer transmission bandwidth is denoted by W_J. The fraction of the communication bandwidth that is jammed is denoted by $\rho = W_J/W$. The jammer one-sided psd over the transmission band W_J is $N_J' = J/W_J = J/(\rho W)$ and the full-band jammer one-sided psd is $N_J = J/W$. Thermal noise with one-sided psd of N_0 also degrades system performance. Therefore, over one part of the transmission band the total noise psd will be $N_0 + N_J'$, and over the remaining part, the total noise psd will be N_0.

Coherent DS Systems. Consider an arbitrary DS system transmitting a signal described by $s_t(t)$ of (6-2), and using a receiver that can be modeled by Figure 6-4. The received signal is

$$r(t) = s_t(t) + n(t) + n_J(t) \tag{6-24}$$

where $n(t)$ represents the thermal noise and $n_J(t)$ represents the partial-band noise jammer. Both noise processes are narrowband relative to the carrier frequency. The receiver despreader output signal is

$$x(t) = \sqrt{2P} \cos[2\pi f_{\mathrm{IF}} t + \theta_d(t)]$$
$$+ n(t)\{2 \cos[2\pi(f_0 + f_{\mathrm{IF}})t + \theta_{\mathrm{SS}}(t)]\} \tag{6-25}$$
$$+ n_J(t)\{2 \cos[2\pi(f_0 + f_{\mathrm{IF}})t + \theta_{\mathrm{SS}}(t)]\}$$

The middle term of this expression is identical to the noise term of (6-4) and its two-sided psd near f_{IF} is $KN_0/2$. This psd is obtained from (6-9). Equation (6-9) applies only to BPSK and QPSK spreading modulation; a similar equation for MSK spreading leads to the same conclusion for the AWGN psd at the despreader output.

The last term of (6-25) is due to the jammer. The psd due to the jammer at the despreader output is the convolution of the jammer psd and the despreading waveform psd. Usually, the jammer will not know the precise center frequency of the communication signal. The jammer center frequency will be denoted by f_J. In general, the despreader output psd is given by

$$S_R(f) * S_J(f) = \int_{-\infty}^{+\infty} S_R(f - \lambda)S_J(\lambda)\, d\lambda \tag{6-26}$$

where $S_J(f)$ is the jammer psd and $S_R(f)$ is the psd of the reference spreading waveform. For BPSK or QPSK spreading, $S_R(f)$ is given by (6-5). Observe that the total power in the reference waveform is 2.0. The jammer two-sided psd $S_J(f)$ is $\frac{1}{2}N'_J$ over the frequency limits $|f \pm f_J| < \frac{1}{2}W_J$ and, for QPSK or BPSK spreading,

$$
\begin{aligned}
S_R(f) * S_J(f) = {} & \frac{N'_J}{W} \int_{f_J-(1/2)W_J}^{f_J+(1/2)W_J} \left\{ \text{sinc}^2\left[(f - \lambda - f_0 - f_{IF})\frac{2}{W} \right] \right. \\
& \left. + \text{sinc}^2\left[(f - \lambda + f_0 + f_{IF})\frac{2}{W} \right] \right\} d\lambda \\
& + \frac{N'_J}{W} \int_{-f_J-(1/2)W_J}^{-f_J+(1/2)W_J} \left\{ \text{sinc}^2\left[(f - \lambda - f_0 - f_{IF})\frac{2}{W} \right] \right. \\
& \left. + \text{sinc}^2\left[(f - \lambda + f_0 + f_{IF})\frac{2}{W} \right] \right\} d\lambda \\
\triangleq {} & S(f)
\end{aligned}
\tag{6-27}
$$

These integrals are depicted graphically in Figure 6-12. A typical spreading waveform psd shown in Figure 6-12a and a jammer psd in Figure 6-12b. The jammer center frequency does not equal the signal center frequency. The reference (i.e., despreading) waveform psd is shown in Figure 6-12c; the same psd is shifted and reversed for the convolution of (6-26) in Figure 6-12d and e. The frequency shifts shown in Figure 6-12d and e are $-f_{IF}$ and $+f_{IF}$, respectively. These shifts are used to evaluate $S(f)$ at $f = \pm f_{IF}$, the primary frequencies of interest. The shaded areas indicate the range of integration for (6-27). Observe that half of the terms of (6-27) are approximately zero for $f = \pm f_{IF}$, and that $S(f)$ is an even function. Thus

$$S(f_{IF}) = S(-f_{IF}) \cong \frac{N'_J}{W} \int_{f_J-(1/2)W_J}^{f_J+(1/2)W_J} \text{sinc}^2\left[(f_0 - \lambda)\frac{2}{W} \right] d\lambda \tag{6-28}$$

A similar result is easily derived for MSK spreading.

The integral of (6-28) may be evaluated numerically for a specific jammer. For extremely narrowband jammers (i.e., $\rho \ll 1$), the $\text{sinc}^2(\cdot)$ function may be assumed constant over the range of integration and

$$
\begin{aligned}
S(\pm f_{IF}) & \cong \frac{N'_J}{W} W_J \, \text{sinc}^2\left[(f_0 - f_J)\frac{2}{W} \right] \\
& = N_J \, \text{sinc}^2\left[(f_0 - f_J)\frac{2}{W} \right]
\end{aligned}
\tag{6-29}
$$

For maximum effectiveness, the very narrowband jammer center frequency must equal the carrier frequency f_0. It may also be deduced from (6-28) for any partial

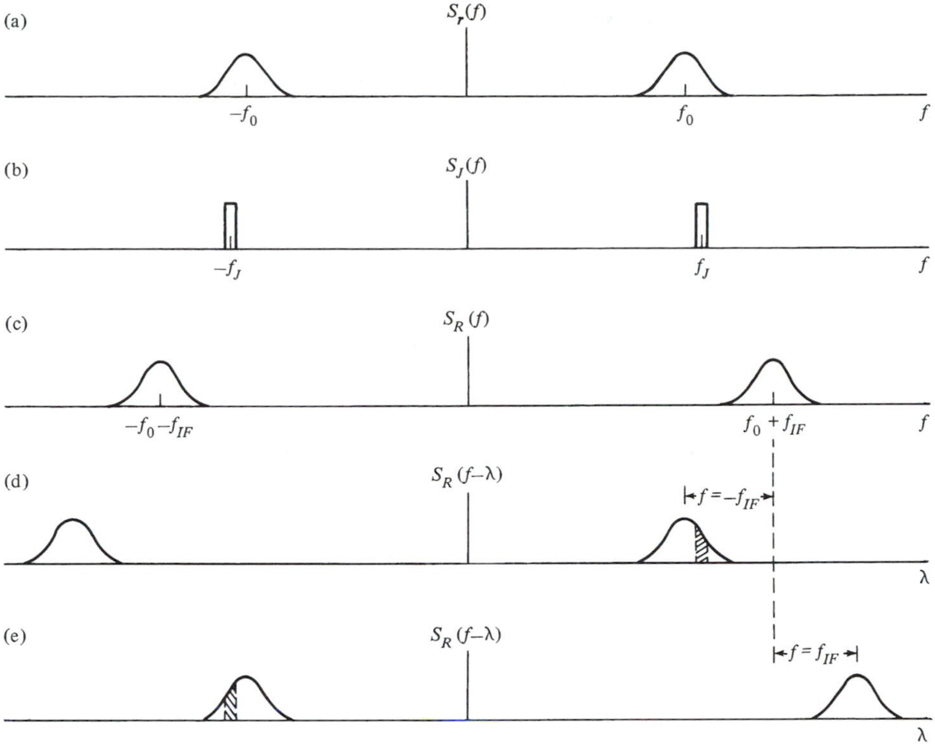

(a)

$S_r(f)$

(b)

$S_J(f)$

(c)

$S_R(f)$

(d)

$S_R(f-\lambda)$

$|\!\leftarrow\! f=-f_{IF}\!\rightarrow\!|$

(e)

$S_R(f-\lambda)$

$|\!\leftarrow\! f=f_{IF}\!\rightarrow\!|$

FIGURE 6-12. Typical power spectral densities used to calculate despreader output density for partial-band jamming: (a) received signal psd; (b) received jammer psd; (c) reference waveform psd; (d) reference psd reversed and shifted by $-f_{IF}$; (e) reference psd reversed and shifted by $+f_{IF}$.

band jammer that the jammer center frequency needs to equal f_0 for maximum effectiveness. When $f_J = f_0$, evaluation of (6-28) as a function of W_J will show that $S(f_{IF})$ is maximized for small W_J; in particular, for $W_J \ll W$, $S(f_{IF}) = N_J$ and for $W_J = W$, $S(f_{IF}) \cong N'_J/2 = N_J/2$. Thus the jammer gains about 3 dB by using a very narrowband signal, but this 3 dB is quickly lost if the jammer center frequency is not properly selected.

The system bit error probability is calculated by assuming that the IF filter output noise may be modeled as bandlimited AWGN with two-sided psd equal to the sum of the thermal noise component and a jammer noise component. For BPSK or QPSK data modulation, the error probability is

$$P_b = Q\left(\sqrt{\frac{E_b}{(N_0/2) + S(f_{IF})}} \right) \tag{6-30}$$

and for very narrowband jamming, with $f_J = f_0$,

$$P_b = Q\left(\sqrt{\frac{2E_b}{N_0 + 2N_J}} \right) = Q\left(\sqrt{\frac{2}{(RN_0/P) + 2(J/P)(R/W)}} \right) \tag{6-31}$$

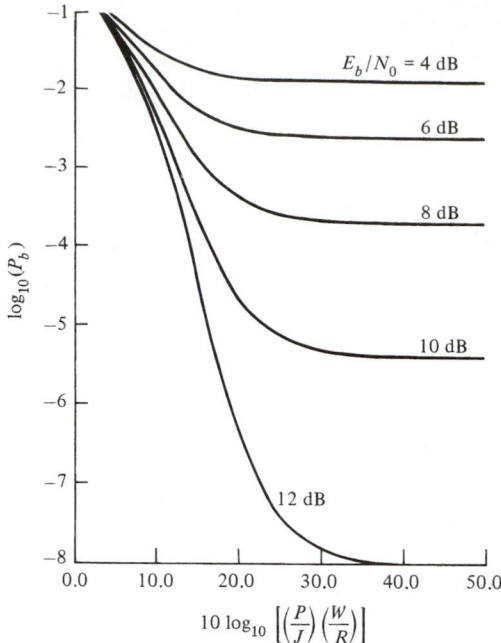

FIGURE 6-13. Performance of coherent direct-sequence spread-spectrum systems in partial-band jamming.

This relationship is plotted in Figure 6-13 using $(P/J)(W/R)$ as the independent variable for various signal-to-thermal noise ratios.

Recall that N_J is the one-sided psd that would be generated by the jammer if the total jammer power J were spread uniformly across the transmission bandwidth W; thus $J = N_J W$. Ignoring thermal noise, the system processing gain is calculated by comparing total jammer power required to increase system error probability to a specified level with and without spectrum spreading. Suppose that the specified error probability is obtained at $\text{SNR}_a = E_b/N_{Ja}$ without spectrum spreading. In this case, the jammer power is uniformly spread over a bandwidth W_d and the total jammer power is $J_1 = N_{Ja} W_d$. With spectrum spreading and narrowband jamming, the same error probability is achieved with $\text{SNR}_b = E_b/N_{Jb}$, and the total jammer power is $J_2 = N_{Jb} W$. SNR_a and SNR_b are related by equating the arguments of the Q-functions for the spread and nonspread cases. Without spreading, the error probability is given by $P_b = Q(\sqrt{2E_b/N_J})$. With spreading the error probability is given by (6-31). Thus equal error probabilities imply that $2\text{SNR}_a = \text{SNR}_b$ and that $2N_{Jb} = N_{Ja}$. Using these relationships, the processing gain is $J_2/J_1 = W/2W_d$, which is half as large as the comparable result for barrage noise jamming. Thus the jammer obtains a 3-dB advantage by using narrowband jamming rather than barrage jamming. Similar results can be derived for MSK spreading modulation.

FH/MFSK. The received signal for the FH/MFSK system is defined by (6-3) and (6-13) and the despreading reference is defined by (6-14). A very simplified block diagram of the receiver was illustrated in Figure 6-6. The system analysis for partial band jamming is the same as the analysis for barrage jamming through (6-15). The received FH signal hops over the entire transmission bandwidth W. On some hops the signal will be at a frequency which is jammed, while on other hops, the signal will be at a frequency where the only interference is thermal noise. When the system is properly synchronized, the despreader output may be represented by

$$x(t) = \sqrt{2P} \sum_{k=-\infty}^{\infty} p_{T_c}(t - kT_c) \cos[2\pi f_{\mathrm{IF}}t + \phi_k + \theta_d(t)] + n_n(t) \quad (6\text{-}32)$$

Because the signal is hopping in and out of the jamming noise, $n_n(t)$ is not stationary.

For slow-frequency-hop systems, the system message error probability is calculated using quasi-static analysis. That is, the error probability is calculated separately for thermal noise interference and thermal plus jamming noise interference, and the two results are averaged. Both error probabilities are calculated using (6-19). For the thermal noise calculation use $N_n = N_0$, and for the thermal plus jamming noise calculation use $N_n = N_0 + N_J'$. Since the fraction of the band that is jammed is ρ, the average bit error probability is

$$\bar{P}_b = (1 - \rho)P_b\left(\frac{E_b}{N_0}\right) + \rho P_b\left(\frac{E_b}{N_0 + N_J'}\right)$$

$$= (1 - \rho)P_b\left(\frac{P}{RN_0}\right) + \rho P_b\left(\frac{1}{(RN_0/P) + (J/P)(R/W)(1/\rho)}\right) \quad (6\text{-}33)$$

where $P_b(\cdot)$ represents (6-19). This equation is plotted in Figure 6-14 for large signal-to-thermal noise ratio and $M = 2$ using the familiar $(P/J)(W/R)$ as the independent variable and using ρ as a parameter.

Observe in Figure 6-14 that for any value of $(P/J)(W/R)$ there is an optimum value of ρ which maximizes the system bit error probability. The jammer has control of ρ and the smart jammer will be able to adjust ρ dynamically to always maximize system bit error probability. The system performance with this worst-case smart jammer has been calculated in Houston [7] for the special case where thermal noise is negligible. In this case, (6-33) simplifies to

$$\bar{P}_b = \rho P_b\left(\frac{E_b}{N_J'}\right) = \rho P_b\left(\rho \frac{E_b}{N_J}\right)$$

$$= \frac{\rho}{2(M - 1)} \exp\left(-\frac{l\rho E_b}{2N_J}\right) \sum_{q=2}^{M} \binom{M}{q}(-1)^q \exp\left[\frac{l\rho E_b(2 - q)}{2N_J q}\right] \quad (6\text{-}34)$$

$$= \frac{\rho}{2(M - 1)} \sum_{q=2}^{M} \binom{M}{q}(-1)^q \exp\left[\frac{l\rho E_b(1 - q)}{N_J q}\right]$$

The jammer wishes to maximize $\bar{P}_b$ by choosing the fractional bandwidth appropriately. To find the maximizing ρ, (6-34) is differentiated with respect to ρ and the

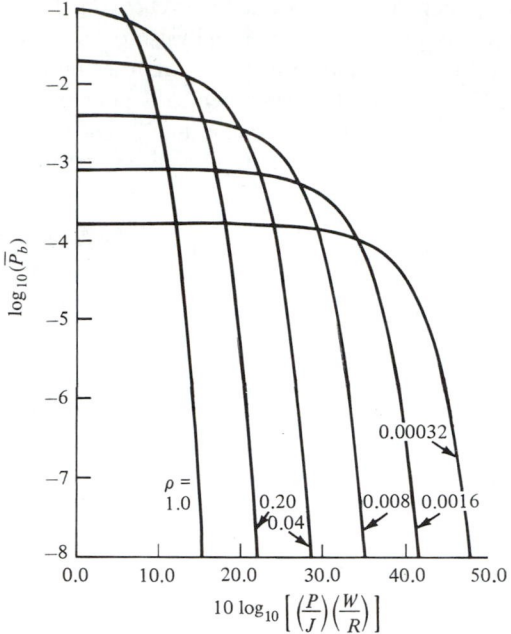

FIGURE 6-14. Performance of FH/BFSK spread spectrum in partial-band jamming.

result is set equal to zero. The result is

$$\frac{d\overline{P_b}}{d\rho} = \frac{\rho}{2(M-1)} \sum_{q=2}^{M} \binom{M}{q} (-1)^q \exp\left[\frac{l\rho E_b(1-q)}{N_J q}\right] \frac{lE_b(1-q)}{N_J q}$$

$$+ \frac{1}{2(M-1)} \sum_{q=2}^{M} \binom{M}{q} (-1)^q \exp\left[\frac{l\rho E_b(1-q)}{N_J q}\right] = 0$$

(6-35)

which can be written [7] as

$$\frac{\displaystyle\sum_{q=1}^{M} \binom{M}{q} \frac{(-1)^q}{q} \exp\left(\frac{y}{q}\right)}{\displaystyle\sum_{q=1}^{M} \binom{M}{q} (-1)^q \exp\left(\frac{y}{q}\right)} = \frac{y-1}{y}$$

(6-36)

where

$$y = l\rho \frac{E_b}{N_J}$$

(6-37)

This equation can be solved numerically for y as a function of M. The results y_0 of this calculation are given in Table 6-1 for several M. Table 6-1 was originally

TABLE 6-1. Solution of (6-36) as a Function of M

M	y_0	k'	$\left(\dfrac{E_b}{N_J}\right)_0$
2	2.00	0.3679	2.000
3	2.19		
4	2.38	0.2329	1.191
5	2.48		
6	2.59		
7	2.70		
8	2.78	0.1954	0.927
16	3.49	0.1812	0.873
32	3.62	0.1759	0.723

Source: Refs. 7 and 8.

presented by Houston [7]; the values in this table have subsequently been corrected by Simon et al. [8]. The numerical values in Table 6-1 are calculated assuming that the thermal noise may be neglected. Different values are found when thermal noise is considered.

The value of y_0 from Table 6-1 is used to calculate the optimum ρ as a function of E_b/N_J using (6-37). Of course, the fractional bandwidth can be no larger than 1.0, so $\rho = 1.0$ is used when $\rho > 1.0$ is found from the calculation. The values of E_b/N_J below which $\rho > 1.0$ are calculated from (6-37) are also given in Table 6-1. For $E_b/N_J > (E_b/N_J)_0$, the worst-case $\overline{P}_b$ is found by substituting $\rho = y_0/(lE_b/N_J)$ into (6-34). The result is

$$(\overline{P}_b)_{\max} = \frac{y_0}{2l(M-1)}\left(\frac{1}{E_b/N_J}\right)\sum_{q=2}^{M}\binom{M}{q}(-1)^q \exp\left(y_0\frac{1-q}{q}\right)$$
$$\triangleq \frac{k'}{E_b/N_J}$$
(6-38)

where k' is a constant which is also given in Table 6-1 for several M. For $E_b/N_J \leq (E_b/N_J)_0$, the average error probability is calculated directly from (6-34) with $\rho = 1.0$. In summary, the worst-case average bit error probability for partial-band jamming of FH/MFSK is given by

$$(\overline{P}_b)_{\max} = \begin{cases} \dfrac{k'}{E_b/N_J} & \dfrac{E_b}{N_J} > \left(\dfrac{E_b}{N_J}\right)_0 \\[3mm] \dfrac{1}{2(M-1)}\displaystyle\sum_{q=2}^{M}\binom{M}{q}(-1)^q \exp\left[\dfrac{lE_b}{N_J}\left(\dfrac{1-q}{q}\right)\right] & \dfrac{E_b}{N_J} \leq \left(\dfrac{E_b}{N_J}\right)_0 \end{cases}$$
(6-39)

where k' and $(E_b/N_J)_0$ are given in Table 6-1. Equation (6-39) is plotted in Figure 6-15 as a function of $E_b/N_J = (P/J)(W/R)$ for $M = 2$, 4, 8, and 16. Comparing Figure 6-8 with Figure 6-15 shows that the jammer can be considerably more effec-

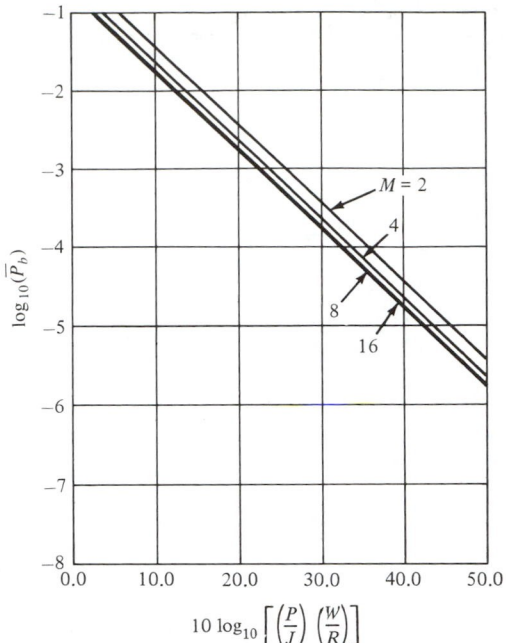

FIGURE 6-15. Performance of FH/MFSK spread spectrum in worst-case partial-band jamming.

tive using partial-band jamming than using barrage jamming. This increased effectiveness is obtained by causing a large degradation to a fraction of the transmitted symbols rather than a little degradation to all symbols. The inverse linear relationship between $(\overline{P_b})_{max}$ and E_b/N_J is typical of the performance of uncoded spread-spectrum communication systems in optimized jamming. The inverse linear relationship is also evident in Figure 6-14 as the envelope of the family of curves shown.

The calculation of the spread-spectrum system processing gain for FH/MFSK in partial-band jamming is somewhat different than the processing gain calculation performed earlier. The reason for this difference is that the mathematical expression for bit error probability has a different form in the jamming environment than it has in the nonjamming environment. Thus the arguments of the error probability expressions can no longer be equated to obtain the same performance with and without spectrum spreading. Processing gain is most easily calculated by plotting the bit error probability expressions with and without spectrum spreading as a function of $(E_b/N_J)_{dB}$ or, equivalently, $(P/J) \times (W/R)$ on the same grid as shown in Figure 6-16 for $M = 2$. At any given $\overline{P_b}$, the difference Δ_{dB} in $(E_b/N_J)_{dB}$ between the two curves of Figure 6-16 can be found, and is attributable to a change in N_J due to spectrum spreading. The N_J required without spectrum spreading is denoted by N_{J1} and the total jammer power required is $J_1 = N_{J1} W_d$. The N_J required with spectrum spreading is denoted by N_{J2} and the total jammer power required is $J_2 = N_{J2} W$. The

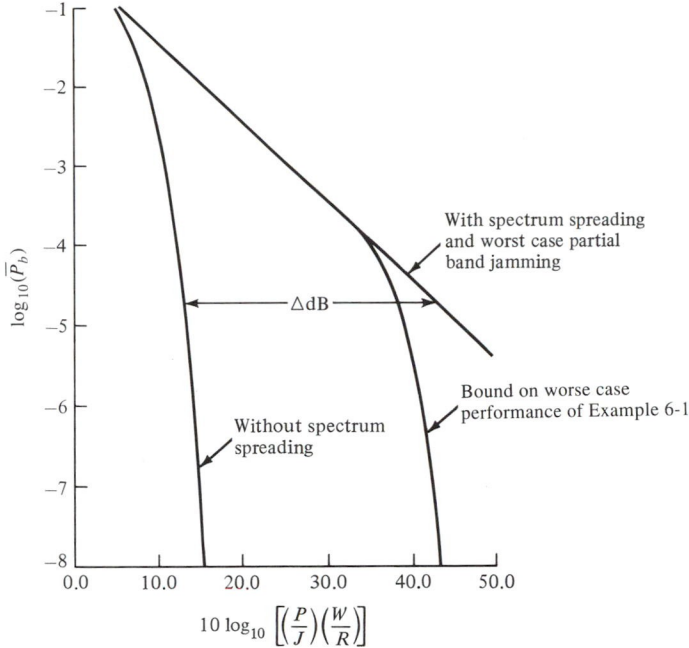

FIGURE 6-16. Calculation of processing gain for FH/MFSK spread spectrum in worst-case partial-band jamming.

relationship between N_{J1} and N_{J2} is

$$\frac{N_{J1}}{N_{J2}} = 10^{0.1\Delta_{\text{dB}}} \tag{6-40}$$

and the system processing gain is

$$\frac{J_2}{J_1} = \frac{N_{J2}W}{N_{J1}W_d} = \left(\frac{W}{W_d}\right) 10^{-0.1\Delta_{\text{dB}}} \tag{6-41}$$

This equation indicates that no matter how large W/W_d is, processing gain can be less than unity for large enough Δ_{dB}. Large Δ_{dB} is seen at low $\overline{P}_b$ on Figure 6-16. Fortunately, however, Δ_{dB} is limited by the fact that the derivation of (6-39) assumes that ρ can be arbitrarily small. The range of useful ρ is bounded below by the fact that no additional performance improvement is obtained by the jammer by using $W_J < W_d$. Thus $W_d/W \le \rho \le 1.0$ is the useful range of ρ, and Δ_{dB} cannot be arbitrarily large.

EXAMPLE 6-1

Consider a FH/MFSK system with $W/W_d = 1000$ and using binary FSK data modulation. The minimum useful $\rho = 0.001$ for this system. The E_b/N_J at which this ρ is required is calculated from (6-37) with $y = y_0$ from Table 6-1; the result is

$$\frac{E_b}{N_J} = \frac{2}{0.001} = 2000 = 33 \text{ dB}$$

Above this value of E_b/N_J, the optimum $\rho = 0.001$ and the complete average error probability expressions is

$$(\overline{P_b})_{\text{max}} = \begin{cases} \dfrac{0.001}{2} \exp\left(-\dfrac{0.001}{2}\dfrac{E_b}{N_J}\right) & 2000 < \dfrac{E_b}{N_J} \\[2ex] \dfrac{0.3679}{E_b/N_J} & 2.0 < \dfrac{E_b}{N_J} < 2000 \\[2ex] \dfrac{1}{2} \exp\left(-\dfrac{1}{2}\dfrac{E_b}{N_J}\right) & \dfrac{E_b}{N_J} < 2.0 \end{cases}$$

This result is also plotted in Figure 6-16. The maximum Δ_{dB} can be read from the figure or calculated from the equation above. The result is about 26 dB.

In all cases, the inverse linear relationship of (6-39) returns to the exponential relationship for very large E_b/N_J. The specific E_b/N_J at which this happens is a function of the bandwidth ratio W/W_d, so that totally general results cannot be derived. The conclusion that the partial-band jammer is considerably more effective than the wideband barrage jammer remains valid despite the bounding of this improvement. It will be demonstrated later that the use of forward error correction coding along with interleaving will largely eliminate the negative effects of the partial-band jammer.

This analysis of partial-band jamming for MFSK has considered only the case where the jammer psd is rectangular, as illustrated in Figure 6-2b. It is possible to obtain similar results [9] assuming that the jammer psd has a Gaussian shape rather than a rectangular shape. Similar performance is obtained when the equivalent noise bandwidth of the Gaussian spectrum is equal to the bandwidth of the rectangular spectrum. This result is important to the jammer since realizing a nearly rectangular output psd is significantly more difficult than realizing a Gaussian output psd.

Finally, this analysis considers only the case where the channel gain is constant. Significantly different results are found for other channels. Omura [10] and Simon et al. [8] have shown that for the Rayleigh-fading channel the partial-band jammer is no more efficient than the full-band jammer. Crepeau [11] has extended these results to the more general Nakagami-m fading channel. The Rayleigh-fading channel and the constant-gain channel are special cases of the Nakagami-m fading channel. Crepeau shows that while the full-band jammer is indeed optimum (from the jammer's viewpoint) for the Rayleigh-fading channel, a small constant-gain component in the channel causes the partial-band jammer again to be optimum for sufficiently large signal-to-noise ratios.

FH/DPSK. Once again consideration will be limited to binary DPSK data modulation. The analysis of the performance of FH/DPSK is identical to that of FH/MFSK except that the data modulation of (6-32) is DPSK rather than MFSK and the data demodulator is modified appropriately. Slow frequency hop is assumed and the noise term $n_n(t)$ of (6-32) is the same as for MFSK; that is, the noise is bandlimited AWGN with psd dependent on whether or not the transmitted carrier frequency is in or out of a jamming band. The average bit error probability is given by

$$\overline{P}_b = (1 - \rho)P_b\left(\frac{E_b}{N_J}\right) + \rho P_b\left(\frac{E_b}{N_0 + N_J'}\right) \tag{6-42}$$

with $P_b(\cdot)$ given by (6-21) when the optimum binary DPSK demodulator is used. Recall that in partial-band jamming, $N_n = N_0$ when the transmitted frequency is not in a jamming band and $N_n = N_0 + N_J' = N_0 + N_J/\rho$ when the transmitted frequency is within a jamming band. Substituting (6-21) into (6-42) yields

$$\overline{P}_b = \tfrac{1}{2}(1 - \rho)\exp\left[-\left(\frac{P}{N_0 R}\right)\right] + \tfrac{1}{2}\rho\exp\left[\frac{-1}{(N_0 R/P) + (J/P)(R/W)(1/\rho)}\right] \tag{6-43}$$

This result is plotted in Figure 6-17 for high signal-to-thermal noise ratio using ρ as a parameter and $(P/J)(W/R)$ as the independent variable. Observe that there is an

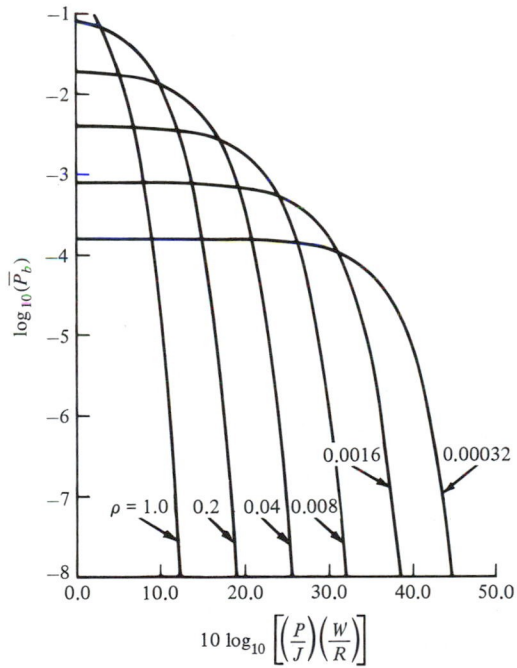

FIGURE 6-17. Performance of FH/DPSK spread spectrum in partial-band jamming.

optimum ρ for any specific value of $(P/J)(W/R)$. The smart jammer will choose ρ to maximize system average bit error probability.

Suppose now that the signal-to-thermal noise ratio is very high so that thermal noise–induced errors can be neglected. Equation (6-43) becomes

$$\overline{P}_b \cong \frac{\rho}{2} \exp\left[-\rho \left(\frac{P}{J}\right)\left(\frac{W}{R}\right) \right] \tag{6-44}$$

and the worst case ρ is found by differentiating with respect to ρ and setting the result equal to zero.

$$\frac{d\overline{P}_b}{d\rho} = \left[-\frac{\rho}{2}\left(\frac{P}{J}\right)\left(\frac{W}{R}\right) + \frac{1}{2} \right] \exp\left[-\rho \left(\frac{P}{J}\right)\left(\frac{W}{R}\right) \right] = 0 \tag{6-45}$$

Solving for the optimum ρ yields

$$\rho_0 = \frac{1}{(P/J)(W/R)} \tag{6-46}$$

and substituting this optimum ρ into (6-44) yields

$$(\overline{P}_b)_{\max} = \frac{1}{2(P/J)(W/R)} \exp(-1) = \frac{0.1839}{(P/J)(W/R)} \tag{6-47}$$

This relationship is only valid for $\rho_0 \leq 1.0$. For cases where (6-46) indicates that $\rho_0 > 1.0$, use $\rho_0 = 1.0$. The worst-case average bit error probability is therefore given by

$$(\overline{P}_b)_{\max} = \begin{cases} \dfrac{e^{-1}}{2(P/J)(W/R)} & \dfrac{P}{J}\left(\dfrac{W}{R}\right) \geq 1.0 \\[3ex] \dfrac{1}{2} \exp\left[-\dfrac{P}{J}\left(\dfrac{W}{R}\right) \right] & \dfrac{P}{J}\left(\dfrac{W}{R}\right) < 1.0 \end{cases} \tag{6-48}$$

This result is plotted in Figure 6-18. The now familiar inverse linear relationship between $(P_b)_{\max}$ and $(P/J)(W/R)$ is seen again. Comparing Figures 6-10 and 6-18 shows that the partial-band jammer can cause significantly worse system performance than the barrage noise jammer. Similar results can be derived for nonbinary DPSK systems.

The system processing gain is calculated using the same procedure used for FH/MFSK. That is, plot $\overline{P}_b$ versus E_b/N_J for the non-spread-spectrum and spread-spectrum cases on the same grid. For a particular $\overline{P}_b$, the difference Δ_{dB} in E_b/N_J between the two curves is found. This difference is due to a change in N_J required to degrade system performance to $\overline{P}_b$. Denote the N_J required without SS by N_{J1} and the associated jammer total power by $J_1 = N_{J1}W_d$. The N_J required with SS is denoted N_{J2} and the associated jammer power by $J_2 = N_{J2}W$. The densities N_{J1} and N_{J2} are related by (6-40) and the processing gain is given by (6-41). Again it appears that the processing gain might be reduced to values less than unity for very low $\overline{P}_b$. This is not true, however, since the jammer fractional bandwidth is practically

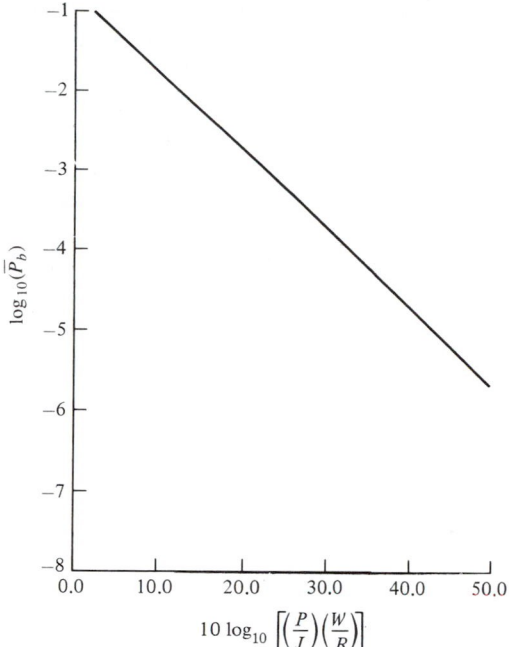

FIGURE 6-18. Performance of FH/DPSK spread spectrum in worst-case partial-band jamming.

limited to $W_d/W \leq \rho$. The particular E_b/N_J above which $\rho \leq W_d/W$ is $E_b/N_J = W/W_d$ calculated from (6-46). Using this lower bound on ρ, the error probability expression can be rewritten

$$(\bar{P}_b)_{\max} = \begin{cases} \dfrac{1}{2} \dfrac{W_d}{W} \exp\left[-\dfrac{P}{J}\left(\dfrac{W}{R}\right)\left(\dfrac{W_d}{W}\right) \right] & \dfrac{W}{W_d} < \dfrac{P}{J}\left(\dfrac{W}{R}\right) \\[4mm] \dfrac{e^{-1}}{2(P/J)(W/R)} & 1.0 \leq \dfrac{P}{J}\left(\dfrac{W}{R}\right) \leq \dfrac{W}{W_d} \\[4mm] \dfrac{1}{2} \exp\left[-\dfrac{P}{J}\left(\dfrac{W}{R}\right) \right] & \dfrac{P}{J}\left(\dfrac{W}{R}\right) < 1.0 \end{cases} \qquad (6\text{-}49)$$

System processing gain should be calculated from this result. These calculations will show that processing gain is significantly reduced by a partial-band jammer.

DS-FH/MFSK and DS-FH/DPSK. The receiver for the hybrid DS-FH systems is illustrated in Figure 6-11 and was described earlier. In partial-band jamming, the noise component at the output of the frequency dehopper may be thermal noise alone or may be the frequency-translated partial-band jammer. When the instantaneous bandwidth of the transmitted DS-FH signal overlaps the spectrum of the partial-band jammer, a portion of the jammer spectrum will appear at the output of the first IF bandpass filter. For simplicity, assume that the partial-band jammer spectrum fills the entire DS bandwidth whenever the two spectra overlap at all. This assumption is equivalent to the assumption that there are a large number of frequency hops across the transmission bandwidth.

In this special case, the DS despreader input noise component is either thermal noise with two-sided psd of $\frac{1}{2}N_0$ or thermal noise plus jamming noise which fills the entire DS bandwidth with two-sided psd of $\frac{1}{2}N_0 + \frac{1}{2}N'_J$. With this input noise, the DS despreader output noise has been calculated in the discussion of barrage noise jamming. The result is that the output noise has a two-sided psd of $KN_0/2$ or $K(N_0 + N'_J)/2$ depending on whether or not the transmission is in a jammed frequency band. The constant K is a function of the details of the DS modulation and is usually near unity. Let $P_b(\cdot)$ denote the bit error probability for the data modulation of interest. Then the average bit error probability is

$$\overline{P}_b = (1 - \rho)P_b\left(\frac{E_b}{KN_0}\right) + \rho P_b\left(\frac{E_b}{KN_0 + KN'_J}\right) \tag{6-50}$$

Except for the constant, K, this equation is the same as the result for FH/MFSK or FH/DPSK. The analysis from this point on is identical to that already performed and will not be repeated. The final result is that (6-39) and (6-49) apply with signal-to-noise ratio increased by a factor $1/K$.

6-3.3 Performance in Pulsed Noise Jamming

The pulsed noise jammer uses essentially the same strategy as the partial band jammer to degrade system performance. This strategy is to degrade performance significantly for a small fraction of the time in order to produce the maximum possible increase in average bit error probability. The pulsed noise jammer is on for a fraction ρ of the time and produces a wideband noise signal with two-sided psd of $N'_J/2 = N_J/2\rho$ when "on." The two-sided psd $N_J/2$ is the density the jammer would produce if "on" continuously. The average jammer power is denoted by J.

It will be shown that the pulsed jammer can produce the same type of degradation for coherent DS systems that the partial-band jammer produced for pure FH systems. A disadvantage of the pulsed noise jammer is that very high peak power may be required from the jammer. Peak power constraints will limit the minimum value of ρ and thereby limit jamming effectiveness.

It is assumed that the jammer pulse repetition frequency is slow enough that a large number of information bits are transmitted between each change of state of the jammer. With this assumption, the situation where the jammer turns "on" or "off" during the transmission of a bit may be ignored, and a quasi-static analysis of the average bit error probability is appropriate. For any of the modulations to be considered, let $P_b(E_b/N_n)$ denote the data demodulator bit error probability, where N_n is the one-sided noise psd at the input to the data demodulator. The magnitude of N_n depends on the particular spreading modulation used and whether or not the jammer is "on." Let N_{n1} and N_{n2} denote the value of N_n when the jammer is "off" and "on," respectively. N_{n1} is not zero because, in general, thermal noise is not neglected. The average bit error probability $\overline{P}_b$ is given by

$$\overline{P}_b = (1 - \rho)P_b\left(\frac{E_b}{N_{n1}}\right) + \rho P_b\left(\frac{E_b}{N_{n2}}\right) \tag{6-51}$$

The jammer selects ρ to maximize $\overline{P}_b$ subject to peak power constraints. The density N_{n2} is also a function of ρ.

Coherent DS Systems. For all the coherent data modulations of interest in this chapter (BPSK, QPSK, OQPSK, and MSK), the bit error probability is given by $P_b(E_b/N_n) = Q(\sqrt{2E_b/N_n})$. The value of N_n is calculated using the same steps used in the barrage noise analysis with the end result that $N_{n1} = KN_0$ and $N_{n2} = K(N_0 + N_J')$, where K is a function of the spreading modulation. For BPSK and QPSK spreading modulations K is given in (6-9). Substituting into (6-51) yields

$$\overline{P}_b = (1 - \rho)Q\left(\sqrt{\frac{2E_b}{KN_0}}\right) + \rho Q\left[\sqrt{\frac{2E_b}{K(N_0 + N_J/\rho)}}\right] \tag{6-52}$$

For the special case where thermal noise is negligible relative to jamming noise, (6-52) may be approximated by

$$\overline{P}_b \cong \rho Q\left(\sqrt{\frac{2\rho E_b}{KN_J}}\right) = \rho Q'\left[\sqrt{\frac{2\rho}{K}\left(\frac{P}{J}\right)\left(\frac{W}{R}\right)}\right] \tag{6-53}$$

This expression is plotted in Figure 6-19 as a function of $(P/J)(W/R)$ using $K = 1.0$ and various ρ. The remainder of this discussion considers only this special case where thermal noise is negligible.

Calculation of the worst-case ρ could be accomplished by differentiating (6-53) with respect to ρ and setting the result equal to zero. Unfortunately, the derivative of the Q-function is not easily calculated so that this method fails. The worst-case $\overline{P}_b$ can, however, be easily determined graphically from the plot of Figure 6-19. The envelope of the family of curves shown represents the worst-case $\overline{P}_b$. This envelope exhibits the usual inverse linear relation between $(P_b)_{max}$ and $(P/J)(W/R)$. An alternative means of calculating the worst-case ρ is to use an upper bound for the Q-function. This bound is

$$\overline{P}_b \leq \frac{\rho}{\sqrt{4\pi\rho y}} \exp(-\rho y) \tag{6-54}$$

where $y = E_b/N_J = (P/J)(W/R)$. Taking the first derivative and setting it equal to zero and solving for ρ yields $\rho = 1/(2y)$. The duty factor can be no larger than unity so that $\rho = 1.0$ is used whenever $y < 0.5$. Substituting the optimum ρ into (6-53) yields

$$(\overline{P}_b)_{max} = \begin{cases} \dfrac{1}{2(P/J)(W/R)} Q\left(\sqrt{\dfrac{1}{K}}\right) & 0.5 < \dfrac{P}{J}\left(\dfrac{W}{R}\right) \\[4mm] Q\left\{\sqrt{\dfrac{2}{K}\left(\dfrac{P}{J}\right)\left(\dfrac{W}{R}\right)}\right\} & \dfrac{P}{J}\left(\dfrac{W}{R}\right) < 0.5 \end{cases} \tag{6-55}$$

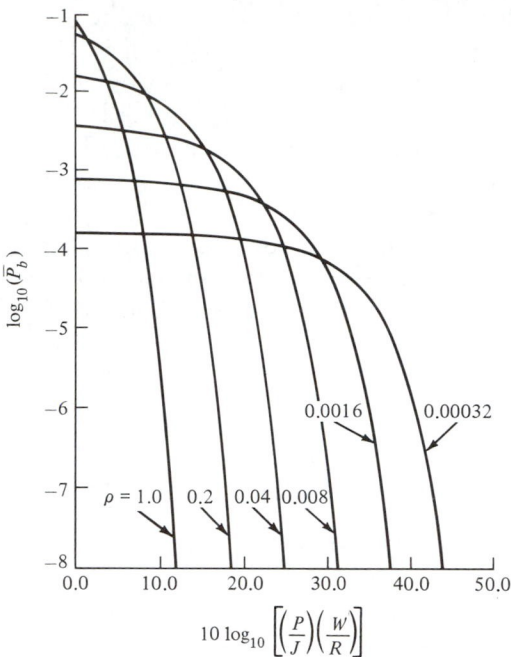

$$10 \log_{10}\left[\left(\frac{P}{J}\right)\left(\frac{W}{R}\right)\right]$$

FIGURE 6-19. Performance of coherent direct-sequence spread spectrum in pulsed noise jamming.

This equation is plotted in Figure 6-20. A similar result is discussed in Viterbi [12] using the bound of (6-54) in the final $(\overline{P}_b)_{\max}$ result.

The value of ρ that maximizes the jammer's effectiveness has been found numerically by Simon et al. [8] using the exact expression (6-53) for average bit error probability. The most effective is $\rho = 0.709/y$ for $y > 0.709$ and $\rho = 1.0$ otherwise. Using this value, the worst-case average bit error probability is

$$(\overline{P}_b)_{\max} = \begin{cases} \dfrac{0.083}{(P/J)(W/R)} & 0.709 \le \dfrac{P}{J}\left(\dfrac{W}{R}\right) \\[4mm] Q\left(\sqrt{\dfrac{2}{K}\left(\dfrac{P}{J}\right)\left(\dfrac{W}{R}\right)}\right) & \dfrac{P}{J}\left(\dfrac{W}{R}\right) \le 0.709 \end{cases}$$

The calculation of the processing gain for coherent direct-sequence systems must take into account the fact that pulsed jamming is also effective against the nonspread system. In fact, the analysis of the pulse-jammed nonspread system is identical to the analysis above except that there is no factor K in the equations comparable to (6-52) and (6-53). The final result is identical to (6-55) with $K = 1.0$ and total bandwidth W_d. The processing gain is then calculated by equating E_b/N_J for the spread and nonspread systems. For both systems the jammer is pulsed with duty factor ρ. The average jammer power without spectrum spreading is $J_1 = \rho N'_J W_d =$

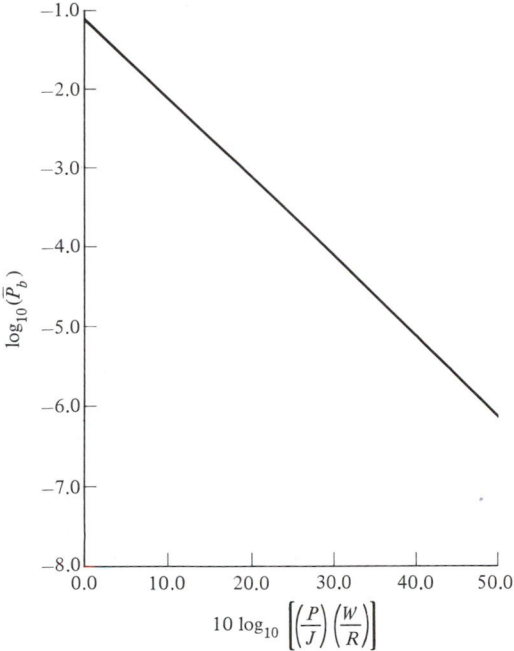

FIGURE 6-20. Performance of coherent direct-sequence spread spectrum in worst-case pulsed noise jamming.

$N_J W_d$ and the average jammer power with spectrum spreading is $J_2 = \rho N'_J W = N_J W$. The processing gain is therefore $J_2/J_1 = W/W_d$.

FH/MFSK. The bit error probability for MFSK is given by (6-19). The data demodulator input noise psd is calculated using the same steps used for barrage noise jamming. The results are $N_{n1} = N_0$ and $N_{n2} = N_0 + N'_J = N_0 + N_J/\rho$ and the average bit error probability is

$$\overline{P}_b = (1 - \rho)P_b\left(\frac{E_b}{N_0}\right) + \rho P_b\left(\frac{E_b}{N_0 + N_J/\rho}\right) \tag{6-56}$$

where $P_b(\cdot)$ is defined in (6-19). This equation is identical to (6-33), which was derived for FH/MFSK in partial-band jamming. Thus the analysis to find the worst case ρ is the same as was done for partial-band jamming and the final result is given by (6-39), which is plotted in Figure 6-15.

The calculation of processing gain for the pulsed jammer is, however, different since the pulsed jammer can also force the nonspread system to an inverse linear relationship. The bit error probability for the pulse jammed nonspread system is also specified by (6-39), and the processing gain is calculated by equating E_b/N_J for the spread and nonspread systems. Once again, jamming resistance is achieved by forcing the jammer to spread its power over a bandwidth W using spread spectrum rather than W_d without spread spectrum. The system processing gain is $J_2/J_1 = W/W_d$.

FH/DPSK. The bit error probability for DPSK is given by (6-21), and the data demodulator input noise psd N_n is calculated using the same steps used for barrage jamming. When the pulse jammer is "on," $N_{n2} = N_0 + N'_J$, and when the pulse jammer is "off," $N_{n1} = N_0$. The average bit error probability is

$$\overline{P}_b = \frac{1 - \rho}{2} \exp\left(-\frac{E_b}{N_0}\right) + \frac{\rho}{2} \exp\left(-\frac{E_b}{N_0 + N_J/\rho}\right) \tag{6-57}$$

which is identical to (6-43). The calculation of the worst-case ρ was described above and the final average bit error probability is given by (6-48) and is plotted in Figure 6-18.

The system processing gain is not the same as calculated for the partial-band jamming case. The pulsed jammer can degrade the nonspread system in the same way as the spread system. Thus the equations describing the bit error probability of the spread and nonspread systems are the same and the processing gain is equal to the bandwidth ratio W/W_d.

DS-FH/MFSK and DS-FH/DPSK. The pulsed jamming performances of DS-FH/MFSK and of DS-FH/DPSK are calculated using (6-51) with N_{n1} and N_{n2} calculated as in the barrage jamming analysis of these hybrid systems. Therefore, when the jammer is "off," $N_n = KN_0$, and when the jammer is "on," $N_n = K(N_0 + N'_J)$, where K is dependent on the details of the DS modulation. The average bit error probability is

$$\overline{P}_b = (1 - \rho)P_b\left(\frac{E_b}{KN_0}\right) + \rho P_b\left(\frac{E_b}{KN_0 + KN'_J}\right) \tag{6-58}$$

where $P_b(\cdot)$ is given by (6-19) for DS-FH/MFSK and by (6-21) for DS-FH/DPSK. From this point on the analysis is the same as the analysis for the pure FH systems pulsed jamming except that the factor K is included. When signal-to-thermal noise ratio is high, thermal noise can be neglected and the worst-case performance is given by (6-39) for DS-FH/MFSK and by (6-48) for DS-FH/DPSK with E_b/N_J in each case increased by $1/K$.

6-3.4 Performance in Single-Tone Jamming

The single-tone jammer is perhaps the easiest of all jamming signals to generate. In early spread-spectrum literature, this jammer was analyzed by assuming that after despreading, the jamming power was equivalent to Gaussian noise. The analysis was simply to compute the jamming power spectrum at the despreader output and to use conventional Gaussian noise techniques thereafter. More recently, a number of authors [13–16] have analyzed the effect of the tone jammer in greater detail and their example is followed here. A single-tone jammer affects only a single hop frequency in a FH system. For this reason, this jammer is not a very effective countermeasure against FH systems and FH systems will be ignored in this section.

Receiver Description and General Analysis. Only coherent direct-sequence SS systems are considered in this section. A generalized receiver model that applies to

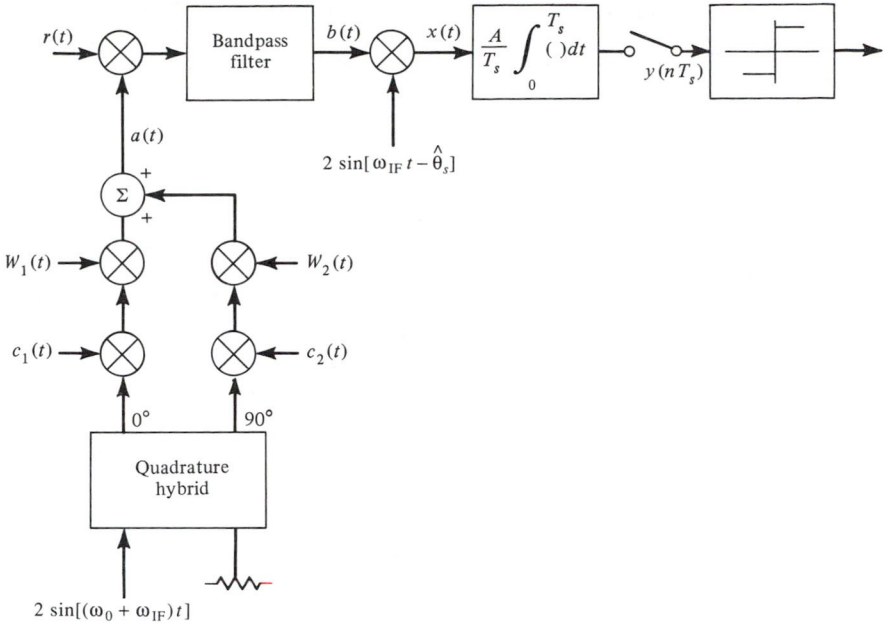

FIGURE 6-21. Conceptual model of coherent direct-sequence spread-spectrum receiver.

all the DS modulations of interest, including BPSK/BPSK, MSK/BPSK, QPSK/BPSK, is illustrated in Figure 6-21. In this figure, $c_1(t)$ and $c_2(t)$ are independent in-phase and quadrature channel spreading waveforms, respectively, and $W_1(t)$ and $W_2(t)$ are in-phase and quadrature weighting functions. For offset-QPSK and MSK spreading modulations, the spreading waveforms $c_1(t)$ and $c_2(t)$ and the weighting functions $W_1(t)$ and $W_2(t)$ are assumed to be appropriately phased. For BPSK/BPSK modulation, set $c_2(t) = 0$ to obtain a single channel. QPSK/QPSK is considered as two independent BPSK/BPSK signals and is therefore a slight modification of the figure. For all modulations except MSK/BPSK, the weighting functions may be set equal to unity while for MSK/BPSK use half-sine functions†

$$W_1(t) = \sqrt{2} \sum_i p_{T_c}(t - iT_c) \sin\left[\frac{\pi}{T_c}(t - iT_c)\right]$$

$$\tag{6-59}$$

$$W_2(t) = \sqrt{2} \sum_i p_{T_c}\left(t - iT_c - \frac{T_c}{2}\right) \sin\left[\frac{\pi}{T_c}\left(t - iT_c - \frac{T_c}{2}\right)\right]$$

where T_c is the spreading waveform chip duration and $p_{T_c}(t)$ is the unit pulse. The

---

†Observe that only the positive half-cycle of the sine-wave weighting function is used here, in contrast to the alternating positive and negative half-cycles of Figure 2-14. In addition, a $\sqrt{2}$ normalization factor is included here which does not appear in Figure 2-14. Either format is valid for MSK.

received spread-spectrum signal is

$$s_t(t) = \sqrt{P}d(t)\{c_1(t)W_1(t)\cos(\omega_0 t + \theta_s) + c_2(t)W_2(t)\sin(\omega_0 t + \theta_s)\} \quad (6\text{-}60)$$

where $d(t)$ is the data modulation, and the received tone jammer is

$$n_J(t) = \sqrt{2J}\cos(\omega_J t + \phi_J) \quad (6\text{-}61)$$

The received signal power is P and the received jammer power is J. The signal frequency ω_0 and the jammer frequency ω_J are not assumed to be equal. The thermal noise component of the received signal is denoted $n(t)$ and the total received signal is $r(t) = s_t(t) + n_J(t) + n(t)$. The purpose of the analysis to follow is to develop an approximate expression for the statistics of the integrator output so that bit error probability can be calculated.

The integrator inputs, $x_1(t)$, $x_2(t)$, and $x_3(t)$, will be calculated separately for the three received signal components. Consider the signal component first. The DS despreader input reference $a(t)$ is

$$a(t) = \sqrt{2}c_1(t)W_1(t)\cos[(\omega_0 + \omega_{IF})t]$$
$$+ \sqrt{2}c_2(t)W_2(t)\sin[(\omega_0 + \omega_{IF})t] \quad (6\text{-}62)$$

Thus the signal component at the bpf output is

$$b_1(t) = [s_t(t)a(t)]_{\text{diff freq}}$$
$$= \frac{\sqrt{2P}}{2}d(t)[W_1^2(t) + W_2^2(t)]\cos(\omega_{IF}t - \theta_s) \quad (6\text{-}63)$$

Perfect synchronization has been assumed. The system is phase coherent so that $\hat{\theta}_s \cong \theta_s$ and the second mixer output due to signal is

$$x_1(t) = \frac{\sqrt{2P}}{2}d(t)\{W_1^2(t) + W_2^2(t)\} \quad (6\text{-}64)$$

The integrator input due to jamming is calculated next. The despreading mixer output component which passes through the bandpass filter is

$$b_2(t) = [n_J(t)a(t)]_{\text{diff freq}}$$
$$= \sqrt{J}c_1(t)W_1(t)\cos[(\omega_{IF} + \Delta\omega)t - \phi_J]$$
$$+ \sqrt{J}c_2(t)W_2(t)\sin[(\omega_{IF} + \Delta\omega)t - \phi_J] \quad (6\text{-}65)$$

where $\Delta\omega = \omega_0 - \omega_J$. The output of the second mixer due to jamming is

$$x_2(t) = \sqrt{J}c_1(t)W_1(t)\cos(\Delta\omega\, t - \phi_J + \theta_s)$$
$$+ \sqrt{J}c_2(t)W_2(t)\sin(\Delta\omega\, t - \phi_J + \theta_s) \quad (6\text{-}66)$$

Finally, the integrator input due to noise is calculated. The received noise is assumed to be bandlimited to approximately the signal bandwidth so that the noise analysis for the barrage noise jammer can be applied directly. The result is that the bandpass filter output is bandlimited white Gaussian noise whose psd is the convolution of the input noise spectrum and the spreading waveform power spectrum evaluated at the IF. For all cases of interest, the two-sided psd is approximately $\frac{1}{2}N_0$, which is identical to the received noise two-sided pdf. Thus

$$b_3(t) = \sqrt{2}\, n_I(t) \cos \omega_{IF}t - \sqrt{2}\, n_Q(t) \sin \omega_{IF}t \tag{6-67}$$

The second mixer output is

$$x_3(t) = \sqrt{2}\, n_I(t) \cos \hat{\theta}_s + \sqrt{2}\, n_Q(t) \sin \hat{\theta}_s \tag{6-68}$$

Since $n_I(t)$ and $n_Q(t)$ are independent, the right-hand side of this equation may be considered a single Gaussian noise process whose two-sided power spectrum is

$$\frac{N_n}{2} = \frac{N_0}{2}[(\sqrt{2} \cos \hat{\theta}_s)^2 + (\sqrt{2} \sin \hat{\theta}_s)^2]$$

$$= N_0 \tag{6-69}$$

over the range of frequencies where it is nonzero.

The integrator output due to noise at the sampling time is

$$y_3(nT_s) = \frac{A}{T_s} \int_0^{T_s} x_3(t)\, dt \tag{6-70}$$

Since $x_3(t)$ is Gaussian and integration is a linear operation, $y_3(nT_s)$ is also Gaussian. The mean of $y_3(T_s)$ is zero and the variance is calculated using [17].

$$\sigma_{y_3}^2 = A^2 \frac{2}{T_s} \int_0^{T_s} \left(1 - \frac{\lambda}{T_s}\right) \{R_{x_3}(\lambda) - E^2[x_3(t)]\}\, d\lambda \tag{6-71}$$

where $R_{x_3}(\lambda)$ is the autocorrelation function of $x_3(t)$. Denote the bandpass filter bandwidth by B; then

$$R_{x_3}(\lambda) = BN_0 \operatorname{sinc} B\lambda \tag{6-72}$$

It is reasonable to assume that B is large relative to the inverse of the integration time since the matched filtering operation is accomplished by the integrator and not the bandpass filter. With this assumption the integral may be approximated by

$$\sigma_{y_3}^2 \cong A^2 \frac{2}{T_s} \int_0^{\infty} BN_0 \operatorname{sinc} B\lambda\, d\lambda = \frac{N_0}{T_s} A^2 \tag{6-73}$$

The total integrator input is $x(t) = x_1(t) + x_2(t) + x_3(t)$. The integrator output due to signal and jamming will be calculated for each modulation of interest.

BPSK/BPSK. To model BPSK/BPSK using the equations above, set $W_2(t) = 0$ and $W_1(t) = \sqrt{2}$. With this substitution, all the signal power is placed in the I channel of both the received signal and the despreading reference. Thus

$$x_1(t) + x_2(t) = \sqrt{2P}\, d(t) + \sqrt{2J}\, c_1(t) \cos(\Delta\omega\, t - \phi_J + \theta_s) \tag{6-74}$$

The integrator output is normalized using $A = 1/\sqrt{2P}$, so that the output data component is ± 1; the result is (signal and jammer only)

$$y(nT_s) = \pm 1 + y_2(nT_s) \tag{6-75}$$

where

$$y_2(nT_s) = \frac{1}{T_s} \sqrt{\frac{J}{P}} \int_0^{T_s} c_1(t) \cos(\Delta\omega\, t - \phi_J + \theta_s)\, dt \tag{6-76}$$

Calculation of the bit error probability requires knowledge of the statistics of y_2. As a first step in evaluating the statistics of y_2, the mean and variance will be calculated. The mean is the ensemble average over all possible spreading code sequences during the integration time,

$$\bar{y}_2 = E\left[\frac{1}{T_s} \sqrt{\frac{J}{P}} \int_0^{T_s} c_1(t) \cos(\Delta\omega\, t + \theta_s - \phi_J)\, dt\right] \tag{6-77}$$

Since $c_1(t)$ takes on values of ± 1 with equal probability, this average is zero for each code chip, and $\bar{y}_2 = 0$.

The evaluation of the variance is considerably more complicated. The first step in this calculation is to evaluate the integral y_2 over each chip interval to express y_2 as a summation. Thus

$$y_2 = \frac{1}{T_s} \sqrt{\frac{J}{P}} \sum_{i=0}^{N-1} c_i \int_{iT_c}^{(i+1)T_c} \cos(\Delta\omega\, t + \theta_s - \phi_J)\, dt$$

$$= \frac{1}{T_s} \sqrt{\frac{J}{P}} \sum_{i=0}^{N-1} c_i \frac{1}{\Delta\omega} \{\sin[\Delta\omega\, (i + 1)T_c + \theta_s - \phi_J]$$
$$- \sin(\Delta\omega i T_c + \theta_s - \phi_J)\} \tag{6-78}$$

where $c_i = \pm 1$ represents the spreading code sequence and N is the number of spreading waveform chips in a data symbol. This expression can be simplified by expressing the sine function in exponential form, factoring common phase angles from all terms, and then recombining. The final result is

$$y_2 = \frac{T_c}{T_s} \sqrt{\frac{J}{P}} \frac{\sin(\Delta\omega\, T_c/2)}{\Delta\omega\, T_c/2} \sum_{i=0}^{N-1} c_i \cos(\Delta\omega\, T_c i + \phi) \tag{6-79}$$

In this equation, $\phi = -\phi_J + \Delta\omega T_c/2 + \theta_s$.

Since the mean of y_2 is zero, the variance of y_2 is the ensemble average over all spreading waveforms of y_2^2. Thus

$$\sigma_{y_2}^2 = E\left[\left(\frac{T_c}{T_s}\right)^2 \left(\frac{J}{P}\right) \frac{\sin^2(\Delta\omega\, T_c/2)}{(\Delta\omega\, T_c/2)^2} \sum_{i=0}^{N-1} c_i \cos(\Delta\omega\, T_c i + \phi) \right.$$

$$\left. \times \sum_{k=0}^{N-1} c_k \cos(\Delta\omega\, T_c k + \phi)\right]$$

$$= \frac{1}{N^2} \frac{J}{P} \frac{\sin^2(\Delta\omega\, T_c/2)}{(\Delta\omega\, T_c/2)^2} \tag{6-80}$$

$$\times E\left[\sum_{i=0}^{N-1}\sum_{k=0}^{N-1} c_i c_k \cos(\Delta\omega\, T_c i + \phi)\cos(\Delta\omega\, T_c k + \phi)\right]$$

$$= \frac{1}{N^2}\frac{J}{P}\frac{\sin^2(\Delta\omega\, T_c/2)}{(\Delta\omega\, T_c/2)^2}\sum_{i=0}^{N-1}\sum_{k=0}^{N-1} E(c_i c_k)\cos(\Delta\omega\, T_c i + \phi)$$

$$\times \cos(\Delta\omega\, T_c k + \phi)$$

Assume now that the spreading code is an infinite sequence of independent equally likely random binary digits so that $E(c_i c_k) = 1$ for $i = k$ and $E(c_i c_k) = 0$ for $i \neq k$. Thus

$$\sigma_{y_2}^2 = \frac{1}{N^2}\frac{J}{P}\frac{\sin^2(\Delta\omega\, T_c/2)}{(\Delta\omega\, T_c/2)^2}\sum_{i=0}^{N-1}\cos^2(\Delta\omega\, T_c i + \phi) \tag{6-81}$$

$$= \frac{1}{N^2}\frac{J}{P}\frac{\sin^2(\Delta\omega\, T_c/2)}{(\Delta\omega\, T_c/2)^2}\sum_{i=0}^{N-1}[\tfrac{1}{2} + \tfrac{1}{2}\cos(2\Delta\omega\, T_c i + 2\phi)]$$

This expression is simplified using the identity [18]

$$\sum_{i=0}^{N-1}\cos(iy + x) = \frac{\cos\{x + [(N-1)/2]\,y\}\sin(Ny/2)}{\sin(y/2)} \tag{6-82}$$

the final result is [14]

$$\sigma_{y_2}^2 = \frac{J}{2PN}\frac{\sin^2(\Delta\omega\, T_c/2)}{(\Delta\omega\, T_c/2)^2}\left\{1 + \frac{\cos[2\phi + (N-1)\,\Delta\omega\, T_c]\sin(N\,\Delta\omega\, T_c)}{N\sin(\Delta\omega\, T_c)}\right\} \tag{6-83}$$

The mean and variance of a probability density do not completely specify the pdf. The exact pdf for the case where the jammer frequency is offset from the received carrier frequency has not been found. However, when $\omega_J = \omega_0$ the pdf can be found exactly, as shown by Levitt [13]. In this special case, (6-76) becomes

$$y_2 = \frac{1}{T_s}\sqrt{\frac{J}{P}}\int_0^{T_s} c_1(t)\cos(-\phi_J + \theta_s)\,dt \tag{6-84}$$

$$= \frac{T_c}{T_s}\sqrt{\frac{J}{P}}\cos(\theta_s - \phi_J)\sum_{i=0}^{N-1} c_i$$

The integral y_2 is a discrete random variable whose value depends on the particular sequence of N spreading code symbols associated with the information symbol being demodulated. Let

$$K = \frac{1}{N}\sum_{i=0}^{N-1} c_i \tag{6-85}$$

and assume that the spreading code is a sequence of independent equally likely binary symbols. Let N_1 be the number of $+1$'s and N_2 be the number of -1's in the spreading code sequence. Then

$$K = \frac{1}{N}(N_1 - N_2) = \frac{1}{N}(2N_1 - N) \qquad (6\text{-}86)$$

All spreading code sequences of length N chips are equally likely. The probability of a particular sequence is $(1/2)^N$ and the probability of a sequence having N_1 ones and N_2 minus ones is $\binom{N}{N_1}(1/2)^N$. For any specific value of K, say k, (6-86) can be solved for N_1. The result is $N_1 = (N/2)(k + 1)$, and the probability that $K = k$ is

$$\Pr(K = k) = \binom{N}{(N/2)(k + 1)}\frac{1}{2^N} \qquad (6\text{-}87)$$

The possible values for K are $k = -1 + 2n/N$, $n = 0, 1, 2, \ldots, N$, and the pdf of K is recognized as the binomial distribution. The mean and variance of K are easily calculated [17]; the results are

$$E(K) = 0 \qquad (6\text{-}88a)$$

$$\sigma_K^2 = \frac{1}{N} \qquad (6\text{-}88b)$$

Since y_2 and K are linearly related, the exact pdf of y_2 in this special case is a discrete binomial pdf with zero mean, and variance $[J \cos^2(\phi_J - \theta_s)/PN]$.

A convenient approximation for the discrete binomial probability density is the continuous Gaussian pdf having the same mean and variance. Figure 6-22 compares the Gaussian pdf and the binomial pdf for the special case where $J \cos^2(\phi_J - \theta_s)/P = 1.0$ and $N = 16$. In this figure, the Gaussian pdf is normalized by the spacing between the discrete lines of the binomial distribution, so that both functions have comparable magnitudes.

Return now to the case where $\omega_J \neq \omega_0$. The discussion of the coherent jammer was intended to make plausible the assumption that the integrator output pdf is Gaussian. The mean of this Gaussian pdf is data dependent and equals ± 1. The variance of the output is the sum of the variance due to thermal noise from (6-73) and the variance due to the spread jammer from (6-83). Assuming that a -1 was transmitted, an error is made whenever the integrator output is positive and

$$P_b = \Pr[y_2(nT_s) + y_3(nT_s) > 1.0]$$

$$= \int_{1.0}^{\infty} \frac{1}{\sqrt{2\pi}\sigma} e^{-\alpha^2/2\sigma^2}\, d\alpha \qquad (6\text{-}89)$$

$$= Q\left(\frac{1}{\sigma}\right)$$

where $\sigma^2 = \sigma_{y_2}^2 + \sigma_{y_3}^2$, and $\sigma_{y_3}^2$ is evaluated using the normalization $A^2 = 1/2P$. An

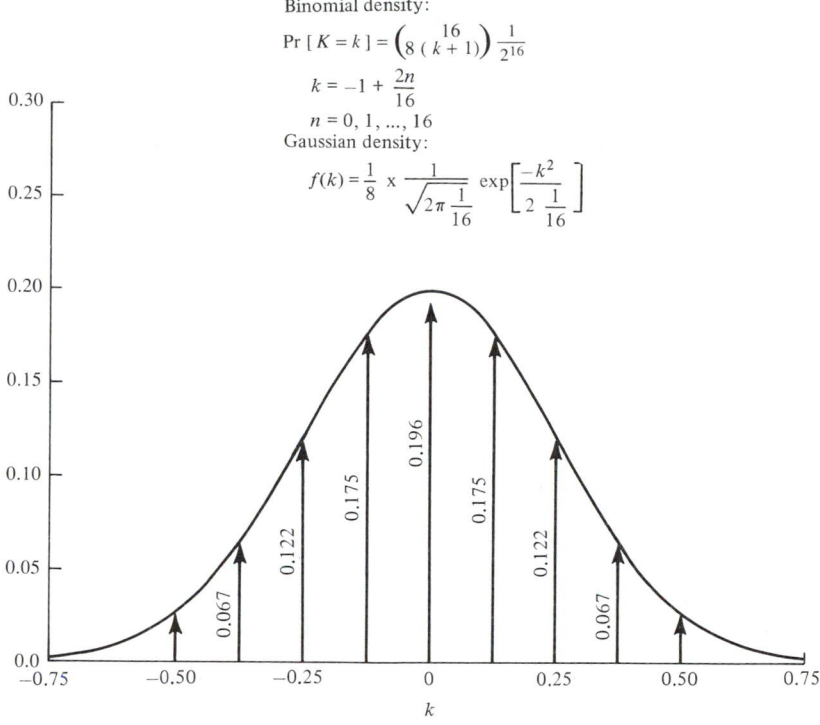

Binomial density:
$$\Pr[K = k] = \left({}_{8}\binom{16}{k+1} \right) \frac{1}{2^{16}}$$
$$k = -1 + \frac{2n}{16}$$
$$n = 0, 1, ..., 16$$
Gaussian density:
$$f(k) = \frac{1}{8} \times \frac{1}{\sqrt{2\pi \frac{1}{16}}} \exp\left[\frac{-k^2}{2 \frac{1}{16}} \right]$$

FIGURE 6-22. Comparison of Gaussian and binomial distribution functions having the same mean and variance.

identical result is obtained for a transmitted $+1$, so that (6-89) represents the total transmission error probability when equally likely symbols are transmitted. For the special case where $\omega_J = \omega_0$,

$$\sigma_{y_2}^2 = \frac{J}{PN} \cos^2(\phi_J - \theta_s) \tag{6-90a}$$

$$\sigma_{y_3}^2 = \frac{N_0}{2PT_s} \tag{6-90b}$$

and

$$P_b = Q\left[\sqrt{\frac{1}{(J/PN) \cos^2(\phi_J - \theta_s) + (N_0/2PT_s)}} \right]$$
$$= Q\left[\sqrt{\frac{2E_b}{2JT_c \cos^2(\phi_J - \theta_s) + N_0}} \right] \tag{6-91}$$

which agrees with Singh [14] and Levitt [13]. Observe that for certain $\phi_J - \theta_s$ the cosine in the denominator equals zero. This occurs when the jammer is phased such that it affects only the quadrature carrier channel. The worst-case jammer phasing results in $\cos^2(\phi_J - \theta_s) = 1.0$. Equation (6-91) is valid only when $\Delta\omega = 0$.

Whenever $\Delta\omega \neq 0$, (6-83) must be used for the variance of y_2 when substituting into (6-89). For high signal-to-thermal noise ratio, the resulting error probability is

$$P_b = Q\left(\sqrt{\dfrac{2PN}{J\dfrac{\sin^2(\Delta\omega\,T_c/2)}{(\Delta\omega\,T_c/2)^2}\left\{1 + \dfrac{\cos[2\phi + (N-1)\,\Delta\omega\,T_c]\,\sin(N\,\Delta\omega\,T_c)}{N\,\sin(\Delta\omega\,T_c)}\right\}}}\right)$$

$$(6\text{-}92)$$

BPSK spreading is being used so that the transmission bandwidth $W = 2/T_c$. Thus $N = T_s/T_c = W/2R$ and $2PN/J = (P/J)(W/R)$. Using $(P/J)(W/R)$ as the independent variable and $\Delta\omega$ as a parameter, (6-92) is plotted in Figure 6-23. Each curve of the figure represents an average over all possible jammer phases. For all curves, $N = 1024$.

QPSK/BPSK. For QPSK spreading modulation and BPSK data modulation, set $W_1(t) = W_2(t) = 1.0$. In this case, the signal component at the integrator input is

$$x_1(t) = \sqrt{2P}\, d(t) \tag{6-93}$$

and the jamming component is

$$x_2(t) = \sqrt{J}\, c_1(t) \cos(\Delta\omega\, t - \phi_J + \theta_s) + \sqrt{J}\, c_2(t) \sin(\Delta\omega\, t - \phi_J + \theta_s) \tag{6-94}$$

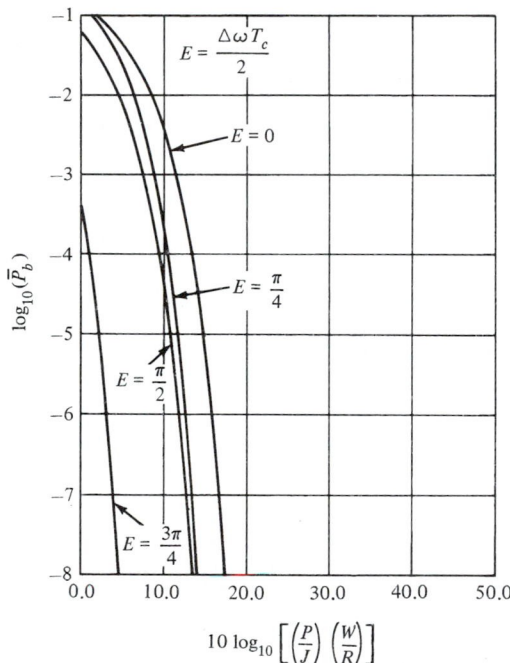

FIGURE 6-23. Performance of coherent BPSK/BPSK spread spectrum in continuous-tone jamming.

The integrator output is again normalized to obtain a data component at the output of ± 1. The normalizing factor is $A^2 = 1/(2P)$ and the normalized integrator output signal and jammer components are

$$y_1(nT_s) + y_2(nT_s) = \pm 1 + \frac{1}{T_s} \sqrt{\frac{J}{2P}} \int_0^{T_s} c_1(t) \cos(\Delta\omega\, t - \phi_J + \theta_s)\, dt$$

$$+ \frac{1}{T_s} \sqrt{\frac{J}{2P}} \int_0^{T_s} c_2(t) \sin(\Delta\omega\, t - \phi_J + \theta_s)\, dt \tag{6-95}$$

The first integral above is identical to the integral of (6-76) and may be approximated by a zero-mean Gaussian random variable with a variance of one-half the magnitude given to (6-83). The one-half factor is due to the jamming power being split between the inphase and quadrature channels. The mean and variance of the second integral of (6-95) are calculated using steps identical to those used to arrive at (6-83). The final result is that the mean is zero and the variance is

$$\sigma_{y_{2b}}^2 = \frac{J}{4PN} \frac{\sin^2(\Delta\omega\, T_c/2)}{(\Delta\omega\, T_c/2)^2} \left\{ 1 - \frac{\cos[2\phi + (N-1)\,\Delta\omega\, T_c]\sin(N\,\Delta\omega\, T_c)}{N\sin(\Delta\omega\, T_c)} \right\} \tag{6-96}$$

Because the spreading waveforms $c_1(t)$ and $c_2(t)$ are independent, the two integrals of (6-95) are independent and their sum may be considered a single Gaussian random variable with zero mean and variance equal to the sum of (6-96) and one-half of (6-83). The variance of the sum is

$$\sigma_{y_2}^2 = \frac{J}{2PN} \frac{\sin^2(\Delta\omega\, T_c/2)}{(\Delta\omega\, T_c/2)^2} \tag{6-97}$$

The integrator output due to noise was calculated earlier. The total integrator output at the sampling time is the sum of a ± 1 signal component and Gaussian noise with zero mean and variance $\sigma^2 = \sigma_{y_2}^2 + \sigma_{y_3}^2$ with $\sigma_{y_3}^2 = N_0/2PT_s$. A transmission error is made when a -1 is transmitted and $y_2 + y_3 > 1.0$. The probability of this event is

$$P_b = \Pr(y_2 + y_3 > 1.0)$$

$$= \int_{1.0}^{\infty} \frac{1}{\sqrt{2\pi}\sigma} e^{-\alpha^2/2\sigma^2}\, d\alpha = Q\left(\sqrt{\frac{1}{\sigma^2}}\right)$$

$$= Q\left[\sqrt{\frac{1}{\dfrac{J}{2PN} \dfrac{\sin^2(\frac{1}{2}\Delta\omega\, T_c)}{(\frac{1}{2}\Delta\omega\, T_c)^2} + \dfrac{N_0}{2PT_s}}} \right] \tag{6-98}$$

Since transmitted information symbols are equally likely and the same error probability expression applies to both, (6-98) also represents the total bit error probability. Observe that the jammer phase does not appear in this equation. Comparing (6-98) with (6-91) when $\Delta\omega = 0$ and for the worst-case jammer phase will show that the performance of QPSK spreading in tone jamming is exactly 3 dB better than the worst-case performance of BPSK spreading.

It is useful to compare the effectiveness of tone jamming to partial-band noise jamming. Ignoring thermal noise, (6-31) for very narrowband partial-band jamming may be written

$$P_b = Q\left(\sqrt{\left(\frac{P}{J}\right)\left(\frac{W}{R}\right)}\right)$$

For the worst-case tone jammer ($\Delta\omega = 0$ and worst-case phase) and BPSK direct-sequence spreading and ignoring thermal noise, (6-91) may be written

$$P_b = Q\left(\sqrt{\frac{E_b}{JT_c}}\right) = Q\left(\sqrt{\frac{P/R}{J(2/W)}}\right) = Q\left(\sqrt{\frac{1}{2}\left(\frac{P}{J}\right)\left(\frac{W}{R}\right)}\right)$$

For the worst-case tone jammer ($\Delta\omega = 0$) and QPSK direct-sequence spreading, and ignoring thermal noise, (6-98) may be written

$$P_b = Q\left(\sqrt{\frac{2PT_s}{J(T_s/N)}}\right) = Q\left(\sqrt{\frac{P}{J}\left(\frac{2T_s}{T_c}\right)}\right) = Q\left(\sqrt{\frac{P}{J}\left(\frac{W}{R}\right)}\right)$$

Thus very narrowband partial-band noise jamming has the same effectiveness as tone jamming for QPSK direct-sequence spreading. However, for BPSK spreading modulation, the tone jammer could be 3 dB more effective than partial-band noise jamming. Fortunately for the communicator, the jammer phase is random, so the jammer can never take advantage of this gain.

MSK/BPSK. Equations (6-59) are used for MSK spreading modulation and BPSK data modulation. The receiver is defined in Figure 6-21. For MSK spreading the two spreading waveforms are offset from one another by $\frac{1}{2}$ chip just as $W_1(t)$ and $W_2(t)$ are offset. Over each spreading waveform chip interval, the weighting function is a half-cycle of a sinusoid whose magnitude is chosen to equal $\sqrt{2}$ so that the total received power is P and the total power in the despreading reference is 2.0. The signal component of the integrator input is calculated using (6-64) and (6-59).

$$x_1(t) = \frac{\sqrt{2P}}{2} d(t)\left\{ 2\sum_i\sum_j p_{T_c}(t - iT_c)p_{T_c}(t - jT_c)\sin\left[\frac{\pi}{T_c}(t - iT_c)\right]\right.$$

$$\times \sin\left[\frac{\pi}{T_c}(t - jT_c)\right]$$

$$+ 2\sum_k\sum_l p_{T_c}\left(t - kT_c - \frac{T_c}{2}\right)p_{T_c}\left(t - lT_c - \frac{T_c}{2}\right)$$

$$\left.\times \sin\left[\frac{\pi}{T_c}\left(t - kT_c - \frac{T_c}{2}\right)\right]\sin\left[\frac{\pi}{T_c}\left(t - lT_c - \frac{T_c}{2}\right)\right]\right\}$$

$$= \sqrt{2P}\, d(t)\left\{\sum_i p_{T_c}(t - iT_c)\sin^2\left[\frac{\pi}{T_c}(t - iT_c)\right]\right\} \tag{6-99}$$

$$+ \sum_k p_{T_c}\left(t - kT_c - \frac{T_c}{2}\right) \sin^2\left[\frac{\pi}{T_c}\left(t - kT_c - \frac{T_c}{2}\right)\right]\Bigg\}$$

$$= \sqrt{2P}\, d(t)\left\{\frac{1}{2} - \frac{1}{2}\sum_i p_T(t - iT_c)\cos\left[\frac{2\pi}{T_c}(t - iT_c)\right]\right.$$

$$\left. + \frac{1}{2} - \frac{1}{2}\sum_k p_{T_c}\left(t - kT_c - \frac{T_c}{2}\right)\cos\left[\frac{2\pi}{T_c}\left(t - kT_c - \frac{T_c}{2}\right)\right]\right\}$$

$$= \sqrt{2P}\, d(t)\left\{1.0 - \frac{1}{2}\cos\left(\frac{2\pi}{T_c}t\right) - \frac{1}{2}\cos\left[\frac{2\pi}{T_c}\left(t - \frac{T_c}{2}\right)\right]\right\}$$

$$= \sqrt{2P}\, d(t)$$

As expected, the despreading operation has completely eliminated the MSK modulation from the data signal. The normalized integrator output due to signal is the usual ± 1.

The integrator input due to jamming is calculated from (6-66) and (6-59). Taking into account that the two spreading waveforms are offset in phase,

$$x_2(t) = \sqrt{2J}\cos(\Delta\omega\, t - \phi_J + \theta_s)\sum_i p_T(t - iT_c)c_{1i}\sin\left[\frac{\pi}{T_c}(t - iT_c)\right]$$

$$+ \sqrt{2J}\sin(\Delta\omega\, t - \phi_J + \theta_s)\sum_j p_{T_c}\left(t - jT_c - \frac{T_c}{2}\right) \qquad (6\text{-}100)$$

$$\times\, c_{2j}\sin\left[\frac{\pi}{T_c}\left(t - jT_c - \frac{T_c}{2}\right)\right]$$

The normalized integrator output is

$$y_2(nT_s) = \sqrt{\frac{J}{P}}\frac{1}{T_s}\sum_{i=0}^{N-1}\int_{iT_c}^{(i+1)T_c} c_{1i}\sin\left[\frac{\pi}{T_c}(t - iT_c)\right]\cos(\Delta\omega\, t - \phi_J + \theta_s)\, dt$$

$$+ \sqrt{\frac{J}{P}}\frac{1}{T_s}\sum_{j=0}^{N-1}\int_{jT_c+T_c/2}^{(j+1)T_c+T_c/2} c_{2j}\sin\left[\frac{\pi}{T_c}\left(t - jT_c - \frac{T_c}{2}\right)\right] \qquad (6\text{-}101)$$

$$\times \sin(\Delta\omega\, t - \phi_J + \theta_s)\, dt$$

Consider the first sum of integrals above. For convenience, let $\phi = -\phi_J + \theta_s$, and expand the trigonometric product to obtain

$$y_{2a}(nT_s) = \sqrt{\frac{J}{P}}\frac{1}{2T_s}\sum_{i=0}^{N-1} c_{1i}\int_{iT_c}^{(i+1)T_c}\left\{\sin\left[\left(\frac{\pi}{T_c} - \Delta\omega\right)t - i\pi - \phi\right]\right.$$

$$\left. + \sin\left[\left(\frac{\pi}{T_c} + \Delta\omega\right)t - i\pi + \phi\right]\right\}dt \qquad (6\text{-}102)$$

Straightforward evaluation of the integrals and trigonometric manipulations result in

$$
y_{2a}(nT_s) = \sqrt{\frac{J}{P}} \frac{1}{T_s} \frac{\pi/T_c}{(\pi/T_c)^2 - (\Delta\omega)^2}
$$

$$
\times \sum_{i=0}^{N-1} c_{1i}\{\cos[(i+1)\,\Delta\omega\,T_c + \phi] + \cos[i\Delta\omega\,T_c + \phi]\}
$$

$$
= \sqrt{\frac{J}{P}} \frac{1}{T_s} \frac{\pi/T_c}{(\pi/T_c)^2 - (\Delta\omega)^2} \sum_{i=0}^{N-1} c_{1i} \cos\left(\frac{\Delta\omega\,T_c}{2}\right)
$$

$$
\times 2\cos\left(\Delta\omega\,T_c i + \phi + \frac{\Delta\omega\,T_c}{2}\right)
$$

(6-103)

The value of $y_{2a}(nT_s)$ is a random variable which depends on the particular sequence of spreading code symbols associated with the data being demodulated. For $\Delta\omega = 0$ this random variable has a binomial distribution which may be approximated accurately by a continuous Gaussian distribution for large N. With this as justification, assume that the random variable $y_{2a}(nT_s)$ may also be approximated by a Gaussian random variable. The mean of $y_{2a}(nT_s)$ is zero and the variance will now be calculated. By definition,

$$
\sigma_{y_{2a}}^2 = E[y_{2a}^2(nT_s)]
$$

$$
= \frac{J}{PT_s^2}\left[\frac{\pi/T_c}{(\pi/T_c)^2 - (\Delta\omega)^2}\right]^2 \sum_{i=0}^{N-1}\sum_{k=0}^{N-1} E[c_{1i}c_{1k}]
$$

$$
\times 4\cos^2\left(\frac{\Delta\omega\,T_c}{2}\right)\cos\left(\Delta\omega\,T_c i + \phi + \frac{\Delta\omega\,T_c}{2}\right)
$$

$$
\times \cos\left(\Delta\omega\,T_c k + \phi + \frac{\Delta\omega\,T_c}{2}\right)
$$

(6-104)

If the spreading code symbols are independent and equally likely, this expression simplifies to

$$
\sigma_{y_{2a}}^2 = \frac{4J}{PT_s^2}\left[\frac{\pi/T_c}{(\pi/T_c)^2 - (\Delta\omega)^2}\right]^2 \sum_{i=0}^{N-1}\cos^2\left(\frac{\Delta\omega\,T_c}{2}\right)
$$

$$
\times \cos^2\left(\Delta\omega\,T_c i + \phi + \frac{\Delta\omega\,T_c}{2}\right)
$$

$$
= \frac{4J}{PT_s^2}\left[\frac{\pi/T_c}{(\pi/T_c)^2 - (\Delta\omega)^2}\right]^2 \cos^2\left(\frac{\Delta\omega\,T_c}{2}\right)
$$

$$
\times \sum_{i=0}^{N-1} \left[\tfrac{1}{2} + \tfrac{1}{2}\cos(2\,\Delta\omega\,T_c i + 2\phi + \Delta\omega\,T_c)\right]
$$

(6-105)

Using (6-82), this equation becomes

$$
\sigma_{y_{2a}}^2 = \frac{4J}{PT_s^2} \left[\frac{\pi/T_c}{(\pi/T_c)^2 - (\Delta\omega)^2} \right]^2 \cos^2\left(\frac{\Delta\omega\, T_c}{2} \right)
$$

$$
\times \left[\frac{N}{2} + \frac{\cos(2\phi + N\,\Delta\omega\, T_c)\,\sin(N\,\Delta\omega\, T_c)}{2\sin(\Delta\omega\, T_c)} \right]
$$

$$
= \frac{J}{4PN} \left(\frac{8}{\pi^2} \right) \left[\frac{\cos(\Delta\omega\, T_c/2)}{1 - (\Delta\omega\, T_c/\pi)^2} \right]^2
$$

$$
\times \left[1.0 + \frac{\cos(2\phi + N\,\Delta\omega\, T_c)\,\sin(N\,\Delta\omega\, T_c)}{N\sin(\Delta\omega\, T_c)} \right]
$$

(6-106)

An identical derivation yields the variance $\sigma_{y_{2b}}^2$ of the second integral of (6-101). The result is

$$
\sigma_{y_{2b}}^2 = \frac{J}{4PN} \left(\frac{8}{\pi^2} \right) \left[\frac{\cos(\Delta\omega\, T_c/2)}{1 - (\Delta\omega\, T_c/\pi)^2} \right]^2
$$

$$
\times \left[1.0 - \frac{\cos(2\phi + N\,\Delta\omega\, T_c)\,\sin(N\,\Delta\omega\, T_c)}{N\sin(\Delta\omega\, T_c)} \right]
$$

(6-107)

Because the two spreading codes are independent, the sum of the two integrals may be assumed to be another Gaussian random variable with variance $\sigma_{y_2}^2 = \sigma_{y_{2a}}^2 + \sigma_{y_{2b}}^2$. Using the equations above gives

$$
\sigma_{y_2}^2 = \frac{J}{2PN} \left(\frac{8}{\pi^2} \right) \left[\frac{\cos(\Delta\omega\, T_c/2)}{1 - (\Delta\omega\, T_c/\pi)^2} \right]^2
$$

(6-108)

The thermal noise component of the integrator output is also a Gaussian random variable with variance $\sigma_{y_3}^2 = N_0/2PT_s$. Bit errors are made when a -1 is transmitted and $y_2 + y_3 > 1.0$ or when $a + 1$ is transmitted and $y_2 + y_3 < -1.0$. These error events have the same probabilities of occurrence and the total bit error probability for this system is

$$
P_b = \Pr[y_2(nT_s) + y_3(nT_s) > 1.0]
$$

$$
= Q\left(\sqrt{\frac{1}{\sigma_{y_2}^2 + \sigma_{y_3}^2}} \right)
$$

(6-109)

$$
= Q\left\{ \sqrt{\frac{1.0}{\dfrac{J}{2PN}\left(\dfrac{8}{\pi^2} \right)\left[\dfrac{\cos(\frac{1}{2}\Delta\omega\, T_c)}{1 - (\Delta\omega\, T_c/\pi)^2} \right]^2 + \dfrac{N_0}{2PT_s}}} \right\}
$$

This expression should be compared to (6-98) for QPSK/BPSK. For $\Delta\omega = 0$ it appears that the MSK/BPSK jammer power is multiplied by $8/\pi^2$. This translates into a 0.91-dB advantage for MSK spreading. Note, however, that the null-to-null transmission bandwidth for MSK spreading is $3.0/T_c$, whereas the same bandwidth for QPSK spreading is $2.0/T_c$. If the transmission bandwidths of the two systems are made equal, the MSK system performs slightly worse than the QPSK system with this jammer.

Although the tone jammer is highly effective, it may be countered by placing a notch filter before the receiver despreading mixer. The notch filter reduces the amount of jamming energy that reaches the despreader and hence the data detector. The notch filter also distorts the signal. Since the receiver does not know the jammer's frequency, adaptive filters are typically used; these may be linear or nonlinear. Extensive research into filter structures, adaption algorithms, and system performance has been done; the reader is directed to the literature [19–25] for additional information on this subject. Significant improvement in communications systems performance has been demonstrated using these techniques.

6-3.5 Performance in Multiple-Tone Jamming

The multiple-tone jammer is the tone equivalent of the partial-band noise jammer and is most effective against FH systems. An estimate of the performance of DS-FH systems may be obtained by considering the output of the DS despreader due to jamming to be Gaussian noise [7]. With this assumption, the DS-FH performance in multiple-tone jamming is the same as the performance in partial-band noise jamming. Only FH systems will be considered in the remainder of this section. The total jamming power is denoted by J, and this power is equally divided between q jamming tones. The jammer has complete knowledge of the transmitted signal structure as well as the exact received signal and jammer power. The jammer does not, however, know the FH pattern. The task of the jammer is to choose q and the tone spacing such that the received bit error probability is maximized.

FH/MFSK. A number of analyses of multiple-tone jamming for FH/MFSK have appeared in the literature [7, 8, 26-33]. The most complicated of these analyses considers MFSK with nonorthogonal frequency spacing and jammer tones which are frequency locked to the communication system MFSK tones. Another analysis presumes that more than one tone can be placed in a single MFSK bandwidth, and the simplest permits only a single tone in each MFSK bandwidth.

First, consider a system that employs orthogonal MFSK with the receiver of Figures 6-6 and 6-7. The jammer has complete knowledge of the signal structure and can therefore select jamming frequencies so that no more than one tone appears in each FH bandwidth. Slow frequency hopping is assumed so that a quasi-static analysis can be used. Orthogonal signaling is achieved using a MFSK tone spacing which is equal to the MFSK symbol rate $R_s = R/\log_2 M$, where R is the system bit rate. There are M tones spaced by R_s hertz, so that the bandwidth for each frequency hop is $W_d = MR/\log_2 M$. The jammer transmits q tones each having power $J_q = J/q$ and spaced in frequency by W_d hertz.

For simplicity, suppose that thermal noise is negligible. Then no symbol errors will be made if $J_q < P$ since even when a jamming tone is within the FH band, the energy detector will still identify the correct signaling tone. When $J_q > P$, a symbol error will be made if the jamming tone is within the FH band but not in the same energy detector filter bandwidth as the desired signaling tone. When $J_q = P$, a symbol error occurs with probability 0.5 under the conditions described for $J_q > P$. The jammer can therefore jam the most FH bands if $J_q = P + \epsilon$, where $\epsilon \ll P$, and the number of jammer tones is approximately given by

$$q \cong \left\lfloor \frac{J}{P} \right\rfloor \tag{6-110}$$

where $[\,\cdot\,]$ denotes the largest integer less than or equal to J/P. Of course, there must be at least one jammer tone so that $q_{min} = 1.0$, and q_{max} is limited by the total number of FH bands or $q_{max} = W/W_d$. When $\lfloor J/P \rfloor \leq 1.0$, no errors will be made since there is insufficient jammer power to jam even a single FH band. Equation (6-110) is approximate only because ϵ has been set to zero. This value of q is optimum since a larger q results in $J_q < P$ and no errors, and since a smaller q results in fewer FH bands being jammed but no larger error probability for any one band.

The total transmission band W contains W/W_d FH bands, so the probability that any one FH band is jammed is

$$p = \frac{q}{W/W_d} \tag{6-111}$$

When a FH band is jammed and q is chosen using (6-110), the symbol error probability is the probability that the jammer and the signal are not in the same band. This probability is $(M - 1)/M$ and the total symbol error probability P_s is

$$P_s = \begin{cases} \dfrac{M-1}{M} & \dfrac{W}{W_d} < \left\lfloor \dfrac{J}{P} \right\rfloor \\[3mm] \dfrac{M-1}{M} \dfrac{qW_d}{W} & 1.0 \leq \left\lfloor \dfrac{J}{P} \right\rfloor \leq \dfrac{W}{W_d} \\[3mm] 0.0 & \left\lfloor \dfrac{J}{P} \right\rfloor < 1.0 \end{cases} \tag{6-112}$$

The symbol error probability is related to the bit error probability by $P_b = MP_s/2(M - 1)$ for orthogonal signals [see (6-19)]. Substituting $q \approx J/P$ and $W_d = MR/\log_2 M$ into (6-112) results in the following expression for bit error probability for worst-case (optimum q) tone jamming:

$$P_b = \begin{cases} 0.5 & \dfrac{P}{J}\left(\dfrac{W}{R}\right) < \dfrac{M}{\log_2 M} \\[3mm] \dfrac{M}{2\log_2 M} \dfrac{1}{(P/J)(W/R)} & \dfrac{M}{\log_2 M} \leq \dfrac{P}{J}\left(\dfrac{W}{R}\right) \leq \dfrac{W}{R} \\[3mm] 0.0 & \dfrac{W}{R} < \dfrac{P}{J}\left(\dfrac{W}{R}\right) \end{cases} \tag{6-113}$$

This relationship is plotted in Figure 6-24 for $M = 2, 4,$ and 8 and $W/R = 1000$. Observe that the usual inverse linear relationship is evident in the midportion of

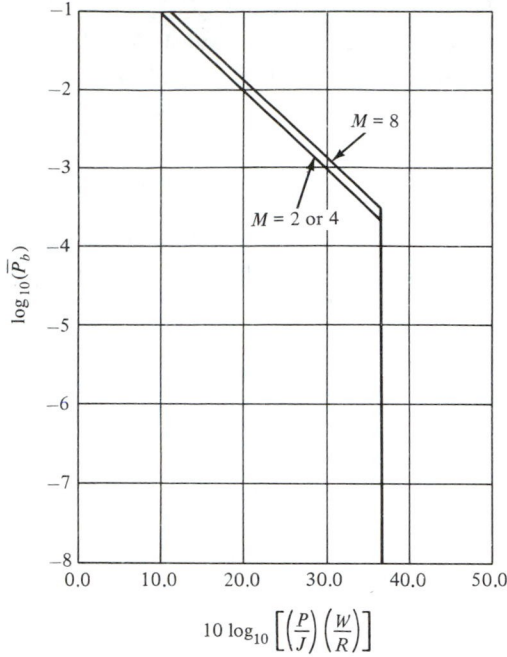

$$10 \log_{10}\left[\left(\frac{P}{J}\right)\left(\frac{W}{R}\right)\right]$$

FIGURE 6-24. Performance of FH/MFSK spread spectrum with worst-case multiple-tone jamming.

these curves. Equation (6-113) should be compared to (6-39) for partial-band noise jamming. In the inverse linear region, it is seen that the same bit error probability is achieved for values of $(P/J)(W/R)$ having a ratio

$$\frac{[(P/J)(W/R)]_{\text{noise}}}{[(P/J)(W/R)]_{\text{tone}}} = \frac{k'}{M/(2 \log_2 M)} \tag{6-114}$$

with k' given in Table 6-1. Thus tone jamming is 4.3, 6.3, 8.3, and 10.5 dB more efficient than noise jamming for $M = 2$, 4, 8, and 16, respectively.

A second type of tone jamming permits the jammer to place more than one tone in each FH band. The jammer will not use this strategy if the precise FH bands are known. Knowledge of the precise FH bands can be denied to the jammer by hopping each symbol frequency independently so that the best that the jammer can do is to jam symbol frequencies randomly across the entire band. The performance of the FH/MFSK system for this case was calculated in Houston [7] and corrected by Levitt [31]. In these references, it is shown that (6-110) is not necessarily optimum because additional error mechanisms are possible. For example, both the signaling tone and one other MFSK tone may be jammed simultaneously. In this case the relative phase of the jammer and signaling tone is a factor and errors may occur even with $J_q < P$. It is concluded in Levitt [8, 31, 32] that the worst-case system bit error probability for this jammer is identical to that specified by (6-113) for large $(P/J)(W/R)$.

FH/DPSK. The analysis of binary and quaternary DPSK with frequency hopping and multitone jamming was first published in Houston [7]. This early work was generalized in Simon [33] to M-ary DPSK with FH. Only the binary case is considered here. Slow frequency hopping is assumed so that a quasi-static analysis may be used. The total jammer power, J, is partitioned into q tones each with power $J_q = J/q$. The signal power is P and each FH band has bandwidth R, where R is the transmitted symbol rate. The transmission bandwidth is W and the total number of FH bands is W/R. The probability that a particular FH band is jammed is

$$\rho = \frac{q}{W/R} \tag{6-115}$$

Assume now that thermal noise is negligible, and consider a single FH band that is jammed. The DPSK demodulator functions by comparing the phase of two successive received symbols. Denote this phase difference by α. When $-\pi/2 < \alpha \leq \pi/2$, the receiver estimates that a 1 has been transmitted, and when $\pi/2 < \alpha \leq 3\pi/2$, the receiver estimates that a zero has been transmitted. The received signal is the phasor sum of the desired tone and the jamming tone as illustrated in Figure 6-25. In Figure 6-25a a 1 has been transmitted and the signal does not change phase between the two symbol times being compared. Thus the phasor sum R_1/θ_1 during the first interval is identical to the phasor sum R_2/θ_2 during the second interval. The receiver calculates $\alpha = \theta_2 - \theta_1 = 0$ and correctly estimates that a 1 has been transmitted regardless of the magnitude of the jammer.

In Figure 6-25b a zero has been transmitted and the desired signal changes phase by π radians between the two signaling intervals being compared. In this case,

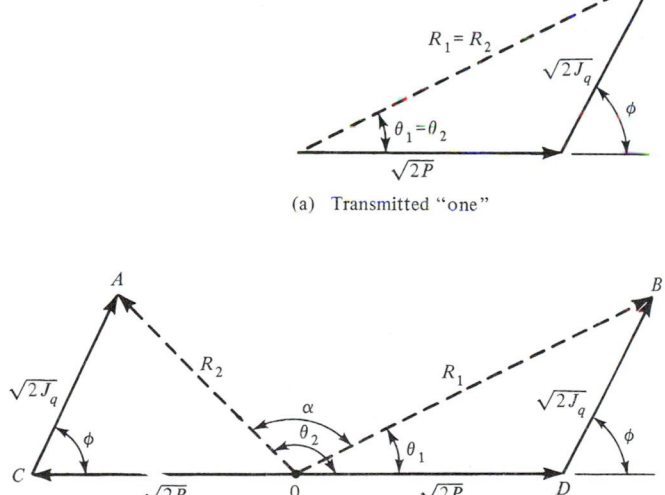

(a) Transmitted "one"

(b) Transmitted "zero"

FIGURE 6-25. Phasor diagrams showing effect of tone jammer on binary DPSK receiver. (From Ref. 7.)

R_1/θ_1 is significantly different from R_2/θ_2 as shown. Using the law of cosines on the triangle OAB of the figure yields

$$\cos \alpha = \frac{R_1^2 + R_2^2 - (2\sqrt{2P})^2}{2R_1R_2} \qquad (6\text{-}116a)$$

Similarly, for triangles OCA and OBD, respectively,

$$\cos \phi = \frac{2P + 2J_q - R_2^2}{2\sqrt{4J_qP}} \qquad (6\text{-}116b)$$

$$\cos(\pi - \phi) = -\cos \phi = \frac{2P + 2J_q - R_1^2}{2\sqrt{4J_qP}} \qquad (6\text{-}116c)$$

Solving (6-116b) and (6-116c) for $R_1^2 + R_2^2$ and substituting into (6-116a) results in

$$\cos \alpha = \frac{2(J_q - P)}{R_1R_2} \qquad (6\text{-}117)$$

When $\pi/2 < \alpha \le 3\pi/2$, the demodulator produces the correct output, but when $-\pi/2 < \alpha \le \pi/2$, an error is made. Equivalently, an error is made whenever $\cos \alpha \ge 0$, which occurs whenever $J_q \ge P$. No error occurs when $J_q < P$.

The optimum jamming strategy is to place just sufficient power in each tone so that an error will always be made when a zero is transmitted using a hop frequency that is jammed. The optimum jamming power is $J_q = P$. The jammer cannot produce any errors when 1's are transmitted. Using this strategy, the optimum number of tones is

$$q = \left\lfloor \frac{J}{P} \right\rfloor \qquad (6\text{-}118)$$

The total bit error probability is

$$P_b = \tfrac{1}{2} \Pr(\text{error}|\text{zero transmitted})$$

$$+ \tfrac{1}{2} \Pr(\text{error}|1 \text{ transmitted}) \qquad (6\text{-}119)$$

$$= \tfrac{1}{2} \Pr(\text{error}|\text{zero transmitted})$$

The conditional error probability given that a zero was transmitted is simply the probability ρ that the hop frequency is jammed. Combining (6-115), (6-118), and (6-119) yields

$$P_b \cong \frac{1}{2} \frac{J/P}{W/R} = \frac{1}{2(P/J)(W/R)} \qquad (6\text{-}120)$$

The usual limits apply to q, that is, $1 \le q \le W/R$. For large jammer power, all FH bands can be jammed and $P_b = \tfrac{1}{2}$. When $J < P$, not even a single FH band can be adequately jammed to create errors and $P_b = 0$. Thus

$$P_b = \begin{cases} 0.5 & \dfrac{P}{J}\left(\dfrac{W}{R}\right) < 1 \\[3ex] \dfrac{1}{2(P/J)(W/R)} & 1 \le \dfrac{P}{J}\left(\dfrac{W}{R}\right) < \dfrac{W}{R} \\[3ex] 0 & \dfrac{W}{R} < \dfrac{P}{J}\left(\dfrac{W}{R}\right) \end{cases} \qquad (6\text{-}121)$$

Once again, the inverse linear relationship is evident in the midrange. Equation (6-121) is plotted in Figure 6-26. Observe that binary FH/DPSK outperforms binary FH/MFSK by exactly 3 dB in the inverse linear region.

6-3.6 Conclusions

It has been demonstrated above that a proper choice of jamming strategy can produce large performance degradations in all the modulation combinations considered. In all instances, the worst-case relationship between bit error probability and $(P/J)(W/R)$ is inverse linear. Values of $(P/J)(W/R)$ in the range $+30$ to $+50$ dB are required to achieve reasonable performance. These large values of $(P/J)(W/R)$ in no way imply that the benefits of spectrum spreading have been nullified by the smart

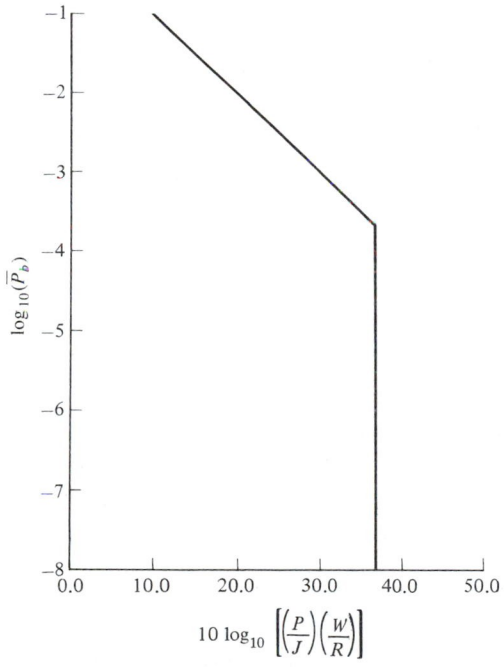

FIGURE 6-26. Performance of FH/DPSK spread spectrum in worst-case multiple-tone jamming.

jammer. The same inverse linear relationships are obtained for the nonspread system without the benefit of the W/R factor. Thus, when comparing the spread and non-spread systems with both optimally jammed, the spread system has a significant power advantage. Meaningful calculations of system processing gains must always include worst-case jamming for both the spread and nonspread systems.

Finally, other analytical techniques are available for the prediction of the performance of spread-spectrum communications systems. For direct-sequence systems, Simon et al. [8, Vol. II, Para. 1.2] present a generalized analysis of the output of the data detection integrator in a BPSK/BPSK system for an *arbitrary* jammer $J(t)$. The technique is very similar to the technique used in this book in the analysis of single-tone jamming of direct-sequence systems. Simon [8] shows that the interference component of integrator output is

$$n = \sqrt{\frac{1}{N}} \sum_{k=1}^{N-1} c_k J_k \qquad (6\text{-}122)$$

where

$$J_k = \sqrt{\frac{2}{T_c}} \int_{kT_c}^{(k+1)T_c} J(t) \cos \omega_0 t \, dt \qquad (6\text{-}123)$$

T_c is the duration of one spreading code chip, N the number of spreading code chips in one information bit, and ω_0 the carrier frequency. For high processing gains (large N), Simon shows that reasonably accurate performance predictions may be made by assuming that the integrator output probability density function is Gaussian. Thus the analysis of the performance with arbitrary jamming reduces to the problem of finding the mean and variance of n in (6-122). Other analysts [34–38] have used (6-122) and (6-123) as a starting point for predicting the performance of multiple-access spread-spectrum system. These analyses of multiple-access systems take the correlation properties of the multiuser spreading code sets into account when calculating link bit error rate. The reader is directed to Nazari [37] or Lehnert [38] for detailed review of these techniques.

6-4 Summary

In this chapter the performance of spread-spectrum systems in jamming environments without error correction coding has been considered. The use of error correction coding to combat the detrimental effects of jamming is considered in Chapter 7. A model of a spread-spectrum communications system for considering the effects of jamming was given in Figure 6-1. Various types of jammers were considered, including barrage (or wideband) noise, partial band noise, and tone jamming. For barrage noise jamming without coding, it was found that the important parameter was $(P/J)(W/R)$, where P is the signal power, J the jammer power, W the spread-spectrum signal bandwidth, and R the data rate. When plotted as a function of this combination of parameters, the bit error probability curves for the communication system approached irreducible values determined by the energy per bit/noise power spectral density ratio, E_b/N_0, for no jamming present. The knee of the curves for each value of E_b/N_0 essentially happened at the same value of $(P/J)(W/R)$. Thus a low value of P/J can be compensated for by a large value of W/R. The latter is approximately the processing gain (recall the definition given in Chapter 2, as being the difference between system performance using spread-spectrum techniques and system performance without spread-spectrum techniques) for the spread-spectrum system. The effects of barrage noise jamming were characterized for coherent DS (Figure 6-5), FH/BFSK (Figure 6-8), and FH/DPSK (Figure 6-10). Similar sets of curves were generated for partial-band jamming, including coherent DS (Figure 6-13), FH/BFSK (Figure 6-14), and FH/DPSK (Figure 6-17). For the latter two modulation schemes, it was found that the bandwidth of the partial-band noise jamming could be optimized so that the bit error probability decreased as the inverse of $(P/J)(W/R)$ (Figures 6-15 and 6-18). A similar result was found for pulsed jamming of DSSS (Figure 6-20). Such jammers were termed *smart jammers* because they could make use of knowledge of the signal structure to impose severe degradations on the system. The effects of tone jamming on DSSS were found to depend on the offset frequency of the tone relative to the spread-spectrum carrier, with zero offset imposing the largest degradation (Figure 6-23). The effects of multiple tone jamming on FH/MFSK (Figure 6-24) and FH/DPSK (Figure 6-26) were also considered. The main conclusion to be gained from this chapter is that smart jammers can inflict severe degradation on spread-spectrum communications systems. Note, however, that the smart jammer can also inflict an equally severe degradation on a nonspread communication system and that the nonspread system does not have the benefit of the spread-spectrum processing gain. What can be done about this serious situation is the subject of the next chapter.

References

[1] R. E. ZIEMER and R. L. PETERSON, *Digital Communications and Spread Spectrum Systems* (New York: Macmillan, 1985).

[2] D. J. TORRIERI, "Fundamental Limitations on Repeater Jamming of Frequency-Hopping Communications," *IEEE J. Selected Areas Commun.,* Vol. SAC-7, pp. 569–575, May 1989.

[3] E. ARTHURS and H. DYM, "On the Optimum Detection of Digital Signals in the Presence of White Gaussian Noise: A Geometric Interpretation and a Study of Three Basic Data Transmission Systems," *IRE Trans. Commun. Syst.,* Vol. CS-10, pp. 336–372, December 1962.

[4] J. H. PARK, "On Binary DPSK Detection," *IEEE Trans. Commun.,* Vol. COM-26, pp. 484–486, April 1978.

[5] T. W. MILLER, *Imperfect Differential Detection of a Biphase Modulated Signal: An Experimental and Analytical Study,* Ohio State University Research Foundation, February 1972. (Available through DTIC: AD 740617.)

[6] W. C. LINDSEY and M. K. SIMON, *Telecommunication Systems Engineering* (Englewood Cliffs, N.J.: Prentice Hall, 1973).

[7] S. W. HOUSTON, "Modulation Techniques for Communication: Part 1. Tone and Noise Jamming Performance of Spread Spectrum M-ary FSK and 2, 4-ary DPSK Waveforms," *Conf. Rec.*, NAECON, pp. 51–58, 1975.

[8] M. K. SIMON, J. K. OMURA, R. A. SCHOLTZ, and B. K. LEVITT, *Spread Spectrum Communications*, Vols. I, II, and III (Rockville, Md.: Computer Science Press, 1985).

[9] H. M. KWON, L. E. MILLER, and J. S. LEE, "Evaluation of a Partial-Band Jammer with Gaussian-Shaped Spectrum Against FH/MFSK," *IEEE Trans. Commun.*, Vol. COM-38, pp. 1045–1049, July 1990.

[10] J. K. OMURA, "Variable Data Bit Rates with a Fixed Hop Rate Noncoherent FH/MFSK System," *Proc. Int. Telem. Conf.*, October 1981.

[11] P. J. CREPEAU, "Performance of FH/BFSK with Generalized Fading in Worst Case Partial-Band Gaussian Interference," *IEEE J. Selected Areas Commun.*, Vol. SAC-8, pp. 884–886, June 1990.

[12] A. J. VITERBI, "Spread Spectrum Communications: Myths and Realities," *IEEE Commun. Mag.*, Vol. 17, pp. 11–18, May 1979.

[13] B. K. LEVITT, "Effect of Modulation Format and Jamming Spectrum on Performance of Direct Sequence Spread Spectrum Systems," *Conf. Rec.*, IEEE Natl. Telecommun. Conf., pp. 3.4.1–3.4.5, 1980.

[14] R. SINGH, "Performance of a Direct Sequence Spread Spectrum System with Long Period and Short Period Code Sequences," *Conf. Rec.*, Int. Conf. Commun., pp. 45.2.1–42.2.5, 1981.

[15] D. L. SCHILLING et al., "Optimization of the Processing Gain of an M-ary Direct Sequence Spread Spectrum Communication System," *IEEE Trans. Commun.*, Vol. COM-28, pp. 1389–1398, August 1980.

[16] L. COUCH, "Performance of DS Spread Spectrum Systems," *Proc. IEEE,* Vol. 68, pp. 298–300, February 1980.

[17] A. PAPOULIS, *Probability, Random Variables and Stochastic Processes,* 3rd ed. (New York: McGraw-Hill, 1991).

[18] I. S. GRADSHTEYN and I. W. RYZHIK, *Tables of Integrals Series and Products* (New York: Academic Press, 1965).

[19] G. J. SOULNIER, P. K. DAS, and L. B. MILSTEIN, "An Adaptive Digital Suppression Filter for Direct-Sequence Spread-Spectrum Communications," *IEEE J. Selected Areas Commun.*, Vol. SAC-3, pp. 676–686, September 1985.

[20] G. J. SOULNIER, P. K. DAS, and L. B. MILSTEIN, "Suppression of Narrowband Interference in a PN Spread-Spectrum Receiver Using a CTD Based Adaptive Filter," *IEEE Trans. Commun.*, Vol. COM-32, pp. 1227–1232, November 1984.

[21] L. Li and L. B. Milstein, "Rejection of Narrow-Band Interference in PN Spread-Spectrum System Using Transversal Filters," *IEEE Trans. Commun.*, Vol. COM-30, pp. 925–928, May 1982.

[22] Y.-C. Wang and L. B. Milstein, "Rejection of Multiple Narrow-Band Interference in Both BPSK and QPSK DS Spread-Spectrum Systems," *IEEE Trans. Commun.*, Vol. 36, pp. 195–204, February 1988.

[23] J. W. Ketchum and J. G. Proakis, "Adaptive Algorithms for Estimating and Suppressing Narrowband Interference in PN Spread-Spectrum Systems," *IEEE Trans. Commun.*, Vol. COM-30, pp. 913–924, May 1982.

[24] R. Vijayan and H. V. Poor, "Nonlinear Techniques for Interference Suppression in Spread-Spectrum Systems," *IEEE Trans. Commun.*, Vol. COM-38, pp. 1060–1065, July 1990.

[25] E. Masry and L. B. Milstein, "Performance of DS Spread-Spectrum Receiver Employing Interference-Suppression Filters Under a Worst-Case Jamming Condition," *IEEE Trans. Commun.*, Vol. COM-34, pp. 13–21, January 1986.

[26] D. J. Torrieri, "*Frequency Hopping, Multiple-Frequency Shift Keying, Coding, and Optimal Partial-Band Jamming*," Tech. Rep. CM/CCM-82-1, Naval Air Systems Command, CM/CCM Center, August 1982 (AD A118585).

[27] M. P. Ristenbatt and J. L. Davis, Jr., "Performance Criteria for Spread-Spectrum Communications," *IEEE Trans. Commun.*, Vol. COM-25, pp. 756–763, August 1977.

[28] L. B. Milstein, R. L. Pickholtz, and D. S. Schilling, "Optimization of the Processing Gain of an FSK-FH System," *IEEE Trans. Commun.*, Vol. COM-28, pp. 1062–1079, July 1980.

[29] H. R. Pettit, *ECM and ECCM Techniques for Digital Communication Systems* (Belmont, Calif.: Lifetime Learning Publications, 1982).

[30] A. J. Viterbi, *A Robust Ratio-Threshold Technique to Mitigate Tone and Partial Band Jamming in Coded Frequency Hopped Communication Links*, Final Report on Contract N00019-81-C-0451, M/A-COM Linkabit, Inc., September 1982 (AD A 123559).

[31] B. K. Levitt, "Use of Diversity to Improve FH/MFSK Performance in Worst Case Partial Band Noise and Multitone Jamming," *Conf. Rec.*, IEEE Mil. Commun. Conf., 1982.

[32] B. K. Levitt, "FH/MFSK Performance in Multitone Jamming," *IEEE J. Selected Areas in Commun.*, Vol. SAC-3, pp. 627–643, September 1985.

[33] M. K. Simon, "The Performance of M-ary FH-DPSK in the Presence of Partial Band Multitone Jamming," *IEEE Trans. Commun.*, Vol. COM-30, pp. 953–958, May 1982.

[34] M. B. PURSLEY, "Performance Evaluation for Phase-Coded Spread-Spectrum Multiple-Access Communications: Part I. Systems Analysis," *IEEE Trans. Commun.*, Vol. COM-25, pp. 795–799, August 1977.

[35] M. B. PURSLEY, D. V. SARWATE, and W. E. STARK, "Error Probability for Direct-Sequence Spread-Spectrum Multiple-Access Communications: Part I. Upper and Lower Bounds," *IEEE Trans. Commun.*, Vol. COM-30, pp. 975–984, May 1982.

[36] J. S. LEHNERT and M. B. PURSLEY, "Error Probability for Binary Direct-Sequence Spread-Spectrum Communications with Random Signature Sequences," *IEEE Trans. Commun.*, Vol. COM-35, pp. 87–98, January 1987.

[37] N. NAZARI and R. E. ZIEMER, "Computationally Efficient Bounds for the Performance of Direct-Sequence Spread-Spectrum Multiple-Access Communications Systems in Jamming Environments," *IEEE Trans. Commun.*, Vol. COM-36, pp. 577–587, May 1988.

[38] J. S. LEHNERT, "Efficient Technique for Evaluating Direct-Sequence Spread-Spectrum Multiple-Access Communications," *IEEE Trans. Commun.*, Vol. 37, pp. 851–858, August 1989.

Problems

(6-1) A communicator uses BPSK direct-sequence spreading modulation and BPSK data modulation. The DS chip rate is 10 Mchips/s, and the data rate is 75 bps. Calculate the ratio of barrage noise average jamming power to pulse noise average jamming power assuming that both jammers increase the communicator's average bit error probability to $P_b = 10^{-2}$. Calculate the same ratio for $P_b = 10^{-5}$.

(6-2) A jammer is to be designed as an ECM against the communicator of Problem 6-1. The jammer will estimate the communicator's carrier frequency with an accuracy of ± 5 MHz. The jammer must increase the communicator's error probability to $P_b = 10^{-3}$ to be effective. What is a good jamming strategy?

(6-3) A communicator uses FH/DPSK modulation to transmit a digitized voice signal using a data rate of 32 kbps. The maximum acceptable transmission bit error probability is $P_b = 10^{-3}$. A smart partial-band noise jammer has a power advantage of 30 dB over the communicator. What spread-spectrum bandwidth must be used to obtain the required bit error probability? How many different frequencies must the FH synthesizer generate?

(6-4) A communicator uses FH/DPSK modulation. The rate of information transmission is 1 Mbps and the total spread transmission bandwidth is 20 MHz. Is a pulse noise jammer more effective or less effective than a partial-band noise jammer against this system? Why?

(6-5) Repeat Problem 6-4 for BPSK/BPSK modulation.

(6-6) Consider the direct-sequence spread-spectrum receiver illustrated below. The receiver input is a BPSK/BPSK signal using chip rate $R_c = 100$ Mchips/s and data rate $R = 1$ Mbps, plus a tone jammer at frequency $f_J = f_0 + \Delta f$. What are reasonable choices for the image reject and IF bandpass filter bandwidths and center frequencies? Assume a signal-to-jammer power ratio of $10 \log(P/J) = -10$ dB. What is the received bit error probability for $\Delta f = 1.5$, 5, 10, and 50 MHz?

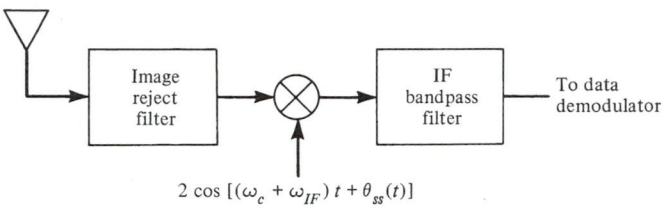

PROBLEM 6-6. Direct-sequence spread-spectrum receiver.

(6-7) Consider the frequency-hop spread-spectrum receiver shown below. Binary FSK data modulation is used with tone spacing of 100 kHz. The data rate is 100 kbps. The FH tone spacing is 200 kHz and there are a total of 1024 tones. Select reasonable image reject and IF bandpass filter bandwidths. The received signal is jammed by a multiple-tone jammer, and the total received signal-to-jammer power ratio is $10 \log(P/J) = -10$ dB. What is the optimal number of jamming tones? What is the optimal jammer tone spacing and tone power level? Select a frequency-hop rate. What is the received bit error probability?

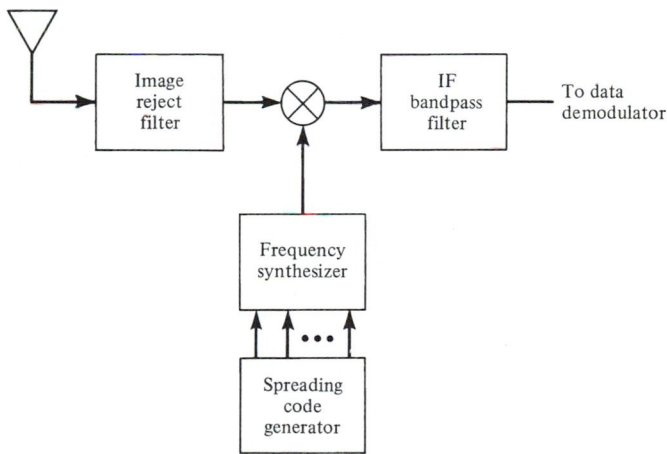

PROBLEM 6-7. Frequency-hop spread-spectrum receiver.

(6-8) Consider the FH communication system described in Problem 6-7.

    (a) Calculate the bit error probability assuming the signal is jammed by a single-tone jammer with $10 \log(P/J) = -10$ dB.

    (b) Calculate the bit error probability assuming that the signal is jammed by a barrage noise jammer with $10 \log(P/J) = -10$ dB.

(6-9) Consider a BPSK/BPSK spread-spectrum system which is jammed by a single-tone jammer whose frequency is identical to the BPSK/BPSK carrier frequency but which is not phase coherent. Suppose that the spreading code is a maximal-length sequence whose period is exactly equal to the information bit period. Calculate the bit error probability as a function of P/J and the jammer phase error θ_J [14].

(6-10) Consider a BPSK/BPSK spread-spectrum system jammed by a single-tone jammer. Assume that there is no thermal noise interference and calculate a lower bound on $10 \log(P/J)$ above which the demodulator makes no errors. (*Hint:* Do not use the Gaussian approximation for the integrator output statistics.)

(6-11) A communicator uses FH/MFSK as an ECCM against a pulsed noise jammer. The MFSK modulator transmits 4-ary symbols. Assume that the received signal-to-thermal noise ratio is $10 \log(E_b/N_0) = +15$ dB and the received signal-to-jammer noise ratio is $10 \log(E_b/N_J) = +3$ dB. Calculate the optimal jammer duty factor and plot the received bit error probability as a function of time. Draw and label the discrete memoryless channel transition diagram representing this channel.

(6-12) A communicator has a transmitter power amplifier whose power output is 10 W. A jammer has a power amplifier whose average power output is 100 W and which can be pulsed using duty factors in the range $0.01 \leq \rho \leq 1.0$. The bandwidth of these two power amplifiers is the same. Propose a spectrum spreading strategy that does not use error-correction coding and which will enable the communicator to achieve a bit error probability of 10^{-5}.

(6-13) Consider a BPSK/BPSK coherent spread-spectrum system operating in a pulsed noise jamming environment. Assume that the information bit rate is 1 kbps and that the direct-sequence chip rate is 10 Mchips/s, and plot the worst-case system bit error probability performance as a function of $10 \log(P/J)$. Calculate and plot the error probability performance for non-spread-spectrum BPSK communications system operating in the same environment. What is the spread-spectrum processing gain?

CHAPTER 7

Performance of Spread-Spectrum Systems with Forward Error Correction

7-1 Introduction

Spectrum spreading by itself produces large communication system performance improvements by effectively spreading the jammer power over the full spread communications bandwidth while simultaneously enabling communications performance to be affected only by the interference in approximately the data bandwidth. The resultant performance was analyzed in Chapter 6. Even with spectrum spreading, worst-case pulse or partial-band jamming was highly effective in degrading communications performance. The effect of worst-case jamming can be mitigated using one or more of the powerful forward error correction (FEC) techniques which have been developed following Shannon's pioneering work. Error correction coding is an extremely complex topic to which entire books are dedicated. In this book a small number of the most basic concepts of FEC are discussed to give the student a preliminary idea of the power of these techniques.

Jamming strategies which concentrate jamming resources on some fraction of the transmitted symbols using either pulsed or partial band techniques cause demodulator output errors to occur in bursts. Because the FEC techniques to be discussed perform best when channel errors are independent from one signaling interval to the next, interleaving is assumed for all FEC schemes. The purpose of the interleaver is to rearrange the order in which coder output symbols are transmitted so that bursts of transmission errors will not appear as bursts at the decoder input because they are reordered by the de-interleaving operation in the receiver. The system model for the coded spread-spectrum system was illustrated in Figure 6-1 and discussed in Section 6-2. This model includes the interleaver and deinterleaver within the discrete memoryless channel (DMC). It is the interleaver/de-interleaver that makes the channel approximately memoryless.

The performance of the FEC schemes in this chapter is calculated as a function of the DMC transition probabilities $p(y_j | x_j, z_j)$ accounting for jammer-state information z_i. These transition probabilities are calculated using the basic techniques discussed in Chapter 6. These techniques must be extended, however, to include instances where the DMC input and output alphabets are not the same. Additional information can be given to the decoder by permitting the DMC input and output alphabets to be different. For example, in addition to outputting a binary 1 or 0, the

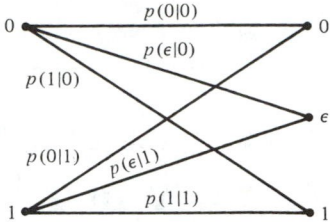

FIGURE 7-1. Transition diagram for soft-decision binary erasure channel.

DMC may also output information indicating the reliability of the 1/0 decision. When additional information is output, the channel is called a soft-decision channel. An example of the transition diagram for a soft-decision channel is shown in Figure 7-1, where a binary input channel has three outputs: (1) one, (2) zero, and (3) erasure. An erasure means that the demodulator is not sure what was transmitted and chooses not to guess. In an additive white Gaussian noise interference environment, decoders using reliability information typically require 1 to 2 dB less transmitter power than decoders not using this information. In the limiting case the DMC output is a real number (i.e., the size of the output alphabet is infinite); in this case, the channel is specified by a conditional (on the input) probability density function.

As discussed in Chapter 6, the DMC output may also include jammer-state information in the form of a variable z_j which in its most basic form, indicates whether or not the jammer is "on" for channel use j. The use of jammer-state information (JSI) is critical for coded systems using soft-decision decoding since, without JSI, the smart jammer can cause complete failure of the communications link using minimal jammer power. For hard-decision decoders JSI is not required, although system performance is always better with JSI. Again, the on/off jammer-state information is the most basic form of JSI; other forms of JSI can include an estimate of jammer received power or other information.

In the remainder of this chapter, the fundamental concepts of both block and convolutional codes are discussed. The student is referred to Lin and Costello [1] or Clark and Cain [2] for detailed introductory discussions of FEC, and to Viterbi and Omura [3], Peterson and Weldon [4], Gallager [5], and Berlekamp [6] for advanced discussions of both information-theoretic and algebraic foundations of FEC.

7-2 Elementary Block Coding Concepts

7-2.1 Block Code Definitions

Block codes can be either linear or nonlinear; discussion here is limited to linear codes. Consider only encoders whose input and output is binary. A block encoder outputs an n-bit binary codeword for each group of k encoder input binary symbols. Each group of k encoder input symbols is called an input message and is identified by a message number m. For binary output codes it can be proven that k must be less than n. The code rate is $R = k/n$. There are 2^k possible input words and each has a unique output codeword associated with it. Since there are 2^n possible output codewords and $k < n$, not all possible output codewords are used.

The error correction capability of any error correction code is due to the fact that not all possible encoder output n-tuples are used. Because of this, it is possible to generate codes with codewords selected so that a number of transmission errors must occur before one codeword will be confused with another. Codewords are represented by binary n-tuples, $\mathbf{x}_m = (x_{m1}, x_{m2}, \ldots, x_{mn})$, where $m = 0, 1, 2, \ldots, 2^k - 1$ is the message associated with the codeword. Any two codewords $\mathbf{x}_m$ and $\mathbf{x}_{m'}$ which differ from one another in d_H places and agree in $n - d_H$ places are said to be separated by *Hamming distance d_H*. For a binary symmetric channel, a total of d_H transmission errors must occur before $\mathbf{x}_m$ is changed into $\mathbf{x}_{m'}$. The minimum Hamming distance between any two of the 2^k codewords in a code is the *minimum distance $d_{\min}$* of the code. An (n,k) binary block code uses a fraction $2^k/2^n = 2^{k-n}$ of all possible output codewords. A low-rate code uses a smaller fraction of the possible output words than a high rate code and codewords can therefore be separated further from one another. Thus low rate codes typically have more error correction capability than high-rate codes.

An important property of linear block codes is that the symbol-by-symbol modulo-2 sum of any two codewords is another codeword. This implies that the all zeros codeword is a codeword in all linear codes. Consider any two codewords $\mathbf{x}_a$ and $\mathbf{x}_b$ separated by Hamming distance d_H, and a third arbitrary codeword $\mathbf{x}_c$. Let $\mathbf{x}_{a'} = \mathbf{x}_a \oplus \mathbf{x}_c$ and $\mathbf{x}_{b'} = \mathbf{x}_b \oplus \mathbf{x}_c$, and notice that the Hamming distance between $\mathbf{x}_{a'}, \mathbf{x}_{b'}$ is also d_H. Using these two properties, it can be demonstrated that the set of Hamming distances between a codeword $\mathbf{x}_m$ and all other codewords $\mathbf{x}_{m'}$, $m' \neq m$, in the code is the same for all m. Denote the number of codewords which are Hamming distance, $d_H = d$ from the all zeros codeword by A_d. The *weight distribution* of the code is the set of all A_d for $d = d_{\min}, \ldots, n$.

Consider the binary block code defined in Table 7-1. This code uses $k = 4$, $n = 7$, and has rate $R = 4/7$ and minimum distance 3. There are a total of $2^k = 16$ codewords. The weight distribution of this code is $A_3 = 7$, $A_4 = 7$, and $A_7 = 1$. Thus there are a total of seven codewords $\mathbf{x}_{m'}$ with distance 3, seven codewords $\mathbf{x}_{m'}$, with distance 4, and one codeword $\mathbf{x}_{m'}$, with distance 7 from any particular codeword $\mathbf{x}_m$. A codeword is transmitted one symbol at a time over the digital channel. Each symbol is corrupted by noise and/or jamming on the channel.

In all cases, the channel can be described by the probability $p(\mathbf{y}|\mathbf{x}_m, \mathbf{z})$ of receiving a particular output n-tuple $\mathbf{y} = (y_1, y_2, \ldots, y_n)$ given that the input was a particular codeword n-tuple $\mathbf{x}_m$ and the jammer-state vector is $\mathbf{z}$. The channel is assumed to be memoryless so that

$$p(\mathbf{y}|\mathbf{x}_m, \mathbf{z}) = \prod_{j=1}^{n} p(y_j | x_{mj}, z_j) \tag{7-1a}$$

EXAMPLE 7-1a _____

Consider a hard-decision coherent BPSK/BPSK direct-sequence spread-spectrum system operating in a worst-case pulsed noise jamming environment. The

TABLE 7-1. Typical Block Code

	Message	Codeword
0	0 0 0 0	0 0 0 0 0 0 0
1	1 0 0 0	1 1 0 1 0 0 0
2	0 1 0 0	0 1 1 0 1 0 0
3	1 1 0 0	1 0 1 1 1 0 0
4	0 0 1 0	1 1 1 0 0 1 0
5	1 0 1 0	0 0 1 1 0 1 0
6	0 1 1 0	1 0 0 0 1 1 0
7	1 1 1 0	0 1 0 1 1 1 0
8	0 0 0 1	1 0 1 0 0 0 1
9	1 0 0 1	0 1 1 1 0 0 1
10	0 1 0 1	1 1 0 0 1 0 1
11	1 1 0 1	0 0 0 1 1 0 1
12	0 0 1 1	0 1 0 0 0 1 1
13	1 0 1 1	1 0 0 1 0 1 1
14	0 1 1 1	0 0 1 0 1 1 1
15	1 1 1 1	1 1 1 1 1 1 1

channel input and output alphabets are $\{0,1\}$ and the channel is a binary symmetric channel (BSC). The BSC may be represented by the transition diagram of Figure 7-2 with transition probability p. When the jammer is "off," $z_j = 0$, transmission errors are caused by thermal noise with a two-sided psd of $N_0/2$. When the jammer is "on," $z_j = 1$, the two-sided noise psd is increased to $(N_0 + N_J)/2$. Assume that the receiver knows both N_0 and N_J. From Chapter 6, the BSC error probability p_i for binary channel use i is

$$p_j = Q\left(\sqrt{\frac{2E_b}{N_{nj}}} \right)$$

where N_{nj} represents the noise psd for channel use j. Let $\mathbf{J}$ be the set of all channel use indices for which $z_j = 1$, and let n_1 be the number of elements of $\mathbf{J}$. Let p_0 and p_1 denote the BSC error probabilities when $N_n = N_0$ and $N_n = N_0 +$

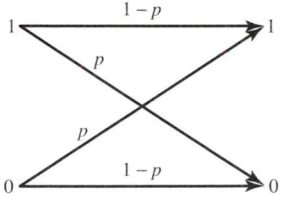

FIGURE 7-2. Binary symmetric channel transition diagram.

N_J, respectively. Assume that perfect JSI is available. Then

$$p(\mathbf{y}\,|\,\mathbf{x}_m,\mathbf{z}) = \prod_{j\in\mathbf{J}} p(y_j\,|\,x_{mj},1)\prod_{j\in\overline{\mathbf{J}}} p(y_j\,|\,x_{mj},0)$$

$$= p_1^{d_1}(1 - p_1)^{n_1 - d_1}p_0^{d_0}(1 - p_0)^{n - n_1 - d_0}$$

where d_0 and d_1 denote the Hamming distances between $\mathbf{y}$ and the codeword $\mathbf{x}_m$ over the channel use indices in $\overline{\mathbf{J}}$ and $\mathbf{J}$, respectively.

When JSI is not available, the probability $p(\mathbf{y}\,|\,\mathbf{x}_m)$ of DMC output $\mathbf{y}$ given that the channel input was $\mathbf{x}_m$ is

$$p(\mathbf{y}\,|\,\mathbf{x}_m) = \prod_{j=1}^{n} p(y_j\,|\,x_{mj}) \tag{7-1b}$$

Since JSI is not available, the decoder does not know the noise psd for each channel use. Therefore, the decoder bases its decisions on average (over all jamming states) transition probabilities.

EXAMPLE 7-1b

Suppose that JSI is not available in the communications system of Example 7-1a. The probability $p(\mathbf{y}\,|\,\mathbf{x}_m)$ is based on the average BSC error rate as given in (6-55). Thus the transition probability p in Figure 7-2 is

$$p = (\overline{P_b})_{\text{max}} \cong \frac{1}{2E_b/N_J}$$

in the range of signal-to-jammer power ratios where the inverse linear relationship of (6-55) is applicable. Then

$$p(\mathbf{y}\,|\,\mathbf{x}_m) = p^{d_H}(1 - p)^{n - d_H}$$

where d_H is the Hamming distance between the received n-vector $\mathbf{y}$ and the codeword $\mathbf{x}_m$ being evaluated.

7-2.2 Optimum Decoding Rule

No Jammer-State Information Available. The decoder inputs are the received vector $\mathbf{y}$, prior knowledge of the entire set of codewords $\mathbf{x}_m$ for $m = 0, \ldots, 2^k - 1$, knowledge of the source output message probabilities $\Pr(m)$, and knowledge of the channel transition probabilities $p(\mathbf{y}\,|\,\mathbf{x}_m)$. Using this information, the decoding rule is designed such the average number of bit errors delivered to the user is minimized. The decoder output bit error probability is denoted P_b.

First consider the much simpler problem of designing a decoding rule which results in the minimum probability of incorrectly estimating the transmitted codeword without regard to the number of information bit errors associated with a particular codeword error. This block decoding error probability is denoted P_B. Denote the decoder estimate of the transmitted codeword by $\hat{\mathbf{x}}$. Given that $\mathbf{y}$ was received and $\hat{\mathbf{x}} = \mathbf{x}_m$, the decoder estimate is correct if $\mathbf{x}_m$ was indeed transmitted. Therefore, P_B is minimized if the decoder chooses as its estimate $\hat{\mathbf{x}}$ the codeword that was most likely to have been transmitted. That is, choose $\hat{\mathbf{x}}$ to be that $\mathbf{x}_m$ with the largest posterior probability $\Pr[\mathbf{x}_m | \mathbf{y}]$.

To calculate the posterior probabilities $\Pr[\mathbf{x}_m | \mathbf{y}]$, Bayes' rule is used. Recall from probability theory that

$$\Pr[\mathbf{x}_m | \mathbf{y}] \Pr[\mathbf{y}] = p(\mathbf{y} | \mathbf{x}_m) \Pr[\mathbf{x}_m] \tag{7-2}$$

Therefore,

$$\Pr[\mathbf{x}_m | \mathbf{y}] = \frac{p(\mathbf{y} | \mathbf{x}_m) \Pr[\mathbf{x}_m]}{\Pr[\mathbf{y}]} \tag{7-3}$$

The denominator on the right side of this equation is

$$\Pr[\mathbf{y}] = \sum_{m'=0}^{2^k-1} \Pr[\mathbf{x}_{m'}] p(\mathbf{y} | \mathbf{x}_{m'}) \tag{7-4}$$

which is positive and independent of the message m. Therefore, the decoding rule may be restated: Choose $\hat{\mathbf{x}}$ to be that $\mathbf{x}_m$ for which

$$p(\mathbf{y} | \mathbf{x}_m) \Pr[\mathbf{x}_m] = p(\mathbf{y} | \mathbf{x}_m) \Pr(m) \tag{7-5}$$

is maximum. This equation is evaluated using knowledge of the channel transition probabilities and the prior probabilities for the messages. When all message probabilities are equal, the decoding rule simplifies further to: Choose $\hat{\mathbf{x}}$ to be that $\mathbf{x}_m$ for which

$$p(\mathbf{y} | \mathbf{x}_m) \tag{7-6}$$

is maximum. The decoder that makes decisions without regard to the message prior probabilities is called a *maximum-likelihood decoder*. This decoding rule is applicable to all discrete memoryless channels, including both hard- and soft-decision channels, which can be characterized by (7-1b).

For the binary symmetric channel $p(\mathbf{y} | \mathbf{x}_m)$ becomes

$$p(\mathbf{y} | \mathbf{x}_m) = p^{d_H}(1 - p)^{n - d_H} \tag{7-7}$$

where d_H is the Hamming distance between $\mathbf{y}$ and $\mathbf{x}_m$ and p is the BSC error probability. Since the logarithm is a monotone increasing function of an increasing argument, the logarithm of $p(\mathbf{y} | \mathbf{x}_m)$ may be used in the maximum-likelihood decoding rule without changing any decoder decisions. Taking the logarithm of (7-7), the decoding rule for the BSC can be stated: Choose $\hat{\mathbf{x}}$ to be that $\mathbf{x}_m$ for which

$$\ln\{p(\mathbf{y} | \mathbf{x}_m)\} = d_H \ln(p) + (n - d_H) \ln(1 - p) \tag{7-8}$$

$$= d_H \ln \frac{p}{1 - p} + n \ln(1 - p)$$

is maximum. Since $\ln[p/(1 - p)] < 0$ for $p < 0.5$, maximizing (7-8) is equivalent to minimizing d_H. Therefore, the decoding rule is a *minimum-distance decoding rule*. For the BSC with $p < 0.5$ the decoder will choose as its estimate the codeword that is closest to the received vector $\mathbf{y}$ in terms of Hamming distance.

The decoding rules just described achieve the minimum possible *block* decoding error probability. The minimization of block decoding error probability is also assumed to minimize decoded *bit* error probability. This will be true with proper assignment of messages to codewords. Messages are assigned such that the most probable block decoding errors cause the minimum number of bit errors. "Good" block codes all have this property.

EXAMPLE 7-2

Consider the $n = 7$, $k = 4$ binary code of Table 7-1. Suppose that the demodulator output is $\mathbf{y} = (1,0,1,1,0,1,0)$. Using the decoding rule just described, the decoder calculates the Hamming distance between $\mathbf{y}$ and all possible codewords. These distances are compared and $\hat{\mathbf{x}}$ is the codeword with the minimum Hamming distance from $\mathbf{y}$. This minimum distance is 1 and $\hat{\mathbf{x}} = (0,0,1,1,0,1,0)$.

Consider next a soft decision memoryless AWGN channel where the channel input is ± 1 and the channel output is a real number with Gaussian statistics. Specifically, the channel is defined by

$$p(y_j | x_j) = \frac{1}{\sqrt{\pi N_0}} \exp\left[-\frac{(y_j - x_j)^2}{N_0} \right] \tag{7-9}$$

This model is based on the assumption that an ideal matched filter receiver is employed and that the channel noise is ideal AWGN. The channel is memoryless, so that

$$p(\mathbf{y} | \mathbf{x}_m) = \prod_{j=1}^{n} p(y_j | x_{mj}) \tag{7-10}$$

The decoder could implement the decoding rule of (7-6) directly using (7-9) and (7-10); however, it is again more convenient to take the logarithm of both sides of (7-6), so that the product of (7-10) becomes a summation. Since the logarithm is a monotone increasing function of its argument, decisions based on $\ln[p(\mathbf{y}|\mathbf{x})]$ will be identical to decisions based on $p(\mathbf{y}|\mathbf{x})$. Using (7-9) and (7-10), and taking the logarithm, (7-6) becomes

$$-\frac{n}{2} \ln(\pi N_0) - \frac{1}{N_0} \sum_{j=1}^{n} (y_j - x_{mj})^2 \tag{7-11a}$$

The first term of (7-11a) is independent of $\mathbf{x}_m$ and may be ignored by the decoder without affecting decoder output decisions. The constant $1/N_0$ of the second term may also be ignored since it affects all channel uses identically. Therefore, the maximum-likelihood decoder chooses its output to be the codeword $\mathbf{x}_m$ for which

$$\sum_{j=1}^{n} (y_j - x_{mj})^2 \tag{7-11b}$$

is *minimum*. The minimum, rather than the maximum, is used due to the negative sign in (7-11a), which has been deleted in (7-11b). The summation above is the square of the Euclidean distance between the received real n-tuple and the codeword real n-tuples, and the decoder chooses the output estimate $\hat{\mathbf{x}}$ to be the codeword closest to $\mathbf{y}$ in Euclidean distance.

EXAMPLE 7-3

Consider the code of Table 7-1 and suppose that message 0 is transmitted. For the hard-decision BSC, the corresponding codeword is $\mathbf{x}_0 = (0,0,0,0,0,0,0)$, and for the real vector AWGN channel, the corresponding codeword is represented by $\mathbf{x}_0 = (+1,+1,+1,+1,+1,+1,+1)$. Suppose further that the received real n-tuple (i.e., the sampled output of the matched filter) is $y = (-0.01,-0.01,1.0,1.0,1.0,1.0,1.0)$. In the hard-decision demodulator, a decision threshold is placed at zero volts and the hard-decision output is $\mathbf{y} = (1,1,0,0,0,0,0)$. This n-tuple is closest to code n-tuple $(1,1,0,1,0,0,0)$ in Hamming distance and the hard-decision decoder outputs message 1 and makes an error. The soft-decision decoder calculates the Euclidean distances shown in Table 7-2. The real vector $\mathbf{y}$ is closest to real vector $(+1,+1,+1,+1,+1,+1,+1)$ in Euclidean distance and the soft-decision decoder outputs message 0 and does not make an error.

The purpose of Example 7-3 is to illustrate a case where a hard-decision decoder makes an error but a soft-decision decoder does not. Because the soft-decision decoder is capable of making correct decisions in cases like this, the soft-decision decoder always performs better than the hard-decision decoder.

Reconsider (7-11b) and expand the square to obtain

$$\sum_{j=1}^{n} \{y_j^2 - 2y_j x_{mj} + x_{mj}^2\} \tag{7-11c}$$

The channel input is binary (i.e., $x_j \in [+1,-1]$), so $x_{mj}^2 = 1$ for all messages m and all channel uses j. Also, y_j^2 is not a function of the codeword m. Therefore, the decoding rule may ignore y_j^2 and x_{mj}^2 and the decoder chooses its output to be the codeword for which

$$-2 \sum_{j=1}^{n} \{y_j x_{mj}\} \tag{7-11d}$$

is minimum. The decoder using this decoding rule correlates the soft-decision demodulator output with all possible transmitted codewords and estimates the transmitted codeword to be the codeword having the maximum correlation with the demodulator output $\mathbf{y}$.

TABLE 7-2. Soft-Decision Decoding Calculation for Example 7-3

Message	Channel Input Vector	Euclidean Distance Between Codeword and $y = (-0.01, -0.01, 1, 1, 1, 1, 1)$
0	(1, 1, 1, 1, 1, 1, 1)	2.04
1	(−1,−1, 1,−1, 1, 1, 1)	5.96
2	(1,−1,−1, 1,−1, 1, 1)	10.00
3	(−1, 1,−1,−1,−1, 1, 1)	14.00
4	(−1,−1,−1, 1, 1,−1, 1)	9.96
5	(1, 1,−1,−1, 1,−1, 1)	14.04
6	(−1, 1, 1, 1,−1,−1, 1)	10.00
7	(1,−1, 1,−1,−1,−1, 1)	14.00
8	(−1, 1,−1, 1, 1, 1,−1)	10.00
9	(1,−1,−1,−1, 1, 1,−1)	14.00
10	(−1,−1, 1, 1,−1, 1,−1)	9.96
11	(1, 1, 1,−1,−1, 1, 1)	10.04
12	(1,−1, 1, 1, 1,−1,−1)	10.00
13	(−1, 1, 1,−1, 1,−1,−1)	14.00
14	(1, 1,−1, 1,−1,−1,−1)	18.04
15	(−1,−1,−1,−1,−1,−1,−1)	21.96

Jammer-State Information Available. In this case, the jammer-state vector **z** is available to the decoder to assist in making decoding decisions. The channel is characterized by (7-1a). Using arguments identical to those used previously, the maximum-likelihood decoder chooses its output to be the codeword $\mathbf{x}_m$ for which

$$p(\mathbf{y} \mid \mathbf{x}_m, \mathbf{z}) \tag{7-12}$$

is maximum. The design of the minimum error probability decoding rule with JSI follows the same steps that were followed without JSI since the JSI is the same for all possible transmitted codewords $\mathbf{x}_m$.

The channel is memoryless, so that (7-1a) is applicable. As before, the output of the decoder is unchanged if the logarithm of (7-12) or (7-1a) is used as the decoding metric. Thus the minimum error probability decoder may use the decoding metric

$$\sum_{j=1}^{n} \ln\{p(y_j \mid x_{mj}, z_j)\} \tag{7-13}$$

By taking the logarithm, the product of (7-1a) has been replaced by a summation in (7-13).

Consider a two-state noise jammer and a binary symmetric channel. Assume that the jammer-state information **z** is perfect. Let **J** denote the set of all channel use indices j for which $z_j = 1$ and denote the BSC error probability for these channel uses by p_1. Let p_0 denote the BSC error probability when $z_j = 0$. Using these definitions, $p(\mathbf{y}|\mathbf{x}_m, \mathbf{z})$ may be written

$$\prod_{j\in\mathbf{J}} p(y_j|x_{mj},1) \prod_{j\in\bar{\mathbf{J}}} p(y_j|x_{mj},0) = p_1^{d_1}(1-p_1)^{n_1-d_1} p_0^{d_0}(1-p_0)^{n-n_1-d_0} \quad (7\text{-}14)$$

where d_1 and d_0 denote the Hamming distances between $\mathbf{y}$ and the codeword $\mathbf{x}_m$ over the channel use indices in $\mathbf{J}$ and $\bar{\mathbf{J}}$, respectively, and n_1 denotes the number of elements in $\mathbf{J}$. Taking the logarithm of the right side of (7-14) yields

$$\ln\{p(\mathbf{y}|\mathbf{x}_m, \mathbf{z})\} = d_1 \ln \frac{p_1}{1-p_1} + n_1 \ln(1-p_1)$$

$$+ d_0 \ln \frac{p_0}{1-p_0} + (n-n_1) \ln(1-p_0) \quad (7\text{-}15)$$

The decoder estimates the transmitted codeword to be the codeword for which (7-15) is largest. The second and fourth terms of the right side of (7-15) are not functions of $\mathbf{x}_m$ and therefore do not have to be considered by the decoder. Ignoring these terms, the maximum-likelihood decoding rule is seen to again be a minimum distance decoding rule. The decoder estimate $\hat{\mathbf{x}}$ is the codeword $\mathbf{x}_m$ which is closest to the received vector $\mathbf{y}$ in terms of the weighted Hamming distance

$$d_1 \ln \frac{p_1}{1-p_1} + d_0 \ln \frac{p_0}{1-p_0} \quad (7\text{-}16)$$

The weighting factor for the jammed channel uses is $\ln[p_1/(1-p_1)]$. When the jammer power is high resulting in $p_1 \cong 0.5$, this weighting factor approaches zero. In contrast, the weighting factor for the nonjammed channel uses, $\ln[p_0/(1-p_0)]$, is a large negative number when p_0 is small. Thus the decoder maximizes (7-16) by choosing the codeword having the smallest d_0 or, equivalently, the codeword closest to the received vector *for the channel uses which are not jammed.* Note that the weighting of the jammed symbols cannot be set to zero since, when all symbols are jammed these symbols are used to make the best possible estimate of the transmitted codeword.

Now consider a binary input soft-decision AWGN channel defined by (7-9) with two-state noise jammer interference. The decoder chooses its output to be the codeword which maximizes $p(\mathbf{y}|\mathbf{x}_m, \mathbf{z})$. Defining $\mathbf{J}$ as above,

$$p(\mathbf{y}|\mathbf{x}_m, \mathbf{z}) = \prod_{j\in\mathbf{J}} p(y_j|x_{mj},1) \prod_{j\in\bar{\mathbf{J}}} p(y_j|x_{mj},0) \quad (7\text{-}17)$$

The one-sided channel noise psd's for jammed and not-jammed channel uses are $N_0 + N_J$ and N_0, respectively. Using (7-9), (7-17) is

$$p(\mathbf{y}|\mathbf{x}_m, \mathbf{z}) = \prod_{j\in\mathbf{J}} \frac{1}{\sqrt{\pi(N_J + N_0)}}$$

$$\times \exp\left[-\frac{(y_j - x_{mj})^2}{N_J + N_0}\right] \prod_{j\in\bar{\mathbf{J}}} \frac{1}{\sqrt{\pi N_0}} \exp\left[-\frac{(y_j - x_{mj})^2}{N_0}\right] \quad (7\text{-}18)$$

Taking the logarithm of (7-18) yields

$$\ln\{p(\mathbf{y}\,|\,\mathbf{x}_m,\mathbf{z})\} = -\frac{n_1}{2}\ln\{\pi(N_J + N_0)\} - \frac{1}{N_J + N_0}\sum_{j\in\mathbf{J}}(y_j - x_{mj})^2$$

$$-\frac{n - n_1}{2}\ln(\pi N_0) - \frac{1}{N_0}\sum_{j\in\bar{\mathbf{J}}}(y_j - x_{mj})^2 \tag{7-19}$$

As before, the first and third terms on the right side of (7-19) are constant for all codewords and therefore may be ignored by the decoder. The minimum error probability decoder therefore chooses its output to be the codeword $\mathbf{x}_m$ for which

$$-\frac{1}{N_J + N_0}\sum_{j\in\mathbf{J}}(y_j - x_{mj})^2 - \frac{1}{N_0}\sum_{j\in\bar{\mathbf{J}}}(y_j - x_{mj})^2 \tag{7-20}$$

is largest. This decoding metric is seen to be a weighted Euclidean distance metric with received symbols that are not jammed, influencing the decoder decision more than the received symbols that are jammed.

Although a general proof has not been given, these examples illustrate an important general concept applicable to all coded communications systems operating with time-varying channels. That concept is: When the channel is time varying, the decoder input metric for each channel use should be weighted so that channel uses with poor reliability (signal-to-noise ratio) have less influence on the decoding decision than channel uses with high reliability.

7-2.3 Calculation of Error Probability

No Jammer State Information Available. The bit and word error probabilities for any specific code are dependent on the details of the code as well as the decoding rule used. In the following, the maximum-likelihood or minimum distance decoding rule is assumed. It is convenient to describe the decoder as a partitioning of the space of all possible received n-tuples $\mathbf{y}$. This space has n dimensions and is partitioned by using the decoding rule on all points in the space, thereby associating a decoder output message with each point. The partitioning results in $M = 2^k$ subspaces Λ_m, where $m = 0, 1, \ldots, M - 1$. When a decoder input $\mathbf{y}$ is within Λ_m, the decoder output is $\hat{m} = m$. A block decoding error occurs whenever message m is transmitted and $\mathbf{y} \in \overline{\Lambda_m}$.

Given that message m was transmitted, the block error probability is given by Viterbi and Omura [3]

$$P_B(m) = \Pr(\mathbf{y} \in \overline{\Lambda}_m\,|\,\mathbf{x}_m)$$

$$= \sum_{\mathbf{y}\in\overline{\Lambda}_m} p(\mathbf{y}\,|\,\mathbf{x}_m) \tag{7-21}$$

The condition on message m is removed by averaging over all messages, yielding

$$P_B = \sum_{m=0}^{M-1} \Pr(m) P_B(m) \tag{7-22}$$

Bit error probability P_b is calculated by considering specific block decoding error events and the number of information bit errors associated with each of these error events. Let $P_B(m,m')$ denote the probability that message m is transmitted and the decoder output is message m'. In terms of the decoding regions $P_B(m,m')$ is the probability that $\mathbf{x}_m$ was transmitted and $\mathbf{y} \in \Lambda_{m'}$. Let $B(m,m')$ denote the number of information bit errors associated with this block decoding error event. For example, when codeword $m = 2$ of the code of Table 7-1 is transmitted and the decoder estimates that codeword $m = 3$ was transmitted, a single bit error occurs in the fourth bit position and $B(2,3) = 1$. The probability $P_B(m,m')$ is

$$P_B(m,m') = \sum_{\mathbf{y} \in \Lambda_{m'}} \Pr(\mathbf{y} \mid \mathbf{x}_m) \tag{7-23}$$

The exact bit error probability of interest is calculated by weighting (7-23) by the associated number of information errors $B(m,m')$ and averaging over all possible transmitted messages m and, for each m, all possible decoding error outputs m'. Specifically,

$$P_b = \frac{1}{k} \sum_{m=0}^{M-1} \Pr(m) \sum_{\substack{m'=0 \\ m' \neq m}}^{M-1} B(m,m') \sum_{\mathbf{y} \in \Lambda_{m'}} \Pr(\mathbf{y} \mid \mathbf{x}_m) \tag{7-24}$$

where the leading $1/k$ factor is due to the fact that k information bits are decoded with each codeword transmission. Without this term the result would be the average number of bit errors per block. Except for the simplest of codes the evaluation of this expression is tedious. For large n the calculation is not possible due to computation time bounds. Fortunately, in most cases good upper bounds on bit error probability exist which are much easier to evaluate. Some of these bounds are discussed next.

The region $\overline{\Lambda}_m$ can be defined by [3]

$$\overline{\Lambda}_m = \{\mathbf{y} \mid \ln[p(\mathbf{y} \mid \mathbf{x}_{m'})] \geq \ln[p(\mathbf{y} \mid \mathbf{x}_m)] \quad \text{for some} \quad m' \neq m\}$$

$$= \bigcup_{m' \neq m} \{\mathbf{y} \mid \ln[p(\mathbf{y} \mid \mathbf{x}_{m'})] \geq \ln[p(\mathbf{y} \mid \mathbf{x}_m)]\} \tag{7-25}$$

$$\equiv \bigcup_{m' \neq m} \Lambda_{mm'}$$

where

$$\Lambda_{mm'} \equiv \left\{ \mathbf{y} \mid \ln\left[\frac{p(\mathbf{y} \mid \mathbf{x}_{m'})}{p(\mathbf{y} \mid \mathbf{x}_m)}\right] \geq 0 \right\}$$

For each m and m', the region $\Lambda_{mm'}$ consists of all $\mathbf{y}$ that are more likely to be due to the transmission of $\mathbf{x}_{m'}$ than to the transmission of $\mathbf{x}_m$ without consideration of any other codewords. All other $\mathbf{y}$ are in the set $\overline{\Lambda}_{mm'}$ and are more likely to be due to the transmission of $\mathbf{x}_m$ than the transmission of $\mathbf{x}_{m'}$. The decoding regions $\Lambda_{mm'}$ and $\overline{\Lambda}_{mm'}$ are the decoding regions for a code which uses only two codewords $\mathbf{x}_m$ and $\mathbf{x}_{m'}$. Given this code and that $\mathbf{x}_m$ was transmitted, the probability of block decoding error is denoted $P_B'(m,m') = \Pr(\mathbf{y} \in \Lambda_{mm'}|\mathbf{x}_m)$. Recall from probability theory that the probability of a union of events is less than or equal to the sum of the probabilities of the component events. Thus block error probability given that $\mathbf{x}_m$ was transmitted can be overbounded by

$$P_B(m) = \Pr(\mathbf{y} \in \overline{\Lambda}_m|\mathbf{x}_m)$$

$$= \Pr\left(\mathbf{y} \in \bigcup_{m \neq m'} \Lambda_{mm'}\Big|\mathbf{x}_m\right)$$

$$\leq \sum_{m \neq m'} \Pr(\mathbf{y} \in \Lambda_{mm'}|\mathbf{x}_m) \tag{7-26}$$

$$= \sum_{m \neq m'} P_B'(m,m')$$

Consider first the evaluation of $P_B'(m,m')$ for a binary symmetric channel. The two codewords $\mathbf{x}_m$ and $\mathbf{x}_{m'}$ are separated by Hamming distance $d_H(\mathbf{x}_m,\mathbf{x}_{m'}) \equiv d(m,m')$. The minimum error probability decoder is a minimum-distance decoder. When $\mathbf{x}_m$ is transmitted a decoding error occurs whenever transmission errors are made that cause $\mathbf{y}$ to be closer in Hamming distance to $\mathbf{x}_{m'}$ than to $\mathbf{x}_m$. Transmission errors in positions where the two codewords are identical increase the distance between $\mathbf{x}_m$ and $\mathbf{y}$ and the distance between $\mathbf{x}_{m'}$ and $\mathbf{y}$ equally and therefore have no effect on the decoding result. Decoding errors are caused by transmission errors in the d symbols where the codewords differ. The specific locations of the error events within the d symbols do not affect the decoder decision and therefore do not affect error probability. Thus the two-codeword error probability is a function only of the Hamming distance $d(m,m')$ between the codewords and the BSC symbol error probability, and is not a function of other details of the codewords. For all codeword pairs $(\mathbf{x}_m,\mathbf{x}_{m'})$ such that $d(m,m') = d$ is even, a decoding error is made whenever $(d/2) + 1$ or more transmission errors occur in the positions where $\mathbf{x}_m$ and $\mathbf{x}_{m'}$ differ. When only $d/2$ errors occur, $\mathbf{y}$ is equidistant from $\mathbf{x}_m$ and $\mathbf{x}_{m'}$, and a decoding error is assumed to be made one-half of the time. The probability of exactly e errors in d symbols is $\binom{d}{e}p^e(1-p)^{d-e}$. Thus, for d even, the two-codeword error probability $P_B'(m,m')$ for any two codewords separated by Hamming distance d is

$$P_B'(m,m') = \sum_{e=(d/2)+1}^{d} \binom{d}{e}p^e(1-p)^{d-e} + \frac{1}{2}\binom{d}{\frac{1}{2}d}p^{d/2}(1-p)^{d/2} \tag{7-27a}$$

For d odd, a decoding error is made whenever $(d + 1)/2$ or more transmission errors occur in the positions where $\mathbf{x}_m$ and $\mathbf{x}_{m'}$ differ, and the two-codeword error probability for d odd is

$$P_B'(m,m') = \sum_{e=(d+1)/2}^{d} \binom{d}{e} p^e (1 - p)^{d-e} \tag{7-27b}$$

Using these relationships, the total block decoding error probability is bounded by

$$P_B \leq \sum_{m=0}^{M-1} \Pr(m) \sum_{\substack{m'=0 \\ m' \neq m}}^{M-1} P_B'(m,m') \tag{7-28}$$

Further simplification of this result is dependent on the properties of linear codes. The specific property of importance is the fact that the set of Hamming distances $d_H(\mathbf{x}_m, \mathbf{x}_{m'})$ between a codeword $\mathbf{x}_m$ and all other codewords $\mathbf{x}_{m'}$ for $m \neq m'$, $m' = 0$, $\ldots$, $M - 1$ is the same for all codewords $\mathbf{x}_m$. Because of this property, the second summation of the bound of (7-28) is the same for any choice of m. Therefore, the error probability is unchanged if $m = 0$ is chosen for all calculations of the second sum. Thus

$$P_B \leq \sum_{m=0}^{M-1} \Pr(m) \sum_{m'=1}^{M-1} P_B'(0,m')$$

$$= \sum_{m'=1}^{M-1} P_B'(0,m') \sum_{m=0}^{M-1} \Pr(m) \tag{7-29}$$

$$= \sum_{m'=1}^{M-1} P_B'(0,m')$$

where $P_B'(0,m')$ is calculated using (7-27).

In principle, the codewords for any specific code can be enumerated to determine the number of codewords A_d that are Hamming distance d from the all-zeros codeword. The two-codeword error probability $P_B'(0,m')$ is the same for all of these codewords and is denoted P_d. Therefore, the block error probability can also be expressed as a sum over values of Hamming distance, specifically,

$$P_B \leq \sum_{d=d_{\min}}^{n} A_d P_d \tag{7-30}$$

For codes with moderate k it is possible to find A_d with the aid of a digital computer. For a small set of codes A_d has been found analytically. Unfortunately, the communications systems engineer is often required to calculate P_B for codes with large k, say $k > 200$, where even a fast digital computer would not be able to enumerate all codewords in a reasonable time. In these cases results can still be obtained by modifying the decoding rules somewhat.

Assume that the decoder will correct up to a maximum of E channel errors and no more. Decoders having this property are called *bounded distance decoders*. Note that a maximum-likelihood decoder is guaranteed to be able to correct t channel errors, but it can, in many cases, correct more than t errors. A bounded distance decoder is guaranteed to make a block decoding error whenever more than E channel errors occur. Therefore, the block error probability is simply the probability of $E + 1$ or more channel errors occurring:

$$P_B = \sum_{i=E+1}^{n} \binom{n}{i} p^i (1 - p)^{n-i} \tag{7-31}$$

This result is an upper bound on P_B for a maximum-likelihood decoder that is guaranteed to correct *at least* E errors.

The results discussed above permit the communication system designer to predict the *block error probability* performance for any linear code. Determining the *bit error probability* requires knowledge of the number of bit errors associated with every possible block decoding error. An exact expression for bit error probability was given in (7-24). Another approach to estimating bit error probability is to use bounds that are a function of block error probability. When a block decoding error is made, a minimum of one bit error will always occur. Since k information bits are associated with each block, a single bit error corresponds to an average bit error probability of P_B/k. Thus a lower bound on bit error probability is

$$\frac{P_B}{k} \leq P_b \tag{7-32}$$

Similarly, not more than k bit errors will occur with each block decoding error. Since k bits are decoded per block, an upper bound on bit error probability is

$$P_b \leq P_B \tag{7-33}$$

These bounds are often adequate for system design purposes.

For linear systematic block codes improved bit error probability estimates can be made. Systematic block codes are a subset of the binary linear codes for which the k information symbols are used directly as part of the codeword [1]. Recall that for a linear code, the number of codewords that are Hamming distance d from any particular codeword is the same for all codewords. Using this fact, the block error probability calculation was simplified to the calculation of the block error probability assuming that codeword $\mathbf{0}$ was transmitted. The brute-force calculation of bit error probability weights each possible block error event by the associated number of bit errors as in (7-24). Using arguments similar to those used to simplify the block error probability calculation, it can be shown that for linear systematic block codes the average bit error probability calculation simplifies to the calculation of bit error probability assuming that the $\mathbf{0}$ codeword was transmitted. This simplification is possible because (1) the number of codewords Hamming distance d from any particular codeword is the same for all codewords, and (2) the number of bit errors caused by any Hamming distance d block decoding error is the same for all codewords.

Consider, for example, the (7,4) Hamming code. If message 7 was transmitted and message 13 was decoded, 2 bit errors would occur. The Hamming distance between the transmitted and decoded codewords is 4. There is a corresponding error event associated with transmission of the **0** codeword. Specifically, if message 0 was transmitted and message 9 was decoded, 2 bit errors would occur. The Hamming distance between the transmitted and decoded codewords is 4.

The probability of information bit error P_b is calculated by assuming that message **0** is transmitted and weighting each block error event by the associated number of information bit errors. Let $B(m')$ denote the number of bit errors that occur when $m = 0$ is transmitted and m' is decoded. The average bit error probability is then bounded by extending the bound of (7-29):

$$P_b \le \sum_{m'=1}^{M-1} \frac{1}{k} B(m') P'_B(0,m') \tag{7-34}$$

$P'_B(0,m')$ has the same value for all codewords $\mathbf{x}_{m'}$, with the same Hamming distance d from **0**; let P_d denote this value. Let B_d denote the total number of bit errors that occur in all block error events involving codewords that are distance d from the all-zeros codeword. Then (7-34) can be rewritten

$$P_b \le \frac{1}{k} \sum_{d=d_{\min}}^{n} B_d P_d \tag{7-35}$$

where $d_{\min}$ is the minimum distance of the code. This equation provides a good bound on bit error probability for any particular code. The values of B_d may be given or may be calculated using a computer for any particular code. For low error rates, the bit error probability is often bounded using only the first term of this equation.

EXAMPLE 7-4 _____

Calculate the bit error probability bound as a function of BSC symbol error probability p for the code of Table 7-1.

Solution: From the list of all codewords, B_3 is the total number of ones in all messages whose codewords are distance 3 from the all-zeros codeword. Thus $B_3 = 12$. Similarly, $B_4 = 16$, and $B_7 = 4$. From (7-27)

$$P_3 = \sum_{e=2}^{3} \binom{3}{e} p^e (1-p)^{3-e}$$

$$P_4 = \frac{1}{2}\binom{4}{2} p^2 (1-p)^2 + \sum_{e=3}^{4} \binom{4}{e} p^e (1-p)^{4-e}$$

$$P_7 = \sum_{e=4}^{7} \binom{7}{e} p^e (1-p)^{7-e}$$

The total average error probability is bounded by

$$P_b < \tfrac{1}{4}(12P_3 + 16P_4 + 4P_7)$$

EXAMPLE 7-5

Calculate a bound on bit error probability for a FH/MFSK-coded spread-spectrum system that is operating in a worst-case partial-band jamming environment. Assume that the error correction code used is the code given in Table 7-1 and that binary FSK data modulation is used.

Solution: In all error probability calculations for coded systems, care must be exercised to attribute the proper amount of energy to information bits and to encoder output symbols. For binary (n,k) block codes, n binary symbols are transmitted for each k binary information bits so that the energy per encoder output symbol is $E_s = kE_b/n$ and E_s is always less than E_b. Symbols with energy E_s are transmitted over the discrete binary channel. The error probability for this channel is given by (6-39) with E_b replaced by kE_b/n. Using Table 6-1, the symbol error probability, which is the BSC crossover probability, is

$$P_s = p = \begin{cases} \dfrac{0.3679}{(4/7)(E_b/N_J)} & \dfrac{4}{7}\dfrac{E_b}{N_J} > 2.00 \\[2ex] \dfrac{1}{2}\exp\left(-\dfrac{4}{7}\dfrac{E_b}{N_J}\dfrac{1}{2}\right) & \dfrac{4}{7}\dfrac{E_b}{N_j} \le 2.00 \end{cases}$$

The error probability bound is calculated using this equation for p in the expression for P_b found in Example 7-4. The result of this calculation is illustrated in Figure 7-3 together with the error probability for the same system without coding.

The usefulness of the techniques just described are limited by the availability of values of B_d for any specific code. For moderate k, B_d can be found with a digital computer. However, for large k other techniques are used. Specifically, once again, limit consideration to systematic codes and the bounded distance decoder. With a bounded distance decoder, the received vector will be changed ("corrected") in at most E positions by the decoder. Odenwalder [7] has argued that the bit error probability for any systematic binary block code decoded using a bounded distance decoder is given approximately by

$$P_b = \frac{1}{n}\sum_{i=E+1}^{n} i\binom{n}{i}p^i(1 - p)^{n-i} \tag{7-36}$$

where n is the code block length, E the number of errors that can be corrected by the decoder, and p the binary symmetric channel transition (error) probability. This relationship is not a true upper bound; however, it provides reasonably accurate results and has been widely used and quoted in the literature.

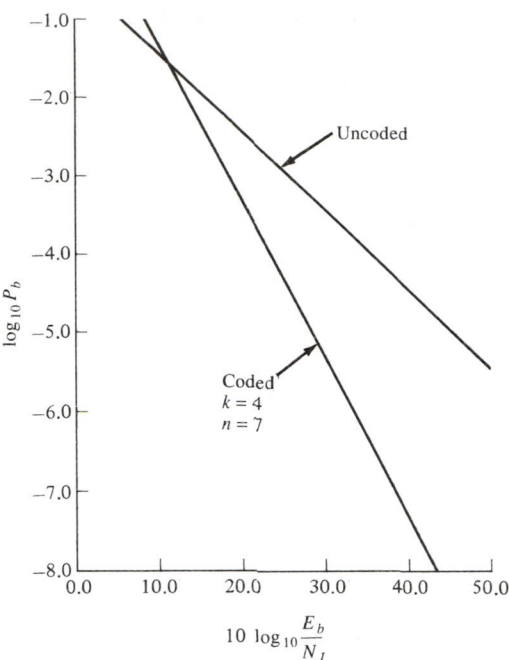

FIGURE 7-3. Comparison of coded and uncoded performance of FH/MFSK in worst-case partial-band jamming.

In summary, a linear binary block code may be thought of as a mapping of encoder input messages into encoder output codewords. The codewords are carefully selected to be separated as far as possible in Hamming distance. The decoder input is a distorted version of the encoder output. The decoder operates by choosing as its estimate of the transmitted codeword the codeword that is "closest" to the received n-tuple **y**. The distance measure used by the decoder is, in general, the a posteriori probability. For the BSC an equivalent distance measure is Hamming distance and for the AWGN channel an equivalent distance measure is Euclidean distance. The design of hardware efficient coders and decoders is the subject of coding theory and is not discussed here. The primary problem of error correction coding is the development of good codes which are, at the same time, reasonably easy to decode. A decoder that operates directly on the principles defined above will be inefficient from a hardware standpoint in almost all cases. The student is referred to Lin and Costello [1] for further discussion of coding and decoding techniques.

Many formulas for block and bit error probability have been given in these discussions. These results are all useful when used prudently by the system designer. These formulas apply when no jammer-state information is available. The modification of these formulas to account for jammer-state information is considered next.

Jammer-State Information Available. The concepts used in the following analysis were first published by Stark [8]. For simplicity it is again assumed that the jammer has only two states, 1 and 0, corresponding to "on" and "off," and that the probability of any specific codeword symbol being jammed is ρ. Assume that ρ is constant. Due to interleaving, the jammer may be assumed to jam each symbol independently (i.e., the channel is memoryless). The DMC transition probability is given by (7-12). For a known jammer state $\mathbf{z} = (z_1, z_2, \ldots, z_n)$, decoding regions $\Lambda_m(\mathbf{z})$ are defined for each possible transmitted message m; $\Lambda_m(\mathbf{z})$ is the set of all DMC output vectors $\mathbf{y}$ for which $p(\mathbf{y} | \mathbf{x}_m, \mathbf{z})$ is larger than $p(\mathbf{y} | \mathbf{x}_{m'}, \mathbf{z})$ for all $m' \neq m$. If message m is transmitted, a decoding error occurs when the received vector $\mathbf{y} \in \overline{\Lambda_m(\mathbf{z})}$, that is, when $\mathbf{y}$ is not in the decoding region for message m. Given $\mathbf{z}$ and that m was transmitted, the block error probability is

$$P_B(m, \mathbf{z}) = \sum_{\mathbf{y} \in \Lambda_m(\mathbf{z})} p(\mathbf{y} | \mathbf{x}_m, \mathbf{z}) \tag{7-37}$$

which is (7-21) with the addition of the condition on JSI.

An upper bound on $P_B(m, \mathbf{z})$ is calculated using steps identical to those described in (7-25) and (7-26) when no JSI was available. The region $\overline{\Lambda_m(\mathbf{z})}$ is defined by

$$\overline{\Lambda_m(\mathbf{z})} = \{\mathbf{y} \,|\, \ln[p(\mathbf{y} | \mathbf{x}_{m'}, \mathbf{z})] \geq \ln[p(\mathbf{y} | \mathbf{x}_m, \mathbf{z})] \quad \text{for some} \quad m' \neq m\} \tag{7-38}$$

$$\equiv \bigcup_{m' \neq m} \Lambda_{mm'}(\mathbf{z})$$

where

$$\Lambda_{mm'}(\mathbf{z}) \equiv \left\{ \mathbf{y} \,\middle|\, \ln\left[\frac{p(\mathbf{y} | \mathbf{x}_{m'}, \mathbf{z})}{p(\mathbf{y} | \mathbf{x}_m, \mathbf{z})}\right] \geq 0 \right\}$$

As before, $\Lambda_{mm'}(\mathbf{z})$ is the set of all $\mathbf{y}$ that are more likely to have been caused by the transmission of $\mathbf{x}_{m'}$ than by the transmission of $\mathbf{x}_m$, ignoring all other codewords. A code with two codewords $\mathbf{x}_m$ and $\mathbf{x}_{m'}$ is defined; the decoding regions for this code are $\Lambda_{mm'}(\mathbf{z})$ and $\overline{\Lambda_{mm'}(\mathbf{z})}$. Given that $\mathbf{x}_m$ was transmitted and the jammer state is $\mathbf{z}$, the probability of block decoding error for this code is denoted

$$P'_B(m, m', \mathbf{z}) = \Pr(\mathbf{y} \in \Lambda_{mm'} | \mathbf{x}_m, \mathbf{z}) \tag{7-39}$$

Application of the union bound to (7-37) gives the overbound

$$P_B(m, \mathbf{z}) = \Pr(\mathbf{y} \in \overline{\Lambda_m(\mathbf{z})} | \mathbf{x}_m, \mathbf{z})$$

$$= \Pr\left(\mathbf{y} \in \bigcup_{m \neq m'} \Lambda_{mm'}(\mathbf{z}) \,\middle|\, \mathbf{x}_m, \mathbf{z}\right) \tag{7-40}$$

$$\leq \sum_{\substack{m'=0 \\ m' \neq m}}^{M-1} \Pr(\mathbf{y} \in \Lambda_{mm'}(\mathbf{z}) | \mathbf{x}_m, \mathbf{z})$$

$$= \sum_{\substack{m'=0 \\ m' \neq m}}^{M-1} P'_B(m, m', \mathbf{z})$$

Consider the evaluation of $P'_B(m,m',\mathbf{z})$ for a binary symmetric channel with error probabilities p_1 and p_0 corresponding to jammer states 1 and 0. The Hamming distance between the two codewords $\mathbf{x}_{m'}$ and $\mathbf{x}_m$ is denoted $d(m,m')$. The maximum-likelihood decoding metric for this channel is the weighted Hamming distance defined by

$$d_1 \ln \frac{p_1}{1-p_1} + d_0 \ln \frac{p_0}{1-p_0} \qquad (7\text{-}16)$$

where d_1 and d_0 denote the Hamming distances between $\mathbf{y}$ and the codeword being tested over the jammed and not-jammed channel use indices, respectively. This metric is evaluated for both codewords, and the decoder output is the codeword for which (7-16) is largest.

Recall from the analysis without JSI that BSC errors for channel uses where $x_{jm} = x_{jm'}$ affect the Hamming distance decoding metric identically for both codewords. Similarly, with JSI the value of (7-16) for both codewords is affected identically by BSC errors for channel uses where $x_{jm} = x_{jm'}$; therefore, these channel use indices may be ignored by the decoder. Let $\mathbf{J}_d(\mathbf{x}_m,\mathbf{x}_{m'},\mathbf{z})$ denote the set of channel uses indices j for which $x_{jm} \neq x_{jm'}$ and $z_j = 1$ and $\overline{\mathbf{J}}_d(\mathbf{x}_m,\mathbf{x}_{m'},\mathbf{z})$ denote the channel use indices for which $x_{jm} \neq x_{jm'}$ and $z_j = 0$. The number of elements in $\mathbf{J}_d(\mathbf{x}_m,\mathbf{x}_{m'},\mathbf{z})$ is $n_{Jd}(\mathbf{x}_m,\mathbf{x}_{m'},\mathbf{z})$ and the number of elements in $\overline{\mathbf{J}}_d(\mathbf{x}_m,\mathbf{x}_{m'},\mathbf{z})$ is $d(m,m') - n_{Jd}(\mathbf{x}_m,\mathbf{x}_{m'},\mathbf{z})$. Let $d'_1(m)$ denote the Hamming distance between $\mathbf{y}$ and $\mathbf{x}_m$ for channel use indices in $\mathbf{J}_d(\mathbf{x}_m,\mathbf{x}_{m'},\mathbf{z})$, and let $d'_0(m)$ denote the Hamming distance between $\mathbf{y}$ and $\mathbf{x}_m$ for channel use indices in $\overline{\mathbf{J}}_d(\mathbf{x}_m,\mathbf{x}_{m'},\mathbf{z})$. Let $d'_1(m') = n_{Jd}(\mathbf{x}_m,\mathbf{x}_{m'},\mathbf{z}) - d'_1(m)$ and $d'_0(m') = d(m,m') - n_{Jd}(\mathbf{x}_m,\mathbf{x}_{m'},\mathbf{z}) - d'_0(m)$ denote the same distances for $\mathbf{x}_{m'}$. Hereafter, the explicit dependence of $d(m,m')$ on m and m' and of $n_{Jd}(\mathbf{x}_m,\mathbf{x}_{m'},\mathbf{z})$, $\overline{\mathbf{J}}_d(\mathbf{x}_m,\mathbf{x}_{m'},\mathbf{z})$, and $\mathbf{J}_d(\mathbf{x}_m,\mathbf{x}_{m'},\mathbf{z})$ on $\mathbf{x}_m$, $\mathbf{x}_{m'}$, and $\mathbf{z}$ will be dropped where the meaning is clear. Using these definitions, the decoder calculates

$$\delta(m) \equiv d'_1(m) \ln \frac{p_1}{1-p_1} + d'_0(m) \ln \frac{p_0}{1-p_0} \qquad (7\text{-}41a)$$

and

$$\delta(m') \equiv d'_1(m') \ln \frac{p_1}{1-p_1} + d'_0(m') \ln \frac{p_0}{1-p_0}$$

$$\qquad (7\text{-}41b)$$

$$= [n_{Jd} - d'_1(m)] \ln \frac{p_1}{1-p_1} + [d - n_{Jd} - d'_0(m)] \ln \frac{p_0}{1-p_0}$$

and outputs m if $\delta(m) > \delta(m')$. If $\delta(m) = \delta(m')$, the decoder chooses m or m' arbitrarily. Ignore for a moment the possibility that $\delta(m) = \delta(m')$. The two-codeword error probability $P'_B(m,m',\mathbf{z})$ is then the probability that

$$d'_1(m) \ln \frac{p_1}{1-p_1} + d'_0(m) \ln \frac{p_0}{1-p_0}$$

$$\qquad (7\text{-}42)$$

$$< [n_{Jd} - d'_1(m)] \ln \frac{p_1}{1-p_1} + [d - n_{Jd} - d'_0(m)] \ln \frac{p_0}{1-p_0}$$

given that message m was transmitted and given $\mathbf{z}$. Rearranging (7-42), the two-codeword error probability is the probability that

$$\delta(m) = d_1'(m) \ln \frac{p_1}{1 - p_1} + d_0'(m) \ln \frac{p_0}{1 - p_0}$$

$$< \frac{n_{Jd}}{2} \ln \frac{p_1}{1 - p_1} + \frac{d - n_{Jd}}{2} \ln \frac{p_0}{1 - p_0} \qquad (7\text{-}43)$$

$$\equiv \alpha(d, n_{Jd})$$

Considering the possibility that $\delta(m) = \delta(m')$, the error probability is

$$P_B'(m, m', \mathbf{z}) = \Pr[\delta(m) < \alpha(d, n_{Jd})] + \tfrac{1}{2}\Pr[\delta(m) = \alpha(d, n_{Jd})] \qquad (7\text{-}44)$$

The Hamming distance between $\mathbf{y}$ and $\mathbf{x}_m$ equals $d_1'(m)$ when $\mathbf{x}_m$ is transmitted, and $d_1'(m)$ transmission errors occur in the n_{Jd} channel uses that are jammed. The probability of $d_1'(m)$ errors in n_{Jd} jammed channel uses is

$$\binom{n_{Jd}}{d_1'} p_1^{d_1'} (1 - p_1)^{n_{Jd} - d_1'} \qquad (7\text{-}45a)$$

where the dependence on m has been dropped since it is no longer necessary. Similarly, the probability of $d_0'(m)$ errors in $d - n_{Jd}$ not-jammed channel uses is

$$\binom{d - n_{Jd}}{d_0'} p_0^{d_0'} (1 - p_0)^{d - n_{Jd} - d_0'} \qquad (7\text{-}45b)$$

Finally, the two-codeword error probability $P_B'(m, m', \mathbf{z})$, given jammer-state vector $\mathbf{z}$, is the sum of the product of the probabilities (7-45) over all d_1' and d_0' where $\delta(m) < \alpha(\mathbf{z})$ or $\delta(m) = \alpha(\mathbf{z})$. Thus

$$P_B'(m, m', \mathbf{z}) = \sum_{d_1'=0}^{n_{Jd}} \sum_{d_0'=0}^{d - n_{Jd}} \beta(d_1', d_0') \binom{n_{Jd}}{d_1'} p_1^{d_1'} (1 - p_1)^{n_{Jd} - d_1'}$$

$$\times \binom{d - n_{Jd}}{d_0'} p_0^{d_0'} (1 - p_0)^{d - n_{Jd} - d_0'} \qquad (7\text{-}46)$$

where $\beta(d_1', d_0') = 1$ if $\delta(m) < \alpha(d, n_{Jd})$, $\beta(d_1', d_0') = \tfrac{1}{2}$ if $\delta(m) = \alpha(d, n_{Jd})$, and $\beta(d_1', d_0') = 0$ otherwise. Observe that the value of $P_B'(m, m', \mathbf{z})$ is a function of the Hamming distance $d(m, m')$ between $\mathbf{x}_m$ and $\mathbf{x}_{m'}$, and the number of jammed symbols $n_{Jd}(\mathbf{x}_m, \mathbf{x}_{m'}, \mathbf{z})$ in channel uses where $x_{jm} \neq x_{jm'}$. Thus

$$P_B'(m, m', \mathbf{z}) = P_B'[d(m, m'), n_{Jd}(\mathbf{x}_m, \mathbf{x}_{m'}, \mathbf{z})]$$

Having evaluated $P_B'[d(m, m'), n_{Jd}(\mathbf{x}_m, \mathbf{x}_{m'}, \mathbf{z})]$, the block coding error probability union bound can now be completed. For a specific message m and jammer state $\mathbf{z}$, the following bound was derived previously:

$$P_B(m, \mathbf{z}) \leq \sum_{\substack{m'=0 \\ m' \neq m}}^{M-1} P_B'[d(m, m'), n_{Jd}(\mathbf{x}_m, \mathbf{x}_{m'}, \mathbf{z})] \qquad (7\text{-}40)$$

The dependence on m and $\mathbf{z}$ are removed via straightforward averaging. The result is

$$P_B = \sum_{\mathbf{z}} \Pr(\mathbf{z})P_B(\mathbf{z})$$

$$= \sum_{\mathbf{z}} \Pr(\mathbf{z})\left[\sum_{m=0}^{M-1} \Pr(m)P_B(m,\mathbf{z})\right] \tag{7-47}$$

$$\leq \sum_{\mathbf{z}} \Pr(\mathbf{z})\left[\sum_{m=0}^{M-1} \Pr(m)\left\{\sum_{\substack{m'=0 \\ m'\neq m}}^{M-1} P_B'[d(m,m'),n_{Jd}(\mathbf{x}_m,\mathbf{x}_{m'},\mathbf{z})]\right\}\right]$$

The summations of (7-47) are finite and may be reordered to yield

$$P_B \leq \sum_{m=0}^{M-1} \Pr(m)\left[\sum_{\mathbf{z}} \Pr(\mathbf{z})\left\{\sum_{\substack{m'=0 \\ m'\neq m}}^{M-1} P_B'[d(m,m'),n_{Jd}(\mathbf{x}_m,\mathbf{x}_{m'},\mathbf{z})]\right\}\right]$$

$$= \sum_{m=0}^{M-1} \Pr(m)\left[\sum_{\substack{m'=0 \\ m'\neq m}}^{M-1}\left\{\sum_{\mathbf{z}} \Pr(\mathbf{z})P_B'[d(m,m'),n_{Jd}(\mathbf{x}_m,\mathbf{x}_{m'},\mathbf{z})]\right\}\right] \tag{7-48}$$

Consider the summation within braces in (7-48). This summation is over all possible jammer-state vectors with m and m' fixed. The function $P_B'[d(m,m'),n_{Jd}(\mathbf{x}_m,\mathbf{x}_{m'},\mathbf{z})]$ has the same value for all $\mathbf{z}$ for which the value of $n_{Jd}(\mathbf{x}_m,\mathbf{x}_{m'},\mathbf{z})$ is the same. Therefore, the summation may be rewritten as a summation over all possible values n_{Jd}. Thus

$$\sum_{\mathbf{z}} \Pr(\mathbf{z})P_B'[d(m,m'),n_{Jd}(\mathbf{x}_m,\mathbf{x}_{m'},\mathbf{z})]$$

$$= \sum_{n_{Jd}=0}^{d(m,m')} \binom{d(m,m')}{n_{Jd}}\rho^{n_{Jd}}(1-\rho)^{d(m,m')-n_{Jd}}P_B'[d(m,m'),n_{Jd}]$$

$$\equiv P_{Jd}(d(m,m'),\rho) \tag{7-49}$$

where

$$\binom{d}{n_{Jd}}\rho^{n_{Jd}}(1-\rho)^{d-n_{Jd}}$$

is the sum of the probabilities of all $\mathbf{z}$ for which $n_{Jd}(\mathbf{x}_m,\mathbf{x}_{m'},\mathbf{z}) = n_{Jd}$. The fraction ρ which now appears in (7-49) is controlled by the jammer and is not a random variable that can be eliminated by statistical averaging.

Rewrite (7-48) using (7-49) to find

$$P_B(\rho) \leq \sum_{m=0}^{M-1} \Pr(m)\sum_{\substack{m'=0 \\ m'\neq m}}^{M-1} P_{Jd}(d(m,m'),\rho) \tag{7-50}$$

For linear codes, as before, the second summation is the same for all m. Thus

$$P_B(\rho) \le \sum_{m'=1}^{M-1} P'_{Jd}(d(0,m'),\rho) = \sum_{d=d_{\min}}^{n} A_d P_{Jd}(\rho) \qquad (7\text{-}51)$$

where $P_{Jd}(\rho)$ denotes the value of $P_{Jd}(d(m,m'),\rho)$ for $d(m,m') = d$, and the summation over m' has been replaced by a summation over all possible Hamming distances d with each term of the summation weighted by the number of codewords A_d, which are Hamming distance d from the all-zeros codeword.

Using the formulas presented above the student is able to calculate an upper bound on block error probability for any specific block code in partial band or partial time (pulsed) jamming environment. The bit error probability of primary interest to the designer is calculated using the bound of (7-33), which is applicable to all codes, or (7-35), which is applicable only to systematic codes. The bound of (7-35) is directly applicable if P_d is replaced by P_{Jd} from the analysis above.

7-3 Elementary Convolutional Coding Concepts

7-3.1 Basic Concepts

The fundamentals of convolutional coding are discussed in this section. The structure of a convolutional code is different from the structure of a block code. Specifically, information sequences are not grouped into distinct independent blocks and encoded. Rather, a continuous† sequence of information bits is mapped into a continuous sequence of encoder output symbols which are then transmitted over a DMC. This mapping is highly structured, enabling a decoding method considerably different from block decoding methods to be applied. Some researchers have argued that convolutional coding can achieve better bit error rate performance on the AWGN channel than a block code with the same complexity. Whether block coding or convolutional coding is preferable for a particular application will depend on the details of that application and the technology available at the time the comparison. In this section the fundamentals of convolutional coding and decoding are described and a method of predicting error probability performance is presented. Discussion is limited to binary-input, binary-output convolutional codes.

The mapping of information sequences into encoder output sequences is done in a manner that improves communication efficiency by enabling the system to correct transmission errors. Convolutionally coded systems process semi-infinite sequences of information symbols. Encoder output sequences are also semi-infinite. The set of encoder output code sequences of a convolutional code is analogous to the set of encoder output codewords of a block code. The elegant structure of convolutional codes enables processing (i.e., coding and decoding) of a semi-infinite sequence by processing a few symbols at a time in both the coder and the decoder when the channel is memoryless. Codeword sequences due to distinct information sequences

---

†Of course, the information sequence is not infinite and eventually terminates. However, for the purpose of analyzing the structure and performance of these codes, the sequence may be considered infinite.

are distinct and are thus separated from one another in Hamming distance. This separation allows some number of transmission errors to be corrected by the decoder. The separation of codeword sequences is made possible by using a small fraction of the possible binary channel input sequences as codeword sequences, just as a small fraction, 2^{k-n}, of possible binary n-vectors were used for block code codewords.

The optimum decoding rule for convolutional codes is the same as for block codes; the decoder will estimate the transmitted code sequence to be the sequence that was most likely to have been transmitted given the known code structure, channel characteristics, and received sequence. For the BSC, the most likely transmitted sequence is the sequence that is closest in Hamming distance to the received sequence. An efficient algorithm will be described for finding this closest sequence. The structure of convolutional codes enables reliability information to be used easily by the decoder. Thus for convolutional codes, soft-decision decoding is often used in modern communications systems.

7-3.2 Definition of a Convolutional Code

Figure 7-4 illustrates a shift-register circuit that generates a simple rate-$\frac{1}{2}$ convolutional code. Input bits are clocked into the circuit from the left. After each input is received the coder output is generated by sampling and multiplexing the outputs of the two modulo-2 adders. For this simple code, two output symbols ($n = 2$) are generated for each input bit ($k = 1$) and the code rate is $k/n = R = \frac{1}{2}$. Observe that a particular input bit influences the output during its own interval as well as the next two input bit intervals. A convolutional code is defined by the number of stages in the shift register, the number of outputs (i.e., the number of modulo-2 adders), and the connections between the shift register and the modulo-2 adders. The state of the encoder is defined to be the contents of the shift register and is completely determined by the previous two information bit inputs. The encoder of Figure 7-4 has four possible states corresponding to all possible contents of the binary two-stage shift register.

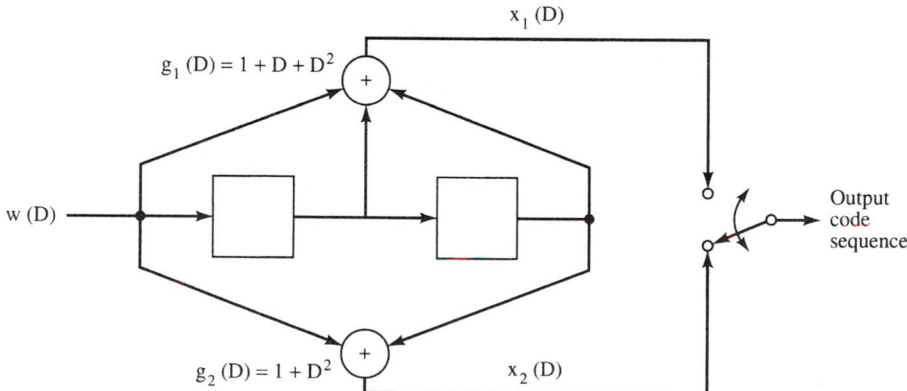

FIGURE 7-4. Rate-$\frac{1}{2}$ convolutional encoder.

The simple convolutional code described above produces two output symbols for each input information bit and has code rate $R = \frac{1}{2}$. In general, convolutional codes produce n output symbols for each k input symbols and have rate $R = k/n$. Although this would appear to make convolutional codes identical to block codes, they are significantly different. Most important, convolutional codes have memory, which causes the rule used to map the k information bits into n code symbols to be a function of past information bits. For example, the encoder of Figure 7-4 maps $k = 1$ information bit into $n = 2$ code symbols using a rule that depends on the previous two information bits. For example, an input 1 produces outputs $11, 00, 01$, or 10 when the previous two inputs, and hence the encoder states, are $00, 10, 01$, or 11, respectively. The number of past inputs that affect the mapping of the current k inputs into n outputs is a critical parameter of convolutional codes. This parameter affects convolutional code performance and complexity much as block length affects block code performance and complexity. The *constraint length* of the code is 1 plus the number of past inputs affecting the current outputs. The constraint length of the code of Figure 7-4 is 3, since the current output pair is a function of the current input plus the two previous inputs. Note that different definitions of constraint length can be found in the literature on convolutional coding. In all cases, however, constraint length is a measure of the memory within the encoder.

The convolutional code of Figure 7-4 can also be represented by a *state-transition diagram,* as shown in Figure 7-5. The state of the encoder is represented by the contents of the circles. The encoder begins in state 00. If the first encoder input bit is a 0, the encoder exits state 00 on the solid branch, outputs the two code symbols

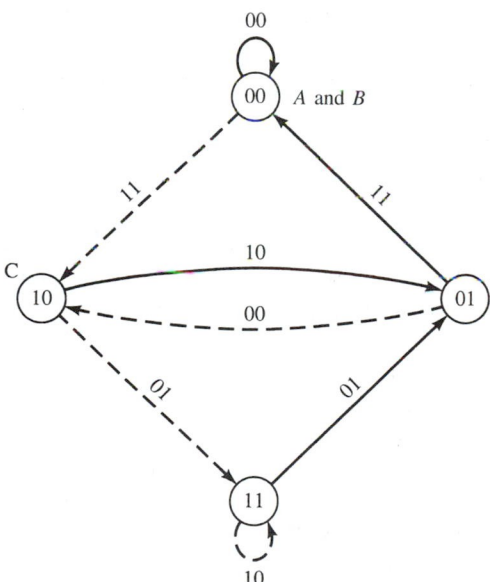

FIGURE 7-5.   State-transition diagram representation for rate-$\frac{1}{2}$ convolutional encoder.

found on this branch (00), and returns to state 00. If the first input bit were 1, the encoder would exit state 00 on the dashed branch, would output the two code symbols found on this branch (11), and would enter state 10. In general, encoder inputs equal to 0 cause the encoder to move along the solid branches to the next state and to output the codeword symbols corresponding to the branch label. Encoder inputs equal to 1 cause the encoder to move along the dashed branches to the next state, again outputting two codeword symbols encountered along the branch. Convolutional coders can therefore be thought of as finite-state machines that change states as a function of the input sequence. The branch labels in Figure 7-5 are calculated directly from the shift-register representation.

A convolutional coder can also be represented by a trellis diagram [9] as illustrated in Figure 7-6 for the code of Figure 7-4. The encoder represented by Figure 7-6 has four states, which are represented by the labels within the circles of the trellis and correspond to the states of the state-transition diagram. The encoding operation starts at point A, which is state 00 on the far left of the trellis. If the first information bit is a 0, the encoder moves along the solid line out of state 00, arriving again at state 00, labeled B. The encoder output is the symbol pair 00, which is the label on the trellis branch between the two states. If the first encoder input were a 1, the encoder would move along the dashed branch out of state 00, arriving at state 10, labeled C. In this case the encoder output is 11, which is the label on the branch connecting states 00 and 10. The second encoder input causes the encoder to move to the right one more branch and to output the associated branch label. This process of moving from left to right through the trellis following the path specified by the information sequence and outputting the branch labels continues as long as desired. A typical input and output sequence for this convolutional code is shown at the bottom of Figure 7-6. For this input sequence, the encoder follows the branches marked with arrows.

Observe that the labels on the branches leaving and entering a state do not change with time in Figure 7-6. This is characteristic of a *time-invariant convolutional code*. In general, the branch labels could vary with time and the resultant convolutional code would be a time-varying convolutional code; in this case the connections to the modulo-2 adders of Figure 7-4 would change after each k input bits. Even more generally the branch labels on the trellis could be selected using any procedure yielding the requited Hamming distance between code sequences. When there are no constraints on the procedure for assigning labels to trellis branches, the resultant code is called a *trellis code*. A trellis code is simply a trellis that defines the structure and memory of the code and an assignment of code symbols to the branches of the trellis. Trellis codes can be linear or nonlinear and can be time varying or time invariant. Convolutional codes are a subset of trellis codes where the assignment of labels to the trellis branches is constrained to follow additional rules. The shift-register encoder (e.g., Figure 7-4) and the state-transition diagram representation (e.g., Figure 7-5) are not always applicable to trellis codes. In this book consideration is limited to time-invariant binary-input, binary-output convolutional code, which may equivalently be represented by a shift-register encoder, a state-transition diagram, or a trellis diagram.

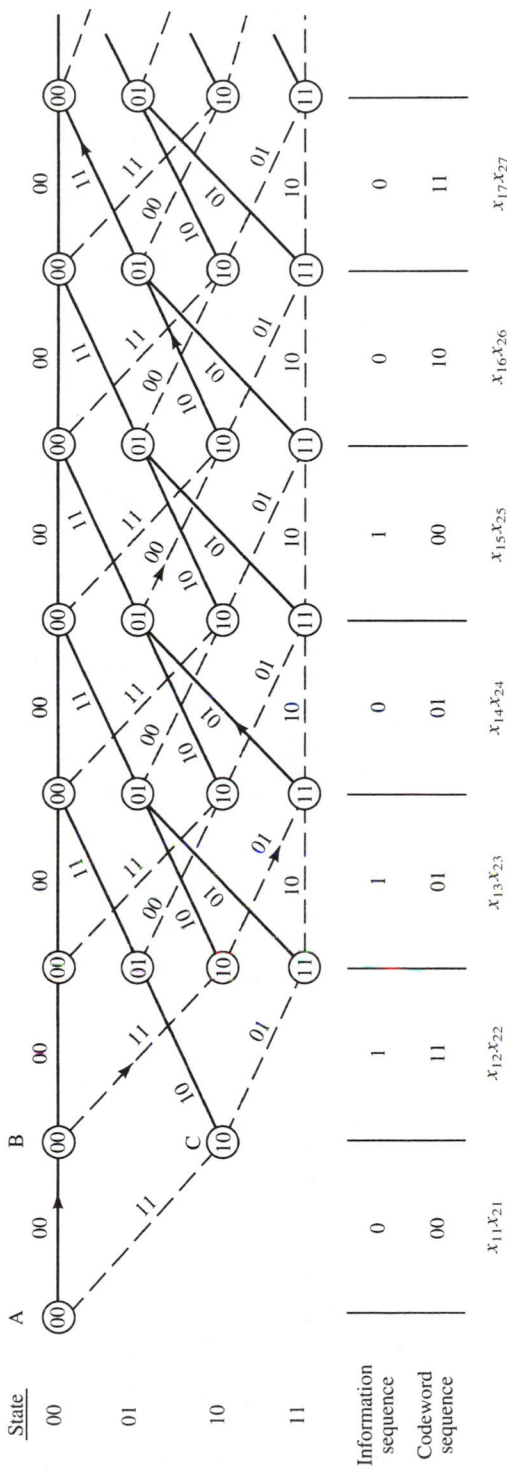

FIGURE 7-6. Trellis representation for rate-$\frac{1}{2}$ convolutional encoder.

EXAMPLE 7-6

A rate $R = \frac{1}{3}$ convolutional code is defined by the shift-register generator illustrated in Figure 7-7. Three outputs are produced for each input bit, and each output is a different modulo-2 sum of the contents of the shift register and the input. The number of stages of the shift register is equal to 3, and the constraint length of the code is 4. Thus an input bit influences the output due to the next three input bits.

Figure 7-8 is a trellis representation of this convolutional code. For a rate-$\frac{1}{3}$ code there are three binary symbols on each of the trellis branches. The trellis labels can be found using the encoder of Figure 7-7. Figure 7-9 is a state-transition-diagram representation of the convolutional code. This diagram was generated directly from the trellis diagram.

For simplicity, for the remainder of this discussion, consideration will be limited to convolutional codes with $k = 1$. The reader is referred to advanced texts such as Lin and Costello [1] for treatment of more complex codes.

7-3.3 Decoding Convolutional Codes

As stated previously, the decoder has knowledge of the code structure (the code trellis), the received sequence, and a statistical characterization (the transition probabilities) of the channel. The function of the decoder is to estimate the encoder input information sequence using a rule or method that results in the minimum possible number of errors being delivered to the information user. There is a one-to-one correspondence between information sequences and encoder output sequences. Further, any information and code sequence pair is uniquely associated with a path through the trellis. Thus the job of the decoder may be viewed as estimating the path through the trellis which was followed by the coder.

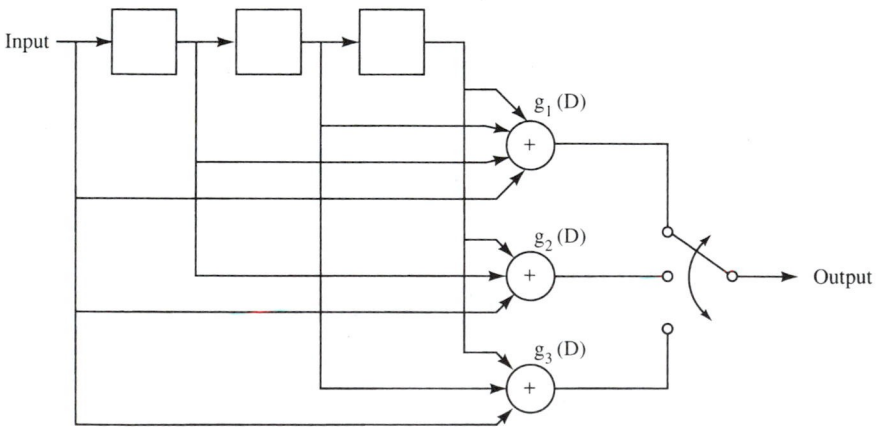

FIGURE 7-7.   Rate-$\frac{1}{3}$, constraint-length-4 convolutional encoder.

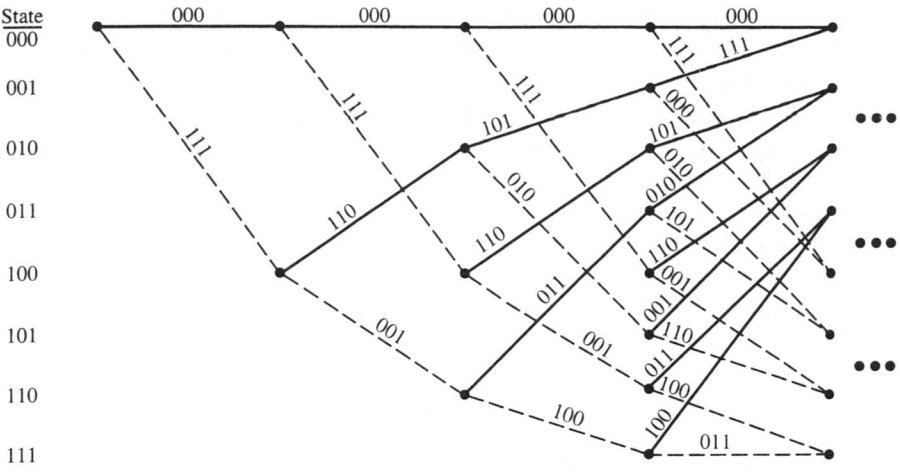

FIGURE 7-8. Trellis representation for $R = \frac{1}{3}$, constraint-length-4 convolutional encoder.

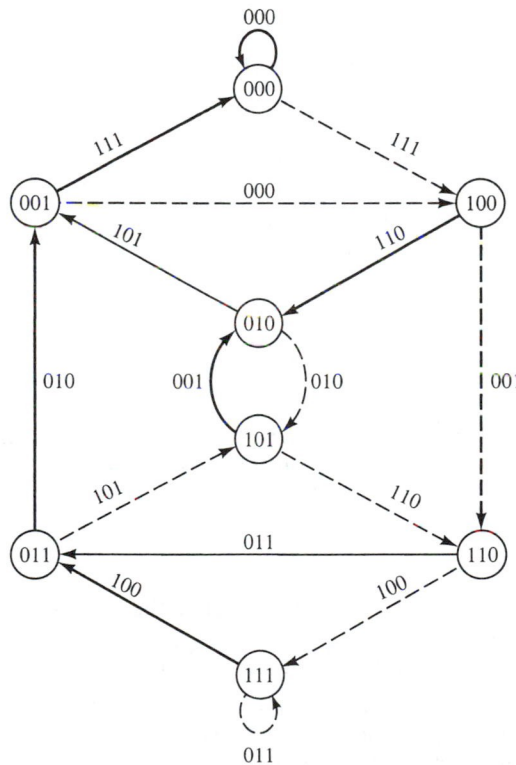

FIGURE 7-9. State-transition-diagram representation of rate-$\frac{1}{3}$, constraint-length-4 convolutional encoder.

The decoding rule for convolutional codes is developed using the same arguments as those used in developing the block decoding rules. Assume that the information source is ideal so that its output symbols are equally likely and independent; thus all paths through the trellis are equally probable and a maximum-likelihood decoding rule is appropriate. The decoder output is the path whose code sequence was most likely to have been input to the channel by the encoder. It is convenient at this point to denote the encoder output semi-infinite sequence corresponding to message sequence or path m by $\mathbf{x}_m = x_{m0}x_{m1}x_{m2}\cdots x_{mj}\cdots$. The DMC output sequence is denoted by $\mathbf{y} = y_0y_1y_2\cdots y_j\cdots$, and the probability of receiving the channel output $\mathbf{y}$ given that the channel input was $\mathbf{x}_m$ is

$$p(\mathbf{y}|\mathbf{x}_m) = \prod_{j=0}^{\infty} p(y_j|x_{mj}) \tag{7-52}$$

Given $\mathbf{y}$, the path most likely to have been followed through the trellis by the coder is the path whose code sequence maximizes $p(\mathbf{y}|\mathbf{x}_m)$. The function $p(\mathbf{y}|\mathbf{x}_m)$ is the *metric* used to compare code sequences $\mathbf{x}_m$ and $\mathbf{x}_{m'}$. For the hard-decision binary symmetric channel, maximization of this metric is equivalent to finding the path through the trellis whose code sequence is closest in Hamming distance to the received sequence. This result corresponds exactly to hard-decision decoding for block codes. One method for finding this closest path is the Viterbi algorithm [9–11], which is discussed in detail later. For now, the decoder for the BSC can be viewed as a processor that searches through all trellis paths to find that single path whose code sequence is closest in Hamming distance to the received sequence.

Since the logarithm is a monotone increasing function of an increasing argument, the decoder could also use the metric $\ln\{p(\mathbf{y}|\mathbf{x}_m)\}$ rather than $p(\mathbf{y}|\mathbf{x}_m)$. In this case the decoder would calculate $\ln\{p(\mathbf{y}|\mathbf{x}_m)\}$ for all paths and would choose the path m with the largest value as the decoder output path. Taking the logarithm converts the product of (7-52) to a summation. Thus the decoding metric is

$$\ln\{p(\mathbf{y}|\mathbf{x}_m)\} = \sum_{j=1}^{\infty} \ln\{p(y_j|x_{mj})\} \tag{7-53}$$

This metric is more commonly used than the metric of (7-52). The sum of terms of (7-53) associated with a single code trellis branch is called the *branch metric* for that branch.

EXAMPLE 7-7

Suppose (for convenience in this example) that the communicator needs to transmit a message that is 3 bits long. The message is convolutionally encoded using the code defined by the trellis of Figure 7-10. To end the message and clear the encoder of message bits, two zeros are appended to the message. Thus a total of 5 bits (three information and two zeros) are encoded by the encoder; 10 code symbols are output. The code symbols are transmitted over a BSC with transition probability $p = 0.1$. The received sequence is 10, 01, 10, 11, 00. What is the decoder output?

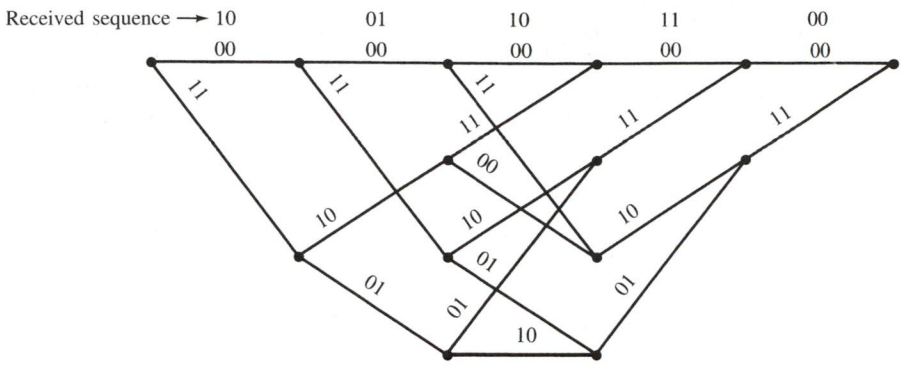

FIGURE 7-10. Decoding trellis.

Solution: Decoding is accomplished using Figure 7-10 by calculating the path metric of (7-53) for all of the eight distinct paths through the trellis and choosing the path with the largest metric. Decoding can also be accomplished by calculating the Hamming distance between the received sequence and the path sequence for all eight paths and choosing the path with the minimum Hamming distance. Both methods will be illustrated. The incremental (i.e., corresponding to one transmitted symbol) log-likelihood symbol metrics for the specified channel are

$$\ln[p(0|0)] = \ln[p(1|1)] = \ln(0.9) = -0.11$$

$$\ln[p(1|0)] = \ln[p(0|1)] = \ln(0.1) = -2.30$$

Consider the all-zero path. The code sequence on this path differs from the received sequence in five positions. Thus the path metric is

$$5(-2.3) + 5(-0.11) = -12.05$$

The Hamming distance between this path and the received sequence is 5. All paths (specified by the encoder input bits) and their path metrics and Hamming distances are listed below.

Received sequence: 10, 01, 10, 11, 00

Path	Code Sequence	Path Metric	Hamming Distance
0, 0, 0, 0, 0	00, 00, 00, 00, 00	−12.05	5
0, 0, 1, 0, 0	00, 00, 11, 10, 11	−14.24	6
0, 1, 0, 0, 0	00, 11, 10, 11, 00	−5.48	2
0, 1, 1, 0, 0	00, 11, 01, 01, 11	−16.43	7
1, 0, 0, 0, 0	11, 10, 11, 00, 00	−14.24	6
1, 0, 1, 0, 0	11, 10, 00, 10, 11	−16.43	7
1, 1, 0, 0, 0	11, 01, 01, 11, 00	−7.67	3
1, 1, 1, 0, 0	11, 01, 10, 01, 11	−9.86	4

> The largest path metric is -5.48, which corresponds to message sequence 0, 1, 0, 0, 0 and path sequence 00, 11, 10, 11, 00. This same path sequence is also closest in Hamming distance to the received sequence, illustrating that for the BSC, both decoding metrics decode the same path. The decoder output is the information sequence 0, 1, 0, 0, 0.

Because of the elegant structure of the codes and the Viterbi Algorithm, convolutional decoders can and do make use of channel output reliability information. As stated previously, this reliability information is generated by allowing the DMC to have a larger number of outputs than inputs. In the limit, the DMC output is permitted to be a continuum of values. Consider, for example, a BPSK system. The BPSK matched-filter output is a real number generated by integrating received signal plus noise over a symbol period. For the AWGN channel the probability density function of the matched filter output is Gaussian, as in (7-9). To create a hard-decision channel, the matched filter output space (the real number line) is partitioned into two regions. If the channel input symbols are equally likely, the boundary between regions is zero, so that matched filter outputs above zero are declared to be 1's while outputs below zero are declared to be 0's. This is illustrated in Figure 7-11, which shows the matched filter output pdf's for channel inputs 0 and 1 (corresponding to channel inputs $+1$ and -1, respectively) on the real number line along with the hard-decision regions. To provide reliability information to the decoder, the matched filter output space is partitioned into more than two regions. Figure 7-11 illustrates further partitioning into four or eight regions. Figure 7-12 illustrates the DMCs resulting from the partitions of Figure 7-11. The transition probabilities for

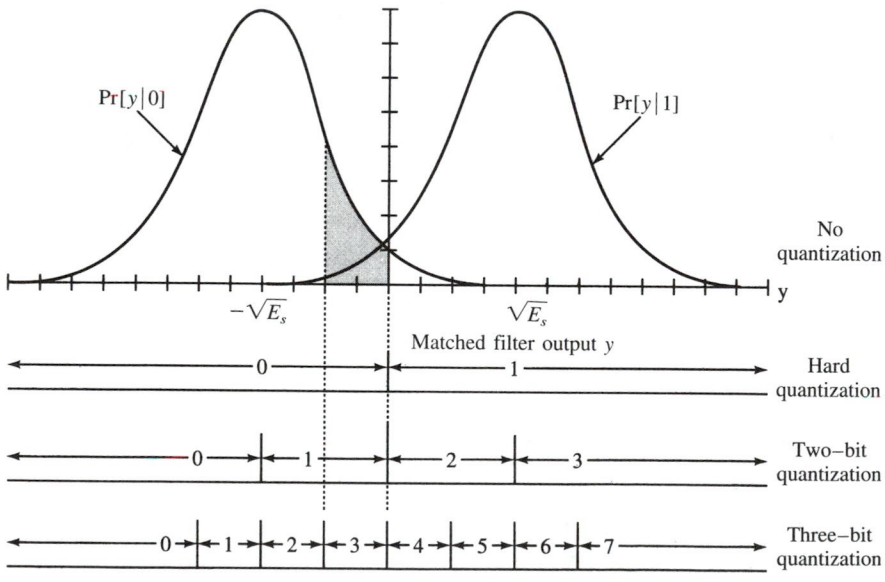

FIGURE 7-11. Soft-decision demodulator thresholds.

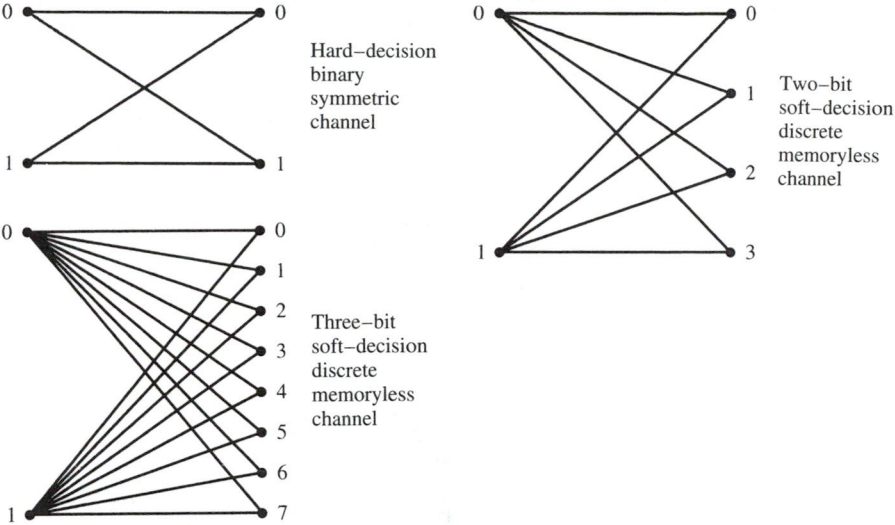

FIGURE 7-12.   Channel models for decoding of convolutional codes.

any of the channels of Figure 7-12 are calculated by calculating the area under the Gaussian density function between the thresholds which define the output symbol. For example, the transition probability $Pr[3|0]$ is the area shown shaded in the figure.

Regardless of whether the channel outputs hard or soft decisions, the decoding rule remains the same. Given the received sequence $\mathbf{y}$, the decoder output is the code sequence $\mathbf{x}_m$ (or path) that maximizes the probability $p(\mathbf{y}|\mathbf{x}_m)$. The Viterbi algorithm can be used to find the correct path through the trellis for hard or soft decisions.

7-3.4 Viterbi Algorithm

The Viterbi algorithm (VA) is an elegant method for performing maximum-likelihood decoding of convolutional codes. The algorithm was first described mathematically by Viterbi [10] in 1967. Since that time the algorithm has been described many times by many authors; most notably Forney [9,12] provides a highly readable and insightful description of the algorithm and its performance. Recall that the function of a maximum-likelihood decoder is to find the code sequence that was most likely to have been transmitted given the received channel output sequence. As discussed previously, this corresponds to finding the path through the trellis whose code sequence has the largest log-likelihood function as defined in (7-53). For the BSC, maximizing this function is equivalent to finding the path through the trellis whose code sequence is closest in Hamming distance to the received sequence. The VA will be explained by example for a BSC using the Hamming distance metric.

Hard-Decision Decoding. Consider the truncated trellis diagram of Figure 7-13a. For convenience in this explanation, the number of trellis branches between

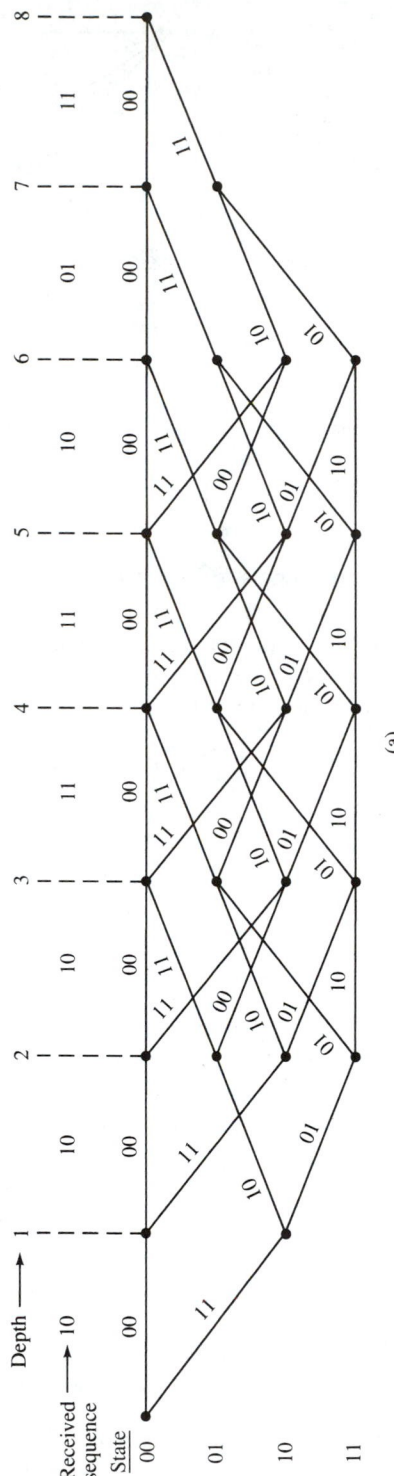

FIGURE 7-13a. Truncated trellis diagram.

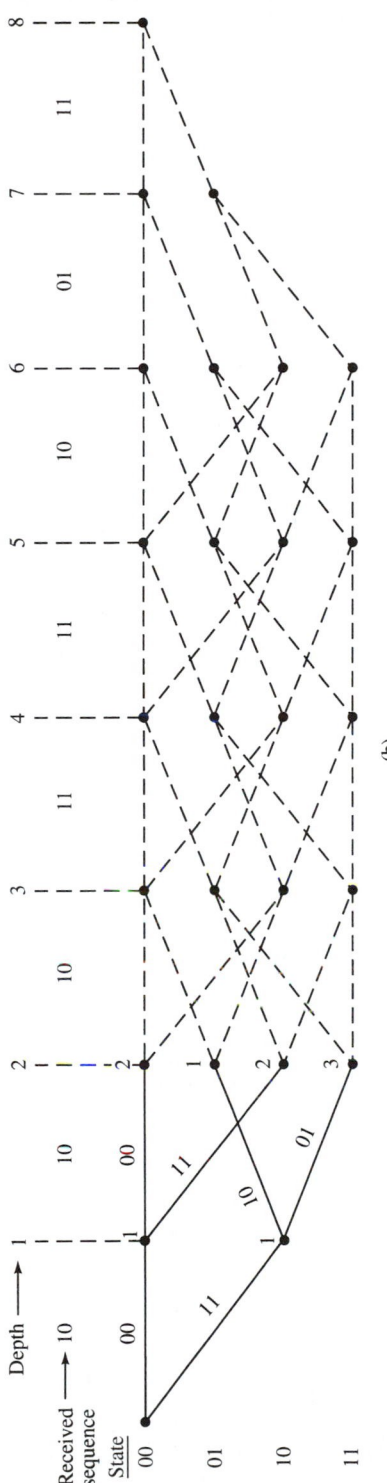

FIGURE 7-13b. Truncated trellis diagram.

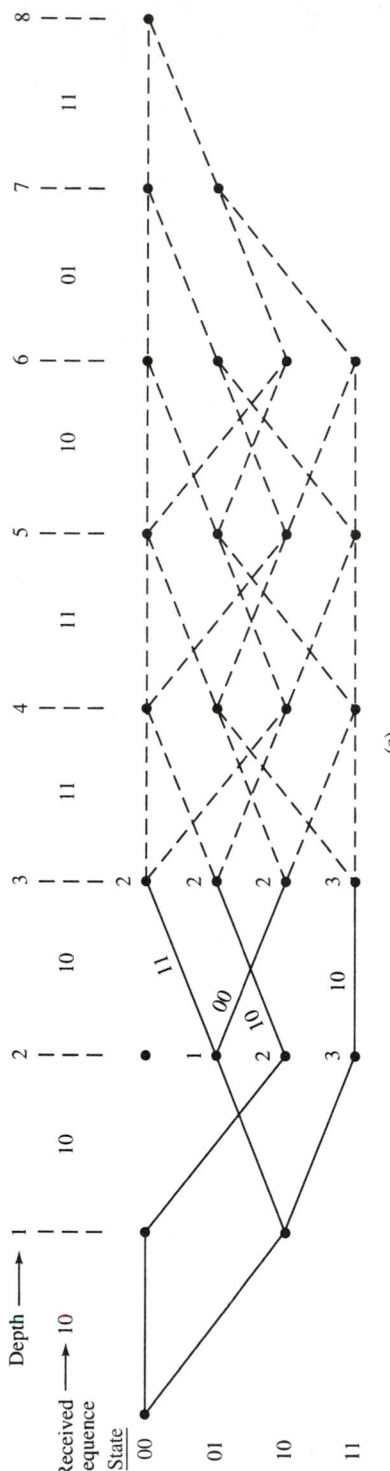

FIGURE 7-13c. Truncated trellis diagram.

(c)

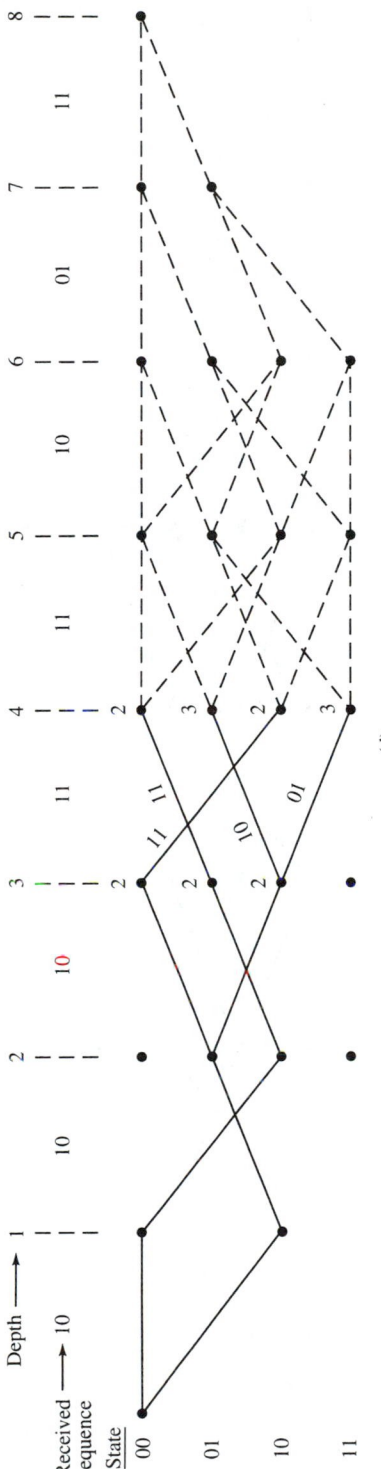

FIGURE 7-13d. Truncated trellis diagram.

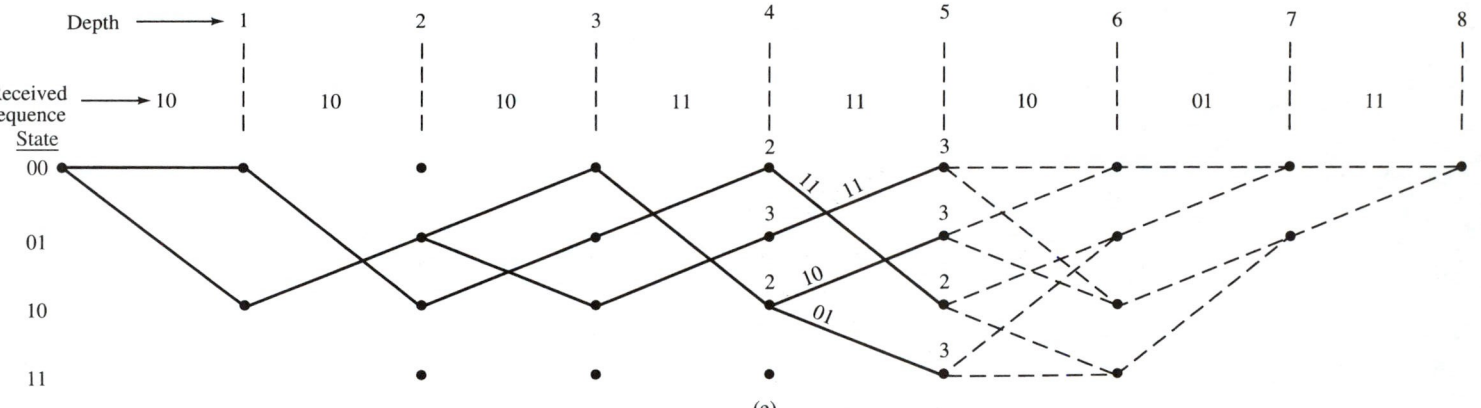

FIGURE 7-13e. Truncated trellis diagram.

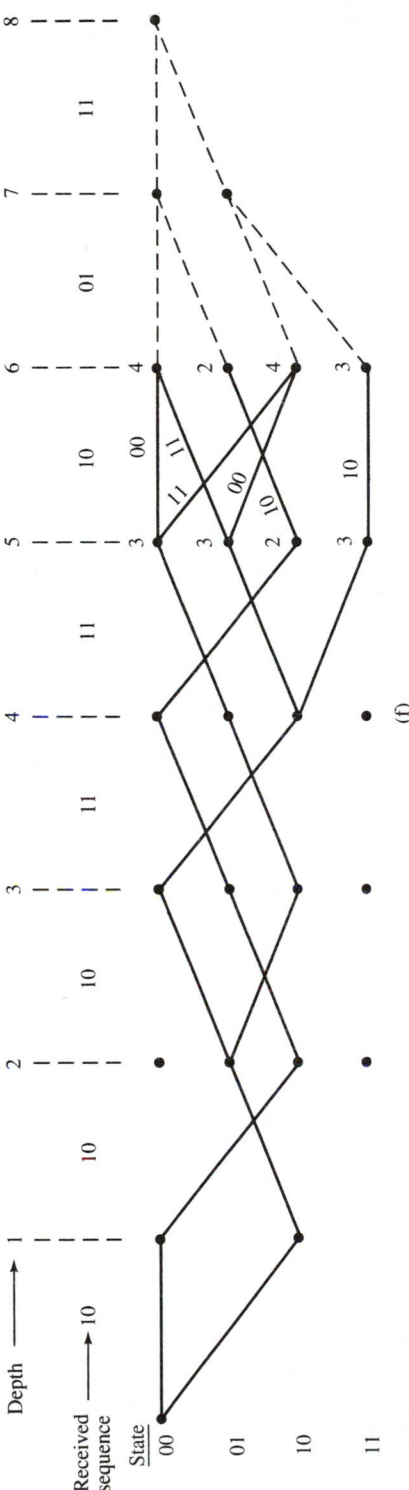

FIGURE 7-13f. Truncated trellis diagram.

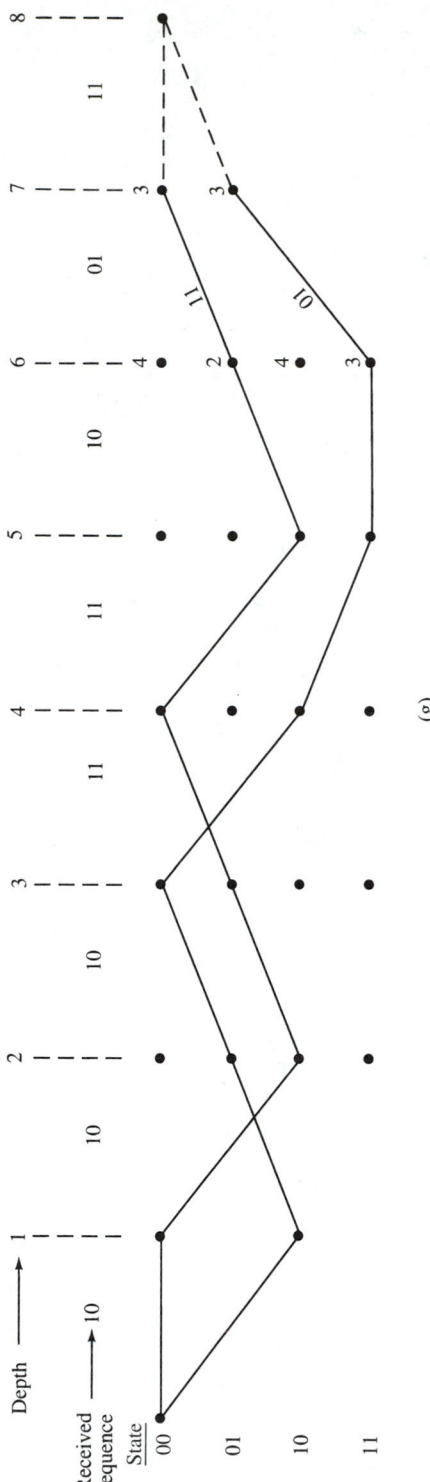

FIGURE 7-13g. Truncated trellis diagram.

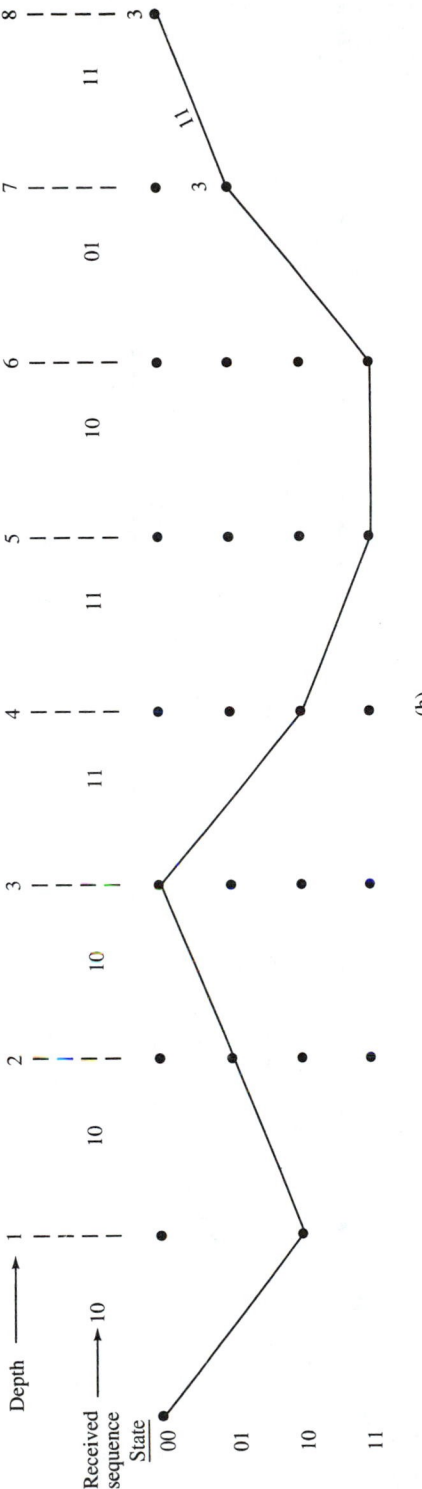

FIGURE 7-13h. Truncated trellis diagram.

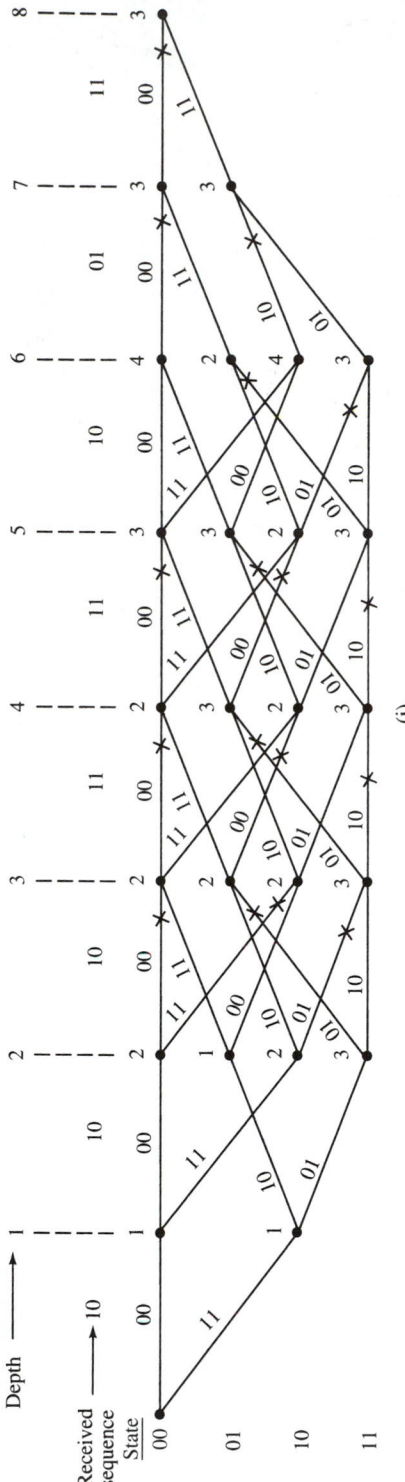

FIGURE 7-13i. Truncated trellis diagram.

the leftmost trellis state and the state being discussed is indicated above the trellis; this number is also referred to as depth into the trellis. Path segments are defined by a sequence of trellis states $\{S_1, S_2, S_3, \ldots\}$. The trellis has been truncated by clearing the encoder by inputting two known zeros at depth 7 and 8. A possible received sequence is written above the trellis; this sequence is decoded in the discussion that follows.

The first decoding step is to calculate the Hamming distance between the first two received symbols and the two symbols on the trellis branches leaving state 00 at depth 0 and leading to the depth 1 states. These Hamming distances are both 1 and are written above the state nodes at depth 1, as illustrated in Figure 7-13b. The decoder stores these Hamming distances and moves to depth 2 of the trellis. Now the decoder calculates the Hamming distance between the first four received symbols and the first four symbols on all four trellis paths leading into depth 2 states. The Hamming distance for the first four symbols of any path is the sum of the Hamming distance for the first two symbols and the Hamming distance accumulated in the branch leading from depth 1 to depth 2. Thus the decoder has only to add the Hamming distances it stored in the first step to the Hamming distance accumulated in the most recent branch. For example, the Hamming distance for the path segment $\{00,10,01\}$ is the Hamming distance written above state 10 plus 0 since no additional Hamming distance is accumulated on the branch $\{10,01\}$. The results of these four calculations are written above the trellis states as before and are stored by the decoder.

The decoding becomes more interesting as depth 3 states are considered. There are two trellis paths leading into each of four states at depth 3. Therefore, eight path segments must be considered. The decoder calculates the accumulated Hamming distance for each path segment by adding the distance noted above the states at depth 2 to the distance accumulated in moving from depth 2 to depth 3. After performing these eight calculations the decoder is able to discard one of the two paths leading into each state. The path discarded is the path with the higher accumulated distance at that depth. The discarded path will never again be considered by the decoder. The reason that a path segment can be discarded is that because it has merged with another path having a lower cumulative Hamming distance, it can never be part of the path through the trellis with the lowest total Hamming distance. Consider, for example, the two paths leading into state 00 at depth 3. The path segment $\{00,00,00,00\}$ has accumulated Hamming distance 3, while the path segment $\{00,10,01,00\}$ has accumulated Hamming distance 2. Suppose that the path with the minimum total Hamming distance passed through state 00 at depth 3. Because this path has the *minimum* Hamming distance of all paths, it must include the path segment $\{00,10,01,00\}$ and cannot include the path segment $\{00,00,00,00\}$. This discarding of one of the two paths leading into each state is the key to the efficiency of the Viterbi algorithm. Figure 7-13c illustrates the surviving path segments after the decoder has discarded half of the paths leading into depth 3 states. Observe that all paths that pass through state 00 at depth 2 have been discarded by the VA. In this and subsequent figures, the surviving paths are shown as solid lines and the paths not yet processed by the decoder are shown as dashed lines.

The decoder now moves to depth 4 states and repeats the process. The Hamming distances for eight path segments are calculated and the path segment with the smallest Hamming distance leading into each state is discarded. The decoding process is illustrated in Figure 7-13b through h. At depth 6 note that two Hamming distance ties occur. In this case no meaningful decision can be made regarding the best path segment and the decoder can either keep both paths leading into the state or select one arbitrarily. Both paths have been retained in Figure 7-13f. In this case both of the pairs of paths retained due to ties are discarded at depth 7. Observe that at depth 7 only two paths remain as potential decoder output paths. At depth 8 a single path is selected by the VA and decoding is complete.

Figure 7-13i illustrates the entire Viterbi decoding procedure on a single trellis. Starting with the encoding trellis, the decoder calculates Hamming distances and discards paths as just described. Rather than redraw the trellis after each step, the discarded paths are denoted by an $\times$ on the discarded branch leading into a state. After reaching depth 8, the decoded path is determined by reversing direction and following the branches that do not have $\times$'s back to depth 0.

Soft-Decision Decoding. The Viterbi algorithm described above using the Hamming distance decoding metric works in exactly the same manner using soft-decision decoding metrics. A minor difference is that for soft-decision decoding, the path with the *largest* log-likelihood metric, $\ln[p(\mathbf{y}\,|\,\mathbf{x}_m)]$, is being found rather than the path with the *smallest* Hamming distance. Thus discarded path segments are the segments with the smaller (rather than the larger) of the two metrics. In both cases the path most likely to have been followed by the encoder is found.

EXAMPLE 7-8

Suppose that the same truncated encoder used for the hard-decision VA description above is used in a soft-decision communication system. Assume that the binary input DMC for this system is defined by Figure 7-14, which illustrates the transition probabilities and the log-likelihood symbol metrics for all transitions. The trellis description of the encoder is illustrated in Figure 7-15a and a possible soft-decision received sequence is given. The branch metrics are calculated from the received sequence and the log-likelihood functions of Figure 7-14. For example, the branch metric for the branch between depth 0 state 00 and depth 1 state 10 is

$$\ln[\Pr(3\,|\,1)] + \ln[\Pr(2\,|\,1)] = -0.22 + (-1.90) = -2.12$$

This branch metric is the number within parentheses on the branch. All branches are similarly labeled.

The trellis is redrawn in Figure 7-15b to illustrate the decoding process. In this figure the cumulative metric for the surviving path segment leading to a node is written immediately above the node. For the nodes at depth 1 the cumulative metric is exactly equal to the first branch metrics. For the nodes at depth 2, the cumulative metric is the sum of the depth 1 metric and the metric accumulated on the branch leading from depth 1 to depth 2. For example, the cumulative metric leading to state 01 at depth 2 is the metric -2.12 at state 10 at depth 1 plus the

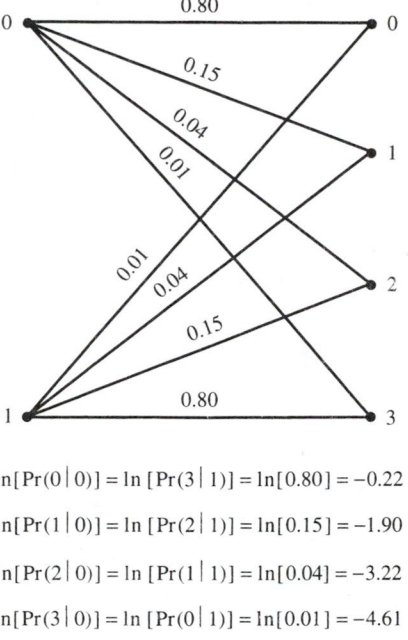

$$\ln[\Pr(0\,|\,0)] = \ln\,[\Pr(3\,|\,1)] = \ln[0.80] = -0.22$$

$$\ln[\Pr(1\,|\,0)] = \ln\,[\Pr(2\,|\,1)] = \ln[0.15] = -1.90$$

$$\ln[\Pr(2\,|\,0)] = \ln\,[\Pr(1\,|\,1)] = \ln[0.04] = -3.22$$

$$\ln[\Pr(3\,|\,0)] = \ln\,[\Pr(0\,|\,1)] = \ln[0.01] = -4.61$$

FIGURE 7-14. Binary input discrete memoryless channel.

branch metric -3.80 accumulated on the branch between depth 1 state 10 and depth 2 state 01.

As before, the decoding becomes more interesting at depth 3, where path segments begin to be discarded. Consider the two path segments leading into state 00 at depth 3. The cumulative metric for the path segment {00,00,00,00} is $(-7.83) + (-5.12) + (-9.22) = -22.17$. The cumulative metric for the path segment {00,10,01,00} is $(-2.12) + (-3.80) + (-0.44) = -6.36$. The path with the smallest metric of -22.17 is discarded by placing an $\times$ on that branch. The metric of the surviving path is written above state 00 at depth 3. Decoding continues in this manner to the end of the trellis. After the decoder has reached depth 8, the decoder reverses direction following the surviving path to determine the correct path through the trellis. Observe that the surviving path is identical to the path found using hard-decision decoding.

Costello [1] has given the following succinct definition, valid for either hard or soft decisions, of the Viterbi algorithm. In this description, L is the total number of information bits processed by the encoder and m is the number of zeros required to clear the decoder.

VITERBI ALGORITHM [1]

Step 1: *Begin at depth $j = 1$; compute the partial metric for the single path entering each state. Store the path (the survivor) and its metric for each state.*

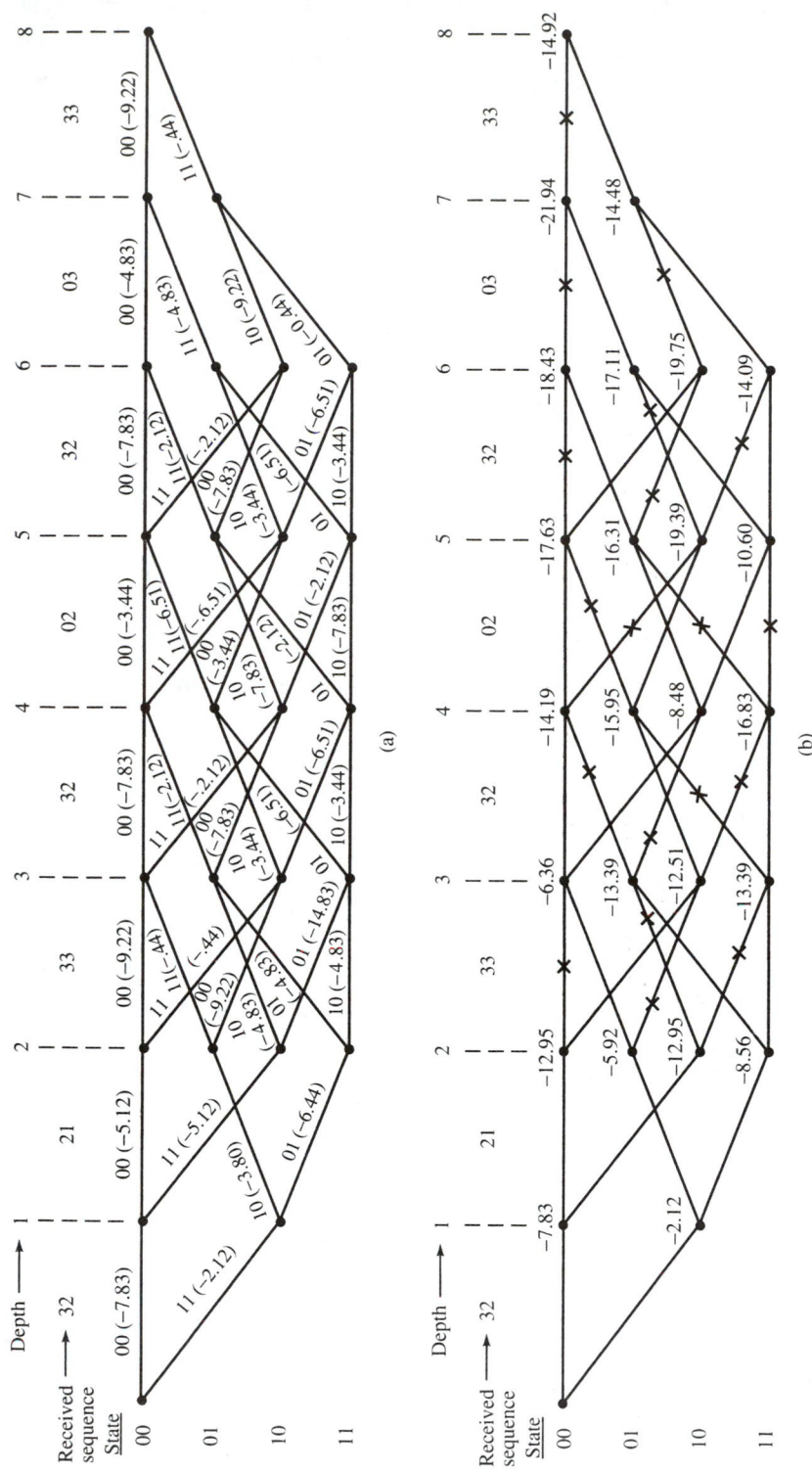

FIGURE 7-15. (a) Soft-decision Viterbi decoding branch metrics; (b) soft-decision Viterbi decoding cumulative path metrics.

Step 2: *Increase j by 1. Compute the partial metric for all the paths entering a state by adding the branch metric entering that state to the metric of the connecting survivor at the preceding depth. For each state, store the path with the largest metric (the survivor), together with its metric, and eliminate all other paths.*

Step 3: *If $j < L + m$, repeat step 2. Otherwise, stop.*

The descriptions of the VA used above presume that after a finite time, the encoder is cleared with a sequence of known zero information bits. This allows the decoder to force all paths to state 00 and complete the decoding. Although this is a valid and useful decoding strategy, it is sometimes desirable to make decoding bit decisions prior to the zeroing of the encoder. The reason for this is that the number of information bits transmitted between zeroing of the encoder is usually very large (perhaps thousands of bits), requiring significant path memory. In addition, reliable bit decisions can be made well in advance of zeroing the decoder. Suppose, for example, that the encoder was never zeroed and that the number of states in the trellis is S. At each depth there are S surviving paths. At any depth the decoder could trace all of the surviving paths backwards through the trellis for some distance. With high probability it would find that all of the S surviving paths would merge into a single surviving path some distance back from the current depth. At the point where a single survivor path is observed, reliable decoding decisions can be output from the decoder. The depth that must be traced backward through the trellis to the point where all paths merge is a random variable.

Rather than actually retrace through the trellis, most decoders include a fixed path memory. When that memory is filled, a decoding decision is forced at that level of the trellis. Typical decoders store all surviving paths for about five times the size of the shift register in the encoder [1,2]. For example, the encoder of Figure 7-6 has a shift register size of 2 bits. Thus a practical decoder would store path segments of length $2 \times 5 = 10$ branches. With high probability all paths would have merged to a single path 10 branches back from the current decoding path and reliable decisions could be output.

The complexity of Viterbi decoding is directly proportional to the number of states in the trellis. The number of states grows exponentially with constraint length, and decoders become impractical for large constraint length. At the time of this writing integrated-circuit Viterbi decoders are available for moderate bit rates for constraint lengths to about nine.

7-3.5 Decoding and Bit Error Probability

Decoding error probability for Viterbi decoding of convolutional codes is calculated using concepts similar to those used to calculate error probabilities for maximum-likelihood decoding of block codes. Error events are enumerated, their probabilities are calculated, and the results used to calculate bounds on decoding error probability. For Viterbi decoding, the appropriate measure of decoding reliability [11,12] is the probability that the correct path through the trellis is discarded at depth j. That is, the correct reliability measure is the probability of making an error at each opportunity for decoding error. An opportunity to discard the correct path exists each time the decoder moves one unit deeper into the trellis. This measure is comparable to the measure used for block codes, where an opportunity for error exists each time a new block is decoded and the appropriate reliability measure is the block decoding error probability. Note that the probability of selecting an incorrect path over the full extent of the trellis is an inappropriate reliability measure. It would not be correct to evaluate the probability that the total semi-infinite decoder output sequence is not continuously merged with the correct sequence. Over all time, the probability that the decoded path diverges from and remerges with the correct path at least once approaches unity.

Since convolutional codes are linear, decoding error probability is the same for all possible transmitted sequences. Thus decoding error probability is calculated assuming that the all-zero sequence is transmitted. The all-zero code sequence corresponds to the all-zero information sequence. The decoding error probability is then the probability that a nonzero path (i.e., a path defined by information sequence having at least one 1) entering state 00 at depth j of the trellis is selected as the survivor. To evaluate decoding error probability, all potential decoding error events at depth j are enumerated and their error probabilities calculated and summed. The summation is the result of a union bounding argument similar to that used for the error probability bound for block codes. The student is referred to Viterbi [11] for the details of the error probability derivation.

Consider the set of all nonzero paths through the code trellis which first merge with the all-zero path at depth j. The number of 1's in a nonzero code sequence is the Hamming distance of that code sequence from the all-zero sequence. The *free distance*, denoted d_f, of the convolutional code is the minimum Hamming distance between any two distinct code sequences; since the code is linear, the free distance is equal to the Hamming weight of the nonzero code sequence with the smallest Hamming weight. Let a_d denote the number of nonzero code sequences that are Hamming distance d from the all-zero path. Because of the definition of free distance, for all $d < d_f$, $a_d = 0$.

Let P_d denote the probability that the decoder discards the all-zero code sequence at depth j in favor of a code sequence having Hamming weight d. This probability is identical to the two-codeword error probability for a block code with minimum distance d; P_d is therefore calculated using (7-27) for the binary symmetric channel with symbol error probability p. In Ref. 11 it is shown that the decoding error probability P_E is overbounded by

$$P_E < \sum_{d=d_{\text{free}}}^{\infty} a_d P_d \tag{7-54}$$

For the continuous output soft-decision additive white Gaussian noise channel it can be shown [11,13] that P_d is given by

$$P_d = \int_{\sqrt{2dRE_b/N_0}}^{\infty} \frac{1}{\sqrt{2\pi}} \exp\left(\frac{-x^2}{2}\right) dx$$

$$= Q\left(\sqrt{\frac{2dRE_b}{N_0}}\right) \tag{7-55}$$

For other soft-decision channels the calculation of P_d is more complex and the student is referred to Refs. 11 and 14 for details.

Equation (7-54) bounds the probability that the Viterbi decoder selects an incorrect path through the trellis. It is not a bound on decoding bit error probability. Bit error probability is calculated using essentially the same techniques as those used above for decoding error probability. To calculate bit error probability, each decoding error event is weighted by the number of bit errors associated with that event. Since the all-zero path is the correct path, the number of bit errors associated with a nonzero path is equal to the number of information 1's associated with that path. In general, there are a total of c_d bit errors associated with all paths that are Hamming distance d from the all-zero path. It can be shown that bit error probability is overbounded by [11].

$$P_b < \sum_{d=d_{\text{free}}}^{\infty} c_d P_d \tag{7-56}$$

where P_d is calculated using (7-27) for the hard-decision binary symmetric channel or (7-55) for the soft-decision AWGN channel. When jammer-state information is available, (7-54) and (7-56) remain valid with appropriate calculation of P_d. Using analysis similar to that used in the development of error probability for block codes with JSI, it can be shown that for the BSC, when JSI is available P_d must be replaced by P_{Jd} of (7-49).

The results presented above for probability of decoding error and probability of bit error are reasonably good upper bounds; they are not exact results. Computer simulations may also be used to determine error probabilities. Extensive computer simulation results are presented in a frequently referenced paper by Heller and Jacobs [15]. The simulation results of this paper show good agreement with the upper bounds for moderate to low bit error rates where the upper bounds presented herein are best. Soft- and hard-decision decoding are also compared in this paper, along with other interesting and practical results. Other practical results are presented in Odenwalder [7].

7-3.6 Other Topics

Space does not permit coverage of many interesting topics related to convolutional coding. Some of these topics are mentioned here briefly along with references.

Sequential Decoding. Prior to the discovery of Viterbi decoding, convolutional codes were decoded using sequential decoding and its variants. Sequential decoding makes use of yet another representation of convolutional codes, the code ''tree.'' Sequential decoding is not precisely maximum-likelihood decoding, although if processing time and decoder memory is not an issue, sequential decoding performance is nearly identical to maximum-likelihood decoding performance. The complexity of sequential decoders is approximately independent of constraint length, so that this decoding method can be used for very-large-constraint-length convolutional codes. Sequential decoding has been used for extremely important deep-space missions. The absence of further discussion of sequential decoding in this chapter is not an indication that this topic is not important. The student is referred to Refs. 10 and 13 for detailed discussions of sequential decoding.

Threshold Decoding. Certain convolutional codes can be decoded using circuitry that is only several times more complex than the convolutional encoder itself. To be decoded so simply the codes must have additional structure is defined in detail by Massey [16] and by Costello [1]. The simplicity of threshold decoders makes possible their implementation at very high speeds. Although the performance of these special convolutional codes is not outstanding for the additive white Gaussian noise channel, the structure is particularly convenient for applications where interleaving is required. For channels where the interference comes in bursts, very large coding gains can be achieved using convolutional coding with threshold decoding.

Concatenated Reed–Solomon/Convolutional Coding. Extremely powerful error correction capability can be obtained by concatenating a Viterbi-decoded convolutional code with a Reed–Solomon [1] block code. Figure 7-16 is a block diagram of a coded communications system that uses both a convolutional and a block code. Information bits from the data source are grouped into blocks of m to create the 2^m-ary symbol alphabet used by the Reed–Solomon (R-S) code. The coding system that appears first in the chain, in this case the R-S code, is called the *outer code* of a concatenated code system. The R-S output symbols are converted to their

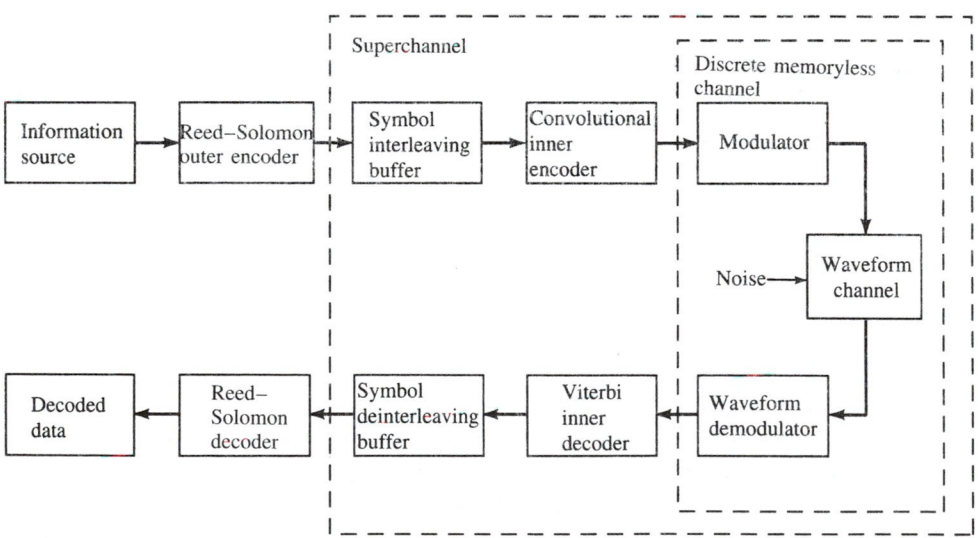

FIGURE 7-16. Concatenated coding system block diagram.

m-bit binary representations and input to a symbol interleaving buffer. The purpose of this buffer is to enable the spreading of bursts of channel symbol errors over R-S codewords as uniformly as possible. The student is referred to Refs. 2, 17, and 18 and to Section 7-5 for additional discussions about interleaver design. For the purpose of this discussion the interleaver/de-interleaver combination may be considered processors that take bursts of channel errors and spread these errors over some number of codewords.

The interleaver output of Figure 7-16 is input to a convolutional coder whose output is sent over the discrete memoryless channel to a Viterbi (in this case) decoder. The convolutional code and Viterbi decoder operate exactly as discussed previously; the convolutional code is called the *inner code* of the system. The output of the Viterbi decoder is an estimate of the binary sequence that was input to the encoder. The Viterbi decoder output is input to the de-interleaver. The portion of the communication system between the interleaver input and the de-interleaver output is called a *superchannel.* The superchannel includes the convolutional coding system as well as the physical channel and can be modeled as a discrete memoryless channel characterized by a set of symbol transition probabilities. Detailed measurements and simulations of Viterbi decoder performance have shown that output errors tend to occur in bursts. It is for this reason that the interleaving scheme is required. Without the interleaver, the superchannel would not be memoryless.

The superchannel output binary symbols are grouped into blocks of m to construct the 2^m-ary symbols for input to the R-S decoder. The R-S decoder performs the processing required to estimate the data source sequence. The R-S decoder is able to correct some of the decoding errors of the Viterbi inner decoder, thus improving communication reliability.

Concatenated coding systems have been used by NASA on recent deep-space missions [19]. Detailed analysis of concatenated coding systems can be found in Refs. 2, 7, 19, and 20.

7-4 Results for Specific Error Correction Codes

The discussions of Sections 7-2 and 7-3 were general and were intended to introduce the fundamental concepts of error correction coding. In this section specific error correction codes are considered. The codes selected for this discussion are a small sample of a very large range of codes which are currently available.

7-4.1 BCH Codes

The family of BCH codes are powerful linear block codes for which excellent decoding algorithms exist. This family of codes contains codes of many rates and a wide range of error correction capability. Note that the Hamming codes are single error-correcting binary BCH codes. All texts on error correction coding contain detailed discussions of this family of codes. The student is referred to Blahut [21] or Lin and Costello [1] for detailed encoder and decoder designs. The spread-spectrum system designer must understand two aspects of any error correction scheme. The first aspect is the complexity of the encoder and decoder. Extremely complex decoders can be implemented only at low speeds and therefore are not usable in many systems. The second aspect is the performance of the code in the anticipated communications environment. Only the latter aspect is considered here. Only binary BCH codes are considered. Most of the decoding algorithms for BCH codes are bounded distance decoders which correct up to E transmission errors but never correct more than E errors. The bit error probability for block codes using bounded distance decoding was given in (7-36). The values of n, k, and E for all known BCH codes up to length $n = 1023$ are given in Lin and Costello [1]. The code minimum distance d_{min} and E are related by

$$E = \begin{cases} \frac{1}{2}d_{min} - 1 & \text{for } d_{min} \text{ even} \\ \dfrac{d_{min} - 1}{2} & \text{for } d_{min} \text{ odd} \end{cases} \tag{7-57}$$

The error probability bound (7-36) is convenient in that knowledge of the code weight distribution is not required.

EXAMPLE 7-9

A BCH code with $n = 63$, $k = 30$, and $E = 6$ exists. This code has a rate of $30/63 = 0.4762$. This code is used in a FH/DPSK spread-spectrum system. Calculate the upper bound on bit error probability when the system is operating in worst-case tone jamming.

Solution: The channel symbol error probability in worst-case tone jamming is calculated using (6-121). When using this equation for coded systems, it gives the symbol (not bit) error probability and R is the symbol rate. That is,

$$p = \begin{cases} 0.5 & \dfrac{P}{J}\left(\dfrac{W}{R_s}\right) < 1 \\ \dfrac{1}{2(P/J)(W/R_s)} & 1 \le \dfrac{P}{J}\left(\dfrac{W}{R_s}\right) < \dfrac{W}{R_s} \\ 0 & \dfrac{W}{R_s} < \dfrac{P}{J}\left(\dfrac{W}{R_s}\right) \end{cases}$$

For the code being considered, the bit rate R and the symbol rate are related by $R_s = 63R/30$. Substituting for R_s in the equation above yields the BSC crossover probability as a function of $(P/J)(W/R)$. Using p in (7-36) yields a bound on information bit error probability. This result is plotted in Figure 7-17 together with the previously calculated result without error correction coding. The student is referred to Refs. 22 and 23 for additional analysis and detailed performance results for spread-spectrum systems using BCH codes in worst-case jamming environments.

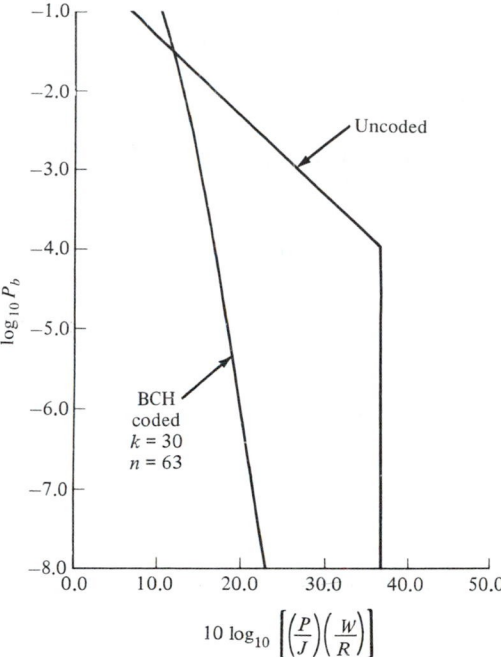

$$10 \log_{10} \left[\left(\frac{P}{J} \right) \left(\frac{W}{R} \right) \right]$$

FIGURE 7-17. Comparison of BCH coded and uncoded performance of FH/DPSK spread-spectrum system in worst-case tone jamming.

7-4.2 Reed–Solomon Codes

The Reed–Solomon (R-S) codes are nonbinary BCH codes which have found numerous applications in spread-spectrum systems. The encoder input is a message k-tuple made of k symbols from an alphabet of $q = 2^m$ symbols. The encoder output is a codeword n-tuple with symbols from the same q-ary alphabet. Observe that since the input–output alphabet size is a power of 2, input and output symbols may be represented by m-ary binary words. That is, the input message may be considered a km-bit binary vector and the output codeword a nm-bit binary vector. The R-S codes are capable of correcting t channel symbol errors where $n - k = 2t$. The block length of standard R-S codes is $n = q - 1$. R-S codes exist for $1 \leq k \leq n - 2$. The R-S codes can also be extended to have block length $n = q$ and $n = q + 1$.

The bit error probability for R-S codes is given in Clark and Cain [2]. The result is an overbound

$$P_b < \frac{2^{k-1}}{2^k - 1} \sum_{j=t+1}^{n} \frac{j + t}{n} \binom{n}{j} p_s^j (1 - p_s)^{n-j} \tag{7-58}$$

where p_s is the channel symbol error probability. The symbol error probability may be the error probability for an actual nonbinary channel such as in FH/MFSK or may be the probability of one or more binary errors in an m-bit word on a binary channel.

When the channel is binary, then

$$p_s = \sum_{e=1}^{m} \binom{m}{e} p^e (1 - p)^{m-e} \tag{7-59}$$

is used for the symbol error probability.

Reed–Solomon codes are often used in concatenated coding systems [7,24] along with a convolutional codes. Simon et al. [24] presents detailed performance results for R-S codes and concatenated R-S/convolutional codes for FH/MFSK systems in worst-case partial-band noise and multitone jamming. Simon's results show that the combination of a good convolutional code and a R-S code can completely eliminate the severe performance degradations due to partial-band jammers. Finally, Stark [8], Pursley and Stark [25], Dou and Milstein [26], Torierri [27], and Ma and Poole [22] all investigate the performance of R-S coding in various spread-spectrum systems.

EXAMPLE 7-10

Consider a R-S code having the parameters $n = 2^m - 1 = 63$, $k = 31$, $t = 16$, and $m = 6$. This code is used in a FH/DPSK spread-spectrum system. Calculate the error probability bound for worst-case tone jamming.

Solution: The worst-case jamming binary symbol error probability for this channel was given in Example 7-9. Since the encoder input and output alphabets are the same, the energy per transmitted information bit E_b is related to the energy per transmitted binary symbol E_s by $E_s = 31 E_b/63$. Therefore, substitute $R_s = 63R/31$ into the equation for p in the last example to obtain the BSC crossover probability as a function of $(P/J)(W/R)$. This p is used in (7-59) to obtain the symbol error probability p_s. The symbol error probability is then used in (7-58) to obtain the desired bound on bit error probability. The result of this calculation is illustrated in Figure 7-18 together with the comparable result without coding.

7-4.3 Maximum Free-Distance Convolutional Codes

Many researchers have used a variety of techniques to search for and find "good" convolutional codes. The quality factor used in these efforts is always a measure of the weight structure of the code. When maximum-likelihood decoding is used, the optimum weight structure is that which has the minimum number of bit errors in the paths through the code trellis which are closest to one another in Hamming distance. The weight structure of the best rate-$\frac{1}{2}$ and rate-$\frac{1}{3}$ codes in this sense were published in Odenwalder [28] and later in Clark and Cain [2]. For convenient reference, these results are presented in Table 7-3 for rate-$\frac{1}{2}$ codes and in Table 7-4 for rate-$\frac{1}{3}$ codes. The constraint length of the code is denoted by v and the number of states in the code trellis is 2^{v-1}. The free distance is the minimum Hamming distance between the codewords on any two paths through the trellis. The code generators are given in octal notation. This notation gives the connections between the encoder shift register stages and the modulo-2 adders. Consider, for example, the constraint length 7 rate-$\frac{1}{2}$

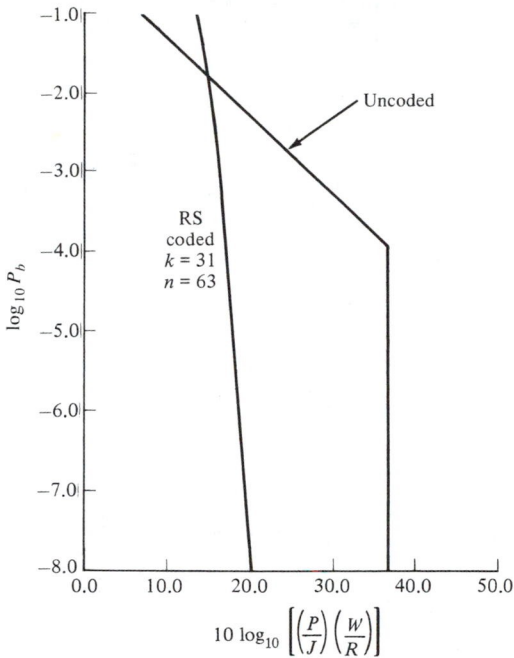

$$10 \log_{10} \left[\left(\frac{P}{J} \right) \left(\frac{W}{R} \right) \right]$$

FIGURE 7-18. Comparison of R-S coded and uncoded performance of FH/DPSK spread spectrum in worst-case tone jamming.

code. The generators are (171,133) in octal or (1111001,1011011) in binary. This means that connections to the first modulo-2 adder are from shift register stages 0, 1, 2, 3, and 6 and to the second modulo-2 adder are from shift register stages 0, 2, 3, 5, and 6. This encoder is illustrated in Figure 7-19.

The performance of these convolutional codes in spread-spectrum systems is found using (7-56), with c_d given in Table 7-3 or 7-4 and P_d given by (7-27) for hard-decision decoding by (7-55) for soft-decision decoding with the continuous AWGN channel, or by (7-49) for binary hard-decision decoding with jammer-state information. The result of these calculations for a FH/DPSK spread-spectrum system in worst-case tone jamming and using a rate-$\frac{1}{2}$ constraint length 5 and constraint length 7 codes are given in Figure 7-20.

7-4.4 Repeat Coding for the Hard-Decision FH/MFSK Channel

Repeat coding for FH/MFSK has been extensively investigated by many authors [8,22–25,27,29–43]. Many more references on this important subject could be cited. The abundance of literature in this area is due to the fact that excellent antijam performance can be obtained with reasonably economic hardware using frequency hopping with noncoherent MFSK modulation. As a result, a number of important military communications systems employ wideband frequency hopping and a great deal of effort was directed to the analysis of system performance during the design

TABLE 7-3. Best Rate-$\frac{1}{2}$ Convolutional Codes and Their Partial Weight Structure

Constraint Length, v	Code Generators	Free Distance, d_f	c_d for $d =$							
			d_f	$d_f + 1$	$d_f + 2$	$d_f + 3$	$d_f + 4$	$d_f + 5$	$d_f + 6$	$d_f + 7$
3	(7,5)	5	1	4	12	32	80	192	448	1,024
4	(17,15)	6	2	7	18	49	130	333	836	2,069
5	(35,23)	7	4	12	20	72	225	500	1,324	3,680
6	(75,53)	8	2	36	32	62	332	701	2,342	5,503
7	(171,133)	10	36	0	211	0	1,404	0	11,633	0
8	(371,247)	10	2	22	60	148	340	1,003	2,642	6,748
9	(753,561)	12	33	0	281	0	2,179	0	15,035	0

Source: Ref. 28.

TABLE 7-4. Best Rate-$\frac{1}{3}$ Convolutional Codes and Their Partial Weight Structure

Constraint Length, v	Code Generators	Free Distance, d_f	c_d for $d =$							
			d_f	$d_f + 1$	$d_f + 2$	$d_f + 3$	$d_f + 4$	$d_f + 5$	$d_f + 6$	$d_f + 7$
3	(7,7,5)	8	3	0	15	0	58	0	201	0
4	(17,15,13)	10	6	0	6	0	58	0	118	0
5	(37,33,25)	12	12	0	12	0	56	0	320	0
6	(75,53,47)	13	1	8	26	20	19	62	86	204
7	(171,145,133)	14	1	0	20	0	53	0	184	0
8	(367,331,225)	16	1	0	24	0	113	0	287	0

Source: Ref. 28.

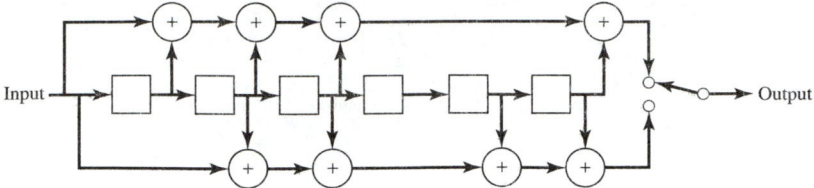

FIGURE 7-19. Rate-$\frac{1}{2}$, constraint-length-7 convolutional encoder.

phase of these systems. The references cited treat nearly every imaginable aspect of coded and uncoded noncoherent MFSK signaling. A few of these topics are reviewed in this section. Throughout this section worst-case two-state partial-band jamming is assumed. Repeat coding is very often referred to in the literature as time diversity. Repeat coding and time diversity are identical.

Consider a system using orthogonal MFSK and assume that the modulator divides each symbol interval into L chips and transmits the symbol L times using a different FH frequency for each chip. Using a different frequency for each chip can be accomplished by using a frequency-hop rate equal to L times the MFSK symbol rate or by using a slower hop rate with interleaving (see Section 7-4.5). Denote the MFSK uncoded symbol rate by R_s and the chip rate by $R_c = LR_s$. The jammer is smart and knows exactly how much power to transmit over a particular bandwidth

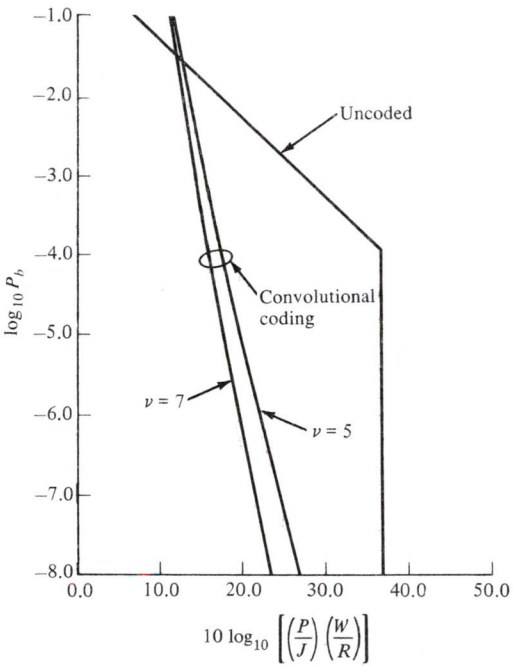

FIGURE 7-20. Comparison of uncoded and convolutionally coded performance of FH/DPSK spread spectrum in worst-case tone jamming.

or over a particular time interval to cause the maximum possible average chip error probability. Assume that the modem employs ideal interleaving so that all chip errors may be considered independent.

The transmitted signal is illustrated conceptually in Figure 7-21. For clarity, the FH pattern is not shown in this figure. The message to be transmitted is one of $M = 2^b$ frequencies, and is repeated L times as shown. The jammer optimizes its bandwidth or duty factor just as it would if normal uncoded FH/MFSK with symbol rate LR_s were being used. The receiver removes the spreading modulation in the normal manner and the despread signal and interference is input to an MFSK demodulator matched to the chip time $T_c = 1/R_c$ as illustrated in Figure 7-22. The demodulator is a bank of noncoherent matched-filter energy detectors; there is one energy detector output e_{ij} for each MFSK symbol i for each repeat code chip j. These energy detector outputs may be processed in many different ways to produce an estimate of the transmitted symbol. Several of these processing methods are reviewed in the following paragraphs.

Hard-Decision Decoding Without JSI. The most straightforward processing makes a hard MFSK symbol decision for each received chip. The hard-decision output for each chip is the MFSK symbol whose energy detector output is largest. In this case the chip error probability for worst-case partial band or pulsed noise jamming of FH/MFSK is calculated using identical steps to those used to calculate the bit error probability of (6-39). With appropriate adjustments to obtain MFSK symbol (as opposed to bit) error probability and to account for the reduced energy per chip due to the repeat coding, (6-39) becomes

$$
P_c = \begin{cases}
\dfrac{2(M-1)}{M}\dfrac{Lk'}{E_b/N_J} & \dfrac{E_b}{N_J} > L\left(\dfrac{E_b}{N_J}\right)_0 \\[3mm]
\dfrac{1}{M}\sum_{m=2}^{M}(-1)^m\binom{M}{m}\exp\left[\dfrac{bE_b}{LN_J}\left(\dfrac{1-m}{m}\right)\right] & \dfrac{E_b}{N_J} \le L\left(\dfrac{E_b}{N_J}\right)_0
\end{cases}
\tag{7-60}
$$

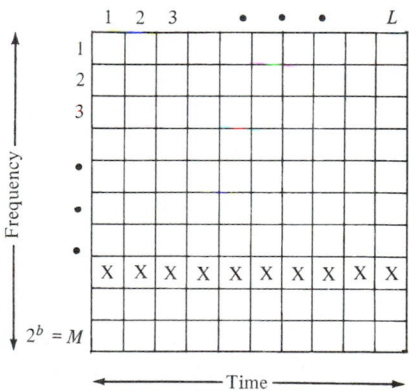

FIGURE 7-21. Transmitted signal for FH/MFSK with repeat coding.

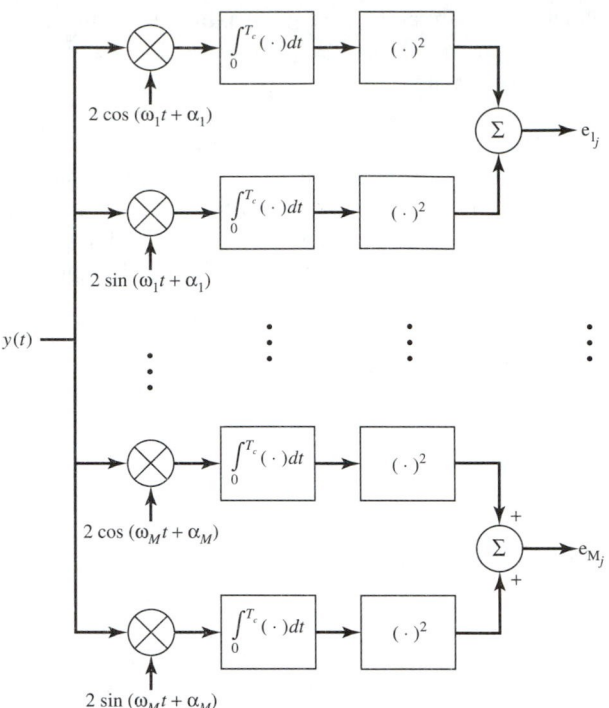

FIGURE 7-22.   MFSK demodulator.

where k' and $(E_b/J_J)_0$ are given in Table 6-1 and P_c denotes the channel chip error probability.

The receiver demodulator output is a sequence of L chip estimates for each information symbol. One possible received symbol sequence is shown in Figure 7-23. The decoding rule used by the receiver is to choose as the decoder output symbol the symbol with the largest number of votes out of L transmissions. For the example of Figure 7-23, the decoded output symbol is a six. Given that P_c is the received chip error probability, the probability p_e of an entry in the correct row and column of the decoding matrix is

$$p_e = 1 - P_c$$

and the probability of l entries in the correct row of the decoding matrix is

$$\Pr(l \text{ correct}) = \binom{L}{l} p_e^l (1 - p_e)^{L-1}$$

$$\equiv P_I(l) \tag{7-61}$$

A symbol error is made whenever the number of entries in *some* incorrect row of the decoding matrix is greater than the number of entries in the correct row. A symbol error *may* be made when the number of entries in an incorrect row equals the number of entries in the correct row.

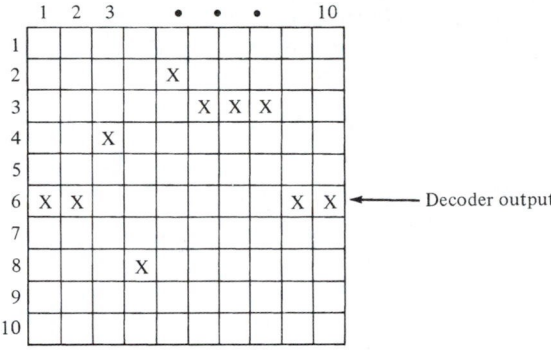

FIGURE 7-23. Possible FH/MFSK decoder input and decoder output estimate.

There are at most L entries in all rows of the decoding matrix. Let $P_{s|l}$ denote the symbol error probability given that there are l entries in the correct row of the decoding matrix. Then the total symbol error probability is

$$P_s = \sum_{l=0}^{L} P_{s|l} P_I(l) \tag{7-62}$$

With l entries in the correct row, there are $L - l$ entries in all incorrect rows. There are $(M - 1)L$ incorrect positions in the decoding matrix of which $(M - 1)l$ cannot contain entries since there are l entries in the correct row. Thus there are a total of $(M - 1)(L - l)$ incorrect positions in the decoding matrix which contain $L - l$ entries. It can be shown that the probability of j entries in a block which has $(L - l)$ elements, given a total of $(L - l)$ entries in $(M - 1)(L - l)$ elements is given by

$$q(j|l) = \frac{\dbinom{L - l}{j}\dbinom{(M - 2)(L - l)}{L - l - j}}{\dbinom{(M - 1)(L - l)}{L - l}} \tag{7-63}$$

Two cases must be considered when calculating total symbol error probability. First consider binary FSK, where whenever a chip error is made an entry appears in the one incorrect row of the decoding matrix. For L even,

$$P_s = \sum_{l=0}^{(L/2)-1} P_I(l) + \frac{1}{2} P_I\left(\frac{L}{2}\right) \tag{7-64}$$

and for L odd,

$$P_s = \sum_{l=0}^{(L-1)/2} P_I(l) \tag{7-65}$$

The first expression is based on the fact that a decoding error always occurs when there is from zero to $(L/2) - 1$ entries in the correct row and occurs with probability $\frac{1}{2}$ when there are $L/2$ entries in the correct row. The second expression is similar but accounts for the fact that there cannot be $L/2$ entries in any row since L is odd. For $M \geq 4$ the error probability expression is, for L even:

$$P_s = P_l(0) + \sum_{l=1}^{L/2-1} (M - 1)P_l(l)\left[\frac{1}{2}q(l|l) + \sum_{j=l+1}^{L-l} q(j|l)\right]$$

$$+ (M - 1)P_l\left(\frac{L}{2}\right)\frac{1}{2}q\left(\frac{L}{2}\middle|\frac{L}{2}\right)$$

(7-66)

and for L odd,

$$P_s = P_l(0) + \sum_{l=1}^{(L-1)/2} (M - 1)P_l(l)\left[\frac{1}{2}q(l|l) + \sum_{j=l+1}^{L-l} q(j|l)\right]$$

(7-67)

The first term in the first equation accounts for the case where there are no entries in the correct row, so some other row (incorrect) must have more entries than the correct row. The last term in the first equation accounts for the case where there are $L/2$ entries in the correct row. An error then occurs with probability $\frac{1}{2}$ only when some incorrect row also has $L/2$ entries. The probability of a particular incorrect row having $L/2$ entries is $q(L/2|L/2)$ and the probability of any one of the $(M - 1)$ incorrect rows having $L/2$ entries is $(M - 1)q(L/2|L/2)$. The first summation in the first equation is over all remaining values of l for which errors can occur. The term in brackets is the probability of a particular incorrect row being decoded given l entries in the correct row. The $(M - 1)$ factor with the middle term accounts for the $(M - 1)$ possible incorrect rows.

Using the equations derived here, the symbol error probability for diversity transmission and worst-case jamming can be calculated. Symbol error probability for orthogonal symbols is related to information bit error probability by

$$P_b = \frac{M}{2(M - 1)}P_s$$

(7-68)

so that bit error probability can also be calculated. The results of this calculation are illustrated in Figure 7-24 for binary FSK. Observe that significant performance improvements are obtained with this simple repeat coding procedure, and that a different L is optimum for each input signal-to-jammer power ratio.

Hard-Decision Decoding with JSI. The availability of jammer-state information complicates the analysis significantly. To facilitate this discussion, consideration is limited to binary FSK so that results derived in Section 7-2.3 will be applicable. As in Section 7-2.3, the jammer-state $\mathbf{z} = (z_1, z_2, z_3, \ldots, z_L)$ and the binary symmetric channel error probabilities p_0 and p_1 corresponding to jammer states 0 and 1 are known to the receiver. The L-repeat code is a two-codeword linear code with block length L. The two codewords are $\mathbf{x}_0 = (0,0,0, \ldots, 0)$ and $\mathbf{x}_1 = (1,1,1, \ldots, 1)$. The Hamming distance between these two codewords is L. In Section 7-2.3

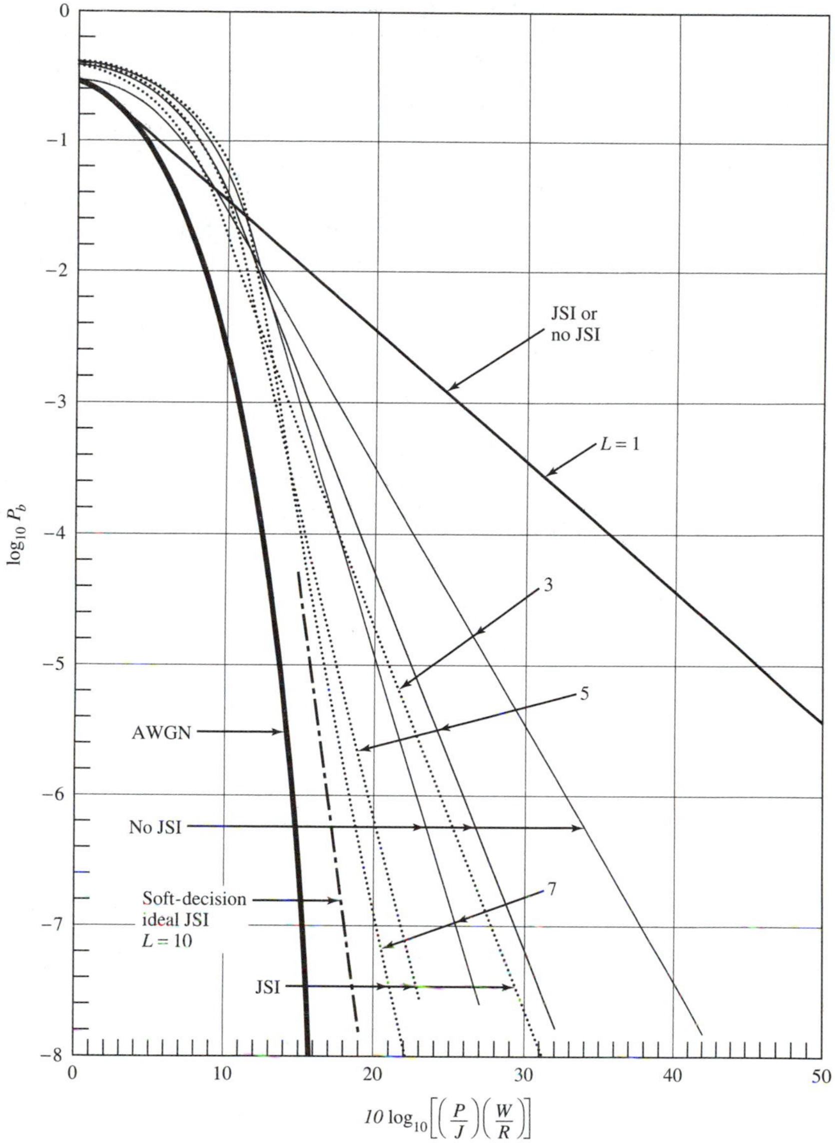

FIGURE 7-24. Performance of FH/BFSK in worst-case partial-band jamming with L-repeat code with and without jammer-state information. All curves are for hard decisions except $L = 10$.

the average block decoding error probability was calculated for any linear binary block code; those general results are directly applicable to this simple binary block code.

The hard-decision demodulator output is the vector $\mathbf{y} = (y_1, y_2, y_3, \ldots, y_L)$ with $y_j \in \{0,1\}$. The block decoder calculates the weighted Hamming distance

$$d_1 \ln \frac{p_1}{1 - p_1} + d_0 \ln \frac{p_0}{1 - p_0} \tag{7-16}$$

between the vector $\mathbf{y}$ and both codewords and estimates the transmitted codeword to be the codeword for which (7-16) is largest. In (7-16) recall that d_1 is the Hamming distance between $\mathbf{y}$ and $\mathbf{x}$ for the symbols that are jammed (i.e., $z_j = 1$) and d_0 is the Hamming distance between $\mathbf{y}$ and $\mathbf{x}$ for the symbols that are not jammed. For this two-codeword code, a block decoding error results in a single bit error, so that bit and block error probabilities are the same. Average block/bit error probability will be calculated.

The desired average block error probability is given by (7-51), which for the repeat code becomes

$$P_B(\rho) = P_{Jd}(L,\rho) \tag{7-69}$$

where there is only one term in the summation of (7-51) and $d(0,1) = L$ for that term. The inequality of (7-51) has been replaced by an equality in (7-69) since there is only one term in the union bound for this degenerate case. $P_{Jd}(L,\rho)$ is calculated using (7-49), which for this case becomes

$$P_{Jd}(L,\rho) = \sum_{n_{Jd}=0}^{L} \binom{L}{n_{Jd}} \rho^{n_{Jd}}(1 - \rho)^{L-n_{Jd}} P_B'[L,n_{Jd}] \tag{7-70}$$

In this equation $P_B'(L,n_{Jd})$ is a special case of (7-46)

$$P_B'(L,n_{Jd}) = \sum_{d_1'=0}^{n_{Jd}} \sum_{d_0'=0}^{L-n_{Jd}} \beta(d_1',d_0') \binom{n_{Jd}}{d_1'}$$

$$\times p_1^{d_1'}(1 - p_1)^{n_{Jd}-d_1'} \binom{L - n_{Jd}}{d_0'} p_0^{d_0'}(1 - p_0)^{L-n_{Jd}-d_0'} \tag{7-71}$$

where d_1' and d_0' are indices here which have the same meaning as in (7-16); $\beta(d_1',d_0') = 1$ if $\delta < \alpha(L,n_{Jd})$, $\beta(d_1',d_0') = \frac{1}{2}$ if $\delta = \alpha(L,n_{Jd})$, and $\beta(d_1',d_0') = 0$ otherwise;

$$\delta = d_1' \ln \frac{p_1}{1 - p_1} + d_0' \ln \frac{p_0}{1 - p_0} \tag{7-72}$$

and

$$\alpha(L,n_{Jd}) = \frac{n_{Jd}}{2} \ln \frac{p_1}{1 - p_1} + \frac{L - n_{Jd}}{2} \ln \frac{p_0}{1 - p_0} \tag{7-73}$$

Using (7-69) through (7-73) the bit error probability for the repeat code may be calculated as a function of the jammer parameter ρ. The parameter ρ is removed by acknowledging that the jammer will choose ρ to maximize bit error probability. Recall from Section 6-3.2 that the partial-band jammer against FH/MFSK will select ρ according to

$$\rho_{opt} = \frac{y_0}{E_b/N_J} \tag{7-74}$$

where $y_0 = 2.0$ from Table 6-1. Assume that the jammer uses this same strategy for the repeat code, with the exception that E_b is replaced by $E_c = E_b/L$, the chip energy for each repetition of the repeat code. Using this optimum (for the jammer) ρ_{opt}, the repeat code bit error probability may be calculated as a function of E_b/N_J. Figure 7-24 illustrates the results of this calculation along with the results of the preceding section for the repeat code without jammer-state information. Observe that the performance with and without JSI is identical for $L = 1$; however, significant performance improvements are seen for other values of L.

The analysis of this section is from Stark [8]. In Ref. 8, however, it is noted that y_0 used in the calculation of the optimum ρ is a function of E_b/N_0, and corrected values of y_0 for several cases are given. The results given above closely approximate those in Figure 4 of Ref. 8 for large E_b/N_0. Stark's results should be used where high accuracy is required. Finally, note that communicator may select L to minimize bit error rate just as the jammer selected ρ to maximize bit error rate. For a particular E_b/N_J and M, there is an optimum L, and bit error rate degrades for L larger or smaller than this optimum.

Soft-Decision Decoding with JSI. Consider the L-repeat code using FH/MFSK with soft-decision decoding and jammer-state information $\mathbf{z} = (z_1, z_2, z_3, \ldots, z_L)$. The soft-decision output of the MFSK demodulator of Figure 7-22 is a vector $\mathbf{y} = (\mathbf{e}_1, \mathbf{e}_2, \mathbf{e}_3, \ldots, \mathbf{e}_L)$, where $\mathbf{e}_j = (e_{1j}, e_{2j}, e_{3j}, \ldots, e_{Mj})$ and e_{mj} is the output of the mth energy detector of Figure 7-22 for the jth code symbol. Given $\mathbf{z}$ and $\mathbf{y}$, the task of the decoder is, as usual, to make the best possible estimate of the transmitted code word. This problem has been analyzed extensively [24, Chap. 2; 30; 33–43]. Many different strategies for processing the soft-decision information are presented in these references. All of these analyses are beyond the scope of this book and therefore results are presented here without analytical justification. The following result is from Simon et al. [24] but was originally from Trumpis [43].

Consider a system operating at very high E_b/N_0 so that thermal noise may be neglected. Thus assume that if any of the symbols of the L-repeat code are received without jamming, the decoder makes a perfect decision (i.e., no decoding error is made). Assume further that $\mathbf{z}$ is generated by the receiver for each received symbol j by examining $\mathbf{e}_j$ and declaring $z_j = 0$ if a single energy detector output is high and $z_j = 1$ if two or more energy detector outputs are high. It is possible for this decoder to make an error only if all L received symbols are jammed. If all L received symbols are jammed, the decoder examines the sums

$$\Lambda_m = \sum_{j=1}^{L} e_{mj} \tag{7-75}$$

for all possible $m = 1, 2, \ldots, M$ and outputs the m for which Λ_m is largest. The exact bit error probability for this system has been calculated by Trumpis [43]. The result is

$$P_b = \beta \left[\frac{LN_J}{(\log_2 M) E_b} \right]^L \tag{7-76}$$

TABLE 7-5. Parameters for the Exact Performance of FH/MFSK Signals with Diversity L in Worst-Case Partial-Band Jamming

L	M = 2 β	M = 2 γ (dB)	M = 4 β	M = 4 γ (dB)	M = 8 β	M = 8 γ (dB)	M = 16 β	M = 16 γ (dB)	M = 32 β	M = 32 γ (dB)
2	0.4168	6.7	0.5959	4.2	0.8575	2.8	1.0796	2.0	1.3659	1.4
3	0.5210	8.8	0.8265	6.1	1.2565	4.6	1.8198	3.7	2.5937	3.0
4	0.6797	10.2	1.1401	7.4	1.8320	5.9	2.8251	4.8	4.1493	4.1
6	1.2392	12.1	2.2245	9.2	3.8464	7.6	6.4370	6.5	10.3502	5.7
8	2.3584	13.5	4.4101	10.5	8.0165	8.9	14.0997	7.7	24.0142	6.8
10	4.6110	14.5	8.8385	11.5	16.5774	9.8	30.2829	8.7	61.6169	7.8

Source: Derived by Trumpis [43] as reported by Simon et al. [24].

provided that $E_b/N_J > \gamma$, where β and γ are given in Table 7-5 for various values of M and L. Bit error probability calculated using (7-76) is plotted in Figure 7-24 for $M = 2$ and $L = 10$. The bit error probability for BFSK in additive white Gaussian noise has also been plotted in Figure 7-24. Observe that the repeat code with soft-decision decoding and jammer-state information has nearly eliminated the advantage that the jammer gained through the use of partial-band techniques. At a bit error rate of 10^{-5}, this simple coding strategy has reduced the required signal power by approximately 30 dB relative to the power needed for binary FSK without coding in worst-case partial-band jamming. Using this technique, the required E_b/N_J is only 2.5 dB more than the E_b/N_J needed for the wideband barrage noise jammer (AWGN). Finally, note that this coding strategy has replaced the inverse-linear relationship between bit error rate and E_b/N_J, which was achieved by the optimum partial-band jammer, by an exponential relationship characteristic of MFSK performance in wideband stationary noise.

It has been noted by Viterbi [29] and Simon et al. [24, Chap. 2] that for FH/MFSK repeat coding using soft-decision decoding with optimum partial-band jamming and with known jammer state, there is an optimum L that is a function of E_b/N_J and the number of MFSK tones per symbol. Levitt [31] demonstrates that there is also an optimum L value for multitone jamming. Thus the jammer and the communicator both optimize their system, making the other's task as difficult as possible. This joint optimization game is discussed in detail in Ref. 24.

7-5 Interleaving

Most forward error correction codes perform well only when the channel errors are completely independent from one signaling interval to the next. All the bounds on bit error probability discussed in Sections 7-2 through 7-4 were based on the assumption that the channel is memoryless. The worst-case jammers analyzed in Chapter 6 are the pulsed noise or pulsed tone jammers for direct-sequence systems or the partial band noise or tone jammers for frequency-hop systems. These jammers produce bursts of errors and therefore the channel is not memoryless. To counter this difficulty, an interleaver is placed between the encoder and the modulator and a de-interleaver is placed between the demodulator and the decoder.

One of two types of interleaving commonly used is called *block interleaving*. A block interleaver uses four N-row by B-column random access memories to randomize errors. Two of these memories are in the transmitter and the others are in the receiver. The transmitter reads encoder output symbols into a memory by columns until it is full. Then the memory is read out to the modulator by rows. While one memory is filling the other is being emptied, so two memories are needed. In the receiver, the inverse operation is effected by reading the demodulator output into a memory by rows and reading the decoder input from the memory by columns.

This operation is illustrated in Figure 7-25 for a $N = 10$, $B = 10$ interleaver. The encoder output symbols are numbered consecutively 1 through 100. These symbols are transmitted over the channel by rows. Thus the order of transmission for the first 20 symbols is 1, 11, 21, 31, 41, 51, 61, 71, 81, 91, 2, 12, 22, 32, 42, 52, 62, 72, 82,

1	11	21	31	41	51	61	71	81	91
2	12	22	32	42	52	62	72	82	92
3	13	23	33	43	53	63	73	83	93
4	14	24	34	44	54	64	74	84	94
5	15	25	35	45	55	65	75	85	95
6	16	26	36	46	56	66	76	86	96
7	17	27	37	47	57	67	77	87	97
8	18	28	38	48	58	68	78	88	98
9	19	29	39	49	59	69	79	89	99
10	20	30	40	50	60	70	80	90	100

$B = 10$

$N = 10$

FIGURE 7-25. Operation of a block interleaver.

92. . . . . A burst of channel errors is assumed to hit channel symbols 41, 51, 61, 71, 81 and symbols 29, 39, 49, 59, 69, as shown by the shaded blocks. The demodulator output is read into the memory by rows resulting in the same array of symbols shown in Figure 7-25. This memory is then read into the decoder by columns. The end result of this operation is that adjacent channel errors are spaced by N symbols at the decoder input. N is chosen to be large enough that errors spaced by N will affect different codewords and may be considered independent.

Another type of interleaving is convolutional interleaving. A convolutional interleaver is depicted in Figure 7-26. All multiplexers in the transmitter and receiver are assumed to operate synchronously. The multiplexer switches change position after each symbol time so that successive encoder outputs enter different rows of the interleaver memory. Each interleaver and de-interleaver row is a shift register mem-

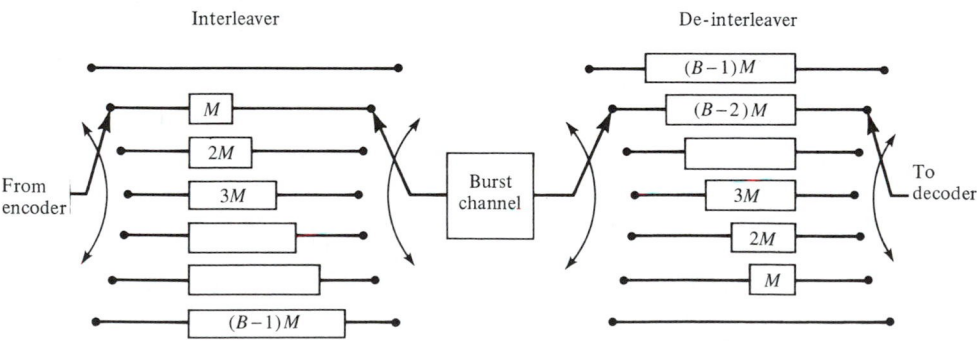

FIGURE 7-26. Operation of a convolutional interleaver.

ory with the number of units of delay shown within the rectangle. The first encoder output enters the top interleaver row, is transmitted over the channel immediately, and enters the de-interleaver memory of $(B - 1)M$ symbols. The second encoder output enters the second row of the interleaver and is delayed M symbol times before transmission. Thus adjacent encoder outputs are transmitted M symbol times apart and are not affected by the same channel error burst. Upon reception, the second encoder symbol is delayed by an additional $(B - 2)M$ symbol time for a total delay of $(B - 1)M$. Observe that all symbols have the same delay after passing through both the interleaver and de-interleaver, so that the decoder input symbols are in the same order as the encoder output symbols.

Interleaving is an essential part of all spread-spectrum systems which are expected to operate in worst-case jamming environments. Additional discussion on the subject of interleaver design can be found in Clark and Cain [2], Forney [18], and Cain and Geist [44].

7-6 Coding Bounds

7-6.1 Error Probability Bounds Using the Channel Parameter D

The estimation of error probability for coded spread-spectrum communications systems may be simplified through the use of well-known bounds on decoded error probability. These bounds are based upon the pioneering work of Shannon [45]. Use of these bounds will yield error rate estimates that are larger than the estimates using, for example, (7-30) with the two-codeword error probability given by (7-27). However, these bounds are easier to evaluate, and yield results that are pessimistic by only a few decibels in signal-to-noise ratio [24,46]. The derivation of these bounds is beyond the scope of this book, so that results will be stated without proof but with references to original source material. In this section, discussion is limited to binary input channels.

As always, the communications path between the encoder output and the decoder input is a memoryless channel that is defined by the transition probabilities $p(y_j | x_{mj}, z_j)$ when jammer-state information is available or $p(y_j | x_{mj})$ when jammer state information is not available. These transition probabilities characterize all of the functions within the dashed box of Figure 6-1; all of these functions define the *coding channel* for the communications system [46]. In the development of the optimum decoding rule for both block and convolutional codes, it was demonstrated that the optimum decoder output is the message (either codeword or convolutional code trellis path) that was most likely to have been transmitted given the received sequence $\mathbf{y}$ and, if available, the jammer-state sequence $\mathbf{z}$. To estimate this most likely message, a decoding metric was employed. For the binary symmetric channel without JSI, this decoding metric was the Hamming distance between $\mathbf{y}$ and a codeword $\mathbf{x}_m$. For the continuous-output binary-input channel without JSI, the decoding metric was the Euclidean distance between $\mathbf{y}$ and $\mathbf{x}_m$. When JSI was available, the decoding metrics were weighted Hamming distance and weighted Euclidean distance. In all instances the metric was a sum related to $\ln\{p(y_j | x_{mj}, z_j)\}$ over all channel use indices j of the codeword or trellis path. Omura and Levitt [46] have generalized this concept to include other, possibly non-maximum-likelihood decoding metrics. Define a general decoding metric $m(y_j, x_{mj}, z_j)$ for each channel use of a memoryless channel. The decoding metric for a particular message is

$$m(\mathbf{y}, \mathbf{x}_m, \mathbf{z}) = \sum_{j=1}^{n} m(y_j, x_{mj}, z_j) \tag{7-77}$$

When $m(y_j, x_{mj}, z_j)$ is the logarithm of the transition probability, the decoding metric is a maximum-likelihood metric.

Consider the block error probability bound (7-30)

$$P_B \leq \sum_{d=d_{\min}}^{n} A_d P_d \tag{7-30}$$

In this result A_d is a function of the particular block code being evaluated and P_d is the two-codeword error probability for any codeword pair separated in Hamming distance by $d(m, m') = d$. It can be shown [3,11,24,46] that for antipodal and orthogonal signal waveforms

$$P_d \leq D^d \tag{7-78}$$

where D is the *channel parameter,* which is a function only of the channel transition probabilities and the decoding metric $m(y_j, x_{mj}, z_j)$. The channel parameter D has been calculated for many different modulation channels and decoding metrics [24,46–49] with and without JSI. Channels and associated decoding metrics having smaller values of D have lower P_d and therefore lower P_B. Through the evaluation of D, the communications system engineer can perform trade-off studies for modulation/demodulation strategies independent of the details of any particular forward error correction code.

Having selected a good modulation and decoding metric, the performance of particular codes may be found by substituting (7-78) into (7-30) to find

$$P_B \leq \sum_{d=d_{\min}}^{n} A_d D^d \tag{7-79}$$

Similarly, the bit error probability for Viterbi decoding of binary rate-$1/n$ convolutional codes (7-56) is overbounded by

$$P_b \leq \sum_{d=d_{\text{free}}}^{\infty} c_d D^d \tag{7-80}$$

For the convenience of the reader, the following formulas for the channel parameter D for binary input channels are presented without proof. The reader is encouraged to consult the references for justification and further detail.

Barrage Noise Jamming. For the binary symmetric hard-decision channel with error probability p using the Hamming distance maximum-likelihood decoding metric, the channel parameter D is [24,46,50]

$$D = \sqrt{4p(1 - p)} \tag{7-81}$$

The Hamming distance decoding metric for this case is

$$m(y_j, x_{mj}) = \begin{cases} 1 & y_j = x_{mj} \\ 0 & y_j \neq x_{mj} \end{cases} \tag{7-82}$$

The Gaussian noise interference for this channel is continuously present at a constant level so that jammer-state information is irrelevant. This result is valid for BPSK, QPSK, and binary-FSK modulation with appropriate calculation of the channel error probability. Channel error probability must be calculated using the correct (i.e., accounting for code rate) channel symbol energy.

For the same AWGN interference, consider BPSK data modulation using a continuous output soft-decision demodulator and the correlation metric (7-11d). In this case, the channel parameter is [24,46]

$$D = e^{-E_c/N_J} \tag{7-83}$$

where E_c/N_J is the symbol energy/jammer power spectral density ratio for the channel. Consider binary-FSK modulation using the soft-decision demodulator illustrated in Figure 7-22. Since binary FSK is being considered, the demodulator has two noncoherent energy detectors and the output for each channel use is a vector $\mathbf{y}_j = (e_{0j}, e_{1j})$. The decoding metrics for message symbols 0 and 1 are $m(\mathbf{y}_j, 0) = e_{0j}$ and $m(\mathbf{y}_j, 1) = e_{1j}$. Using this metric, the channel parameter is [24]

$$D = \min_{0 \leq \lambda \leq 1} \left\{ \frac{1}{1 - \lambda^2} \exp\left[-\frac{\lambda}{1 + \lambda}\left(\frac{E_c}{N_J}\right) \right] \right\} \tag{7-84}$$

The minimization over λ must, unfortunately, be done for each value of signal-to-noise ratio.

Pulse or Partial-Band Jamming Without JSI. Consider a two-state binary symmetric hard-decision channel due to pulse or partial-band jamming. Ignore thermal noise so that the BSC error probability is zero when the channel use is not jammed. Denote the BSC error probability when the channel use is jammed by p_1. As usual, denote the probability that a channel use is jammed by ρ. Since the demodulator does not know which received symbols are jammed, the Hamming distance decoding metric (7-82) is used. Hamming distance is calculated over all channel uses. The channel parameter D is [24]

$$D(\rho) = \sqrt{4\rho p_1(1 - \rho p_1)} \tag{7-85}$$

The channel parameter is a function of ρ both directly and through the calculation of p_1. As always, the jammer controls ρ so that calculation of worst-case performance must include the calculation of this worst-case ρ for each signal-to-interference ratio. The calculation of the worst-case ρ may be done numerically.

Soft-decision decoding without jammer-state information is not recommended since if the jammer is permitted to use the worst-case ρ, the jammer is always able to degrade error probability to unacceptable levels. The student is referred to [24] for the calculation of the channel parameter and for further analysis.

Pulse or Partial-Band Jamming with JSI. Consider the same two-state hard-decision BSC considered immediately above. In this case, jammer-state information is used to enable weighting the Hamming distance for the jammed symbols differently from the nonjammed symbols as in (7-16). Note, however, that the Hamming distance accumulated on the jammed symbols cannot be ignored. If all jammed symbols were ignored, the jammer could disable all communications merely by being "on" at any transmit power J. Thus the demodulator is assumed to give Hamming distance accumulated on unjammed symbols significantly more weight than Hamming distance accumulated on jammed symbols. Then the channel parameter is [24]

$$D(\rho) = \rho\sqrt{4p_1(1 - p_1)} \tag{7-86}$$

This result is valid for BPSK, QPSK, and binary-FSK modulation with appropriate calculation of the channel error probability for the jammed symbols. The channel parameter is a function of ρ both directly and through p_1. The worst-case ρ maximizes $D(\rho)$ and may be determined numerically.

Now consider BPSK data modulation with soft-decision demodulation and perfect jammer-state information. Assume that thermal noise on the nonjammed symbols is negligible. The decoding metric is a weighted correlation as in (7-20), with much more weight given to the unjammed symbols than to the jammed symbols. As in the hard-decision case, the jammed symbols may not be given zero weight since this would permit the jammer to disable communications using any transmit power J. With these assumptions, the channel parameter [24] is

$$D(\rho) = \rho \exp\left(-\rho\frac{E_c}{N_J}\right) \tag{7-87}$$

where again maximization of $D(\rho)$ is required to determine worst-case performance. In this case the worst-case ρ may be calculated by taking the derivative of (7-87), setting the result equal to zero, and then solving for ρ. The result is $\rho_{wc} = (E_c/N_J)^{-1}$ for $E_c/N_J \geq 1$, and substituting this value into (7-87) yields

$$D(\rho_{wc}) = \left(\frac{E_c}{N_J}\right)^{-1} e^{-1} \tag{7-88}$$

Finally, consider binary FSK modulation using the soft-decision demodulator illustrated in Figure 7-22. As before, the decoding metrics for message symbols 0 and 1 are $m(\mathbf{y}_j, 0) = e_{0j}$ and $m(\mathbf{y}_j, 1) = e_{1j}$; a weighted sum of these metrics is used as the message decoding metric. The channel parameter is [24]

$$D = \min_{0 \leq \lambda \leq 1}\left\{\frac{\rho}{1 - \lambda^2}\exp\left[-\frac{\lambda\rho}{1 + \lambda}\left(\frac{E_c}{N_J}\right)\right]\right\} \tag{7-89}$$

7-6.2 Computational Cutoff Rate R_0

The computational cutoff rate [3,13,51] (also called the exponential bound parameter) R_0 is another indicator of the quality of the memoryless coding channel defined by the transition probabilities $p(y_j \mid x_{mj})$. Using random coding arguments it can be shown that block codes with block length n and code rate $R = k/n$ exist for which the decoded bit error probability P_b using maximum-likelihood decoding satisfies [13,52]

$$P_b \leq 2^{-n(R_0 - R)} \tag{7-90}$$

where R_0 is the computational cutoff rate. For binary input channels, the code rate R is the number of information bits that are transmitted over the channel for every binary channel use. Of course, the communicator would like R to be as large as possible. The upper bound (7-90) states that the code rate R can be no larger than R_0 if the communicator expects to find a block code that will be effective in reducing decoded bit error probability. Thus the value of R_0 is a valuable measure of the goodness of the coding channel. For convolutional codes, random coding arguments can be used to demonstrate the existence of time-varying codes whose decoded bit error probability satisfies [3,11,52]

$$P_b < \frac{2^{-\nu R_0 / R}}{\{1 - 2^{-[(R_0/R) - 1]}\}^2} \tag{7-91}$$

where ν is the constraint length of the code. This bound is valid only for $R < R_0$. Again, the code rate R can be no longer than R_0 if the communicator expects to find a convolutional code that will be effective in reducing bit error probability.

For a stationary binary symmetric channel with error probability p, the computational cutoff rate is [13]

$$R_0 = -\log_2 [\tfrac{1}{2} + \sqrt{p(1 - p)}] \tag{7-92}$$

For BPSK data modulation using a continuous-output soft-decision demodulator, the computational cutoff rate is [13]

$$R_0 = 1 - \log_2 \left[1 + \exp\left(-\frac{E_c}{N_J} \right) \right] \tag{7-93}$$

For both the hard- and soft-decision demodulators the cutoff rate increases as signal-to-noise ratio increases to a maximum of 1 information bit per binary channel use. The transmission of a larger number of information bits per channel use requires that the channel input alphabet be larger [13].

The computational cutoff rate for binary channels is related to the channel parameter by [24,47]

$$R_0 = 1 - \log_2(1 + D) \tag{7-94}$$

Using this formula, the cutoff rate may be found for all the cases considered in the preceding section. When D is a function of the parameter ρ, R_0 is also a function of ρ and worst-case performance is determined by *minimization* of R_0 over all possible

ρ. Often, the computation cutoff rate, rather than the channel parameter D, is used as a figure of merit for the comparison of modulation and demodulation strategies. Omura and Levitt [46] present plots of R_0 as a function of received signal-to-noise ratio for hard- and soft-decision demodulation of BPSK with barrage and partial-time jamming with and without jammer-state information. These plots show the relative performance gained by the additional system complexity required for soft decisions and/or the use of jammer-state information.

7-7 Summary

Many results have been presented in this chapter. These results enable the system designer to evaluate the performance of coded spread-spectrum communications systems for a wide variety of conditions. These conditions include communications in the presence of worst-case partial-band jamming for FH systems and worst-case partial-time jamming for DS systems. It has been shown that using the very simple repeat code with interleaving, the worst-case partial-band or partial-time jammer can be largely defeated. The repeat code may be viewed as transmitting the energy associated with a single information bit a little at a time using different frequencies and/or different time periods. That is, the repeat code enables diversity transmission. The use of diversity transmission in turn forces the jammer to transmit over a wider bandwidth and/or to transmit a larger fraction of the time. In summary, the use of diversity moves the optimum jamming strategy from a partial band or time strategy toward a full-band or full-time strategy. A goal of the spread-spectrum system designer is to develop a modulation and coding strategy for which the worst-case jammer is a wideband barrage noise jammer. Using coding techniques that are more powerful than the repeat code, the system designer can not only defeat the worst-case jammer but can achieve a communications reliability (i.e., bit error rate) superior to that which would be achieved in wideband barrage noise jamming. An extremely important conclusion of Chapters 6 and 7 is that forward error correction coding and interleaving are essential functions in spread-spectrum communications systems operating in the presence of smart jammers.

In conventional (i.e., non-spread-spectrum) communications systems, the employment of forward error correction (FEC) requires either that the transmission bandwidth be increased or that the information bit rate be decreased to accommodate the redundancy of the code. In contrast, the employment of FEC in spread-spectrum systems does not require bandwidth expansion or data rate reduction. Further, the processing gain due to the spreading modulation is not decreased by the use of forward error correction. Viterbi showed that this fact is true for both direct-sequence and frequency-hop systems in his famous paper [52] "Spread Spectrum Communications: Myths and Realities."

Finally, the subjects addressed in this chapter describe only a small portion of the research results regarding coded spread-spectrum communications. A comprehensive study of this subject may be found in Simon et al. [24], from which some of the material for this chapter has been extracted. Much additional information is available for the references cited in this chapter and in Chapter 6. In addition to presenting theoretical background, some of these references [8,22–25] present comparisons of the signal-to-jammer power ratio necessary to achieve a particular communications bit error rate for specific codes and jamming strategies.

References

[1] S. LIN and D. J. COSTELLO, JR., *Error Control Coding: Fundamentals and Applications* (Englewood Cliffs, N.J.: Prentice Hall, 1983).

[2] G. C. CLARK and J. B. CAIN, *Error-Correction Coding for Digital Communications* (New York: Plenum, 1981).

[3] A. J. VITERBI and J. K. OMURA, *Principles of Digital Communication and Coding* (New York: McGraw-Hill, 1979).

[4] W. W. PETERSON and E. J. WELDON, *Error-Correcting Codes* (Cambridge, Mass.: MIT Press, 1972).

[5] R. G. GALLAGER, *Information Theory and Reliable Communication* (New York: Wiley, 1968).

[6] E. R. BERLEKAMP, *Algebraic Coding Theory* (New York: McGraw-Hill, 1968).

[7] J. P. ODENWALDER, "Error Control," in *Data Communications, Networks, and Systems,* Thomas Bartee, ed. (Indianapolis, Ind.: Howard W. Sams, 1985).

[8] W. E. STARK, "Coding for Frequency-Hopped Spread-Spectrum Communication with Partial-Band Interference: Part II. Coded Performance," *IEEE Trans. Commun.,* Vol. COM-33, pp. 1045–1057, October 1985.

[9] G. D. FORNEY, JR., "The Viterbi Algorithm," *Proc. IEEE,* Vol. 61, pp. 268–278, March 1973.

[10] A. J. VITERBI, "Error Bounds for Convolutional Codes and an Asymptotically Optimum Decoding Algorithm," *IEEE Trans. Inf. Theory,* Vol. IT-13, pp. 260–269, April 1967.

[11] A. J. VITERBI, "Convolutional Codes and Their Performance in Communications Systems," *IEEE Trans. Commun. Technol.,* Vol. COM-19, pp. 751–772, October 1971.

[12] G. D. FORNEY, JR., "Convolutional Codes: II. Maximum Likelihood Decoding," *Inf. Control,* Vol. 25, pp. 222–226, July 1974.

[13] J. M. WOZENCRAFT and I. M. JACOBS, *Principles of Communications Engineering* (New York: Wiley, 1965).

[14] A. J. VITERBI and J. K. OMURA, *Principles of Digital Communication Engineering* (New York: McGraw-Hill, 1979).

[15] J. A. HELLER and I. M. JACOBS, "Viterbi Decoding for Satellite and Space Communications," *IEEE Trans. Commun. Technol.,* Vol. COM-19, pp. 835–848, October 1971.

[16] J. L. MASSEY, *Threshold Decoding* (Cambridge, Mass.: MIT Press, 1963).

[17] J. B. CAIN and J. M. GEIST, "Modulation, Coding, and Interleaving Tradeoffs for Spread Spectrum Systems," *Conf. Rec., IEEE Natl. Telecommun. Conf.,* 1981.

[18] G. D. FORNEY, "Burst-Correcting Codes for the Classic Bursty Channel," *IEEE Trans. Commun. Technol.,* Vol. COM-19, pp. 772–781, October 1971.

[19] J. H. YUEN, ed., *Deep Space Telecommunications Systems Engineering* (New York: Plenum Press, 1983).

[20] G. D. FORNEY, *Concatenated Codes* (Cambridge, Mass.: MIT Press, 1967).

[21] R. BLAHUT, *Theory and Practice of Error Control Codes* (Reading, Mass.: Addison-Wesley, 1983).

[22] H. H. MA and M. A. POOLE, "Error-Correcting Codes Against Worst-Case Partial Band Jammer," *IEEE Trans. Commun.,* Vol. COM-32, pp. 124–133, February 1984.

[23] J. S. LEE, R. H. FRENCH, and L. E. MILLER, "Error-Correcting Codes and Nonlinear Diversity Combining Against Worst Case Partial-Band Noise Jamming of Frequency-Hopping MFSK Systems," *IEEE Trans. Commun.,* Vol. COM-36, pp. 471–478, April 1988.

[24] M. K. SIMON, J. K. OMURA, R. A. SCHOLTZ, and B. K. LEVITT, *Spread Spectrum Communications,* Vols. I, II and III (Rockville, Md.: Computer Science Press, 1985).

[25] M. B. PURSLEY and W. E. STARK, "Performance of Reed-Solomon Coded Frequency-Hop Spread-Spectrum in Partial Band Interference," *IEEE Trans. Commun.,* Vol. COM-33, pp. 767–774, August 1985.

[26] R. H. DOU and L. B. MILSTEIN, "Erasure and Error-Correction Decoding Algorithm for Spread-Spectrum Systems with Partial-Time Interference," *IEEE Trans. Commun.,* Vol. COM-33, pp. 858–862, August 1985.

[27] D. J. TORRIERI, "Frequency Hopping with Multiple Frequency Shift Keying and Hard Decisions," *IEEE Trans. Commun.,* Vol. COM-32, pp. 574–582, May 1984.

[28] J. P. ODENWALDER, *Optimal Decoding of Convolutional Codes,* Ph.D. dissertation, University of California, Los Angeles, 1970.

[29] A. J. VITERBI and I. M. JACOBS, "Advances in Coding and Modulation for Noncoherent Channels Affected by Fading, Partial Band, and Multiple-Access Interference," in *Adv. Commun. Syst.,* McGraw-Hill, 1975.

[30] A. J. VITERBI, "A Robust Ratio-Threshold Technique to Mitigate Tone and Partial Band Jamming in Coded MFSK Systems," *Conf. Rec.,* 1982 IEEE Military Communications Conference, pp. 22.4-1–22.4-5, October 1982.

[31] B. K. LEVITT, "Use of Diversity to Improve FH/MFSK Performance in Worst Case Partial Band Noise and Multitone Jamming," *Conf. Rec.*, 1982 IEEE Military Communications Conference, pp. 28.2-1–28.2-5, October 1982.

[32] W. E. STARK, "Coding for Frequency-Hopped Spread-Spectrum Communications with Partial-Band Interference: Part I. Capacity and Cutoff Rate," *IEEE Trans. Commun.*, Vol. COM-33, pp. 1036–1044, October 1985.

[33] J. S. LEE, R. H. FRENCH, and L. E. MILLER, "Probability of Error Analyses of a BFSK Frequency-Hopping System with Diversity Under Partial-Band Jamming Interference: Part I. Performance of Square-Law Linear Combining Soft Decision Receiver," *IEEE Trans. Commun.*, Vol. COM-32, pp. 645–653, June 1984.

[34] J. S. LEE, L. E. MILLER, and Y. K. KIM, "Probability of Error Analysis of a BFSK Frequency-Hopping System with Diversity Under Partial-Band Jamming Interference: Part II. Performance of Square-Law Nonlinear Combining Soft Decision Receiver," *IEEE Trans. Commun.*, Vol. COM-32, pp. 1243–1250, December 1984.

[35] J. S. LEE, L. E. MILLER, and R. H. FRENCH, "The Analysis of Uncoded Performance for Certain ECCM Receiver Design Strategies for Multihop/Symbol FH/MFSK Waveforms," *J. Selected Areas Commun.*, Vol. SAC-3, pp. 611–621, September 1985.

[36] L. E. MILLER, J. S. LEE, and A. P. KADRICHU, "Probability of Error Analysis of a BFSK Frequency-Hopping System with Diversity Under Partial-Band Jamming Interference: Part III. Performance of a Square-Law Self-Normalizing Soft Decision Receiver," *IEEE Trans. Commun.*, Vol. COM-34, pp. 669–675, July 1986.

[37] B. K. LEVITT, "Strategies for FH/MFSK Signaling with Diversity in Worst-Case Partial-Band Noise," *J. Selected Areas Commun.*, Vol. SAC-3, pp. 622–626, September 1985.

[38] B. K. LEVITT, "FH/MFSK Performance in Multitone Jamming," *J. Selected Areas Commun.*, Vol. SAC-3, pp. 627–643, September 1985.

[39] C. M. KELLER and M. B. PURSLEY, "Diversity Combining for Channels with Fading and Partial-Band Interference," *J. Selected Areas Commun.*, Vol. SAC-5, pp. 248–260, February 1987.

[40] C. M. KELLER and M. B. PURSLEY, "Clipped Diversity Combining for Channels with Partial-Band Interference: Part I. Clipped-Linear Combining," *IEEE Trans. Commun.*, Vol. COM-35, pp. 1320–1328, December 1987.

[41] C. M. KELLER and M. B. PURSLEY, "Clipped Diversity Combining for Channels with Partial-Band Interference: Part II. Ratio Statistic Combining," *IEEE Trans. Commun.*, Vol. 37, pp. 145–151, February 1989.

[42] R. VISWANATHAN and K. TAGHIZADEH, "Diversity Combining in FH/BFSK Systems to Combat Partial Band Jamming," *IEEE Trans. Commun.*, Vol. 36, pp. 1062–1069, September 1988.

[43] B. D. TRUMPIS, "On the Optimum Detection of Fast Frequency Hopped MFSK Signals in Worst Case Jamming," TRW internal memorandum, June 1981.

[44] J. B. CAIN and J. M. GEIST, "Modulation, Coding, and Interleaving Tradeoffs for Spread Spectrum Systems," *Conf. Rec.*, IEEE Natl. Telecommun. Conf., 1981.

[45] C. E. SHANNON, "A Mathematical Theory of Communication," *Bell Syst. Tech. J.*, Vol. 27, pp. 379–423, 623–656, July and October 1948.

[46] J. K. OMURA and B. K. LEVITT, "Coded Error Probability Evaluation for Antijam Communication Systems," *IEEE Trans. Commun.*, Vol. COM-30, pp. 896–903, May 1982.

[47] J. K. OMURA and B. K. LEVITT, "A General Analysis of Anti-jam Communication Systems," *Conf. Rec.*, Natl. Telecommun. Conf., November 1981.

[48] P. J. LEE, "Performance of a Normalized Energy Metric Without Jammer State Information for an FH/MFSK System in Worst Case Partial Band Jamming," *IEEE Trans. Commun.*, Vol. COM-33, pp. 869–877, August 1985.

[49] F. EL-WAILLY, *Convolutional Code Performance Analysis of Jammed Spread-Spectrum Channels Without Side Information*, Ph.D. dissertation, Illinois Institute of Technology, Chicago, 1982.

[50] R. G. GALLAGER, *Information Theory and Reliable Communication* (New York: Wiley, 1968).

[51] J. M. WOZENCRAFT and R. S. KENNEDY, "Modulation and Demodulation for Probabilistic Coding," *IEEE Trans. Inf. Theory*, Vol. IT-12, pp. 291–297, July 1966.

[52] A. J. VITERBI, "Spread-Spectrum Communications: Myths and Realities," *IEEE Commun. Mag.*, Vol. 12, pp. 11–18, May 1979.

Problems

(7-1) Consider an error correction code whose codewords are generated using a four-stage maximal-length shift register whose generator is $g(D) = 1 + D + D^4$ and which is illustrated in Figure 8-16. A codeword is generated by loading the shift register with the information word and clocking the register 15 times. The codeword appears serially in the right stage of the shift register. What is the output message estimate of a maximum-likelihood hard decision decoder whose input 15-tuple is $\mathbf{y} = (101100101011000)$?

(7-2) Repeat Problem 7-1 for a Gaussian channel with a continuous output using a soft-decision decoder with input $y = (+0.8, -0.1, +1.1, +0.25, +1.0, -1.0, -0.35, +0.95, +0.87, -0.01, +0.20, +1.0, +1.7, -0.82, +0.13)$.

(7-3) Consider the convolutional code defined by the trellis diagram of Figure 7-4. What is the output message estimate for a hard-decision maximum-likelihood decoder whose input is $y = (00101011000100)$?

(7-4) Repeat Problem 7-3 for a Gaussian channel with a continuous output using a soft-decision maximum-likelihood decoder whose input is $\mathbf{y} = (+1.1, +0.95, -0.78, +1.0, -1.0, +0.95, +0.98, -0.82, -0.90, -1.0, +0.01, -0.90, -1.0, -1.0)$.

(7-5) Consider a BPSK/BPSK direct-sequence spread-spectrum system designed for a data rate of 1 Mbps and a direct sequence chip rate of 100 Mchips/s. An $n = 15$, $k = 5$, $t = 3$ BCH code together with a block interleaver are used to improve system performance. The system operates in a pulsed noise environment and the maximum acceptable average bit error probability is $P_b = 10^{-5}$.
   (a) What is the maximum BSC crossover probability that will yield the required system P_b?
   (b) What is the minimum required signal-to-jammer power ratio?
   (c) What is the processing gain of the system relative to a nonspread, noncoded BPSK system operating in the same environment?

(7-6) Repeat Problem 7-5 using the best constraint-length-4, rate-$\frac{1}{3}$ convolutional code of Table 7-4.

(7-7) Draw a single stage of the convolutional code trellis for the best constraint-length-4, rate-$\frac{1}{3}$ code of Table 7-4. Draw a block diagram of this convolutional encoder.

(7-8) Consider a FH/DPSK spread-spectrum system using a data rate of 1 Mbps, a frequency-hop tone spacing of 2 MHz, and using 64 tones. This system is jammed by a smart partial-band jammer.
   (a) Calculate the computational cutoff rate R_0 as a function of $10 \log(P/J)$ and plot the result.
   (b) Calculate and plot the upper bound on bit error probability of (7-90) assuming a block code having block length 7 and rate $R = 0.571$.

(7-9) Derive an expression for decoded bit error probability for a FH/DPSK spread-spectrum system that uses repeat coding. Assuming that the system operates in a worst-case jamming environment.

(7-10) Calculate the error probability upper bound for the best constraint-length-7, rate-$\frac{1}{2}$ convolutional code as a function of E_b/N_J using (7-56) and using (7-91) and compare the results. Assume FH/DPSK modulation and barrage noise jamming.

CHAPTER 8

Introduction to Fading Channels

8-1 Introduction

In previous chapters the channel model employed in analyzing and designing spread-spectrum communication systems was an additive white Gaussian noise (AWGN) channel model in which the primary impairment was thermal noise from the receiver front end. In this chapter the concept of fading channels is introduced. The motivation for studying fading channels is to provide the background necessary to understand the design and performance of code-division multiple-access (CDMA) digital cellular communication systems, the topic of Chapter 9. Central to the discussion of the performance of *any* cellular radio system, however, is a description of the land mobile radio channel, channel impairments, and methods of overcoming channel impairments.

The chapter begins with an overview of fading channels in which a mathematical model of general fading channels is introduced along with the terminology of fading channels. Next, mobile radio channels are characterized in terms of time and frequency selectivity. As will be seen, fading channels, and in particular, mobile radio channels, can cause a significant degradation in the performance of a communication system. The concept of diversity methods for fading channels is therefore introduced. Diversity methods will be shown to help overcome the performance degradation of fading channels. Several diversity methods are discussed here, including the methods that are employed in the example CDMA digital cellular systems of Chapter 9.

8-2 Statistical Model of Fading

In many analyses of the performance of various types of communication systems, the communication channel is modeled as a linear time-invariant system whose transfer function consists of a frequency-independent magnitude less than unity proportional to the propagation loss, and a delay term proportional to the propagation delay between the channel modulator and the channel demodulator. In addition, the channel is usually considered to be corrupted by AWGN. Although this simple additive white Gaussian noise model is quite accurate for channels such as deep-space communication channels, it is an overly simplified model for a number of radio channels, including high-frequency (HF) long-distance communications achieved via the ionosphere, microwave communications beyond the horizon achieved via the use of tropospheric scatter, and communications to a mobile platform. In the latter three channels, the received signal has been shown experimentally to undergo a process known as fading. A fading channel may exhibit such properties as selective frequency response, intersymbol interference in digital communications, spreading of signals in the frequency domain, time-varying amplitude response, or any combination of these attributes. A comprehensive model for a fading channel is given in this section.

In addition to the three examples of fading channels given above, there exist several other types of channels that exhibit fading. These include very-high-frequency (VHF) communication channels between an aircraft and a synchronous satellite relay [1]; artificially created communications channels temporarily created in an interesting experiment known as the West Ford Project† [2]; line-of-sight microwave communication links, which occasionally undergo severe fading due to the formation of tropospheric inversion layers, permitting multiple transmission paths between the transmitting and receiving antennas [3]; and communication at millimeter to optical wavelengths in line-of-sight paths through the nonionized atmosphere [4]. Further examples of the effects of fading over HF channels may be found in the text by Goodman [5]; details of fading over mobile VHF and UHF radio channels are given in Sections 8-3 and 8-4.

Fading encountered over either a mobile radio channel or an HF ionospheric channel, for example, has been verified experimentally to be of two types: short-duration rapid fading over time spans of less than 1 s and long-duration slow fading over time spans from 1 s to 1 h or longer. The statistics of the two fading processes are different; hence these two types of fading must be accounted for in the channel model.

The origin of the fading mechanism for most of the fading channels mentioned above may be traced to the scattering of an electromagnetic wave by a random medium. To see how this leads to fading, consider the following: Let a single continuous sine wave be scattered by a random medium. The scattered components may be resolved into in-phase and quadrature components. The instantaneous amplitudes of the two types of components may be shown to be uncorrelated. Using the central limit theorem, as the number of in-phase and quadrature components becomes large, the sum of the in-phase component approaches a Gaussian random variable. Similarly, the quadrature components add to form an identically distributed Gaussian random process. Hence the in-phase and quadrature random processes collectively form a zero-mean complex Gaussian random process [6]. If the random medium is a single surface and is time-invariant, the received signal, after scattering,

---

† In the West Ford Project, 20 kg of 2-cm-long copper dipoles were injected into a 3600-km-high orbit around the earth, and trans-horizon communications were conducted at 8 GHz using the orbiting dipole belt as a scattering mechanism. Fading effects due to the scattering were predicted prior to the experiment and were confirmed experimentally during the course of the experiment. Because of the shape of the scatterers, this project was also known as the ''Needles'' project. The *Proceedings of the IEEE* dedicated the May 1964 issue to a discussion of this project.

can be shown to have a Rayleigh-distributed amplitude and uniformly distributed phase (i.e., the signal is undergoing fading). Further discussion of the modeling of scattering by random medium as a complex Gaussian random process may be found in Ref. 7; in the following it will be accepted that scattering results in a zero-mean complex Gaussian random process, provided that a "sufficient number" of random scatterers exists for the particular geometry under consideration.

8-2.1 General Fading Channel Model

In this section a general model of the fading channel is developed. Since the signals under consideration are complex signals (i.e., narrowband signals centered at some frequency ω_0), the use of complex envelope notation is a convenient method of describing these signals. To clarify the notation used below, several standard properties [6–8] of bandpass signals are given in complex envelope notation.

The transmitted signal may be expressed in complex envelope notation as

$$s_0(t) = \text{Re}[u_0(t)e^{j2\pi f_0 t}] \tag{8-1}$$

where $u_0(t)$ is a lowpass signal having a Fourier transform $U_0(f)$,

$$u_0(t) = \int_{-\infty}^{\infty} U_0(f)e^{j2\pi ft}\, df \tag{8-2}$$

and f_0 is the center (reference) frequency of $s_0(t)$. By *narrowband signals* it is meant that if the normalized bandwidth B_n of $u_0(t)$ is defined as

$$B_n = \frac{\int_{-\infty}^{\infty} |U_0(f)|^2\, df}{|U_0(0)|^2} \tag{8-3}$$

then

$$B_n \ll f_0 \tag{8-4}$$

Given a linear, time-invariant system with an impulse response $h(t)$ and a transfer function $H(f)$, where

$$h(t) = \int_{-\infty}^{\infty} H(f)e^{j2\pi ft}\, df = \text{Re}[2h_e(t)e^{j2\pi f_0 t}] \tag{8-5}$$

and $H(f)$ is a bandpass function around f_0, the output $y(t)$ due to an input $s_0(t)$ given by (8-1) is

$$y(t) = \text{Re}[y_e(t)e^{j2\pi f_0 t}] \tag{8-6}$$

where

$$y_e(t) = \int_{-\infty}^{\infty} h_e(t - \sigma)u_0(\sigma)\, d\sigma \tag{8-7}$$

Equations (8-6) and (8-7) illustrate the advantage of using complex envelope notation for bandpass signals and systems; to evaluate the output of a bandpass system due to a bandpass signal input, simply convolve the lowpass equivalent impulse response with the lowpass equivalent input signal, multiply by $e^{j2\pi f_0 t}$, and take the real part of the resulting product.

For a linear time-varying bandpass system with impulse response $h(t,\tau)$ where

$$h(t,\tau) = \text{Re}[2h_e(t,\tau)e^{j2\pi f_0 t}] \tag{8-8}$$

the expression corresponding to (8-7) is

$$y_e(t) = \int_{-\infty}^{\infty} h_e(t,u)u_0(u)\ du \tag{8-9}$$

In (8-8), $h(t,\tau)$ denotes the output at time t due to an impulse at time τ.

Finally, the following algebraic identity is needed: Given two complex numbers X and Y,

$$\text{Re}[X]\text{Re}[Y] = \tfrac{1}{2}\text{Re}[XY] + \tfrac{1}{2}\text{Re}[XY^*] \tag{8-10}$$

Other properties of complex envelope notation will be developed as needed.

The fading channel model will now be developed. Consider the propagation model shown in Figure 8-1. In the figure, a signal $s_0(t)$ given by (8-1) is transmitted and scattered by a *moving* random medium which is assumed to be modeled as a layered scatterer, where each layer has an incremental thickness dr. Associated with each layer is a propagation delay τ, which is in addition to the nominal propagation delay t_0 between the transmitter and the receiver. The received signal due to scattering from the layer whose incremental propagation delay is τ is given by

$$s(\tau,t) = \text{Re}[\beta(\tau,t-t_0)u_0(t-t_0-\tau)e^{j2\pi f_0(t-t_0-\tau)}] \tag{8-11}$$

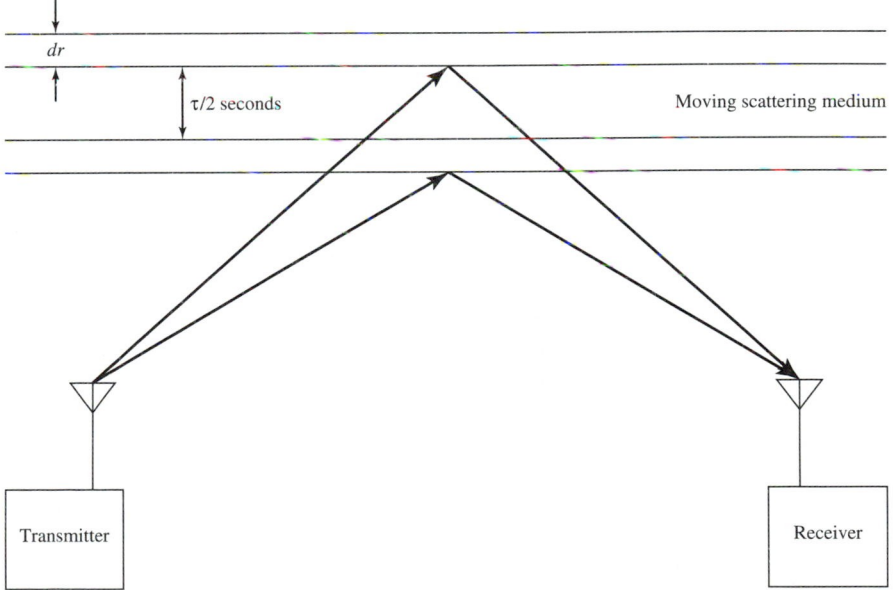

FIGURE 8-1. Propagation model of a fading channel.

where $\beta(\tau,t)$ is the time-varying transmission coefficient for a wave scattering off a layer whose incremental propagation delay is τ. Several comments about the form of (8-11) are in order. First, (8-11) assumes that the thickness dr and its corresponding delay $d\tau$ may be chosen small enough such that $\beta(\tau,t)$ is constant in its first argument over the interval $d\tau$. This allows the effect of scattering off a single layer to be modeled as a simple time-varying multiplicative factor as opposed to some form of superposition of responses. Second, it is important to note that the received signal $s(t,\tau)$ is a function of *two* arguments: the first denotes time and the second denotes the incremental propagation delay due to the layer from which the transmitted signal is being scattered. Hence the received signal is both time varying due to the fact that the scattering volume was assumed to be moving, and layer dependent since the scattering medium was assumed to be able to be modeled as consisting of differential layers. Third, the time origins of the various terms of (8-11) should be noted. For a received signal at time t due to scattering from a layer having a *total* propagation delay of $t_0 + \tau$ associated with it, the signal would have to be transmitted at time $t - t_0 - \tau$. Thus the arguments of $u_0(\cdot)$ and $\exp[j2\pi f(\cdot)]$ are the transmitted time origin. The second argument of $\beta(\tau,t - t_0)$ refers to the nominal time at which the received signal was transmitted. Finally, the statistical nature of $\beta(\tau,t)$ must be determined. From the discussion above, $\beta(\tau,t)$ is assumed to be a zero-mean complex Gaussian random process.† Inherent in this assumption is the fact that $\beta(\tau,t)$ has a Rayleigh-fading envelope and a uniformly distributed phase, both of which are time varying due to the presence of a moving random medium. Hence the term $\exp(j2\pi f_0\tau)$ may be absorbed into $\beta(\tau,t)$ in (8-11).

The total received signal is a superposition of the responses due to all the scattered layers:

$$s(t) = \int_{-\infty}^{\infty} s(\tau,t)\, d\tau = \mathrm{Re}\left[\int_{-\infty}^{\infty} \beta(\tau,t - t_0)u_0(t - t_0 - \tau)e^{j2\pi f_0(t-t_0)}\, d\tau\right] \quad (8\text{-}12)$$

Equation (8-12) may also be written in the form

$$s(t) = \mathrm{Re}[u(t - t_0)e^{j2\pi f_0(t-t_0)}] \quad (8\text{-}13)$$

where

$$u(t) = \int_{-\infty}^{\infty} \beta(\tau,t)u_0(t - \tau)\, d\tau = \int_{-\infty}^{\infty} \beta(t - \tau,t)u_0(\tau)\, d\tau \quad (8\text{-}14)$$

Comparing (8-13) and (8-14) with (8-8) and (8-9), it is readily seen that $\beta(\tau,t)$ represents a time-varying equivalent low-pass impulse response for the general fading channel. Applying an impulse at time α as the input function in (8-14) yields

$$u_I(t) = \int_{-\infty}^{\infty} \beta(\tau,t)\delta(t - \alpha - \tau)\, d\tau = \beta(\tau - \alpha,t) \quad (8\text{-}15)$$

That is, $\beta(\tau - \alpha,t)$ is the impulse response at time t to an impulse applied at time α.

---

†A complex Gaussian random process $n(t)$ is a complex random process whose real $[n_R(t)]$ and imaginary $[n_I(t)]$ parts are each real Gaussian random processes [i.e., $n(t) = n_R(t) + jn_I(t)$]. Complex Gaussian random processes are frequently used to model bandpass-filtered white Gaussian noise.

Equation (8-12) represents the most general model of a fading channel that will be discussed here. Before proceeding further, experimental evidence justifying the use of this model as well as suggesting other models will be examined briefly. To begin with, if the random medium is slowly varying with time and can be modeled as consisting of a single layer, (8-12) reduces to

$$s(t) = \text{Re}[\beta u_0(t - t_0)e^{j2\pi f_0(t-t_0)}] \tag{8-16}$$

where β is a zero-mean complex Gaussian random *variable*. The assumptions necessary for the transition from (8-12) to (8-16) are examined in greater detail below. Letting the transmitted signal be a continuous sine wave at frequency f_0, (8-16) becomes

$$s(t) = \text{Re}[\beta e^{j2\pi f_0(t-t_0)}] \tag{8-17}$$

It can be shown that the magnitude of a zero-mean complex Gaussian random variable is Rayleigh distributed and the phase uniformly distributed. Hence (8-17) predicts that the amplitude of the received sine wave is Rayleigh distributed. Experimental evidence [9–11] for the VHF and UHF mobile channel and the HF ionospheric channel tends to support this prediction; at the same time, experimental data for some fading channels exhibit a Rician amplitude distribution. Rician fading is usually considered to be due to a specular component in the received path (i.e., the presence of a fixed nonrandom scatterer in the transmitter–receiver propagation path). In this case, (8-12) may be modified to include a deterministic component in the received signal:

$$s(t) = \text{Re}\Bigg[Au_0(t - t_0)e^{j2\pi f_0(t-t_0)} \\ + \int_{-\infty}^{\infty} \beta(\tau,t - t_0)u_0(t - t_0 - \tau)e^{j2\pi f_0(t-t_0)}\, d\tau \Bigg] \tag{8-18}$$

where A is the transmission coefficient associated with the specular path.

The comments above are not meant to suggest that all fading is Rician or Rayleigh in nature; on the contrary, experimental data for the HF ionospheric channel, the mobile radio channel, and the millimeter-to-optical wave line-of-sight channel indicate the amplitude of the received signal is lognormally distributed in many instances. In Ref. 7 it is suggested that deviations from a Rayleigh amplitude distribution may be due to an insufficient number of scatters in a scattering "layer" to support the complex Gaussian random process hypothesis stated above. For the situation where the direct line-of-sight path encounters a large number of obstacles (as is often the case for the mobile radio channel), the lognormal statistics that are observed may readily be explained. For a mobile user, the obstacles may be modeled as a large number of randomly moving planes with respect to a moving vehicle. The transmitted signal is then modulated by the product of the random transmission coefficients of each of the planes. Assuming the transmission coefficients to be independent of each other, the logarithm of the overall channel amplitude transmission function is the sum of the logarithms of the plane transmission functions and, using a central limit theorem argument, is thus normally distributed. The channel transmission function is then seen to be lognormally distributed in amplitude. Despite the fact that some channels may be modeled as exhibiting a lognormal amplitude distribution, for the remainder of this section it is assumed that the fading channels of concern exhibit predominately Rayleigh fading.

In determining the performance of communication systems over fading channels, frequent use is made of second-order statistical properties of the channel model. In particular, assuming $\beta(\tau,t)$ to be a zero-mean complex Gaussian process, given $u_0(t)$, $u(t)$ given by (8-14) is also a zero-mean complex Gaussian process which is completely characterized statistically by its correlation function. For a bandpass process represented by complex envelope notation, *two* correlation functions are needed to completely characterize the second-order properties of the process. To see this, the autocorrelation function of $s(t)$ given by (8-12) is defined in the usual manner as

$$R_s(t_1,t_2) = E[s(t_1)s(t_2)] \tag{8-19}$$

Using (8-12) and (8-10) in (8-19) gives

$$
\begin{aligned}
R_s(t_1,t_2) = \text{Re}&\left\{ \int_{-\infty}^{\infty} \int_{-\infty}^{\infty} \tfrac{1}{2}E[\beta(\tau_1,t_1 - t_0)\beta(\tau_2,t_2 - t_0)]u_0(t_1 - t_0 - \tau_1)u_0 \right. \\
&\left. \times (t_2 - t_0 - \tau_2)e^{j2\pi f_0(t_1 + t_2 - 2t_0)}\, d\tau_1\, d\tau_2 \right\} \\
+ \text{Re}&\left\{ \int_{-\infty}^{\infty} \int_{-\infty}^{\infty} \tfrac{1}{2}E[\beta(\tau_1,t_1 - t_0)\beta^*(\tau_2,t_2 - t_0)]u_0 \right. \\
&\left. \times (t_1 - t_0 - \tau_1)u_0^*(t_2 - t_0 - \tau_2)e^{j2\pi f_0(t_1 - t_2)}\, d\tau_1\, d\tau_2 \right\}
\end{aligned}
\tag{8-20}
$$

Note that to evaluate (8-20) *two* correlation functions are needed,

$$\tilde{\Lambda}(\tau_1,\tau_2;t_1,t_2) \equiv \tfrac{1}{2}E[\beta(\tau_1,t_1)\beta(\tau_2,t_2)] \tag{8-21}$$

and

$$\Lambda(\tau_1,\tau_2;t_1,t_2) \equiv \tfrac{1}{2}E[\beta(\tau_1,t_1)\beta^*(\tau_2,t_2)] \tag{8-22}$$

$\Lambda(\cdot)$ given by (8-22) is defined to be the space-time cross-correlation function (or simply the correlation function) of the fading process [7]. In most applications, the narrowband process $\beta(\tau,t)$ is so constituted that [6,12]†

$$\tilde{\Lambda}(\tau_1,\tau_2;t_1,t_2) = 0 \tag{8-23}$$

One other correlation function associated with the general fading channel model

---

†Examples of processes that do not satisfy (8-23) are given in Refs. 13 and 14. It is easily seen that (8-23) is a *necessary* condition for stationary bandpass processes. For by representing the equivalent bandpass impulse response in complex envelope notation as

$$B(\tau,t) = \text{Re}[\beta(\tau,t)e^{j2\pi f_0 t}]$$

as in (8-8), the autocorrelation function of $B(\tau,t)$ is given by

$$
\begin{aligned}
R_B(\tau_1,t_1;\tau_2,t_2) &= \text{Re}\{\tfrac{1}{2}E[\beta(\tau_1,t_1)\beta(\tau_2,t_2)]e^{j2\pi f_0(t_1 + t_2)}\} + \text{Re}\{\tfrac{1}{2}E[\beta(\tau_1,t_1)\beta^*(\tau_2,t_2)]e^{j2\pi f_0(t_1 + t_2)}\} \\
&= \text{Re}[\tilde{\Lambda}(\tau_1,\tau_2;t_1,t_2)e^{j2\pi f_0(t_1 + t_2)}] + \text{Re}[\Lambda(\tau_1,\tau_2;t_1,t_2)e^{j2\pi f_0(t_1 - t_2)}]
\end{aligned}
$$

For $R_B(\cdot)$ to be a function only of the time difference $t_1 - t_2$, $\tilde{\Lambda}(\cdot)$ must vanish above. A different proof of this result utilizing the spectra of the bandpass processes may be found in Ref. 6. Grettenberg [15] has proven that a necessary and sufficient condition for (8-23) to hold is that $\beta(\tau,t)$ and $e^{j\theta}\beta(\tau,t)$ be identically distributed for all real θ. A more direct proof of the latter result may be found in Ref. 16.

given by (8-12) is the frequency–time cross-correlation function [7] defined by

$$R(f_1,f_2;t_1,t_2) \equiv \tfrac{1}{2}E[H(f_1;t_1)H^*(f_2;t_2)] \tag{8-24}$$

where

$$H(f;t) = \int_{-\infty}^{\infty} \beta(\tau,t)e^{-j2\pi f\tau}\, d\tau \tag{8-25}$$

is the spatial Fourier transform of the equivalent lowpass impulse response $\beta(\tau,t)$ for the channel. Using (8-25) in (8-24) yields

$$R_F(f_1,f_2;t_1t_2) = \int_{-\infty}^{\infty}\int_{-\infty}^{\infty} \Lambda(\tau_1,\tau_2;t_1,t_2)e^{j2\pi(f_1\tau_1-f_2\tau_2)}\, d\tau_1\, d\tau_2 \tag{8-26}$$

which is recognized to be the double Fourier transform of the space-time cross-correlation function. Because of the dependency on both frequency and time, the function $R_F(f_1,f_2;t_1,t_2)$ is also known as the spaced-frequency spaced-time correlation function of the channel.

In practice, the very general fading channel model developed above is difficult to use in the performance analysis of communication systems due to the mathematical complexities involved. Furthermore, as has been shown in Ref. 12, a simpler model is warranted for most radio channels, as will now be discussed.

8-2.2 WSSUS Fading Channels

The wide-sense stationary uncorrelated scattering (WSSUS) fading channel model will now be developed from the general fading channel model (8-12) presented above. In this development, the wide-sense stationary (WSS) and the uncorrelated scattering (US) channel models will also be defined.

In the above it was noted that some channels exhibit two types of fading: long-term fading and short-term fading. Short-term fading over these channels is often such that the short-term fading statistics are approximately stationary over time. Hence it is convenient to define a subclass of the general fading channel model known as wide-sense stationary (WSS) channels. Bello [12] defines the WSS channel as a channel whose correlation functions $R_F(\cdot)$ and $\Lambda(\cdot)$ are invariant under a time translation. The space-time cross-correlation function given by (8-22) satisfies

$$\Lambda(\tau_1,\tau_2;t_1,t_2) = \Lambda(\tau_1,\tau_2;t_1 - t_2) \tag{8-27}$$

while the frequency-time cross-correlation function given by (8-24) satisfies

$$R(f_1,f_2;t_1,t_2) = R(f_1,f_2;t_1 - t_2) \tag{8-28}$$

for the WSS channel.

While developing the general fading channel model above, it was assumed that the scattering medium could be modeled as consisting of differential layers. A reasonable assumption for many channels is to assume that the complex Gaussian process $\beta(\tau,t)$ is independent of $\beta(\sigma,t)$ for $\tau \neq \sigma$. Note that this is equivalent to assuming that the effect of scatterers in one differential layer is independent of the effect of the scatterers in all other differential layers. The space-time cross-correlation function for such a channel is given by

$$\Lambda(\tau_1,\tau_2;t_1,t_2) = \Lambda(\tau_1;t_1,t_2)\delta(\tau_1 - \tau_2) \qquad (8\text{-}29)$$

Channels whose space-time cross-correlation functions satisfy (8-29) are known as uncorrelated scattering (US) channels [12]. Note that the US channel is the wide-sense dual of the WSS channel using the definitions of duality given by Bello [17].

Channels that exhibit both WSS channel characteristics and US channel characteristics are known as wide-sense-stationary uncorrelated scattering (WSSUS) channels [12]. Using (8-27) and (8-29), the space-time cross-correlation function for a WSSUS channel is given by

$$\Lambda(\tau_1,\tau_2;t_1,t_2) = \rho(\tau_1;t_1 - t_2)\delta(\tau_1 - \tau_2) \qquad (8\text{-}30)$$

where $\rho(\tau_1; t_1 - t_2) \equiv \Lambda(\tau_1,\tau_2;t_1,t_2)$ for the special case of a WSSUS channel. Note that for WSSUS channels, two assumptions on $\Lambda(\cdot)$ are being made: (1) the scattering processes due to different layers are statistically uncorrelated; and (2) the scattering processes in each layer are wide-sense stationary. Since $\beta(\tau,t)$ is a zero-mean complex Gaussian process that satisfies (8-23), wide-sense stationarity of $\beta(\tau,t)$ implies strict-sense stationarity of $\beta(\tau,t)$ [16].

Although the channel models above simplify the determination of the performance of communication channels considerably, they would be of little value unless these models correspond to actual fading channels. Fortunately, most radio channels appear to exhibit WSSUS channel properties [7,12]. As noted above, radio channel fading is often characterized by the superposition of short-term fading on long-term fading. The short-term fading is usually found to exhibit stationary characteristics while the long-term fading is often highly nonstationary, depending on the interval of interest. Bello [12] has introduced the term quasi-wide-sense-stationary uncorrelated scattering (QWSSUS) to describe such a channel. As might be expected, a QWSSUS channel has WSSUS channel characteristics over time intervals on the order of the duration of short-term fading. In light of this, for the remainder of this section, the WSSUS channel model, and, in particular, the Gaussian WSSUS (GWSSUS) channel model will be considered only.

8-2.3 Doubly Spread Channels

In the preceding section, the WSSUS fading channel was developed as a general fading channel model. In this section the most general class of WSSUS channels, known either as doubly spread channels [6] or doubly dispersive channels [18]. The goal is to characterize the various parameters of doubly spread channels and to introduce the notation that is often used in conjunction with fading channels.

A word about terminology is first in order. Doubly spread channels are so called because they spread the time *and* frequency waveforms of a signal transmitted through the channel. Demonstration of this spreading in both domains must wait until discussion of the singly spread degenerate channels that exhibit spreading in only one domain.

For the doubly spread channel, the space-time cross-correlation function of the channel is given by (8-30). The scattering function $S_{DR}(\tau,f)$ is defined to be

$$S_{DR}(\tau,f) \equiv \int_{-\infty}^{\infty} e^{-j2\pi f t} \rho(\tau,t)\, dt \tag{8-31}$$

where $\rho(\tau,t)$ is defined implicitly by (8-30). Note that the scattering function is the temporal Fourier transform of $\rho(\tau,t)$.

The scattering function provides a means of characterizing the Doppler spread of the channel, as will be seen shortly. It should be noted that most of the results up to this point and throughout the rest of this section could just as well have been developed in terms of the channel scattering function instead of the channel correlation function; see Ref. 18 for such an approach. Because the channel scattering function and the channel correlation function are Fourier transform pairs, such parallelism is to be expected.

If the scattering function is concentrated in one region of the (t,f) plane, it may be characterized grossly in terms of its moments. Consider the following definitions [6]: Define the *mean delay* to be

$$m_R \equiv \frac{1}{2\sigma_b^2} \int_{-\infty}^{\infty}\int_{-\infty}^{\infty} \tau S_{DR}(\tau,f)\, d\tau\, df \tag{8-32}$$

The *mean-square delay spread* is defined to be

$$L \equiv \frac{1}{2\sigma_b^2} \int_{-\infty}^{\infty}\int_{-\infty}^{\infty} \tau^2 S_{DR}(\tau,f)\, d\tau\, df - m_R^2 \tag{8-33}$$

The *mean Doppler shift* is defined to be

$$m_D \equiv \frac{1}{2\sigma_b^2} \int_{-\infty}^{\infty}\int_{-\infty}^{\infty} f S_{DR}(\tau,f)\, d\tau\, df \tag{8-34}$$

The *mean-square Doppler spread* is defined to be

$$B \equiv \frac{1}{2\sigma_b^2} \int_{-\infty}^{\infty}\int_{-\infty}^{\infty} f^2 S_{DR}(\tau,f)\, d\tau\, df - m_D^2 \tag{8-35}$$

In (8-32) through (8-35),

$$2\sigma_b^2 = \int_{-\infty}^{\infty}\int_{-\infty}^{\infty} S_{DR}(\tau,f)\, d\tau\, df \tag{8-36}$$

In addition, define the duration T and the bandwidth W of a narrowband bandpass signal given by (8-1) to be

$$T = \frac{1}{E_t} \int_{-\infty}^{\infty} t^2 |u_0(t)|^2\, dt \tag{8-37}$$

and

$$W = \frac{1}{E_t} \int_{-\infty}^{\infty} f^2 |U_0(f)|^2\, df \tag{8-38}$$

where

$$E_t = \int_{-\infty}^{\infty} |u_0(t)|^2 \, dt \qquad (8\text{-}39)$$

and $U_0(f)$ is defined by (8-2). The definitions given above are given only for completeness. For strictly time- or bandlimited scattering functions and/or bandpass signals, simpler definitions will be used for L, B, T, and W.

An underspread channel is defined to be one for which [6]

$$BL < 1 \qquad (8\text{-}40)$$

Similarly, an overspread channel is defined to be one for which [6]

$$BL > 1 \qquad (8\text{-}41)$$

In reviewing the literature on doubly spread channels, the terms correlation (or coherence) time and correlation (or coherence) bandwidth are often encountered in characterizing the channel [18]. The correlation time of a fading channel is defined as the time separation τ_c beyond which samples of the received signal are independent. Since the channel scattering process is assumed to be modeled as a zero-mean complex Gaussian process, given the transmitted signal, the received signal is also a zero-mean complex Gaussian process. Thus independence of time samples is implied if the correlation function of the envelope is zero. Using (8-30) and (8-14), the correlation between time samples of the received signal envelope is

$$R_{TE}(t_1,t_2) \equiv \tfrac{1}{2}E[u(t_1)u^*(t_2)] = \int_{-\infty}^{\infty} \rho_T(\tau, t_1 - t_2)u_0(t_1 - \tau)u_0^*(t_2 - \tau) \, d\tau \qquad (8\text{-}42)$$

By convention, the correlation time for the channel is chosen to be the smallest time separation $\tau_c = t_1 - t_2$ for which

$$R_{TE}(t_1,t_2) = 0 \qquad (8\text{-}43)$$

In a similar fashion, the correlation bandwidth of a fading channel is defined to be the frequency separation W_c beyond which samples of the Fourier transform of the received complex envelope are independent. From the comments above, the received signal envelope is a complex Gaussian process and it therefore follows that the Fourier transform of the received envelope is also a complex Gaussian process. Thus independence of frequency samples is implied if the correlation function of the Fourier transform of the received envelope is zero. From (8-14), the Fourier transform of the received envelope is

$$U(f) = \int_{-\infty}^{\infty}\int_{-\infty}^{\infty} \beta(\tau,t)u_0(t - \tau)e^{-j2\pi ft} \, d\tau \, dt \qquad (8\text{-}44)$$

Using (8-44), the correlation between frequency samples of the Fourier transform of the received signal envelope is given by

$$R_{FE}(f_1,f_2) \equiv \tfrac{1}{2}E[U(f_1)U^*(f_2)\}]$$

$$= \int_{-\infty}^{\infty}\int_{-\infty}^{\infty}\int_{-\infty}^{\infty} \rho(\tau,t_1 - t_2)u_0(t_1 - \tau)u_0^*(t_2 - \tau)e^{j2\pi(f_2t_2 - f_1t_1)} \, d\tau \, dt_1 \, dt_2 \qquad (8\text{-}45)$$

By convention, the correlation bandwidth for the channel is chosen to be the *smallest* frequency separation $W_c = f_1 - f_2$ for which

$$R_{FE}(f_1, f_2) = 0 \tag{8-46}$$

Finally, in the above, the doubly spread channel was characterized in terms of its channel correlation functions and the temporal Fourier transform of this quantity, the channel scattering function. Alternatively, the channel could have been characterized by the spatial Fourier transforms of these two quantities:

$$R_{DR}(\nu, t) \equiv \int_{-\infty}^{\infty} \rho(\tau, t) e^{-j2\pi\nu\tau} \, d\tau \tag{8-47}$$

and

$$P_{DR}(\nu, f) \equiv \int_{-\infty}^{\infty} S_{DR}(\tau, f) e^{-j2\pi\nu\tau} \, d\tau \tag{8-48}$$

The quantity $R_{DR}(\nu, t)$ defined in (8-47) and the quantity $P_{DR}(\nu, f)$ in (8-48) are known as the two-frequency correlation function [6] and the Doppler cross-power spectral density [12], respectively. The usefulness of having the four quantities $\rho(\tau, t)$, $S_{DR}(\tau, f)$, $R_{DR}(\nu, t)$, and $P_{DR}(\nu, f)$ comes in characterizing doubly spread channels that have correlation functions that are concentrated in one or more of the variables time t, Doppler spread f, delay spread τ, or delay frequency ν. The two-frequency correlation function is useful in parameterizing frequency-selective fading channels.

In this section the most general WSSUS fading channel, the doubly spread channel, has been defined and characterized. Often, fading radio channels exhibit spreading predominately in either the time or frequency domains only. In still other cases, the fading effects are such that they may be modeled by a random *variable* instead of a random process. In such cases it is convenient to define subclasses of doubly spread channels having specific characteristics. In Sections 8-2.4 through 8-2.5, models for three subclasses of doubly spread channels, also known as degenerate channels [6], will be developed.

8-2.4 Time-Selective Fading Channels

At the beginning of this section, the general model of a fading channel was developed by assuming that the scattering medium could be modeled as a randomly *moving, layered* volume of scatterers, each layer of which could be modeled as a complex Gaussian process. In this section it will be assumed that the scattering medium can be modeled as a *single* layer of randomly moving scatterers. As will be seen, such an assumption leads to the development of a class of channels known as time-selective fading channels.

To begin with, assume that a signal $s_0(t)$ given by (8-1) is transmitted through a scattering medium that can be modeled as a single layer. The received signal is

$$s(t) = \text{Re}[\beta(t - t_0)u_0(t - t_0)e^{j2\pi f_0(t - t_0)}] \tag{8-49}$$

where t_0 is the propagation time between the transmitter and receiver and $\beta(t)$ is the time-varying transmission coefficient due to the scattering medium. From the comments above it will be assumed that $\beta(t)$ is a sample function from a zero-mean complex Gaussian process. Equation (8-49) may also be written as

$$s(t) = \text{Re}[u(t - t_0)e^{j2\pi f_0(t-t_0)}] \tag{8-50}$$

where

$$u(t) = \beta(t)u_0(t) \tag{8-51}$$

For the WSSUS channel model, $\beta(t)$ is a stationary random process. It is important to note that since $\beta(t)$ is a complex-valued process it influences both the amplitude and phase of the transmitted signal $s_0(t)$.

Equations (8-49) or (8-50) and (8-51), together with the condition that $\beta(t)$ is a stationary process, constitute what is known as the time-selective fading channel model [19]. Other adjectives often used to describe this channel model include channels dispersive only in frequency [18], frequency-flat fading channels [19], or Doppler-spread channels [6].

Equation (8-49) could just as easily have been derived directly from the doubly spread fading channel model by noting that (8-49) is identical in form to (8-11), except for the argument of β. Hence the general fading channel model (8-12) reduces to (8-11), where the range variable τ of β is no longer important. In terms of correlation functions of the channel, the channel correlation function for the time-selective fading channel is

$$\rho(\tau; t - s) = \rho(0; t - s)\delta(\tau) \tag{8-52}$$

In the following, (8-52) will be used for the channel correlation function for time-selective fading channels rather than define a new time autocorrelation function that directly characterizes $\beta(t)$ in (8-49).

Some comments on (8-51) are in order. First, since $u_0(t)$, the transmitted envelope, is being multiplied by $\beta(t)$, which is independent of frequency, the received envelope $u(t)$ in (8-51) is easily seen to undergo fading that is independent of frequency [i.e., the various frequency components of $u_0(t)$ fade identically (frequency-flat fading)]. Nevertheless, $\beta(t)$ in (8-51) is a time-varying function and as such, acts to modulate the transmitted envelope of $u_0(t)$. The effect of modulation of the envelope of a transmitted pulse signal is shown in Figure 8-2 for two different on–off pulse durations. For short-duration pulses, no effects are observed, while for longer-duration pulses, the random modulation effects are readily apparent. This modulation process leads to spreading of the Fourier transform of the transmitted envelope in the frequency domain—hence the origin of the term Doppler-spread fading. Finally, since $\beta(t)$ in (8-51) is independent of τ, the uncorrelated scattering condition of WSSUS channels is not really necessary in characterizing time-selective fading channels.

Using (8-52) in (8-31), the scattering function for a time-selective fading channel is

$$S_{DR}(\tau, f) = S_D(0, f)\delta(\tau) = \delta(\tau) \int_{-\infty}^{\infty} e^{-j2\pi f t} \rho(0, t) \, dt \tag{8-53}$$

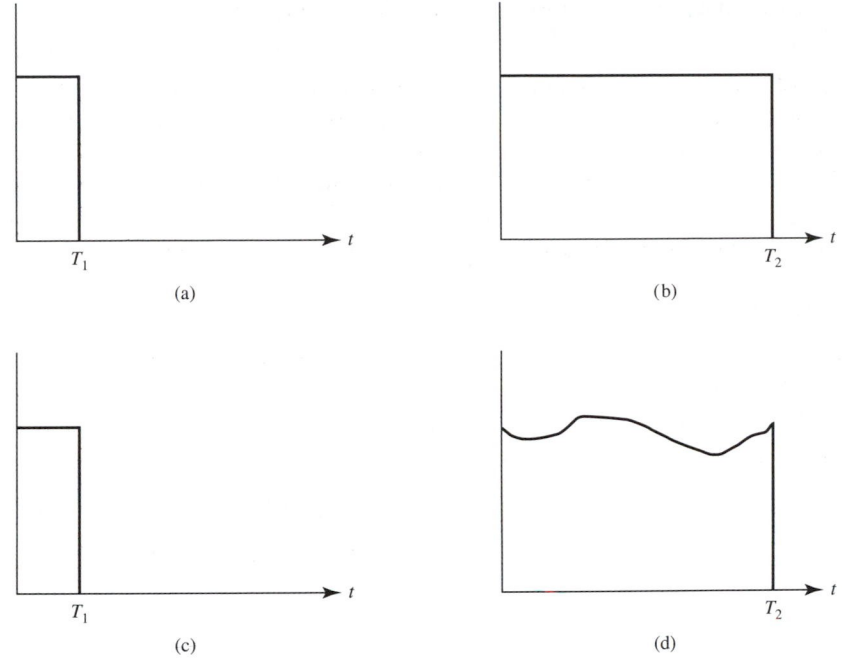

FIGURE 8-2. Effect of transmitting a pulse over a time-selective fading channel: (a) transmitted short pulse; (b) transmitted long pulse; (c) received short pulse; (d) received long pulse.

The term $S_D(0,f)$ in (8-51) is known as the Doppler scattering function or the Doppler power spectrum of the channel.

From (8-53) and (8-34), it can be seen that the mean Doppler shift for a time-selective fading channel is

$$m_D = \frac{1}{2\sigma_b^2} \int_{-\infty}^{\infty} fS_D(0,f)\, df \qquad (8\text{-}54)$$

Similarly, from (8-53) and (8-35), it can be seen that the mean-square Doppler spread is

$$B = \frac{1}{2\sigma_b^2} \int_{-\infty}^{\infty} f^2 S_D(0,f)\, df - m_D^2 \qquad (8\text{-}55)$$

where

$$2\sigma_b^2 = \int_{-\infty}^{\infty} S_D(0,f)\, df \qquad (8\text{-}56)$$

The mean-square Doppler spread characterizes the Doppler spread around the mean Doppler shift due to the time-selective properties of the fading channel.

Using (8-53) in (8-32) and (8-33), it is readily seen that the mean delay and mean-square delay for a time-selective channel are identically zero:

$$m_R = L = 0 \tag{8-57}$$

Thus the time-selective fading channel exhibits spreading in frequency but not in delay and is therefore often called a singly spread channel. Another singly spread channel will be encountered in the next section.

To clarify the relationship between correlation time, the channel correlation functions, and the correlation between time samples of the received signal envelope for the time-selective fading channel, consider the following example.

EXAMPLE 8-1

In this example the relationship between the channel correlation function and the correlation time of the channel will be examined. Note that by using (8-52) in (8-42), the correlation between time samples of the received signal envelope for a time-selective fading channel is

$$R(t_1,t_2) = \rho(0,t_1 - t_2)u(t_1)u_0^*(t_2) \tag{8-58}$$

For this example, let the Doppler scattering function for the channel be of the form

$$S_D(0,f) = \begin{cases} 1 & \dfrac{-B}{2} < f < \dfrac{B}{2} \\ 0 & \text{elsewhere} \end{cases} \tag{8-59}$$

The corresponding channel correlation function $\rho(0,\Delta t)$ is therefore

$$\rho(0,\Delta t) = B\frac{\sin \pi B\, \Delta t}{\pi B\, \Delta t}$$

The Doppler scattering function and the resulting channel correlation function are depicted in Figure 8-3. The correlation time as defined by (8-43) is given by

$$\tau_c = \frac{1}{B} \tag{8-60}$$

provided that

$$u_0(t)u_0^*(t_1 - \Delta t) \neq 0 \qquad \text{for } 0 < \Delta t < \frac{1}{B} \tag{8-61}$$

Thus time samples of the received signal separated by $1/B$ seconds will be uncorrelated.

Suppose, however, that a pulse of duration T is transmitted through the channel, where

$$T \ll \frac{1}{B} \tag{8-62}$$

In this case it is the pulse itself that determines τ_c and the correlation time is equal to the duration of the pulse.

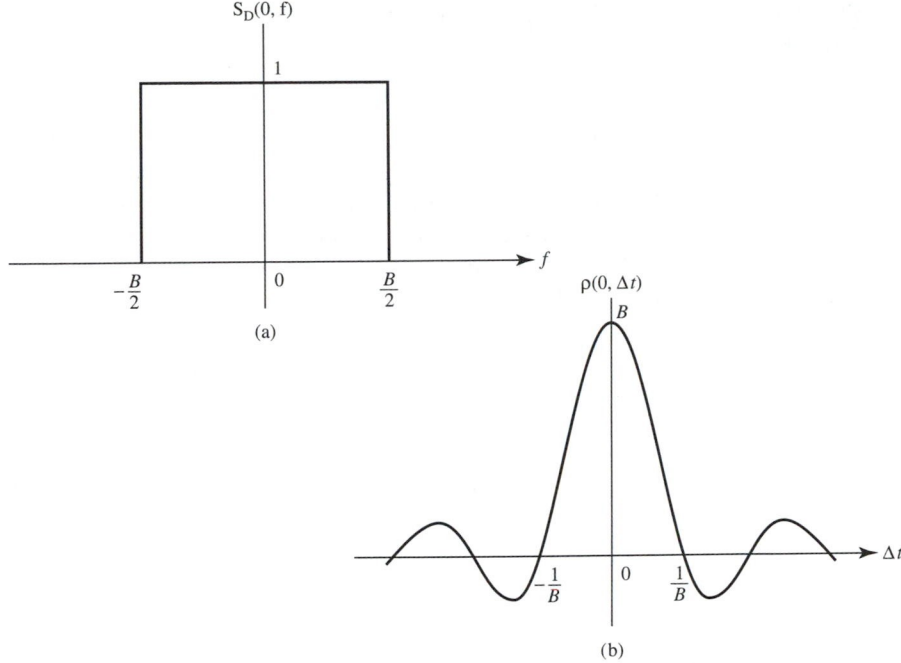

FIGURE 8-3. (a) Doppler scattering function $S_D(0,f)$ and (b) channel correlation function $\rho(0,\Delta t)$ for a time-selective fading channel.

From (8-58) and this example, it may be concluded that provided that (8-62) holds, time samples of the received envelope separated by less than the pulse duration T will be correlated. Below it will be seen that (8-62) is a necessary condition for the development of the nondispersive channel model.

8-2.5 Frequency-Selective Fading Channels

The second subclass of doubly spread channels that will be examined are known as frequency-selective fading channels [19]. As is the case for time-selective fading channels, frequency-selective fading channels result from doubly spread channels by making different assumptions about characteristics of the scattering medium. For frequency-selective fading channels, it is assumed that the scattering medium may be modeled as a *fixed* (nonmoving) volume consisting of differential layers. For a transmitted signal $s_0(t)$ given by (8-1), the received signal scattered from a single layer is

$$s(t,\tau) = \text{Re}[\beta(\tau)u_0(t - t_0 - \tau)e^{j2\pi f_0(t-t_0-\tau)}\,d\tau] \qquad (8\text{-}63)$$

where t_0 is the nominal propagation delay due to the scattering layer and $\beta(\tau)$ is the transmission coefficient for a wave scattering off a layer whose propagation delay is τ. Note in particular that $\beta(\tau)$ is a random *variable* that is indexed by the additional delay τ. It will be assumed that $\beta(\tau)$ is a zero-mean complex random variable.

The total received signal is a superposition of all the responses due to the individual layers:

$$s(t) = \text{Re}\left[\int_{-\infty}^{\infty} \beta(\tau)u_0(t - t_0 - \tau)e^{j2\pi f_0(t-t_0-\tau)}\, d\tau\right] \tag{8-64}$$

where $\beta(\tau)$ is assumed to be zero except for the finite region where scatterers exist. Since $\beta(\tau)$ is a zero-mean complex Gaussian random variable with a Rayleigh-distributed amplitude and a uniformly distributed phase, the factor $e^{-j2\pi f_0\tau}$ may be absorbed into $\beta(\tau)$. In this case, (8-64) becomes

$$s(t) = \text{Re}\left[\int_{-\infty}^{\infty} \beta(\tau)u_0(t - t_0 - \tau)e^{j2\pi f_0(t-t_0)}\, d\tau\right] \tag{8-65}$$

or, equivalently,

$$s(t) = \text{Re}[u(t - t_0)e^{j2\pi f_0(t-t_0)}] \tag{8-66}$$

where

$$u(t) = \int_{-\infty}^{\infty} \beta(\tau)u_0(t - \tau)\, d\tau = \int_{-\infty}^{\infty} \beta(t - \tau)u_0(\tau)\, d\tau \tag{8-67}$$

Equations (8-65) or (8-66) and (8-67) together with the condition that the individual scattering layers are uncorrelated constitute the frequency-selective fading channel model. Observe that (8-65) could have been derived directly from (8-12) simply by dropping the time dependence of $\beta(\tau,t)$. Thus frequency-selective fading channels are properly designated a subclass of doubly spread channels. Note that (8-67) describes the input–output characteristics for a linear time-invariant filter with complex impulse response $\beta(t)$. The effect of a frequency-selective fading channel on the envelope of an on–off pulse signal is shown in Figure 8-4. The transmitted signal shown at the top is of duration T. The envelope of the received signal is shown at the bottom. Not only is the amplitude distorted by the random fading process, but the signal is spread in time to duration $T_1 + \Delta t$. For reasons that will become apparent, frequency-selective fading channels are also known as time-flat fading channels [19], channels dispersive only in time [18], and range or delay-spread fading channels [6].

The channel correlation function for a frequency-selective fading channel is

$$g(\tau) \equiv \rho(\tau,0) \tag{8-68}$$

The term $g(\tau)$ is also known as the range scattering function, the multipath intensity profile, or the power delay spectrum for a frequency-selective fading channel. The range of values for which $g(\tau)$ is nonzero is known as the multipath spread of the channel.

The channel scattering function for a frequency-selective fading channel may be found using (8-68) in (8-31):

$$S_{DR}(\tau,f) = g(\tau)\delta(f) \tag{8-69}$$

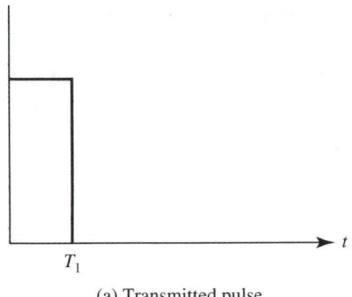

(a) Transmitted pulse

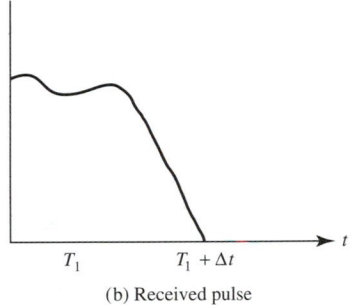

(b) Received pulse

FIGURE 8-4. Effect of transmitting a pulse over a frequency-selective fading channel.

Using (8-69), the mean delay for a frequency-selective fading channel is

$$m_R = \frac{1}{2\sigma_b^2} \int_{-\infty}^{\infty} \tau g(\tau) \, d\tau \tag{8-70}$$

while the mean-square delay spread is given by

$$L = \frac{1}{2\sigma_b^2} \int_{-\infty}^{\infty} \tau^2 g(\tau) \, d\tau - m_R^2 \tag{8-71}$$

where

$$2\sigma_b^2 = \int_{-\infty}^{\infty} g(\tau) \, d\tau \tag{8-72}$$

Using (8-69) in (8-33) and (8-35), the mean Doppler shift and the mean-square Doppler spread for a frequency-selective fading channel are found to be identically zero.

EXAMPLE 8-2

The relationship between the two-frequency correlation function, the transmitted envelope, and the correlation between frequency samples of the received signal envelope will be demonstrated in this example. To do this, the correlation between frequency samples of the Fourier transform of the received signal envelope when a bandlimited signal is transmitted must first be computed. Using (8-68) in (8-45), the frequency correlation function $R_{FE}(f_1,f_2)$ becomes

$$R_{FE}(f_1,f_2) = \int_{-\infty}^{\infty}\int_{-\infty}^{\infty}\int_{-\infty}^{\infty} g(\tau)u_0(t_1 - \tau)u_0^*(t_2 - \tau)e^{j2\pi(f_2 t_2 - f_1 t_1)}\, d\tau\, dt_1\, dt_2 \quad (8\text{-}73)$$

Equation (8-73) may be rewritten as

$$R_{FE}(f_1,f_2) = U_0(f_1)U_0^*(f_2)\int_{-\infty}^{\infty} g(\tau)e^{j2\pi\tau(f_2 - f_1)}\, d\tau \quad (8\text{-}74)$$

where $U_0(f)$ is the Fourier transform of $u_0(t)$ as defined in (8-2). Note that the quantity

$$R_R(f,0) \equiv \int_{-\infty}^{\infty} g(\tau)e^{-j2\pi\tau f}\, d\tau \quad (8\text{-}75)$$

appearing in (8-74) is the two-frequency correlation function $R_{DR}(f,t)$ for zero time separation as defined in (8-47). Using (8-75), (8-74) may be written as

$$R_{FE}(f_1,f_2) = U_0(f_1)U_0^*(f_2)R_R(f_1 - f_2,0) \quad (8\text{-}76)$$

Now let $g(\tau)$ be of the form

$$g(\tau) = \begin{cases} 1 & -\dfrac{L}{2} \leq \tau \leq \dfrac{L}{2} \\ 0 & \text{elsewhere} \end{cases} \quad (8\text{-}77)$$

Figure 8-5 shows the channel correlation function and the corresponding two-frequency correlation function for this example.

Using (8-75), the two-frequency correlation function for the channel correlation function given by (8-77) is

$$R_R(f_1 - f_2,0) = L\frac{\sin \pi L(f_1 - f_2)}{\pi L(f_1 - f_2)} \quad (8\text{-}78)$$

Assume that a signal is transmitted whose envelope function has a Fourier transform given by

$$U_0(f) = \begin{cases} 1 & -\dfrac{W}{2} \leq f \leq \dfrac{W}{2} \\ 0 & \text{elsewhere} \end{cases} \quad (8\text{-}79)$$

Using (8-78) and (8-79) in (8-76), it is readily seen that if

$$W > \frac{1}{L} \quad (8\text{-}80)$$

then the frequency components of $U_0(f)$ separated by multiples of $1/L$ hertz will

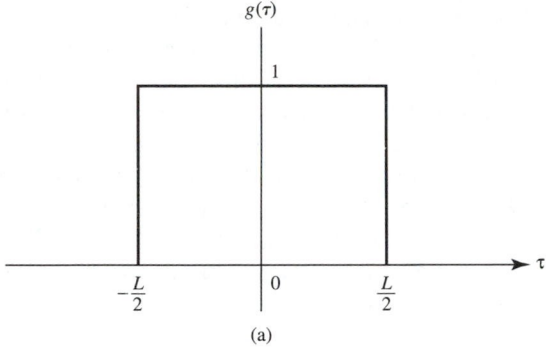

$$g(\tau)$$

(a)

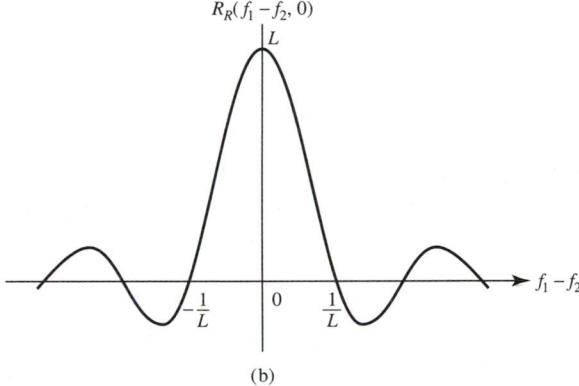

$$R_R(f_1 - f_2, 0)$$

(b)

FIGURE 8-5. (a) Range scattering function $g(\tau)$ and (b) two-frequency correlation function $R_R(f_1 - f_2, 0)$ for a frequency-selective fading channel.

be uncorrelated. From (8-46) it is seen that the correlation bandwidth of the channel for this case is

$$W_c = \frac{1}{L} \tag{8-81}$$

If, however, instead of condition (8-80) the condition is

$$W \ll \frac{1}{L} \tag{8-82}$$

then using (8-78) and (8-79) in (8-76), the samples of the Fourier transform are correlated for all frequencies in the bandwidth of the signal.

Below it will be seen that (8-82) is a necessary condition for the development of the nondispersive fading channel model.

Before leaving this section, mention should be made of a concept frequently encountered in the literature on fading channels—namely, that of time-frequency duality [6,17]. A certain similarity should have been observed between the equations characterizing a time-selective channel and the equations characterizing a frequency-selective channel. As an example of this, compare the expressions for the correlation between time samples of the received envelope for a time-selective fading channel given by (8-58) and the correlation between frequency samples of the Fourier transform of the received signal envelope for a frequency-selective fading channel given by (8-76). Except for the fact that time appears in the arguments of (8-58) while frequency appears in the arguments of (8-76) and that the Fourier transforms of the terms of (8-58) are used in (8-76), the two expressions are identical in form. To emphasize this "duality" concept, by taking a Fourier transform, (8-67) may be expressed as

$$U(f) = H(f)U_0(f) \qquad (8\text{-}83)$$

where $H(f)$ is the Fourier transform of a sample function of $\beta(\tau)$:

$$H(f) = \int_{-\infty}^{\infty} \beta(\tau)e^{-j2\pi f\tau}\, d\tau \qquad (8\text{-}84)$$

Comparing (8-51) with (8-83), it may be seen that except for the fact that Fourier transforms are used in (8-83), the defining equations for the two channel models are identical in form. As was noted above, even the conditions of wide-sense stationarity for the time-selective fading channel model and uncorrelated scattering for the frequency-selective fading channel model are dual concepts. Bello [17] has thoroughly discussed the various concepts associated with time-frequency duality. Using Bello's definitions, Van Trees [6] has proved that a time-selective fading channel is the dual of a frequency-selective fading channel. The principal advantages of recognizing this duality are in deriving equivalent-circuit models of fading channels, designing optimal receivers for the two types of channels [6], and in evaluating the performance of communication systems over time-selective and frequency-selective fading channels [19,20].

8-2.6 Nondispersive Fading Channels

In the preceding two sections, two subclasses of the doubly spread fading channel model have been discussed. To derive the channel models for these two subclasses, it has been assumed either that the scattering medium is moving but may still be modeled as a single layer, or that the scattering layer is fixed but consists of differential layers. In this section it will be assumed that the scattering medium is both fixed (nonmoving) and can be modeled as a single layer.

For a transmitted narrowband signal given by (8-1), the received signal, after passing through a fixed, single-layer scattering medium, is given by

$$s(t) = \mathrm{Re}[\beta u_0(t - t_0)e^{j2\pi f_0(t-t_0)}] \qquad (8\text{-}85)$$

where t_0 is the propagation time between the transmitter and the receiver and β is assumed to be a zero-mean complex Gaussian *random variable*. Equation (8-85) constitutes the fading channel model for a nondispersive fading channel [18]. Note that (8-85) could have been derived directly from the channel model for a doubly spread fading channel (8-12) by dropping the time dependency of $\beta(\tau,t)$ and by noting that scattering is due to a single layer. A more precise derivation of the nondispersive fading channel model starting with the doubly spread fading channel model follows below.

The channel correlation function for a nondispersive fading channel is given by

$$\rho(\tau,\Delta t) = \rho(0,0)\delta(\tau) \qquad -\infty < \tau < \infty; \quad -\infty < \Delta t < \infty \qquad (8\text{-}86)$$

Using (8-86) in (8-31), the scattering function for the nondispersive fading channel is given by

$$S_{DR}(\tau,f) = \rho(0,0)\delta(\tau)\delta(f) \qquad (8\text{-}87)$$

From (8-87) and (8-32) to (8-35), the mean delay, mean-square delay spread, mean Doppler shift, and mean-square Doppler shift for a nondispersive fading channel are all identically zero. The nondispersive fading channel thus exhibits no spreading in frequency or time—hence the name *nondispersive*. Another term often used to describe the nondispersive fading channel is flat-flat fading channel [19]. Since β in (8-85) has a Rayleigh-distributed amplitude and a uniformly distributed phase, the nondispersive fading channel is also known simply as a Rayleigh fading channel.

As mentioned above, a nondispersive fading channel is a special case of a doubly spread channel. More generally, it is a special case of both time-selective fading and frequency-selective fading channels. Above it was observed that for a time-selective fading channel satisfying (8-62), time samples of the received envelope would be correlated over the duration of the transmitted envelope. It was also noted that for a frequency-selective fading channel and a signal satisfying (8-82), frequency samples of the received envelope would be correlated for all frequencies in the bandwidth of the transmitted signal. Since a nondispersive fading channel does not exhibit spreading in either time or frequency, (8-62) and (8-82) combined must be the condition for a doubly spread channel to exhibit nondispersive fading:

$$BL \ll \frac{1}{WT} \qquad (8\text{-}88)$$

Since the time–bandwidth product for any signal must be greater than unity [21],

$$TW > 1 \qquad (8\text{-}89)$$

(8-89) combined with (8-88) indicates that nondispersive fading will occur provided that

$$BL \ll 1 \qquad (8\text{-}90)$$

From (8-40) it is seen that nondispersive fading will occur in a doubly spread channel provided that the channel is underspread.

In Figure 8-6 the hierarchy of fading channels that have been considered up to this point are shown with the relationships between each of the channel models indicated by arrows.

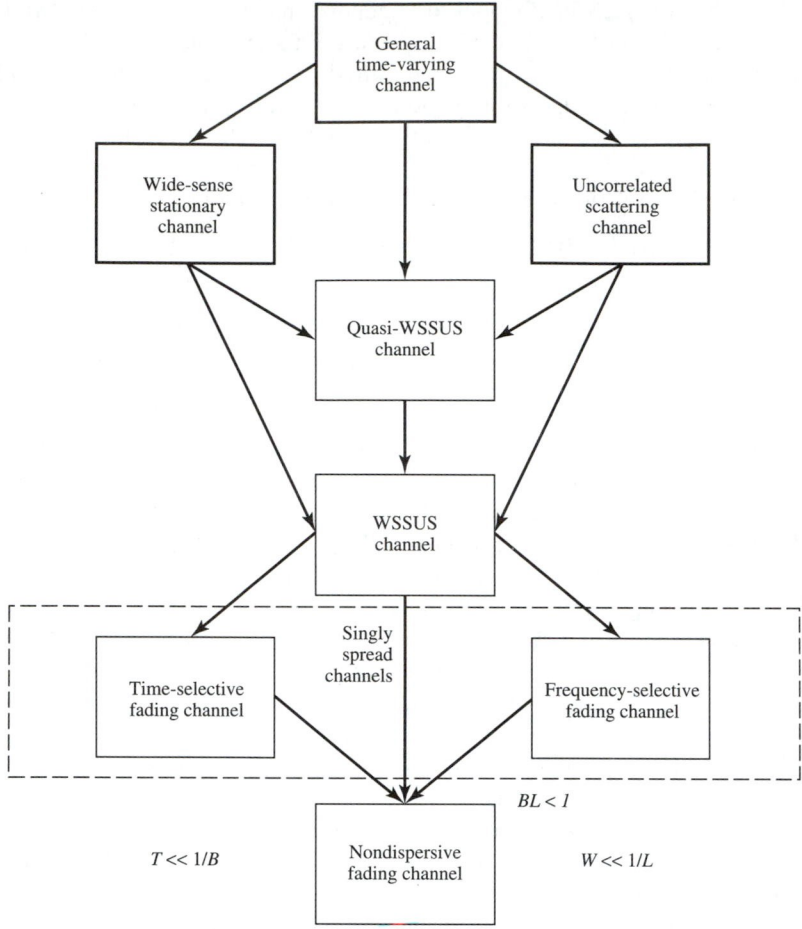

FIGURE 8-6. Summary of hierarchy of fading channel models.

8-3 Characterization of the Mobile Radio Channel

The discussion in Section 8-2 concentrated on determining the statistical parameters of a *general* fading channel. In this section a specific example of the general fading channel model, the mobile radio channel, will be characterized in terms of its statistical properties. In addition, the fundamentals of propagation in the mobile radio environment will be introduced. The mobile radio channel exhibits the short-term statistics of a doubly spread GWSSUS channel described in the preceding section. In addition, due to the specific environment in which a mobile radio operates, the mobile radio channel is often characterized by long-term statistics that are distributed on a lognormal basis. This long-term statistic is caused by *shadowing* by various obstacles in the direct path of the mobile radio wave. Because the presence or absence of these obstacles varies from location to location, the ability to communicate with any given mobile radio station may be statistically characterized in terms of a *coverage reliability* parameter. In the following subsections, each of the foregoing properties of the mobile radio channel will be described further. The material discussed here is intended to be an overview of the mobile radio channel with particular emphasis on those aspects that affect the performance of spread-spectrum mobile radio systems operating in the cellular ($\approx$800 to 900 MHz) bands and in the personal communication system (PCS) bands ($\approx$1.7 to 2.2 GHz). Because of the growing interest in personal communication systems, cellular radio systems, and wireless data devices, there have a number of recent texts that treat the mobile radio channel in more detail. These include the texts by Jakes [9], Lee [22], Parsons [23], Hess [24], and Linnartz [25]. The interested reader should consult these texts for more information on the mobile radio channel, particularly in the subject area of mean path attenuation.

8-3.1 Time-Selective Fading

As Section 8-2 noted, a doubly spread GWSSUS channel model exhibits both time- and frequency-selective fading. The earliest studies of the mobile radio channel (see, e.g., Refs. 26 to 28) indicated the presence of time-selective fading on both narrowband and wideband radio channels. The time-selective fading manifests itself as rapid variations of the received signal envelope as the mobile receiver moves through a field of local scatterers. Figures 8-7 and 8-8 illustrate the situation in which the envelope of the received signal is plotted as a function of time for a vehicle traveling at 10 and 25 mph, respectively, with a transmitted sinusoidal carrier of 845 MHz. As will be noted shortly, the duration of the fades from the median signal level is dependent on vehicle speed and the operating frequency.

From Section 8-2, a time-selective fading channel may be characterized in terms of its Doppler power spectrum, correlation time, and so on. As Gans [28] has noted, the statistical characteristics of the received signal for a mobile radio channel are functions of the polarization of the antenna with respect to the received signal. While this would lead to three separate characterizations of time-selective fading for the mobile radio channel, in the following, only one such characterization (the ''vertical monopole'' case) is considered, with the other two cases of polarization described in the literature (see, e.g., Ref. 22). From consideration of the angle of arrival and the polarization of the antenna, the autocorrelation function of the signal received by a mobile radio receiver corresponding to a constant, unmodulated carrier transmitted signal is given by

$$a(\tau) = J_0(2\pi f_m \tau) \qquad (8\text{-}91)$$

where $J_0(\cdot)$ is the zero-order Bessel function of the first kind and f_m is the maximum Doppler frequency shift given by

$$f_m = \frac{v}{\lambda} = \frac{v f_0}{c} \qquad (8\text{-}92)$$

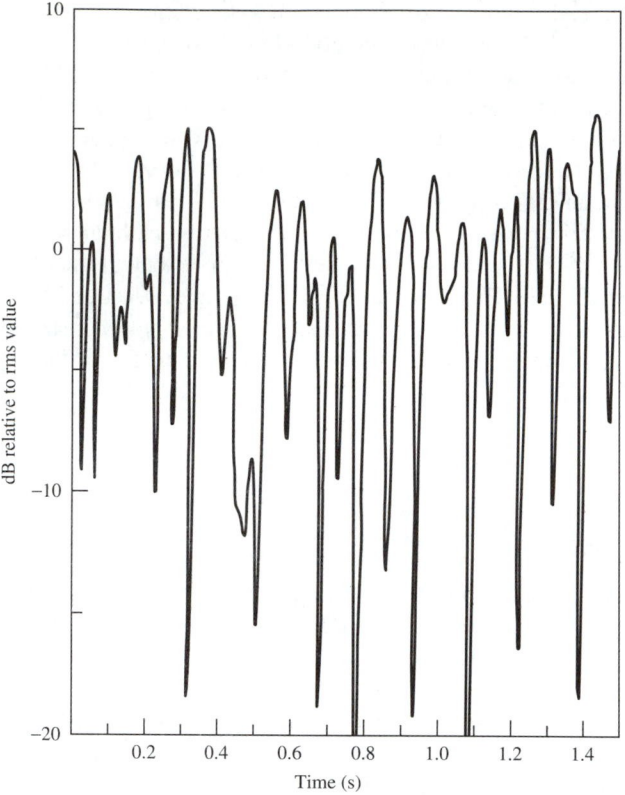

FIGURE 8-7. "Typical" Rayleigh fading at 845 MHz; vehicle speed is 10 mph.

in which v is the vehicle speed (in m/s), λ the wavelength of the transmitted signal, c the speed of light ($= 3 \times 10^8$ m/s in a vacuum), and f_0 the frequency of the transmitted signal (in hertz). The Doppler spectrum is the Fourier transform of the autocorrelation function and is given by

$$
S(f) = \begin{cases} \dfrac{1}{\pi f_m} \dfrac{1}{\sqrt{1 - (f/f_m)^2}} & |f| \le f_m \\ 0 & \text{elsewhere} \end{cases}
\tag{8-93}
$$

Equation (8-93) is valid only for vertical monopole antennas and scatterers uniformly distributed around the antenna. A plot of the Doppler spectrum for a mobile radio channel is shown in Figure 8-9. Note the abrupt frequency cutoff of the Doppler spectrum as a function of the maximum Doppler frequency f_m. The correlation (or coherence) time (see Section 8-2) of this channel is usually assumed to be given by

$$
\tau_c = \frac{1}{2f_m}
\tag{8-94}
$$

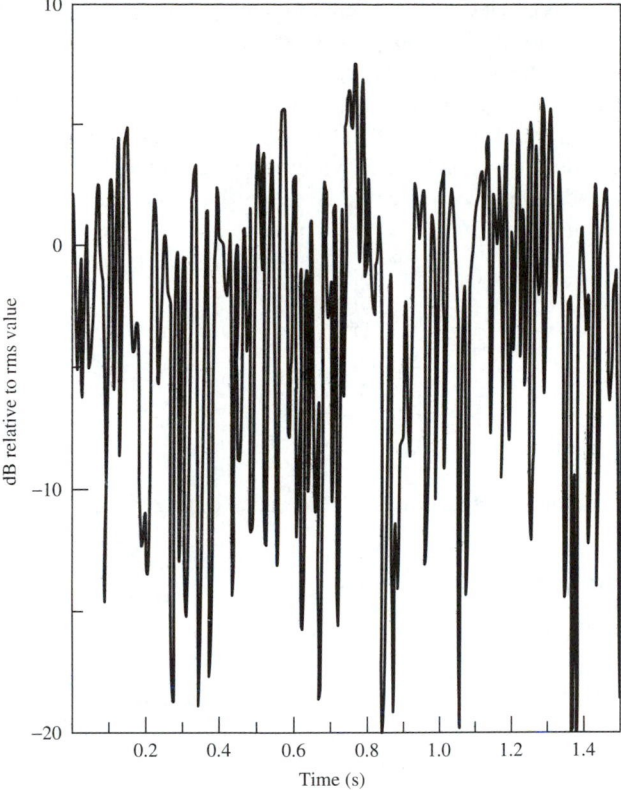

FIGURE 8-8. Example of typical Rayleigh-faded signal at 845 MHz; vehicle speed is 25 mph.

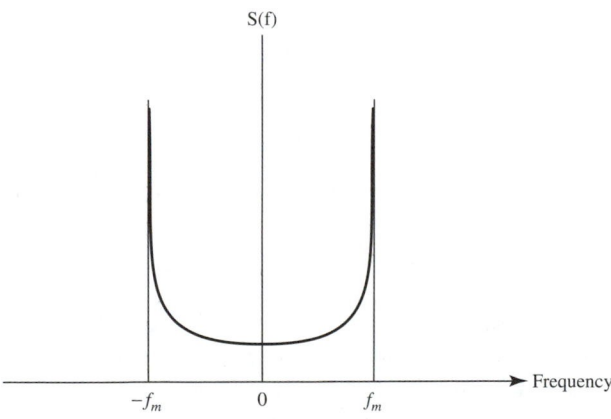

FIGURE 8-9. Plot of Doppler spectrum for a mobile radio channel.

In addition to the correlation time of the channel, in the digital communications literature, the term *normalized Doppler* appears. The normalized Doppler is defined to be the product of the maximum Doppler frequency and the symbol duration of the transmitted signal. As indicated above, the duration of a fade is a function of the vehicle speed and the mobile radio frequency. It can be shown [9, 22] that for a vertically polarized monopole in a Rayleigh-fading mobile radio environment, the duration of a fade, defined to be the average duration of a fade below a given threshold A, is given by

$$\bar{t} = \frac{\lambda}{\sqrt{2\pi}\, vR} (e^{R^2} - 1) \tag{8-95}$$

where

$$R = \frac{A}{\sqrt{2\sigma^2}} \tag{8-96}$$

in which $2\sigma^2$ is the average power of the received signal.

8-3.2 Frequency-Selective Fading

In addition to exhibiting time-selective fading, the mobile radio channel also exhibits frequency-selective fading, as characterized in Section 8-2. In that section it was noted that frequency-selective fading may be characterized in terms of the mean delay and the mean delay spread of the channel. For discrete multipath arrival times, a number of related parameters are also often used, including mean excess delay, rms delay spread, and maximum excess delay. The terms are defined as follows.

Let $P_m(\tau_k)$ denote the average power delay profile, where τ_k is the propagation delay from the transmitter to the receiver. Then

$$\text{mean excess delay} \equiv d_m = \frac{\Sigma\, \tau_k P_m(\tau_k)}{\Sigma\, P_m(\tau_k)} - \tau_A \tag{8-97}$$

where τ_A is the first arrival delay. Similarly,

$$\text{rms delay spread} \equiv s_m = \left[\frac{\Sigma(\tau_k - d_m - \tau_A)^2 P_m(\tau_k)}{\Sigma\, P_m(\tau_k)}\right]^{1/2} \tag{8-98}$$

and maximum excess delay $= \max(\tau_k - \tau_A)$.

In Table 8-1, some "typical" delay spread profile parameters from various sources are listed. Observe from the table that the rms delay spread values are on the order of 1 to 3 μs for urban terrain, 0.1 to 1 μs for suburban terrain, and 6 to 7 μs for rural mountainous terrain, respectively. Note that while the values cited are, for the most part, for 900 MHz, the same values are expected to hold for 2 GHz.

In addition to the various delay profile parameters noted above, in characterizing spread-spectrum systems operating over mobile radio channels it is often necessary to specify the power delay profile of the channel. Because the power delay profile is location dependent, this is often a very difficult task. Nonetheless, in 1986, a scientific study group in Europe specified several "average" power delay profiles that

TABLE 8-1. Typical Delay Spread Results Reported in the Literature

Source	Location	Terrain	Mean Excess Delay (μs)	RMS Delay Spread (μs)	Maximum Excess Delay (μs)	Frequency (MHz)
Cox and Leck [29–31]	New York City	Urban	1.1	1.3	4	910
Van Rees [32]	The Hague, The Netherlands	Urban			3	910
Cox [33, 34]	New Jersey	Suburban	0.03–1.9	0.1–2.05	4.5	910
Nielson [35–37]	San Francisco and Palo Alto	Urban–Suburan		2.1–2.5	5–6	1370
Turin [38]	San Francisco and Berkeley	Urban–Suburban			1–4	488, 1280, 2920
Hata [39]	Tokyo	Urban		3	5	920
Nilson [40]	Stockholm	Urban			6	900
Bajwa [41]	Birmingham, England	Urban Suburban	1.5 0.6	1 0.8		450
Zogg [42]	Berne, Switzerland	Rural with Mountains	4.8	6.8	35	
Bang [43]	Orkanger, Norway	Rural with Mountains		7	20	949
Young [44]	New York City	Urban			10	450

modeled the mobile radio channel at 900 MHz quite well and were in fact used later in specifying the performance of the GSM digital cellular system, to be discussed in Section 9-4.5 [45]. Figure 8-10 plots two of these profiles for the typical urban environment and for the hilly terrain environment. Although both of these average power delay profiles are decaying exponentials in delay indicating a minimum phase channel characteristic (i.e., the channel impulse response decays as time increases), non-minimum-phase channels *can* exist in the mobile radio environment and must be taken into account in designing digital communication systems.

8-3.3 Mobile Radio Path Loss

In free space it is well known that radio waves propagate with a path loss characteristic of the form $A(f)d^{-2}$, where $A(f)$ is a function of frequency, antenna gains, reference distance, and so on, and d is the distance (in meters, feet, miles, and so on, consistent with the units of the reference distance) between the transmitter and the receiver. In the mobile radio environment, propagation is no longer in free space and is often not line of sight. Factors affecting the mean path loss include base and

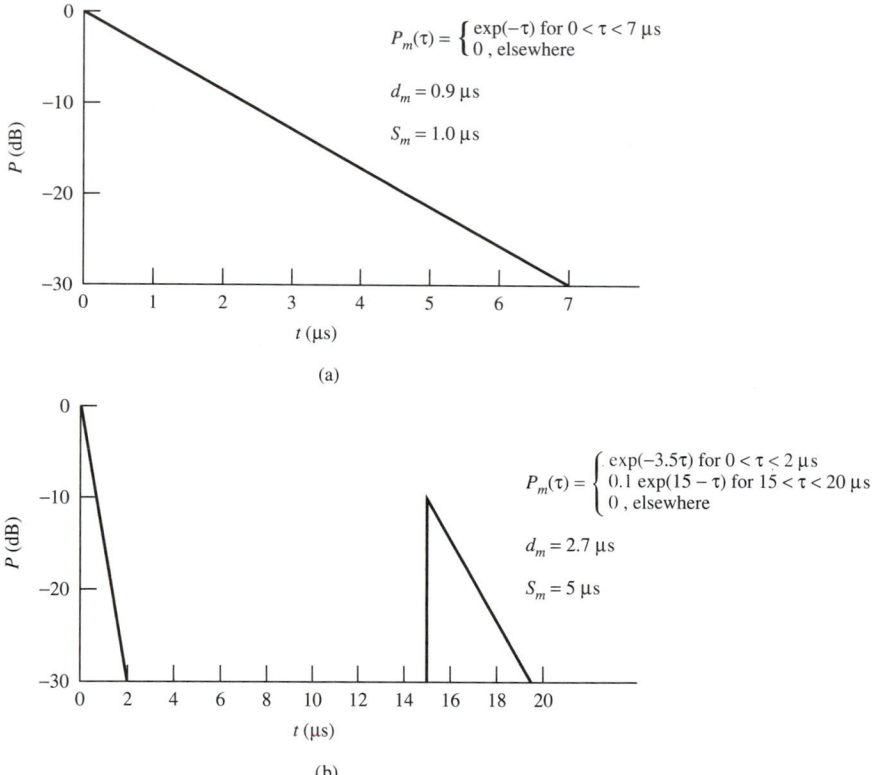

FIGURE 8-10. Cost 207 average power delay profiles: (a) typical delay profile for suburban and urban areas; (b) typical "bad"-case delay profile for hilly terrain. (From Cost 207 Document 207 TD (86)51 rev 3.)

mobile antenna heights, operating frequency, the presence (or absence) of localized reflectors or absorbers, the presence (or absence) of buildings, trees, and foliage between the transmitter and the receiver, and various ducting effects dependent on atmospheric conditions. Over the period of many years, a number of attempts at predicting the mean path loss in the land mobile radio environment have been made. These include the Egli model, the Longley–Rice model, the Okumura method, and the Hata model, all of which are described in the text by Parsons [23]. Of these methods, the Okumura method [46] has often been used for land mobile radio channels. The Okumura method for mean path loss prediction is based on an extensive experimental study of propagation path loss in various terrain for frequencies from 100 MHz to 3 GHz. In an attempt to simplify the results of Okumura, Hata [47] fitted a set of lines to the Okumura data resulting in a path loss model of the form $A + B \log_{10} R$ (in dB), where A and B are functions of frequency and height and R is the distance between the transmitter and the receiver. The Hata approximation for a large city in an urban area is shown in Table 8-2.

The Hata model thus predicts mean path loss as a functional of the form Kd^{-r}, where r ranges from 2 to 4. This form of the propagation loss model can simplify a number of calculations that predict cell coverage reliability and co-channel coverage reliability, as will be seen in subsequent sections. As Table 8-2 indicates, the Hata approximation to the Okumura results are valid only for relatively tall base station antennas (30 to 300 m) and for frequencies less than 1.5 GHz. In Ref. 48, a modified Hata path-loss prediction model is provided for use with low base station antenna heights and over the frequency range 1500 to 2000 MHz. The modified version of the Hata model is given by

$$\text{Path loss (dB)} = 46.3 + 33.9 \log_{10}(f(\text{MHz})) - 13.82 \log_{10}(h_{\text{base an}}(m))$$
$$- a(h_{\text{mobile}}(m)) + [44.0 - 6.55 \log_{10}(h_{\text{base}}(m))] \cdot \log_{10}(R(\text{km}))$$

where

$$a(h_{\text{mobile}}(m)) = [1.1 \log_{10}(f(\text{MHz})) - 0.7] \cdot h_{\text{mobile}}(m)$$
$$- [1.56 \log_{10}(f(\text{MHz})) - 0.8]$$

Again this propagation path-loss model is of the form Kd^{-r}.

TABLE 8-2. Hata Model for a Large City in an Urban Area

$$L(\text{dB}) = 69.55 + 26.16 \log_{10} f - 13.82 \log_{10} h_t - a(h_r)$$
$$+ (44.9 - 6.55 \log_{10} h_t) \log_{10} R$$

$$\text{where } a(h_r) = \begin{cases} 8.29(\log_{10} 1.54 h_r)^2 - 1.1 & f \leq 200 \text{ MHz} \\ 3.2(\log_{10} 11.75 h_r)^2 - 4.97 & f \geq 400 \text{ MHz} \end{cases}$$

f = operating frequency in MHz $(150 < f < 1500)$
h_t = transmitter antenna height in meters $(30 < h_t < 200)$
h_r = receiver antenna height in meters $(1 < h_r < 10)$
R = distance between the base station and the mobile station in km $(1 < R < 20)$

8-3.4 Shadowing

As indicated at the start of this section, in addition to the short-term statistics of the mobile radio channel, which lead to time- and frequency-selective fading, the presence of location-dependent obstacles leads to long-term fading or shadowing. Although shadowing results in a time-varying received signal, this fading phenomenon is unlike the time-selective fading considered above, as vehicle speed is *not* a factor in determining the fading statistics. Instead, the nature of the terrain surrounding the base and the mobile antennas as well as the respective antenna heights with respect to the terrain determines the extent of shadowing. Figure 8-11 illustrates a typical shadowing situation. Since obstacles in the propagation path between the base and the mobile antennas lead to shadowing, frequently simply moving the mobile location will change the effects of shadowing.

Shadowing may be modeled as a multiplicative, slowly time-varying random process. Hence if the lowpass-equivalent transmitted signal is given by $u_0(t)$, the received signal $r(t)$ *in the absence of time- and frequency-selective fading* will be given by

$$r(t) = Am(t)u_0(t) \tag{8-99}$$

where A accounts for the antenna gains and the mean path loss between the transmitting antenna and the receiving antenna and $m(t)$ is the random process due to shadowing. In the presence of time and/or frequency-selective fading, (8-99) still holds true with $u_0(t)$ replaced by a suitable term that includes the short-term fading effects.

The shadowing random process is generally assumed to be lognormally distributed (i.e., the distribution function associated with the long-term fading process is a normal distribution function when the values are measured in decibels). If Y is a Gaussian random variable with mean μ and variance σ^2, and if X is related to Y by $Y = \ln X$, then X is lognormally distributed with the density function

$$p(x) = \begin{cases} \dfrac{1}{\sqrt{2\pi\sigma^2}\,x} e^{-(\ln x - \mu)^2/2\sigma^2} & x \geq 0 \\[2mm] 0 & x < 0 \end{cases} \tag{8-100}$$

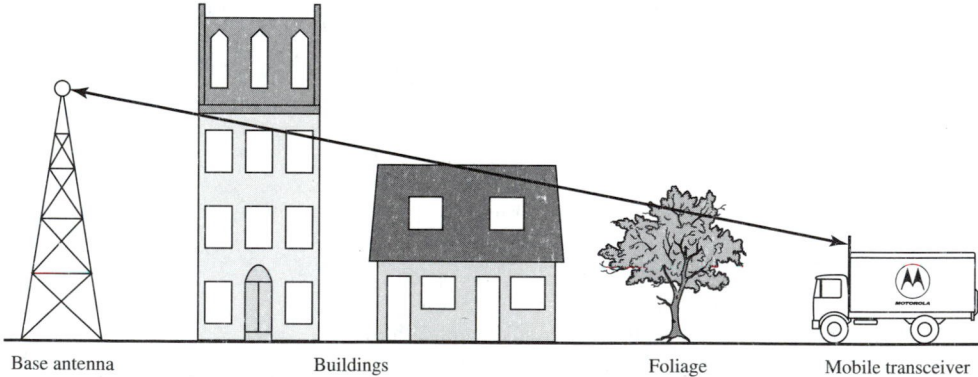

| Base antenna | Buildings | Foliage | Mobile transceiver |

FIGURE 8-11.   Example of shadowing in a mobile radio system.

It may be readily shown that the first and second moments of X are given by

$$E\{X\} = e^{(\mu+\sigma^2/2)}$$
$$E\{X^2\} = e^{(2\mu+2\sigma^2)} \tag{8-101}$$

A plot of the density function of X is shown in Figure 8-12. Provided that all parameters are measured in decibels (as is customary in propagation measurements), the density function usually dealt with is the Gaussian density function of Y, that is,

$$p(y) = \frac{1}{\sqrt{2\pi\sigma^2}} e^{-(y-\mu)^2/2\sigma^2} \tag{8-102}$$

In this case μ is the mean path loss measured in decibels (either measured directly or predicted using a method described in Section 8-3.3) and σ is the standard deviation of the path loss. In cellular systems the value of σ is a critical parameter, as the choice of σ determines both the reliability of coverage and the co-channel interference reliability, as will be seen shortly. The standard deviation σ is a function of terrain and antenna heights. For cellular and microcellular environments, σ may range between 4 and 12 dB.

The lognormal shadowing process accounts for path-loss variations in the terrain that cannot readily be predicted using straightforward mean path-loss predictions. It has been argued by several authors (see, e.g., Linnartz [25]) that the goal of path-loss prediction methods in the mobile radio environment should be to reduce σ to zero. This is an admirable goal but is usually difficult to impossible to accomplish in practice.

8-3.5 Coverage Reliability

Using the lognormal shadowing process characterization developed in Section 8-3.4 it is possible to determine the reliability of signal coverage in a given area. Two measures of coverage reliability are commonly used. The first measure is the frac-

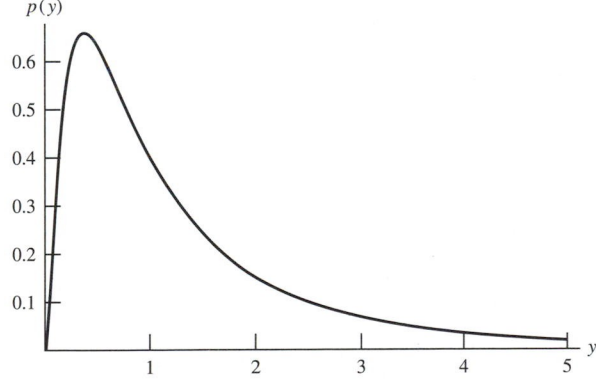

FIGURE 8-12. Plot of density function of lognormal random variable, $\mu = 0$, $\sigma = 1$.

tion of locations at a distance $r = R$ where the received signal has a signal strength above some threshold A measured in dBm, also known as the contour reliability. The threshold may be any value, although usually it is associated with a given level of performance. Let μ be the mean path loss up to the distance $r = R$. Then the probability that the received signal power Z (in dBm) at a distance R exceeds the threshold A is given by

$$
P[Z \geq A] = \int_{A}^{\infty} \frac{1}{\sqrt{2\pi\sigma^2}} e^{-(x-\mu)^2/2\sigma^2} \, dx
$$

$$
= Q\left(\frac{A - \mu}{\sigma}\right)
$$

(8-103)

where $Q(x)$ is the complementary normal distribution function and σ is the lognormal standard deviation in dB.

EXAMPLE 8-3 _____

Assume that, via measurement, the mean received signal strength at a location at distance R from a transmitter is -100 dBm and that the lognormal standard deviation in the area is 6 dB. Assume that the system threshold is -107 dBm. Then the probability that the received signal strength at all locations $r = R$ from the transmitter exceeds the threshold $A = -107$ dBm is given by

$$
Q\left(\frac{-107 - (-100)}{6}\right) = Q\left(\frac{-7}{6}\right) = 87.7\%
$$

The second commonly used measure of coverage reliability is the percentage of locations within a circle of radius R from the transmitter in which the received signal strength exceeds a given threshold value, also known as the area reliability. Let ACP denote the area coverage probability, the fraction of the useful area within a circle of radius R for which the signal strength exceeds a given threshold x_0. Assume that the mean signal strength $\bar{x}$ (in dBm) within the circle of radius R is given by

$$
\bar{x} = \alpha - 10n \log_{10}\frac{r}{R}
$$

(8-104)

where α and n are constants and that shadowing is given by a lognormal distribution with a standard deviation of σ decibels. Jakes [9] (as corrected by Hess [24]) has shown that the ACP is given by

$$
\text{ACP} = \frac{1}{2}\left\{1 - \text{erf}(a) + \exp\left(\frac{1 - 2ab}{b^2}\right)\left[1 - \text{erf}\left(\frac{1 - ab}{b}\right)\right]\right\}
$$

(8-105)

where

$$
a = \frac{x_0 - \alpha}{\sigma\sqrt{2}}
$$

$$b = \frac{10n \log_{10} e}{\sigma \sqrt{2}}$$

$$\text{erf}(z) = \frac{2}{\sqrt{\pi}} \int_0^z e^{-t^2} \, dt$$

For a given propagation environment, σ, n, and hence b are fixed. Often, from system design constraints, the ACP is fixed to a certain percentage. Hence it is necessary to solve (8-105) for a, given the ACP and b. Then given a, the excess path margin required to achieve ACP = 0.9 is given by

$$\text{excess path margin} = a\sigma\sqrt{2} \qquad (8\text{-}106)$$

The set of equations above has been solved numerically and the results are shown in Figure 8-13. This figure plots the excess path loss required for 90% coverage as a function of the lognormal standard deviation σ, with the propagation law n as a parameter. For example, with a propagation law of $n = 4$ (corresponding to the propagation of the form d^{-4}) and a lognormal standard deviation of $\sigma = 6$ dB, 3 dB of additional power above the threshold value of $x_0 = \bar{x}$ into the receiver is required to achieve a 90% ACP within the circle of radius R.

8-4 Requirement for Diversity in Fading Channels

In Sections 8-2 and 8-3 the nature of fading was first described in general and then specialized to the mobile radio case. Although the statistical characterization of fading should now be apparent to the reader, the deleterious effects of fading on the performance of digital communication systems may not be so obvious. Two simple examples will now demonstrate these effects.

EXAMPLE 8-4. Time-Selective Fading ————————————————————

Consider the transmission of a binary PSK signal over a slowly time-selective fading channel. Assume that the fading is sufficiently slow in time such that the frequency spreading of the received equivalent lowpass signal may be neglected. Then the received signal may be written as

$$r(t) = \alpha b(t) + n(t) \qquad (8\text{-}107)$$

where α is a Rayleigh random variable, $b(t)$ is the lowpass equivalent of the BPSK signal, and $n(t)$ represents the complex white Gaussian noise due to the channel. The signal $b(t)$ is assumed to take on the values ± 1, depending on the value of the transmitted data. From Chapter 1 the conditional probability of error of the received signal is given by

$$P_e = Q\left(\sqrt{\frac{2\alpha^2 E_b}{N_0}} \right) \qquad (8\text{-}108)$$

Let γ_0 denote the quantity $E\{\alpha^2\}(E_b/N_0)$. Then the probability density function of

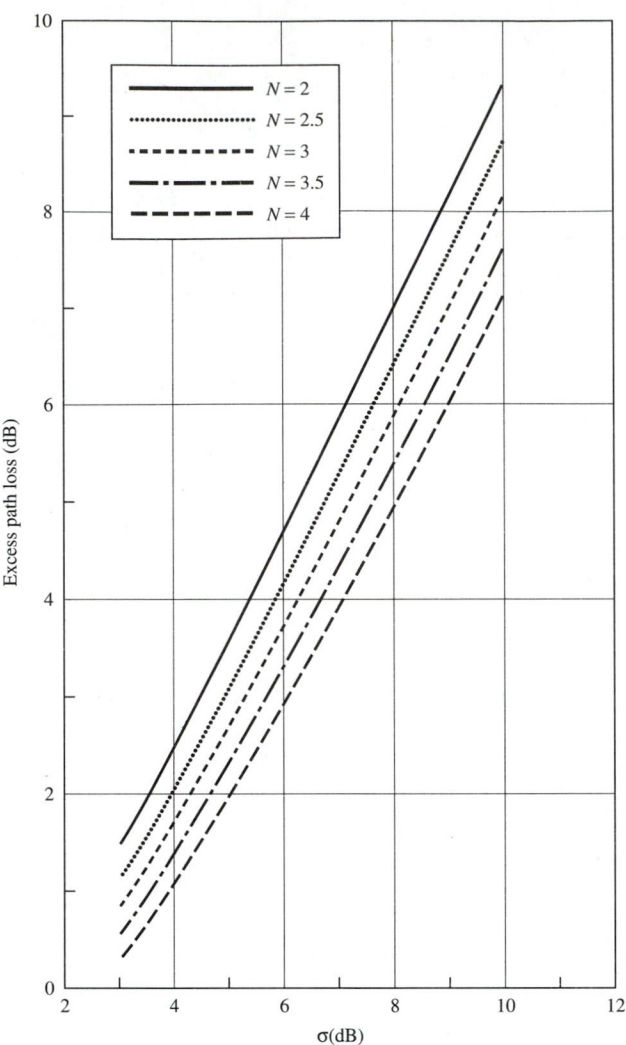

FIGURE 8-13. Excess path loss required for 90% coverage for various propagation laws.

$\gamma = \alpha^2(E_b/N_0)$ is given by

$$f_\gamma(\gamma) = \frac{1}{\gamma_0}e^{-\gamma/\gamma_0} \qquad \gamma \geq 0 \tag{8-109}$$

The (unconditional) average probability of error $\overline{P}_e$ is then given by

$$\overline{P}_e = \int_0^\infty Q(\sqrt{2\gamma})\frac{1}{\gamma_0}e^{-\gamma/\gamma_0}\,d\gamma \tag{8-110}$$

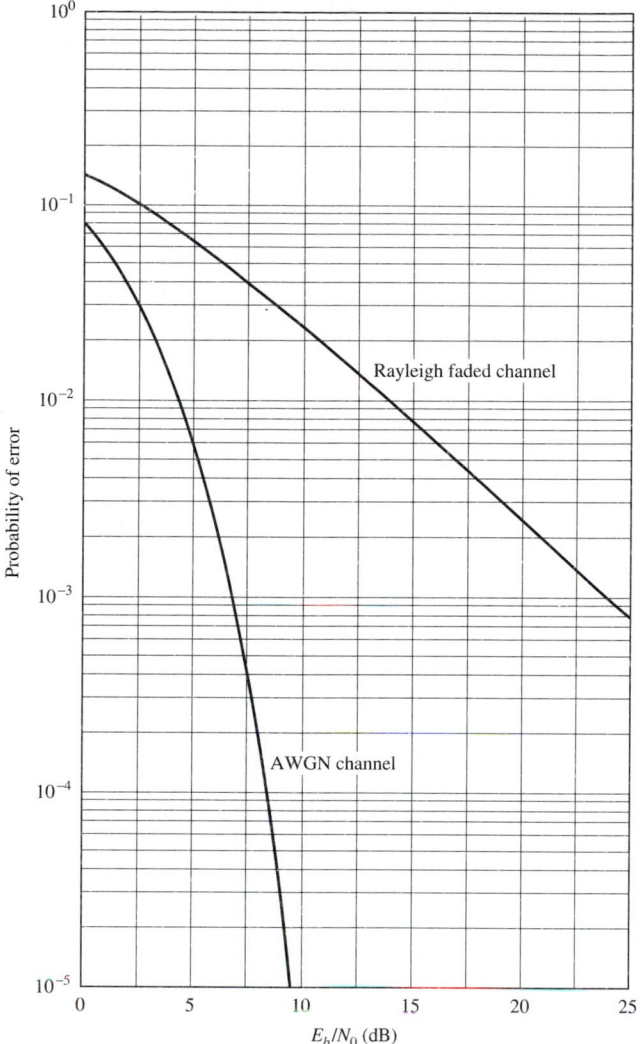

FIGURE 8-14. Probability of error for BPSK over AWGN and Rayleigh-faded channels.

which, after integration yields

$$\overline{P}_e = \frac{1}{2}\left(1 - \sqrt{\frac{\gamma_0}{1 + \gamma_0}}\right) \tag{8-111}$$

In Figure 8-14, the average probability of error $\overline{P}_e$ is plotted along with the probability of error for BPSK transmitted over an AWGN nonfaded channel. Observe the significant degradation in performance due to the Rayleigh-fading amplitude distribution. For example, at a value of $\overline{P}_e = 10^{-3}$, the faded signal requires 17.2 dB more power than the corresponding signal transmitted over an AWGN channel.

EXAMPLE 8-5. Impairment Due to Multipath ————————————

In this example, assume that a BPSK signal with bit period T_b is transmitted over a channel with impulse response $h(t) = \delta(t) + \lambda\delta(t - \tau)$, where τ is limited to the interval $[0, T_b)$. The receiver is assumed to be a correlation receiver matched to the transmitted signal (i.e., the receiver is "mismatched" to the received signal). Letting the transmitted signal be represented by

$$s(t) = Ab(t)\cos\omega_0 t \qquad (8\text{-}112)$$

where A is the amplitude of the signal and $b(t)$ takes on the values ± 1 depending on the transmitted data value, after mixing the received signal with $2\cos\omega_0 t$, the signal into the integrator over the time interval $[0, T_b)$ is given by

$$x(t) = \begin{cases} b_0 A + b_{-1} A\lambda + n(t) & 0 \le t \le \tau \\ b_0 A + b_0 A\lambda + n(t) & \tau \le t \le T_b \end{cases} \qquad (8\text{-}113)$$

where $n(t)$ represents AWGN with a two-sided power spectral density of $N_0/2$ due to the receiver front-end noise and b_{-1} and b_0 represent the values of $b(t)$ during $[-T_b, 0)$ and $[0, T_b)$, respectively. The output of the integrator is given by

$$Z = b_0 A T_b + b_0 A\lambda(T_b - \tau) + b_{-1} A\lambda\tau + \int_0^{T_b} n(t)\, dt \qquad (8\text{-}114)$$

An error occurs when either $Z < 0$ and $b_0 = +1$ or $Z > 0$ and $b_0 = -1$. To determine the probability of error, it is necessary to evaluate the probability of error $P\{\text{error} \mid b_0, b_{-1}\}$ conditioned on the value of the *two* bits b_0 and b_{-1} and then average over the four conditional error probabilities. From symmetry considerations, it may be shown that

$$\begin{aligned} P\{\text{error} \mid +1, +1\} &= P\{\text{error} \mid -1, -1\} \text{ and} \\ P\{\text{error} \mid +1, -1\} &= P\{\text{error} \mid -1, +1\} \end{aligned} \qquad (8\text{-}115)$$

and hence

$$P\{\text{error}\} = \tfrac{1}{2} P\{\text{error} \mid +1, +1\} + \tfrac{1}{2} P\{\text{error} \mid +1, -1\} \qquad (8\text{-}116)$$

Z is a conditional Gaussian random variable with variance $N_0 T_b$. The conditional probability of error given $b_0 = +1$, $b_{-1} = +1$ is then given by

$$\begin{aligned} P\{\text{error} \mid +1, +1\} &= Q\left(\frac{AT_b(1 + \lambda)}{\sqrt{N_0 T_b}}\right) = Q\left(\sqrt{\frac{A^2 T_b(1 + \lambda)}{N_0}}\right) \\ &= Q\left(\sqrt{\frac{2E_b(1 + \lambda)}{N_0}}\right) \end{aligned} \qquad (8\text{-}117)$$

where E_b is the energy per *transmitted* data bit.

Similarly, the conditional probability of error given $b_0 = +1$, $b_{-1} = -1$ is given by

$$P\{\text{error} \mid +1, -1\} = Q\left(\frac{AT[1 + \lambda - 2\lambda(\tau/T_b)]}{\sqrt{N_0 T_b}}\right)$$

$$= Q\left(\sqrt{\frac{A^2 T_b[1 + \lambda - 2\lambda(\tau/T_b)]}{N_0}} \right) \qquad (8\text{-}118)$$

$$= Q\left(\sqrt{\frac{2E_b[1 + \lambda - 2\lambda(\tau/T_b)]}{N_0}} \right)$$

The (unconditional) probability of error is then

$$P_e = \frac{1}{2} Q\left(\sqrt{\frac{2E_b(1 + \lambda)}{N_0}} \right) + \frac{1}{2} Q\left(\sqrt{\frac{2E_b[1 + \lambda - 2\lambda(\tau/T_b)]}{N_0}} \right) \qquad (8\text{-}119)$$

In Figure 8-15, P_e is plotted for several values of $A = \lambda$ and $B = \tau/T_b$†. Note that as τ approaches T_b, P_e is degraded significantly from the single-path case.

Examples 8-4 and 8-5 indicate the rather significant degradation in performance that may occur over Rayleigh-faded or multipath channels. A further degradation in performance will occur if the channel exhibits frequency-selective fading of the type demonstrated in previous sections. For mobile radio applications, this degradation due to fading is often unacceptable, as it will result in requiring that the transmitters involved transmit with the excess power required to overcome the deleterious fading effects. For portable transceivers, this requirement can have a significant impact on the design.

8-4.1 Diversity Approaches

One method commonly employed to overcome the degradation in performance due to fading is the use of *diversity.* The goal of diversity is to reduce the depth of the fades and/or the fade duration by supplying the receiver with multiple replicas of the transmitted signal that have passed over independently fading channels. Given that the channels are independent, the probability that all the channels fade below a certain threshold level at the same time is significantly lower than the probability that one channel will fade below the threshold level.‡ Diversity reception may be achieved by a number of means. In mobile radio applications, the most common approaches employed are time diversity, antenna diversity, multipath diversity, path diversity, and frequency diversity [7,9,49]. Figure 8-16 depicts each of these methods.

In *time diversity,* identical replicas of the transmitter signal are sent spaced in time with the time spacing between transmissions (hopefully) exceeding the coherence time of the channel. The most common method of implementing time diversity is through the use of forward error correction coding with or without interleaving.

---

†The apparent *gain* over the single-path case for *certain* combinations of A and B in the two-path case is due to the fact that the figure plots P_e as a function of the transmitted energy per bit E_b. In the two-path case, the receiver can sometimes take advantage of the energy in the second path, resulting in an increase in the *received* energy per bit.

‡Let p denote the probability $p = \Pr\{x < T\}$, where x is the received signal at some time and T denotes a fixed threshold. Then for L independent channels, the probability that all the channels are below the threshold T is p^L.

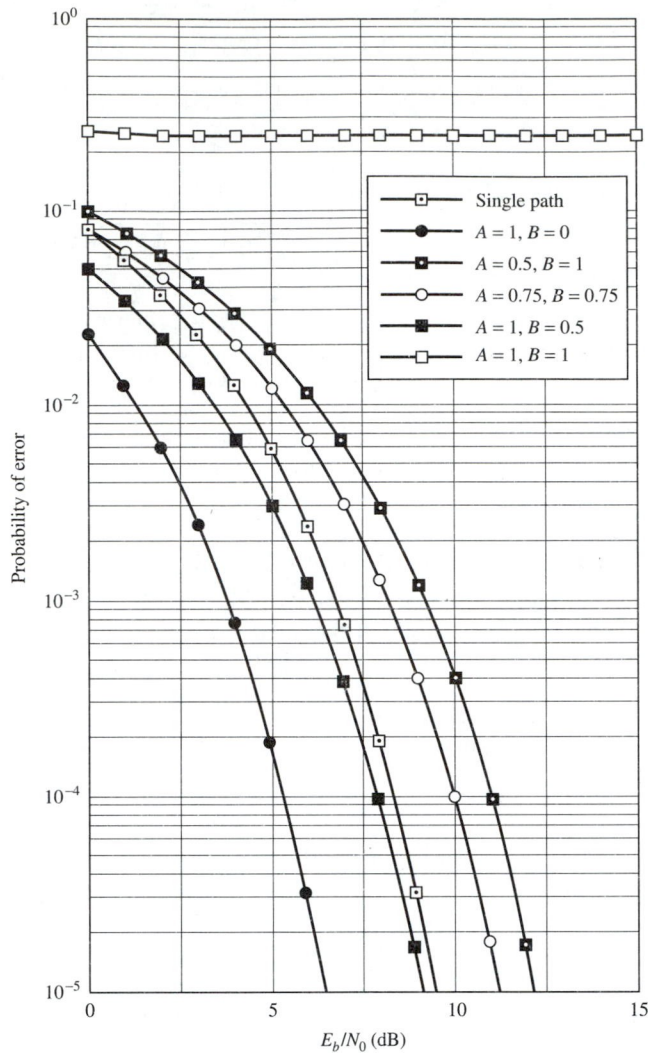

FIGURE 8-15. Probability of error for BPSK transmission over a two-path channel.

For short-duration fades, the error-correcting capability of the code can correct the symbols corrupted during the fade. For longer-duration fades, interleaving is employed to further spread out in time both the information and parity symbols.

In *antenna diversity,* also known as space diversity, multiple antennas spaced sufficiently far apart are employed at the receiver (and sometimes at the transmitter, too). In a Rayleigh-scattering field, the separate antennas will encounter different ''scattering volumes''; hence the received signal will undergo independent fading.

Another diversity approach that is sometimes available is *multipath diversity.* On frequency-selective fading channels, the received signal will consist of multiple

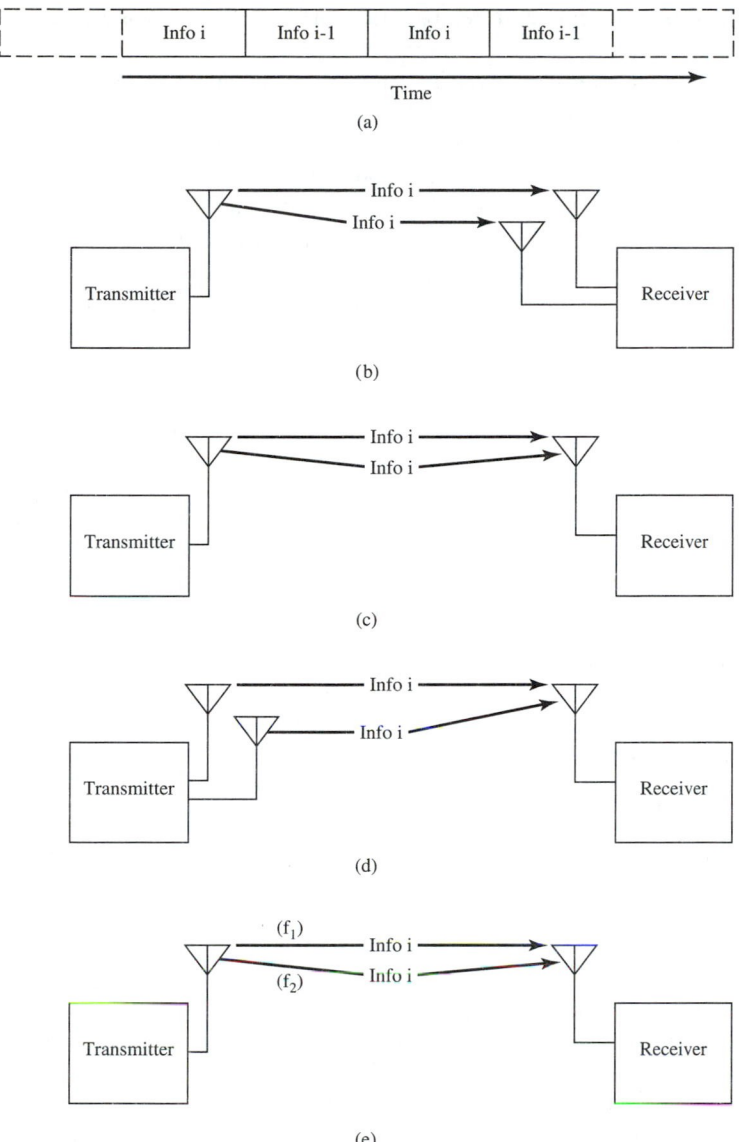

FIGURE 8-16. Methods of diversity in radio systems: (a) time diversity; (b) antenna diversity; (c) multipath diversity; (d) path diversity; (e) frequency diversity.

copies of the transmitted signal, each experiencing a different path delay and amplitude which is dependent on the instantaneous power delay profile of the channel. If the signal bandwidth is sufficiently wide, a receiver can resolve the multipath components and combine the multiple copies in an advantageous manner. A receiver structure that performs this operation is known as a Rake receiver, so-called by its inventors, Price and Green [50]. Because spread-spectrum signals are often wideband, the Rake concept can often be used. The Rake receiver is further developed below.

In some cases, multipath diversity can be achieved in the absence of a frequency-selective fading channel. By intentionally creating different transmitter paths through transmission at different transmitter sites, *path diversity* is created. Although such an approach may initially appear to be wasteful in terms of transmitter resources, it can often overcome the effects of fading, thereby justifying its use.

Finally, in *frequency diversity*, multiple copies of the information signal are sent over independent fading channels, where the frequency spacing between channels exceeds the coherence bandwidth of the channel. Frequency diversity may either be explicit, as in the case of fast-frequency hopping, or implicit in the case of direct-sequence spread spectrum.

These five approaches do not exhaust the list of diversity methods. Other diversity approaches include polarization diversity, angle (of arrival) diversity, and implicit multipath diversity via channel equalization. The interested reader should consult the references given above for more information on the last three approaches.

8-4.2 Diversity Combining Methods

To make use of any of the diversity approaches outlined above, methods of combining the multiple branches in a manner that improves the performance of the communication system must be considered. If the combining process is to be effective, the combiner should not degrade the performance of the system relative to the performance achieved by using any *single* branch available and should improve the performance of the receiver relative to the *best* branch. In this section, two combining methods are considered: maximal ratio combining and equal gain combining. Figure 8-17 shows a block diagram representation of the channel, receiver, and combiner for both of these approaches for Lth-order diversity. For the purposes of this discussion, it will be assumed that binary PSK is sent at the transmitter, and therefore coherent combining is done at the receiver. It is also assumed that predetection combining is being done (i.e., combining is done prior to final detection).

Assume that the transmitter equivalent lowpass signal is given by

$$s(t) = u(t) \tag{8-120}$$

The signal on each of the L branches at the input to the receiver $x_k(t)$ is given by

$$x_k(t) = \beta_k(t)u(t) + n_k(t) \qquad k = 1, ..., L \tag{8-121}$$

where $\beta_k(t)$ is the complex channel gain of the kth path and $n_k(t)$ is the additive noise in the kth receiver branch. The most general linear combining rule is

$$c(t) = \sum_{k=1}^{L} \alpha_k x_k(t) \tag{8-122}$$

For the maximal ratio combining case, it is desirable to maximize the instantaneous signal-to-noise (SNR) ratio and therefore, hopefully, minimize the probability of

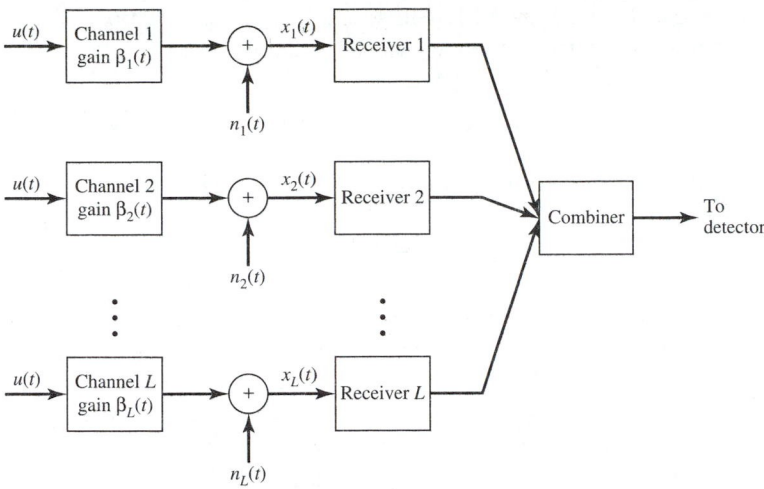

FIGURE 8-17. Block diagram representation of combining approach used in equal gain and maximal ratio combining.

error at the output of the combiner. This may be done by proper selection of the combiner coefficients, as will now be demonstrated.

The signal $s(t)$ and noise $n(t)$ components at the output of the combiner are, respectively,

$$s(t) = u(t) \sum_{k=1}^{L} \alpha_k \beta_k \qquad (8\text{-}123)$$

$$n(t) = \sum_{k=1}^{L} \alpha_k n_k(t) \qquad (8\text{-}124)$$

where it has been assumed that $\beta_k(t)$ is slowly time varying and may be represented by β_k.

Assuming that the $n_k(t)$ are independent, the instantaneous SNR is then given by†

$$\gamma = \frac{\left| \sum_{k=1}^{L} \alpha_k \beta_k \right|^2}{\sum_{k=1}^{M} |\alpha_k|^2 N_k} \qquad (8\text{-}125)$$

where N_k denotes the variance of $n_k(t)$. To maximize the instantaneous SNR γ, the Schwarz inequality for complex quantities may be employed, which is of the form

$$\left| \sum_{k=1}^{L} a_k b_k \right|^2 \le \left(\sum_{k=1}^{L} |a_k|^2 \right) \left(\sum_{k=1}^{L} |b_k|^2 \right) \qquad (8\text{-}126)$$

†Note that $u(t)$ does *not* appear explicitly in (8-125) since $u(t) = \pm 1$.

for complex quantities a_k and b_k. Equality, and hence maximization, is achieved if $b_k = Ka_k$ for an arbitrary complex constant K. Letting

$$a_k = \frac{\beta_k^*}{\sqrt{N_k}} \qquad b_k = \alpha_k\sqrt{N_k} \qquad (8\text{-}127)$$

then the Schwarz inequality is maximized if and only if

$$\alpha_k = K\frac{\beta_k^*}{N_k} \qquad (8\text{-}128)$$

for some arbitrary complex constant K. Referring back to Figure 8-17, the maximal ratio combining rule (8-122) is just a weighted summation of each of the L branches, whereby each branch is multiplied by the complex conjugate of the channel gain β_k and divided by the noise variance of that channel. Hence the receiver must be capable of determining the channel gain and phase as well as the noise variance for each branch. If the N_k consist only of thermal noise, a good receiver design will have equal values of N_k for all k. In the latter case, the optimal linear combining receiver must determine the path phase and gain for each channel. The decision statistic at the output of the matched-filter detector following the maximal ratio combiner in this case is given by

$$Z = \text{Re}\left\{2E\sum_{k=1}^{L}|\beta_k|^2 + \sum_{k=1}^{L}\beta_k N_k\right\} \qquad (8\text{-}129)$$

where E is the energy per transmitted bit.

An alternative combining rule, known as the equal-gain combining rule, simply neglects the channel gain and combines each of the L channels with equal gain (but conjugate phase). This can greatly simplify the receiver structure at the expense of performance. For the equal-gain combining rule, the decision statistic at the output of the maximal ratio combiner in this case is given by

$$Z = \text{Re}\left\{2E\sum_{k=1}^{L}|\beta_k| + \sum_{k=1}^{L}N_k\right\} \qquad (8\text{-}130)$$

8-4.3 Performance of Maximal Ratio Combining

The performance of the maximal ratio combining rule for BPSK may now be determined. In the following it will be assumed that each path of the receiver undergoes independent Rayleigh fading and that each branch exhibits independent thermal noise with equal variance. In this case the decision statistic at the output of the maximal ratio combiner is

$$Z = 2E\sum_{k=1}^{L}|\beta_k|^2 + \sum_{k=1}^{L}\text{Re}\{\beta_k\}N_0 \qquad (8\text{-}131)$$

where the β_k are independent, identically distributed Rayleigh random variables, and $N_0/2$ is the two-sided noise power spectral density. For BPSK, the decision statistic is compared to a threshold of zero to determine if a $+1$ or -1 was transmitted.

To determine the probability of error at the output of the combiner decision process, the conditional probability of error given a fixed set of β_k and N_0 (and hence a fixed SNR) will be evaluated first. The conditional probability of error will then be averaged over the density function of the set of β_k to yield the average probability of error.

Following from the first example in this section, the conditional probability of error given the set of β_k is given by

$$P_e(\gamma_b) = Q(\sqrt{2\gamma_b}) \tag{8-132}$$

where, following Proakis [49], the SNR per bit γ_b is given by

$$\gamma_b = \frac{E}{N_0} \sum_{k=1}^{L} \beta_k^2$$
$$= \sum_{k=1}^{L} \gamma_k \tag{8-133}$$

and γ_k is the instantaneous SNR on the kth branch. Noting that γ_b is the sum of the square of L independent Rayleigh random variables, if it is assumed that all the channels have equal average SNRs, it may be shown [7,49] that the probability density function of γ_b is just that of a chi-squared random variable with $2L$ degrees of freedom and is therefore given by

$$p(\gamma_b) = \frac{\gamma_b^{L-1} e^{-\gamma_b/\gamma_c}}{(L-1)! \, \gamma_c^L} \tag{8-134}$$

where

$$\gamma_c = \frac{E}{N_0} E\{\beta_k^2\} \tag{8-135}$$

is the average SNR per channel.

The unconditional probability of error is thus given by

$$P_e = \int_0^\infty P_e(\gamma_b) p(\gamma_b) \, d\gamma_b \tag{8-136}$$

With some degree of difficulty,† this integral may be evaluated to yield

$$P_e = \frac{(2L-1)! \, _2F_1(L,L+\frac{1}{2},L+1,-1/\gamma_c)}{2^{2L} \gamma_c^L (L-1)! \, L!} \tag{8-137}$$

† This integral does *not* appear in any of the standard tables of integrals, such as M. Abramowitz and I. Stegun, *Handbook of Mathematical Functions* (New York: Dover, 1970) or I. S. Gradshteyn and I. M. Ryshik, *Table of Integrals, Series and Products* (New York: Academic Press, 1965). It may, however, be evaluated readily using one of the advanced symbolic mathematics programs such as Mathematica.

where $_2F_1(a,b,c,d)$ is the (Gauss) hypergeometric function.† Using the special properties of the Gauss hypergeometric function [52], the expression for P_e above may be expressed as

$$P_e = \left(\frac{1}{2}\right)^L \left(1 - \sqrt{\frac{\gamma_c}{1 + \gamma_c}}\right)^L \sum_{k=0}^{L-1} \binom{L + k - 1}{k} \left(\frac{1}{2}\right)^k$$
$$\times \left(1 + \sqrt{\frac{\gamma_c}{1 + \gamma_c}}\right)^k \tag{8-133}$$

In Figure 8-18, P_e is shown plotted as a function of the average SNR per bit $\gamma_b = L\gamma_c$ for various values of the diversity order L. Note that increasing the value of L results in a significant reduction in transmit power required to achieve a given probability of error.

Under conditions of high SNR per channel ($\gamma_c \gg 1$), it may readily be seen that

$$\left(1 + \sqrt{\frac{\gamma_c}{1 + \gamma_c}}\right)^k \longrightarrow 2^k \tag{8-139}$$

By the properties of binomial coefficients, it may be shown [53] that

$$\sum_{k=0}^{M} \binom{r + k}{k} = \binom{M + r + 1}{M} \tag{8-140}$$

Letting $r = M = L - 1$ and noting that $\binom{2L - 1}{L - 1} = \binom{2L - 1}{L}$, it can be seen that for high-SNR conditions,

$$\sum_{k=0}^{L-1} \binom{L + k - 1}{k} \left(\frac{1}{2}\right)^k \left(1 + \sqrt{\frac{\gamma_c}{1 + \gamma_c}}\right)^k$$
$$\longrightarrow \sum_{k=0}^{L-1} \binom{L + k - 1}{k} = \binom{2L - 1}{L} \tag{8-141}$$

It may also be shown that under high-SNR conditions,

$$\left(\frac{1}{2}\right)^L \left(1 - \sqrt{\frac{\gamma_c}{1 + \gamma_c}}\right)^L \longrightarrow \left(\frac{1}{2}\right)^L \left(\frac{1}{2\gamma_c}\right)^L = \left(\frac{1}{4\gamma_c}\right)^L \tag{8-142}$$

Thus under high-SNR conditions, the probability of error P_e for maximal ratio

---

† The Gauss hypergeometric function $_2F_1(a,b,c,d)$ is defined as [see, e.g., Abramowitz and Stegun (1970)]

$$_2F_1 = \frac{\Gamma(c)}{\Gamma(a)\Gamma(b)} \sum_{n=0}^{\infty} \frac{\Gamma(a + n)\Gamma(b + n)}{\Gamma(c + n)} \frac{d^n}{n!}$$

where $\Gamma(x)$ is the gamma function. A C program for $_2F_1(a,b,c,d)$ may be found in L. Baker, *C Mathematical Function Handbook* (New York: McGraw-Hill, 1992).

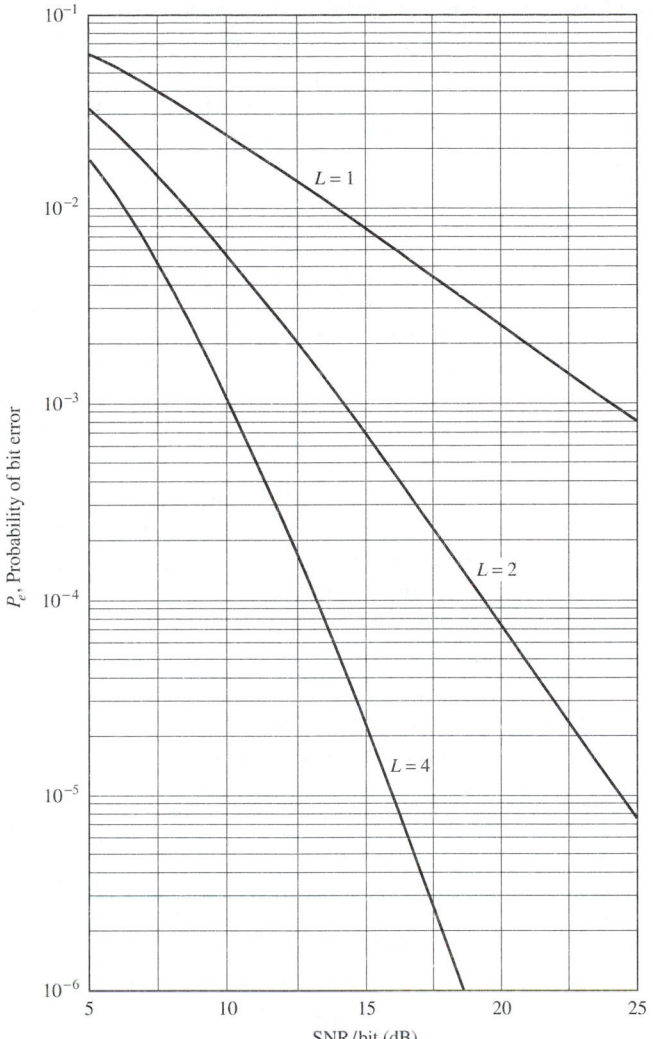

FIGURE 8-18. Performance of binary PSK with L-fold diversity using maximal ratio combining.

combining may be approximated by

$$P_e \approx \binom{2L-1}{L}\left(\frac{1}{4\gamma_c}\right)^L \tag{8-143}$$

This asymptotic approximation to P_e is a very important result. The formula above indicates that for large SNR values, the probability of error over a Rayleigh-fading channel as a function of the SNR can change from an inverse linear relationship for $L = 1$ (no diversity) to an inverse-squared relationship for $L = 2$, to an inverse Lth relationship for arbitrary L. Thus the performance loss resulting from transmission over a Rayleigh-fading channel versus transmission over a AWGN channel can gradually be overcome as the diversity order increases. Furthermore, although the asymptotic result above was derived assuming BPSK and maximal ratio coherent predetection combining, the general relationship between the probability of error and the SNR as a function of the diversity order L for other diversity combining methods may be shown [49] to be of the form

$$P_e \cong K(L)\left(\frac{1}{\gamma_c}\right)^L \qquad (8\text{-}144)$$

where $K(L)$ is a function of L and the particular diversity combining method.

Before leaving the subject of the performance of maximal ratio combining, the reader may question why a similar result for the performance of equal-gain combining could not also be derived. The problem lies in obtaining the probability density function of the coefficients of the combiner $p(\gamma_b)$ in a neat closed-form solutions, as both Brennan [51] and Stein [7] note.† The performance of the equal-gain combiner may best be evaluated using either a numerical integration approach or a Monte Carlo simulation approach. Finally, although the results derived above hold only for the case of equal SNR channels, further results for the case of unequal SNR channels may be found in Stein [7] and Proakis [49].

8-4.4 Other Diversity Combining Methods

Besides the maximal ratio and equal-gain combining approaches toward combining noted above, other methods of achieving diversity exist. In the instances where noncoherent modulation methods are used, *noncoherent* combining methods may be employed together with any of the diversity approaches noted previously. Frequently, square-law noncoherent combining is employed together with noncoherent modulation methods such as FSK or orthogonal modulation methods. The performance of noncoherent combining is analyzed in Ref. 49.

Another approach toward diversity combining is selection combining. Selection combining is defined as the method of choosing as the system decision variable the channel output that yields the largest instantaneous SNR. As such, selection com-

---

†The problem here is that, from (8-130), the decision statistic is, in part, the sum of L independent Rayleigh random variables. This is in contrast to (8-129), where the corresponding sum consisted of the sum of L squares of Rayleigh random variables. As was shown in the latter case, the sum of L independent squared Rayleigh random variables is a chi-squared random variable. The sum of L Rayleigh random variables requires $L - 1$ convolutions of the Rayleigh density function. For $L = 2$, the resulting density function for the sum of just two independent Rayleigh random variables, each having a density function given by

$$f_X(x) = \frac{2x}{A}e^{-x^2/A}$$

is given by

$$f_Y(y) = \frac{y}{Ae^{y^2/A}} - \frac{\sqrt{(\pi/2)}(A - y^2)\,\mathrm{erf}[y/\sqrt{2}\sqrt{A}]}{A^{3/2}e^{y^2/2A}}$$

which is definitely *not* a "neat" result.

bining is not actually a combining method in the sense of maximal ratio combining or equal-gain combining; rather, it is a selection approach toward achieving diversity. Selection diversity may often achieve the performance of a maximal ratio combiner without the expense of complex combining hardware. However, its application to spread-spectrum systems appears to be limited. Further discussion of selection combining may be found in Stein [7] and Jakes [9].

8-4.5 The Rake Receiver

In Section 8-4.1 it was noted that one way of achieving a diversity improvement on a frequency-selective fading channel was through use of a Rake receiver. The Rake receiver concept is based on the notion that a wideband channel may be represented in terms of a discrete-time model, as will now be shown.

Discrete-Time Channel Model. A transmitted signal having a lowpass equivalent representation $u(t)$ is assumed to be transmitted over a frequency-selective fading channel with a time-varying lowpass equivalent impulse response given by $\beta(\tau,t)$. The signal $u(t)$ is assumed to be bandlimited to the frequency range $|f| < W/2$. The sampling theorem will show that the signal $y(t)$ at the output of the channel may be expressed as

$$y(t) = W \frac{\sin \pi Wt}{\pi Wt} * \sum_{n=-\infty}^{\infty} u\left(\frac{n}{W}\right) \delta\left(t - \frac{n}{W}\right) * \beta(\tau,t)$$

$$= W \frac{\sin \pi Wt}{\pi Wt} * \sum_{n=-\infty}^{\infty} u\left(\frac{n}{W}\right) \beta\left(t - \frac{n}{W}, t\right) \qquad (8\text{-}145)$$

$$= W \frac{\sin \pi Wt}{\pi Wt} * \sum_{n=-\infty}^{\infty} u\left(t - \frac{n}{W}\right) \beta\left(\frac{n}{W}, t\right)$$

where $*$ denotes convolution and the term $W(\sin \pi Wt/\pi Wt)$ is the Fourier transform of an ideal lowpass filter centered at $f = 0$ with bandwidth $W/2$. In (8-145) the last sum is indicative of passing the signal $u(t)$ through an infinite-length tapped-delay line [or finite impulse response (FIR) filter] with tap coefficients given by $\{\beta(n/W, t)\}$. From Section 8-2 the time spread of a frequency-selective fading channel will be limited to T_m seconds. Then the total number of nonzero taps in the tapped-delay line model will be limited to at most $L = T_m W$ taps. If it is understood that $u(t)$ is strictly bandlimited to $[-W/2, W/2]$, (8-145) implies that an equivalent model for a frequency-selective channel model limited to L taps is as shown in Figure 8-19. The tapped-delay line channel model has been studied extensively (see, e.g., Refs. 54 and 55).

The received signal $y(t)$ at the output of the tapped-delay line channel model may be expressed as

$$y(t) = \sum_{n=1}^{L} \beta\left(\frac{n}{W}, t\right) u\left(t - \frac{n}{W}\right) \qquad (8\text{-}146)$$

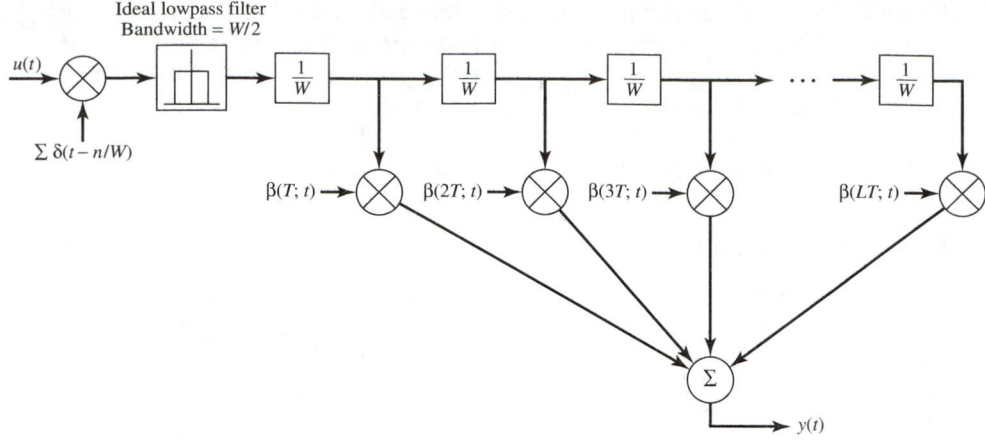

FIGURE 8-19. Tapped-delay line model of a frequency-selective fading channel. The boxes marked by $1/W$ correspond to delays of value $1/W$ seconds.

From Section 8-2 it is assumed that the frequency-selective fading process is a Gaussian WSSUS process. Hence the coefficients of the tapped-delay line are uncorrelated zero-mean complex Gaussian random processes and are therefore statistically independent.

Rake Receiver. In Section 8-4.1 it was noted that an equivalent representation for a bandlimited signal transmitted over a frequency-selective fading channel was given by (8-146). The signal at the receiver input is then given by

$$r(t) = \sum_{n=1}^{L} \beta\left(\frac{n}{W}, t\right) u\left(t - \frac{n}{W}\right) + z(t) \qquad (8\text{-}147)$$

where $z(t)$ is a zero-mean Gaussian white noise random process with a one-sided power spectral density of N_0. In the absence of intersymbol interference considerations (the so-called ''one-shot'' approach), the optimal receiver for detection of binary antipodal signals transmitted over a frequency-selective fading channel is a complex matched-filter receiver (or equivalently, a correlator receiver) with a decision statistic given by

$$V = \text{Re}\left\{ \int_0^T r(t) h^*(t)\, dt \right\}$$

$$= \text{Re}\left\{ \sum_{n=1}^{L} \int_0^T r(t) \beta^*\left(\frac{n}{W}, t\right) u^*\left(t - \frac{n}{W}\right) dt \right\} \qquad (8\text{-}148)$$

where $h(t)$ is the matched-filter impulse response and T is the bit duration. The decision rule is then given by

$$\text{decide } u(t) = \begin{cases} +1 & \text{if } V > 0 \\ -1 & \text{if } V \le 0 \end{cases} \tag{8-149}$$

Figure 8-20 shows a block diagram of the receiver corresponding to this receiver structure.

With a change of variable, (8-148) may also be rewritten as

$$V = \text{Re}\left\{ \sum_{n=1}^{L} \int_0^T r\left(t - \frac{n}{W}\right) \beta^*\left(\frac{n}{W}, t\right) u^*(t)\, dt \right\} \tag{8-150}$$

This yields an alternative receiver structure as shown in Figure 8-21. Although these receiver structures are equivalent, there are often implementation issues to be considered when choosing one approach over the other. For obvious reasons, the former receiver structure is called a delayed reference structure, while the latter structure is called a delayed signal version. Both of these receiver structures are examples of Rake receivers. Rake receivers are so named because their regular structure is analogous to the projecting teeth of an ordinary garden rake [50,56].

The performance of the optimal receiver will now be evaluated. The following assumptions will be made at the outset:

1. The channel is assumed to be time *in*variant over the decision interval $[0,T]$. This implies that the channel, although perhaps doubly spread, may be considered to be purely frequency selective over the decision interval and that the

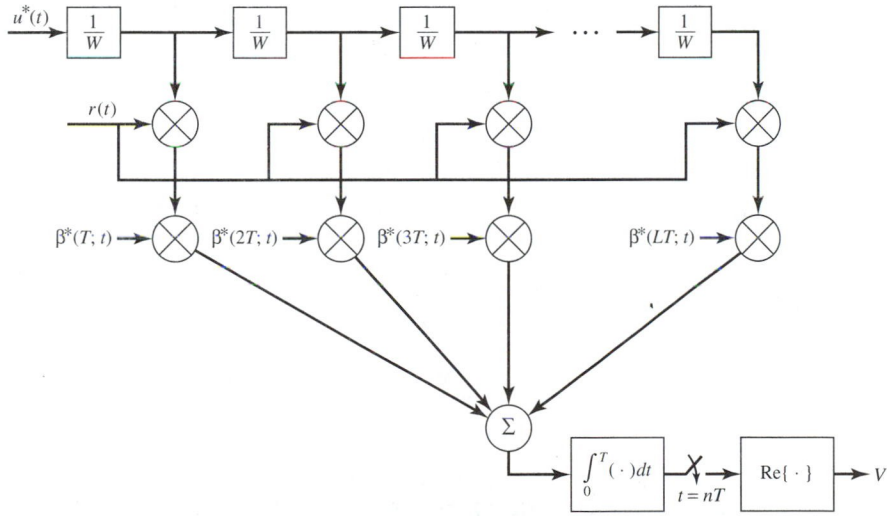

FIGURE 8-20. Matched-filter receiver for binary antipodal signals transmitted over a frequency-selective fading channel (Rake), delayed reference version. The boxes marked by $1/W$ correspond to delays of value $1/W$ seconds.

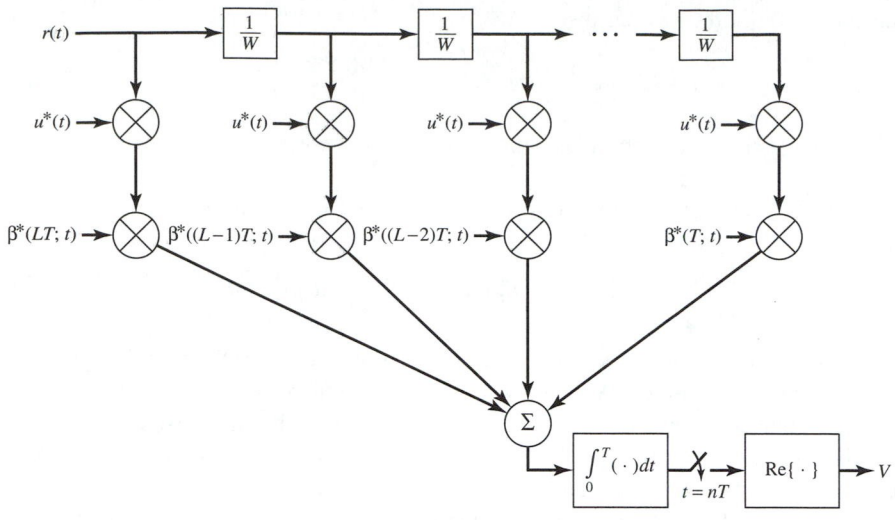

FIGURE 8-21. Matched-filter receiver for binary antipodal signals transmitted over a frequency-selective fading channel (Rake), delayed signal version. The boxes marked by $1/W$ correspond to delays of value $1/W$ seconds.

channel coefficients may be expressed simply as

$$\beta\left(\frac{n}{W},t\right) = \beta\left(\frac{n}{W}\right) \tag{8-151}$$

2. The channel gain estimates are assumed to be known *exactly*. Any estimation errors in channel gains are therefore assumed to be negligible.

3. The signal $u(t)$ is assumed to have zero autocorrelation at nonzero time offsets, that is,

$$\int_0^T u\left(t - \frac{n}{W}\right)u^*\left(t - \frac{k}{W}\right) dt \equiv 0 \qquad \text{for } k \neq n \tag{8-152}$$

Then inserting (8-151) into (8-150), the decision variable becomes

$$V = \text{Re}\left\{\sum_{n=1}^{L} \beta\left(\frac{n}{W}\right)\int_0^T r(t)u^*\left(t - \frac{n}{W}\right) dt\right\}$$

$$= \text{Re}\left\{\sum_{n=1}^{L} \beta^*\left(\frac{n}{W}\right)\int_0^T \left[\sum_{k=1}^{L} \beta\left(\frac{k}{W}\right)u\left(t - \frac{k}{W}\right) + z(t)\right]u^*\left(t - \frac{n}{W}\right) dt\right\}$$

$$= \text{Re}\left\{\sum_{n=1}^{L} \beta^*\left(\frac{n}{W}\right)\sum_{k=1}^{L} \beta\left(\frac{k}{W}\right)\int_0^T u\left(t - \frac{k}{W}\right)u^*\left(t - \frac{n}{W}\right) dt\right\} \tag{8-153}$$

$$+ \text{Re}\left\{\sum_{n=1}^{L} \beta^*\left(\frac{n}{W}\right)\int_0^T z(t)u^*\left(t - \frac{n}{W}\right) dt\right\}$$

From the assumption on the autocorrelation of $u(t)$, (8-152), this reduces to

$$V = \text{Re}\left\{\sum_{n=1}^{L}\left|\beta\left(\frac{n}{W}\right)\right|^2 \int_0^T u\left(t - \frac{n}{W}\right)u^*\left(t - \frac{n}{W}\right)dt\right\}$$

$$+ \text{Re}\left\{\sum_{n=1}^{L}\beta^*\left(\frac{n}{W}\right)\int_0^T z(t)u^*\left(t - \frac{n}{W}\right)dt\right\} \qquad (8\text{-}154)$$

$$= \text{Re}\left\{2E\sum_{n=1}^{L}c_n^2 + \sum_{n=1}^{L}c_n N_n\right\}$$

where $c_n = |\beta(n/W)|$ and

$$N_n = e^{j\theta_n}\int_0^T z(t)u^*\left(t - \frac{n}{W}\right)dt \qquad (8\text{-}155)$$

Careful comparison of (8-154) with the corresponding equation (8-130) for the maximal ratio combining case reveals that for the case of coherent detection of BPSK over a frequency-selective fading channel *with identical tap weights,* the two decision variables are identical. Hence the performance of an L-branch Rake receiver in frequency-selective fading is identical to that of an L-branch maximal ratio combiner in time-selective Rayleigh fading. The performance of the Rake receiver is therefore also shown in Figure 8-18.

Before proceeding, some additional remarks about the Rake receiver are in order:

1. Despite the comments made above in regard to the similarity between the performance of the Rake receiver and the maximal ratio combiner, there remains a subtle difference in performance *if* antenna diversity is used with the maximal ratio combiner. In the analysis above for both the maximal ratio combiner and the Rake receiver, it was assumed that the received signal power was the sum of the received signal power in each of the branches [see (8-133)]. This is indeed the case for all diversity systems. However, the required *transmitter* power required to achieve a given level of performance is a different situation. In the frequency-selective fading multipath case considered above in conjunction with the Rake receiver, the transmitter power is *divided* among all the possible channels. Hence assuming that the multipath is completely resolvable and neglecting the propagation loss, the received signal power considered in the Rake analysis is equal to the transmitted power. In the antenna diversity situation, again neglecting the propagation loss, the received signal power in *each* of the branches is equal to the power transmitted. Thus in the antenna diversity situation in which maximal ratio combining is employed, the received signal power (and hence the overall performance) is $10 \log_{10}(L)$ decibels greater than the Rake receiver case, assuming identical orders of diversity L in both situations.†

---

† Needless to say, in practice, all diversity approaches are not equivalent because, for example, antennas may be correlated, the multipath may not be resolvable, and so on. Engineering judgment is required in selecting the type of diversity to be employed.

2. The tap spacing $1/W$ in the Rake receiver is determined not by the channel but, rather, by the bandwidth of the transmitted signal. There is no direct correspondence, then, between the power delay profile of a frequency-selective fading channel and the corresponding tap weights and spacing of a Rake receiver designed to operate over that channel.

3. The cross-correlation assumption required above [see (8-152)] in deriving the Rake receiver structure requires proper selection of the spreading sequences in a direct-sequence system. As several authors have noted [49,57], nonorthogonality results in a degradation of the performance of the Rake receiver.

4. The Rake receiver requires that the transmitted signal have a bandwidth W in excess of the coherence bandwidth of the channel. The greater the bandwidth, the higher the order of diversity that could be obtainable with the Rake receiver. The most common application of the Rake receiver is thus naturally in conjunction with a direct-sequence spread-spectrum system since, with such a system, the transmitted signal bandwidth is a function of the chip rate and not (just) the information rate.

5. The analysis above considered only ideal channel gain estimation and equal tap weights in the Rake receiver. Consideration of nonequal tap weights and channel gain estimation errors is given in Proakis [49].

Although the analysis given here examined only the Rake receiver structures for coherent detection of binary antipodal signals, other Rake receivers structures exist for other forms of modulation and noncoherent detection (and hence combining). Further examples of these structures may be found in Refs. 49 and 56 to 58. Reference 58 provides some useful information on the performance of several of these structures operating over identical channels.

8-4.6 Summary: The Benefits of Diversity

At the start of this section, two examples were given of transmission of BPSK over fading and multipath channels in which a significant degradation of the performance of the communication system was noted relative to the performance that would be obtainable over an AWGN channel. In the remainder of this section, methods of combating this degradation through the use of diversity in the receiver were developed and characterized in terms of performance. In the absence of channel impairments, the probability of error of BPSK as a function of SNR may be overbounded by the relationship $P_e \leq \frac{1}{2}e^{-\text{SNR}}$ (see, e.g., Ref. 59). Without diversity, the relationship between the probability of error and SNR is asymptotically an inverse linear one [see (8-143)]. With diversity, this relationship goes from an inverse linear relationship to an inverse Lth-order relationship: that is, the exponential relationship for an AWGN channel is gradually approached as the order of diversity increases.

The performance improvement due to diversity may be achieved by more than one form of diversity. For example, time diversity via channel coding may be used together with antenna diversity to yield a higher order of diversity than either approach used separately. It will be seen in Chapter 9 that this approach toward combining diversity methods has been used by the designers of the example CDMA digital cellular systems considered in this book.

8-5 Summary

A number of different concepts were introduced in this chapter related to fading channels. The concept of fading channels was introduced and the parameters and statistical description of such channels were described. The general class of wide-sense-stationary uncorrelated scattering (WSSUS) fading channels was introduced. The category of WSSUS fading channels may be further subdivided into three classes: time-selective fading channels, frequency-selective fading channels, and doubly spread fading channels. The latter class of fading channels exhibits both time and frequency selectivity.

Mobile radio channels can exhibit the characteristics of a doubly spread channel. Propagation over a mobile radio channel is a function of many parameters, including antenna heights, operating frequency, and terrain. Terrain dependence can result in a lognormal fading process that needs to be taken into account in planning the design of a cellular system.

Fading, both time selective and frequency selective, will significantly degrade the performance of a communication system. Introducing diversity can change the probability of bit error performance from an $(SNR)^{-1}$ relationship to a $(SNR)^{-L}$ relationship, where L is the order of diversity. Diversity may take many forms, including antenna, time, space, and frequency. The Rake receiver was shown to be one implementation of a diversity receiver for a frequency-selective fading channel.

References

[1] F. E. Bond and H. F. Meyer, "Fading and Multipath Considerations in Aircraft/Satellite Communications Systems," in *Communication Satellite Systems Technology,* R. B. Marsten, ed. (New York: Academic Press: 1966).

[2] I. L. Lebow, P. R. Drouilhet, N. L. Dagget, J. N. Harris, and J. N. Nagy, Jr., "The West Ford Belt as a Communications Medium," *Proc. IEEE,* Vol. 52, pp. 543–563, May 1964.

[3] W. C. Jakes, "An Approximate Method to Estimate an Upper Bound on the Effects of Multipath Delay Distortion on Digital Transmission," *IEEE Int. Conf. Commun.,* pp. 47.1.1–47.1.5, June 1978.

[4] J. W. Strohbehn, "Line-of-Sight Wave Propagation Through the Turbulent Atmosphere," *Proc. IEEE,* Vol. 56, pp. 1301–1318, August 1968.

[5] J. M. Goodman, *HF Communications: Science and Technology* (New York: Van Nostrand Reinhold, 1992).

[6] H. L. Van Trees, *Detection, Estimation, and Modulation Theory,* Part III (New York: Wiley, 1971).

[7] S. Stein, "Fading Communication Media," Chap. 9 of *Communication Systems and Techniques* (New York: McGraw-Hill, 1966).

[8] S. STEIN and J. J. JONES, *Modern Communication Principles* (New York: McGraw-Hill, 1967).

[9] W. C. JAKES, JR., ed., *Microwave Mobile Communications* (New York: Wiley, 1974). Also reprinted by IEEE Press, 1994.

[10] R. W. E. MACNICOL, "The Fading of Radio Waves of Medium and High Frequencies," *Proc. IEE,* Vol. 96, pp. 517–524, November 1949.*

[11] G. L. GRISDALE, J. G. MORRIS, and D. S. PALMER, "Fading of Long-Distance Radio Signals and a Comparison of Space- and Polarization-Diversity Reception in the 6–18 Mcs Range," *Proc. IEE,* Vol. 104, pp. 39–51, January 1957.*

[12] P. A. BELLO, "Characterization of Randomly Time-Variant Linear Channels," *IEEE Trans. Commun. Syst.,* Vol. CS-11, pp. 360–393, December 1963.

[13] P. A. BELLO, "On the Approach of a Filtered-Pulse Train to a Stationary Gaussian Process," *IRE Trans. Inf. Theory,* Vol. IT-7, pp. 144–149, July 1961.

[14] W. M. BROWN and R. B. CRANE, "Conjugate Linear Filtering," *IEEE Trans. Inf. Theory,* Vol. IT-15, pp. 462–465, July 1969.

[15] T. L. GRETTENBERG, "A Representation Theorem for Complex Normal Processes," *IEEE Trans. Inf. Theory,* Vol. IT-11, pp. 305–306, April 1965.

[16] K. S. MILLER, *Complex Stochastic Processes* (Reading, Mass.: Addison-Wesley, 1974).

[17] P. A. BELLO, "Time-Frequency Duality," *IEEE Trans. Inf. Theory,* Vol. IT-10, pp. 18–33, January 1964.

[18] R. S. KENNEDY, *Fading Dispersive Communication Channels* (New York: Wiley, 1969).

[19] P. A. BELLO and B. D. NELIN, "The Effect of Frequency Selective Fading on the Binary Error Probabilities of Incoherent and Differentially Coherent Matched Filter Receivers," *IEEE Trans. Commun. Syst.,* Vol. CS-11, pp. 170–186, June 1963. See corrections in *IEEE Trans. Commun. Syst.,* Vol. COM-12, pp. 230–231, December 1964.*

[20] P. A. BELLO and B. D. NELIN, "The Influence of Fading Spectrum on the Binary Error Probabilities of Incoherent and Differentially Coherent Matched Filter Receivers," *IRE Trans. Commun. Syst.,* Vol. CS-10, pp. 160–168, June 1962.*

[21] A. PAPOULIS, *The Fourier Integral and Its Applications* (New York: McGraw-Hill, 1962).

[22] W. C. Y. LEE, *Mobile Communications Engineering* (New York: McGraw-Hill, 1982).

---

*References marked with an asterisk have been reprinted in K. Brayer, ed., *Data Communications via Fading Channels* (New York: IEEE Press, 1975).

[23] J. D. PARSONS, *The Mobile Radio Propagation Channel* (New York: Halsted Press, 1992).

[24] G. C. HESS, *Land-Mobile Radio System Engineering* (Boston: Artech House, 1993).

[25] J.-P. LINNARTZ, *Narrowband Land-Mobile Radio Networks* (Boston: Artech House, 1993).

[26] R. H. CLARKE, ''A Statistical Theory of Mobile Radio Reception,'' *Bell Syst. Tech. J.,* Vol. 47, pp. 957–1000, 1968.

[27] E. N. GILBERT, ''Energy Reception for Mobile Radio,'' *Bell Syst. Tech. J.,* Vol. 44, pp. 1779–1803, 1965.

[28] M. J. GANS, ''A Power Spectral Theory of Propagation in the Mobile Radio Environment,'' *IEEE Trans. Veh. Technol.,* Vol. VT-21, pp. 27–38, 1972.

[29] D. C. COX and R. P. LECK, ''Distributions of Multipath Delay Spread and Average Excess Delay for 910-MHz Urban Mobile Radio Paths,'' *IEEE Trans. Antennas Propagation,* Vol. AP-23, pp. 206–213, March 1975.

[30] D. C. COX and R. P. LECK, ''Correlation Bandwidth and Delay Spread Multipath Propagation Statistics for 910-MHz Urban Mobile Radio Channels,'' *IEEE Trans. Commun.,* Vol. COM-23, pp. 1271–1280, November 1975.

[31] D. C. COX, ''910 MHz Urban Mobile Radio Propagation: Multipath Characteristics in New York City,'' *IEEE Trans. Commun.,* Vol. COM-21, pp. 1188–1194, November 1973.

[32] J. VAN REES, ''Measurements of Impulse Response of a Wideband Radio Channel at 910 MHz from a Moving Vehicle,'' *Electron. Lett.,* Vol. 22, No. 5, pp. 246–247, February 27, 1986.

[33] D. C. COX, ''Time- and Frequency-Domain Characterizations of Multipath Propagation at 910 MHz in a Suburban Mobile-Radio Environment,'' *Radio Sci.,* Vol. 7, pp. 1069–1077, December 1972.

[34] D. C. COX, ''Delay Doppler Characteristics of Multipath Propagation at 910 MHz in a Suburban Mobile Radio Environment,'' *IEEE Trans. Antennas Propagation,* Vol. AP-20, pp. 625–635, September 1972.

[35] D. L. NIELSON, ''Microwave Propagation Measurements for Mobile Digital Radio Application,'' *Proc. EASCON '77,* pp. 14-2A–14-2L, 1977.

[36] D. L. NIELSON, *''Microwave Propagation and Noise Measurements for Mobile Digital Radio Application,''* Packet Radio Note 4, SRI Project 2325, ARPA Contract DAHC15-73-C-0187, SRI International, January 1975.

[37] S. C. FRALICK and J. C. GARRETT, ''Technological Considerations for Packet Radio Networks,'' *Proc. AFIPS Natl. Comput. Conf.,* pp. 233–243, 1975.

[38] G. L. TURIN, F. D. CLAPP, T. L. JOHNSTON, S. B. FINE, and D. LAVRY, "A Statistical Model of Urban Multipath Propagation," *IEEE Trans. Veh. Technol.,* Vol. VT-21, pp. 1–9, February 1972.

[39] M. HATA and T. MIKI, "Performance of MSK High-Speed Digital Transmission in Land Mobile Radio Channels," *Proc. GLOBECOM '84,* pp. 16.3.1–16.3.6, 1984.

[40] M. NILSON, S. NORSIN, and M. MIZUNO, "Radio Wave Propagation Aspects for Future Digital Mobile Radio Systems," *Proc. Eurocon '88,* pp. 295–296, June 1988.

[41] A. S. BAJWA and J. D. PARSONS, "Small Area Characterisation of UHF Urban and Suburban Mobile Radio Propagation," *IEE Proc.,* Part F, Vol. 129, No. 2, pp. 102–109, April 1982. See also P. W. Huish and E. Gurdenli, "Propagation Measurement and Planning Requirements for Digital Cellular Systems," *Proc. 2nd Nordic Seminar Digital Land Mobile Radio Commun.,* Paper 47, October 1986.

[42] A. ZOGG, *Reflection Measurement of TV-Test Lines: Echo Diagrams of the First Measurement Campaign* (in German), Directorate General of the Swiss PTT, Report VD 11.1045 U, 1985. Cited in R. W. Lorenz, "Variation of Multipath Spread in Mobile Radio and Its Impact on Digital Transmission," *Proc. 2nd Nordic Seminar Digital Land Mobile Radio Commun.,* Paper 49, October 1986.

[43] S. BANG, T. MASENG, S. T. OLSEN, and O. TRANDEM, *Multipath Measurements at 949 MHz,* GSM WP2 document 209/87, ELAB, Norwegian Institute of Technology, June 4, 1987.

[44] W. R. YOUNG, JR., and L. Y. LACY, "Echoes in Transmission at 450 Megacycles from Land-to-Car Radio Units," *Proc. IRE,* Vol. 38, pp. 255–258, March 1950.

[45] COST 207 WG1, *Proposal on Channel Transfer Functions to Be Used in GSM Tests Late 1986,* COST 207 TD (86) 51 Rev 3, Paris, September 29–30, 1986.

[46] Y. OKUMURA, E. OHMORI, T. KAWANO, and K. FUKUDA, "Field Strength and Its Variability in VHF and UHF Land-Mobile Radio Service," *Rev. Tokyo Elec. Commun. Lab.,* Vol. 16, pp. 825–873, September–October 1968.

[47] M. HATA, "Empirical Formula for Propagation Loss in Land Mobile Radio Services," *IEEE Trans. Veh. Technol.,* Vol. VT-29, pp. 317–325, August 1980.

[48] P. E. MOGENSEN, P. EGGERS, C. JENSEN, AND J. B. ANDERSEN, "Urban Area Radio Propagation Measurements at 955 and 1845 MHz for Small and Micro Cells," *Proc. Globecom'91,* pp. 36.4.1–36.4.6, December 1991.

[49] J. G. PROAKIS, *Digital Communications,* 2nd ed. (New York: McGraw-Hill, 1989).

[50] R. PRICE and P. E. GREEN, JR., "A Communication Technique for Multipath Channels," *Proc. IRE,* Vol. 46, pp. 555–570, March 1958.

[51] D. G. BRENNAN, "Linear Diversity Combining Techniques," *Proc. IRE,* Vol. 47, pp. 1075–1102, June 1959.*

[52] Y. L. LUKE, *The Special Functions and Their Approximations* (2 vols.) (New York: Academic Press, 1969).

[53] D. E. KNUTH, *The Art of Computer Programming,* Vol. 1: *Fundamental Algorithms,* 2nd ed. (Reading, Mass.: Addison-Wesley, 1973).

[54] J. C. HANCOCK and P. A. WINTZ, *Signal Detection Theory* (New York: McGraw-Hill, 1966).

[55] T. KAILATH, "Channel Characterization: Time-Variant Dispersive Channels," Chap. 6 of *Lectures on Communication Theory,* E. J. Baghdady, ed. (New York: McGraw-Hill, 1961).

[56] G. L. TURIN, "Introduction to Spread-Spectrum Antimultipath Techniques and Their Application to Urban Digital Radio," *Proc. IEEE,* Vol. 68, pp. 328–353, March 1980.

[57] P. A. BELLO, "Performance of Some RAKE Modems over the Nondisturbed Wide Band HF Channel," *Proc. 1989 IEEE Military Commun. Conf.* (MILCOM), pp. 39.3.1–39.3.4, September 1989.

[58] J. M. WOZENCRAFT, "Sequential Reception of Time-Variant Dispersive Transmissions," Chap. 12 of *Lectures on Communication Theory,* E. J. Baghdady, ed. (New York: McGraw-Hill, 1961).

[59] J. M. WOZENCRAFT and I. M. JACOBS, *Principles of Communication Engineering* (New York: Wiley, 1965).

Problems

(8-1) Given a bandpass filter at center frequency f_0 with lowpass equivalent impulse response

$$h_e(t) = \alpha e^{-\alpha t} u(t)$$

where $u(t)$ is the unit step, find its response to the input

$$s_0(t) = \begin{cases} A \cos 2\pi f_0 t & 0 \le t \le T \\ 0 & \text{otherwise} \end{cases}$$

(8-2) Prove (8-10).

(8-3) The Doppler scattering function of a certain time selective channel is

$$S_D(0,f) = \frac{e^{-(f-f_0)^2/2\sigma_f^2}}{\sqrt{2\pi\sigma_f^2}}$$

where f_0 and σ_f^2 are constants.
(a) Obtain the channel correlation function.
(b) What is the mean-square Doppler spread?

(8-4) The channel correlation function for a frequency-selective fading channel is

$$g(\tau) = \rho(\tau,0) = Ae^{-a|\tau-\tau_0|}$$

where A, a, and τ_0 are constants.
(a) What is the channel scattering function?
(b) What is the mean delay?
(c) Find the mean-square delay spread.

(8-5) As a communication engineer, you are asked to design a system that will provide radio communications within a circular "cell" area. A cell is being designed that will have a 4-mile radius. Zero-decibel gain antennas are assumed to be employed in both the base and the mobile. The communication system being proposed requires a power level of -107 dBm into the receiver to achieve a specified minimum level of performance. The radio propagation loss is assumed to be given by a Hata model, which is given by

$$\text{path loss (dB)} = A + B \log_{10} R$$

where R is the distance in miles from the transmitter, $A = 127$ dB, and $B = 35$. Determine the required transmit power needed to achieve a received signal level of -107 dBm at the mobile receiver. Assume that the mobile receiver is the only user and neglect lognormal shadowing effects.

(8-6) Derive (8-142); that is, show that for large SNR, (8-142) becomes

$$\left(\frac{1}{2}\right)^L \left(1 - \sqrt{\frac{\gamma_c}{1 + \gamma_c}}\right)^L \rightarrow \left(\frac{1}{2}\right)^L \left(\frac{1}{2\gamma_c}\right)^L = \left(\frac{1}{4\gamma_c}\right)^L$$

(8-7) Consider Problem 8-5, taking into account lognormal shadowing effects. Assume that propagation is of the form $Kd^{-3.5}$ and the lognormal standard deviation is $\sigma = 10$ dB. Compute the transmit power required to achieve a contour reliability of 90% for the 4-mile cell and a single user. Compute the transmit power required to achieve an area coverage probability of 90% for the 4-mile cell and a single user.

(8-8) A channel has a lowpass equivalent transfer function given by

$$H(f) = A\frac{\sin \pi Af}{\pi Af}$$

Suppose that a transmitter exhibits a transmission bandwidth of $W = 10/A$. How many taps are required in the Rake receiver matched to this system/ channel? What is the equivalent order of diversity? Write an expression for the asymptotic probability of error for the system incorporating a Rake receiver.

(8-9) Without any channel coding, a binary antipodal communication system requires 1 W of transmitted power to achieve a 10^{-4} probability of error in the receiver when the system operates over a Rayleigh-fading channel. The receiver employs no diversity. Estimate the order of diversity required to achieve the same probability of error given that the transmitter can actually transmit only 100 mW of power. Assume that the gains of the transmitting and receiving antennas are the same for both cases.

CHAPTER 9

Code-Division Multiple-Access Digital Cellular Systems

9-1 Introduction

In previous chapters we considered the analysis and design of spread-spectrum communication systems; these discussions concentrated heavily on the various communication subsystems. In this chapter *complete* spread-spectrum communication *systems* are introduced and analyzed. In particular, this chapter considers a system that has recently become very timely: code-division multiple-access (CDMA) digital cellular communication systems. As the reader will observe, the analysis and design of CDMA digital cellular systems is quite complex, since the analysis requires knowledge of mobile radio channels, methods of implementing diversity, and cellular system topology, as well as the information presented throughout this book. A complete analysis of these systems will generally require full systems simulation, a topic not discussed further here. Instead, the analytical tools required to analyze and design a CDMA digital cellular radio system are presented.

The chapter begins with an overview of the principles of cellular radio systems. Next, the fundamental relationships between cellular system topology and the performance of the modulation/coding/diversity subsystems on cellular system designs are noted. The chapter concludes with a description and analysis of six different CDMA digital cellular mobile radio systems that have either been proposed or have actually been implemented and tested in the field. Given the rapid and continuing growth of cellular radio systems throughout the world, CDMA digital cellular radio systems will probably be the widest deployed form of spread-spectrum systems for voice communications.†

9-2 Cellular Radio Concept

In this section the fundamentals of the cellular radio concept will be developed. The goal of this section is to present the underlying ideas behind cellular radio systems

---

† This statement may also hold true for data communications. However, a contender for the leadership position for data communications may be the Electronic Industries Association (EIA) Interim Standard-60 (IS-60) Home Automation System (CEBus), a power-line control system for appliances in the home that employs chirp spread-spectrum communications.

without studying any one particular cellular radio system in detail. The ideas presented here are applicable to both wide area (macrocellular) cellular radio systems and small area (microcellular) personal communication systems (PCS).

The concepts of cellular radio may be traced back to the 1960s [1,2]. Up to the introduction of cellular radio in the early 1980s, the common form of mobile radio telephone system operated in the 150- or 450-MHz common carrier band using a limited set of frequency channels (<20) in a given geographical area. The geographical areas were associated with metropolitan areas throughout the United States. Each frequency was used only once in the geographical area, leading to a high demand for a very limited channel resource.

In the United States, docket 18262 [3] was introduced by the Federal Communications Commission (FCC) in 1968 with the goal of changing this situation by introducing a paired band of 666 channels in the 800-MHz spectrum. Users of this new band were to introduce new technology that would improve on the spectral efficiency of the earlier radiotelephone systems. The concept that was introduced that would lead to improved spectral efficiency is that of frequency reuse within the serving area. Frequency reuse was made possible by the introduction of *cells* within the geographical area. Within the United States the concept of cellular radio was developed primarily by AT&T and Motorola Inc., with contributions by others [4–6]. As a result of the widespread deployment and acceptance of cellular radio systems both in the United States and throughout the world (see Tables 9-1 and 9-2), a number of texts have appeared that provide further details of the design and implementation of cellular systems [7–9].

9-2.1 Fundamentals of Cellular Radio Systems

Cellular radio systems—macro- *and* microcellular, analog *and* digital, conventional *and* CDMA—have a number of fundamental characteristics, as summarized in the following:

1. As shown in Figure 9-1, the geographical area is broken up into smaller geographic areas. To simplify the analysis, uniform-area polygons that tessellate the plane (i.e., cover the area without overlap or missed areas) are often assumed to represent the cells. Although triangles and squares have been used in some analyses, the hexagon is most frequently employed in the literature, as has been used in the figure. Base stations are located either in the middle of the cell (''center illumination'') or at the corner of the cell (''corner illumination'').

2. By the cellular construction in characteristic 1, it is assumed that communication with a mobile in a given cell is made to the base station that serves the cell. This implies that the signal strength of the serving cell exceeds the signal strength of the base station in surrounding cells. The power of both the base station transmitter and the mobile station transmitter is intentionally limited to that which is required to communicate within the cell area. Direct mobile-to-mobile communications are not permitted in the system.

3. Because of the power limitation described in characteristic 2 and propagation

TABLE 9-1. Comparison of First- and Second-Generation Cellular Radio Systems: Analog

Parameter	System					
	AMPS[a] (U.S., Canada)	NAMPS[b] (U.S.)	NMT[c] (Scandinavia)	MCS-L1 MCS-L2[d] (Japan)	C450 (Germany)	TACS ETACS[e] (U.K.)
Transmission frequency (MHz)						
Base	869–894	869–894	935–960	870–885	461–466	917–933, 935–960
Mobile	824–849	824–849	890–915	925–940	451–456	872–888, 890–915
Duplexing method	FDD	FDD	FDD	FDD	FDD	FDD
Multiple-access method	FDMA	FDMA	FDMA	FDMA	FDMA	FDMA
Channel bandwidth (kHz)	30.0	10.0	25[f]	25.0 12.5	20.0 10.0	25.0
Total channels	832	832 × 3	1999	600 1200	222 444	1000
Voice	Analog	Analog	Analog	Analog	Analog	Analog
Modulation	PM	PM	PM	PM	PM	PM
Peak deviation (kHz)	± 12	± 5	± 5	± 5	± 4	± 9.5
Control	Digital	Digital	Digital	Digital	Digital	Digital
Modulation	FSK	FSK	FFSK	FSK	FSK	FSK
Peak deviation (kHz)	± 8	± 8	± 3.5	± 4.5	± 2.5	± 6.4
Channel rate (kbps)	10.0	10.0	1.2	0.3	5.3	8.0

[a] Advanced Mobile Phone System (AMPS). This is described further in Refs. 5 and 8, and Chapter 2 of Ref. 70.

[b] Narrowband AMPS. See M. P. Metroka, "An Introduction to Narrowband AMPS," *Globecom '91*, pp. 41.2.1-41.2.6, December 1991 for further details.

[c] Nordic Mobile Telephone (NMT) System. This is described further in Chapter 3 of Ref. 70.

[d] Mobile Control Station (MCS)-L2. This is a system designed by NTT in Japan and is described in Chapter 5 of Ref. 70.

[e] Total Access Communication System (TACS). ETACS stands for Extended TACS. This is described further in Chapter 4 of Ref. 70.

[f] 12.5 kHz interleaved.

TABLE 9-2. Comparison of First- and Second-Generation Cellular Radio Systems: Digital

Parameter	System				
	GSM[a] (Europe)	PCN[b] (U.K.)	IS-54[c] (U.S., Canada)	PDC[d] (Japan)	IS-95[e] (U.S.)
Transmission frequency (MHz)					
Base	890–915	1710–1785	869–894	810–826	869–894
Mobile	935–960	1805–1880	824–849	940–956	824–849
Duplexing method	FDD	FDD	FDD	FDD	FDD
Multiple-access method	TDMA	TDMA	TDMA	TDMA	CDMA
Channel bandwidth (kHz)	200.0	200.0	30.0	25.0	1.23 MHz
Traffic channels per RF channel	8	16	3	3	60
Total channels	125×8	375×8	832×3		60×5
Voice coder	RPE/LTP-LPC	RPE/LTP-LPC	VSELP	VSELP	QCELP
Speech rate (kbps)	13.0	13.0	8.0	11.2	8
Modulation	GMSK	GMSK	$\pi/4$-DQPSK	$\pi/4$-DQPSK	DS/CDMA
Channel rate (kbps)	270.8	270.8	48.6	42	—
Channel coding	Conv.	Conv.	Conv.	Conv.	Conv.
Base–mobile	1/2	1/2	1/2	1/2	1/2
Mobile–base	1/2	1/2	1/2	1/2	1/3

[a]Global System for Mobile Communications (GSM).
[b]Personal Communication Network (PCN).
[c]Telecommunications Industry Association (TIA) Interim Standard (IS) 54.
[d]Personal Digital Cellular (PDC).
[e]Telecommunications Industry Association (TIA) Interim Standard (IS) 95.

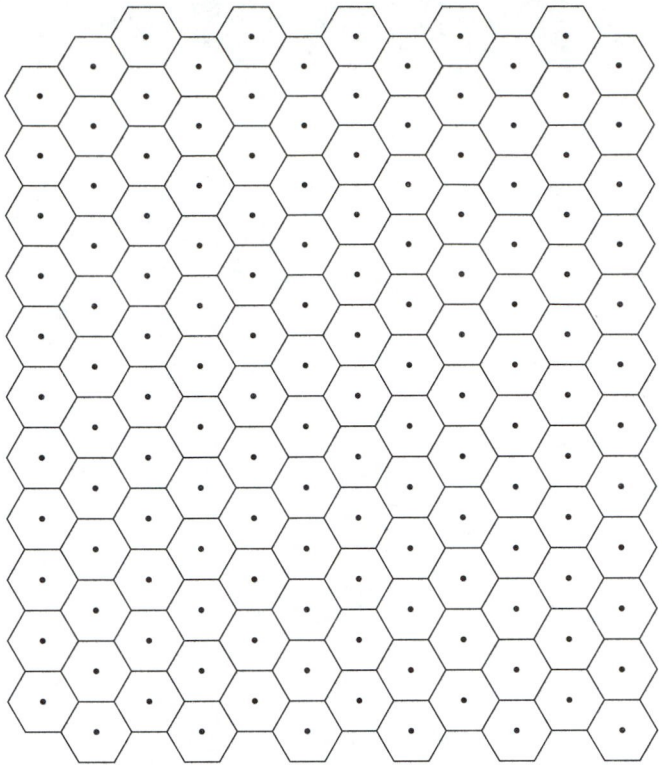

FIGURE 9-1. Hexagonal cell pattern for cellular radio systems. Center dot denotes base site antenna.

characteristics of radio waves, as studied in Section 8-3, frequencies may be *reused* within the overall geographic area. The spectral efficiency of the system is therefore increased by a factor equal to the number of times a frequency may be reused within a geographical area. To minimize cochannel interference, the total available allocation of channels is divided into N disjoint subsets, with each subset assigned to one cell of a group of N cells. The number N is the *frequency-reuse factor* of the cellular system and is dependent on the cochannel interference immunity of the modulation, diversity, and coding methods employed, the antenna pattern characteristics, and the characteristics of radio propagation in the area, as will be seen below. Figure 9-2 shows a cellular system frequency reuse of 7 (a so-called seven-cell pattern).

4. As a mobile proceeds from one cell to another during the course of a call, a central controller automatically reroutes the call from the old cell to the new cell without noticeable interruption, in a process known as *handoff*.

5. As demand for the radio channels within a given cell increases beyond the capacity of the cell, measured in terms of the number of supportable calls that may occur simultaneously, the overloaded cell is ''split'' into smaller cells, each with its own base station and central controller, as shown in Figure 9-3.

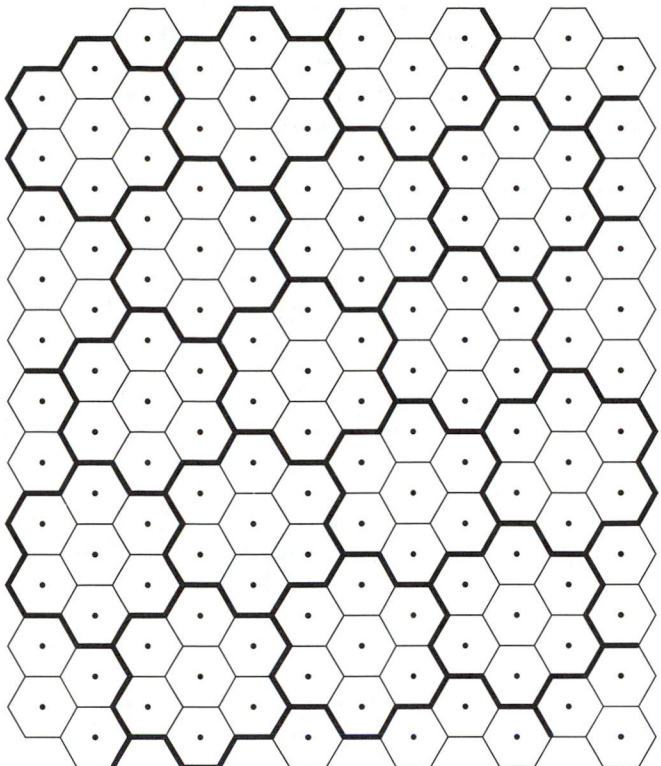

FIGURE 9-2. Seven-cell reuse pattern.

The radio-frequency allocations of the original cellular system are then reallocated to account for the smaller cells.

6. For cellular systems, the base-to-mobile and mobile-to-base communication paths are allocated in pairs of frequencies separated by a fixed bandwidth.† Such an approach is known as frequency-division duplex (FDD).

Specific cellular systems often exhibit additional system characteristics. One additional characteristic that is often employed is the concept of sectors in a cell. Rather than using omnidirectional antennas at the base station, a cell may be split into equal-area sectors through the use of directional antennas. Directional antennas often permit the use of smaller frequency-reuse factors and therefore a reduction in geographic separation between sites. For example, a 12-cell-site repeat pattern employing omnidirectional antennas can be shown to be equivalent to a seven-cell pattern employing 120° sectors in terms of co-channel rejection performance [10].

---

†This statement is true for all existing terrestrial cellular systems. For PCS systems, however, several approaches time-multiplex a single frequency for both up and down links. Such systems are known as time-division duplex (TDD) systems. Examples of TDD PCS systems include the Digital European Cordless Telephone (DECT), the second-generation cordless telephone (CT-2), and the Japanese Personal Handyphone (PHP) system.

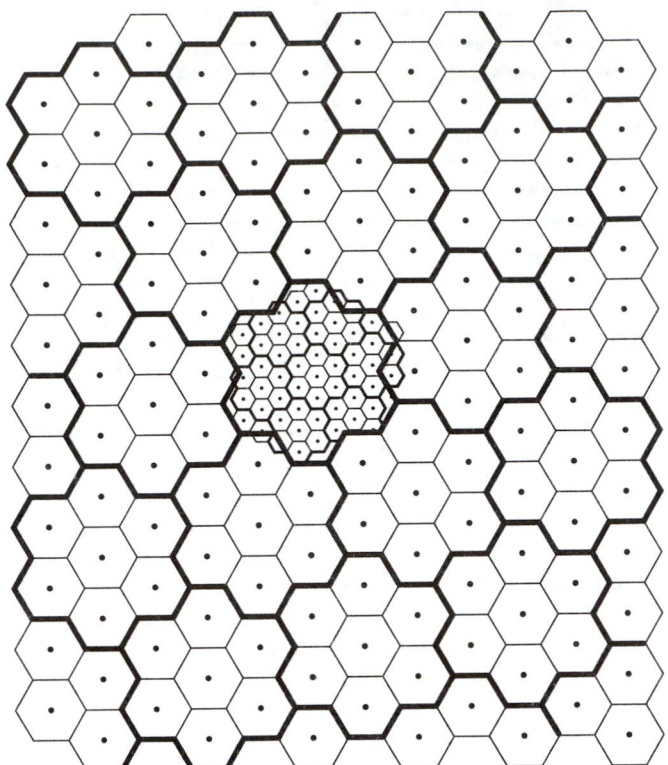

FIGURE 9-3. Cell splitting in seven-cell pattern.

As will be seen below, for spread-spectrum systems, a sectorized cell pattern can lead to a significant increase in system capacity.

Another system concept that is often employed in some cellular systems is that of power control of the transmitter in the uplink (mobile to base) as well as sometimes in the downlink (base to mobile). Power control in the uplink tends to minimize co-channel interference within a cellular system. Because mobiles are restricted to transmit at the minimum amount of power (or a quantized step size thereof) required to achieve acceptable performance with the serving base station, interference into surrounding cells is minimized. Furthermore, power control is absolutely required for any degree of operation in certain types of CDMA digital cellular systems.

9-2.2 Co-channel Interference Protection Prediction

As suggested above, the cell repeat pattern N is a function of a number of parameters, which in turn determine the co-channel interference (C/I) properties of a particular cell system design. In this subsection two approaches toward estimating the C/I protection factor of a particular cell system design are considered.

Method 1: Geometrical Approach to C/I Protection Prediction. Define an area equivalent hexagon (AEH) to be a hexagon whose area is equal to the area of a group of N cells in a reuse pattern. Let R be defined to be the distance from the center of the AEH to a corner of the AEH. Then by the geometrical properties of hexagons, it may be shown that

$$\text{area (AEH)} = \frac{3\sqrt{3}}{2} R^2 \tag{9-1}$$

Similarly, it may be readily seen that if D is the distance between similar cells in adjacent patterns, D is related to R by

$$D = \sqrt{3}R \tag{9-2}$$

and hence

$$\text{area (AEH)} = \frac{\sqrt{3}}{2} D^2 \tag{9-3}$$

Letting r be the diameter of each of the (small) cells that constitute the AEH, the area of the AEH in terms of the cell diameter r is

$$\text{area (AEH)} = N \cdot \frac{3\sqrt{3}}{2} r^2 = \frac{\sqrt{3}}{2} D^2 \tag{9-4}$$

or

$$\frac{D}{r} = \sqrt{3N} \tag{9-5}$$

Now assuming a propagation path loss of the form kd^{-4} and using as a worst-case interference mobile, a mobile located at the vertex of the central cell, the signal power received at a target mobile from its base station is proportional to Akr^{-4} for some transmitter power A. To a first approximation, the distance between the target mobile and an interfering base station is $D - r$; hence the interference power is proportional to $Ak(D - r)^{-4}$. The co-channel interference protection ratio for the target mobile is thus

$$\frac{C}{I} = \left(\frac{r}{D - r}\right)^{-4} \tag{9-6}$$

or, in terms of decibels,

$$\frac{C}{I}(\text{dB}) = 40 \log_{10} \frac{D - r}{r} = 40 \log_{10}(\sqrt{3N} - 1) \tag{9-7}$$

where (9-5) has been used. For example, an $N = 12$-cell pattern should achieve a C/I protection ratio of approximately 28 dB.

Note that this approach completely neglects the effects of lognormal shadowing. Nonetheless, it does yield a simple rule of thumb in relating cell-reuse pattern size to C/I protection ratio. As shown here, the co-channel interference protection ratio increases as a function of the cell-reuse pattern size. To optimize the communication link quality, N should therefore be large. However, as N increases, the number of available frequencies in each frequency set decreases, and the overall capacity of the system measured in terms of users per area per megahertz decreases.

Method 2: Monte Carlo Approach to C/I Protection Prediction. As has been noted, the simple geometrical approach to C/I protection prediction in a cellular system neglects the lognormal shadowing effects described in Section 8-3. Hence the geometrical approach is commonly used only in first-pass designs of cellular systems. More commonly, a Monte Carlo approach is employed, as will now be described.

In the Monte Carlo approach to C/I protection prediction, a probability distribution function of the required C/I over some fraction of the area, also known as the location reliability, is calculated via a simulation approach. For calculation of the uplink C/I protection ratio, a target mobile is randomly placed in the serving cell. Interfering mobiles are randomly located in each of the possible interfering cells. Shadowing is accounted for by assigning a lognormal random variable with a fixed σ to each of the randomly placed mobiles. Using a fixed propagation path-loss equation, the signal received at the base from the target mobile is computed where the path loss is equal to the path loss predicted from the propagation equation plus the path loss due to the lognormal random variable. Next, the interference power at the base due to each of the interfering mobiles is computed in a similar manner. Then the C/I ratio is computed from the simulated signal power and sum of the interference power received at the base. The target mobile is then randomly moved and the interferers are again randomly placed with new lognormal shadowing losses drawn randomly. A new C/I value is computed and the process is repeated many more times. A histogram of the C/I values obtained is computed and a probability distribution function is then calculated from the histogram. A ''typical'' distribution resulting from the Monte Carlo approach is shown in Figure 9-4, which plots the distribution function of C/I as a function of the cell pattern size for a lognormal shadowing σ of 6 dB, and the Hata model described in Section 8-3 is employed as the propagation model. An approach similar to the above may be used to predict the downlink location reliability.

The Monte Carlo method can be made quite exact by properly accounting for the shadowing model, the propagation model, path correlation models [11], power control, handoff algorithms, and the antenna patterns employed. There are a number of subtleties in the Monte Carlo approach; some of these are described in the literature [6,12–14]. Complete understanding of the assumptions made is required to accurately reproduce the results from a Monte Carlo simulation.

9-2.3 Cellular Concept Revisited

The goal of cellular systems design is thus to maximize spectral efficiency through the use of frequency reuse. In designing a cellular system, a trial design might begin with a choice of a multiple-access method, a choice of a modulation/coding method, and a choice of diversity method(s). The performance of the radio link subsystem as

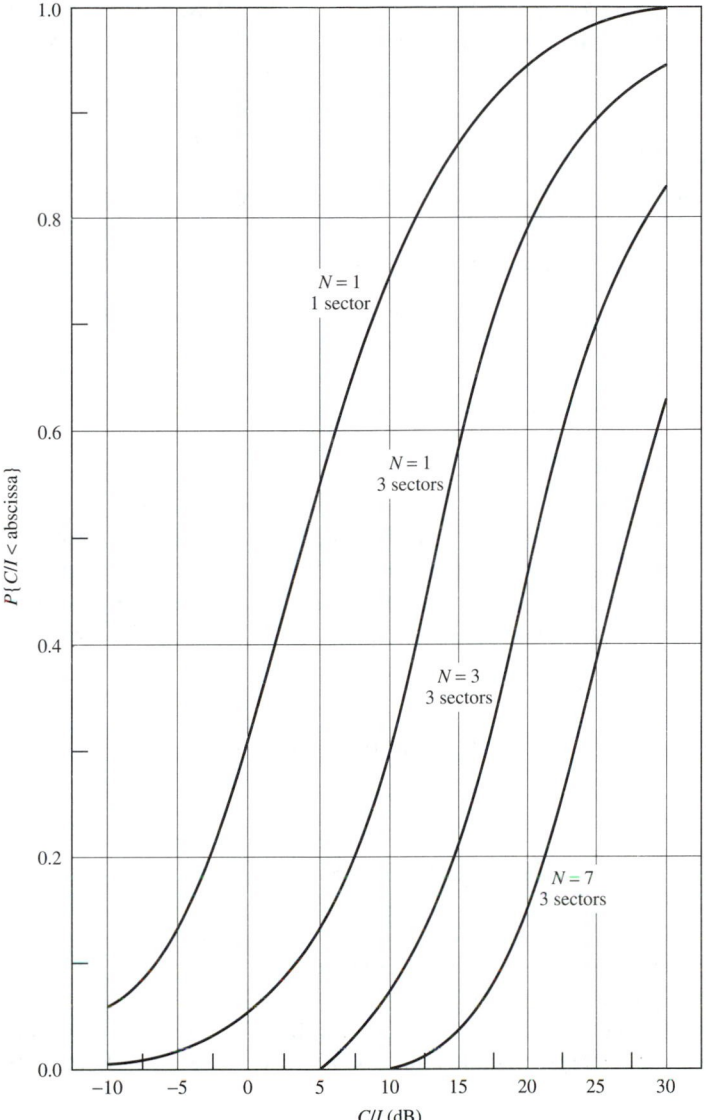

FIGURE 9-4. C/I location reliability for several cellular system topologies.

a function of co-channel interference is then evaluated using analysis and simulation with the tools that have been presented in this chapter and elsewhere in this book. A minimum performance requirement as a function of C/I is established based on the system requirements—for a digital system this may be a required decoded probability of error or a required block error probability. Next, a trial cellular system design is attempted by selection of the reuse factor N, the sectorization factor, and the system coverage reliability goals. Using either the geometric approach or the Monte Carlo simulation approach, the resulting C/I coverage protection is computed, for a given reliability in the latter case. Next, the radio subsystem performance is compared to the C/I protection requirement prediction. If the radio subsystem performance exceeds the C/I protection requirement, the system design is deemed permissible. Finally, system capacity goals and frequency allocation goals must be compared with the corresponding calculated parameters. Often, a number of iterations must be done to complete the system design. Clearly, the design process is a complex engineering task. A simple example can help solidify this design process.

EXAMPLE 9-1.   Cellular System Design ——————————————————

Assume that it is desired to use Gaussian minimum shift keying (GMSK) without diversity and without coding in a digital cellular system design. In the absence of coding and diversity, GMSK requires an 18-dB C/I protection ratio to meet a required 10^{-2} bit error rate (BER) goal. Using either the geometrical approach or extrapolating from the results shown in Figure 9-4, a 12-cell system employing omnidirectional antennas at the base station will yield an 18-dB C/I protection ratio over 90% of the area. If the total (one-way) available system bandwidth is 20 MHz and the transmitted GMSK signal requires a channel spacing of 25 kHz, then, neglecting the overhead required for signaling channels in a real cellular system, the total number of channels available within a reuse pattern is 20 MHz/ 25 kHz = 800 channels. This will permit a total of 800/12 = 66 channels to be allocated to each cell.

Other examples of cellular system design may be found in Ref. 15.

9-3 CDMA Digital Cellular Systems

In Section 9-2 the fundamentals of cellular radio systems were discussed without regard to the type of system. In this section the discussion of cellular systems is extended to the specific case of code-division multiple-access (CDMA) digital cellular systems. The goal here is to introduce additional tools that will permit the study of the example CDMA digital cellular systems considered in Section 9-4.

9-3.1 General Aspects of CDMA Digital Cellular Systems

CDMA digital cellular systems share *most* of the attributes of analog and narrowband digital cellular systems; that is, they are governed by the same basic principles of cellular systems that were discussed in   Section 9-2. Hence the modulation/ coding/diversity methods employed determine the cochannel interference immunity of the system, and the cell system topology/propagation model/sectorization determines the cochannel protection available to the system. To the extent that the modulation method combined with coding and diversity can overcome the cochannel interference provided in the cell system, measured in terms of operation above some threshold, communication within a CDMA digital cellular system is possible.

CDMA digital cellular systems also have some unique attributes. From the beginning, CDMA digital cellular systems were designed to offer improvements over analog cellular system designs in the following two areas:

1. An improvement in capacity/spectral efficiency over analog cellular systems.
2. An improvement in quality, measured in terms of improved speech quality and/or improved system reliability, which in turn is measured in terms of number of "dropped" calls, interference rejection capability, and so on.

Details associated with the first area of improvement are treated here. Because improvements in quality are very system specific, the aspects of the second topic are deferred until Section 9-4. A number of different measures for the system capacity/spectral efficiency of a cellular system exist in the literature. Two such system capacity measures are developed here.

Approach 1: Relative Spectral Efficiency. In Ref. 16, the relative spectral efficiency η (relative to a 30-kHz-channel-spacing AMPS system) of a cellular radio system was defined to be

$$\eta = \frac{U}{F_r} \cdot \frac{30 \text{ kHz}}{W} \tag{9-8}$$

where U is the average number of users/cell, W the bandwidth required for one-way transmission, and F_r the frequency-reuse factor, defined as N in the preceding section. The relative spectral efficiency thus provides a measure of the number of users per cell per 30 kHz of bandwidth. The relative spectral efficiency metric has been employed by a number of authors, including the designers of several of the systems described in the next section. For reference purposes, the relative spectral efficiency of an analog AMPS system employing a 12-cell reuse design with omnidirectional antennas is given by

$$\eta_{\text{AMPS}} = \tfrac{1}{12} = 0.0833\% \tag{9-9}$$

Approach 2: CDMA System Capacity. For CDMA digital cellular systems that (1) employ power control such that all uplink (mobile-to-base) signals are received at the same power level, and (2) all users are spread (via either direct-sequence or frequency-hop spread spectrum) over the total available bandwidth W, a simple equation for capacity may be developed as follows. Assuming that a cell has a total of N mobile users, the composite uplink received signal at the base site will consist of the desired signal with power S and $N - 1$ interfering users, each also having power S, by the power control constraint. The signal-to-interference ratio at the base is thus $S/I = S/(N - 1)S = 1/(N - 1)$. The energy/bit to interference power spectral density E_b/I_0 may be derived similarly. The energy/bit is just the signal power S divided by the information bit rate R (i.e., $E_b = S/R$). The interference power spectral density is the interference power divided by the spreading bandwidth W [i.e., $I_0 = S(N - 1)/W$]. Hence the expression for E_b/I_0 becomes

$$\frac{E_b}{I_0} = \frac{S/R}{(N - 1)S/W} = \frac{W/R}{N - 1} \tag{9-10}$$

Solving (9-10) for N yields the capacity of the cell in terms of user channels/bandwidth W:

$$N = 1 + \frac{W/R}{E_b/I_0} \tag{9-11}$$

which for N large is

$$N \cong \frac{W/R}{E_b/I_0} \tag{9-12}$$

The derivation above for the number of users N per bandwidth W assumes that each interferer is transmitting continuously. In fact, for voice (and some data) information sources, in a two-way conversation, the percentage of time that a speaker is active (i.e., talking) ranges from 35 to 50% [17]. The interference in (9-12) is thus reduced by the voice activity factor d. In addition, *provided that a single cell-reuse pattern yields acceptable performance,* the number of users in (9-12) may be increased by the sectorization factor g. This implies that a cell may be "split" into g cells determined by the sectorization approach used. Finally, because the interference at a base station will come from both within a cell, as has already been noted, as well as *outside* the cell, a term f accounting for an increase in interference power due to interfering users outside the target cell must be factored into the capacity equation.

Taking all of these elements into account, the modified equation for capacity is given by

$$N = \frac{W/R}{E_b/I_0}\left(\frac{1}{d}\right)gf \tag{9-13}$$

where E_b/I_0 is the required energy/bit to interference spectral density necessary to achieve a given level of performance. This equation was first developed by Cooper and Nettleton [18] and was subsequently embraced by Gilhousen et al. [19].

Equation (9-13) was derived for the uplink (mobile to base); a similar calculation for the downlink (base to mobile) would yield the same result.† As the papers referenced above indicate, (9-13) should only be used as a first-order approximation to capacity. In particular, Gilhousen et al. [19] reveals many of the more subtle details of analyzing the capacity of a CDMA digital cellular system. Several of the assumptions made in deriving (9-13) deserve more attention. The issues of power control, voice activity, and sectorization will now be addressed further.

Power Control. In the derivation of the equation for capacity, (9-13), it was assumed that the mobile transmitter power was controlled in such a manner that the received power at the base from each of the users was identical. Note that this requirement on power control is stronger than the one made in the preceding section in discussing cellular systems. Without *very* good power control, in direct-sequence

---

† But this analysis neglects the location of the distribution of interferers. In the uplink case, the interferers are always moving; in the downlink case the interferers remain stationary. This leads to certain difficulties in applying the concept of "interference averaging" to the downlink case.

(and certain types of frequency hopping) CDMA cellular systems, multiple access communications is impossible. Assume, for example, that a cellular system has two mobiles each trying to communicate with the base with the same transmit power. One mobile is located near the base station, the other near the cell boundary. From considerations of the path-loss equations in Section 8-3, a 60-dB or more signal power difference at the base station between the two mobiles is quite possible. At the base station, the near-in mobile appears as a wideband jammmer with 60 dB more power than the far-out jammer. Unless very wideband spreading is used, the far-out mobile will never be received by the base station. This is the familiar *near-far* problem of spread-spectrum multiple-access communication systems. The solution to overcoming the near-far problem is the use of stringent power control in the uplink. This is a challenging engineering problem; several solutions are discussed in the references given for the examples in Section 9-4. Note that the near-far problem does not exist in the downlink path as the base station can transmit all signals with equal power, thereby ensuring that none of the multiple access signals are dominant jammers.

Power control may also be used in the downlink. In this case, power control is usually used for *allocation* of power to users in fringe areas (near cell boundaries, for example). Power control of the downlink requires significantly less dynamic range than that required in the uplink. Note that perfect downlink power control *creates* a near-far problem at the mobile.

Voice Activity Factor. The modified equation for capacity given above includes a term $(1/d)$ to accommodate a voice activity factor. This equation is correct if it is assumed that all the interference comes from a single interferer with voice activity factor d. In fact, as has been noted by Levitt [20], the interference due to the other users is a binomially distributed random variable. In a CDMA digital cellular system with $N + 1$ users, each having a voice activity factor d, a given link experiences interference from n other channels at a given instant of time, where n is distributed according to

$$\Pr\{n\} = \binom{N}{n} d^n (1 - d)^{N-n} \qquad n = 0, 1, 2, ..., N \qquad (9\text{-}14)$$

which has a mean of $E\{n\} = dN$. In Figure 9-5, the probability density function of n is shown plotted for $N = 60$ for three different values of d. As expected, the mode of each of these curves occurs near the mean of n, but there is a wide distribution of values around the mean. This indicates that the simple $(1/d)$ factor in (9-13) must be *decreased* in value to accommodate a higher level of interference than that predicted in the simple analysis given above.

Sectorization. The modified equation for capacity above includes a multiplier term for the capacity which accounts for sectorization. Just as interference from adjacent cells is accounted for in the capacity equation, so must interference from adjacent *sectors* be accounted for in the capacity equation. Interference from adjacent sectors is due to both (1) radio propagation from adjacent cells and (2) nonideal sectorization antennas. In Ref. 21, the nonideal sectorized antenna patterns are ac-

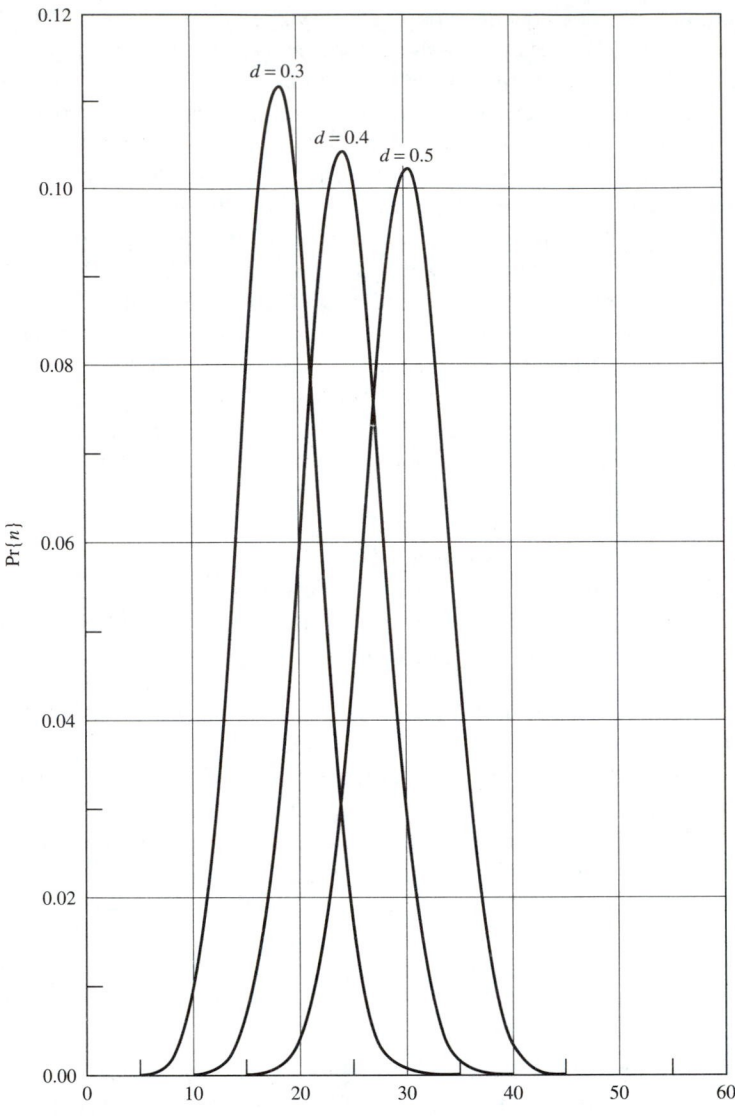

FIGURE 9-5.   Envelope of discrete density function for various voice activity factors, $N = 60$.

counted for by defining the equivalent uniform beamwidth of the antenna as the beamwidth of a rectangular function enclosing the same area as that of the original antenna. For the simplified capacity equation given above, the loss from ideal sectorization may be accounted for in a reduced value for g (or f).

9-3.2 Special Aspects of CDMA Digital Cellular Systems

Several of the CDMA digital cellular systems to be studied below have certain additional attributes in common. These attributes are noted here.

Direct-Sequence CDMA Digital Cellular Systems. In direct-sequence CDMA digital cellular systems and in certain frequency-hopped CDMA digital cellular systems, the signals are assumed to occupy the full allocated bandwidth all the time. Interferers are therefore assumed to come from all directions, not just a few directions as is the case with narrowband cellular systems. The large number of interferers leads to the concept of *interference averaging* in treating interference. That is, interference is assumed to obey the law of large numbers—hence the interference is equated to the mean level of interference observed with no dominant interferers. The interference averaging concept then allows, for example, interference to be treated as a Gaussian random process, large code cross-correlation peaks to be ignored, and so on. Further examples of the effects of interference averaging are given in Ref. 22.

Slow-Frequency-Hop CDMA Digital Cellular Systems. Slow-frequency-hopped (SFH) CDMA digital cellular systems can benefit from two means of performance improvements over nonfrequency hopped systems. First, at slow vehicle speeds, slow frequency hopping provides a form of frequency diversity through exploitation of frequency selectivity over the system bandwidth. From hop to hop the fading process is decorrelated either over the span of an interleaver or over the duration of long fades [23]. This frequency diversity results in improved performance, which is independent of vehicle speed, provided that the hopping dwell period is sufficiently small.

Second, by careful assignment of small correlation frequency-hopping patterns, a given source of interference can affect the interference only a small portion of time. This interference reduction effect has been termed *interferer diversity* in Refs. 24 and 25. Interferer diversity causes the interference experienced by any user to be reduced, as the interference experienced by the user comes not from a single dominant interferer, as is the case in analog cellular systems, but from the aggregate of all users, each sampled one at a time. Provided that coding with adequate interleaving is employed in the SFH system, individual samples of extreme interference may be corrected in the decoding process.

Several systems that take advantage of both of these attributes of SFH systems are discussed next.

9-4 Specific Examples of CDMA Digital Cellular Systems

The concepts developed throughout this chapter are perhaps best understood by looking at some specific examples of CDMA digital cellular systems. In this section a number of examples of CDMA digital cellular systems are given. As these systems vary in degree of development, the treatment of each of these examples will vary in scope. Furthermore, because several of these systems are still under development and/or specific implementations may be proprietary to a given manufacturer, certain details of implementation and performance will not be given.

Six examples of CDMA systems are provided below. The first system is a direct-sequence implementation; all the remaining systems are frequency-hopped implementations.

9-4.1 North American DS-CDMA Digital Cellular System (IS-95)

As noted earlier, in the United States and in many other parts of the world, an ever-increasing demand has been placed on the resources of the existing analog cellular systems. In many cities it has been projected that the existing analog systems would run out of capacity (i.e., excessive blocked calls and/or blocked handoff attempts) in less than 10 years. In June 1989, Qualcomm Inc. proposed the use of DS-CDMA technology as a means of overcoming the expected capacity limits of analog cellular systems. In the years following, Qualcomm has further developed the idea of a DS-CDMA digital cellular system into an actual implementation that has now been adopted as an Interim Standard (IS-95) [26] by the Telecommunications Industry Association (TIA), an ANSI-approved standards body, as one of two† digital standards accepted for deployment in North America by the existing analog cellular carriers within their allocated spectrum at 800 MHz. The IS-95 DS-CDMA system (with possible modifications) has also been proposed to be used in the personal communication systems (PCS) frequency band of 1850 to 1970 MHz which the Federal Communications Commission released for PCS applications in September 1993. At the present time, trial systems for the IS-95 system have been deployed in San Diego, Spokane, and in Washington, D.C., with other trial systems planned in the near future. Several equipment manufacturers will provide base and subscriber unit equipment for the IS-95 system; at the present time these include Qualcomm, OKI, Motorola, Nokia, and Sony for subscriber equipment and Motorola and AT&T for base equipment.

The IS-95 system has been designed to replace a number of existing analog cellular radio channels with a single DS-CDMA carrier. Asymmetric modulation methods have been employed in the forward link (base-to-mobile) and the reverse link (mobile-to-base), although both links are spread at a 1.2288-Mchips/s rate, thereby permitting both links to occupy the same bandwidth. Table 9-3 summarizes many of the parameters of the IS-95 system. Conceptual block diagrams of the forward and reverse link are shown in Figure 9-6a and b, respectively. Because of the differences between the forward link and the reverse link, each link will be described separately below.

---

†The other North American digital cellular standard adopted by the TIA is a digital three-slot TDMA system designed to be retrofitted one-for-one into the existing 30-kHz AMPS channels. This system has been commercially deployed in several areas of the United States and Canada by Southwestern Bell, McCaw, and others.

TABLE 9-3. IS-95 DS-CDMA Digital Cellular System Parameters

Parameter	Forward Link	Reverse Link
Frequency band (MHz)	869–894	824–849
Access method	DS/CDMA-FDMA	DS/CDMA-FDMA
Cell-reuse pattern	Single-sector reuse	Single-sector reuse
Modulation	BPSK with Walsh orthogonal covering	64-ary orthogonal signaling
DS spreading	QPSK spread, period $= 2^{15}$ with modulation shaping	OQPSK spread, period $= 2^{15}$ with modulation shaping
Chip rate (Mchips/s)	1.2288	1.2288
FEC	Rate $\frac{1}{2}$, k = 9 convolutional code	Rate $\frac{1}{3}$, k = 9 convolutional code
Interleaving	Block interleaver, duration = 20 ms	Block interleaver, duration = 20 ms
Scrambling	Decimated $2^{42} - 1$ length user sequence	$2^{42} - 1$ length user sequence
Receiver structure	Three-branch Rake receiver	Four-branch Rake receiver
Speech coder Method Data rates	QCELP 9.6, 4.8, 2.4, 1.2 kbps, rate determined by VAD	QCELP 9.6, 4.8, 2.4, 1.2 kbps, rate determined by VAD
Frame duration (ms)	20	20
Diversity methods	Frequency diversity: wideband signal Time diversity: interleaving Path diversity: multipath Rake	Frequency diversity: wideband signal Time diversity: interleaving Path diversity: multipath Rake + antenna diversity

Forward Link Description. As Figure 9-6a indicates, the forward link employs coherent modulation/demodulation [19,22,27]. A variable-rate speech coder is employed based on Qualcomm's QCELP (Qualcomm code-excited LPC) vocoder method. The variable-rate speech coder forms an integral part of the VAD approach employed in the IS-95 system. As the table indicates, the speech coder outputs data at one of four rates—9.6, 4.8, 2.4, or 1.2 kbps, depending on the speech activity at a given instant in time. Assume for the moment that the highest data rate of 9.6 kbps is used. The output of the speech coder is then rate-$\frac{1}{2}$ coded with a constraint-

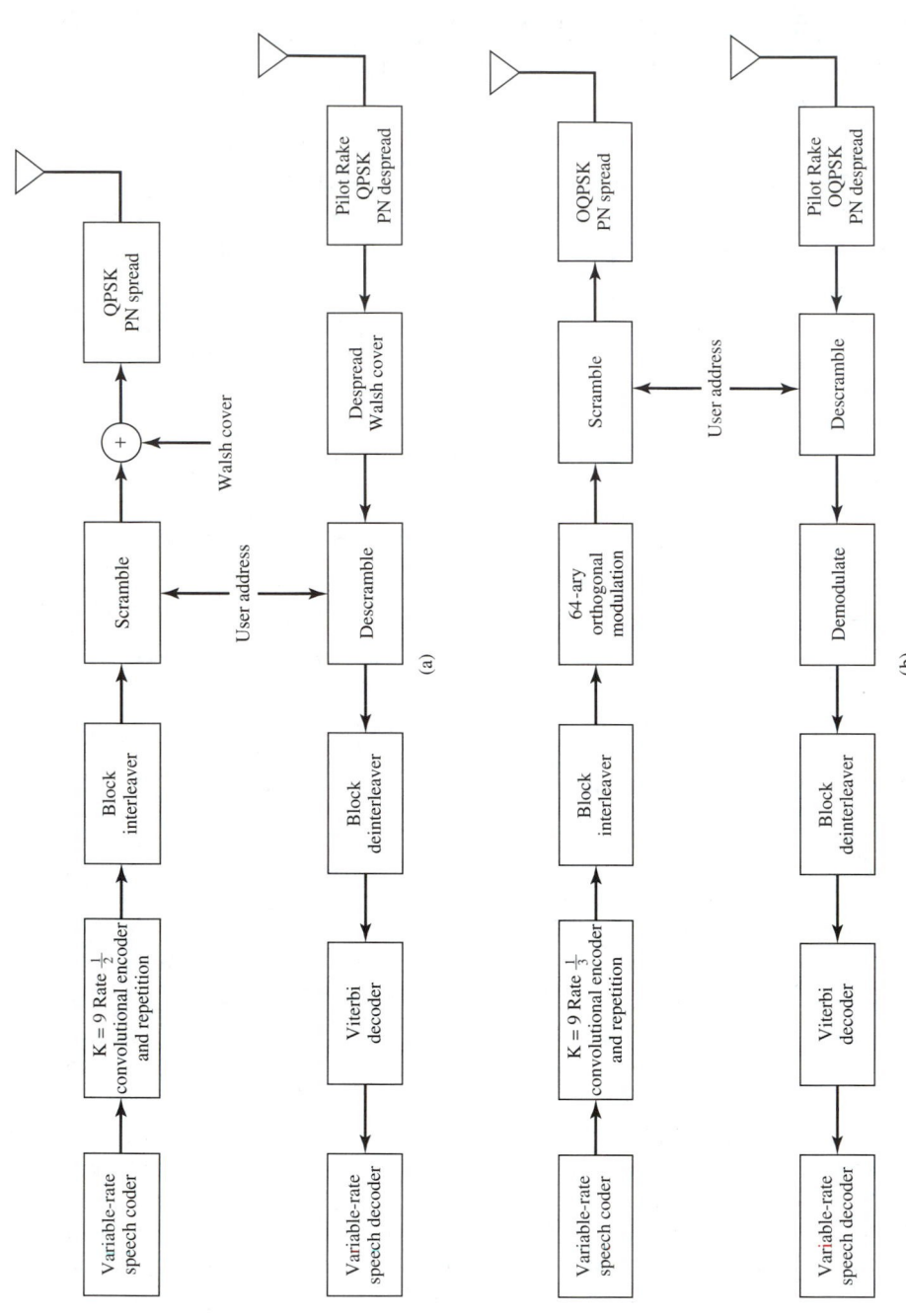

FIGURE 9-6. Conceptual block diagram of IS-95 (a) forward link and (b) reverse link.

length-9 convolutional code, using the generator polynomials

$$g_1(x) = 1 + x + x^2 + x^3 + x^5 + x^7 + x^8$$

$$g_2(x) = 1 + x^2 + x^3 + x^4 + x^8$$

thereby increasing the bit rate to 19.2 ksymbols/s. The output of the convolutional coder is then block interleaved using a 24×16 block interleaver over a 384-symbol (20-ms) interval. The coded symbols are scrambled, for the purpose of privacy, by exclusive-ORing the coded symbols with a long code sequence that consists of a decimated $(2^{42} - 1)$-length PN sequence. The scrambled, coded symbols are then exclusive-ORed with a row of a dimension-64 Hadamard matrix. This process known as *Walsh covering* ensures that each user within a given cell (or sector, if sectors are used) is (strictly) orthogonal to every other user within the cell (or sector), assuming that different rows of the Hadamard matrix are used for each user. As is well known, the orthogonality of Walsh codes breaks down in the presence of multipath; nonetheless, in the absence of multipath, orthogonality within the cell can be maintained. After the Walsh covering process, the covered, scrambled, coded symbol stream is at 64×19.2, or 1.2288 Msymbols/s. The symbol stream then modulates the 800-MHz carrier using BPSK modulation *with* QPSK spreading. Each of the two branches of the QPSK spreader uses a separate PN code of period 2^{15} (32,768) chips with a chip rate of 1.2288 Mchips/s. The two period-2^{15} PN sequences, unique to a given cell (sector), are distinguished from other cells (sectors) by time offsets from the base code. Thus other cells or sectors will have effectively low PN code cross-correlations with a given cell or sector, permitting multiple-access operation over the entire DS-CDMA cellular system. Following PN spreading, each of the two branches is filtered separately with a finite impulse response (FIR) filter prior to modulating a quadrature carrier and combining. Figure 9-7 shows further details of the transmitter operations.

Some comments about the signal structure of the forward path are in order. First, even though the information is conveyed via BPSK modulation, QPSK spreading is employed. For a single-user DS system, either form of modulation would yield equivalent performance. However, in the multiple-access environment, the use of QPSK spreading effectively randomizes the phase of the desired user relative to that of the other users, thereby ensuring that the other users do not introduce a significant phase alignment degradation factor into the desired user's receiver [28]. Second, although the Hadamard matrix of dimension 64 contains 64 rows, thus permitting up to 64 users within a cell (or sector), only 61 Walsh codes are available to individual users; the remaining three Walsh codes are reserved for the pilot, paging, and sync channels, which are common to all the users in the cell. Specifically, the pilot channel is simply a constant-level signal that is exclusive-ORed with the all-zeros Walsh cover signal and sent over the air after QPSK spreading (see Figure 9-7). It is the presence of this fixed, known signal in the forward channel that permits *coherent* demodulation at the mobile receiver. The fixed-level pilot is also used in performing open-loop power control in the reverse link. Third, although only the full-rate (9.6-kbps) traffic channel was considered in the description above, the other three traffic rates are transmitted in a similar fashion, with the following

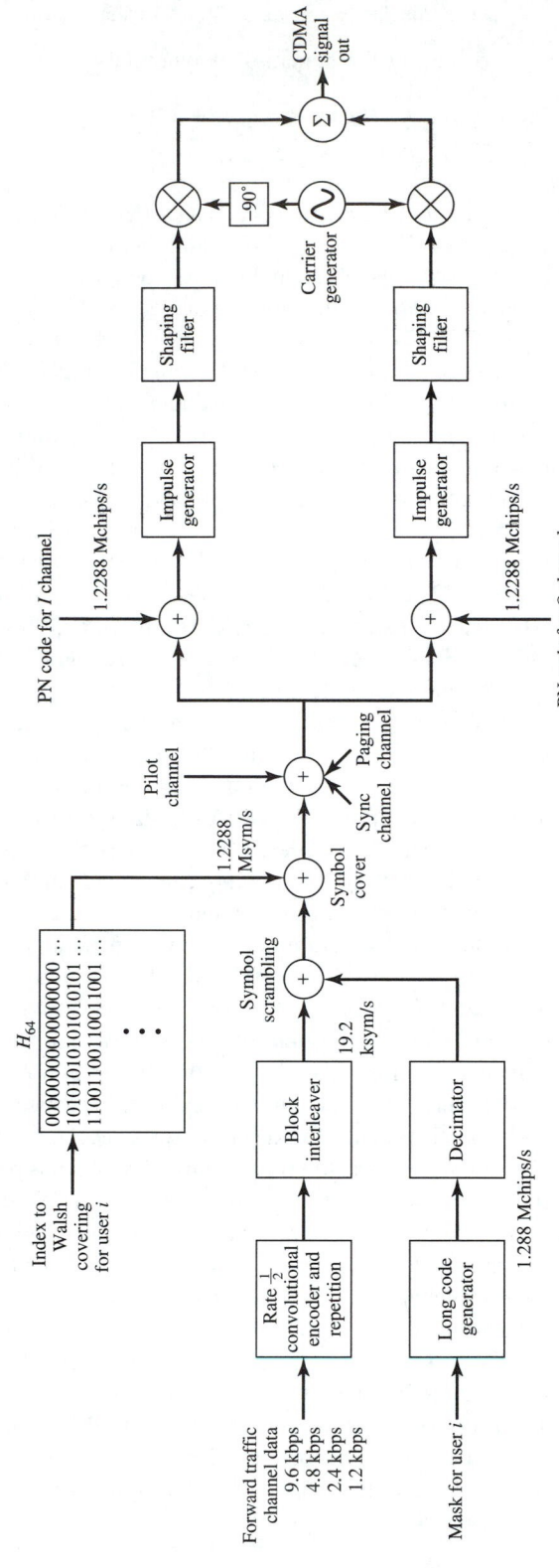

FIGURE 9-7. Detailed IS-95 transmitter structure for forward link.

modifications. Since voice activity is affected in the IS-95 system by transmitting speech at lower rates, for the forward link, the power of the forward path is reduced by $10 \log_{10}(9.6 \text{ kbps/speech coder rate})$ dB for the lower rates (i.e., the transmitter power is reduced by 3, 6, and 9 dB for speech coder rates of 4.8, 2.4, and 1.2 kbps, respectively). Hence low speech activity will result in lower transmit power and a subsequent reduction in interference to other users. To maintain a constant E_b/N_0 into the receiver, the output symbols of the conventional coder are repeated two, four, or eight times, corresponding to speech coder rates of 4.8, 2.4, and 1.2 kbps, respectively, prior to interleaving.

As indicated in Figure 9-6a, in the forward link receiver, the operations at the base transmitter are essentially reversed. The received signal is initially despread, with the quadrature period 2^{15} PN sequences. The resulting signal will then consist of the signals intended for all users in the desired cell, the pilot, paging and sync signals for the desired cell, and an attenuated version of interference from other cells. The pilot signal is initially recovered from the despread signal and is used to perform coherent demodulation of the desired user's signal as well to perform multipath searching and tracking, as will be described shortly. The despread, coherently detected signal is further despread by the desired user's Walsh cover. In the absence of multipath, the Walsh-despread signal will contain only the desired user's signal and an attenuated version of signals from other cells/sectors. The Walsh despread signal is then descrambled with the decimated version of the user's long ($2^{42} - 1$ length) code, deinterleaved, convolutionally decoded with a soft-decision Viterbi decoder, and then applied to the speech decoder.

In practice, the received signal must be synchronized in time at the receiver. This is accomplished through the use of a search processor that is part of the pilot signal recovery. In addition, in the presence of multipath, the simple coherent receiver structure for the desired user must be replaced by a variable-time-delay coherent Rake receiver structure. In the latter case, the search processor is employed to identify likely multipath components for the Rake receiver.

Reverse Link Description. As noted at the beginning of this subsection, the modulation method employed in the reverse link of the IS-95 system differs from that employed in the forward link. In particular, the reverse link employs *noncoherent* modulation. From Figure 9-6b, in the mobile transmitter of the reverse link, the QCELP speech coder output is supplied to a rate-$\frac{1}{3}$ constraint-length-9 convolutional encoder with generator polynomials

$$g_1(x) = 1 + x^2 + x^3 + x^5 + x^6 + x^7 + x^8$$

$$g_2(x) = 1 + x + x^3 + x^4 + x^7 + x^8$$

$$g_3(x) = 1 + x + x^2 + x^5 + x^8$$

Assume for the moment that the speech coder is operating at full rate (9.6 kbps). After the convolutional coder, the symbol rate is now 28.8 ksym/s. The output of the convolutional coder is applied to a 32×18 block interleaver of length 576 symbols (20 ms). The coded, interleaved symbols are then block coded using a 64-ary orthogonal modulator. The 64-ary orthogonal modulator uses six sequential symbols from the block interleaver to form an index into a Hadamard matrix of dimension 64. The index is then used to select a row of the Hadamard matrix whose 64 elements are then transmitted *in place of* the 6-bit index.

Note that the 64-ary orthogonal modulation process is identical to encoding the symbols using a first-order Reed–Muller code [29,30]. Recognizing this relationship, Figure 9-8 shows a simple method of encoding the interleaved symbols. In the figure, the 6-bit index into the Hadamard matrix is represented by the binary vector $u_6 u_5 u_4 u_3 u_2 u_1$. The synchronous 5-bit binary counter is assumed to be initially reset. All the counters are then clocked from a common clock as shown. The outputs of the clock and each of the counter stages are weighted by one of the symbol values in the binary vector $u_6 u_5 u_4 u_3 u_2 u_1$ as indicated. The clock is then clocked 64 times and the six weighted bit streams are modulo-2 summed. The resultant summed stream is simply the row of the Hadamard matrix that corresponds to the index $u_6 u_5 u_4 u_3 u_2 u_1$.

At the output of the 64-ary modulation process, the symbol rate is now 28.8 ksym/s × 64/6 = 307.2 ksym/s. The coded symbol stream is then spread and scrambled (again for privacy) by mixing with the 1.2288-Mbit/s long code (period $2^{42} - 1$) from the long-code generator. As was the case with the forward link, the scrambled data stream is spread further by simultaneously spreading the data stream in quadrature with two short-length (period 2^{15}) sequences. Now, however, the resulting quadrature channels are applied to an offset-QPSK modulator for *I/Q* filtering and for mixing up to the output carrier frequency. Figure 9-9 shows further details of the reverse link transmitter.

As was the case for the forward link, for the lower data rates (4.8, 2.4, and 1.2 kbps), the convolutionally coded symbols are repeated two, four, and eight times, respectively, prior to interleaving. Now, however, the voice activity function is affected by transmitting only one of the code symbol repetitions at the full-rate power. This yields a constant E_b/N_0 at the receiver *for a given data rate*, but does result in a variable-rate transmission. To reduce interference, the time position of the

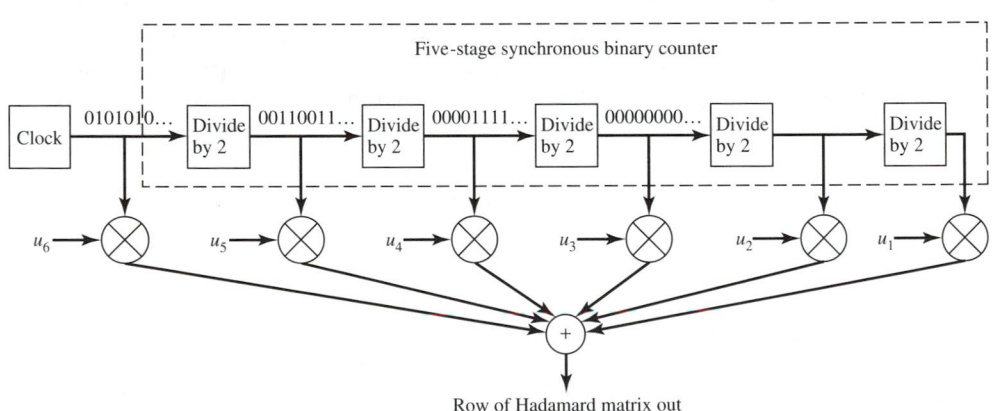

FIGURE 9-8. 64-ary sequence generator.

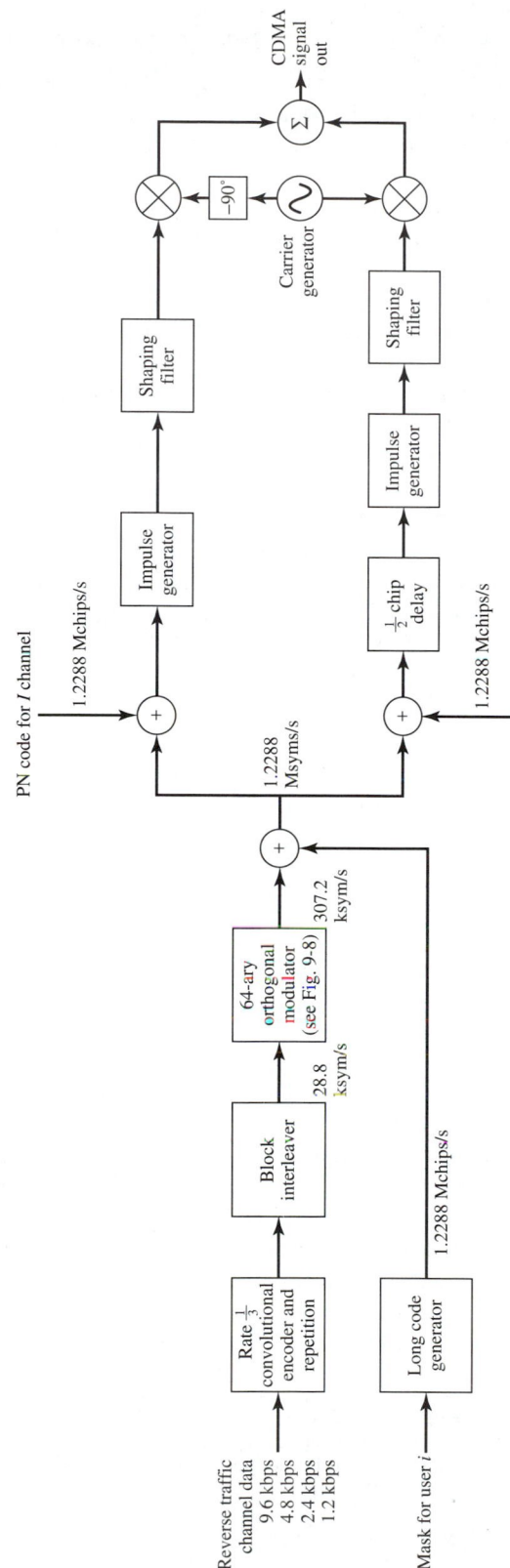

FIGURE 9-9. Detailed IS-95 transmitter structure for reverse link.

transmitted code symbol repetition is randomized so that at the lower data rate, consecutive transmitted bursts do not occur evenly spaced in time.

The reader may question the need for a change in signaling format between the forward and reverse links. In the forward link, many users are naturally multiplexed together in the process of transmitting to the multiple mobile receivers. In this link, a common pilot channel may be inserted that will permit coherent demodulation for *all* users. In the reverse link, however, insertion of a dedicated pilot channel for each mobile transmitter to permit coherent demodulation at the base site receiver will result in a significant increase in co-channel interference, thereby resulting in a decrease in system capacity. Hence coherent demodulation is *not* employed in the reverse link; the most power-efficient noncoherent modulation method is employed instead (*M*-ary noncoherent modulation). As will be seen below, the block code employed admits to a soft-decision decoding technique that is also computationally efficient. In addition, offset QPSK modulation is employed in the reverse link to minimize the mobile transmitter's peak-to-average excursions.†

As indicated in Figure 9-6b, at the reverse link receiver, the operations at the mobile transmitter are essentially reversed. The received signal is initially despread with the quadrature period 2^{15} PN sequences. The resulting signal is then further despread/descrambled with the long user code before being applied to the 64-ary demodulator/noncoherent Rake receiver.

The 64-ary noncoherent demodulator/block code decoder takes as an input, the length-64 received signal *real-valued* vector $\{s_i\}$, $i = 1, 64$. The block decoder then performs a correlation with each row of the dimension 64 Hadamard matrix, where the binary elements of the Hadamard matrix are mapped to ± 1 using the mapping

$$
\begin{aligned}
1 &\rightarrow -1 \\
0 &\rightarrow +1
\end{aligned}
\tag{9-15}
$$

That is, the block decoder computes

$$
r_j = \sum_{i=1}^{64} s_i H_i^j \qquad j = 1,64
\tag{9-16}
$$

where H_i^j denotes the jth row of the dimension-64 Hadamard matrix after performing the mapping on the elements given in (9-15). In the absence of soft decisions to the convolutional decoder (and deinterleaver), the block demodulator determines the index j corresponding to the maximum of all the r_j. The binary value of the maximum index is then mapped to a corresponding 6-bit convolutionally coded symbol for application to the deinterleaver/convolutional decoder. Straightforward computation of the correlation process in the block decoder requires 64×64 real multiplications and 64×64 real additions for every six convolutionally coded symbols. By

---

† The peak-to-average ratio of a transmitted signal is simply the ratio of the instantaneous peak power to the average transmitted power. The link performance is determined by the average transmitted power, but the transmitter power amplifier capacity is determined by the peak power. For a given transmit filter shaping, OQPSK modulation results in a lower peak-to-average power requirement than QPSK.

taking advantage of the symmetries of the elements of the Hadamard matrix, the number of computations required in performing the 64 correlations above may be reduced from the order of 642 computations to $64 \log_2 64 = 384$ operations. The algorithm that enables the reduction in the number of operations is known as the fast Hadamard transform (FHT), an algorithm that was employed in some early deep-space probes (see, e.g., Refs. 29 and 31). The FHT has a structure very similar to that of the fast Fourier transform (FFT); hence it exhibits the computational reduction properties of the FFT. Figure 9-10 shows the FHT structure for 8-ary orthogonal modulation—the FHT demodulator for 64-ary modulation exhibits the same form of this structure. The FHT shown in this figure corresponds to a radix 2 decimation in time FFT [32], and hence the numerical algorithm follows this FFT structure except that the "twiddle factors" $e^{j2\pi nk/N}$ are replaced with real multiplies,† and no bit-reversal reordering of the indices is necessary.

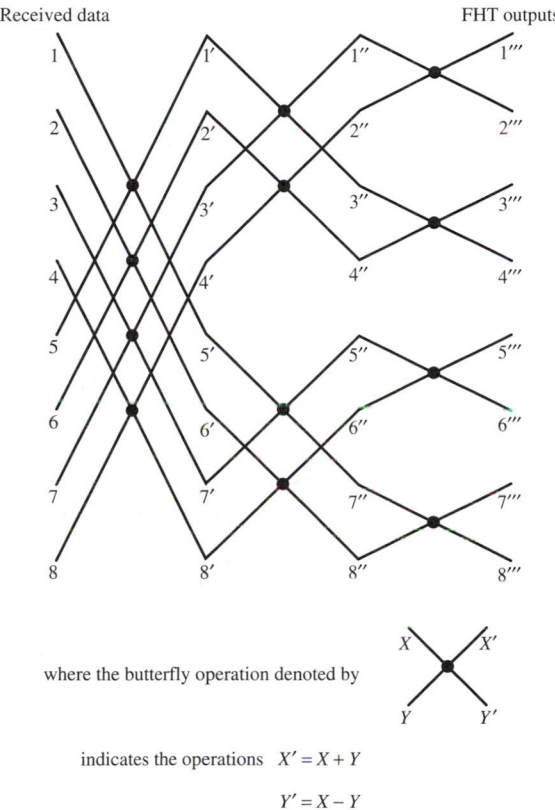

where the butterfly operation denoted by

indicates the operations $X' = X + Y$

$$Y' = X - Y$$

FIGURE 9-10. FHT signal flow diagram for $M = 8$ FHT, indicating a regular algorithm structure. $M = 64$ FHT would have six stages, each stage having 32 butterflies.

† In fact, because soft-decision samples are supplied to the convolutional decoder, four FHT operations are required for every 6 convolutionally coded symbols.

The output of the block code demodulator is then fed to the block deinterleaver, the Viterbi convolutional decoder, and the speech decoder. Not described above is the search processor required in the Rake receiver for tracking multipath as well as the noncoherent Rake combiner. These were described in detail earlier in the book. Further details of VLSI implementations of the IS-95 system may be found in Refs. 33 to 36.

Performance. The performance of the IS-95 system has been analyzed by a number of authors [19,27,37,38]. Because the system incorporates power control, multiple modulation/coding methods, and various forms of diversity, a single approach to system analysis will not adequately predict the performance of the system. As a result, several of the published analyses seem to contradict other analyses. For this system, Monte Carlo simulations appear to the best method of analyzing system performance, especially over fading channels. In the balance of this subsection, a first-order analysis of the performance of the system will be given.

For both the forward and reverse channels, the capacity of the system, as given by (9-13), is a function of the modulation/coding/diversity subsystem as well as of the cellular system design. For the forward channel, analytical probability of bit error bounds may readily be obtained under certain channel conditions. For the AWGN channel, the probability of bit error, assuming soft-decision decoding, may be upper bounded by the expression [39]

$$P_b \leq \sum_{d=d_{\text{free}}}^{\infty} \beta_d Q\left(\sqrt{\frac{2dRE_b}{N_0}} \right) \qquad (9\text{-}17)$$

where β_d is the number of information 1's in all nonzero paths through the trellis with Hamming weight d of the forward link convolutional code which merge with the all-zero path for the first time, R is the code rate, and d_{free} is the free distance of the convolutional code. The parameters β_d and d_{free} of a number of convolutional codes, including those employed in the IS-95 system, may be found in Ref. 40. For the Rayleigh-fading channel, several upper bounds may be employed with the assumption that the interleaver is of infinite length.†

Assuming that hard-decision decoding is employed, the probability of error bound for Viterbi decoding on a binary symmetric channel holds for the Rayleigh-fading channel and is given by [39] (also see Chapter 7)

$$P_b \leq \sum_{d=d_{\text{free}}}^{\infty} \beta_d P_d \qquad (9\text{-}18)$$

---

† While an infinite-length interleaver is not physically realizable, provided that the channel is sufficiently decorrelated using the finite-length interleaver actually employed in the system, the bounds obtained with the infinite-length interleaver will also hold for the finite-length interleaver. Sufficient decorrelation may imply high Doppler rates and hence high vehicle speeds (see Section 8-3).

where

$$P_d = \begin{cases} \displaystyle\sum_{k=(d+1)/2}^{d} \binom{d}{k} p^k (1-p)^{d-k} & d \text{ odd} \\[4mm] \displaystyle\frac{1}{2}\binom{d}{d/2} p^{k/2}(1-p)^{k/2} + \sum_{k=d/2+1}^{d} \binom{d}{k} p^k (1-p)^{d-k} & d \text{ even} \end{cases} \qquad (9\text{-}19)$$

and

$$p = \frac{1}{2}\left(1 - \sqrt{\frac{1}{1 + 1/(RE_b/N_0)}}\right) \qquad (9\text{-}20)$$

For soft-decision decoding on the Rayleigh-fading channel, it may be shown that an upper bound on bit error performance, assuming perfect channel-state information, is given by (9-18), where P_d is given by (8-138) [41–43].

Figure 9-11 plots each of these performance bounds as a function of E_b/N_0. The actual performance of the forward link over a Rayleigh-fading channel will probably lie between the hard- and soft-decision bounds. Assuming that a 6-dB E_b/I_0 is required to meet a 10^{-3} probability of decoded bit error rate, (9-13) indicates that for a 50% VAD factor, $f = 0.6$, and $g = 3$ sectors, the forward link capacity is given by

$$N = \frac{1.2288 \text{ MHz}/9.6 \text{ kHz}}{3.981 \ (= 6 \text{ dB})} \cdot \frac{1}{0.5} \cdot 3 \cdot 0.6 = 115.7 \text{ users} \qquad (9\text{-}21)$$

Further refinement of the value of forward link capacity requires simulation of both the modem and the cellular system design.

Determination of the performance of the reverse link is complicated by the concatenated coding scheme employed. Hence bounds for performance only over an AWGN channel will be given. For the reverse link, the probability of bit error may be upper bounded by (9-18) with P_d given approximately by the hard-decision probability of error for noncoherent detection of 64-ary orthogonal signals [44,45], (and Chapter 7)

$$P_d = [2\sqrt{p(1-p)}]^d \qquad (9\text{-}22)$$

where

$$p \cong \frac{1}{2}\sum_{n=1}^{63} (-1)^{n+1}\binom{63}{n}\frac{1}{n+1}e^{-6n(E_b/N_0)/(n+1)} \qquad (9\text{-}23)$$

In Figure 9-12, this bound on the probability of bit error on an AWGN channel is plotted as a function of E_b/N_0. From this graph the required E_b/N_0 to achieve a 10^{-3} decoded probability of error is approximately 6 dB. Again using (9-13) and assuming the same values of d, f, and g as for the forward channel, the capacity of the reverse channel is

$$N = \frac{1.2288 \text{ MHz}/9.6 \text{ kbps}}{3.981 \ (= 6 \text{ dB})} \cdot \frac{1}{0.5} \cdot 3 \cdot 0.6 = 115.7 \text{ users} \qquad (9\text{-}24)$$

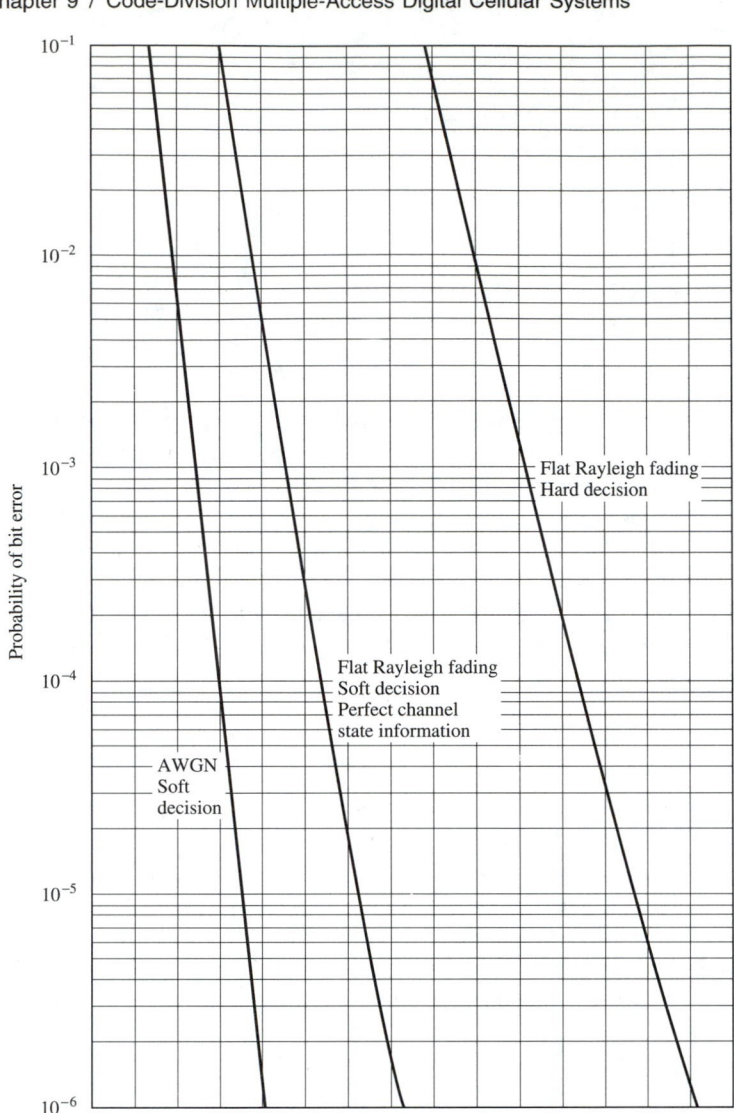

FIGURE 9-11. BER bounds for forward channel of IS-95 system.

Other results on the performance of the reverse link may be found in Refs. 37, 46, and 47.

The capacity calculations above have made various assumptions about the channel characteristics and the factors $d, f,$ and g in (9-13). To achieve a 10^{-3} probability of error at the stated values of E_b/N_0 for other channel conditions, additional diver-

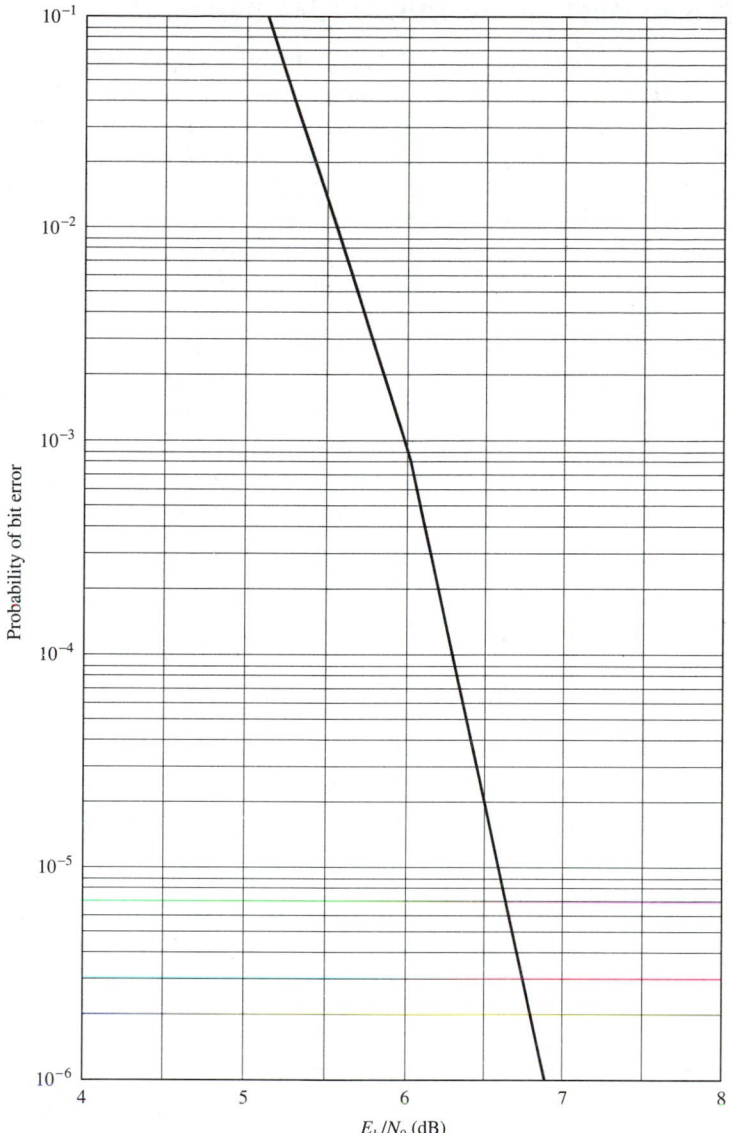

FIGURE 9-12. Upper bound on probability of error for reverse channel of IS-95 system, AWGN channel.

sity will be required in many instances. In the IS-95 system, the additional diversity is present in the forward path in the form of multipath diversity (via multipath combining in the Rake receiver) and path diversity in which multiple cell sites are used to transmit to a given mobile user. For the reverse path, multipath diversity is again employed along with three-branch antenna diversity.

9-4.2 Cooper and Nettleton DPSK-FHMA System

In a series of papers beginning in 1977 [18,21,48,49], Cooper and Nettleton, then both at Purdue University, described what has become recognized as the first serious approach to applying spread-spectrum code-division multiple-access methods to the design of a digital cellular system. Indeed, many of the concepts employed in the other CDMA digital cellular systems that have been developed since that time can trace their origins to the Cooper–Nettleton proposal. This proposal involves a frequency-hopped multiple-access (FHMA) cellular system employing differential PSK (DPSK) as the modulation method. Hence this system will subsequently be called a DPSK-FHMA system.

The Cooper–Nettleton proposal actually predated the commercial deployment of *analog* cellular systems. It was, in fact, an initial attempt to displace the analog FM modulation technology that was then being discussed as the modulation choice for future cellular systems. Table 9-4 summarizes the parameters of the DPSK-FHMA system. Figure 9-13 shows a conceptual block diagram of the DPSK-FHMA transmitter and receiver pair. In the following, a brief description of the operation of the system is given.

Transmitter. As Table 9-4 indicates, the DPSK-FHMA system is a fast-frequency-hop CDMA system. Speech data coded at approximately 32 kbps is fed to a Walsh or Hadamard block orthogonal channel coder of dimension 5. Thus data are encoded with a rate (5/32) block code which increases the symbol rate by 32/5 relative to the information rate. The output of the channel coder is then binary differentially encoded using a conventional DPSK encoder *but with a delay element equal to one Walsh symbol time* (5 information bits or 32 coded symbols). DPSK is used because the multipath nature of the channel (as well as the fast-hopping rate) destroys any phase coherence in the received signal. *Each* coded symbol is then mixed to a frequency determined by an M-ary FSK generator (or FH synthesizer), where $M = 32$. For a 31.25-kbps information rate, the frequency-hopping rate of the

TABLE 9-4. Parameters of DPSK-FHMA System

Frequency	825–845 MHz mobile transmit
	870–890 MHz base transmit
Total occupied bandwidth, W	20 MHz
Speech coder rate, R	31.25 kbps
Channel coding	Hadamard orthogonal code, $k = 5$
Modulation method	DPSK
Access method	Fast-frequency-hop CDMA
Diversity approach	Frequency diversity
Number of branches, L	32
Channel spacing	200 kHz
Cell reuse pattern	Single cell

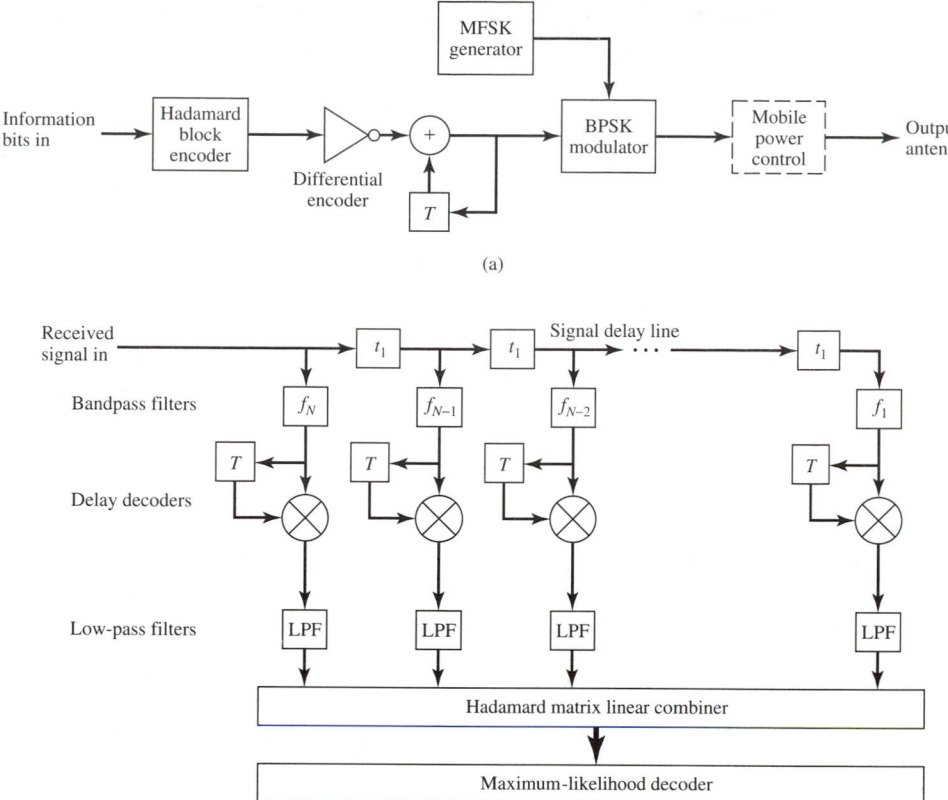

FIGURE 9-13. Block diagram of DPSK-FHMA (a) transmitter and (b) receiver.

synthesizer is

$$f_{\text{hop}} = 31.25 \text{ kbps} \cdot \frac{32 \text{ coded symbols}}{5 \text{ information bits}} = 200 \text{ khops/s} \qquad (9\text{-}25)$$

Note that this is truly a *fast*-hopping system since there are $32/5 = 6.4$ hops per information bit, each hop of duration $t_1 = 1/200$ khops/s $= 5 \, \mu$s. The fast-hopping rate allows the system to use frequency diversity to overcome the effects of Rayleigh fading, as well as permitting the use of "good" FH sequences. Because of the fast-hopping rate, Cooper and Nettleton termed each dwell period a "chip," by analogy with direct-sequence spread spectrum.

The frequency-hopping patterns are assumed to be selected from a much larger set of frequencies in a manner such that the signals have a low aperiodic correlation among themselves. Hence no two frequency sequences are identical. Specifically,

Cooper and Nettleton proposed the use of the Yates–Cooper [50] sequences, a set of sequences that has the one-coincidence property; that is, when a frequency-hop sequence is shifted in time by one dwell period with respect to any other frequency-hop sequence in the set, there will be at most one dwell period for which two sequences will occupy the same frequency. The Yates–Cooper sequences are all of the form

$$f_i^k = f_0 + a_i^k f_1 \qquad i = 1, 2, ..., m \qquad k = 1, 2, ..., m \qquad (9\text{-}26)$$

where f_i^k is the ith frequency for user k, the set $\{a_i^k\}$ the set of integers associated with the kth FH sequence, f_1 the minimum frequency shift, and f_0 the nominal carrier frequency.

The desired property of the sets of frequency-hopping sequences is that collectively, the sequences will occupy the entire $W = 20$ MHz bandwidth with minimal cross-correlation between sequences of different users. To minimize the interference level between users, the mobile transmitter is assumed to have power control on the uplink such that at the base station, each user is received with equal power.

Receiver. The receiver must frequency dehop the signal, differentially decode the dehopped signal, and perform frequency diversity combining to overcome the effects of multipath. Figure 9-13 shows a receiver structure to perform all these operations. Rather than constructing a receiver to perform all these operations in serial, Cooper and Nettleton advocated the use of a parallel receiver architecture. Specifically, the receiver shown employs 32 bandpass filters, corresponding to the 32 frequencies used in the frequency-hop sequence. Each bandpass filter is assumed to be matched to a rectangular chip of duration t_1 and thus has a noise equivalent bandwidth of $1/t_1$. A tapped delay line is used to align in time the chips of the received sequence for each Walsh codeword. Thus all of the 32 chips in the FH sequence (or Walsh codeword) pass through each of the filters simultaneously.† The output of each of the matched bandpass filters is then followed by a DPSK decoder and a low-pass filter (to remove second harmonic terms).

The outputs of the lowpass filters are then fed to a combiner and decoder circuit. The function of the combiner is to combine the appropriate outputs of the low-pass filter in a manner that will correspond to a possible transmitted codeword. Since the transmitter employed a Walsh–Hadamard block encoder, the receiver may use the corresponding decoding matrix to perform the combining function. Alternatively, from the previous discussion of the IS-95 system, a fast Hadamard transform could be used in the receiver to perform the combining function in order to minimize combiner complexity. After combining, the decoder selects the combiner output that yields the greatest amplitude, performs a mapping from the combiner output index to decoded bits, and outputs the decoded bits.

---

† Despite the similarity of the receiver structure to that of the receivers shown in Section 8-4, the DPSK-FHMA receiver is *not* a Rake receiver, as detection is done at different frequencies, not at one frequency.

Performance. The performance of the DPSK-FHMA system was originally analyzed by Cooper and Nettleton [18] and analyzed further by Henry [51]†. In the following, Henry's results will be used to comment on the capacity and performance of the DPSK-FHMA system. Details of the analysis are contained in the original paper.

Henry's analysis began with the observation that for a system consisting of M mobile users occupying a bandwidth W, assuming that the transmitters were uncorrelated, for large M, the received interference to a single user was approximately equivalent to white Gaussian noise with a one-sided power density of

$$S(f) = P\frac{M - 1}{W} \tag{9-27}$$

where P is the (power-controlled) power of each mobile, as received at the base station. Assuming that the combiner outputs are independent *and* Gaussian, in the absence of thermal noise and Rayleigh fading, Henry was able to show that the bit error probability as a function of a signal-to-noise parameter ρ is given by

$$P_b \cong \frac{N - 1}{2}Q(\sqrt{\rho}) \tag{9-28}$$

where

$$\rho = \frac{N}{2}\frac{\rho_f}{1 + 1/2\rho_f} \tag{9-29}$$

and $\rho_f = N/(M - 1)$, where N is the number of chips/code symbol. Henry was also able to derive a similar result for the bit error probability for the Rayleigh-fading channel. In Figure 9-14, the probability of bit error is shown plotted for both the nonfaded and faded cases as a function of ρ_f.

Assuming that the required bit probability of error for acceptable communication is 10^{-3}, then from Figure 9-14 and the definition of efficiency from Section 9-3, the relative spectral efficiency of the DPSK-FHMA system may be calculated. For the faded case, a bit error probability of 10^{-3} is achievable for $\rho_f = 2.5$ dB. For $N = 32$ this corresponds to $M = 19$ users yielding a relative spectral efficiency for the one-cell case of

$$\eta = 19 \times \frac{30\text{ kHz}}{20\text{ MHz}} = 0.029 \tag{9-30}$$

9-4.3 Bell Labs Multilevel FSK Frequency-Hop System

The small relative spectral efficiency of the Cooper–Nettleton DPSK-FHMA digital cellular system prompted a number of researchers at Bell Labs during the period

---

† In Ref 51, Henry noted a 3-dB discrepancy between his performance results and those of the first papers of Cooper and Nettleton. In their later papers [52,53], Cooper and Nettleton acknowledged this discrepancy and presented results in general agreement with those of Henry.

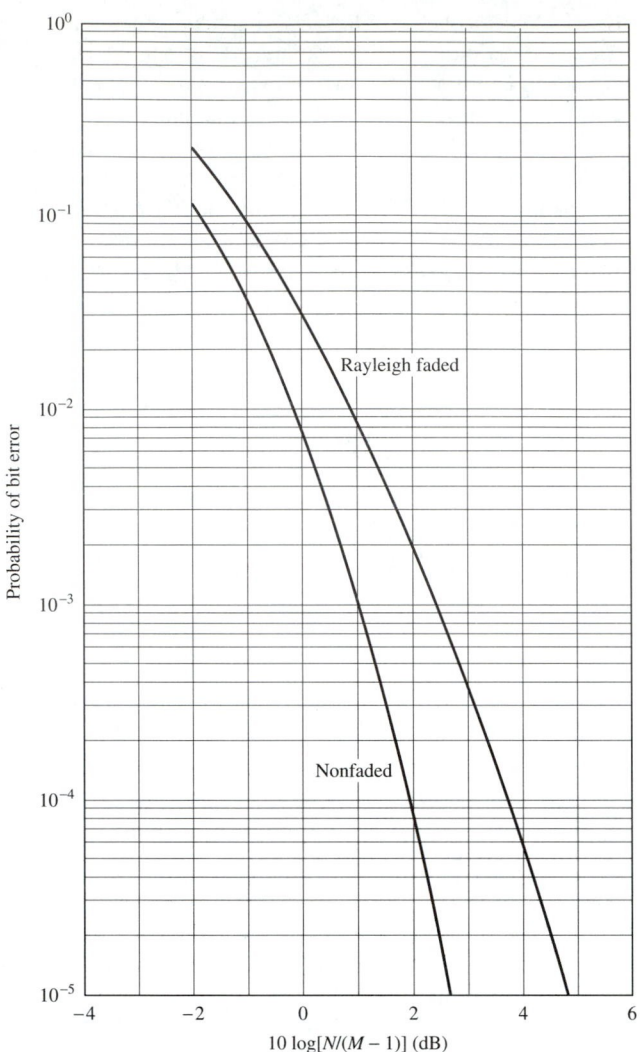

FIGURE 9-14. Probability of bit error for DPSK-FHMA, single-cell case.

1979–1982 [54] to investigate other forms of spread-spectrum multiple-access methods suitable for digital cellular systems. In particular, a team of researchers designed (on paper), a frequency-hopped spread-spectrum multiple-access system that employed multilevel (*M*-ary) frequency-shift keying as the modulation method. The capacity of this system was shown to be roughly three times the capacity of the Cooper–Nettleton proposal. The original basis of the Bell Labs proposal was a paper by Viterbi [55] in which he proposed the use of multilevel FSK-FHMA in a processing satellite transponder which communicated with low-data-rate mobile users.

The specifications of the Bell Labs proposal are shown in Table 9-5. Figure 9-15

TABLE 9-5. Parameters of the FH-FSK System

Frequency band	825–845 MHz mobile transmit
	870–890 MHz base transmit
Access method	Fast-frequency-hop spread-spectrum multiple access
Modulation method	M-ary FSK
Cell design	Single-cell reuse
Occupied bandwidth	20 MHz
Source data rate	32.895 kbps
Forward error correction	Repetition coding, rate K/L

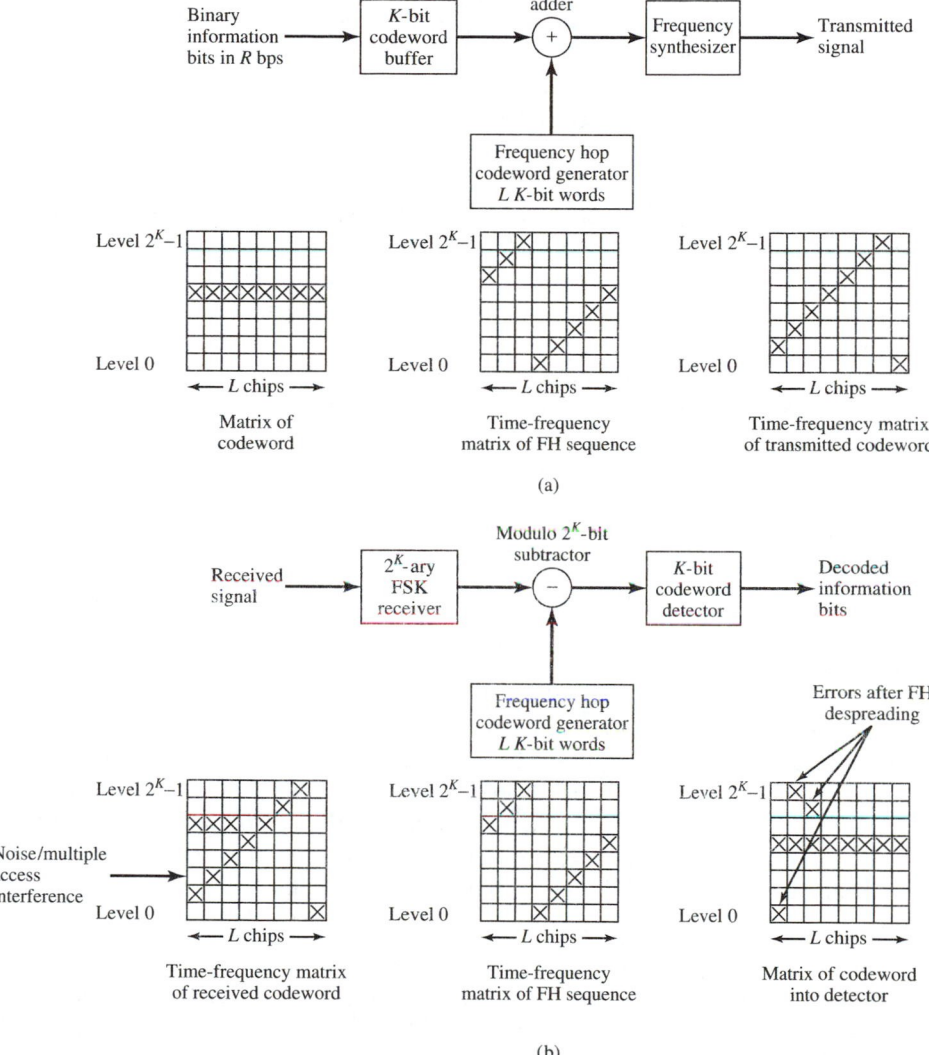

FIGURE 9-15. Multilevel FSK-FH transmitter (a) and receiver (b) block diagrams.

shows a conceptual block diagram of the transmitter and the receiver for this system. Operation of this system will now be considered in more detail.

Transmitter and Receiver. As shown in Figure 9-15, the transmitter consists of a buffer, a frequency-hop sequence generator (''address'' generator), and a frequency-hop synthesizer (or ''tone'' generator). The buffer groups K information bits/symbol and forms a code symbol or chip based on the binary representation of the K bits (which ranges from 0 to $2^K - 1$). The code symbol is then repeated $L - 1$ times to form a codeword consisting of L chips.

The address generator generates a frequency-hop sequence or address that is unique to a given user, generating a value in the range $[0, 2^K - 1]$ every chip time and repeating after every L chips. At each chip time, the output of the codeword buffer and the address buffer are added together modulo 2^K. The resulting M-ary FSK/frequency hop address is then sent to the frequency-hop synthesizer, which generates a hop frequency every chip time according to the equation

$$f_h = f_0 + (\text{composite address}) \cdot f_{\text{step}} \tag{9-31}$$

where f_0 is the nominal carrier frequency, f_{step} the minimum frequency step size of the synthesizer, and the composite address take on values in the range $[0, 2^K - 1]$.

The receiver, which is assumed to be chip and codeword synchronous to the transmitter, performs a spectrum analysis of the received signal and outputs a detected frequency index (or indices) in the range $[0, 2^K - 1]$ every chip time. The address of the frequency-hopping sequence is then subtracted from the detected index modulo-2^K every chip time. In the absence of noise and multiple-access interference, the resulting value following the modulo 2^K subtraction would be a value in the range $[0, 2^K - 1]$ repeated L times, corresponding to the repetition coding in the transmitter. The decoder would then output the K-bit sequence corresponding to the received codeword.

In the presence of noise and/or multiple-access interference, the spectrum analyzer will output *multiple* values in the range $[0, 2^K - 1]$ every chip time. Following subtraction of the desired user's frequency-hop address modulo-2^K, the received value sequence will now consist of multiple values every chip time, as represented by a detection *matrix* in Figure 9-15. A majority logic decision rule is used in this situation which picks the codeword associated with the row of the detection matrix that contains the greatest number of entries. Errors will occur when entries due to noise and/or multiple-access interference cause a row to be formed with more entries than that corresponding to the correct codeword. The frequency-hopping sequences must therefore be selected such that correlation between users is minimized and sufficient frequency diversity is achieved to overcome the effects of Rayleigh fading.

Performance. The original paper by Goodman et al. [54] provides details of the analysis of the performance of the system. The supporting equations of this analysis are contained in Table 9-6.

In Figure 9-16, a plot of the bit error probability is plotted as a function of the

TABLE 9-6. Summary of Supporting Equations for Performance Analysis of FH-FSK System

Upper bound on bit error rate:

$$P_B \cong \frac{2^{K-1}}{2^K - 1} \left\{ 1 - \sum_{i=1}^{L} P_C(i) \left[P(i,0) + \frac{1}{2}P(i,1) \right] \right\} \tag{9-32}$$

where

$$P_C(i) = \text{probability of } i \text{ entries in correct row}$$

$$= \binom{L}{i}(1 - p_D)^i p_D^{L-i} \tag{9-33}$$

and

$$P(n,0) = \text{Probability that no unwanted row has } n \text{ or more entries}$$

$$= \left[\sum_{m=0}^{n-1} P_S(m) \right]^{2^K-1} \tag{9-34}$$

$P(n,1)$ = probability that n is the maximum number of entries in an unwanted row and only one unwanted row has n entries

$$= (2^K - 1)P_S(n) \left[\sum_{m=0}^{n-1} P_S(m) \right]^{2^K-2} \qquad n = 1, 2, \dots, L \tag{9-35}$$

where

$$P_S(m) = \text{probability of } m \text{ entries in a spurious row}$$

$$= \binom{L}{m} p_i^m (1 - p_i)^{L-m} \tag{9-36}$$

and p_i is the probability of insertion due to interference or falsing and is given by

$$p_i = p + p_F - pp_F \tag{9-37}$$

p is the probability of insertion due to interference and is given by

$$p = [1 - (1 - 2^{-K})^{M-1}](1 - p_D) \tag{9-38}$$

For a Rayleigh-fading channel, the probability of false alarm p_F and the probability of detection p_D is given by [56]

$$p_F = e^{-\beta^2/2} \tag{9-39}$$

$$p_D = 1 - e^{-\beta^2/2(1+\text{SNR})} \tag{9-40}$$

where β is the normalized threshold and SNR is the signal-to-thermal-noise ratio.

---

number of users M for various values of SNR for a value of $\beta = 2.75$, $K = 8$, and $L = 19$. If it is assumed that a 10^{-3} probability of bit error yields acceptable performance, in the absence of thermal noise (SNR $= \infty$), the system exhibits a capacity of 212 users for the one-cell case. Assuming an occupied bandwidth of 20 MHz, this

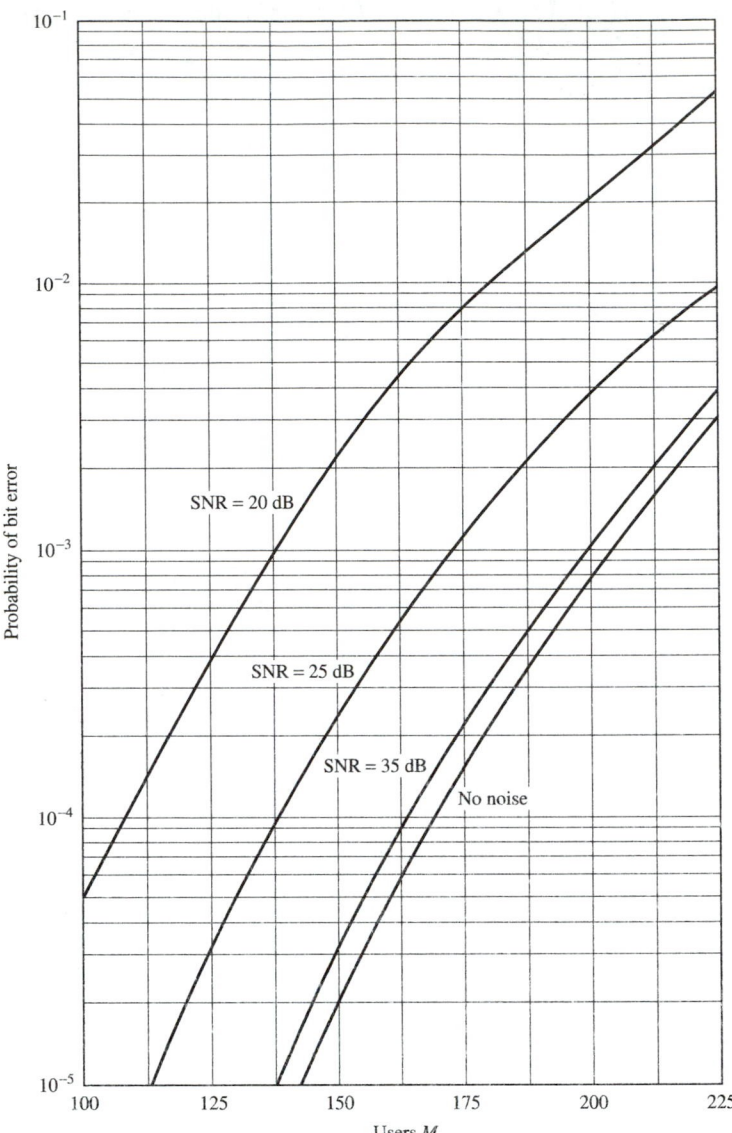

FIGURE 9-16. Probability of bit error versus number of users M for various SNR values.

yields a relative spectral efficiency of

$$\eta = (212 \text{ users}) \frac{30 \text{ kHz}}{20 \text{ MHz}} = 0.318 \qquad (9\text{-}41)$$

or about ten times the relative spectral efficiency of the DPSK-FHMA system described in the preceding subsection.

Related Work. Subsequent to the publication of the original paper on the FH-FSK system, there have appeared a number of publications that offered enhancements to the proposed multilevel FH-FSK system. The latter results are summarized briefly below.

1. *Einarsson (1980) [57]:* In place of the simple frequency-hopping patterns that were used in the original paper, Einarsson proposed the use of shortened Reed–Solomon codes, in the process deriving a set of frequency-hopping codes that guaranteed minimal interference. In addition, Einarsson analyzed the performance of the system for the nonsynchronous case. An excellent overview of Einarsson's sequence construction as well as the design of frequency-hopping sequences based on Reed–Solomon codes is contained in Ref. 58.

2. *Haskell (1981) [59]:* Haskell compared the use of random frequency-hop sequences (as used by Goodman et al.), the finite field construction frequency-hop sequences considered by Einarrson, and linearly increasing (chirp) frequency-hop sequences via computer simulation. He showed that both the chirp and finite-field frequency-hop sequence construction yielded improved performance over the random sequence approach for the noiseless, nonfaded case. By incorporating a simple checking procedure into the decoding process for the chirp frequency-hop sequence case, relative spectral efficiencies of $\eta = 0.55$ were achieved for the noiseless, nonfaded case with frame synchronization among all users assumed.

3. *Timor (1980, 1981) [60, 61]:* Timor considered a technique for improved decoding of FHMA-FSK signals employing Einarsson's construction for frequency-hop sequences. He later extended his results by making use of the algebraic structure of the finite-field frequency-hop sequences for all M users. The additional information required for the latter technique is generated in decoding all M users, and hence this technique is generally suitable for use at the base station only, where such information is readily available. With this improved method, a relative spectral efficiency of $\eta = 0.72$ was achieved for the frame synchronous, nonfaded, noiseless case.

4. *Agusti and Junyent (1983) [62]:* All of the previous analyses of the FH multilevel FSK system used an isolated cell model. Agusti and Junyent generalized the previous results to account for unsynchronized users, power control, and interference from other cells. They found that when co-channel interference from adjacent cells was taken into account, the relative spectral efficiency of the FH multilevel FSK system was reduced by a factor of $\frac{1}{2}$ from that of the isolated cell analysis.

9-4.4 SFH900 System

Prior to the development of the GSM system in Europe, described in the next section, several proposals were put forth as trial systems to be used in selecting the access and channelization approaches for GSM. One of these systems, the SFH900 system, was developed by Laboratoire Central de Télécommunications (LCT) in France [63]. This system is interesting because it combined slow frequency hopping, concatenated coding, and channel equalization to combat the effects of both time- and frequency-selective fading. An experimental prototype of this system was con-

TABLE 9-7. System Parameters of the SFH900 System

Frequency band	900-MHz GSM band
Modulation	GMSK, $B_b T = 0.3$
Access method	Intracell-TDMA
	Intercell-SFH CDMA
TDMA structure	3 slots/frame
SFH dwell period	4 ms
Cell-reuse pattern	1 cell/3 sectors
Channel coding	Interhop: Reed–Solomon
	Intrahop: cyclotomicaly shortened Reed–Solomon
Information data rate	16 kbps
Channel equalization	MLSE equalizer
Gross channel bit rate	200 kbps

structed and was demonstrated in field tests in Paris during late 1986, which were used to test various component parts of the GSM system.† The important parameters of the SFH900 system are shown in Table 9-7.

System Description. As Table 9-7 notes, the SFH900 system employs TDMA within a cell and SFH-CDMA between cells as the access methods. The TDMA frame format consists of a transmit slot, a receive slot, and a frequency switching slot, denoted by T, R, and S slots, respectively. All slots and frames are synchronized within the system to a common clock. The use of T-R-S slot structure within a cell permits easy implementation of a SFH system without requiring a fast-switching synthesizer. Between cells, SFH is employed as the multiple-access method.

A concatenated coding scheme is employed in the system to provide both frequency and interferer diversity and to combat random errors that would occur on the mobile channel. The concatenated coding scheme employs a Reed–Solomon code as an interhop, or outer, code. The coding/hopping is designed to transmit a different Reed–Solomon symbol on each hop so as to minimize the effect of multiple-access frequency collisions among the various users. This is the classical approach to channel coding for frequency-hopping communications using Reed–Solomon codes [64]. In addition to the interhop code, within code dwell periods, the information content of each slot is coded further using cyclotomicaly shortened Reed–Solomon codes (CSRS), shortened Reed–Solomon codes in which all computations (error syndrome evaluation, error locator polynomial computation, and error value determination) are done over the binary field GF(2) rather than the extension field GF(2^m), as is the case for Reed–Solomon codes. Further details of CSRS codes may be found in Refs. 25 and 65. The intrahop code is used to correct random errors that occur during the dwell period.

† Other prototype systems that were built for the Paris field trials include the SEL/AEG CD-900 system, the ANT/Bosch S-900-D system, the Philips/TRT Mats-D system, the Ericsson DMS-90 system, the Televerket MAX system, the Mobira narrowband TDMA system, and the ELAB ADPM system.

The modulation method employed in the system is Gaussian minimum shift keying with a low-pass filter parameter of $B_b T = 0.3$. A 16-state maximum-likelihood sequence estimator equalizer is used to combat the effects of intersymbol interference due to frequency-selective fading. Frequency hopping occurs at the TDMA frame rate of 250 hops/s. Within a cell, all hopping is coordinated such that two users never occupy the same frequency at the same time ("orthogonal" hopping). Between cells, a different frequency-hopping code is employed for all nearby cells, resulting in uncorrelated interference from hop to hop. The channel coding is matched to the frequency-hopping pattern so that frequency collisions are corrected. Within the SFH900 system, frequency hopping is therefore used to obtain both frequency diversity and interferer diversity, the latter being a benefit of the CDMA aspects of SFH.

Performance An analysis of the performance of the SFH900 system is contained in Ref. 25. The net result of this analysis is that the SFH900 system is able to achieve a symbol erasure rate on the order of 4×10^{-2} (which corresponds to a decoded bit error rate of 5×10^{-3}) for a channel capacity of 3.5 users per sector per megahertz.

9-4.5 GSM-SFH Digital Cellular System

The GSM-SFH digital cellular system represents another slow-frequency-hopping approach to digital cellular systems. The GSM system has now been deployed commercially throughout western Europe and will be deployed in other countries as well. Its commercial success throughout the world led the European Telecommunications Standards Institute (ETSI), the standards body in charge of the GSM system, to change the meaning of the GSM initials from the original "Groupe Spécial Mobile," the designated name for the Pan European digital cellular system, to "Global System for Mobile Communications"† in 1992. The GSM system is described in a formal set of documents consisting of some 5200 pages that is available directly from ETSI [66]. In addition, there have been a number of journal articles [67], trade press articles [68], commercial journals, trade/technical conferences [69], and even a complete book [9] or portions of a book [70] directed toward the GSM system. Because of the availability of information on the GSM system, coverage here will be concise.

As already noted, the GSM system has been deployed at 800/900 MHz in many western European nations, including the United Kingdom, France, Germany, Denmark, and Sweden, among others. In addition, an extension of the GSM standard, defined by ETSI as DCS1800 for the 1800-MHz band, is currently being deployed in several European countries [71]. A modification of the latter system has been proposed for PCS frequency bands in the United States. Manufacturers of GSM/DCS1800 equipment include Motorola, Alcatel, Siemens, Ericsson, and others.

† Both GSM and Global System for Mobile Communications are registered trademarks of the European Telecommunications Standards Institute.

The GSM system was designed to supplant the wide variety of analog cellular systems within Europe with a common digital cellular system. It was designed at the outset to support a wide range of digital services, including both bearer services and teleservices. The system is an 8 slot/frame TDMA system operating at a data rate of 270.8333 kbps over the air. The duration of the basic TDMA frame is 4.615 ms†和 and the modulation format employed is $B_b T = 0.3$ GMSK. Speech information is convolutionally coded using a rate-$\frac{1}{2}$ constraint-length-5 convolutional code. At the channel data rates employed, the transmitted signal is expected to undergo moderate to severe frequency-selective fading of up to four symbol times (≈ 15 μs). Hence the GSM specifications implicitly require that some form of equalizer be incorporated into both the mobile and base receivers. To combat slow time-selective fading and to provide interference diversity (see Section 9-3), slow frequency hopping at the frame rate of 217 hops/s is specified in the GSM specifications. Table 9-8 summarizes many of the parameters of the GSM/DCS1800 specifications. In addition, Figure 9-17 shows a conceptual block diagram of a GSM transceiver that is based on an experimental system [72]. In the following, a brief description of the signal-processing aspects of one specific implementation of a GSM transceiver is given. Further details of this implementation are given in Refs. 72 and 73.

Signal Processing Description. In the transmitter, the speech signal is transformed into a digital representation at 13 kbps using a residual pulse-excited, linear prediction coder with long-term prediction (RPE-LPC/LTP). The speech coder, which also implements a voice activity detector (VAD), is then used to turn off the transmitter during low-speech-activity periods.‡ The most significant bits of the speech coder output are then rate-$\frac{1}{2}$ coded, and the resulting (coded plus uncoded bits) are block diagonal interleaved. The interleaved symbols are then formatted with training symbols inserted and applied to a GMSK modulator where the carrier is modulated with the coded, interleaved information stream and transmitted. Every frame time, the carrier is frequency hopped, using one of two algorithms, as described below.

In the receiver, the signal is dehopped and demodulated to baseband using a quadrature demodulator. The demodulated signal is then phase and frequency corrected, equalized, deinterleaved, convolutionally decoded, and applied to a speech decoder. The speech decoder then converts the signal back to analog speech.

Not shown in Figure 9-17 is the significant amount of control structure required to support the GSM system. There exists a framing hierarchy whereby 8 slots form a TDMA frame, 26 frames form a multiframe, 51×26 multiframes form a superframe, and 2048 superframes (2,715,648 TDMA frames) form a hyperframe (see

---

† Although at first glance the parameters of the GSM system appear to be arbitrary quantities, in fact, all frequencies and times within the system may be derived from a common 13-MHz clock: for example, 13 MHz/48 = 270.8333 kHz, the system data rate; 13 MHz/60,000 = 216.666 Hz, the TDMA frame rate; and so on.

‡ Although no speech is transmitted during the silence periods, periodically a parametric estimate of the acoustic background noise is completed and transmitted during one of the silence periods. The acoustic background noise parameters are then synthesized to prevent complete silence from being heard. GSM terms this background noise insertion "comfort noise," for obvious reasons.

TABLE 9-8. GSM/DCS1800 System Specifications

	GSM System	**DCS1800 System**
Frequency bands		
Mobile transmit	890–915 MHz	1710–1785 MHz
Base transmit	935–960 MHz	1805–1880 MHz
Channels per carrier	8	
Channel bit rate	270.83 kbps	
Channel spacing	200 kHz	
Modulation	$B_bT = 0.3$ GMSK	
Symbol alphabet size	Binary, differentially coded	
Co-channel interference protection	≤ 12 dB	
Time slot duration	0.58 ms	
Frame duration	4.6 ms	
Frequency-hopping rate	216.68 hops/s	
Channel coding		
Type	Convolutional	
Rate		
Speech channel	$\frac{1}{2},1$	
Signaling channels	$\frac{1}{2}$	
Interleaving depth	Eight-block diagonal interleaving for speech channel	
Maximum channel delay to be equalized	16 μs	
Speech coder		
Type	RPE/LTP-LPC	
Rate	13.0 kbps	
Frame length	20 ms	
Block length	260 bits	
Classes	Class I: 182 bits	
	Class II: 78 bits	
Gross speech rate with FEC coding	22.8 kbps	
Frequency accuracy	Base station: 0.05 ppm	
	Mobile station: 0.1 ppm	

Figure 9-18). The framing hierarchy is used for synchronization, monitoring of other carriers, frequency-hopping control, and encryption of the TDMA slots. In addition to the information-bearing slots, other synchronization information and both common and dedicated control channel information are transmitted periodically between the base site and the subscriber unit.

Slow-Frequency-Hopping Algorithms. Early in the design of the GSM system, the system architects realized the benefits of including slow frequency hopping in the system. At the present time, two algorithms for hopping the channels are avail-

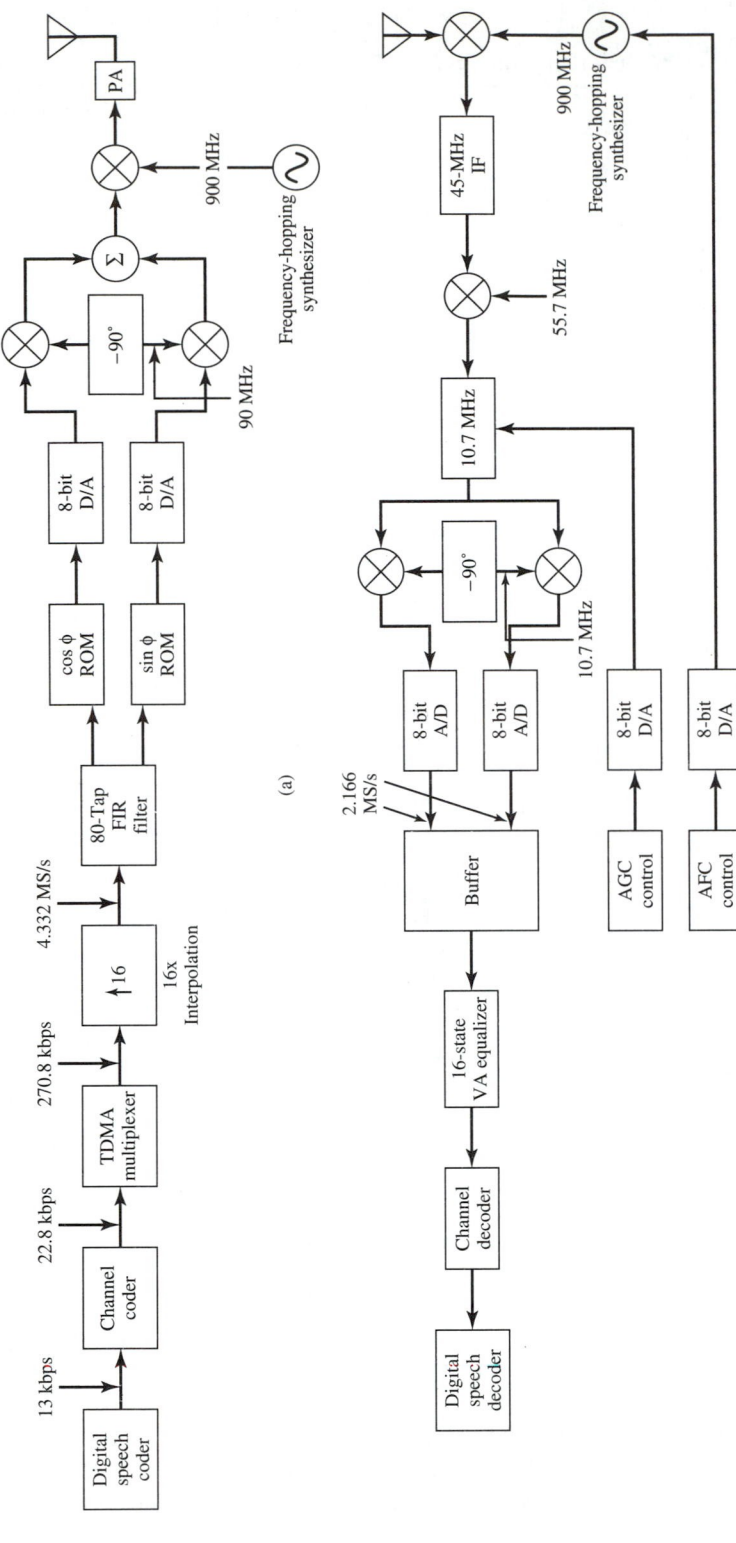

(a)

FIGURE 9-17. Block diagram of a GSM transceiver: (a) transmitter; (b) receiver.

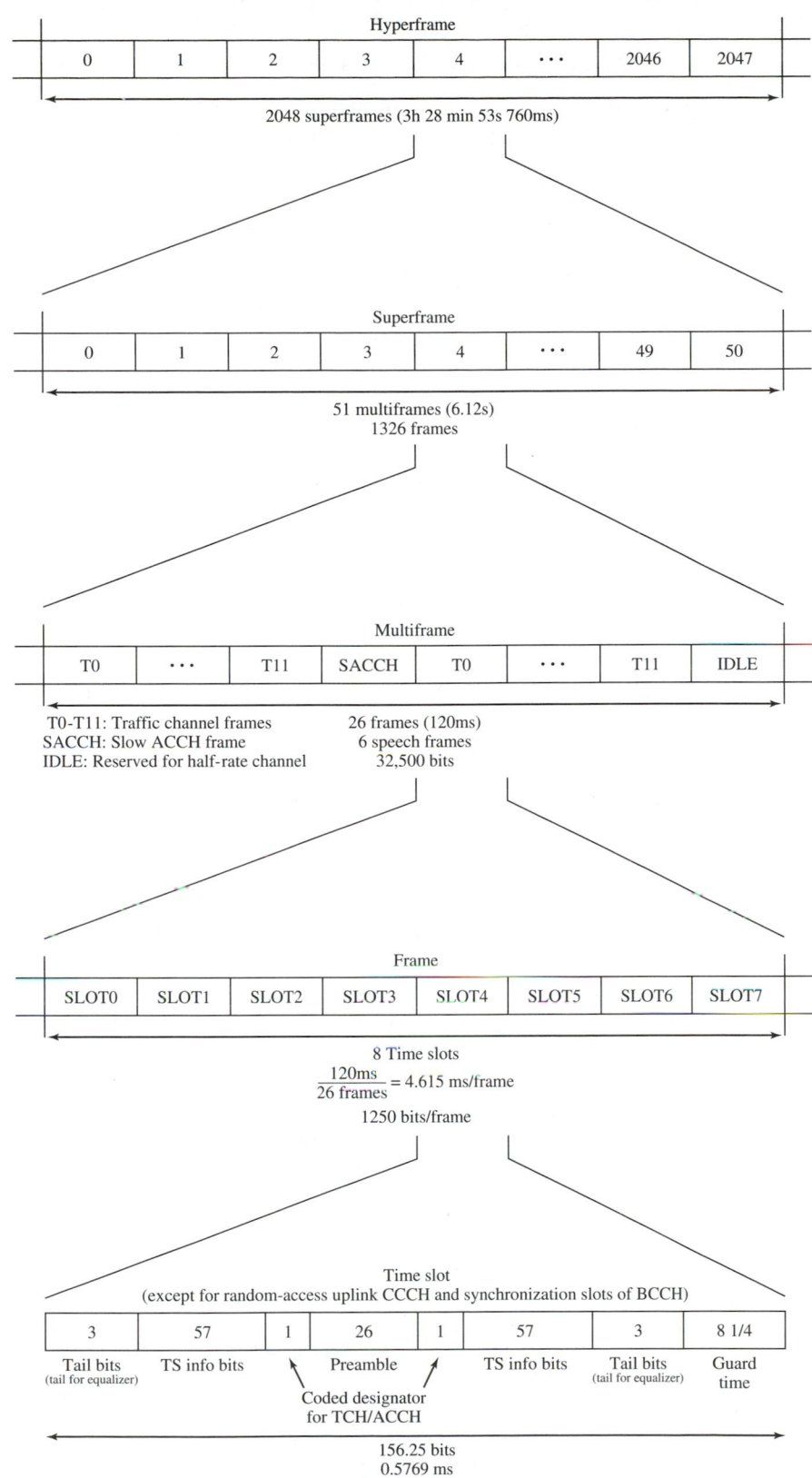

FIGURE 9-18. Hierarchy of GSM framing.

able to the system operators. The first, designed for systems employing a small number of RF carriers, is simply a cyclic permutation of the frequencies available within a cell. This approach will not result in interference diversity but will improve the performance of the system at low vehicle speeds via randomization of time-selective fading. The second approach employs a pseudo-random sequence generator to generate pseudo-random sets of carrier frequencies to be used throughout a cellular system. With this approach, all users within a cell are orthogonal in frequency and users in immediately adjacent cells will interfere at most once per n frames, where n is the number of available frequencies.

Performance. As noted in the beginning of this section, the GSM system is currently commercially deployed. As a result, except for information contained in the literature that describes precommercial systems (see, e.g., Ref. 74), information on the performance of the GSM system is difficult to obtain, as such information is usually tightly controlled by the equipment manufacturers and/or the system operators, for competitive reasons. However, some indication of the performance of the GSM system may be obtained from the GSM specifications themselves. By way of example, an early version of the GSM specifications indicated that for the typical urban channel model (see Section 8-3) and an equivalent 50-km/h vehicle speed, with a signal power input to the receiver of -104 dBm, the word error rate was to be less than 3% without frequency hopping and 6% with frequency hopping. Other performance specifications are contained in the current GSM specifications.

9-4.6 Hybrid SFH TDMA/CDMA System for PCS Applications

In October 1993, the U.S. Federal Communications Commission (FCC) released 140 MHz of spectrum in the 1.85 to 1.99 GHz region for use in PCS applications [75]†. Anticipating the release of this spectrum, during 1992–1994, one of Motorola's R&D laboratories designed, constructed, and field-tested a prototype PCS system that incorporated a hybrid form of multiple access—in particular, slow-frequency-hopping CDMA combined with time-division multiple access (TDMA). This system was designed to permit true portable/vehicular mobility with large-radius cells (1 to 4 miles) while allowing for coexistence with lower-cost PCS systems that operate in pedestrian-only environments. In contrast to several other systems proposals for PCS bands (including two that have already been discussed: the IS-95 DS-CDMA system and the GSM system), it was deemed desirable to support the full range of PCS services, including medium- to high-data-rate voice and data services, by incorporating these services into the initial system design. The hybrid SFH TDMA/CDMA system is similar in many ways to several of the CDMA systems that have been discussed as the reader will note in the following.

---

† 120 MHz of the PCS spectrum release was allocated to licensed PCS services. The licensed PCS bands were allocated on a paired basis with the assignments given by 1850–1910/1930–1990 MHz. 20 MHz of the PCS spectrum release was allocated to unlicensed PCS devices in the band at 1910 to 1930 MHz.

From the beginning it was desired to design a system that would both work in the large-cell mobile environment and would exhibit minimum complexity. Accordingly, a study was conducted in which all possible trade-offs between coding/equalization/diversity were investigated. In particular, the study looked at constraint-length-6 to constraint-length-9 convolutional codes, maximum-likelihood sequence estimation (MLSE) equalization, frequency hopping, and antenna diversity. The study determined that for a wide range of the types of channels that were expected to be encountered by the system, the combination of a convolutional code of constraint length 6 or 7, plus slow frequency hopping plus antenna diversity, yielded the best performance overall at relatively low complexity. Reference 76 gives further details of this study. Surprisingly enough, channel equalization via MLSE yielded little to no performance improvement over even strongly frequency-selective fading channels (one- to two-symbol delay, equal-gain paths) when the other diversity/coding elements were also present.

Although not obvious, the hybrid multiple-access approach does offer several advantages over pure TDMA or SFH-CDMA schemes. SFH-CDMA is employed as the *inter*cell multiple-access method in the hybrid system. Within a cell, orthogonality is strictly maintained by requiring that no two users occupy the same frequency at the same time. In contrast to the DS-CDMA system as well as the two fast FH systems discussed earlier, this requirement results in absolutely *no* intracell interference under *all* channel conditions.† The SFH-CDMA approach allows the system to incorporate frequency diversity, which minimizes the performance degradation due to slow fading channels, as well as interference diversity, which ensures that the system is not subject to a worst-case interference problem, as discussed in Section 9-3.

TDMA is employed as the *intra*cell multiple-access method in this system. TDMA readily permits the introduction of such system concepts as ''bandwidth on demand'' (via concatenation of time slots to support higher data rates) and asymmetric uplink and downlink data rates (via assignment of different numbers of slots to both paths). From a base site implementation viewpoint, TDMA allows a reduction in the total number of RF transceivers required. Finally, from a portable/mobile transceiver implementation viewpoint, TDMA inherently permits ''reasonable'' synthesizer switching times and handoff measurement during the times when the transceiver is neither transmitting or receiving.

The parameters of the hybrid SFH TDMA/CDMA system (as implemented in the prototype) are shown in Table 9-9. A block diagram of the system appears in Figure 9-19, and a diagram of the frame structure employed appears in Figure 9-20. In the following, a concise description of the operation of the transmitter and receiver is given. Further details of transmitter/receiver operation and actual implementation are given in Ref. 77.

Transmitter. Refer to Figure 9-19. As Table 9-9 indicates, a 32-kbps ADPCM speech coder utilizing the CCITT (now ITU-T) G.721 algorithm is employed as the

---

†Recall that in the IS-95 system, the orthogonality of intracell interference for the downlink is degraded in the presence of multipath channels. For the two fast-FH systems, intracell users are treated identically to intercell users in terms of co-channel interference.

TABLE 9-9. Hybrid SFH TDMA/CDMA PCS System Specifications

Frequency band	1850–1970 MHz
Slots/frame	10
Channel bit rate	500 kbps
Frequency hopping rate	500 hops/S
Duplex method	FDD
Intracell multiplexing	TDM/FDM (orthogonal SFH)
Intercell multiplexing	SFH/CDMA
Modulation	QPSK
Pulse shaping	Raised cosine, $\alpha = 0.5$
Time slot duration	0.2 ms
Frame duration	2.0 ms
Gross data rate/slot	34 kbps
Channel coding	
Type	Rate $\frac{1}{2}$, $K = 6$ convolutional, soft decision
Interleaving span	40 ms
Maximum differential multipath delay	16 μs
Speech coder	32 kbps ADPCM

speech coder in the prototype system. This speech coding algorithm permits "toll-quality" speech at moderate complexity. Other submultiple speech coding rates are also supported in the system and yield correspondingly higher spectral efficiency. The output of the speech coder is supplied to a rate-$\frac{1}{2}$ constraint-length-6 convolutional coder with generator polynomials

$$g_1(D) = 1 + D + D^2 + D^3 + D^5$$

$$g_2(D) = 1 + D^2 + D^4 + D^5$$

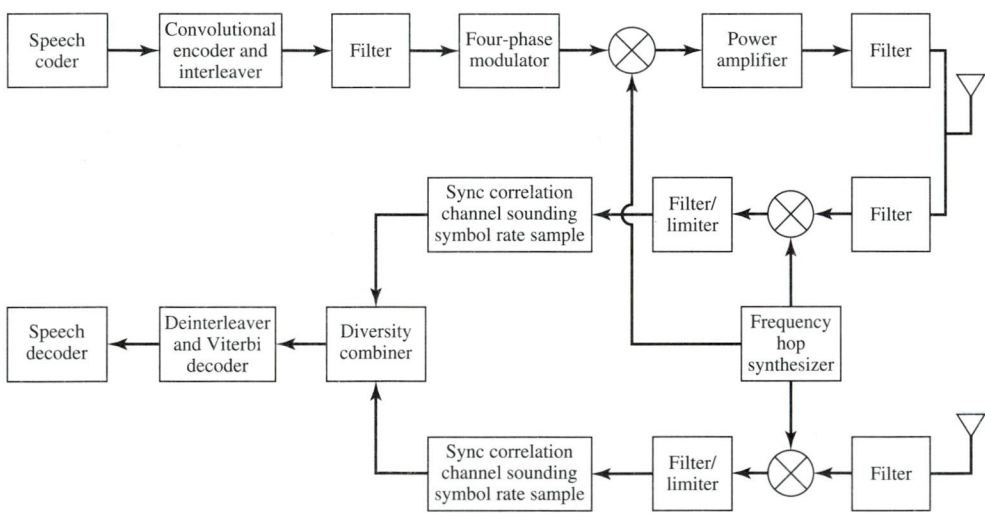

FIGURE 9-19. Hybrid SFH TDMA/CDMA PCS system block diagram.

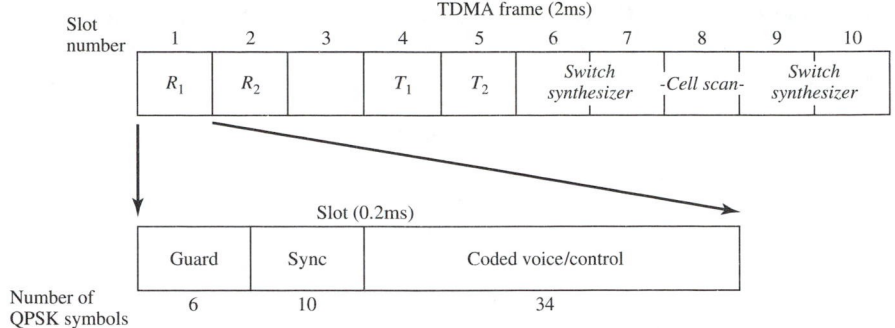

FIGURE 9-20. TDMA frame and slot structure for hybrid SFH TDMA/CDMA PCS system.

The output of the convolutional coder is supplied to a convolutional interleaver with parameters $(I, j) = (18, 9)$ (using Odenwalder's [78] definitions for I and j) which has an end-to-end delay corresponding to 40 ms of 32-kbps speech. The interleaved symbols are then fed to a filtered QPSK modulator whose output is upconverted with a frequency-hop synthesizer, amplified, and transmitted.

Receiver. The receiver employs two diversity antennas and two independent receiver branches which are subsequently combined. In each of the branches, the signal is dehopped using a synthesizer time-locked to the transmit synthesizer and filtered/hard-limited prior to undergoing sync acquisition, channel sounding, and symbol rate downconversion. Hard-limiting is used in the receiver to simplify the receiver signal processing in several ways. First, hard-limiting inherently reduces the dynamic range of the arithmetic required in subsequent signal processing stages. Second, hard-limiting removes the requirement for an automatic gain control (AGC) subsystem in each of the diversity branches. Finally, because AGC is not required in the receiver, initial acquisition of the frequency-hopped transmitted signal is easier, since dynamic range searching is eliminated as a parameter in the initial time-frequency search process that is necessary to acquire the signal. Although hard-limiting any signal (in particular, multiple-access signals) generally results in a degradation in performance, Ref. 76 indicates that the performance degradation due to hard limiting is only about 1.2 dB for this system.

Following hard-limiting, sync acquisition, and downsampling, the two diversity branches are combined in a "near" maximal ratio combining fashion. The diversity combining process is no longer strictly maximal ratio combining, due to the presence of the hard limiters in each of the branches. In Ref. 77 it is shown that the optimal channel gain ρ to interference power ρ_n^2 diversity weighting coefficient is given by

$$\frac{\rho}{\sigma_n^2} = \frac{R_{rx}}{1 - R_{rx}^2} \tag{9-42}$$

where R_{rx} is the synchronization word cross-correlation. The combiner also tracks timing and frequency offsets in the receiver. Following combining, the received signal is sent to a convolutional deinterleaver and a soft-decision Viterbi decoder. The decoded data are then transferred to the ADPCM speech decoder.

Performance. Figure 9-21 plots the probability of bit error versus SNR with and without coding for this system. Observe that a 10^{-3} probability of bit error is achieved at a 1-dB SNR.

Because this system is designed to work in a cellular environment, the performance of the radio subsystem, in co-channel interference is of significant interest. Figure 9-22 plots the probability of decoded error versus C/I ratio for several different fading channel models. The ideal frequency-selective fading channel model assumes that from hop to hop, the channel characteristics are statistically independent. The typical urban channel model is as described in Section 8-3. Finally, the two-ray, one-symbol-delay model assumes equal-amplitude paths with independent Rayleigh fading on each path. From the discussion in Section 9-3, assuming that the

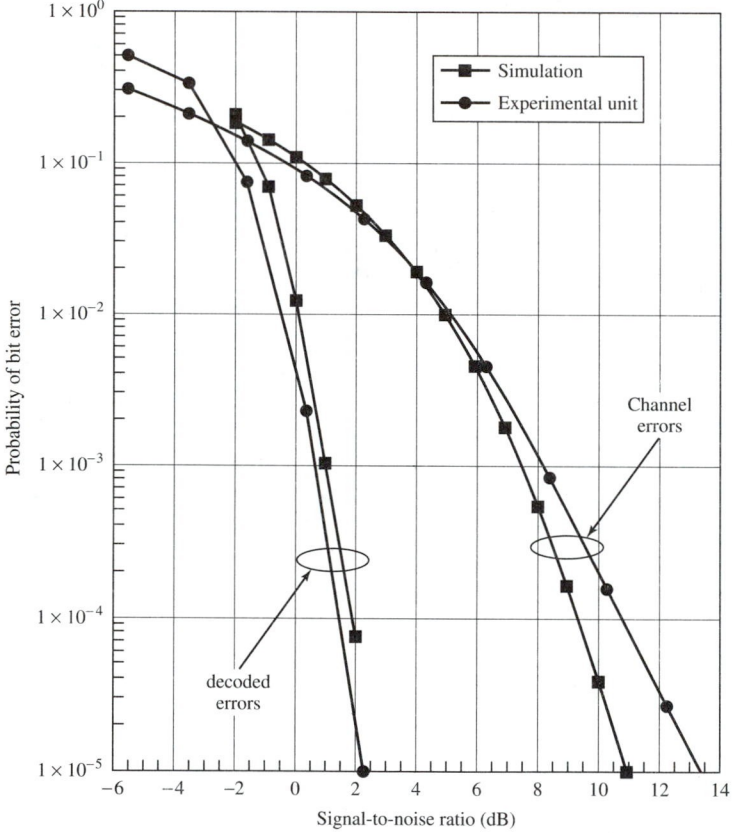

FIGURE 9-21. AWGN channel performance of the hybrid SFH TDMA/CDMA experimental prototype as compared to simulation.

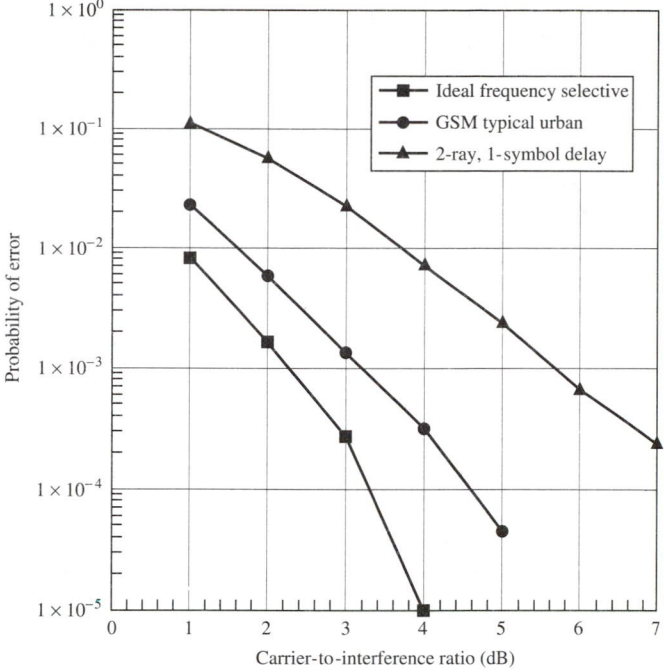

FIGURE 9-22. Probability of decoded bit error as a function of C/I for different fading channel models.

speech coder can operate at a 10^{-3} probability of error, the hybrid SFH TDMA/CDMA system can support a one-cell, three-sector-reuse pattern with a 90% coverage reliability, as the 10^{-3} probability of error occurs at or below a C/I ratio of 5 dB or less for the three frequency-selective fading channels shown.

In Figures 9-23 and 9-24, the location reliability is plotted for the hybrid system for uplink and downlink, respectively, for a one-cell, three-sector design. These plots were obtained using a Monte Carlo simulation that incorporates both the cellular system model *and* the modulation/coding/diversity model for the system. The propagation model employed in both of these plots assumed a $d^{-3.6}$ propagation law and a lognormal standard deviation of $\sigma = 8$ dB. For the uplink, the energy/bit to thermal noise power spectral density was 13 dB, indicative of realistic power limits placed on a mobile transmitter. For the downlink, this power limit was increased to 40 dB E_b/N_0. Thirty-six carriers were assumed to be available for the system. In each of the figures a number of curves are plotted as a function of the system loading L. The system loading is a measure of the percentage of channels within a sector's carrier subset that is actually used. Hence $L = 1.0$ indicates that every carrier frequency is used in every sector. Values of L greater than 1.0 are permissible by borrowing carrier frequencies from adjacent sectors. Other parameters of the simulation are indicated in the figures and in Ref. 79. These two figures indicate that using a system threshold of 10^{-3} probability of bit error, up to $1.1 \times 36 \approx 40$ carriers/cell may be supported with a location reliability of greater than 95%. As

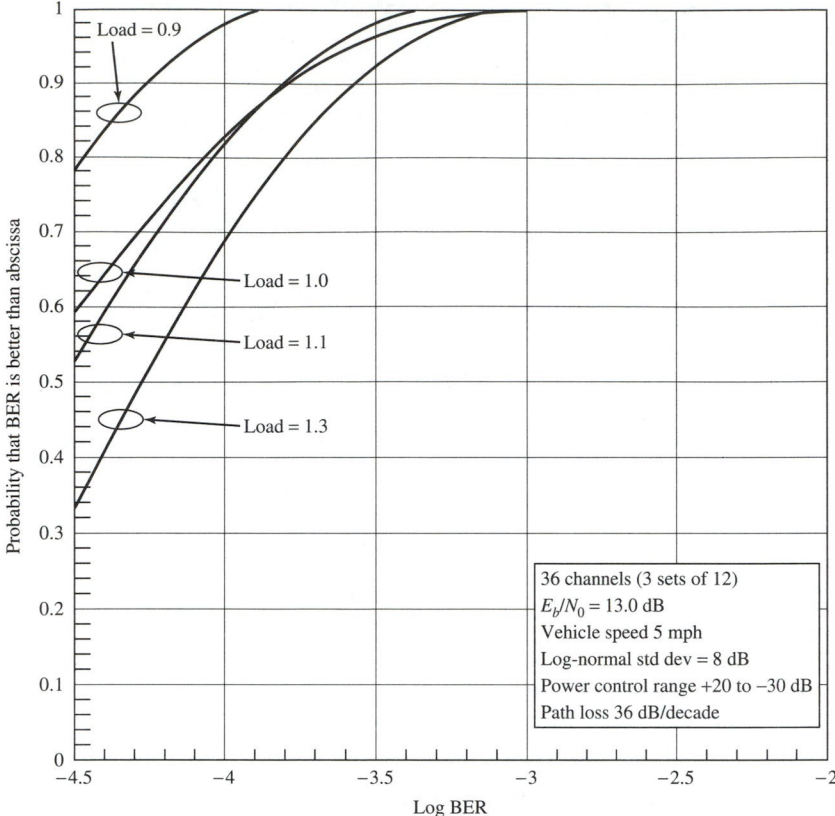

FIGURE 9-23. Uplink location reliability with load as a parameter for a system with 36 available RF carriers for the hybrid SFH TDMA/CDMA PCS system.

each carrier supports 10 time slots at the 16-kbps speech coder rate, 400 channels may be supported by an allocation of 36×400 kHz = 14.4 MHz. Thus the system supports 27 channels per cell per megahertz. Taking into account VAD, the system could support up to 54 speech users per cell per megahertz.

9-5 Summary

A number of concepts were introduced in this chapter related to the analysis and design of code-division multiple-access digital cellular communication systems. Cellular system designs are functions of both the topology of the cell and the modulation/coding/diversity methods employed. Code-division multiple-access digital cellular radio systems can exhibit improved capacity over existing single-channel-per-carrier analog cellular systems. Several different CDMA digital cellular systems designs were described in Section 9-4. These designs included a direct-sequence CDMA system, two fast-frequency-hopped CDMA systems, and three slow-

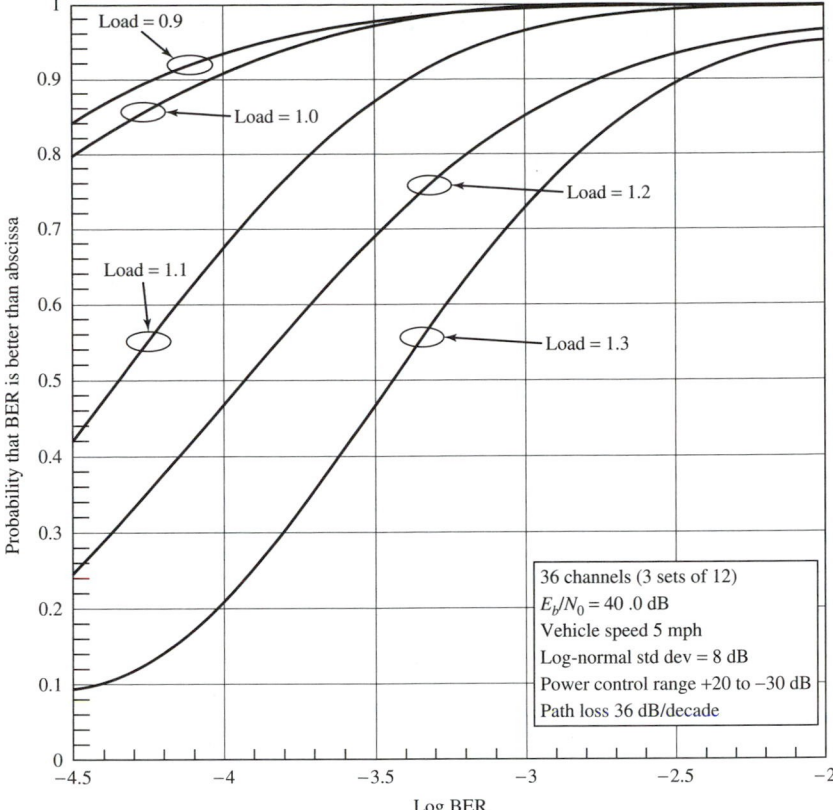

FIGURE 9-24. Downlink location reliability with load as a parameter for a system with 36 available RF carriers for the hybrid SFH TDMA/CDMA PCS system.

frequency-hopped systems. The important aspects of each of these systems were described in detail.

From the material presented in this chapter, the reader should now have gained an appreciation for the level of detail necessary in designing a CDMA digital cellular system. Not only is an understanding of the "modem" functions (channel coder, modulator, receiver, channel decoder) and the associated subsystems needed, but knowledge of the nature of the mobile radio channel, cellular system topology, and design of diversity subsystems must also be mastered to design and/or analyze a CDMA digital cellular system successfully.

In addition to the CDMA digital cellular system concepts covered here, the reader should be aware of the large body of literature that exists that discusses spread-spectrum multiple-access communication systems in general. While this material is generally beyond the scope of this book, several key references to the latter body of material will be made briefly here. The interested reader should examine both the paper cited and the references cited within each paper for more details on spread-spectrum multiple-access communication systems.

1. *Direct-sequence spread-spectrum multiple-access (DS-SSMA) systems:* A key paper for spreading code design for DS-SSMA systems is Ref. 80. DS-SSMA systems design for the AWGN channel is considered in Refs. 81 to 84. DS-SSMA systems designs for fading channels are given in Refs. 85 to 88. Performance bounds and limits for DS-SSMA communication systems are provided in Refs. 89 to 91.

2. *Frequency-hopped spread-spectrum multiple-access (FH-SSMA) systems:* Several representative papers on FH-SSMA communication systems are Refs. 92 to 94.

3. *Spread-spectrum packet radio systems:* Although spread-spectrum packet radio systems differ somewhat from the classical SSMA systems, there is also considerable overlap between these two classes of communication systems. A good reference to spread-spectrum packet radio systems is Ref. 95.

4. *Multiuser detection:* For DS-SSMA systems, the detectors employed in the receivers are single-user detectors. However, usually much is known about the nature of the other users in a DS-SSMA, so that multiuser detection is *theoretically* possible in some cases. Reference 96 is a recent survey paper on this fascinating subject.

References

[1] H. J. Schulte, Jr., and W. A. Cornell, "Multi-area Mobile Telephone System," *IRE Trans. Veh. Commun.,* Vol. VC-9, pp. 49–53, May 1960.

[2] J. S. Engel, "The Effects of Cochannel Interference on the Parameters of a Small-Cell Mobile Telephone System," *IEEE Trans. Veh. Technol.* Vol. VT-18, pp. 110–116, November 1969.

[3] Federal Communications Commission, Docket 18262, *FCC Report 14,* pp. 311–321, July 17, 1968.

[4] P. A. Bello and B. D. Nelin, "The Effect of Frequency Selective Fading on the Binary Error Probabilities of Incoherent and Differentially Coherent Matched Filter Receivers," *IEEE Trans. Commun. Sys.,* Vol. CS-11, pp. 170–186, June 1963.

[5] AT&T, "Advanced Mobile Phone Service," *Bell Syst. Tech. J.,* Vol. 58, No. 1, January 1979.

[6] J. J. Mikulski, "A System Plan for a 900-MHz Portable Radio Telephone," *IEEE Trans. Veh. Technol.,* Vol. VT-26, pp. 76–81, February 1977.

[7] W. C. Jakes, Jr., ed., *Microwave Mobile Communications* (New York: Wiley, 1974). Also reprinted by IEEE Press, 1994.

[8] W. C. Y. Lee, *Mobile Cellular Telecommunications Systems* (New York, McGraw-Hill, 1989).

[9] M. Mouly and M. B. Pautet, *The GSM System for Mobile Communications* (Palaiseau, France: Mouly-Pautet, 1992).

[10] M. STERN, "Analysis Shows Advantages of Four-Cell Site Repeat Patterns," *Mobile Radio Technol.,* pp. 42–48, January 1984.

[11] V. GRAZIANO, "Propagation Correlations at 900 MHz," *IEEE Trans. Veh. Technol.,* Vol. VT-27, pp. 182–189, November 1978.

[12] J. E. STJERNVALL, "Calculation of Capacity and Co-channel Interference in a Cellular System," *First Nordic Seminar Digital Land Mobile Radiocommun.,* Espoo, Finland, February 5–7, 1985.

[13] D. C. COX, "Cochannel Interference Considerations in Frequency Reuse Small-Coverage-Area Radio Systems," *IEEE Trans. Commun.,* Vol. COM-30, pp. 135–42, January 1982.

[14] Y. S. YEH and S. C. SCHWARTZ, "Outage Probability in Mobile Telephony Due to Multiple Log-Normal Interferers," *IEEE Trans. Commun.,* Vol. COM-32, pp. 380–388, April 1984.

[15] D. E. BORTH, M. J. McLAUGHLIN, and J. J. MIKULSKI, "Implementation of a Digital Mobile Radio Incorporating Combined Modulation/Coding," *Second Nordic Seminar Digital Land Mobile Radio Commun.,* Stockholm, Sweden, pp. 85–89, October 14–16, 1986.

[16] O. C. YUE, "Spread Spectrum Mobile Radio, 1977–1982," *IEEE Trans. Veh. Technol.,* Vol. VT-32, pp. 98–105, February 1983.

[17] P. T. BRADY, "A Statistical Analysis of On–Off Patterns in 16 Conversations," *Bell Syst. Tech. J.,* Vol. 47, pp. 73–91, January 1968.

[18] G. R. COOPER and R. W. NETTLETON, "A Spread Spectrum Technique for High Capacity Mobile Communications," *IEEE Trans. Veh. Technol.,* Vol. VT-27, pp. 264–275, November 1978.

[19] K. S. GILHOUSEN, I. M. JACOBS, R. PADOVANI, A. J. VITERBI, L. A. WEAVER, JR., and C. E. WHEATLEY III, "On the Capacity of a Cellular CDMA System," *IEEE Trans. Veh. Technol.,* Vol. 40, pp. 303–312, May 1991.

[20] B. K. LEVITT, "The Effect of Activity Factor on Performance of CDMA Voice Communication Links," *MSAT-X Quart.,* No. 24, Jet Propulsion Laboratory, pp. 13–17, July 1990.

[21] G. R. COOPER and R. W. NETTLETON, *"Spectral Efficiency in Cellular Land-Mobile Communications: A Spread-Spectrum Approach,"* Final Report, TR-EE 78-44, Purdue University, West Lafayette, Ind., October 31, 1978.

[22] A. SALMASI and K. S. GILHOUSEN, "On the System Design Aspects of Code Division Multiple Access (CDMA) Applied to Digital Cellular and Personal Communications Networks," *Proc. 41st IEEE Veh. Technol. Conf.,* pp. 57–62, May 19–22, 1991.

[23] M. MIZUNO, ''Randomization Effects of Errors by Frequency-Hopping Techniques in a Fading Channel,'' *IEEE Trans. Commun.*, Vol. COM-30, pp. 1052–1056, May 1982.

[24] D. VERHULST, M. MOULY, and J. SZPIRGLAS, ''Slow Frequency Hopping Multiple Access for Digital Cellular Radiotelephone,'' *IEEE Trans. Veh. Technol.*, Vol. VT-33, pp. 179–190, August 1984.

[25] J. L. DORNSTETTER and D. VERHULST, ''Cellular Efficiency with Slow Frequency Hopping: Analysis of the Digital SFH900 Mobile System,'' *IEEE J. Selected Areas Commun.*, Vol. SAC-5, pp. 835–848, June 1987.

[26] TIA/EIA/IS-95 Interim Standard, *Mobile Station–Base Station Compatibility Standard for Dual Mode Wideband Spread Spectrum Cellular System*, Telecommunications Industry Association, Washington, D.C., July 1993.

[27] A. M. VITERBI and A. J. VITERBI, ''Erlang Capacity of a Power Controlled CDMA System,'' *IEEE J. Selected Areas Commun.*, Vol. 11, pp. 892–900, August 1993.

[28] A. J. VITERBI, *CDMA: Principles of Spread Spectrum Multiple Access Communication* (Reading, MA: Addison-Wesley, 1995).

[29] F. J. MACWILLIAMS and N. J. A. SLOANE, *The Theory of Error Correcting Codes* (New York: North-Holland, 1977).

[30] W. C. LINDSEY and M. K. SIMON, *Telecommunication Systems Engineering* (Englewood Cliffs, N.J.: Prentice Hall, 1973; also New York: Dover, 1991).

[31] E. C. POSNER, ''Combinatorial Structures in Planetary Reconnaissance,'' in *Error Correcting Codes*, H. B. Mann, ed. (New York: Wiley, 1969).

[32] A. V. OPPENHEIM and R. W. SCHAFER, *Discrete-Time Signal Processing* (Englewood Cliffs, N. J.: Prentice Hall, 1989).

[33] R. KERR, K. GILHOUSEN, B. WEAVER, R. PADOVANI, and H. DEHESH, ''The CDMA Digital Cellular System: An ASIC Overview,'' *Proc. IEEE 1992 Custom Integrated Circuits Conf.*, pp. 10.1.1–10.1.7, May 3–6, 1992.

[34] J. HINDERLING, T. RUETH, K. EASTON, D. EAGLESON, J. LEVIN, and R. KERR, ''CDMA Mobile Station Modem ASIC,'' *Proc. IEEE 1992 Custom Integrated Circuits Conf.*, pp. 10.2.1–10.2.5, May 3–6, 1992.

[35] J. K. HINDERLING, T. RUETH, K. EASTON, D. EAGLESON, D. KINDRED, R. KERR, and J. LEVIN, ''CDMA Mobile Station Modem ASIC,'' *IEEE J. Solid-State Circuits*, Vol. 28, pp. 253–260, March 1993.

[36] R. KERR, ''CDMA Digital Cellular: An ASIC Overview,'' *Appl. Microwave Wireless*, pp. 30–41, Fall 1993.

[37] A. J. VITERBI, A. M. VITERBI, and E. ZEHAVI, "Performance of Power-Controlled Wideband Terrestrial Digital Communication," *IEEE Trans. Commun.,* Vol. 41, pp. 559–569, April 1993.

[38] J. W. KETCHUM, "Down-Link Capacity of Direct Sequence CDMA for Application in Cellular Systems," *IEEE Symp. Spread Spectrum Tech. Appli.,* King's College, London, pp. 151–157, September 1990.

[39] A. J. VITERBI, "Convolutional Codes and Their Performance in Communication Systems," *IEEE Trans. Commun.,* Vol. COM-19, pp. 751–772, October 1971.

[40] J. CONAN, "The Weight Spectra of Some Short Low-Rate Convolutional Codes," *IEEE Trans. Commun.,* Vol. COM-32, pp. 1050–1053, September 1984. See also correction in *IEEE Trans. Commun.,* Vol. COM-33, p. 570, June 1985.

[41] J. HAGENAUER, "Viterbi Decoding of Convolutional Codes for Fading- and Burst-Channels," *Proc. 1980 Int. Zurich Seminar Digital Commun.,* pp. G2.1–G2.7, March 4–6, 1980.

[42] J. HAGENAUER, N. SESHADRI, and C.-E. W. SUNDBERG, "The Performance of Rate-Compatible Punctured Convolutional Codes for Digital Mobile Radio," *IEEE Trans. Commun.,* Vol. 38, pp. 966–980, July 1990.

[43] C. C. CHAN, "Error Protection for Tactical Voice Modems," *Proc. Tactical Commun. Conf.,* pp. 481–498, April 24–26, 1990.

[44] A. J. VITERBI and J. K. OMURA, *Principles of Digital Communication and Coding,* (New York: McGraw-Hill, 1979).

[45] J. G. PROAKIS, *Digital Communications,* 2nd ed. (New York: McGraw-Hill, 1989).

[46] L. F. CHANG and N. SOLLENBERGER, "Comparison of Two Interleaving Techniques for CDMA Radio Communications Systems," *Proc. 42nd Veh. Technol. Conf.,* pp. 275–278, May 10–13, 1992.

[47] L. F. CHANG, F. LING, D. D. FALCONER, and N. SOLLENBERGER, "Comparison of Two Orthogonal Coding Techniques for CDMA Radio Communications Systems," *IEEE Trans. Commun.,* to appear, 1995.

[48] G. R. COOPER and R. W. NETTLETON, "A Spread Spectrum Technique for High Capacity Mobile Communications," *Proc. 27th IEEE Veh. Technol. Conf.,* pp. 98–103, March 16–18, 1977.

[49] G. R. COOPER, R. W. NETTLETON, and D. P. GRYBOS, "Cellular Land-Mobile Radio: Why Spread Spectrum?" *IEEE Commun. Mag.,* Vol. 17, pp. 17–24, March 1979.

[50] R. D. Yates and G. R. Cooper, *Design of Large Signal Sets with Good Aperiodic Correlation Properties,* Technical Report TR-EE 66-13, Purdue University, West Lafayette, Ind., September 1966.

[51] P. S. Henry, "Spectrum Efficiency of a Frequency-Hopped-DPSK Spread-Spectrum Mobile Radio System," *IEEE Trans. Veh. Technol.,* Vol. VT-28, pp. 327–332, November 1979.

[52] R. W. Nettleton and G. R. Cooper, "Performance of Frequency-Hopped Differentially Modulated Spread-Spectrum Receiver in a Rayleigh Fading Channel," *IEEE Trans. Veh. Technol.,* Vol. VT-30, pp. 14–29, February 1981.

[53] M. Matsumoto and G. R. Cooper, "Multiple Narrow-Band Interferers in an FH-DPSK Spread-Spectrum Communication System," *IEEE Trans. Veh. Technol.,* Vol. VT-30, pp. 37–42, February 1981.

[54] D. J. Goodman, P. S. Henry, and V. K. Prabhu, "Frequency-Hopped Multilevel FSK for Mobile Radio," *Bell Syst. Tech. J.,* Vol. 59, pp. 1257–1275, September 1980.

[55] A. J. Viterbi, "A Processing Satellite Transponder for Multiple Access by Low Rate Mobile Uses," *Proc. Digital Satellite Commun. Conf.,* Montreal, Canada, pp. 166–174, October 1978.

[56] S. Stein, "Fading Communication Media," Chap. 9 of *Communication Systems and Techniques* (New York: McGraw-Hill, 1966).

[57] G. Einarsson, "Address Assignment for a Time-Frequency-Coded, Spread-Spectrum System," *Bell Syst. Tech. J.,* Vol. 59, pp. 1241–1255, September 1980.

[58] D. V. Sarwate, "Reed–Solomon Codes and the Design of Sequences for Spread-Spectrum Multiple-Access Communications," in *Reed-Solomon Codes and Their Applications,* S. B. Wicker and V. K. Bhargava, eds. (New York: IEEE Press, 1994).

[59] B. G. Haskell, "Computer Simulation Results on Frequency-Hopped MFSK Mobile Radio: Noiseless Case," *IEEE Trans. Commun.,* Vol. COM-29, pp. 125–132, February 1981.

[60] U. Timor, "Improved Decoding of Frequency-Hopped Multilevel FSK System," *Bell Syst. Tech. J.,* Vol. 59, pp. 1839–1855, December 1980.

[61] U. Timor, "Multistage Decoding of Frequency-Hopped FSK System," *Bell Syst. Tech. J.,* Vol. 60, pp. 471–483, April 1981.

[62] R. Agusti and G. Junyent, "Performance of an FH Multilevel FSK for Mobile Radio in an Interference Environment," *IEEE Trans. Commun.,* Vol. COM-31, pp. 840–846, June 1983.

[63] J. L. Dornstetter and D. Verhulst, "The Digital Cellular SFH 900 System," *Proc. IEEE Int. Conf. Commun.,* pp. 36.3.1–36.3.5, June 1986.

[64] M. B. PURSLEY, "Reed–Solomon Codes in Frequency-Hop Communications," in *Reed-Solomon Codes and Their Applications,* S. B. Wicker and V. K. BHARGAVA, eds. (New York: IEEE Press, 1994).

[65] J. L. DORNSTETTER, U.S. Patent 4,754,458, "Method of Transmission, with the Possibility of Correcting Bursts of Errors, of Information Messages and Encoding and Decoding Devices for Implementing This Method," June 28, 1988.

[66] EUROPEAN TELECOMMUNICATIONS STANDARDS INSTITUTE, *GSM Specifications* (Sophia Antipolis, France: ETSI TC-SMG, 1991).

[67] D. J. GOODMAN, "Second Generation Wireless Information Networks," *IEEE Trans. Veh. Technol.,* Vol. 40, pp. 366–374, May 1991.

[68] Alcatel Alsthom Publications S.A., *Electrical Communication,* Key topic: Global System for Mobile Communications, 2nd Quarter, 1993.

[69] CEPT/GSM Operators, *Digital Cellular Mobile Communication Seminar,* October 16–18, 1990.

[70] C. WATSON, "Radio Equipment for GSM," Chap. 7 of *Cellular Radio Systems,* D. M. BALSTON and R. C. V. MACARIO, eds. (Boston: Artech House, 1993).

[71] A. HADDEN and P. KNIGHT, "The Birth of Personal Communications Networks," Chap. 8 of *Cellular Radio Systems,* D. M. BALSTON and R. C. V. MACARIO, eds. (Boston: Artech House, 1993).

[72] D. E. BORTH and P. D. RASKY, "Signal Processing Aspects of Motorola's Pan European Digital Cellular Validation Mobile," *Proc. 1991 IEEE Int. Phoenix Conf. Comput. Commun.,* pp. 416–423, March 27–30, 1991.

[73] J. M. NOWACK, D. E. BORTH, and P. D. RASKY, "Soft-Output MLSE Equalization Methods for the Mobile Radio Channel," *Proc. 29th Annual Allerton Conf. Commun. Control Comput.,* pp. 11–20, October 2–4, 1991.

[74] D. E. BORTH and P. D. RASKY, "An Experimental RF Link System to Permit Evaluation of the GSM Air Interface Standard," *Proc. 3rd Nordic Seminar Digital Land Mobile Radio Commun.,* Paper 6.3, Copenhagen, Denmark, September 12–15, 1988.

[75] FEDERAL COMMUNICATIONS COMMISSION, "In the Matter of Amendment of the Commission's Rules to Establish New Personal Communications Services," *Second Report and Order,* GEN Docket 90-314, FCC 93-451, October 22, 1993.

[76] P. D. RASKY, G. M. CHIASSON, and D. E. BORTH, "Hybrid Slow Frequency-Hop/CDMA-TDMA as a Solution for High-Mobility, Wide-Area Personal Communications," *Proc. 4th Winlab Workshop Third Generation Wireless Inf. Networks,* East Brunswick, N.J., pp. 199–215, October 19–20, 1993.

[77] P. D. RASKY, G. M. CHIASSON, and D. E. BORTH, "An Experimental Slow Frequency-Hopped Personal Communication System for the Proposed U.S. 1850–1990 MHz Band," *Proc 2nd Int. Conf. Universal Personal Commun.,* Ottawa, Canada, pp. 931–935, October 12–15, 1993.

[78] J. P. ODENWALDER, "Error Control," Chap. 10 of *Data Communications, Networks, and Systems,* T. C. Bartee, ed. (Indianapolis, Ind.: Howard W. Sams, 1985).

[79] P. D. RASKY, G. M. CHIASSON, D. E. BORTH, and R. L. PETERSON, "Slow Frequency-Hop TDMA/CDMA for Macrocellular Personal Communication Systems," *IEEE Personal Commun.,* Vol. 1, Second Quarter, 1994.

[80] D. V. SARWATE and M. B. PURSLEY, "Crosscorrelation Properties of Pseudo-random and Related Sequences," *Proc. IEEE,* Vol. 68, pp. 593–619, May 1980.

[81] M. B. Pursley, "Performance Evaluation for Phase-Coded Spread-Spectrum Multiple Access Communication: Part I. System Analysis," *IEEE Trans. Commun.,* Vol. COM-25, pp. 795–799, August 1977.

[82] M. B. PURSLEY and D. V. SARWATE, "Performance Evaluation for Phase-Coded Spread-Spectrum Multiple Access Communication: Part II. Code Sequence Analysis," *IEEE Trans. Commun.,* Vol. COM-25, pp. 800–803, Aug. 1977.

[83] M. B. PURSLEY, F. D. GARBER, and J. S. LEHNERT, "Analysis of Generalized Quadriphase Spread-Spectrum Communications," *Proc. 1980 IEEE Int. Conf. Commun.,* pp. 15.3.1–15.3.6, June 1980.

[84] F. D. GARBER and M. B. PURSLEY, "Performance of Offset Quadriphase Spread-Spectrum Multiple-Access Communications," *IEEE Trans. Commun.,* Vol. COM-29, pp. 305–314, March 1981.

[85] D. E. BORTH and M. B. PURSLEY, "Analysis of Direct-Sequence Spread-Spectrum Multiple-Access Communication over Rician Fading Channels," *IEEE Trans. Commun.,* Vol. COM-27, pp. 1566–1577, October 1979.

[86] L. B. MILSTEIN and D. L. SCHILLING, "Performance of a Spread Spectrum Communication System Operating over a Frequency-Selective Fading Channel in the Presence of Tone Interference," *IEEE Trans. Commun.,* Vol. COM-30, pp. 240–247, January 1982.

[87] J. S. LEHNERT and M. B. PURSLEY, "Multipath Diversity Reception of Direct-Sequence Spread-Spectrum Multiple-Access Communications," *IEEE Trans. Commun.,* Vol. COM-35, pp. 1189–1198, November 1987.

[88] E. A. GERANIOTIS and M. B. PURSLEY, "Performance of Coherent Direct-Sequence Spread-Spectrum Communications over Specular Multipath Fading Channels," *IEEE Trans. Commun.,* Vol. COM-33, pp. 502–508, June 1985.

[89] J. S. LEHNERT, "Efficient Technique for Evaluating Direct-Sequence Spread-Spectrum Multiple-Access Communications," *IEEE Trans. Commun.,* Vol. 37, pp. 851–858, August 1989.

[90] N. NAZARI and R. E. ZIEMER, "Computationally Efficient Bounds for the Performance of Direct-Sequence Spread-Spectrum Multiple-Access Communications Systems in Jamming Environments," *IEEE Trans. Commun.,* Vol. 36, pp. 577–587, May 1988.

[91] H. E. ROWE, "Bounds on the Number of Signals with Restricted Cross Correlation," *IEEE Trans. Commun.,* Vol. COM-30, pp. 966–974, May 1982.

[92] C. D. FRANK and M. B. PURSLEY, "On the Statistical Dependence of Hits in Frequency-Hop Multiple Access," *IEEE Trans. Commun.,* Vol. 38, pp. 1483–1494, September 1990.

[93] M. V. HEGDE and W. E. STARK, "On the Error Probability of Coded Frequency-Hopped Spread Spectrum Multiple-Access Systems," *IEEE Trans. Commun.,* Vol. 38, pp. 571–573, May 1990.

[94] E. GERANIOTIS, "Multiple-Access Capability of Frequency-Hopped Spread-Spectrum Revisited: An Analysis of the Effect of Unequal Power Levels," *IEEE Trans. Commun.,* Vol. 38, pp. 1066–1077, July 1990.

[95] M. B. PURSLEY, "The Role of Spread Spectrum in Packet Radio Networks," *Proc. IEEE,* Vol. 75, pp. 116–134, January 1987.

[96] S. VERDU, "Recent Progress in Multiuser Detection," in *Advances in Communication and Signal Processing* (Berlin: Springer Verlag, 1989).

Problems

(9-1) Consider the downlink portion of the IS-95 system described in Section 9-3.1. A cell is being designed that will have a 4-mile radius. Zero-dB-gain antennas are assumed to be employed in both the base and the mobile. The one-sided thermal noise power spectral density is -174 dBm/Hz. Receivers are assumed to have a 6-dB noise figure. The radio propagation loss is assumed to be given by a Hata model, which is given by

$$\text{path loss (dB)} = A + B \log_{10} R$$

where R is the distance in miles from the transmitter, $A = 127$ dB, and $B = 35$. Assume that all filters are ideal in the sense that the noise bandwidth of the filters are equal to the data rates.

(a) Determine the required transmit power needed to achieve a 10^{-3} probability of error in a mobile receiver located on the cell boundary assuming that 6 dB E_b/N_0 is required to achieve this error rate performance. Assume that the mobile receiver is the only user and neglect lognormal shadowing effects.

(b) A second mobile user is now added to the system, which is also located on the cell boundary. What is the transmit power now required to achieve a 10^{-3} probability of bit error for either of the two users? Assume that multiple-access interference may be treated as independent Gaussian noise with zero mean and variance equal to the power of the multiple-access interference power.

(c) Repeat part (b) for the case of 20 users, all located on the cell boundary.

(9-2) Consider the downlink portion of the prototype SFH system described in Section 9-4.6. A cell is being designed that will have a 4-mile radius. Zero-dB-gain antennas are assumed to be employed in both the base and the mobile. The one-sided thermal noise power spectral density is -174 dBm/Hz. Receivers are assumed to have a 6-dB noise figure. The radio propagation loss is assumed to be given by a Hata model, which is

$$\text{path loss (dB)} = A + B \log_{10} R$$

where R is the distance in miles from the transmitter, $A = 127$ dB, and $B = 35$. Assume that all filters are ideal in the sense that the noise bandwidth of the filters is equal to the data rates.

(a) Determine the required *average* transmit power needed to achieve a 10^{-3} probability of error in a mobile receiver located on the cell boundary assuming that 6 dB E_b/N_0 is required to achieve this error rate performance. Assume that the mobile receiver is the only user and neglect lognormal shadowing effects.

(b) A second mobile user is now added to the system, which is also located on the cell boundary. What is the average transmit power now required to achieve a 10^{-3} probability of bit error for *either* of the two users?

(9-3) Assume that the capacity equation (9-13) holds for all values of information rate R.

(a) Compute the relative spectral efficiency of the downlink portion of the IS-95 system. Compare your result with the relative spectral efficiencies of the systems described in Sections 9-4.2 and 9-4.3.

(b) Repeat part (a) assuming a 32-kbps information rate R.

(9-4) Use the FHT shown in Figure 9-10 to compute the FHT of the received noisy sequence 0.2., 0.4, -0.2, -0.8, -0.8, -0.2, 0.4, -0.2. What is the most likely transmitted binary sequence assuming that the mapping $0 \rightarrow +1$, $1 \rightarrow -1$ was employed at the transmitter?

(9-5) Draw the combining network required in Figure 9-13 assuming 4 chips/2 information bits.

(9-6) As a communication systems engineer, you are asked to comment on the relative transmitted power between two system choices: Approach A uses a DS-CDMA system operating over a frequency-selective fading channel. The channel is such that five independent equal-gain paths are supported that are resolvable by the chip rate of the transmitter. Approach B uses a DS-CDMA system operating over a time-selective fading channel. Five uncorrelated antennas are combined with a maximal ratio combiner in the receiver. Assume that the outputs of each of the antennas are of equal gain.

CHAPTER 10

Low-Probability-of-Intercept Methods

10-1 Introduction

In previous chapters the generation and reception of spread-spectrum signals was discussed. The reception process was assumed to be cooperative—that is, the receiver knew the essential properties of the transmitted signal to allow the implementation of an optimized receiver in the sense of minimum probability of error in AWGN and jamming. What about unintended receivers? Is it possible for an interceptor to detect the presence of a spread-spectrum signal and, if detected, to extract certain signal parameters? The answer to both of these questions is "Yes, to some degree." Methods that unintended interceptors may use to detect the presence of spread-spectrum signals and to extract certain signal parameters, mainly code chip rate, are explored in this chapter. After discussing the nature of the covert communications problem, the subject of energy detection of unknown signals is discussed. This was first explored in Chapter 5 in relation to acquisition of the spectrum spreading code in the receiver of a cooperative system. After reviewing energy detection, the optimum receiver for detection of a direct-sequence spread-spectrum signal is discussed. Certain more practical approximations to the optimum receiver are hypothesized and their performances analyzed. Next, the optimum detector for frequency-hop spread-spectrum signals is discussed and its performance relative to energy detection shown. A simplified approximation to the optimum frequency-hop spread-spectrum signal detector is examined next. Finally, the extraction of spread-spectrum signal parameters is discussed and the design of spread-spectrum signals to minimize the extraction of these parameters is examined. In addition to several papers that will be referenced in the sequel, a number of books have been published in the past few years that deal with the subject of spread-spectrum signal detection and parameter extraction [1–3].

10-2 Nature of Covert Communications

In any communications operation, there may be several ingredients or "players," as illustrated in Figure 10-1. First, there are the intended communicators, which make use of the transmitter and receiver shown in the figure. Second, there may be unin-

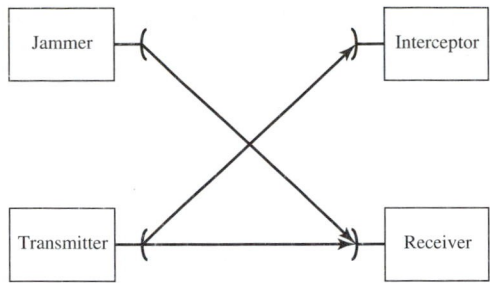

FIGURE 10-1.   Covert communications process.

tentional sources or jammers. Third, there may be an eavesdropper illustrated by the intercept receiver in Figure 10-1.

As a possible example that could happen in ordinary life, consider a conversation with a friend on a cordless telephone. The jammer could be a radio or television set playing in the background. The intercept receiver could be a snoopy neighbor with a scanning receiver, available in most electronics stores.

In situations as depicted in Figure 10-1, it is desirable for the intended communicators (hereafter referred to as the *communicators* or the *communications link*) to use minimum transmit power and the highest-sensitivity receiver possible to minimize the possibility of intercept. On the other hand, to minimize the effects of jamming, the communications should use the greatest transmit power at their disposal and the least sensitive receiver that will still allow the communication to take place. These are clearly conflicting conditions. In addition, there may be several variables present and options available in any communications game, such as propagation conditions, antenna gains, modulation types, employment of special signal processors, and so on. This discussion can be put on a more quantitative basis in terms of a measure of communications quality, or *quality factor* [4], which is the ratio of the intercept range to the range in which the communicators can communicate. The performance of the communications link is determined by the desired bit error performance, which implies a certain required E_b/N_0 depending on modulation type, background conditions, and processing at the receiver as reviewed in Chapter 1. As will be shown later, the interceptor is characterized by its probabilities of detection and false alarm. The desired probability of false alarm determines the threshold to be used for detection in the intercept receiver; to achieve a certain desired probability of detection, the intercept receiver uses this threshold to determine its required received signal-to-noise ratio. If this ratio cannot be achieved under the assumed conditions, other processing methods, geometries, or receiver implementations (e.g., a higher-gain antenna or a lower-noise front-end amplifier) must be explored. This illustrates to some degree the many trade-offs to be explored in the covert communications problem and the desire on the part of the interceptor to extract information from the transmission process.

A term that is used in an almost offhand fashion in discussions relating to uncooperative detection of spread-spectrum signals is *LPI* (low probability of intercept) *signal*. It is useful to define this term before beginning the consideration of intercept receivers. According to Polydoros and Weber [5], an LPI signal is a spread-spectrum waveform whose code is unknown (unexploitable) from the viewpoint of an interceptor. Thus any potential interceptor is forced to use wideband detection techniques because he or she cannot exploit the correlation detection option of the intended receiver who knows the spectrum spreading code used by the transmitter and must only synchronize the code generated locally at the receiver with the code spreading the incoming signal.

10-3 Energy Detection in AWGN

Before spread-spectrum signal parameters can be estimated, the presence of the signal must be detected. A longstanding method for detecting the presence of an unknown signal in noise is energy detection [6,7]. The block diagram of an energy detector is shown in Figure 10-2. It consists of a filter to eliminate noise outside the bandwidth W of interest (perhaps the signal bandwidth plus an additional amount for uncertainities), a square-law device, a finite-time integrator, a sampler that samples the integrator output at the end of the integration interval T (which might be the expected signal duration, if known), and a threshold comparison device. If the sampler output at time T is greater than the threshold K, the decision is made that signal plus noise was present; if it is less than K, the decision is made that noise alone was present. From the communicators' standpoint, T should be as short as possible to minimize detection. Early applications for energy detectors were in radio astronomy, so they are also known as radiometers.

It was shown in Section 5-3.4 that the probability density function of the samples at the integrator output with AWGN of spectral density N_0 alone into the energy detector is approximately chi-square (see Figure 5-10). For signal plus noise at the input, the probability density function of the sampler output is noncentral chi-square [see (5-66)]. The performance of the detector can be characterized by its probabilities of false alarm and detection. The probabilities of false alarm and detection, assuming the chi-square statistics at the integrator output are sufficiently accurate, are given by [6]

$$P_F = \int_{K_0}^{\infty} \frac{u^{TW-1} e^{-u}}{\Gamma(TW)} \, du \qquad (10\text{-}1)$$

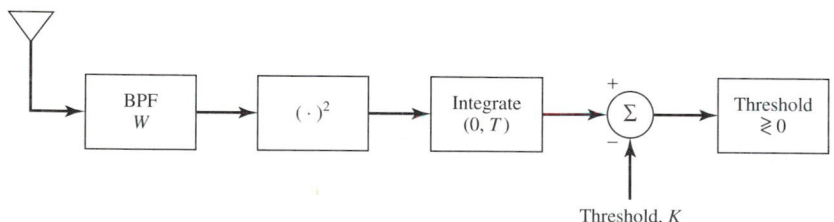

FIGURE 10-2.   Energy detector for detection of an unknown signal in AWGN.

and

$$P_D = \int_{K_0}^{\infty} \left(\frac{u}{S}\right)^{(TW-1)/2} e^{-(u+S)} I_{TW-1}(2\sqrt{uS})\, du \qquad (10\text{-}2)$$

respectively, where

$$S = \frac{E_s}{N_0} \qquad (10\text{-}3)$$

is the signal energy-to-noise spectral density ratio, $K_0 = K/2$, $\Gamma(x)$ is the gamma function, and $I_n(z)$ is the modified Bessel function of the first kind and order n. Equations (10-1) and (10-2) completely characterize the operation of the detector. One way to use the detector parameters is to choose an acceptable false-alarm probability and solve for the corresponding K_0. This threshold value is then substituted into (10-2), which can then be solved for the required value of S to give the desired probability of detection.† This is a difficult task, as seen in Chapter 5, because even though (10-1) can be integrated for TW an integer by repeated integration by parts [9] (this is not feasible for typical values of TW for spread-spectrum signals), (10-2) requires series evaluation or numerical integration. The nomogram of Figure 5-11 was given in Chapter 5 for the purpose of hand calculation of P_F and P_D, and a program for numerical series approximation [10–12] for the noncentral chi-squared cumulative probability density function is discussed in Appendix F.

Urkowitz [6] also gave an approximation of the noncentral chi-squared cumulative probability density function by a central chi-square having a different number of degrees of freedom ($2TW$) and a modified threshold level. The new number of degrees of freedom, as mentioned in Chapter 5, is given by

$$D = 2\frac{(TW + S)^2}{TW + 2S} \qquad (10\text{-}4)$$

and the threshold is produced by dividing the old threshold by the factor‡

$$G = \frac{TW + 2S}{TW + S} \qquad (10\text{-}5)$$

Tables for the central chi-squared cumulative distribution can then be used [13].

For large values of $2TW$, the central limit theorem applies and the integrator output can be approximated as Gaussian for both noise alone and signal plus noise present at the input. In terms of the Gaussian Q-function, the Gaussian approximation results in the expression

$$P_F = Q\left(\frac{K_0 - TW}{\sqrt{TW}}\right) \qquad TW \gg 1 \qquad (10\text{-}6)$$

---

† Called the Neyman–Pearson criterion for detection.
‡ In terms of the notation in Example 5-2, $\lambda = 2S$ and $n = 2TW$.

for the probability of false alarm, and

$$P_D = Q\left(\frac{K_0 - TW - S}{\sqrt{TW + 2S}}\right) \qquad TW \gg 1 \tag{10-7}$$

for the probability of detection.

A plot of P_D versus P_F is called the operating characteristic of the receiver. Operating characteristics, calculated using the Gaussian approximation, are shown in Figure 10-3 for several values of TW and S.

To determine the signal-to-noise ratio required for a desired P_F and P_D, the inverse Q-function of (10-6) and (10-7) results in

$$Q^{-1}(P_F) = \frac{K_0 - TW}{\sqrt{TW}} \tag{10-8}$$

$$Q^{-1}(P_D) = \frac{K_0 - TW - S}{\sqrt{TW}} \tag{10-9}$$

respectively, where $2S$ has been neglected with respect to the TW (in keeping with the large time–bandwidth approximation) in the denominator of the right-hand side of (10-9). Substitution of (10-8) into (10-9) and rearranging results in

$$d = \frac{S}{\sqrt{TW}} = Q^{-1}(P_F) - Q^{-1}(P_D) \tag{10-10}$$

as a normalized minimum signal-to-noise ratio, referred to as the deflection, required to achieve the specified values of P_F and P_D on the right-hand side. The actual minimum required signal-to-noise ratios are

$$S\big|_{\text{req'd}} = \frac{E_s}{N_0}\bigg|_{\text{req'd}} = \sqrt{TW}\,d \tag{10-11}$$

in terms of signal energy-to-noise spectral density at the integrator output or, in terms of signal-to-noise power ratio at the receiver input,

$$\frac{P_s}{P_n}\bigg|_{\text{req'd}} = \frac{E_s}{N_0 WT} = \frac{d}{\sqrt{TW}} \tag{10-12}$$

The assumed large TW character of the signal (i.e., large signal bandwidth and long duration, thus allowing a long integration time) allowed the chi-squared statistics of the integrator output to be replaced by Gaussian statistics. If TW is not large, (10-10) results in a value of d that is too small. A correction factor can be computed for given values of P_F and P_D as a function of TW [14], which has been reproduced in Ref 1. Sample values are extracted from the curves given in Ref. 1 and are listed in Table 10-1.

FIGURE 10-3. Operating characteristics for an energy detector receiver for $TW = 500$, 1000, 2500, and 5000: (a) $S = 10$ dB; (b) $S = 15$ dB.

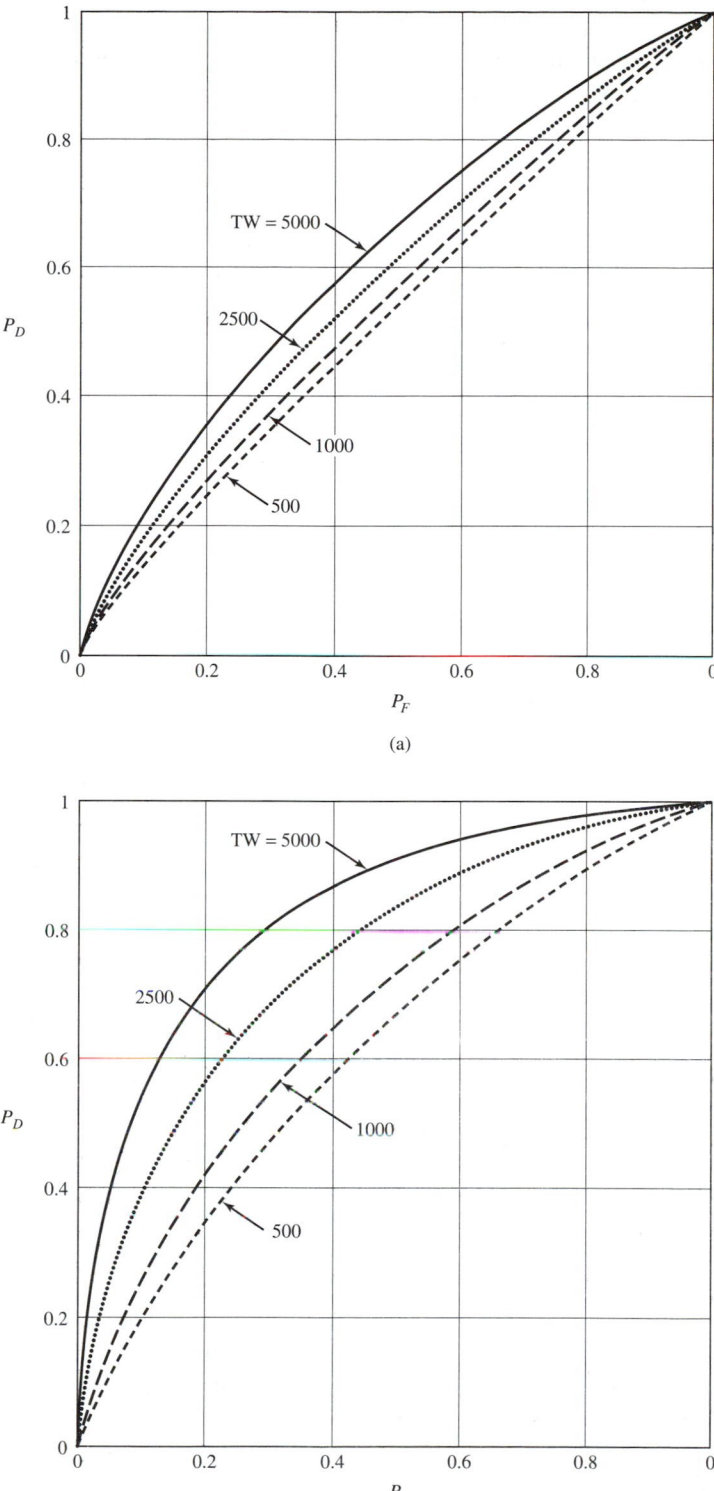

(a)

(b)

TABLE 10-1. Correction Factor η (dB) for Gaussian Approximation in Energy Detection; $P_D = 0.9$

TW	$P_F = 10^{-4}$	$P_F = 10^{-8}$	$P_F = 10^{-12}$
10,000	0.25	0.25	0.25
1,000	0.4	0.5	0.6
100	0.8	1.2	1.3
10	2.2	2.8	3.2
1	4.8	5.8	6.5

EXAMPLE 10-1

In this example, the results of the Gaussian approximation are compared to computations using the nomogram given in Figure 5-11. Assume that $P_F = 10^{-3}$ and $P_D = 0.8$ are desired for $TW = 125$. Find the threshold K_0 and the energy-to-noise-spectral density ratio S that will give these values.

Solution: A table of Q-functions or a rational approximation (see Appendix B) used to solve (10-6) and (10-7) results in

$$\frac{K_0 - TW}{\sqrt{TW}} = 3.09 \quad \text{and} \quad \frac{K_0 - TW - S}{\sqrt{TW + 2S}} = -0.842$$

Simultaneous solution of these equations for K_0 and S (a computer mathematics package is handy here) yields $K_0 = 159.6$ and $S = 47 = 16.72$ dB. If the more approximate equations (10-11) and (10-10) are used, $S = 16.43$ dB.

For solution using the nomogram of Figure 5-11, one first solves for the threshold value $V_T = 2K_0$, given $P_D = 0.001$ and $n = 2TW = 250$. This yields $V_T = 325$ from the nomogram of Figure 5-11 (compare this to $2K_0 = 319.2$ found above). Solution for S requires trial and error. The modified number of degrees of freedom, n', is given by (5-68) or (10-4) and the modified threshold is given by (5-69) or using (10-5). However, the value of S is to be found and appears in both equations. By trying several values for $\lambda = 2S$ (again, a computer mathematics package is handy), one arrives at the modified threshold and degrees of freedom $n' = 268.8$ and $V_T' = 257$ for $\lambda = 90$, which line up with $P_D = 0.8$ on the nomogram (note that the line for n' has to be extended somewhat). This corresponds to $S = \lambda/2 = 45 = 16.5$ dB, which is surprisingly close to the value of 16.72 dB found using the Gaussian approximation.

This design is for a particular performance in terms of P_D and P_F. What happens if this design point is not achieved? For sufficiently large TW, (10-6) and (10-7) can be used to find the change in performance if K_0 and/or S vary from their ideal values. For example, a 1% low value of S yields $P_D = 0.79$ and a 10% low value yields $P_D = 0.70$.

10-4 Optimum Intercept Receivers for Spread-Spectrum Signals

10-4.1 Introduction

In this section, optimum receiver structures for detecting spread-spectrum signals in AWGN backgrounds are examined. This not only gives an idea of how much degradation results from using the simple energy detector discussed in the preceding section, but also provides guidelines as to how to approximate the optimum receiver with suboptimum but simpler structures. By *optimum* is meant a receiver that implements a likelihood ratio test (LRT) [15], which is a well-known procedure based on statistical testing of hypotheses. The Neyman–Pearson detection criterion mentioned in relation to the energy detector can be shown to be an LRT in which the threshold is set by the desired probability of false alarm. The particular form that the LRT takes in signal interception depends on what is presumed known beforehand. For example, in the case of the energy detector, very little was assumed known about the signal—only that it occupied a bandwidth of W and existed for a time duration T. Two general cases are considered in this section: (1) the signal to be intercepted is a direct-sequence spread-spectrum signal, and (2) it is a frequency-hop spread-spectrum signal. The latter case can also be generalized to hybrid spread-spectrum signals. The material of this section is based on the work of several investigators and has been reported in Refs. 1, 5, 16, and 17.

10-4.2 Optimum Intercept Receiver for Direct-Sequence Spread Spectrum

The results summarized in this section are based on the work of Polydoros and Weber [5]. They consider a direct-sequence spread-spectrum signal. The receiver must decide between the two hypotheses of signal plus noise or noise alone, expressed as

$$r(t) = \begin{cases} \sqrt{2P_s}c(t)\cos(\omega_0 t + \phi) + n(t) & (H_1) \\ n(t) & (H_0), \quad 0 \le t \le T \end{cases} \tag{10-13}$$

where

$$c(t) = \sum_{n=-\infty}^{\infty} c_n p(t - nT_c - \epsilon T_c) \tag{10-14}$$

is the spreading code. The various quantities in the equations above are defined as follows:

$n(t)$ is AWGN of single-sided spectral density N_0, W/Hz.

P_s is the average signal power.

$\omega_0 = 2\pi f_0$ is the carrier frequency, rad/s.

ϕ is the carrier phase (to be dealt with in more detail later).

T_c is the chip period.

$T = NT_c$ is the observation time in seconds, assumed to be an integer number of chip periods.

$p(t)$ is a unit-amplitude pulse of duration T_c seconds.

ϵ specifies the signal epoch, usually modeled by a random variable uniformly distributed in $[0,1)$.

$\{c_n\}$ is a sequence of independent identically distributed chip values equally likely to take on the values $+1$ or -1.

The signal, if present, is assumed to occupy the entire observation interval. Several specific cases are considered in Ref 5. Likelihood ratio tests (LRTs) for them are given below without derivation. The reader is referred to Ref. 5 for more details.

Synchronous Coherent Detectors. If both the carrier phase ϕ and signal epoch ϵ are assumed known, the likelihood ratio test for deciding between H_0 and H_1 can be based on the synchronously detected in-phase signal

$$r'(t) = r(t) \cos(\omega_0 t + \phi) \tag{10-15}$$

so that the LRT becomes

$$\Lambda[r'(t)] = \prod_{j=1}^{N} \exp(-\gamma_c) \cosh\left(\frac{2\sqrt{P_s}}{N_0} r_j\right) \overset{H_1}{\underset{H_0}{\gtrless}} \Lambda_0 \tag{10-16}$$

where

$$r_j = \int_{(j-1)T_c}^{T_c} r'(t)\, dt \tag{10-17a}$$

and

$$\gamma_c = \frac{P_s T_c}{N_0} = \frac{E_c}{N_0} \tag{10-17b}$$

is the chip energy/noise spectral density ratio. For a Neyman–Pearson test, the threshold of the test, Λ_0, would be set by the allowed probability of false alarm. It is often more convenient to implement a log-LRT (LLRT), which for this case is

$$l = -N\gamma_c + \sum_{j=1}^{N} \ln \cosh\left(\frac{2\sqrt{P_s}}{N_0} r_j\right) \overset{H_1}{\underset{H_0}{\gtrless}} \ln \Lambda_0 \tag{10-18}$$

For low values of E_c/N_0 (-10 dB or less), the approximation $\ln \cosh x \approx x^2/2$ is valid, so that (10-18) can be simplified to

$$\lambda = \sum_{j=1}^{N} r_j^2 \overset{H_1}{\underset{H_0}{\gtrless}} \lambda_0 \tag{10-19}$$

which is asymptotically optimum as γ_c approaches zero.

If N is large enough so that the Gaussian approximation holds, P_D for the LLRT is

$$P_D = Q\left[\frac{Q^{-1}(P_F) - a\sqrt{N}\gamma_c}{\sqrt{1 + b\gamma_c}}\right] \tag{10-20a}$$

where, for the synchronous coherent detector considered here,

$$a = \sqrt{2} \quad \text{and} \quad b = 4 \tag{10-20b}$$

Typically, $b\gamma_c \ll 1$, so that (10-20a) can be solved to yield

$$d_\lambda = \sqrt{2\,N}\gamma_c = Q^{-1}(P_F) - Q^{-1}(P_D) \tag{10-21}$$

for the minimum deflection required to give the specified values of P_F and P_D on the right-hand side. As pointed out in Ref. 5, the performance of the test is dictated by the product $N^{1/2}\gamma_c$, which is typical of tests depending on postdetection integration as opposed to the product $(N\gamma_c)^{1/2}$, as is the case for coherent integration of the whole signal. The latter case requires knowledge of the spreading code (the intended receiver has this knowledge).

In terms of the overall observation time, T, and direct-sequence spread-spectrum null-to-null bandwidth, $W = 2T_c^{-1}$, the minimum required signal-to-noise ratio at the detector input to give the desired P_F and P_D is

$$\left.\frac{P_s}{P_n}\right|_{\text{req'd}} = \frac{P_s}{N_0 W} = \frac{1}{2}\frac{d_\lambda}{\sqrt{TW}} \tag{10-22a}$$

or, in terms of total integrated signal energy,

$$\left.\frac{E_s}{N_0}\right|_{\text{req'd}} = \frac{1}{2}\sqrt{TW}d_\lambda \tag{10-22b}$$

Performance of the optimum LLRT (10-18) and the suboptimum rule (10-19) are shown in Figure 10-4. These results, taken from Ref. 5, are for $N = 50$ and $N = 1000$, to contrast conditions under which the Gaussian assumption is valid with a situation where it is not.

EXAMPLE 10-2 _____

In Example 10-1 it was found that for a signal of time–bandwidth product 125, an energy detector provided the operating point $P_F = 0.001$ and $P_D = 0.8$ for a total signal energy-to-noise-spectral density ratio $S = E_S/N_0$ of 16.43 dB at the intercept receiver (Gaussian approximation for large TW). Suppose that this signal is actually direct-sequence spread-spectrum and the suboptimum detector of (10-19) is used. What is the required value of S for this detector?

Solution: From Example 10-1 and using (10-21), it is found that $d_\lambda = 3.09 + 0.84 = 3.93$. From (10-22b), the required value of S is $0.5(125)^{1/2}(3.93) = 21.97 = 13.42$ dB, which is exactly 3.01 dB better than the value found for the energy detector in Example 10-1 using the Gaussian approximation.

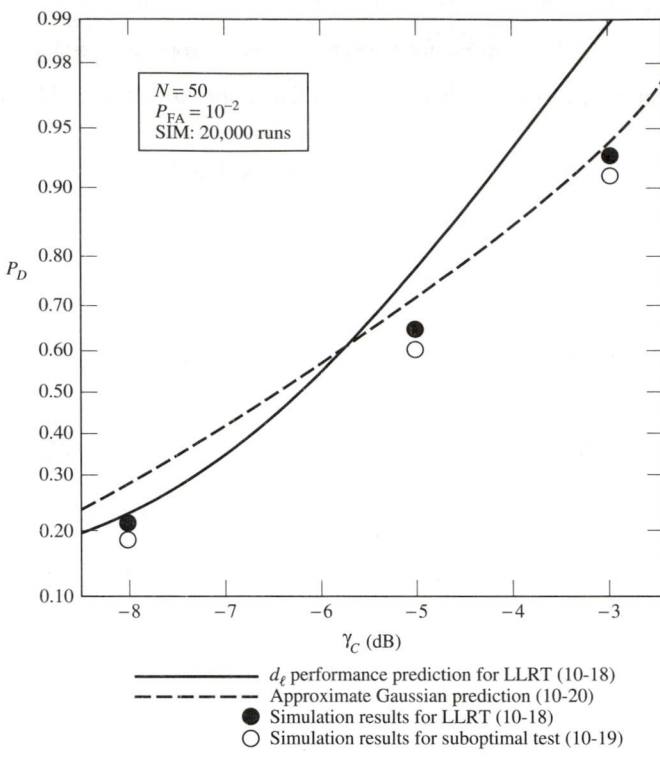

d_ℓ performance prediction for LLRT (10-18)
--- Approximate Gaussian prediction (10-20)
● Simulation results for LLRT (10-18)
○ Simulation results for suboptimal test (10-19)

(a)

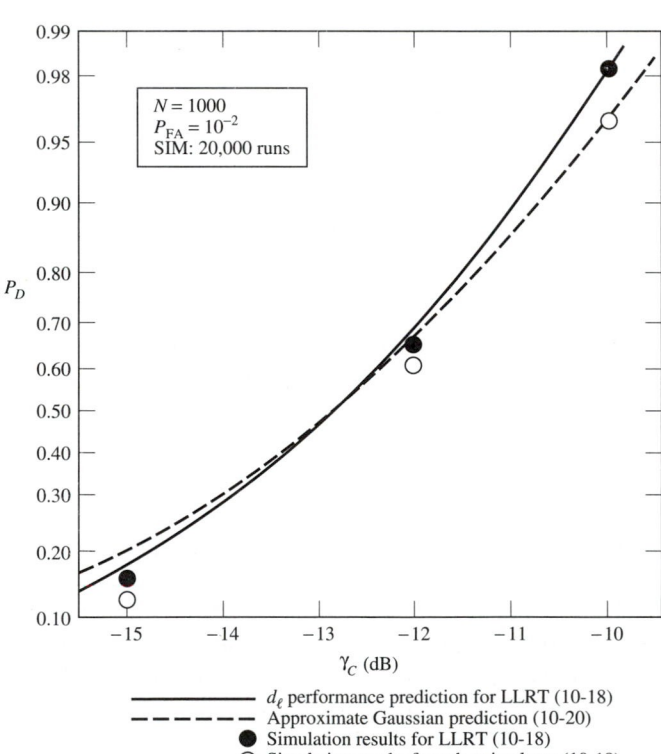

d_ℓ performance prediction for LLRT (10-18)
--- Approximate Gaussian prediction (10-20)
● Simulation results for LLRT (10-18)
○ Simulation results for suboptimal test (10-19)

(b)

Synchronous Noncoherent Detectors. For this case, the carrier phase is modeled as a uniform random variable in $[0, 2\pi)$. The LRT is

$$\Lambda[r(t)] = \exp(-N\gamma_c) \prod_{i=1}^{2^N} I_0\left(\frac{2\sqrt{P_s}}{N_0} R_i\right) \underset{H_0}{\overset{H_1}{\gtrless}} \Lambda_0 \qquad (10\text{-}23)$$

where $I_0(\cdot)$ is the zero-order modified Bessel function. The quantity R_i is the ith correlation envelope, defined as

$$R_i = \sqrt{e_{I_i}^2 + e_{Q_i}^2} \qquad i = 1, 2, \ldots, 2^N \qquad (10\text{-}24)$$

where

$$e_{\alpha_i} = \sum_{j=1}^{N} r_{\alpha_i} c_{ij} \qquad \alpha = I, Q \qquad (10\text{-}25\text{a})$$

with

$$\begin{bmatrix} r_{I_j} \\ r_{Q_j} \end{bmatrix} = \sqrt{2} \int_{(j-1)T_c}^{jT_c} r(t) \begin{bmatrix} \cos \omega_0 t \\ \sin \omega_0 t \end{bmatrix} dt \qquad j = 1, 2, \ldots, N \qquad (10\text{-}25\text{b})$$

This decision rule is hard to mechanize. Polydoros and Weber [5] propose the replacement of the received waveform under H_1 with

$$r(t) = \sqrt{2P_s} \sum_{j=-\infty}^{\infty} c_j p(t - jT_c) \cos(\omega_0 t + \phi_j) + n(t) \qquad (10\text{-}26)$$

where $\{\phi_j\}$ is a sequence of identically distributed, independent random variables, uniform in the interval $[0, 2\pi)$ (i.e., a system where the carrier phase is independent chip to chip). In this case the LLRT becomes

$$l[r(t)] = -N\gamma_c + \prod_{i=1}^{N} \ln I_0\left(\frac{2\sqrt{P_s}}{N_0} r_i\right) \underset{H_0}{\overset{H_1}{\gtrless}} l_0 \qquad (10\text{-}27)$$

where r_j is the envelope of the jth chip, given by

$$r_j = \sqrt{r_{I_j}^2 + r_{Q_j}^2} \qquad (10\text{-}28)$$

Using the approximation $\ln I_0(x) \cong x^2/4$, it follows that (10-27) reduces to the suboptimum rule

$$\lambda = \sum_{j=1}^{N} r_j^2 \underset{H_0}{\overset{H_1}{\gtrless}} \lambda_0 \qquad (10\text{-}29)$$

FIGURE 10-4 Probability of detection versus chip signal-to-noise ratio for the synchronous coherent detector; optimum and suboptimum tests; $P_F = 10^{-2}$. (a) 50 chip observation interval; (b) 1000 chip observation interval.

which is identical to (10-19) except that here r_j corresponds to a noncoherent chip integration.

Performance under the low-signal-to-noise assumption [i.e., (10-29) holds] and assuming $N \gg 1$, so that the Gaussian approximation is valid, results in an expression like (10-20a) but with $a = 1$ and $b = 2$, so that in the chip noncoherent case

$$\left.\frac{E_s}{N_0}\right|_{\text{req'd}} = \frac{1}{\sqrt{2}}\sqrt{TW_s}\, d_\lambda \tag{10-30}$$

This shows a 1.5-dB loss with respect to the coherent detector. Since the carrier-noncoherent detector, characterized by the decision rule (10-27), is bracketed between the coherent and the chip-noncoherent detector, one concludes that any implementational complexity beyond that of (10-29) is unjustified from practical considerations.

Asynchronous Detectors. In most practical cases, the code timing epoch, ϵ, is a random variable. Krasner [16] has considered the implementation of configurations to optimally average the likelihood ratio over the epoch uncertainty. Polydoros and Weber [5] have considered reduced-complexity suboptimum asynchronous detectors and have compared them with the performance of the optimum detector. They have shown that the loss of chip synchronism in the coherent case costs between 0.63 dB (lower bound) and 1.42 dB (upper bound). They conclude that a rough figure of 1 dB loss results due to lack of synchronism in the coherent case, with a similar figure in the noncoherent case.

Two Independent Observations. Polydoros and Weber [5] hypothesize a coherent receiver that contains two independent channels and makes two observations of the same signal with independent noise in each observation. If both observations could be combined coherently, a 3-dB gain in signal-to-noise ratio over the single-observation case would be obtained. In the more practical case where coherent combining is not possible but two separate detections are used, the gain is 1.5 dB over the single-observation case.

Comparisons. Any discussion of interception of spread-spectrum signals is incomplete without bounds between the best and worst cases. Clearly, the best one could hope for is if the spreading code were known exactly by the intercept receiver. This would be the performance of the correlation detector employed by the intended receiver of the spread-spectrum signal. The worst one can do is to use a radiometer, where the detector has no knowledge of the signal structure. The options discussed above fall somewhere between these two bounds. Figure 10-5 shows the performance of these various receiver implementations in terms of probability of detection versus chip signal-to-noise ratio. The assumed probability of false alarm is 10^{-2} and the number of chips observed is 1000.

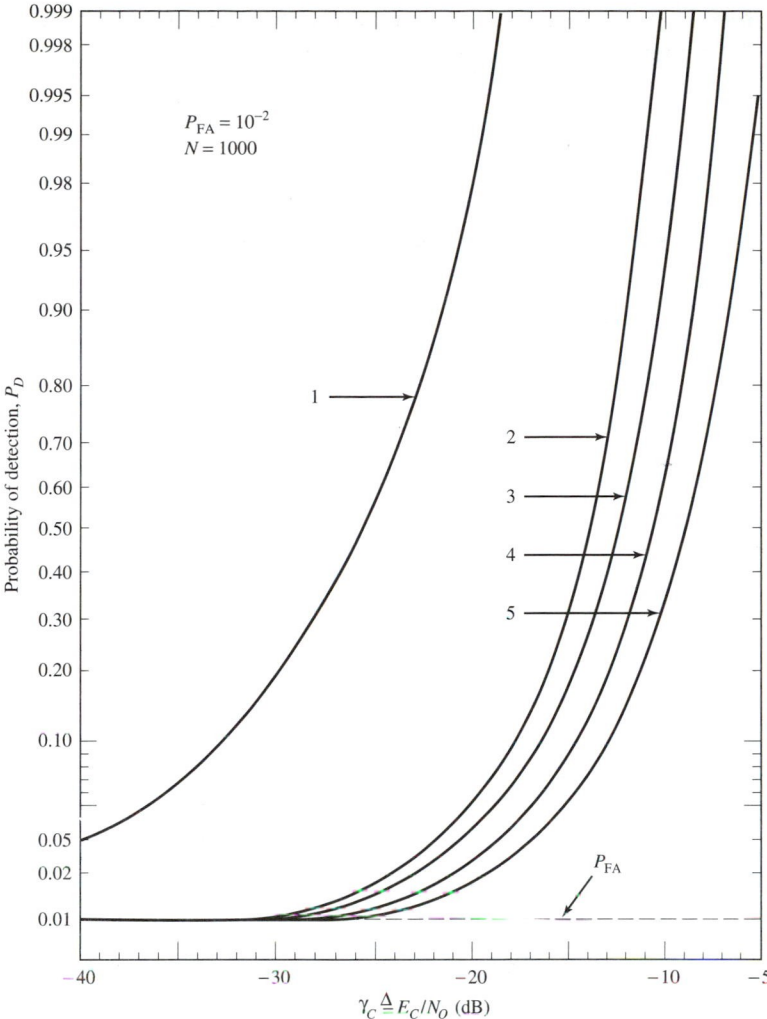

FIGURE 10-5. Performance of various intercept receiver options: 1, Completely known waveform; 2, two independent receptions, synchronous coherent correlation detector; 3, synchronous coherent detector; 4, synchronous chip-noncoherent detector; 5, radiometer. (From Ref. 5. Copyright © 1985 IEEE. Reprinted with permission.)

EXAMPLE 10-3

A direct-sequence spread-spectrum BPSK signal with chip rate of 10 Mchips/s is observed for 0.1 ms. An operating performance point for an intercept receiver of $P_F = 10^{-4}$ and $P_D = 0.9$ is desired. Find the required value of $S = E_s/N_0$, where E_s is the total observed signal energy, to provide this operating point for (a) a radiometer; (b) a synchronous coherent detector; (c) a noncoherent chip detector. If the observation interval changes to 1 ms, what are the new values?

Solution: (a) For the given modulation, $W \cong 2/T_c = 2/10^{-7} = 20$ MHz. Since the observation interval is 0.1 ms, $WT = 2000$. Solution of (10-10) gives

$$d = Q^{-1}(P_F) - Q^{-1}(P_D) = 3.72 + 1.28 = 5$$

Equation (10-11) then gives

$$S_{1a} = d\sqrt{TW} = 5\sqrt{2000} = 223.6 = 23.5 \text{ dB}$$

The corresponding predetection signal-to-noise power ratio from (10-12) is

$$\left(\frac{P_s}{P_n}\right)_{1a} = \frac{d}{\sqrt{TW}} = \frac{5}{\sqrt{2000}} = 0.112 = -9.5 \text{ dB}$$

If the observation interval is a factor of 10 longer, then

$$S_{2a} = d\sqrt{TW} = 5\sqrt{20,000} = 707.2 = 28.5 \text{ dB}$$

and

$$\left(\frac{P_s}{P_n}\right)_{2a} = \frac{d}{\sqrt{TW}} = \frac{5}{\sqrt{20,000}} = 0.035 = -14.5 \text{ dB}$$

(b) Equations (10-22) give

$$S_{1b} = \frac{1}{2}\sqrt{2000} \times 5 = 111.8 = 20.5 \text{ dB}$$

and

$$\left(\frac{P_s}{P_n}\right)_{1b} = \frac{5}{2\sqrt{2000}} = 0.056 = -12.5 \text{ dB}$$

for the 0.1-ms observation interval. For the 1-ms observation interval, the results are 5 dB more for the energy/noise spectral density ratio and 5 dB less for the predetection signal-to-noise power ratio.

(c) These results are 1.5 dB more than the corresponding results for case (b). All of these results are summarized in Table 10-2.

TABLE 10-2. Summary of Results for Example 10-3

Detector	$T = 0.1$ ms		$T = 1$ ms	
	S (dB)	P_s/P_n (dB)	S (dB)	P_s/P_n (dB)
Radiometer	23.5	-9.5	28.5	-14.5
Noncoherent chip	22.0	-11.0	27.0	-16.0
Coherent direct-sequence	20.5	-12.5	25.5	-17.5

Note the effect of the observation interval on the energy/noise spectral density ratio and signal-to-noise power ratio. An increase of a factor of 10 in observation interval length means that more signal energy is integrated, which explains the increase of 5 dB in going from 0.1 ms to 1 ms in observing the signal. However, since more energy is gathered, the signal-to-noise power ratio decreases by 5 dB because signal energy is converted to average signal power by dividing by the time interval (i.e., a factor of 10 or 10 dB).

10-4.3 Intercept Receivers for Frequency-Hop Spread Spectrum

In detecting frequency-hop spread-spectrum signals without knowledge of the hopping code, the best the inteceptor can hope to do is integrate energy in each hop-frequency cell and combine this energy optimally to make an overall decision. For typical numbers of hop frequencies used in frequency-hop spread spectrum, this can result in impractical receiver structures. However, it does provide a comparison by which suboptimum receiver structures can be judged. In this section, this optimum receiver structure is presented along with its performance. A suboptimum receiver structure will be investigated and compared in performance with the optimum structure in the next section. The discussion presented in this section follows that given in Ref. 1.

Optimum Receiver Structure for Frequency-Hop Spread Spectrum. The notation to be used in this discussion of frequency-hop spread-spectrum signal interception is defined in Figure 10-6. The total frequency-hop spread-spectrum signal bandwidth is denoted by W and its duration by T, as before. Each hop is of duration T_h and the bandwidth of each hopped signal element is denoted by W_h. In addition,

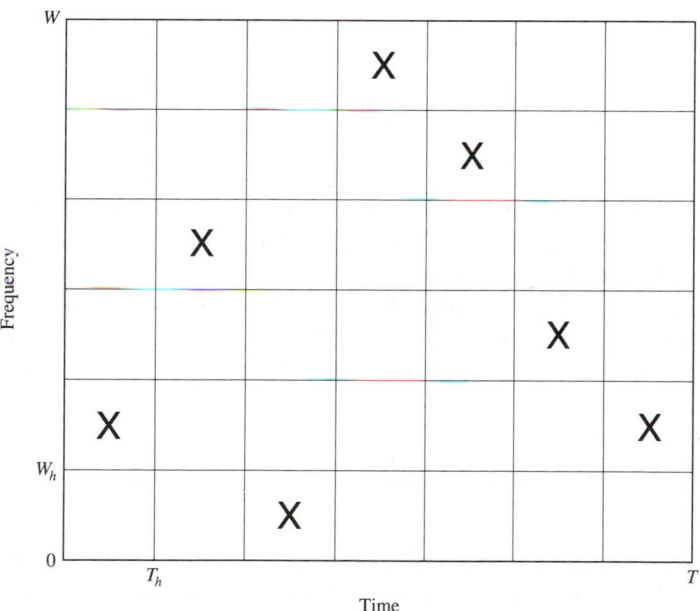

FIGURE 10-6. Signal structure for considering interception of frequency-hop spread-spectrum signals with typical hop pattern.

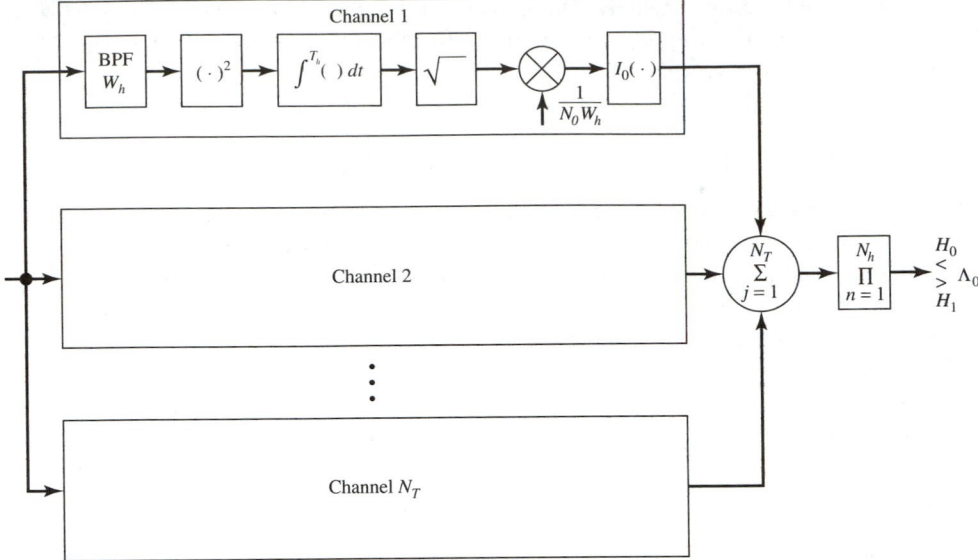

FIGURE 10-7. Optimum detector for FHSS signals, assuming knowledge of the hopping times and frequency intervals. (From Ref. 1.)

there are a total of

$$N_T = \frac{W}{W_h} \qquad (10\text{-}31a)$$

hopping bands and

$$N_h = \frac{T}{T_h} \qquad (10\text{-}31b)$$

hops in the observation interval. The hopping times and frequency slots for the hops are initially assumed known by the inteceptor. Although unrealistic from a practical standpoint, the performance of the receiver for this idealized case will give a best possible case with which to compare the performances of receivers for which less is assumed known about the intercepted signal.

The optimum receiver structure,† shown in Figure 10-7, corresponds to the LRT

$$\Lambda[r(t)] = \prod_{n=1}^{N_h} \sum_{j=1}^{N_T} I_0\left(\frac{1}{N_0 W_h} \sqrt{\int_{(j-1)T_h}^{jT_h} r_j^2(t)\, dt} \right) \qquad (10\text{-}32)$$

where $r_j(t)$ is the output of the filter matched to the jth hopping pulse and $I_0(\cdot)$ is the modified zero-order Bessel function, as before. Although this receiver structure

---

† Van Trees [15, Prob. 4.5.11] considers a problem similar to the present case. See also Appendix A of Ref. 3.

is highly nonlinear, it is approximately valid for large N_h to use Gaussian statistics. Figure 10-8 shows the signal-to-noise ratio required of a wideband radiometer to provide a certain level of detection divided by the signal-to-noise ratio required by the optimum receiver for frequency-hop spread spectrum, assuming that Gaussian statistics are valid, versus the product WT. These are computed by solving the transcendental equation [1]

$$\frac{P_s}{P_n} = \frac{1}{2N_T} I_0^{-1} \left\{ 1 - N_T \, \exp\left[N_T \left(\frac{P_s}{P_n} \right)^2 \right] \right\} \tag{10-33}$$

where $I_0(\cdot)$ is the modified Bessel function of order zero, N_T the number of frequency slots, and P_s/P_n the signal-to-noise power ratio. For example, for $N_T = 10^5$

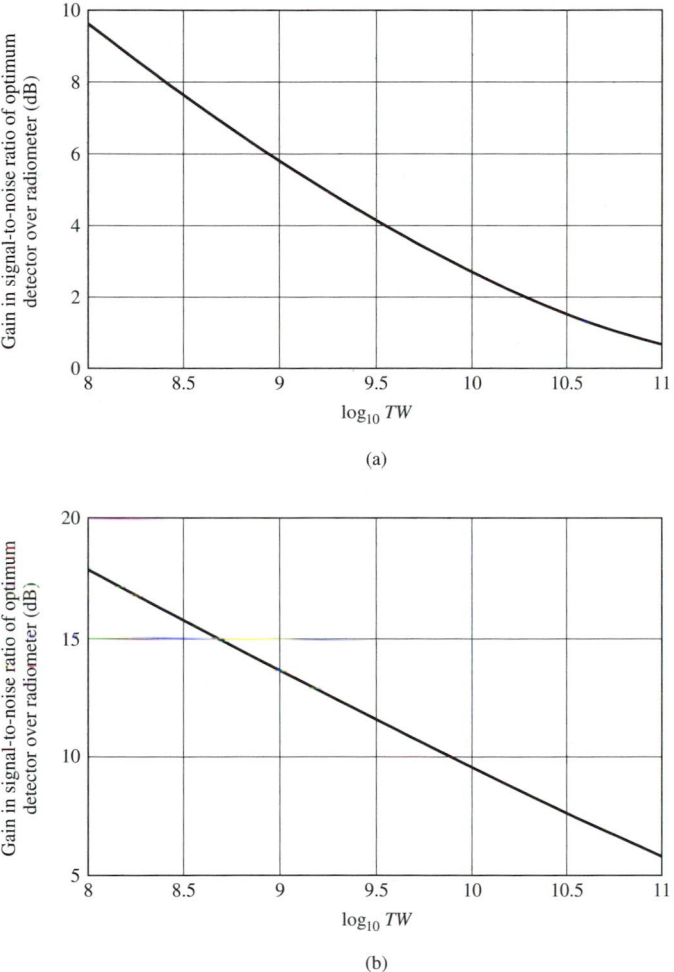

FIGURE 10-8. Signal-to-noise ratio gain of the optimum FHSS detector over the radiometer: (a) $N_T = 10^5$; (b) $N_T = 10^6$. (Adapted from Ref. 1.)

and $WT = 10^8$, the optimum receiver shows approximately a 9.5-dB advantage over the radiometer. This improvement is gained by assuming complete knowledge of the signal structure discussed above and by using a detector with 10,000 parallel channels! This is clearly a difficult, if not impossible task, even with very-large-scale integrated-circuit technology. Because of this complex structure, an approximation to the optimum receiver is considered in the next section.

10-5 The OR/BMWD: Approximately Optimum Spread-Spectrum Signal Detector

The suboptimum detection of spread-spectrum signals in AWGN from the standpoint of an unwanted interceptor will now be considered [3,8]. Through intelligence means, the interceptor is assumed to have knowledge of T_h and W_h for the spread-spectrum system, but does not know the spreading code being used. The general case of a frequency-hopped/direct-sequence spread-spectrum signal (i.e., hybrid modulation) will be considered. Thus the diagram of Figure 10-6 holds with the understanding that now it is possible for $W_h T_h$ to be much greater than 1. It will also be assumed that the interceptor knows the frequency band, hopping subbands, and hopping intervals. This is in keeping with the desire to obtain an upper bound for the performance of any interceptor (i.e., the best that it can do). Note that the case of hybrid modulation can be specialized to direct-sequence spread spectrum by using one hopping interval or to frequency-hop spread spectrum by assuming no direct-sequence spreading code is used during the hopping.

The receiver structure is shown in Figure 10-9. Since the interceptor does not know the hopping sequence or the spreading code, the best it can do is to perform energy detection in the frequency-hopping subbands over each hop dwell and to combine these separate detections. The means for combining these separate detections will be called a binary moving-window detector (BMWD), in which detections

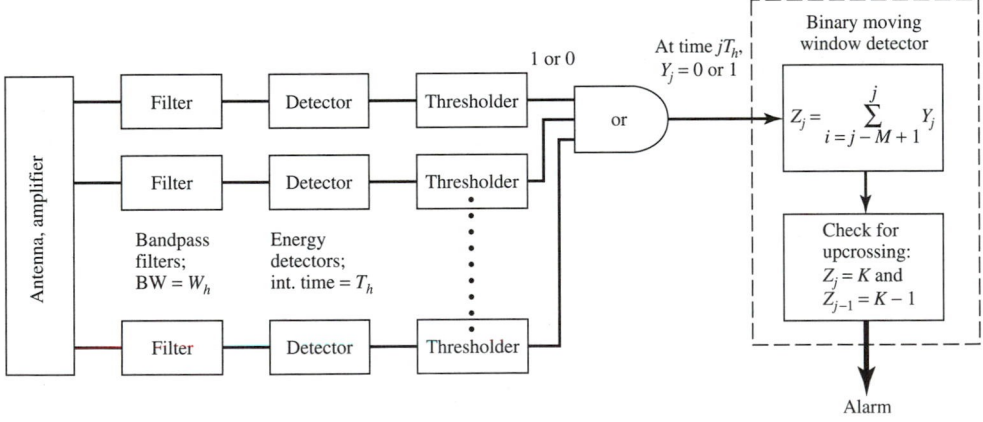

FIGURE 10-9. Channelized receiver with an OR/BMWD for combining detections. (Adapted from Ref. 8.)

in the separate frequency slots at time intervals jT_h, where j is an integer and T_h is the thresholding interval (assumed ideally to be at the end of each hop dwell), are combined by an OR operation and a moving window of ORed detections is summed and compared with a threshold. It is denoted as an OR/BMWD [3,8]. In Figure 10-9 the BMWD is shown inside the dashed box.

Analysis of the OR/BMWD proceeds initially with the results of the energy detector performance analysis, except that they are applied to each pulse in the signal structure (i.e., each small rectangle). Let there be N_T frequency slots of bandwidth W_h as shown in Figure 10-6, for a total bandwidth of $W = N_T W_h$. If noise alone is present at the receiver input, the probability that a 1 enters the BMWD at any thresholding time jT_h is

$$P_0 = 1 - (1 - Q_F)^{N_T} \tag{10-34}$$

where Q_F is the probability of false alarm on a single cell in Figure 10-6. The probability that a 1 enters the BMWD at any thresholding time jT_h when signal plus noise is present at the receiver input is

$$P_1 = 1 - (1 - Q_D)(1 - Q_F)^{N_T - 1} \tag{10-35}$$

where Q_D is the probability of detection on a single cell. (This assumes that only one cell is occupied per time slot of duration T_h.) Note that the signal energy to be used in Q_D is that of a single frequency-hop pulse, or E_p.

At each time jT_h, the BMWD compares the sum Z_j of the M most recent inputs with a threshold K. It is assumed that there are N_h cells observed during the observation interval duration T; it follows therefore that $M = N_h$ will make use of all the information available. The system's alarm is triggered when $Z_{j-1} = K - 1$ and $Z_j = K$. For noise alone present, the probability of this false alarm event is [3]

$$P_F = \binom{N_h - 1}{K - 1} P_0^K (1 - P_0)^{N_h - K + 1} \tag{10-36a}$$

The false-alarm rate (FAR) is defined as

$$\text{FAR} = \frac{P_F}{T_h} \tag{10-36b}$$

Either P_F or FAR will usually be specified, and P_0 must be found.

The overall probability of detection is approximately equal to the probability that there are at least K ones in the window at the instant of coincidence in time of the window with the transmission. This probability is given by

$$P_D = \sum_{i=K}^{N_h} \binom{N_h}{i} P_1^i (1 - P_1)^{N_h - i} \tag{10-37}$$

Since an energy detector is assumed for each cell, the tasks of computing P_0 and P_1 are the same as computing P_F and P_D for the radiometer. Now, however, the time–bandwidth product is that per cell and not that of the signal in the entire observation interval.

To summarize, the calculational procedure for determining the signal-to-noise ratio required by the OR/BMWD to produce a desired P_F and P_D is as follows:

1. Guess a value for K and calculate P_0 for a desired P_F or FAR from (10-36).
2. Calculate Q_F from (10-34).
3. Calculate P_1 for a desired P_D from (10-37).
4. Calculate Q_D from (10-35).
5. Find E_p/N_0, the per cell signal energy-to-noise spectral density ratio, for the Q_D found above (note that the time–bandwidth product per cell may not allow the Gaussian approximation to be used).
6. Assume another value for the second threshold, K, and repeat steps 1 to 5.
7. Choose as the optimum value of K the one that gives the minimum E_p/N_0.
8. Convert the value of E_p/N_0 found to $E_s/N_0 = N_h E_p/N_0$ or $P_s/P_n = E_p/(T_h N_0 W)$.

Note that steps 1 and 3 involve the solution of polynomial equations in P_0 and P_1, respectively. This can be done with a root finder on a calculator or computational package for a computer.

EXAMPLE 10-4

Consider detection of hybrid spread-spectrum signals with an OR/BMWD detector according to the foregoing algorithm. Several sets of signal parameters are given in Table 10-3. Values of the minimum signal-to-noise ratio required to give $P_F = 10^{-3}$ and $P_D = 0.9$ divided by the signal-to-noise ratio required by a radiometer to give the same performance, or gain, are given in decibels.

TABLE 10-3. Summary of Performance Results for an OR/BMWD Compared to a Radiometer; $T = 0.01$ s, $P_D = 0.9$, $P_F = 0.001$

$T_h W_h$	TW	N_T	N_h	Gain (dB)
100	100	1	1	0
100	500	1	5	−0.645
100	500	2	2	0.213
100	5000	4	12	0.91
100	5×10^4	25	20	3.387
100	10^5	25	40	2.739
100	10^6	10^3	10	10.765
100	10^7	10^4	10	15.289
100	10^7	10^5	1	22.57
100	10^8	10^5	10	19.894
1000	10^9	10^5	10	19.707
1000	$10^{9.5}$	10^5	31	17.92
100	10^8	10^6	1	27.337
100	10^9	10^6	10	24.445
1000	10^{10}	10^6	10	24.37
500	10^{10}	10^6	20	23.34

Note that for $N_T N_h$ small, the radiometer actually performs better than the more complex OR/BMWD. Large TW signals with large values for N_T generally give increasingly larger gains as these parameters increase.

Clearly, it becomes more and more infeasible to build a channelized receiver as the number of hopping intervals increases much beyond 100 or 1000. Thus even if a frequency-hop spread-spectrum signal hops over 10,000 bands, practical considerations would demand a channelized receiver with far fewer channels, say 1000. The optimization of hybrid spread-spectrum signals to force a potential inteceptor to use a radiometer has been considered in Refs. 18 and 19.

10-6 Estimation of Spread-Spectrum Signal Parameters

Once the presence of a spread-spectrum signal has been detected by an intercept receiver, it is of interest to determine certain signal parameters if possible. One of the simplest signal parameters to determine in the case of DSSS is the spreading code clock rate. This is examined briefly in this section. The details may be found in Refs. 20 and 21.

For simplicity, the received signal is represented as

$$r(t) = s(t) + n(t) \tag{10-38}$$

where

$$s(t) = A \sum_{k=-\infty}^{\infty} a_k p(t - kT_c) \tag{10-39}$$

In (10-39), $\{a_k\}$ is the spreading code chip sequence, each member of which takes on the values ± 1, $p(t)$ a pulse function with unit energy, A the signal amplitude, T_c the chip period, and $n(t)$ is white Gaussian noise with two-sided power sepctral density $N_0/2$. Note that a baseband signal has been assumed for simplicity, the implication being that the interceptor has been able to estimate the carrier frequency and demodulate the signal. The case of a modulated carrier can be handled by assuming complex signals.

It is well known that the clock rate of a signal can be estimated with the delayand-multiply circuit shown in Figure 10-10. Let the output of the bandpass filter at the delay-and-multiply circuit input be denoted as $\hat{r}(t)$. The output of the multiplier is then

$$y(t) = \hat{r}(t)\hat{r}(t - T_d) \tag{10-40}$$

where the parameter T_d can be chosen to optimize circuit performance.

Consider the power spectral density of $y(t)$, which under suitable stationarity assumptions can be written as

$$S_y(f) = S_s(f) + S_{s \times s}(f) + S_{s \times n}(f) + S_{n \times n}(f) \tag{10-41}$$

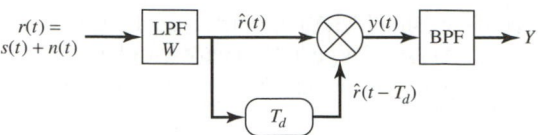

FIGURE 10-10. Delay-and-multiply circuit for estimating code chip rate of a DSSS signal.

where $S_s(f)$ consists of discrete spectral lines at harmonics of the chip rate, $S_{s \times s}(f)$ is the signal self-noise spectrum, $S_{s \times n}(f)$ is the signal × noise spectrum, and $S_{n \times n}(f)$ is the noise × noise spectrum. In writing this form for the power spectrum at the multiplier output, it is implied that the signal can be observed for a sufficiently long time so that stationarity and steady-state conditions are approximately satisfied at the multiplier output. An alternative approach to estimating signal parameters based on the theory of cyclostationary processes has been promoted by Gardner [22–25].

With the steady-state point of view adopted here, the rate line at the chip rate or its harmonic can be found by passing $y(t)$ through a bandpass filter centered at $f_c = 1/T_c$ (or one of its harmonics). Let the equivalent noise bandwidth of the filter, B, be small enough so that the various noise spectra in (10-41) (i.e., the last three terms) can be assumed constant in the vicinity of the rate line. Then the output signal-to-noise ratio is defined as

$$\text{snr}_o = \frac{S_s(f_c)}{B[S_{s \times s}(f_c) + S_{s \times n}(f_c) + S_{n \times n}(f_c)]} \tag{10-42}$$

Assuming $\{a_k\}$ to be a sequence of identically distributed, independent random variables, the various spectra appearing in (10-41) and (10-42) can be expressed as [21]

$$S_s(f) = A^4 f_c^2 \sum_{n=-\infty}^{\infty} |\hat{P}(nf_c) * \hat{P}(nf_c) e^{-j2\pi n T_d f_c}|^2 \, \delta(f - nf_c) \tag{10-43}$$

$$S_{s \times s}(f) = A^4 f_c^2 |\hat{P}(nf_c) * \hat{P}(nf_c) e^{-j2\pi f(T_d - T_c)}|^2 \tag{10-44}$$

$$S_{s \times n}(f) = A^2 N_0 f_c \int_{|f|-W}^{W} |\hat{P}(|f| - \lambda)|^2 \{1 + \cos[2\pi(|f| - 2\lambda)]\} \, d\lambda \tag{10-45}$$

and

$$S_{n \times n}(f) = \frac{1}{2} N_0^2 W \left(1 - \frac{|f|}{2W}\right) \left\{1 + \text{sinc}\left[4WT_d\left(1 - \frac{|f|}{2W}\right)\right]\right\} \tag{10-46}$$

where $P(f)$ and $\hat{P}(f)$ are the Fourier transforms of $p(t)$ and $\hat{p}(t)$, respectively, the asterisk denotes convolution, and $\text{sinc } x = (\sin \pi x)/(\pi x)$. The expression for the self noise (10-44) is obtained under the assumption that the overlap between pulses shifted by more than T_c can be neglected. Under low input signal-to-noise ratio conditions, it will be seen later that the self-noise can be neglected.

In terms of the input signal-to-noise ratio, defined as

$$\frac{A^2 T_c}{N_0} = \frac{E_c}{N_0} = \frac{E_c}{N_0 T_c f_c} = \frac{P_c}{N_0 f_c} \tag{10-47}$$

the output signal-to-noise ratio from the bandpass filter can be written as

$$\text{snr}_o = \left(\frac{E_c}{N_0}\right)^2 \frac{k_s}{B_n[k_{nn} + (E_c/N_0)k_{sn} + (E_c/N_0)^2 k_{ss}]} \tag{10-48}$$

where k_s, k_{ss}, k_{sn}, and k_{nn} are normalized power spectrum coefficients related to the spectra defined by (10-43)–(10-46) (see the problems) and $B_n = BT_c$ is a normalized bandwidth for the bandpass filter. It is apparent from (10-48) that for large E_c/N_0 the self-noise term in the denominator (i.e., the third term) dominates, whereas for small E_c/N_0 it can be neglected. Figure 10-11 shows normalized snr_o versus $W_n = WT_c$ with E_c/N_0 as a parameter for a rectangular pulse shape function. For these curves, T_d is fixed at $\frac{1}{2}$ chip. The relatively flat region between 1 and 2 for W_n indicates that a prefilter bandwidth between about f_c and $2f_c$ is optimum. Figure 10-12 again shows normalized snr_o but now plotted as a function of $T_{dn} = T_d/T_c$ with $W_n = 1.5$. These curves indicate that $T_d \approx 0.5$ is about optimum.

It is of interest to inquire about pulse shapes that will minimize the detectability of the chip-rate line [21]. It is sufficient, in general, to minimize k_s in (10-48), which is proportional to the square of

$$R = f_c \left| \int_{-\infty}^{\infty} \hat{P}(\lambda)\hat{P}(f_c - \lambda)e^{-j2\pi\lambda T_d} \, d\lambda \right| \tag{10-49}$$

Since the integrand is the product of the spectrum of the prefiltered pulse shape and its translate by f_c, it follows that any pulse shape function strictly bandlimited to $0.5 f_c$ hertz will result in the absence of a chip-rate line. Of course, such pulses

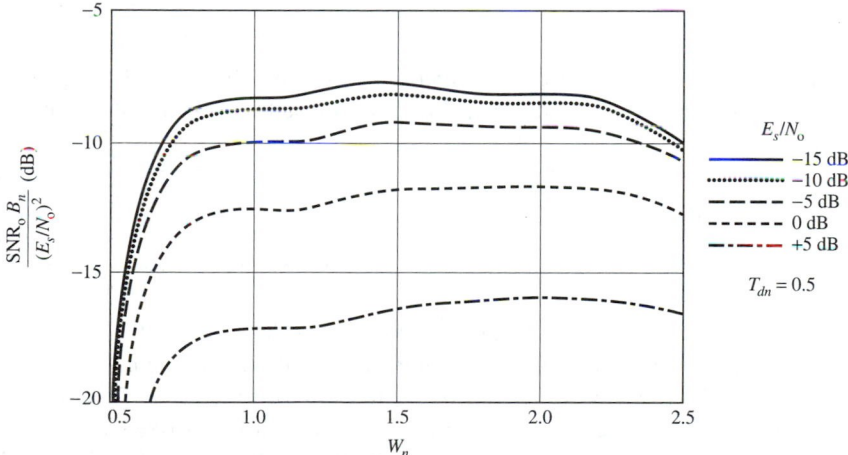

FIGURE 10-11. Normalized snr_o versus $W_n = WT_c$ for a rectangular pulse-shape function. (From Ref. 21. Copyright © 1988 IEEE. Reprinted with permission.)

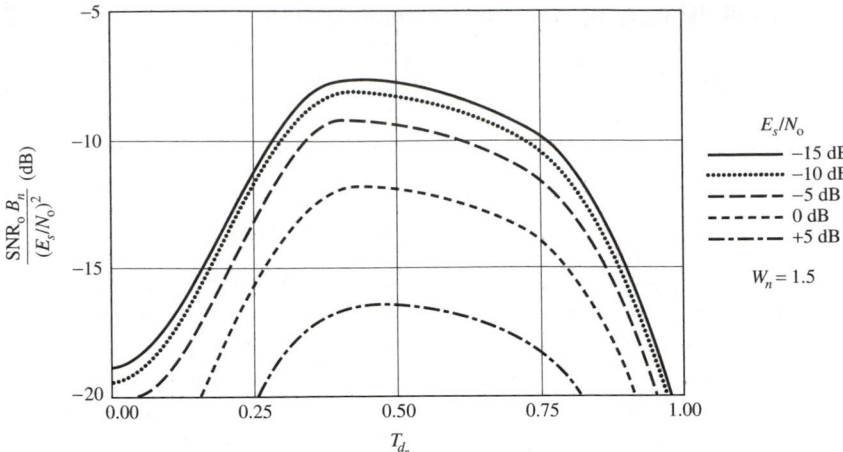

FIGURE 10-12. Normalized snr_o versus T_d/T_c for a delay-and-multiply receiver. (From Ref. 21. Copyright © 1988 IEEE. Reprinted with permission.)

are noncausal. A pulse-shape function that is more nearly realizable with a spectrum that approximates a rectangle is the Nyquist pulse with a raised cosine spectrum given by

$$
P(f) = \begin{cases}
0 & |f| > \dfrac{1 + \alpha}{2T_c}, \ |\alpha| \leq 1 \\[2ex]
a & |f| < \dfrac{1 - \alpha}{2T_c} \\[2ex]
\dfrac{a}{2}\left[1 + \sin\left(\dfrac{\pi}{\alpha}(|f|T_c - 0.5)\right)\right] & \text{otherwise}
\end{cases}
\tag{10-50}
$$

This spectrum is rectangular for $\alpha = 0$ and rolls off smoothly to zero for α increasing to unity. For this pulse spectrum, a closed-form result for R is possible, which is

$$
R = \left| \frac{\alpha}{8(1 - \alpha/4)} \text{sinc } \alpha T_d f_c + 0.5 \text{ sinc}(1 + \alpha T_d f_c) \right.
$$

$$
\left. + 0.5 \text{ sinc}(1 - \alpha T_d f_c) \right|^2 \tag{10-51}
$$

for $W \geq 0.5(1 + \alpha)f_c$, with a similar expression for $0.5f_c < W < 0.5(1 + \alpha)f_c$. If $W < 0.5f_c$, the pulses are bandlimited to the extent that no rate line exists. Figures 10-13 through 10-15 show results analogous to those of Figures 10-11 and 10-12, except for a Nyquist pulse [21]. In Figures 10-13 and 10-14, α is fixed at 1 and 0.5, respectively, $T_d = 0.5\ T_c$, and E_c/N_0 is a parameter. In Figure 10-15, $T_d = 0.5T_c$, $E_c/N_0 = -10$ dB, and α is a parameter. Figure 10-14 indicates that for $\alpha = 0.5$, it

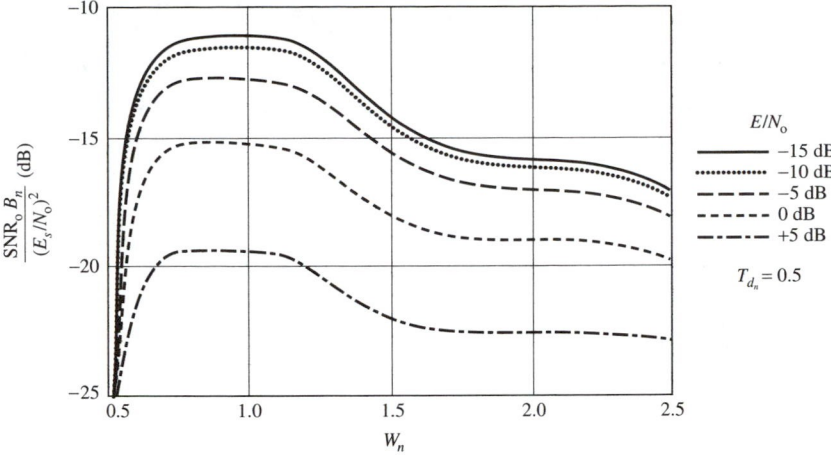

FIGURE 10-13. Normalized snr_o versus WT_c, $\alpha = 1$, and E_c/N_0 as a parameter for a delay-and-multiply receiver with Nyquist pulse shape used. (From Ref. 21. Copyright © 1988 IEEE. Reprinted with permission.)

would be very difficult for an inteceptor to optimize its predetection bandwidth due to the narrow peak in the snr_o curves at about $WT_c = 0.65$. Figure 10-15 reinforces this point, where it is seen that the snr_o curves become more peaked as α decreases— that is, as the spectrum of the pulse-shape function becomes more nearly rectangular. This is only a brief look at extraction of spread-spectrum signal parameters. For further analyses and discussion, the reader may consult Refs. 20 to 28.

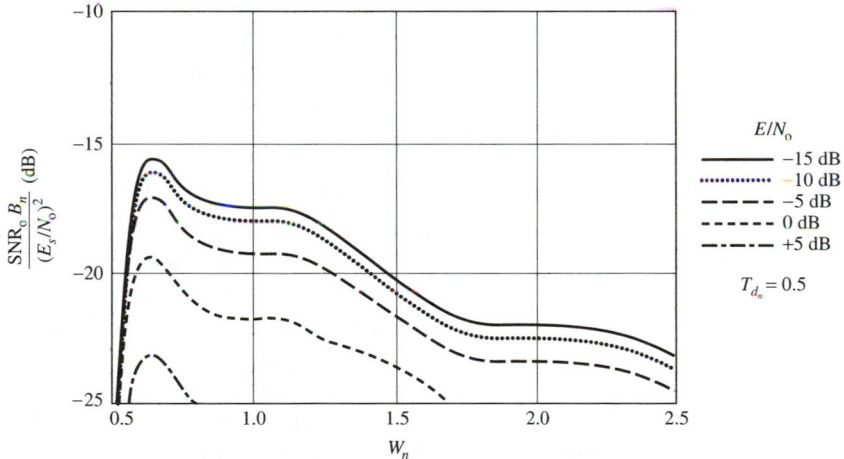

FIGURE 10-14. Normalized snr_o versus WT_c, $\alpha = 0.5$, and E_c/N_0 as a parameter for a delay-and-multiply receiver with Nyquist pulse shape used. (From Ref. 21. Copyright © 1988 IEEE. Reprinted with permission.)

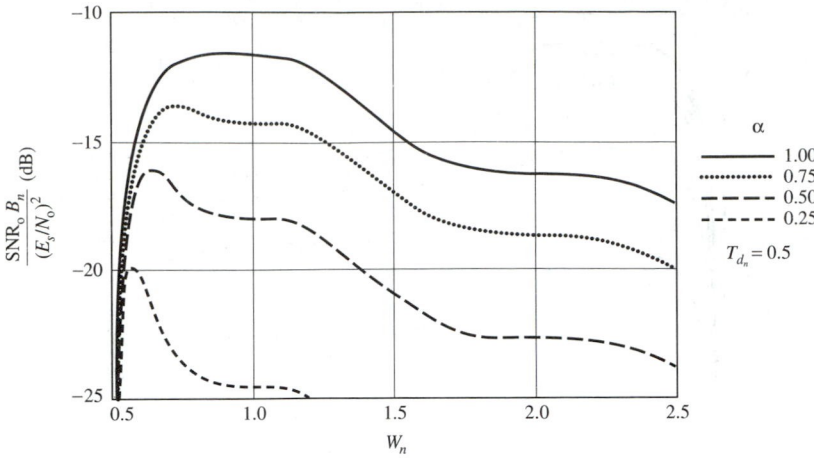

FIGURE 10-15. Normalized snr_o versus WT_c, $E_c/N_0 = -10$ dB, and α as a parameter for a delay-and-multiply receiver with Nyquist pulse shape used. (From Ref. 21. Copyright © 1988 IEEE. Reprinted with permission.)

10-7 Summary

The purpose of this chapter was to investigate the detection and parameter estimation of spread-spectrum signals from the viewpoint of an interceptor that has only partial knowledge of the signal parameters. After discussing the general nature of the covert communications problem, the simplest detector for unknown signals, the energy detector or radiometer, was described and characterized in terms of probabilities of detection, P_D, and false alarm, P_F. An important signal parameter for evaluating these probabilities is the time–bandwidth product, TW. For large TW, Gaussian statistics at the detector output are approximately valid, and P_D and P_F can be expressed in terms of Gaussian Q-functions. A plot of P_D versus P_F, termed the receiver operating characteristic (ROC), can be used to completely characterize the performance of the receiver; the ROC of the radiometer was given for large time–bandwidth product signals. For smaller values of TW (say, 1000 or less), cumulative central and noncentral chi-squared distributions are required to evaluate the ROC. For sufficiently small values of TW (say 125 or less) the nomogram of Chapter 5 can be used. Intermediate values of TW require the numerical evaluation of the central and noncentral chi-squared distributions. A numerical routine is described in Appendix F for evaluating chi-squared distributions for any value of TW.

After discussion of the energy detector, optimum intercept receivers for DSSS signals were discussed. To achieve an upper bound for the performance of such receivers, it was assumed that both carrier phase and chip clock were available but that the actual chip sequence was unknown (termed the synchronous coherent detector). This detector shows a 3-dB loss relative to the radiometer. Next the receiver that did not require carrier coherence was considered but did require a chip-coherent clock (termed the synchronous noncoherent detector), and it was found to operate at a 1.5 dB loss relative to the synchronous coherent detector. Finally, an estimate of between 0.63 and 1.42 dB loss if a nonsynchronous detector was used was given on the basis of results quoted in Ref. 5.

The optimum intercept receiver for FHSS was discussed next. It consists of energy detection over each hop time for each hopping frequency slot. The results of these detections are summed and the product of all hopping intervals in the observation interval taken. For large TW signals, this received structure is clearly infeasible. However, for large TW signals with a large number of hop intervals, the performance of this receiver is greatly superior to that of a radiometer. As a more practical approximation to the optimum intercept receiver for FHSS, the binary moving-window detector with input from ORed energy detectors was considered. Although its analysis is complex, it can give performance approaching that of the optimum FHSS intercept receiver for sufficiently large TW and number of hopping intervals.

The last topic discussed was the estimation of spread-spectrum signal parameters. Only the case of code clock rate was discussed. A standard circuit for performing the estimate of code clock rate is a delay-and-multiply. It was characterized in terms of output signal-to-noise ratio. By suitably choosing the pulse shape function, the performance of this detector can be minimized (i.e., the communicator can defeat this estimation scheme by optimizing its pulse shape). Since the delay-and-multiply circuit is really a square-law device, it is conceivable that the intercept detector could counteract any pulse shaping used on the part of the communicator to defeat it by using a higher-order detector law. This indeed can be demonstrated but is left to the references [20,29].

References

[1] M. K. Simon, J. K. Omura, R. A. Scholtz, and B. K. Levitt, *Spread Spectrum Communications*, Vol. III (Rockville, Md.: Computer Science Press, 1985), Chap. 4.

[2] D. L. Nicholson, *Spread Spectrum Signal Design* (Rockville, Md.: Computer Science Press, 1987).

[3] R. A. Dillard and G. M. Dillard, *Detectability of Spread-Spectrum Signals* (Norwood, Mass.: Artech House, 1989).

[4] L. L. Gutman, "System Quality Factors for LPI Communications," *IEEE AES Mag.*, pp. 25–28, December 1989.

[5] A. Polydoros and C. L. Weber, "Detection Performance Considerations for Direct-Sequence and Time-Hopping LPI Waveforms," *IEEE J. Selected Areas Commun.*, Vol. SAC-3, pp. 727–744, September 1985.

[6] H. Urkowitz, "Energy Detection of Unknown Deterministic Signals," *Proc. IEEE*, Vol. 55, pp. 523–531, April 1967.

[7] R. C. Emerson, "First Probability Densities for Receivers with Square Law Detectors," *J. Appl. Phys.*, Vol. 24, pp. 1168–1176, September 1953.

[8] R. A. DILLARD, "Detectability of Spread-Spectrum Signals," *IEEE Trans. Aerosp. Electron. Syst.,* Vol. AES-15, pp. 526–537, July 1979.

[9] D. P. MEYER and H. A. MAYER, *Radar Target Detection: Handbook of Theory and Practice* (New York: Academic Press, 1973), p. 21 ff.

[10] G. M. DILLARD, "Recursive Computation of the Generalized Q-Function," *IEEE Trans. Aerosp. Electron. Syst.,* Vol. AES-9, pp. 614–615, July 1973.

[11] C. W. HELSTROM, "Approximate Evaluation of Detection Probabilities in Radar and Optical Communications," *IEEE Trans. Aerosp. Electron. Syst.,* Vol. AES-14, pp. 630–640, July 1978.

[12] S. PARL, "A New Method of Calculating the Generalized Q-Function," *IEEE Trans. Inf. Theory,* Vol. IT-26, pp. 121–124, January 1980.

[13] M. ABRAMOWITZ and I. STEGUN, eds., *Handbook of Mathematical Functions* (New York: Dover, 1972).

[14] D. G. WOODRING, *Performance of Optimum and Suboptimum Detectors for Spread Spectrum Waveforms,* Naval Research Laboratory Technical Report 8432, December 1980.

[15] H. L. VAN TREES, *Detection, Estimation, and Modulation Theory* (New York: Wiley, 1968).

[16] N. F. KRASNER, "Optimal Detection of Digitally Modulated Signals," *IEEE Trans. Commun.,* Vol. COM-30, pp. 885–895, May 1982.

[17] A. POLYDOROS and C. L. WEBER, "Optimal Detection Considerations for Low Probability of Intercept," *Milcom '82 Proc.,* pp. 2.1.1–2.1.5, October 1982.

[18] R. E. ZIEMER and R. L. PETERSON, "Low Probability of Detection Tradeoff Design in DS/FH Modems," *Milcom '82 Proc.,* pp. 24.6.1–24.6.6, October 1982.

[19] R. E. ZIEMER and J. M. LIEBETREU, "Spread-Spectrum Signal Design for Low Probability of Intercept Communications," *IEEE Int. Commun. Conf. Rec.,* pp. F6.81–F6.85, June 1983.

[20] D. E. REED, "The Performance of Rate-Line Generation Circuits for Determining the Symbol Rate of Weak and Bandlimited Digitally Modulated Signals," Ph.D. dissertation, University of Colorado at Colorado Springs, 1988.

[21] D. E. REED and M. A. WICKERT, "Minimization of Detection of Symbol-Rate Spectral Lines by Delay and Multiply Receivers," *IEEE Trans. Commun.,* Vol. COM-36, pp. 118–120, January 1988.

[22] W. A. GARDNER, "A Unifying View of Second-Order Measures of Quality for Signal Classification," *IEEE Trans. Commun.,* Vol. COM-28, pp. 807–816, June 1980.

[23] W. A. GARDNER, "Structural Characterization of Locally Optimum Detectors in Terms of Locally Optimum Estimators and Correlators," *IEEE Trans. Inf. Theory,* Vol. IT-28, pp. 924–932, November 1982.

[24] W. A. GARDNER, "Signal Interception: A Unifying Theoretical Framework for Feature Detection," *IEEE Trans. Commun.,* Vol. COM-36, pp. 897–906, November 1982.

[25] W. A. GARDNER and C. M. SPOONER, "Signal Interception: Performance Advantages of Cyclic-Feature Extraction," *IEEE Trans. Commun.,* Vol. COM-40, pp. 149–159, January 1992.

[26] S. HINEDI and A. POLYDOROS, "DS/LPI Autocorrelation Detection in Noise Plus Random-Tone Interference," *IEEE Trans. Commun.,* Vol. COM-38, pp. 805–817, June 1990.

[27] N. C. BEAULIEU, W. L. HOPKINS, and P. J. MCLANE, "Interception of Frequency-Hopped Signals," *IEEE J. Selected Areas Commun.,* Vol. SAC-8, pp. 853–870, June 1990.

[28] J. F. KUEHLS and E. GERANIOTIS, "Presence Detection of Binary-Phase-Shift-Keyed and Direct-Sequence Spread-Spectrum Signals," *IEEE J. Selected Areas Commun.,* Vol. SAC-8, pp. 915–933, June 1990.

[29] D. E. REED and M. A. WICKERT, "Nonstationary Moments of a Random Binary Pulse Train," *IEEE Trans. Inf. Theory,* Vol. IT-35, pp. 700–703, May 1989.

Problems

(10-1) Compare the values of K_0 and S found by using the nomogram of Figure 5-11 and the approximation given by (10-6) and (10-7) for $P_D = 0.7$, $P_F = 0.01$, and $TW = 10$ and 100.

(10-2) Derive the simplification of (10-18) given by (10-19). Give an explicit expression for λ_0.

(10-3) Obtain (10-20a) assuming that N is sufficiently large in (10-19) to allow λ to be approximated as Gaussian.

(10-4) A direct-sequence spread-spectrum BPSK signal with a chip rate of 100 Mchips/s is observed for 1 ms. A receiver operating point of $P_D = 0.8$ and $P_F = 0.01$ is desired. Find the required P_s/P_n, where P_n is measured in the DSSS spread bandwidth, to provide this perfomance level if the intercept receiver is **(a)** a radiometer; **(b)** a synchronous coherent detector for DSSS; **(c)** a noncoherent chip detector for DSSS. Repeat for an observation interval of 0.1 ms. Make up a table like Table 10-2, summarizing the results.

(10-5) Derive (10-33). Use a computer mathematics package to verify the curves of Figure 10-8.

(10-6) Referring to (10-36a), plot P_F versus P_0 for $N_h = 4, 10, 100,$ and 1000 and $K = N_h/2 - 0.1N_h$, N_h, and $N_h/2 + 0.1N_h$ (for $N_h = 4$, use $K = 1, 2,$ and 3). Use Stirling's approximation for the factorials for $N_h = 1000$. [*Note:* Stirling's approximation is $n! \cong (2\pi n)^{1/2}(n/e)^n$.]

(10-7) **(a)** Referring to (10-37), plot P_D versus P_1 for $N_h = 4$ and $K = 1, 2,$ and 3.
(b) Repeat for $N_h = 10$ and $K = 4, 5,$ and 6.

(10-8) **(a)** Show that (10-34) and (10-35) when solved for Q_F and Q_D yield

$$Q_F = 1 - (1 - P_0)^{1/N_T}$$

$$Q_D = 1 - \frac{1 - P_1}{(1 - P_0)^{1 - 1/N_T}}$$

(b) What are the conditions on P_0 and P_1 such that Q_F and Q_D are proper probabilities?

(10-9) Write a computer mathematics package program to verify the gain values given in Table 10-3.

(10-10) **(a)** Derive (10-48), relating the constants k_s, k_{nn}, k_{sn}, and k_{ss} to the spectral densities in (10-42).
(b) Obtain specific expressions for these constants for a rectangular pulse-shape function and thereby verify the curves shown in Figure 10-11.

(10-11) Show that (10-51) follows from (10-49) for the Nyquist pulse with spectrum given by (10-50).

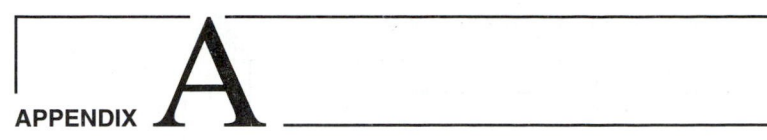

Summary of Phase-Locked Theory

A-1 Introduction

In Chapter 4 it was seen that code tracking loops had the structure of a feedback control system similar to a phase-locked loop. This appendix gives a summary of the features of phase-locked loops important to considering code tracking loop fundamentals. In addition, phase-locked loops play important roles in establishing coherent references in digital communication systems. References 1 to 4 are comprehensive treatments of phase-locked-loop theory, and Refs. 5 and 6 address coherent reference generation.

A-2 Phase-Locked-Loop Models and Characteristics of Operation

Since phase-locked loops play important roles in establishing coherent references in digital communication systems, their properties are described here.

A-2.1 Synchronized Mode: Linear Operation

The block diagram for a phase-locked loop of arbitrary order is shown in Figure A-1. It consists of a phase detector whose output is a monotonic function of the phase difference between the input signal and the reference input, a loop filter with transfer function $F(s)$, and a voltage-controlled oscillator which produces the reference signal, $e_o(t)$. The input signal is represented as

$$x_c(t) = A_c \cos(2\pi f_0 t + \phi) \tag{A-1}$$

and the voltage-controlled oscillator (VCO) output, or reference signal, is represented as†

$$e_o(t) = -A_v \sin(2\pi f_0 t + \theta) \tag{A-2}$$

The frequency deviation of the VCO output is proportional to its input, $e_v(t)$;

† It is initially assumed that the loop is operating in the frequency-synchronized mode; that is, only the phase of the VCO must be synchronized with the input signal phase.

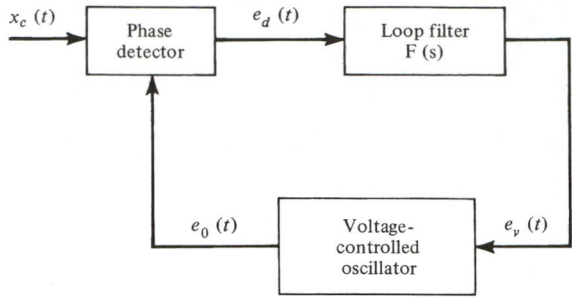

FIGURE A-1. General-order phase-locked-loop block diagram.

that is

$$\frac{d\theta}{dt} = K_v e_v(t) \tag{A-3}$$

where K_v is the VCO constant in rad/s/V.

If the phase detector is assumed to be an ideal multiplier followed by a lowpass filter whose sole effect is to remove the double-frequency component at the multiplier output, the phase detector output is

$$e_d(\psi) = K_d \sin \psi \tag{A-4}$$

where

$$\psi = \phi - \theta \tag{A-5}$$

is the phase error and K_d is a proportionality constant. For the sinusoidal phase detector, $K_d = \frac{1}{2} A_c A_v K_m$, where K_m is the multiplier constant. The phase detector characteristic given by (A-4) is illustrated in Figure A-2 together with several other possible phase detector characteristics. If the phase error is small, the sinusoidal phase detector characteristic shown in Figure A-2a is linear to a good approximation. Therefore, if $|\psi| \ll 1$, all the phase detectors with characteristics shown in Figure A-2 have approximately the same effect on loop operation. If the phase error is large, all impose nonlinear effects on system operation. Such nonlinear behavior will be discussed in detail later, but for now loop operation is assumed to be entirely within one of the linear regions with positive slope. These can be shown to be stable lock-point regions for a first-order loop which has $F(s) = 1$ by employing phase-plane arguments [7].

With the foregoing definitions and assuming operation in the linear mode, the equations describing loop operation will now be obtained. It is convenient to do so using Laplace transform notation and by considering the signal phase as the signal of interest. A loop model using Laplace-transformed quantities and assuming linear operation is shown in Figure A-3. The Laplace-transformed loop equations are

$$E_d(s) = K_d[\Phi(s) - \Theta(s)] = K_d \Psi(s) \tag{A-6}$$

$$E_v(s) = F(s)E_d(s) \tag{A-7}$$

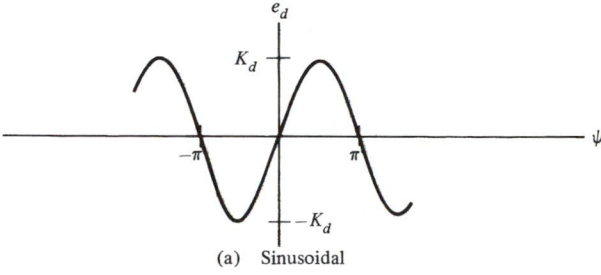

(a) Sinusoidal

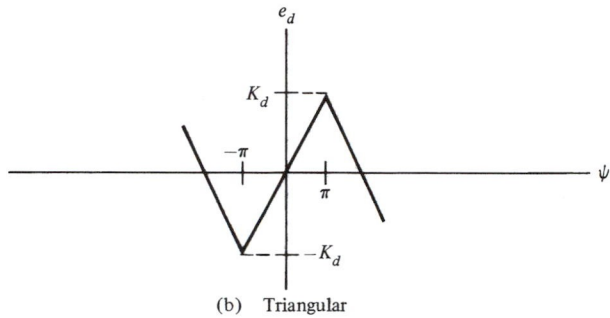

(b) Triangular

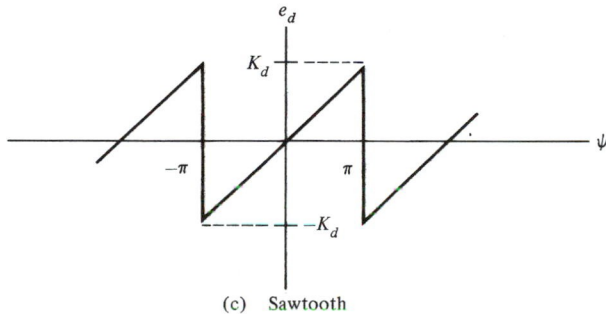

(c) Sawtooth

FIGURE A-2. Phase detector characteristics.

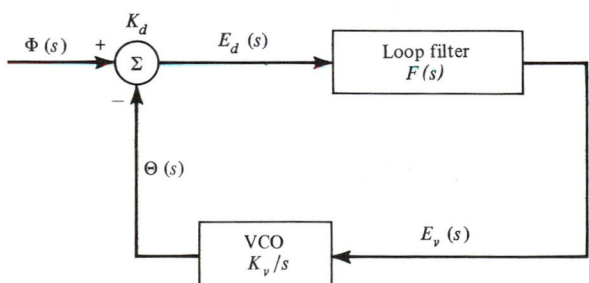

FIGURE A-3. Laplace-transformed phase-locked-loop model for operation in linear mode.

$$\theta(s) = \frac{K_v E_v(s)}{s} \qquad (A\text{-}8)$$

The following ratios of Laplace-transformed quantities, or transfer functions, relating to loop operation in the synchronized mode may be solved for, and are frequently used:

1. The closed-loop transfer function:

$$H(s) \triangleq \frac{\Theta(s)}{\Phi(s)} = \frac{K_v K_d F(s)}{s + K_v K_d F(s)} \qquad (A\text{-}9)$$

2. The phase error transfer function:

$$H_e(s) \triangleq \frac{\Phi(s) - \Theta(s)}{\Phi(s)} = \frac{\Psi(s)}{\Phi(s)}$$

$$= 1 - \frac{\Theta(s)}{\Phi(s)} \qquad (A\text{-}10)$$

$$= 1 - H(s) = \frac{s}{s + K_v K_d F(s)}$$

3. The VCO control-voltage/input-phase transfer function:

$$H_v(s) = \frac{v_c(s)}{\Phi(s)}$$

$$= \frac{sH(s)}{K_v} \qquad (A\text{-}11)$$

$$= \frac{K_d s F(s)}{s + K_v K_d F(s)}$$

It is convenient to write the closed-loop transfer function in terms of the open-loop transfer function, which is defined as

$$G(s) = \frac{K_v K_d F(s)}{s} \qquad (A\text{-}12)$$

Substituting (A-12) into (A-9) results in

$$H(s) = \frac{G(s)}{1 + G(s)} \qquad (A\text{-}13)$$

The open-loop dc gain is defined as

$$K = K_v K_d \qquad (A\text{-}14)$$

which is a generalization of the total effective loop gain of the first-order loop.

By appropriate choice of $F(s)$, any order closed-loop transfer function can be obtained. Consideration here will be restricted to first- and second-order loops. Various types of loop filters for second-order loops are employed. Circuit diagrams for two of these types are illustrated in Figure A-4. For second-order loops, it is customary to express the denominator of the closed-loop transfer function in terms of the damping factor, ζ, and natural frequency, ω_n, as

$$D(s) = s^2 + 2\zeta\omega_n s + \omega_n^2 \tag{A-15}$$

The closed-loop transfer functions and noise equivalent bandwidths for the first- and second-order loops are summarized in Table A-1.

The closed-loop frequency response for a second-order loop with active filter is shown in Figure A-5 for several values of damping factor, ζ. The frequency response corresponding to its phase-error transfer function is shown in Figure A-6 for $\zeta = 0.707$. In terms of its effect on input phase, Figure A-5 shows that a phase-locked loop performs a lowpass filtering operation. In the application to FM demodulation, the loop bandwidth is made large in order that $\theta(t)$ closely tracks $\phi(t)$, thus making the VCO input proportional (or nearly so) to the modulating signal. This follows from the defining equation for the VCO (A-3). Conversely, when establishing a coherent reference for digital data demodulation, it is desirable to have the loop bandwidth narrow to minimize the effects of input noise to the loop on $\phi(t)$ in terms of phase jitter. The limitation on how narrow the loop bandwidth can be made is determined by the amount of phase-jitter noise on the carrier to which the loop is being locked. These points are examined further in the following sections.

A-2.2 Effects of Noise

The input to the linear phase-locked loop of Figure A-3 will now be assumed to be signal plus stationary, bandlimited, Gaussian noise,

$$x_r(t) = x_c(t) + n(t) \tag{A-16}$$

where $x_o(t)$ is given by (A-1) and $n(t)$ is represented in phase/quadrature form as

$$n(t) = n_c(t) \cos 2\pi f_o t - n_s(t) \sin 2\pi f_0 t \tag{A-17}$$

Using the VCO output signal representation (A-2) and the loop parameters defined previously, it can be shown that the output of an ideal multiplier-type phase detector

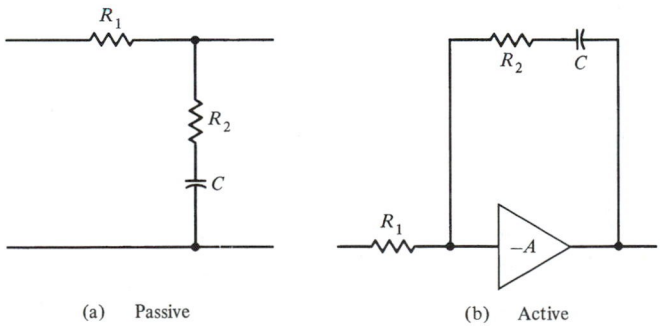

(a) Passive   (b) Active

FIGURE A-4. Second-order phase-locked-loop filters.

TABLE A-1. Transfer Functions and Parameters for First- and Second-Order Phase-Locked Loops

Loop Filter, $F(s)$	Natural Frequency,[a] ω_n (rad/s)	Damping Factor ζ	Closed-Loop Transfer Function, $H(s)$	Error Transfer Function, $1 - H(s)$	Single-Sided Noise/Equivalent Bandwidth[b,c] (Hz)
1 (first order)	K	—	$\dfrac{K}{s + K}$	$\dfrac{s}{s + K}$	$\dfrac{K}{4}$
$\dfrac{s\tau_2 + 1}{s\tau_1 + 1}$ (passive, second order)	$\sqrt{\dfrac{K}{\tau_1}}$	$\dfrac{\omega_n}{2}(\tau_2 + K^{-1})$	$\dfrac{(2\zeta\omega_n - \omega_n^2/K)s + \omega_n^2}{D(s)}$	$\dfrac{s^2 + \omega_n^2 s/K}{D(s)}$	$\dfrac{K\tau_2(1/\tau_2^2 + K/\tau_1)}{4(K + 1/\tau_2)}$
$\dfrac{s\tau_2 + 1}{s\tau_1}$ (active, second order)	$\sqrt{\dfrac{K}{\tau_1}}$	$\dfrac{\tau_2\omega_n}{2}$	$\dfrac{2\zeta\omega_n s + \omega_n^2}{D(s)}$	$\dfrac{s^2}{D(s)}$	$\dfrac{1}{2}\omega_n\left(\zeta + \dfrac{1}{4\zeta}\right)$
$\dfrac{1}{s\tau + 1}$ (lag, second order)	$\sqrt{\dfrac{K}{\tau}}$	$\dfrac{1}{2\sqrt{K\tau}}$	$\dfrac{\omega_n^2}{D(s)}$	$\dfrac{s^2 + 2\zeta\omega_n}{D(s)}$	$\dfrac{K}{4}$

[a] $K = K_v K_d$.

[b] The noise equivalent bandwidth of a filter with transfer function $H(f)$ and maximum gain H_0 is given by $B_N = (1/H_0^2)\displaystyle\int_0^\infty |H(f)|^2 \, df$.

[c] For a second-order loop with $\zeta = 0.5$, $B_L = 0.5\omega_n$; with $\zeta = 1/\sqrt{2}$, $B_L = 0.53\omega_n$. B_L is the single-sided noise equivalent bandwidth in hertz, and the dimensions of ω_n are rad/s.

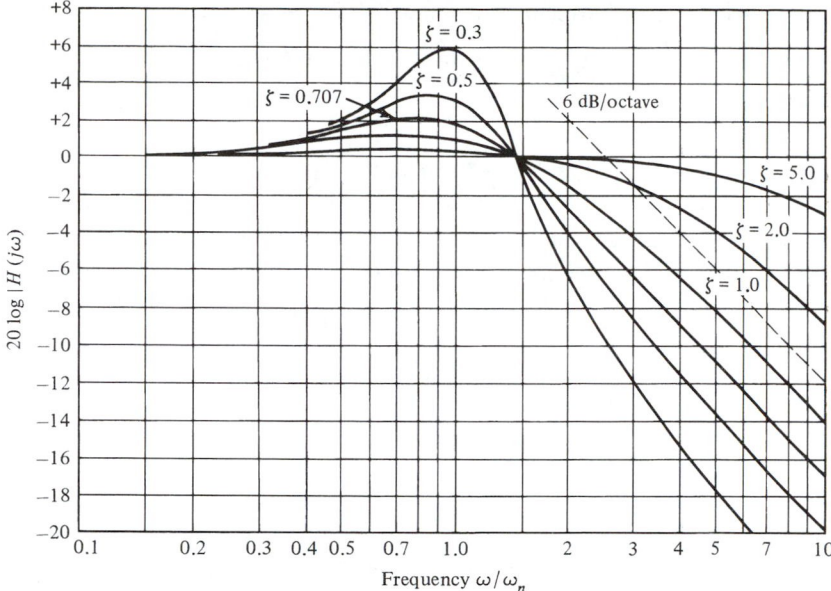

FIGURE A-5.   Frequency response of a high-gain second-order loop. (Reproduced from Ref. 1 with permission.)

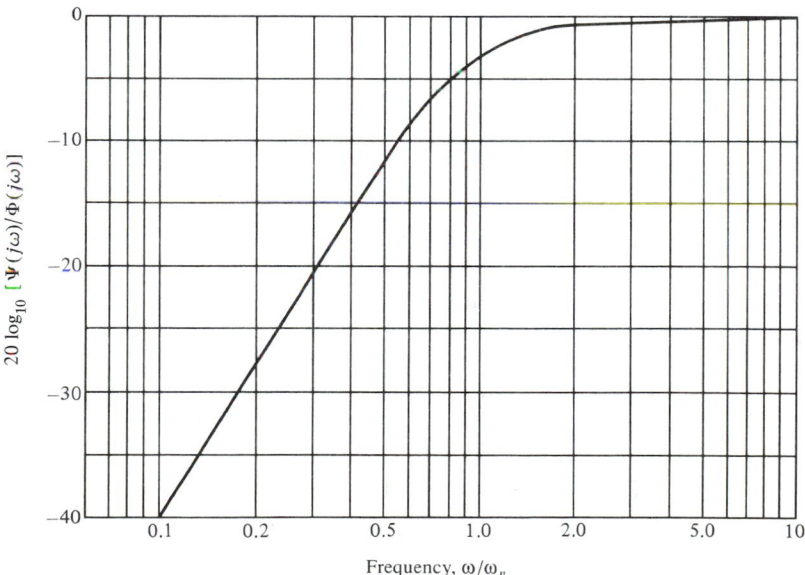

FIGURE A-6.   Error response of high-gain loop, $\zeta = 0.707$. (Reproduced from Ref. 1 with permission.)

(sinusoidal characteristic as shown in Figure A-2a is [1]

$$e_d(t) = K_d[\sin(\phi - \theta) - n'(t)] \tag{A-18}$$

where

$$n'(t) = \frac{n_c(t)}{A_c} \cos\theta - \frac{n_s(t)}{A_c} \sin\theta \tag{A-19}$$

Thus the noise-equivalent model for an ideal multiplier-type phase detector is as shown in Figure A-7. Furthermore, if the single-sided noise spectral density of $n(t)$ is N_0, the single-sided noise spectral density of $n'(t)$ can be shown to be

$$S_{n'}(f) = \frac{2N_0}{A_c^2} \qquad f \le \frac{B}{2} \tag{A-20}$$

If the noise bandwidth of $n(t)$ is B hertz (single sided), the variance of the input noise is

$$\sigma_n^2 = N_0 B \tag{A-21}$$

and that of $n'(t)$, with single-sided bandwidth $B/2$, is

$$\sigma_{n'}^2 = \frac{N_0 B}{A_C^2} = \frac{\sigma_n^2}{A_c^2} \tag{A-22}$$

The only assumption made in deriving $n'(t)$ is that the VCO phase, $\theta(t)$, is very slowly varying (ideally time-invariant, but arbitrary). Linearity of the phase detector has not been imposed. Thus, since it is linearly dependent on $n_c(t)$ and $n_s(t)$, which are Gaussian, $n'(t)$ is Gaussian, assuming that θ is constant or very slowly varying.

If the input noise to the loop is sufficiently small, the $\sin(\phi - \theta)$ operation in Figure A-7 can be replaced by $(\phi - \theta)$, and the appropriate closed-loop model with noise at the input is then as shown in Figure A-8. Since the equivalent noise, $n'(t)$, is additive at the input, the variance of the VCO output phase is

$$\sigma_\theta^2 = S_{n'}(0)B_L$$
$$= \frac{2N_0 B_L}{A_c^2} \tag{A-23}$$

where B_L is the single-sided equivalent noise bandwidth of the closed loop. For the

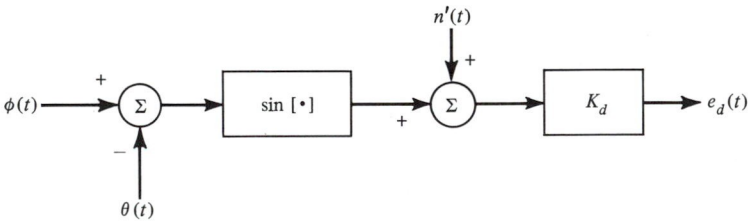

FIGURE A-7. Noise-equivalent model for sinusoidal phase detector.

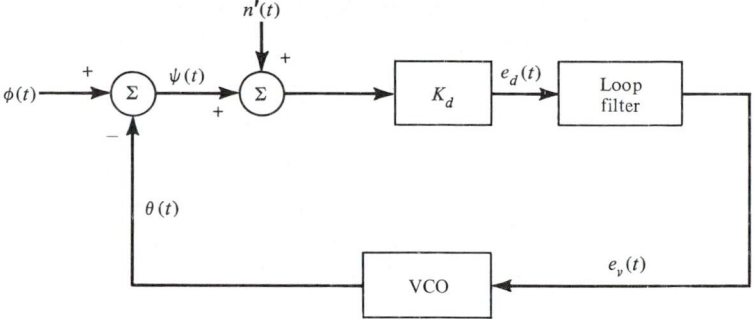

FIGURE A-8. Linear model for phase-locked loop with additive noise present at the input.

second-order loop with active filter, the ratio of equivalent noise bandwidth to natural frequency from Table A-1 is

$$\frac{B_L}{\omega_n} = \frac{\zeta + \zeta/4}{2} \tag{A-24}$$

which has a minimum value of $\frac{1}{2}$ for $\zeta = \frac{1}{2}$, giving a minimum VCO phase variance due to additive input noise of

$$\sigma_{\theta,\min}^2 = \frac{N_0 \omega_n}{A_c^2} = \frac{2N_0 B_L}{A_c^2} \qquad \text{(second-order loop, active filter, } \zeta = \frac{1}{2}\text{)} \tag{A-25}$$

A damping factor of $\zeta = 1/\sqrt{2} = 0.707$, which is often used due to transient response considerations, gives a VCO phase variance due to noise which differs from (A-25) by only 6%.

The signal-to-noise ratio at the loop input, with noise measured in a loop bandwidth, is

$$\rho = (\text{SNR})_L = \frac{A_c^2}{2N_0 B_L} \tag{A-26}$$

In terms of ρ, σ_θ^2 is given by

$$\sigma_\theta^2 = \frac{1}{\rho} \tag{A-27}$$

a result that was derived assuming operation in the linear region of the phase detector characteristic.

An exact analysis for the variance of the VCO phase due to noise has been carried out only for the first-order phase-locked loop assuming no frequency offset and no modulation on the carrier. This result is shown in Figure A-9 together with (A-25) from the linearized analysis. The method used to solve the nonlinear problem, known as the Fokker–Planck technique, gives a probability density function for the phase error of the form

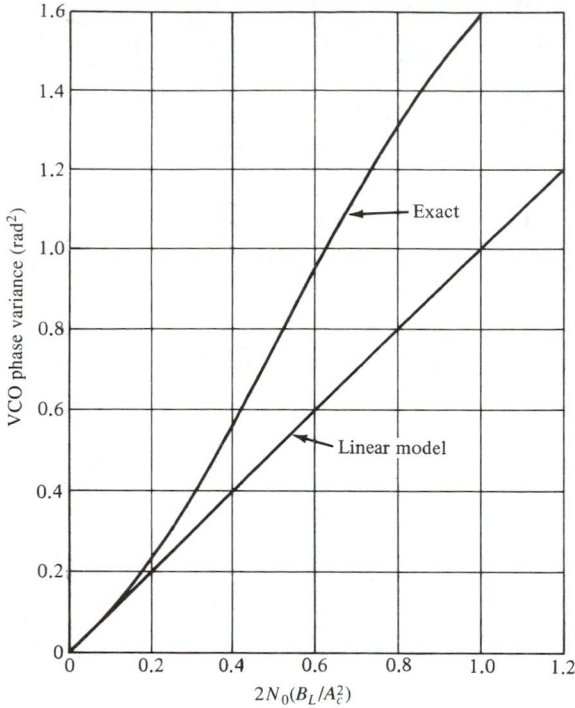

FIGURE A-9. Comparison of exact and approximate values for first-order PLL phase error variance. (Reproduced from Ref. 2 with permission.)

$$p(\psi) = \frac{\exp(\rho \cos \psi)}{2\pi I_0(\rho)} \qquad |\psi| \le \pi \qquad \text{(A-28)}$$

where $I_0(\cdot)$ is the modified Bessel function of order zero. For ρ large, it can be shown by using the asymptotic formula $I_0(\rho) \cong \exp(\rho)/\sqrt{2\pi\rho}$ that $p(\psi)$ tends to a Gaussian density function with zero mean and variance σ_θ^2 [2].

Other approximate noise analysis methods have been devised for analyzing the behavior of phase-locked loops operating into the nonlinear region. These methods deal only with second-order statistics such as variance, and do not take into account cycle slipping. Again, appealing to the Fokker-Planck analysis results for the first-order phase-locked loop, it has been shown [2] that the average time between cycle slips (i.e., the average time for the loop phase error to reach $\pm 2\pi$ after starting initially at zero) is given by

$$T_{\text{av}} = \frac{\pi^2 \rho I_0^2(\rho)}{2B_L}$$

$$\cong \frac{\pi}{4B_L} \exp(2\rho) \qquad \rho \gg 1 \qquad \text{(A-29)}$$

In addition, the probability distribution of the time between slips is exponential;

that is,

$$P(T) = 1 - \exp\left(\frac{-T}{T_{av}}\right) \tag{A-30}$$

where T is the time to a slip given the loop started with zero error.

These results apply exactly to a first-order loop with a sinusoidal phase detector characteristic operating in additive white Gaussian noise. Exact Fokker–Planck solutions for the second-order loop have not been obtained. However, experimental measurements and approximate nonlinear analyses for the second-order loop show that the exact nonlinear analysis results for the first-order loop are in close agreement with the second-order loop results for $(SNR)_L > 0$ dB.

A-2.3 Phase-Locked-Loop Tracking of Oscillators with Phase Noise

The effect of oscillator phase jitter on the phase error of a second-order phase-locked loop with active filter and damping factor $\zeta = 1/\sqrt{2}$ will now be analyzed. To find the variance of the loop phase error, σ_ψ^2, due to the phase-locked loop tracking an oscillator with phase-jitter power spectral density $G_\phi(f)$, the relationship for the output noise variance of a linear system in terms of input noise spectral density and system frequency response function will be used. In the present context, however, the input noise spectral density is $G_\phi(f)$ and the system frequency response function is the loop-phase-error frequency response, $1 - \tilde{H}(f)$, where the tilde denotes $H(s)|_{s=j2\pi f}$. Thus the loop-phase-error variance is†

$$\sigma_\psi^2 = \int_0^\infty G_\phi(f)|1 - \tilde{H}(f)|^2 \, df \tag{A-31}$$

where the lower limit is zero since $G_\phi(f)$ is a single-sided power spectral density. From Table A-1 it follows that

$$|1 - \tilde{H}(f)|^2 = |1 - H(s)|^2_{s=j2\pi f, \zeta=1\sqrt{2}}$$

$$= \left| \frac{(j\pi f)^2}{(j2\pi f)^2 + 2\zeta\omega_n(j2\pi f) + \omega_n^2} \right|^2_{\zeta=1/\sqrt{2}} \tag{A-32}$$

$$= \frac{(f/f_n)^4}{1 + (f/f_n)^4}$$

with $f_n \triangleq \omega_n/2\pi$. To carry out the evaluation of (A-31), $G_\phi(f)$ is represented in terms of the asymptotic expression

$$G_\phi(f) = \begin{cases} \dfrac{k_1}{f^3} + \dfrac{k_2}{f^2} + k_4 & f \le f_m \\[2ex] \dfrac{k_1}{f^3} + \dfrac{k_2}{f^2} & f > f_m \end{cases} \tag{A-33}$$

† The phase-noise spectra are assumed to be single-sided and single-sideband in this appendix.

which is based on the theory of oscillator noise [5,6]. The loop-phase-error variance due to phase jitter on the input signal can be written

$$\sigma_\psi^2 = \int_0^\infty \frac{f_n^{-2} k_1 x}{1 + x^4} \, dx + \int_0^\infty \frac{f_n^{-1} k_2 x^2}{1 + x^4} \, dx + \int_0^{f_m/f_n} \frac{k_1 x^4 f_n}{1 + x^4} \, dx \qquad \text{(A-34a)}$$

Carrying out the integration, the phase-error variance can be expressed as [7]

$$\sigma_\psi^2 = \frac{k_1 \pi^3}{\omega_n^2} + \frac{k_2 \pi^2}{\sqrt{2}\omega_n} + k_4 f_m \qquad \text{(A-34b)}$$

EXAMPLE A-1

Evaluate σ_ψ^2 using the values $k_1 = 10^{-3}$, $k_2 = 10^{-5}$, and $k_4 = 10^{-16}$ for a loop bandwidth of 1 Hz.

Solution: From Table A-1, if $\zeta = 1/\sqrt{2}$, the loop bandwidth is

$$B_L = \frac{1}{2}\omega_n\left(\zeta + \frac{1}{4\zeta}\right) = 0.53\omega_n \qquad \text{(A-35)}$$

For $B_L = 1$ Hz, $\omega_n = 1.89$ rad/s. Therefore,

$$\sigma_\psi^2 = \frac{(10^{-3})\pi^3}{(1.89)^2} + \frac{(10^{-5})\pi^2}{\sqrt{2}(1.89)} + (10^{-16})(10^9)$$

$$= 8.7 \times 10^{-3} + 3.71 \times 10^{-5} + 10^{-7} = 8.72 \times 10^{-3} \text{ rad}^2$$

or $\sigma_\psi = 0.094$ rad.

A-2.4 Phase Jitter Plus Noise Effects

From (A-23) it is seen that the VCO phase variance due to additive noise at the loop input increases linearly with B_L. For a constant input phase, this translates directly to a linear increase in phase-error variance with B_L due to noise. On the other hand, (A-34b) shows that the phase-error variance due to phase jitter on the input signal decreases with increasing ω_n and B_L. Therefore, an optimum value of B_L exists which provides a minimum in the phase-error variance due to both additive input noise and phase jitter on the input signal. This optimum value is illustrated by Figure A-10.

EXAMPLE A-2

Find the optimum loop bandwidth (in the sense of minimum phase-error variance) for an active filter second-order loop with $\zeta = 1/\sqrt{2}$ which is tracking an oscillator with the phase-noise characteristics given in Example A-1 assuming that the loop operates in an additive noise background with

$$10 \log_{10}\left(\frac{A_c^2}{2N_0}\right) = 40 \text{ dB-Hz}$$

Solution: Total mean-square phase error, or variance, due to both phase noise and background noise from (A-23) and (A-34b) is

$$\sigma_{\psi,T}^2 = \frac{2N_0 B_L}{A_c^2} + \frac{k_1 \pi^3}{\omega_n^2} + \frac{k_2 \pi^2}{\sqrt{2}\omega_n} + k_4 f_m$$

where $\omega_n = B_L/0.53 = 1.89 B_L$. Substituting previously obtained values for constants, this can be written as

$$\sigma_{\psi,T}^2 = 10^{-4} B_L + \frac{8.7 \times 10^{-3}}{B_L^2} + \frac{3.71 \times 10^{-5}}{B_L} + 10^{-7}$$

Differentiation of this with respect to B_L and setting the result equal to zero results in

$$10^{-4} - \frac{17.4 \times 10^{-3}}{B_{L,opt}^3} - \frac{3.71 \times 10^{-5}}{B_{L,opt}^2} = 0$$

or

$$B_{L,opt} \cong 5.6 \text{ Hz}$$

For this optimum bandwidth, the total phase-error standard deviation is

$$\sigma_{\psi,T,opt} \cong 0.03 \text{ rad}$$

Now consider a phase-locked loop with phase noise, $\theta_n(t)$, on the VCO output. It can be shown† that the phase-error variance due to the noisy VCO is

$$\sigma_{\psi}^2 VCO = \int_0^\infty |1 - H(j2\pi f)|^2 G_{\theta_n}(f) \, df \qquad (A-36)$$

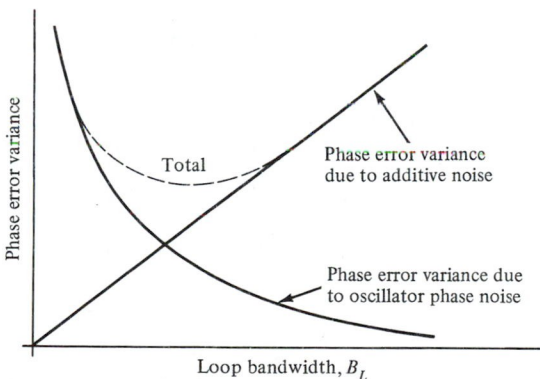

FIGURE A-10. Optimization of phase-error variance.

---

†For example, see Blanchard [3]. Note that Blanchard's closed-loop VCO phase noise and instantaneous phase error due to VCO phase noise are the same.

where $G_{\theta_n}(f)$ is the *single-sided* power spectral density of the VCO phase noise. (Note that additive phase noise at the VCO output is no different from additive phase noise at the loop input.) Suppose that the VCO phase-noise spectral density is random phase walk with

$$G_{\theta_n}(f) = \frac{K_{\text{VCO}}}{f^2} \tag{A-37}$$

For a second-order loop with $\zeta = 1/\sqrt{2}$ and natural frequency ω_n, it follows from (A-34b) that

$$\sigma_{\psi,\text{VCO}}^2 = \frac{K_{\text{VCO}}\pi^2}{\sqrt{2}\omega_n} = \frac{K_{\text{VCO}}\pi^2}{\sqrt{2}(1.89)B_L} \qquad \left(\zeta = \frac{1}{\sqrt{2}}\right) \tag{A-38}$$

If the loop also has additive white noise present at its input, an optimum bandwidth exists, as in Example A-2, that will minimize the total phase-error variance due to VCO jitter and input noise.

One final comment needs stressing in regard to tracking an oscillator with phase jitter. If frequency multiplication or division by N is used prior to the loop, the phase noise is multiplied or divided by N and the phase-noise spectral density is multiplied or divided by N^2, respectively. This follows because the instanteneous output phase, $\theta_{\text{out}}(t)$, of a frequency multiplier is $\theta_{\text{out}}(t) = N\theta_{\text{in}}(t)$, where $\theta_{\text{in}}(t)$ is the input phase, with division by N for a divider.

A-2.5 Transient Response

The tracking error, $\psi(t)$, of the loop for various input phase functions, $\phi(t)$, can be determined by obtaining the inverse Laplace transform of

$$\psi(s) = [1 - H(s)]\phi(s) \tag{A-39}$$

where $1 - H(s)$ is the phase-error transfer function [see (A-10)]. Typical transient input phase functions are

1. A step, $\phi(t) = \Delta\phi\, u(t)$, for which

$$\phi_s(s) = \frac{\Delta\phi}{s} \tag{A-40}$$

2. A ramp (frequency step), $\phi(t) = \Delta\omega\, tu(t)$, for which

$$\phi_r(s) = \frac{\Delta\omega}{s^2} \tag{A-41}$$

3. A parabola (frequency ramp), $\phi(t) = \frac{1}{2}\Delta\dot{\omega}\, t^2u(t)$, for which

$$\phi_p(s) = \frac{\Delta\dot{\omega}}{s^3} \tag{A-42}$$

4. A parabola in frequency, $\phi(t) = \frac{1}{6} \Delta\ddot{\omega}\, t^3 u(t)$, for which

$$\phi_{fp}(s) = \frac{\Delta\ddot{\omega}}{s^4} \tag{A-43}$$

The phase-error responses of a first-order loop to each of the first three inputs, respectively, are

$$(1)\ \psi_s(t) = \Delta\phi\, e^{-Kt} u(t) \tag{A-44}$$

$$(2)\ \psi_r(t) = \frac{\Delta\omega}{K}(1 - e^{-Kt}) u(t) \tag{A-45}$$

$$(3)\ \psi_p(t) = \frac{\Delta\dot{\omega}}{K^2}(Kt + e^{-Kt} - 1) u(t) \tag{A-46}$$

Note that $\psi_r(t)$ is the indefinite integral of $\psi_s(t)$ with $\Delta\phi$ replaced by $\Delta\omega$, and $\psi_p(t)$ is the indefinite integral of $\psi_r(t)$ with $\Delta\omega$ replaced by $\Delta\dot{\omega}$. Also note that only for the phase step is the steady-state VCO phase error zero. For the frequency step (phase ramp), the steady-state phase error is

$$\psi_{r,\mathrm{ss}} = \frac{\Delta\omega}{K} \tag{A-47}$$

which approaches zero as the loop gain approaches infinity. However, the loop bandwidth also goes to infinity with increasing loop gain so that phase error variance due to additive noise at the input becomes progressively larger with decreasing steady-state phase error. From (A-46), it is seen that the frequency ramp (phase parabola) results in an essentially linearly increasing phase error.

All these comments apply to the case where loop components do not saturate; that is, (A-39) is predicated under the assumption of linear loop operation. For a sinusoidal phase detector, the VCO input is really $K \sin\psi$, so that (A-47) becomes

$$\left|\frac{\Delta\omega}{K}\right| = |\sin(\psi_{r,\mathrm{ss}})| \le 1$$

This establishes the *hold-in range* of a first-order loop as

$$-K \le \Delta\omega_H \le K \ \mathrm{rad/s} \tag{A-48}$$

For a high-gain second-order loop, the phase-error transfer function is given by

$$1 - H(s) = \frac{s^2}{s^2 + 2\zeta\omega_n s + \omega_n^2} \tag{A-49}$$

(see Table A-1). The steady-state phase error can be found from

$$\lim_{t\to\infty} \psi(t) = \lim_{s\to 0} \{s[1 - H(s)]\phi(s)\} \tag{A-50}$$

For the phase step and phase ramp (frequency offset) inputs it is seen that the steady-state phase error for a second-order loop is zero.

The Laplace transform inversion of (A-39) in response to a frequency ramp, (A-42), and parabola in frequency, (A-43), yields, respectively, the following transient response for $\zeta < 1$:

$$\psi_p(t) = \frac{\Delta\dot{\omega}}{\omega_n^2}\left\{1 - e^{-\zeta\omega_n t}\left[\cos(\omega_n\sqrt{1-\zeta^2}t)\right.\right.$$

$$\left.\left. + \frac{\zeta}{\sqrt{1-\zeta^2}}\sin(\omega_n\sqrt{1-\zeta^2}t)\right]\right\}u(t) \qquad \text{(A-51)}$$

$$\qquad\qquad\qquad\qquad\qquad\qquad\qquad\qquad \text{(frequency ramp)}$$

$$\psi_{fp}(t) = \frac{\Delta\ddot{\omega}}{\omega_n^3}\left\{\omega_n t - 2\zeta + 2\zeta e^{-\zeta\omega_n t}\left[\cos(\omega_n\sqrt{1-\zeta^2}t)\right.\right.$$

$$\left.\left. - \frac{1-2\zeta^2}{2\zeta\sqrt{1-\zeta^2}}\sin\left(\omega_n\sqrt{1-\zeta^2}t\right)\right]\right\}u(t) \qquad \text{(A-52)}$$

$$\qquad\qquad\qquad\qquad\qquad\qquad\qquad\qquad \text{[frequency parabola]}$$

Figure A-11 shows the transient phase error due to an input ramp in frequency, and Figure A-12 shows the transient phase error due to a parabolic frequency input. For the frequency ramp, it is is seen that the steady-state phase error is

$$\psi_{ss,p} = \frac{\Delta\dot{\omega}}{\omega_n^2} \qquad \text{(A-53)}$$

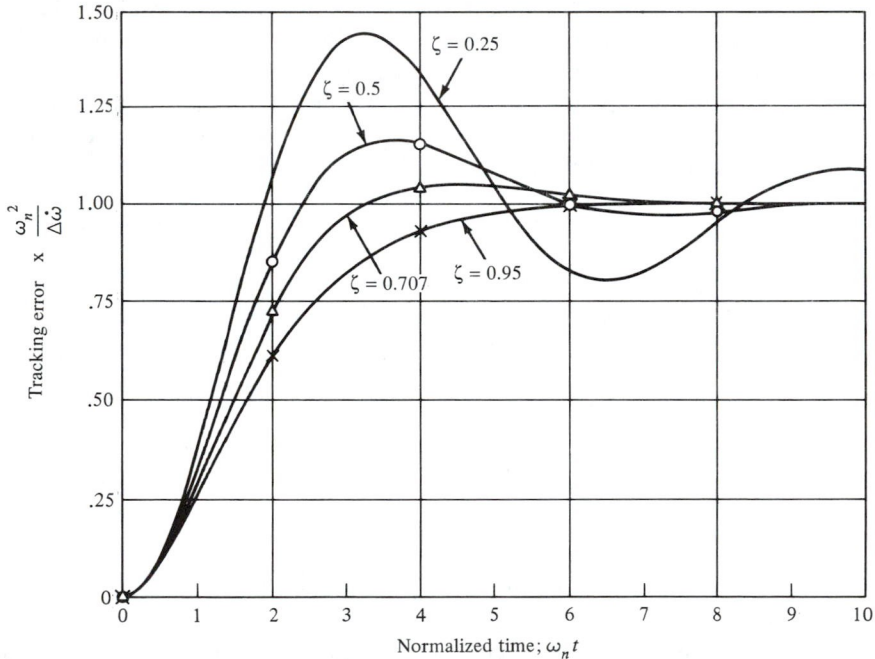

FIGURE A-11. Second-order phase-locked-loop tracking error for frequency ramp input.

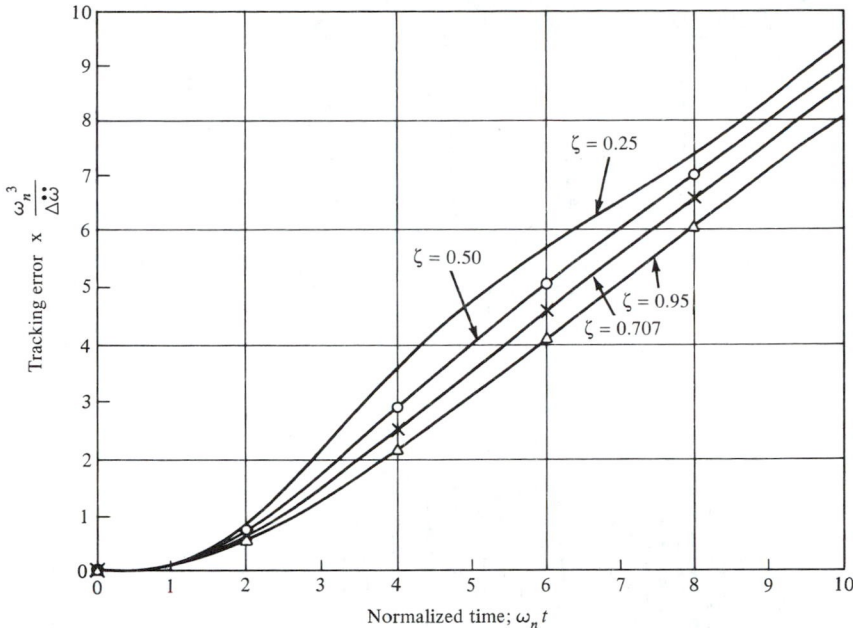

FIGURE A-12.   Second-order phase-locked-loop tracking error for frequency parabola input.

Again, use of a sinusoidal phase detector would have resulted in

$$\sin \psi_{\mathrm{ss},p} = \frac{\Delta\dot\omega}{\omega_n^2} \leq 1$$

which established that the maximum permissible rate of change of input frequency to a second-order loop as

$$\Delta\dot\omega_{\max} = \omega_n^2 \ \mathrm{rad/s^2} = 0.53B_L^2 \qquad \zeta = \frac{1}{\sqrt{2}} \tag{A-54}$$

provided that no loop components saturate.

Theoretically, a second-order loop with infinite dc gain can never permanently lose lock. Its response to a large enough frequency offset will be a temporary loss of lock, resulting in cycle slipping, after which it will relock. The frequency-step limit below which the loop does *not* slip cycles is called the *pull-out frequency*. It has been established from phase-plane portraits for the second-order loop with sinusoidal phase detector that the pull-out frequency satisfies the empirical relation [1]

$$\Delta\omega_{\mathrm{po}} = 1.8\omega_n(\zeta + 1) \qquad \mathrm{rad/s} \tag{A-55}$$

A-2.6 Phase-Locked-Loop Acquisition

The initial application of a sinusoidal signal to a phase-locked loop with carrier frequency different from the quiescent frequency of the VCO results in a beat

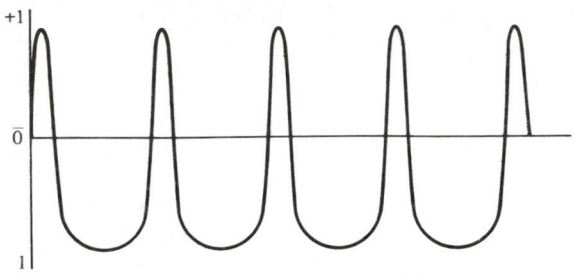

(a) Typical beat-note wave shape, first-order loop,
$\Delta\omega/K = 1.10$ [1]

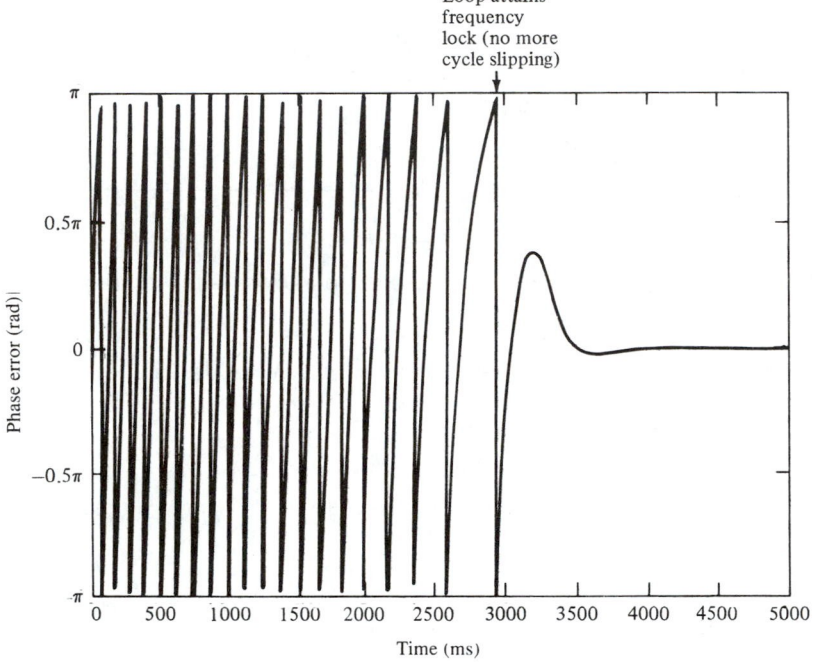

(b) Transient response of a second-order phase-locked
loop with an initial frequency offset of $\Delta f = 10$ Hz
and a noise bandwidth of $B_n = 5$ Hz [7]

FIGURE A-13. Phase-error signals for phase-locked loops in acquisition.

frequency at the VCO output (i.e., the loop slips cycles as illustrated in Figure A-13. The beat-frequency waveform has a nonzero average value, which tends to drive the VCO toward *frequency lock* (i.e., the VCO frequency matches that of the input carrier frequency). The dc component of the phase detector output is called the *pull-in voltage*. In a second-order loop, which includes an integrator prior to the VCO, the pull-in voltage is integrated and the loop will eventually reach frequency lock provided that saturation of a loop component does not occur first.

For a first-order loop, an integrator is not present, and the loop will acquire lock only if the frequency offset is within the *lock-in range* of the loop. The inequalities

$$-K < 2\pi(f_c - f_0) < K \tag{A-56}$$

establish the frequency offset limit of VCO quiescent frequency from input carrier frequency within which a first-order loop may acquire lock. If the magnitude of the frequency offset in rad/s exceeds K, it is impossible for a first-order loop to have a static phase error which will drive the VCO frequency to match the input frequency. If (A-56) is satisfied, the loop will acquire lock, but the time to lock depends on the initial phase error between the input and VCO. Figure A-14 illustrates phase error transients for a first-order phase-locked loop for an initial frequency offset of zero. It is seen that if the initial phase error is near 180°, an extremely long transient can take place. This phenomenon, known as *hang-up,* is not unique to the first-order loop.

Returning to the second-order loop, the duration of the initial beat frequency transient illustrated in Figure A-13b is the *pull-in time*. If $\Delta\omega - 2\pi(f_c - f_0) \gg K$, the pull-in time of a second-order loop is approximately [1]

$$T_p \cong \frac{(\Delta\omega)^2}{2\zeta\omega_n^3} \qquad \text{seconds} \tag{A-57}$$

For a high-gain loop with $\zeta = 1/\sqrt{2}$, this becomes

$$T_p \cong \frac{4(\Delta f)^2}{B_L^3} \qquad \text{seconds} \qquad \left(\zeta = \frac{1}{\sqrt{2}}\right) \tag{A-58}$$

where $\Delta f = f_c - f_0$ hertz and B_L is the single-sided loop bandwidth in hertz. Once $\Delta\omega \leq K$, the loop ceases to skip cycles and quickly snaps into lock. The additional

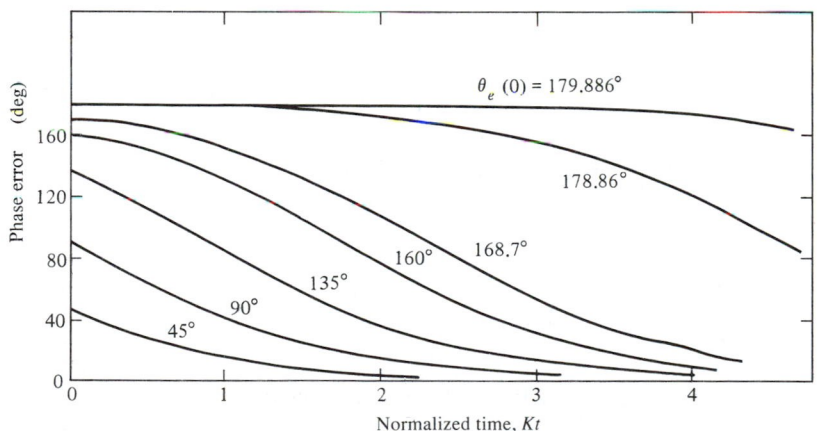

FIGURE A-14. Transient phase errors in first-order PLL. (Reproduced from Ref. 1 with permission.)

time required for the loop to settle, T_S, is approximately [7]

$$T_S \cong \frac{1.5}{B_L} \tag{A-59}$$

where the final phase error is 0.1 rad or less.

Because the pull-in time for a second-order loop can be exceedingly long, an *acquisition aid* is often used. This usually takes the form of a ramp applied to the VCO input or a square wave applied to the integrator input. The maximum rate of change for the VCO frequency is given by (A-54) under noise-free conditions. (Note that there is no difference if a frequency ramp is placed on the input or the VCO output.) With noise present, the sweep rate must be reduced. Empirical data suggest that the maximum sweep rate should be limited to†

$$\Delta\dot{\omega}_{\max} = \omega_n^2[1 - (\text{SNR})_L^{-1/2}] \tag{A-60}$$

where $(\text{SNR})_L$ is the carrier-to-noise ratio with noise measured in a loop bandwidth. Once lock is acquired, the sweep can be removed. If injected as a square wave into the integrator of an active loop filter, the removal of the sweep does not have to be particularly rapid under normal conditions since once the loop is locked the sweep voltage is compensated by the phase detector output.

Removal of the sweep requires the use of a lock detector, which can be implemented by means of the *coherent amplitude detector* illustrated in Figure A-15. The output of such a detector, once the loop is locked, is proportional to signal amplitude at the loop input. Therefore, it can also be used to control open-loop gain, which depends on input signal amplitude. Another way to remove the effect of variations of the input signal amplitude on loop gain is to precede the loop with a limiter.

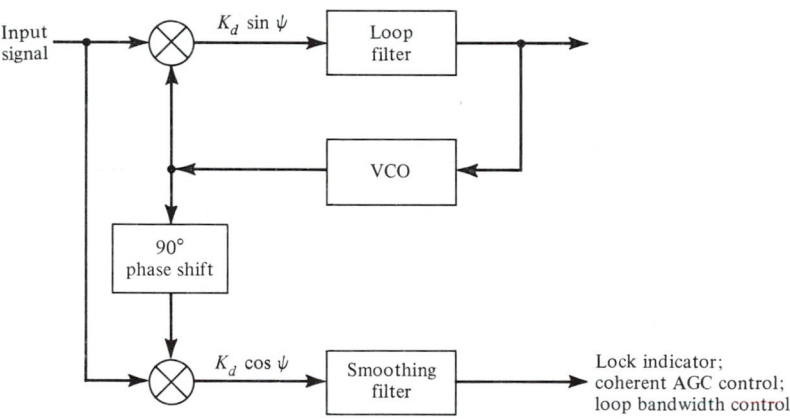

FIGURE A-15. Coherent amplitude detector.

†See Gardner [1, p. 81]. Note that Gardner's $(\text{SNR})_L$ is one-half of the definition used here.

Another way in which acquisition can be speeded up is to employ a wider loop bandwidth during acquisition. This is shown by (A-58). Narrowing the loop bandwidth once acquisition has been achieved is facilitated by means of a coherent amplitude detector.

Finally, note that the settling time for a phase-locked loop is improved with the addition of noise to the loop, either as external noise or as a *dithering signal*. This is illustrated by Figure A-16 for a second-order loop. Note that an optimum value of the order of 20 dB apparently exists for loop signal-to-noise ratio to provide minimum settling time. The use of phase acquisition aids for reducing settling time has been studied with the conclusion that the phase acquisition time can be reduced for loop signal-to-noise ratios above 12 dB. Below 12 dB, no signifiant advantage from the acquisition aid was realized [9].

A-2.7 Effects of Transport Delay

At sufficiently high frequencies, the delay associated with the phase-locked-loop layout can effectively add additional poles to the loop transfer function. Thus a loop designed to be a second-order loop is, in essence, a higher-than-second-order loop. Such delays may cause a loop to operate in a totally different manner from the one for which it was designed. In particular, a loop designed to be second order and therefore thought to be unconditionally stable may effectively be third order or higher and therefore be only conditionally stable.

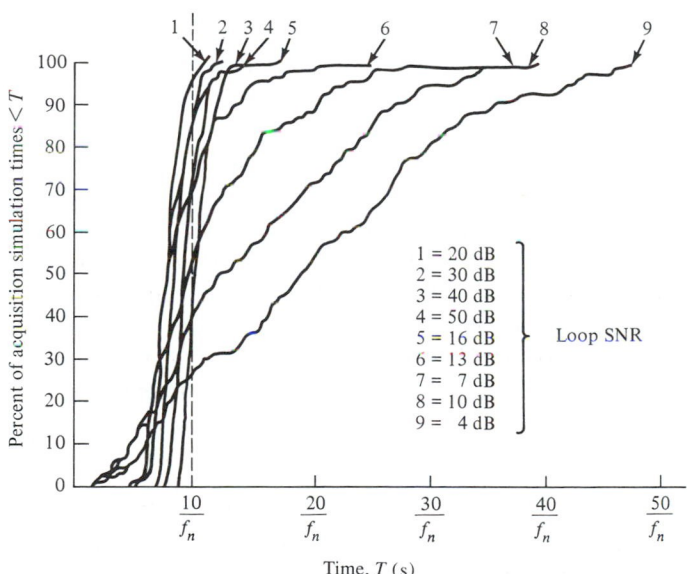

FIGURE A-16. Computer simulations of second-order phase-locked-loop acquisition time for zero-frequency offset and an initial phase error of $\epsilon_T = \pi$ rad. (From Ref. 8. Copyright © IEEE, 1973. Reprinted with permission.)

A-3 Frequency Synthesis

A-3.1 Introduction

A frequency synthesizer is a device for generating several possible output frequencies from a single, highly stable reference frequency. Systems applications to communications include HF radio, frequency-division multiple-access satellite communications, and spread-spectrum communications systems. There are three main techniques used for frequency synthesizer implementation, although combinations of these may be used as well as variants of these techniques. The three methods of frequency synthesis are

1. Digital (or table lookup).
2. Direct (or mix and divide).
3. Phase-locked (or indirect).

The function of a synthesizer is described mathematically by the equation

$$f_2 = \frac{n_2 f_1}{n_1} \tag{A-61}$$

where f_1 is the reference frequency and n_1 and n_2 are integers.

Recalling that frequency multiplication of a sinusoid multiplies both the nominal frequency and the phase deviation by the multiplication factor, n_2/n_1, it is seen from the following model of a reference signal

$$r(t) = A[1 + a(t)] \cos \left[\omega_0 t + \phi(t) + \frac{\alpha t^2}{2} \right] \tag{A-62}$$

that the long-term stability of a frequency synthesizer is that of the stable reference multiplied by n_2/n_1. That is, the term $\alpha t^2/2$ in the argument of (A-62), which reflects long-term drift, is multiplied by n_2/n_1, as are the terms $\omega_0 t$ and $\phi(t)$. It is tempting at this point to say that the short-term stability, or phase noise, of the synthesizer frequency is determined from $n_2 \phi(t)/n_1$. However, short-term stability depends on the manner in which the output frequencies are synthesized.

Each of the synthesis techniques listed has advantages and disadvantages. The reader is encouraged to consult more detailed discussion of synthesizer design (such as Refs. 5, 6, and 10) before embarking on any synthesizer design.

A-3.2 Digital Synthesizers

The basic idea of a digital synthesizer is illustrated by Figure A-17. With each clock pulse, which occurs at frequency f_1, the accumulator increments a phase variable, θ, by the amount $a \Delta \theta$ where a is a proportionality constant. The value of the phase variable, θ, serves as the address to a memory containing N-bit numbers proportional to $\cos \theta$, quantized to 2^N levels. The memory output is converted to an analog voltage by a digital-to-analog (D/A) converter.

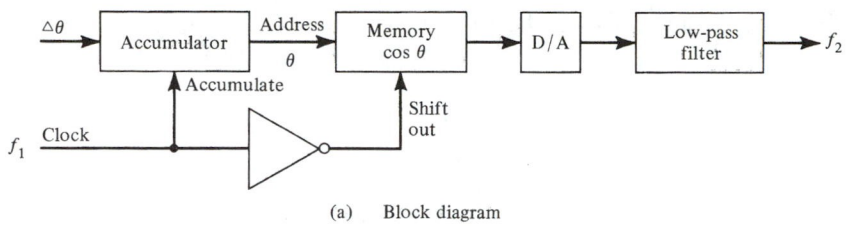

(a)   Block diagram

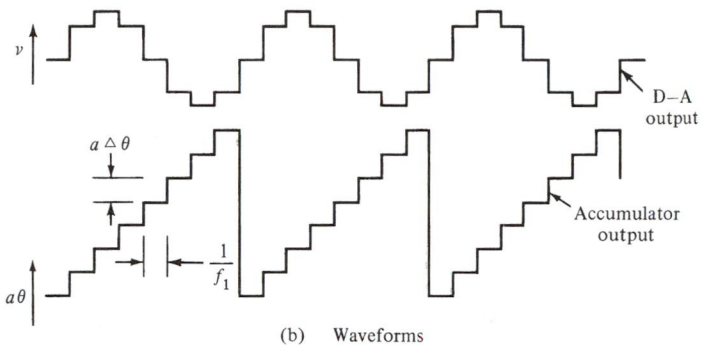

(b)   Waveforms

FIGURE A-17.   Principle of operation for a digital frequency synthesizer. (Reproduced from Ref. 5 with permission.)

The capacity of the accumulator corresponds to one complete cycle of cos θ. Let n_1 be the capacity of the accumulator and let n_2 be the increment in the accumulator value for each clock cycle. Then the number of clock cycles required to cycle the accumulator is n_1/n_2, and the frequency of the accumulator cycle, which is also the frequency of the D/A converter output, is given by (A-62). The resolution of the synthesizer is the change in frequency that occurs when n_2 changes by one. This is

$$\Delta f = \left(\frac{n_2 + 1}{n_1} - \frac{n_2}{n_1}\right)f_1 = \frac{f_1}{n_1} \tag{A-63}$$

Because the structure of the digital synthesizer implies no fewer than two phase values for each cycle of the output, the theoretical maximum value for f_2 is $f_1/2$. However,

$$f_{2,\max} = \frac{f_1}{4} \tag{A-64}$$

is more practical to allow reasonable lowpass output filters.

EXAMPLE A-3

Consider a digital synthesizer for which the capacity of the accumulator is 2^8. Obtain (a) the clock frequency required to produce a 32-kHz resolution, (b) the increment in accumulator contents at each clock pulse to produce a 160-kHz output frequency, and (c) the maximum synthesizer output frequency, $f_{2,\max}$.

Solution: (a) From (A-63) the clock frequency is

$$f_1 = n_1 \, \Delta f = (2^8)(2^5 \times 10^3)$$

$$= 2^{13} \text{ kHz}$$

$$= 8.192 \text{ MHz}$$

(b) To produce $f_2 = 160$ kHz, n_2 is calculated from (A-61) to be

$$n_2 = \frac{n_1 f_2}{f_1}$$

$$= \frac{(2^8)(2^4 \times 10^4)}{2^{13} \times 10^3}$$

$$= 5$$

(c) From (A-64) $f_{2,\max} = 2^{11} \times 10^3$ Hz $= 2.048$ MHz.

Advantages of digital synthesizers are that frequencies can be changed very rapidly and that fine resolution is relatively easy to attain. In addition, digital synthesizers may be used to directly generate modulated waveforms. A disadvantage is that the maximum synthesized frequency is limited, by the speed of the digital logic and memory, to several hundred megahertz. The spurious frequency components near the generated frequency depend on the quantization accuracy used to generate $\cos \theta$. Spurious sidelobe levels of -50 to -60 dB relative to the desired component are possible [10].

A-3.3 Direct Synthesis

Configurations. In the direct frequency synthesis process the desired frequency is built up by multiplication, mixing (summation or subtraction of a reference), and division of a single reference frequency. Many combinations obviously can be used to produce a desired frequency in this way. For example, 7381 kHz can be produced as the 7381th harmonic of 1 kHz or as the 7th harmonic of 1000 kHz plus the 3rd harmonic or 100 kHz plus the 8th harmonic of 10 kHz plus the first harmonic of 1 kHz.

Figure A-18 shows a direct synthesizer for producing frequencies over a 10-MHz range with a resolution of 1 Hz. The reference frequencies of 3 MHz and 27 to 36 MHz could be derived by multiplication of a 1-MHz frequency, which in turn is derived from a stable oscillator, say, of 5 MHz by division. Note that at least one input to each mixer overlaps the output frequency range. Thus it is impossible to eliminate the mixer feed through of this frequency from the output by a fixed output filter. A more practical arrangement which avoids this problem is shown in Figure A-19.

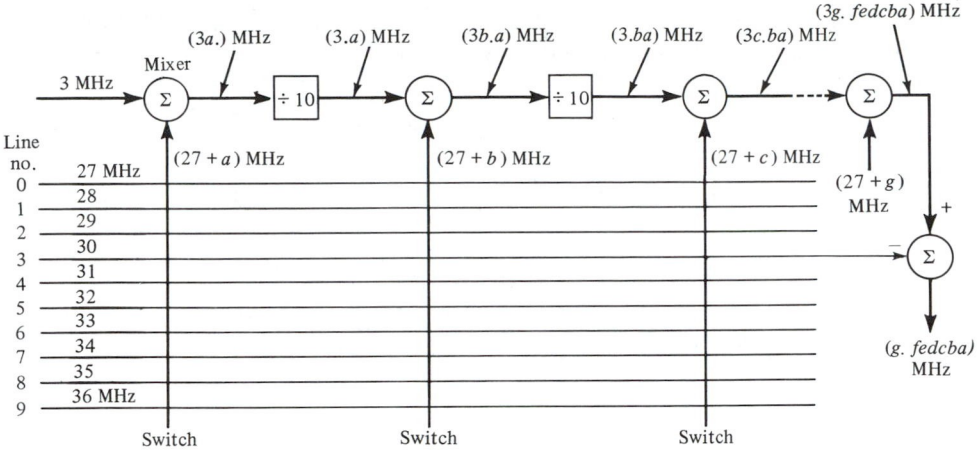

FIGURE A-18. Principle of operation for a direct frequency synthesizer. (Reproduced from Ref. 5 with permission.)

EXAMPLE A-4

Synthesize the frequency 7.123456 MHz with the direct-synthesis scheme of Figure A-19.

Solution: Figure A-20 shows the mathematical construction of the desired frequency.

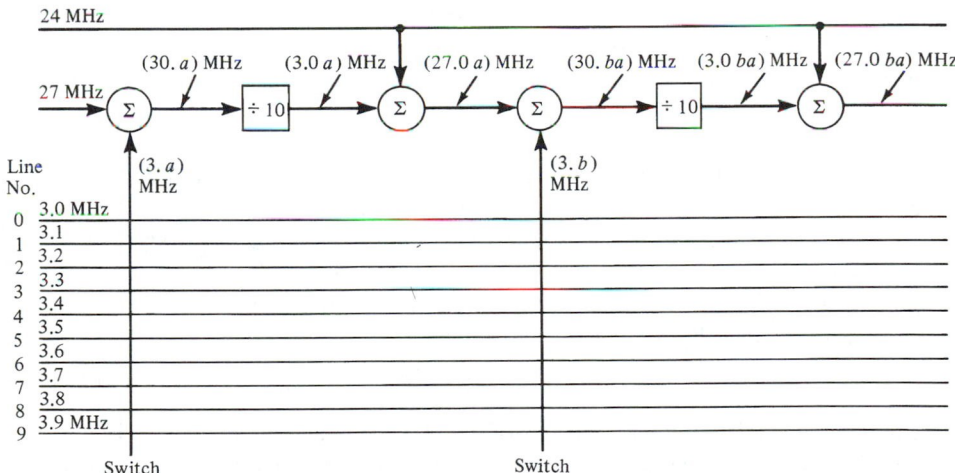

FIGURE A-19. Modified direct synthesizer that avoids mixer feedthrough problem.

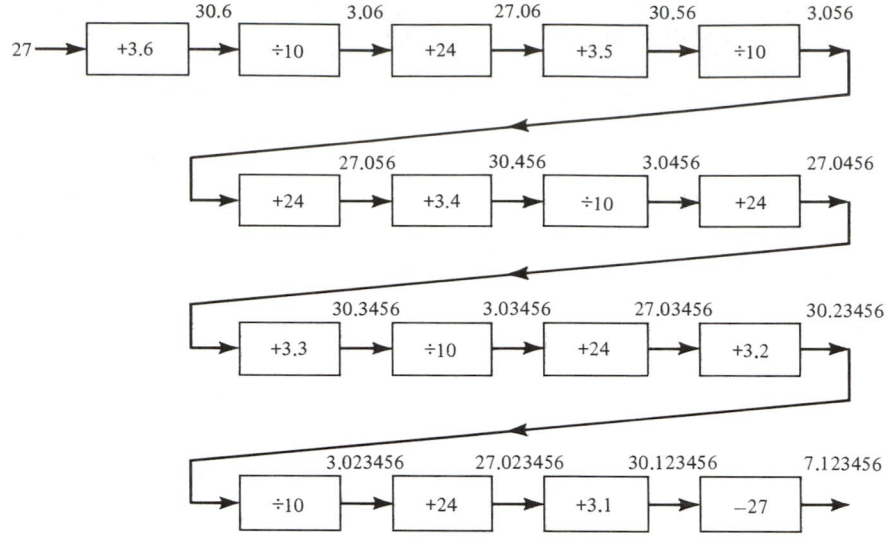

FIGURE A-20. Direct synthesis of the frequency 7.123456 MHz (all frequencies are in units of MHz).

An advantage of direct synthesizers is that the output frequency can be changed rapidly—essentially at the speed of the switches, although some allowance must be made for delay through the system. Another advantage is that the output spectrum can be as clean as the reference oscillator spectrum with FM sidebands increased by the effective multiplication ratio from input to output. In addition, direct synthesizers can have very fine resolution. Disadvantages of direct synthesizers are that they require considerable power because of the large number of LO signals required, and they are bulky. Direct synthesizers with spurious sidelobe levels of -100 dB have been reported.

Spurious Frequency Component Generation in Direct Synthesizers. As an example of spurious frequency component generation in direct synthesizers, consider the synthesizer example of Figure A-20 and the potential spurious responses (or "spurs") in the output of the top chain of mixers and dividers.

The spurious response or spurs at a mixer output, with inputs of frequencies f_s and f_L, are defined by the relationship

$$nf_s + mf_L = f_I \tag{A-65}$$

where f_I is the mixer output frequency of interest and m and n are integers. Equation (A-65) results from the fact that no mixer is a perfect product device but, rather, is more accurately modeled as producing cross-products of integer powers of each input at its output. Only one of the resultant output frequencies, usually

$$f_{I_1} = f_s + f_L \tag{A-66}$$

or

$$f_{I_2} = |f_s - f_L| \tag{A-67}$$

is of interest; others are referred to as *spurs*. The condition $m = 1$, $n = 0$ identifies signal port feedthrough and $m = 0$, $n = 1$ local oscillator feedthrough. The relative amplitudes of these various spurs are a function of the particular mixer design and drive level. Further, where possible, the designer will filter out undesired spurs.

A-3.4 Phase-Locked Frequency Synthesizers

Configurations. The block diagram of a simple phase-locked synthesizer is shown in Figure A-21. The condition (A-61) is satisfied by virtue of the fact that the VCO output frequency divided by n_2 is locked to the reference frequency divided by n_1. Output frequency selection is provided by changing the divider integers, n_1 and n_2. Usually, digital counters are used to provide the desired divider integers.

The minimum increment in output frequency is given by (A-63). The loop bandwidth must be smaller than this minimum increment in order to suppress ripple and ensure loop stability. Therefore, small increments in output frequency demand small loop bandwidths. On the other hand, output phase jitter is dominated by the VCO if loop bandwidth is small. In addition, loop acquisition time is inversely proportional to loop bandwidth [see (A-58) and (A-59)], so that small loop bandwidth implies slow switching between synthesized frequencies.

These conflicting requirements present significant challenges in the design of phase-locked synthesizers. One simple solution to this problem is illustrated in Figure A-22, where the final synthesizer output frequency is

$$f_2 = \frac{n_2 f_1}{n_1 m} \tag{A-68}$$

so that frequency increments of

$$\Delta f = \frac{f_1}{n_1 m} \tag{A-69}$$

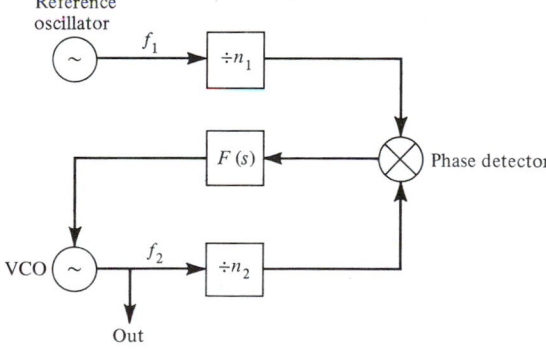

FIGURE A-21. Basic phase-locked synthesizer.

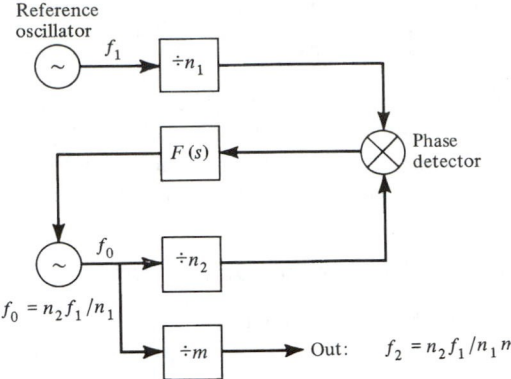

FIGURE A-22. Modified basic phase-locked synthesizer with divider at output.

are obtained. However, phase comparison occurs at frequency f_1/n_1, which alleviates the loop bandwidth problem by a factor of m through operation of the dividers at the VCO output at m times the frequency required for the basic configuration of Figure A-21.

Output Phase Noise. To consider the output phase noise of a phase-locked synthesizer, the model of Figure A-23 will be used and following quantities are defined:

$G_\phi(f)$ = single-sided phase-noise power spectrum of the reference oscillator

$G_\theta(f)$ = single-sided phase-noise power spectrum of the VCO

$\dfrac{2N_0}{A_c^2} = $ equivalent single-sided power spectral level of the additive white input noise *referred to the closed loop input* (see Figure A-8)

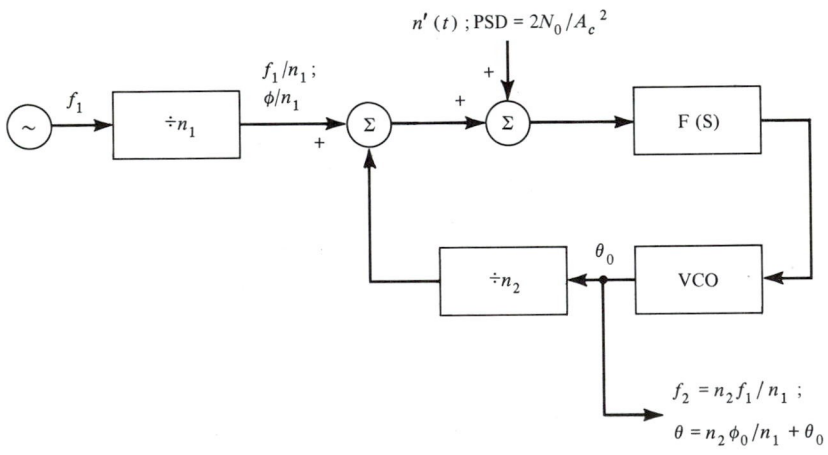

FIGURE A-23. Model for computing phase noise at output of a phase-locked synthesizer (zero subscripts on output phases denote loop filtered quantities).

With these definitions, the phase-noise variance of the phase-locked synthesizer output is

$$\sigma_0^2 = \frac{2N_0 B_L}{A_c^2} + \left(\frac{n_2}{n_1}\right)^2 \int_0^\infty G_\phi(f)|H(f)|^2 \, df$$

$$+ \int_0^\infty G_\theta(f)|1 - H(f)|^2 \, df$$

$$(A-70)$$

where $H(f)$ is the closed-loop frequency response. Thus, with negligible additive noise, the output phase noise variance is the same as the VCO for small loop bandwidths and that of the reference source multiplied by $(n_2/n_1)^2$ if loop bandwidth is large.

Spur Generation in Indirect Synthesizers. The spur problem also exists in indirect synthesizers and is particularly troublesome in synthesizers where a large tuning range and fast acquisition are desired. The latter implies a wideband loop, although the problem can sometimes be alleviated by using preset tuning of the VCO, which permits a narrower bandwidth loop.

References

[1] F. M. GARDNER, *Phaselock Techniques,* 2nd ed. (New York: Wiley, 1979).

[2] A. J. VITERBI, *Principles of Coherent Communication* (New York: McGraw-Hill, 1966).

[3] A. BLANCHARD, *Phase-Locked Loops* (New York: Wiley, 1976).

[4] H. MEYR and G. ASCHIED, *Synchronization of Digital Communications* (New York: Wiley, 1990).

[5] W. F. EGAN, *Frequency Synthesis by Phase Lock* (New York: Wiley, 1981).

[6] V. MANASSEWITSCH, *Frequency Synthesizers Theory and Design,* 2nd ed. (New York: Wiley, 1981).

[7] J. J. SPILKER, JR., *Digital Communications by Satellite* (Englewood Cliffs, N.J.: Prentice Hall, 1977), Chap. 12.

[8] S. L. GOLDMAN, "Second-Order Phase-Lock-Loop Acquisition Time in the Presence of Narrow-Band Gaussian Noise," *IEEE Trans. Commun.,* Vol. COM-21, pp. 297–300, April 1973.

[9] H. MEYR, "Phase Acquisition Statistics for Phase-Locked Loops," *IEEE Trans. Commun.,* Vol. COM-28, pp. 1365–1372, August 1980.

[10] J. GORSKI-POPIEL, ed., *Frequency Synthesis: Techniques and Applications* (New York: IEEE Press, 1975).

[11] M. ABRAMOWITZ and I. STEGUN, eds., *Handbook of Mathematical Functions,* (New York: Dover, 1972).

Gaussian Probability Function†

The Gaussian probability function of unit variance and zero mean is

$$Z(x) = \frac{e^{-x^2/2}}{\sqrt{2\pi}} \tag{B-1}$$

and the corresponding cumulative distribution function is

$$P(x) = \int_{-\infty}^{x} Z(t)\, dt \tag{B-2}$$

The Q-function is defined as

$$Q(x) = 1 - P(x) = \int_{x}^{\infty} Z(t)\, dt \tag{B-3}$$

An asymptotic expansion for $Q(x)$ valid for large x is

$$Q(x) = \frac{Z(x)}{x}\left[1 - \frac{1}{x^2} + \frac{1 \cdot 3}{x^4} + \cdots + \frac{(-1)^n 1 \cdot 3 \ldots \cdot (2n-1)}{x^{2n}}\right] + R_n \tag{B-4}$$

where

$$R_n = (-1)^{n+1} 1 \cdot 3 \ldots \cdot (2n+1) \int_{x}^{\infty} \frac{Z(t)}{t^{2n+2}}\, dt \tag{B-5}$$

which is less in absolute value than the first neglected term. For moderate values of x, several rational approximations are available. One such approximation is

$$1 - Q(x) = P(x) = 1 - Z(x)(b_1 t + b_2 t^2 + b_3 t^3 + b_4 t^4 + b_5 t^5) + \epsilon(x)$$

$$t = \frac{1}{1 + px} \tag{B-6}$$

$$|\epsilon(x)| < 7.5 \times 10^{-8}$$

$$p = 0.2316419$$

---

† The notation used here is that of Abramowitz and Stegun [1, pp. 931 ff.].

TABLE B-1. Abbreviated Table of Values for $Q(x)$ and $Z(x)$

x	$Q(x)$	$Z(x)$	x	$Q(x)$	$Z(x)$
0.0	0.50000	0.39894	2.0	0.02275	0.05399
0.1	0.46017	0.39695	2.1	0.01786	0.04398
0.2	0.42074	0.39104	2.2	0.01390	0.03547
0.3	0.38209	0.38138	2.3	0.01072	0.02833
0.4	0.34458	0.36827	2.4	0.00820	0.02239
0.5	0.30854	0.35206	2.5	0.00621	0.01753
0.6	0.27425	0.33322	2.6	0.00466	0.01358
0.7	0.24196	0.31225	2.7	0.00347	0.01042
0.8	0.21186	0.28969	2.8	0.00256	0.00792
0.9	0.18406	0.26608	2.9	0.00187	0.00595
1.0	0.15866	0.24197	3.0	0.00135	0.00443
1.1	0.13567	0.21785	3.1	0.00097	0.00327
1.2	0.11507	0.19419	3.2	0.00069	0.00238
1.3	0.09680	0.17137	3.3	0.00042	0.00723
1.4	0.08076	0.14973	3.4	0.00034	0.00123
1.5	0.06681	0.12952	3.5	0.00023	0.00087
1.6	0.05480	0.11092	3.6	0.00016	0.00061
1.7	0.04457	0.09405	3.7	0.00011	0.00042
1.8	0.03593	0.07895	3.8	7.24×10^{-5}	0.00029
1.9	0.02872	0.06562	3.9	4.81×10^{-5}	0.00020
			4.0	3.17×10^{-5}	0.00013

$$b_1 = 0.319381530 \qquad b_4 = -1.821255978$$

$$b_2 = -0.356563782 \qquad b_5 = 1.330274429$$

$$b_3 = 1.781477937$$

The error function can be related to the Q-function by

$$\text{erf}(x) \triangleq \frac{2}{\sqrt{\pi}} \int_0^x e^{-t^2} \, dt = 1 - 2Q(\sqrt{2}x) \tag{B-7}$$

The complementary error function, $\text{erfc}(x) = 1 - \text{erf}(x)$, can be approximated similarly to the Q-function.

A short table of values for $Q(x)$ and $Z(x)$ is given in Table B-1. Extensive tables of $P(x)$, $Z(x)$ and its derivatives can be found in Abramowitz and Stegun [1].

Reference

[1] M. ABRAMOWITZ and I. STEGUN, Eds., *Handbook of Mathematical Functions* (New York: Dover, 1972) (originally published in 1964 as NBS Applied Mathematics Series 55).

Power Spectral Densities for Sequences of Random Binary Digits and Random Tones

Calculation of the transmitted spread-spectrum power spectral density (psd) requires knowledge of the power spectral density of the spreading function. The psd of a random binary sequence is usually calculated from the time autocorrelation function as was done in the text. The psd of a random sequence of tones is normally approximated by a sum of delta functions at the tone frequencies, with each weighted by the probability of that tone being transmitted. The exact psd for the special case where the tones are phase coherent from one transmission to the next was given previously (Lindsey and Simon [1]). In this appendix, the ensemble autocorrelation function for a sequence of random binary symbols is calculated and is used in the calculation of the psd of a sequence of noncoherent tones.

Consider an infinite random sequence of binary symbols $\ldots, a_{-1}, a_0, a_1, \ldots,$ where $a_n \in \{+1, -1)$ and the random process generated using these symbols:

$$s(t,\mathbf{a},T) = \sum_{n=-\infty}^{\infty} a_n p(t + T - nT_c) \tag{C-1}$$

In this equation, **a** represents a random vector, T is a random phase required to make $s(t)$ stationary, T_c is the sequence chip duration, and $p(t)$ is the unit pulse of duration T_c. One possible sample function of this random process is illustrated in Figure C-1. The a_n's above are independent and it is equally likely that any a_n is $+1$ or -1. The phase T is uniformly distributed over the interval $(0,T_c)$.

The ensemble autocorrelation function is defined by

$$R_s(t_1,t_2) = E[s(t_1)s(t_2)] \tag{C-2}$$

where the expected value is over all **a** and all T. Substituting (C-1) into (C-2) yields

$$R_s(t_1,t_2) = E\left[\sum_{n=-\infty}^{\infty} \sum_{n'=-\infty}^{\infty} a_n a_{n'} p(t_1 + T - nT_c) p(t_2 + T - n'T_c) \right] \tag{C-3}$$

$$= \sum_n \sum_{n'} E_1[a_n a_{n'}] E_2[p(t_1 + T - nT_c)p(t_2 + T - n'T_c)]$$

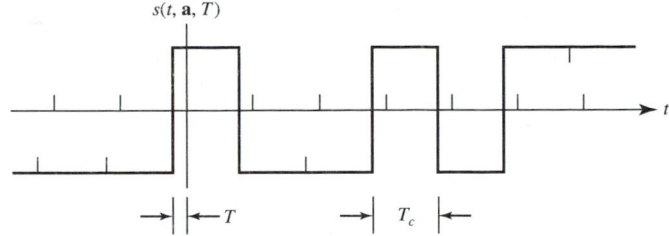

FIGURE C-1. Sample function of the random process $s(t, \mathbf{a}, T)$.

since the expectation is linear and the random variables a_n and T are independent. In (C-3), expectation E_1 is over a_n and E_2 is over T. It is easily shown that $E_1(a_n a'_n) = 0$ for $n \neq n'$ and $E_1(a_n a_{n'}) = 1$ for $n = n'$, so that

$$R_s(t_1, t_2) = \sum_n E_2[p(t_1 + T - nT_c)p(t_2 + T - nT_c)] \tag{C-4}$$

At this time it is convenient to return the summation to within the expected value and to write the expected value explicitly. This yields

$$R_s(t_1, t_2) = \frac{1}{T_c} \int_0^{T_c} \sum_n p(t_1 + T - nT_c)p(t_2 + T - nT_c) \, dT \tag{C-5}$$

The product within the summation is illustrated in Figure C-2 for a particular $t_1 - t_2 < T_c$ and n. For $t_1 - t_2 > T_c$, the product is zero since the pulses are non-overlapping. The area under the product pulse is $T_c - |t_1 - t_2|$. The complete time function under the integral of (C-5) is a sum of time translations of the function shown in Figure C-2d as illustrated in Figure C-2e. The integral of (C-5) can now be calculated by inspection. The result is the shaded area of Figure C-2e.

$$R_s(t_1, t_2) = \begin{cases} \dfrac{1}{T_c}(T_c - |t_1 - t_2|) & \text{for } |t_1 - t_2| < T_c \\[2mm] 0 & \text{for } |t_1 - t_2| > T_c \end{cases} \tag{C-6}$$

This function is a function only of $|t_1 - t_2| = \tau$ and not the absolute value of t_1 and t_2 as expected. The result is identical to the result obtained previously using the time autocorrelation, so that the power spectral density, the Fourier transform of $R_s(\tau)$, is given by

$$S_s(f) = T_c \, \text{sinc}^2 f T_c \tag{C-7}$$

as in Example 2-1.

Consider next a random sequence of tones having frequencies chosen from a set of 2^k possible frequencies and having a random phase each time a new frequency is

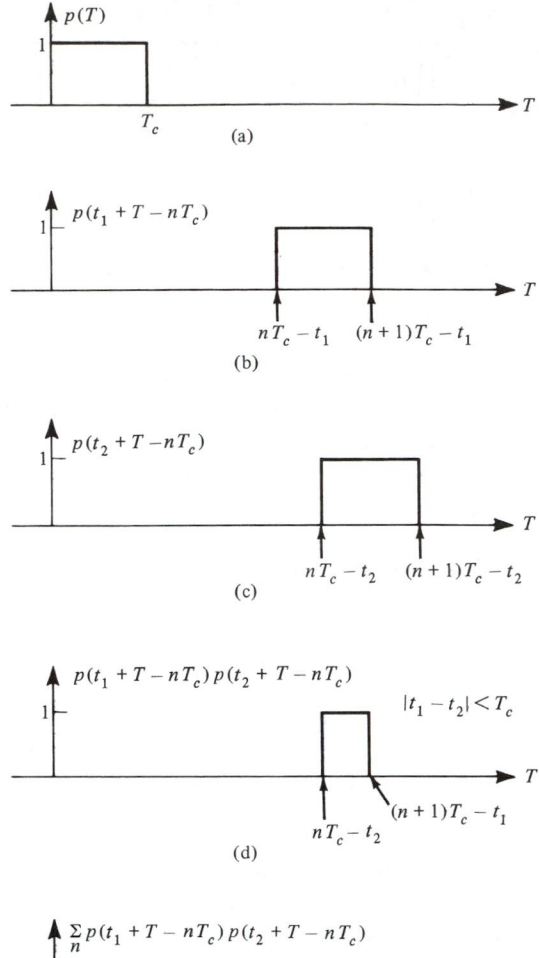

FIGURE C-2. Functions used to calculate $R_s(t_1, t_2)$.

selected. Assume that the random phase is uniformly distributed over $(0, 2\pi)$ and that the phase of the time of frequency change is uniformly distributed over $(0, T_c)$. Thus the random process being considered is

$$s(t) = 2 \sum_{n=-\infty}^{\infty} p(t + T - nT_c) \cos(\omega_n t + \phi_n) \qquad (\text{C-8})$$

The ensemble autocorrelation function is given by

$$R_s(t_1,t_2) = E[s(t_1)s(t_2)]$$

$$= 4E\left[\sum_n \sum_{n'} p(t_1 + T - nT_c)p(t_2 + T - n'T_c)\right. \qquad \text{(C-9)}$$

$$\left. \times \cos(\omega_n t_1 + \phi_n) \cos(\omega_{n'} t_2 + \phi_{n'})\right]$$

when the expectation is over all ω_n, ϕ_n, and T. The linearity of the ensemble average, and the independence of ω_n, ϕ_n, and T, are used to obtain

$$R_s(t_1,t_2) = 2 \sum_n \sum_{n'} E_1[p(t_1 + T - nT_c)p(t_2 + T - n'T_c)]$$

$$\times \{E_2[\cos(\omega_n t_1 + \omega_{n'} t_2 + \phi_n + \phi_{n'})] \qquad \text{(C-10)}$$

$$+ E_2[\cos(\omega_n t_1 - \omega_{n'} t_2 + \phi_n - \phi_{n'})]\}$$

where E_1 is over all T and E_2 is over all ω_n and ϕ_n. For $n \neq n'$, the average over all ϕ_n and ϕ'_n in E_2 implies that these terms are zero. This is also true for the sum frequency term even when $n = n'$. Thus

$$R_s(t_1,t_2) = 2 \sum_n E_1[p(t_1 + T - nT_c)p(t_2 + T - nT_c)]$$

$$\times E_2[\cos(\omega_n[t_1 - t_2])] \qquad \text{(C-11)}$$

All frequencies are assumed to be discrete and equally probable, so that the second expected value for any n is simply a summation over all 2^k frequencies, yielding

$$R_s(t_1,t_2) = \frac{1}{2^{k-1}} \sum_{m=1}^{2^k} \cos(\omega_m[t_1 - t_2]) \sum_n E_1[p(t_1 + T - nT_c)$$

$$\times p(t_2 + T - nT_c)] \qquad \text{(C-12)}$$

The remaining expected value was evaluated above so that the final result can be written directly and is

$$R_s(t_1,t_2) = R_s(\tau) = \begin{cases} \dfrac{1}{2^{k-1}} \displaystyle\sum_{m=1}^{2^k} \cos \omega_m \tau \left(1 - \dfrac{|\tau|}{T_c}\right) & \text{for } |\tau| < T_c \\[2em] 0 & \text{for } |\tau| > T_c \end{cases} \qquad \text{(C-13)}$$

The power spectral density of $s(t)$ is the Fourier transform of (C-13). The Fourier transform is a linear operation, so that each term of the sum can be independently evaluated. Each term of (C-13) is a product of terms each of whose Fourier transform is known. The frequency convolution theorem is invoked, yielding

$$S_s(f) = \frac{T_c}{2^k} \sum_{m=1}^{2^k} \{\mathrm{sinc}^2[(f - f_m)T_c] + \mathrm{sinc}^2[(f + f_m)T_c]\} \qquad \text{(C-14)}$$

This expression is equivalent to the continuous frequency term of (2-41). In this case none of the signal power is in discrete components as was anticipated.

Reference

[1] W. C. LINDSEY and M. K. SIMON, *Telecommunication Systems Engineering* (Englewood Cliffs, N.J.: Prentice Hall, 1973).

APPENDIX **D**

Calculation of the Power Spectrum of the Product of Two *M*-Sequences

The following development is a minor modification of the development by Gill [1]. The power spectrum of $c(t)c(t + \epsilon)$ is calculated by first determining the autocorrelation function and then using the Wiener–Khintchine theorem. Denote the product $c(t)c(t + \epsilon)$ by $b(t,\epsilon)$. The autocorrelation function of this periodic function is

$$R_b(\tau,\epsilon) = \frac{1}{T} \int_0^T b(t,\epsilon)b(t + \tau,\epsilon) \, dt \qquad (D\text{-}1)$$

The functions $c(t)$, $c(t + \epsilon)$, and $b(t + \epsilon)$ are illustrated in Figure D-1 for a 7-bit m-sequence. For any particular m-sequence and any ϵ, $R_b(\tau,\epsilon)$ can be calculated directly. This calculation is facilitated, however, by recognizing [1] that $b(t,\epsilon)$ can be represented by the sum of two functions $p(t,\epsilon) + q(t,\epsilon)$, which are also illustrated in Figure D-1. The function $p(t,\epsilon)$ is a binary-valued function with period T_c, the code chip duration. The function $q(t,\epsilon)$ is a three-valued function whose nonzero values are the same as the original m-sequence shifted in phase. Using this decomposition, the autocorrelation function can be written as the sum

$$R_b(\tau,\epsilon) = R_p(\tau,\epsilon) + R_{pq}(\tau,\epsilon) + R_{qp}(\tau,\epsilon) + R_q(\tau,\epsilon) \qquad (D\text{-}2)$$

Each of these autocorrelations can be separately evaluated and separately Fourier transformed to obtain the desired power spectrum. The following discussion is limited to values of $|\epsilon| \le T_c$.

Consider first the autocorrelation of $p(t,\epsilon)$. This autocorrelation

$$R_p(\tau,\epsilon) = \frac{1}{T_c} \int_0^{T_c} p(t,\epsilon)p(t + \tau,\epsilon) \, dt \qquad (D\text{-}3)$$

is illustrated in Figure D-2a for $0 \le |\epsilon| \le T_c/2$ and in Figure D-2b for $T_c/2 < |\epsilon| \le T_c$. These functions are calculated by inspection from Figure D-1d. Each of the functions is a periodic sequence of triangular pulses. For $0 \le |\epsilon| \le T_c/2$.

$$R_p(\tau,\epsilon) = \left(1 - 2\frac{|\epsilon|}{T_c}\right) + \frac{|\epsilon|}{T_c} \sum_{n=-\infty}^{\infty} \Lambda(\tau - nT_c, |\epsilon|) \qquad (D\text{-}4a)$$

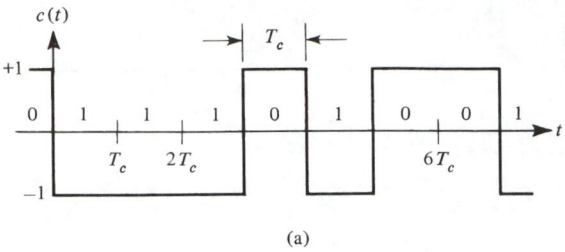

(a)

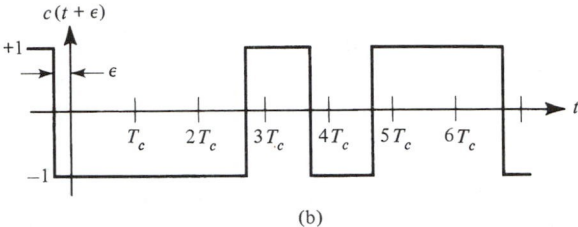

(b)

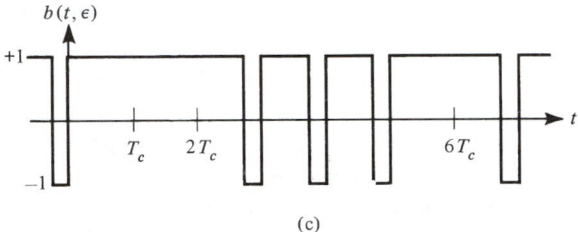

(c)

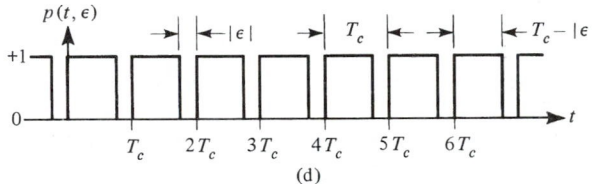

(d)

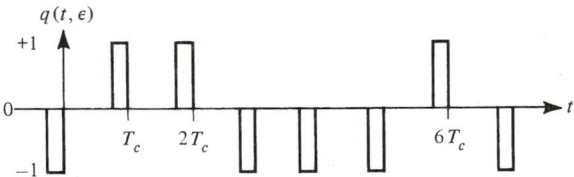

FIGURE D-1. Waveform used in the calculation of the power spectrum of $c(t)c(t + \epsilon)$. (From Ref. 1.)

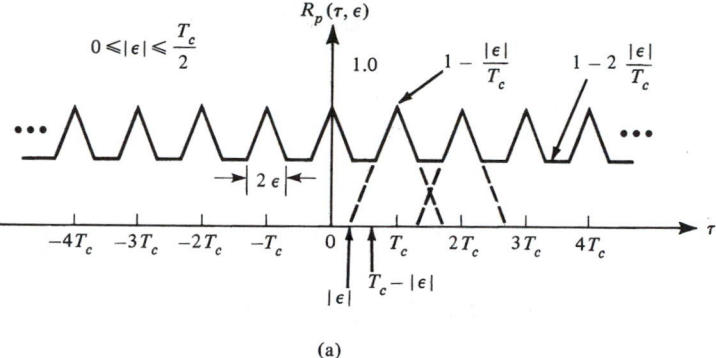

(a)

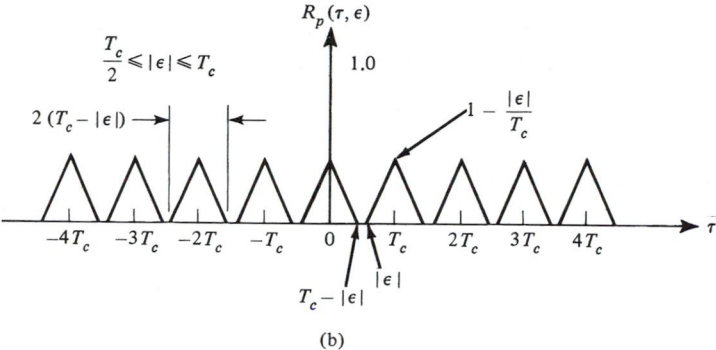

(b)

FIGURE D-2. Autocorrelation function of $p(t,\epsilon)$: (a) $0 \leq |\epsilon| < T_c/2$; (b) $T_c/2 \leq |\epsilon| \leq T_c$. (From Ref. 1.)

and for $T_c/2 < |\epsilon| \leq T_c$,

$$R_p(\tau,\epsilon) = \left(1 - \frac{|\epsilon|}{T_c}\right) \sum_{n=-\infty}^{\infty} \Lambda(\tau - nT_c, T_c - |\epsilon|) \qquad \text{(D-4b)}$$

In (D-4) the function $\Lambda(\tau,B)$ is a triangular pulse of height 1.0 and width $2B$, as illustrated in Figure D-3. When $\epsilon \leq T_c/2$ the triangular pulses rest on a constant pedestal of height $(1 - 2|\epsilon/T_c|)$.

The Fourier transform of a periodic waveform $x(t)$ with period T, for example,

$$x(t) = \sum_{m=-\infty}^{\infty} y(t - mT) \qquad \text{(D-5a)}$$

is

$$X(f) = \frac{1}{T} \sum_{n=-\infty}^{\infty} Y(nf_0)\delta(f - nf_0) \qquad \text{(D-5b)}$$

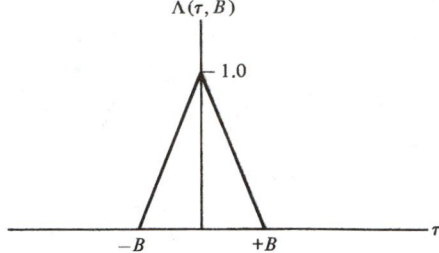

FIGURE D-3. Triangular pulse waveform.

where $Y(f)$ is the Fourier transform of a single pulse $y(t)$ and $f_0 = 1/T$. The Fourier transform of the triangular pulse $\Lambda(\tau,B)$ is $S_\Lambda(f,B) = B \, \text{sinc}^2 fB$. Therefore, substituting the triangular pulse for $y(t)$ in (D-5) yields the power spectrum of $p(t,\epsilon)$. The result for $0 \le |\epsilon| \le T_c/2$ is

$$S_p(f,\epsilon) = \left(1 - 2\frac{|\epsilon|}{T_c}\right)\delta(f) + \left(\frac{|\epsilon|}{T_c}\right)^2 \sum_{n=-\infty}^{\infty} \text{sinc}^2(nf_c|\epsilon|)\delta(f - nf_c) \tag{D-6a}$$

and for $T_c/2 < |\epsilon| \le T_c$ is

$$S_p(f,\epsilon) = \left(1 - \frac{|\epsilon|}{T_c}\right)^2 \sum_{n=-\infty}^{\infty} \text{sinc}^2[nf_c(T_c - |\epsilon|)]\delta(f - nf_c) \tag{D-6b}$$

where $f_c = 1/T_c$.

The autocorrelation function of $q(t,\epsilon)$ is illustrated in Figure D-4a for $0 \le |\epsilon| < T_c/2$ and Figure D-4b for $T_c/2 \le |\epsilon| \le T_c$. These autocorrelation functions are calculated by inspection of Figure D-1e and from the fact that the values of $q(t,\epsilon)$, where it is nonzero, are the same as the original *m*-sequence values shifted in phase. The functions of Figure D-4a or D-4b can be decomposed into two periodic sequences of triangular functions, that is, $R_q(\tau,\epsilon) = R_{qa}(\tau,\epsilon) + R_{qb}(\tau,\epsilon)$, where, for $0 \le |\epsilon| \le T_c/2$,

$$R_{qa}(\tau,\epsilon) = -\frac{|\epsilon|}{NT_c} \sum_{n=-\infty}^{\infty} \Lambda(\tau - nT_c,|\epsilon|) \tag{D-7a}$$

and for $T_c/2 < |\epsilon| \le T_c$,

$$R_{qa}(\tau,\epsilon) = -\frac{1}{N}\left(2\frac{|\epsilon|}{T_c} - 1\right) - \frac{1}{N}\left(1 - \frac{|\epsilon|}{T_c}\right) \sum_{n=-\infty}^{\infty} \Lambda(\tau - nT_c, T_c - |\epsilon|) \tag{D-7b}$$

and for all $0 \le |\epsilon| \le T_c$,

$$R_{qb}(\tau,\epsilon) = \left(1 + \frac{1}{N}\right)\frac{|\epsilon|}{T_c} \sum_{m=-\infty}^{\infty} \Lambda(\tau - mNT_c,|\epsilon|) \tag{D-7c}$$

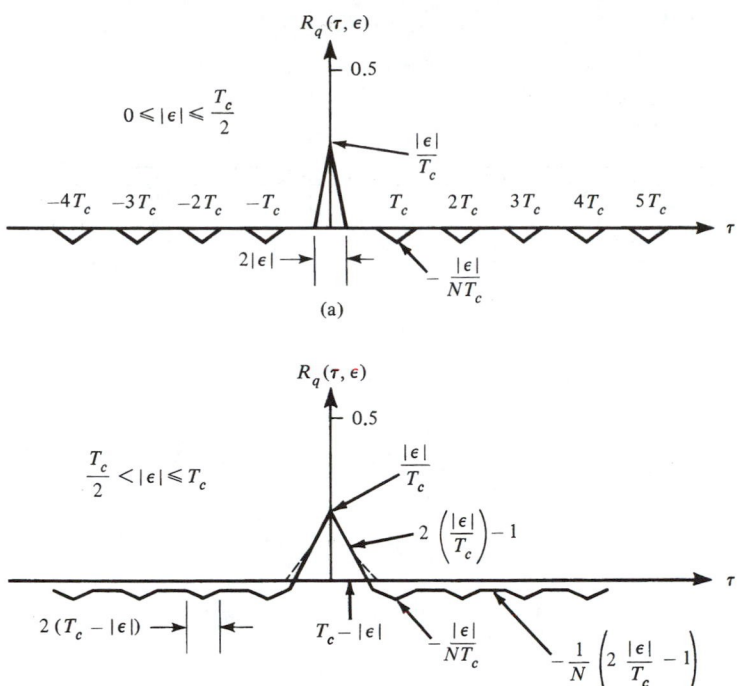

FIGURE D-4. Autocorrelation function of $q(t,\epsilon)$: (a) $0 < |\epsilon| \le T_c/2$; (b) $T_c/2 < |\epsilon| \le T_c$. (From Ref. 1.)

Now consider the cross-correlation functions $R_{pq}(\tau,\epsilon)$ and $R_{qp}(\tau,\epsilon)$. By definition

$$R_{qp}(\tau,\epsilon) = \frac{1}{NT_c} \int_0^{NT_c} q(t,\epsilon)p(t + \tau,\epsilon)\, dt \qquad \text{(D-8)}$$

Using the substitution $\lambda = t + \tau$, this equation becomes

$$R_{qp}(\tau,\epsilon) = \frac{1}{NT_c} \int_\tau^{\tau + NT_c} q(\lambda - \tau,\epsilon)p(\lambda,\epsilon)\, d\lambda \qquad \text{(D-9)}$$

$$= R_{pq}(-\tau,\epsilon)$$

since both $p(t,\epsilon)$ and $q(t,\epsilon)$ are periodic with period NT_c. The function $R_{pq}(\tau,\epsilon)$ is found by inspection of Figure D-1 and knowledge that values of $q(t,\epsilon)$ follow the original *m*-sequence. The function is illustrated in Figure D-5a for $0 \le |\epsilon| \le T_c/2$ and in Figure D-5b for $T_c/2 < |\epsilon| < T_c$. These functions can be written as a sum of triangle functions on a constant pedestal yielding for $0 \le |\epsilon| \le T_c/2$,

$$R_{pq}(\tau,\epsilon) = -\frac{|\epsilon|}{NT_c} + \frac{|\epsilon|}{NT_c} \sum_{n=-\infty}^{\infty} \Lambda(\tau - nT_c, |\epsilon|) \qquad \text{(D-10a)}$$

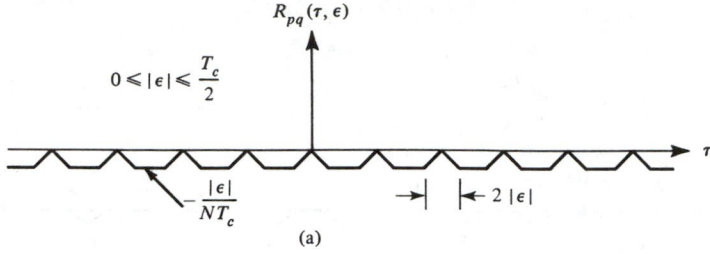

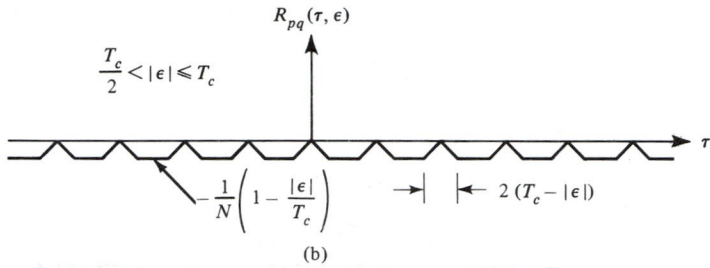

FIGURE D-5. Cross-correlation function of $p(t,\epsilon)$ and $q(t,\epsilon)$: (a) $0 < |\epsilon| \leq T_c/2$; (b) $T_c/2 < |\epsilon| \leq T_c$. (From Ref. 1.)

and for $T_c/2 \leq |\epsilon| \leq T_c$,

$$R_{pq}(\tau,\epsilon) = -\frac{1}{N}\left(1 - \frac{|\epsilon|}{T_c}\right) + \frac{1}{N}\left(1 - \frac{|\epsilon|}{T_c}\right) \sum_{n=-\infty}^{\infty} \Lambda(\tau - nT_c, T_c - |\epsilon|) \qquad \text{(D-10b)}$$

The sum of the final three terms $R_{\Sigma}(\tau,\epsilon)$ of (D-2) can now be written for $0 \leq |\epsilon| \leq T_c/2$,

$$R_{\Sigma}(\tau,\epsilon)$$

$$= R_{pq}(\tau,\epsilon) + R_{qp}(\tau,\epsilon) + R_q(\tau,\epsilon)$$

$$= 2R_{pq}(\tau,\epsilon) + R_{qa}(\tau,\epsilon) + R_{qb}(\tau,\epsilon)$$

$$= -\frac{2|\epsilon|}{NT_c} + \frac{2|\epsilon|}{NT_c} \sum_{n=-\infty}^{\infty} \Lambda(\tau - nT_c, |\epsilon|)$$

$$\qquad\qquad\qquad\qquad\qquad\qquad\qquad\qquad\qquad\qquad\qquad\text{(D-11)}$$

$$\quad - \frac{|\epsilon|}{NT_c} \sum_{n=-\infty}^{\infty} \Lambda(\tau - nT_c, |\epsilon|) + \left(1 + \frac{1}{N}\right)\frac{|\epsilon|}{T_c} \sum_{m=-\infty}^{\infty} \Lambda(\tau - mNT_c, |\epsilon|)$$

$$= -\frac{2|\epsilon|}{NT_c} + \frac{|\epsilon|}{NT_c} \sum_{n=-\infty}^{\infty} \Lambda(\tau - nT_c, |\epsilon|)$$

$$\quad + \left(1 + \frac{1}{N}\right)\frac{|\epsilon|}{T_c} \sum_{m=-\infty}^{\infty} \Lambda(\tau - mNT_c, |\epsilon|)$$

Observe that the triangle functions have been combined in the last line. For $T_c/2 < |\epsilon| \leq T_c$ this expression becomes

$$R_\Sigma(\tau,\epsilon) = R_{pq}(\tau,\epsilon) + R_{qp}(\tau,\epsilon) + R_q(\tau,\epsilon)$$

$$= 2R_{pq}(\tau,\epsilon) + R_{qa}(\tau,\epsilon) + R_{qb}(\tau,\epsilon)$$

$$= -\frac{2}{N}\left(1 - \frac{|\epsilon|}{T_c}\right) + \frac{2}{N}\left(1 - \frac{|\epsilon|}{T_c}\right) \sum_{n=-\infty}^{\infty} \Lambda(\tau - nT_c, T_c - |\epsilon|)$$

$$- \frac{1}{N}\left(2\frac{|\epsilon|}{T_c} - 1\right) - \frac{1}{N}\left(1 - \frac{|\epsilon|}{T_c}\right) \sum_{n=-\infty}^{\infty} \Lambda(\tau - nT_c, T_c - |\epsilon|)$$

$$+ \left(1 + \frac{1}{N}\right)\frac{|\epsilon|}{T_c} \sum_{m=-\infty}^{\infty} \Lambda(\tau - mNT_c, |\epsilon|)$$

$$\text{(D-12)}$$

$$= -\frac{1}{N} + \frac{1}{N}\left(1 - \frac{|\epsilon|}{T_c}\right) \sum_{n=-\infty}^{\infty} \Lambda(\tau - nT_c, T_c - |\epsilon|)$$

$$+ \left(1 + \frac{1}{N}\right)\frac{|\epsilon|}{T_c} \sum_{m=-\infty}^{\infty} \Lambda(\tau - mNT_c, |\epsilon|)$$

Equation (D-5) is employed to find the Fourier transforms and complete the calculation of the power spectrum of $c(t)c(t + \epsilon)$. Using (D-5) and $S_\Lambda(f,B) = B \, \text{sinc}^2 fB$, for $0 < |\epsilon| \leq T_c/2$,

$$S_\Sigma(f,\epsilon) = -\frac{2|\epsilon|}{NT_c}\delta(f) + \frac{1}{N}\left(\frac{|\epsilon|}{T_c}\right)^2 \sum_{n=-\infty}^{\infty} \text{sinc}^2(nf_c|\epsilon|)\delta(f - nf_c)$$

$$\text{(D-13a)}$$

$$+ \frac{1}{N}\left(1 + \frac{1}{N}\right)\left(\frac{|\epsilon|}{T_c}\right)^2 \sum_{m=-\infty}^{\infty} \text{sinc}^2\left(m\frac{f_c}{N}|\epsilon|\right)\delta\left(f - m\frac{f_c}{N}\right)$$

and for $T_c/2 < |\epsilon| \leq T_c$,

$$S_\Sigma(f,\epsilon) = -\frac{1}{N}\delta(f) + \frac{1}{N}\left(1 - \frac{|\epsilon|}{T_c}\right)^2 \sum_{n=-\infty}^{\infty} \text{sinc}^2[nf_c(T_c - |\epsilon|)]\delta(f - nf_c)$$

$$\text{(D-13b)}$$

$$+ \frac{1}{N}\left(1 + \frac{1}{N}\right)\left(\frac{|\epsilon|}{T_c}\right)^2 \sum_{m=-\infty}^{\infty} \text{sinc}^2\left(m\frac{f_c}{N}|\epsilon|\right)\delta\left(f - m\frac{f_c}{N}\right)$$

Finally, to simplify the final expression for the power spectral density, examine the function $\text{sinc}^2[nf_c(T_c - |\epsilon|)]$ which appears in both (D-6) and (D-13). For $n \neq 0$,

$$\text{sinc}^2[nf_c(T_c - |\epsilon|)] \triangleq \frac{\sin^2[\pi nf_c(T_c - |\epsilon|)]}{[\pi nf_c(T_c - |\epsilon|)]^2}$$

$$= \frac{\sin^2[\pi n - \pi n f_c|\epsilon|]}{[\pi n f_c|\epsilon|]^2} \frac{[\pi n f_c|\epsilon|]^2}{[\pi n f_c(T_c - |\epsilon|)]^2}$$

$$= \frac{\cos^2(\pi n) \sin^2(\pi n f_c|\epsilon|)}{[\pi n f_c|\epsilon|]^2} \left(\frac{|\epsilon|}{T_c - |\epsilon|}\right)^2 \qquad (D\text{-}14)$$

$$= \text{sinc}^2(n f_c|\epsilon|) \frac{(|\epsilon|/T_c)^2}{(1 - |\epsilon|/T_c)^2}$$

and for $n = 0$,

$$\text{sinc}^2[n f_c(T_c - |\epsilon|)] = 1 = \text{sinc}^2 n f_c|\epsilon|$$

At this point the final expression for the power spectrum can be written by adding (D-6) and (D-13) and using (D-14). This result is, for $0 \le |\epsilon| < T_c/2$,

$$S_b(f,\epsilon) = \left(1 - 2\frac{|\epsilon|}{T_c}\right)\delta(f) + \left(\frac{|\epsilon|}{T_c}\right)^2 \sum_{n=-\infty}^{\infty} \text{sinc}^2(n f_c|\epsilon|)\delta(f - n f_c)$$

$$- 2\frac{|\epsilon|}{NT_c}\delta(f) + \frac{1}{N}\left(\frac{|\epsilon|}{T_c}\right)^2 \sum_{n=-\infty}^{\infty} \text{sinc}^2(n f_c|\epsilon|)\delta(f - n f_c)$$

$$+ \frac{1}{N}\left(1 + \frac{1}{N}\right)\left(\frac{|\epsilon|}{T_c}\right)^2 \sum_{m=-\infty}^{\infty} \text{sinc}^2\left(\frac{m f_c}{N}|\epsilon|\right)\delta\left(f - \frac{m f_c}{N}\right)$$

$$= \left[1 - \left(1 + \frac{1}{N}\right)\left(\frac{|\epsilon|}{T_c}\right)\right]^2 \delta(f) \qquad (D\text{-}15)$$

$$+ \left(1 + \frac{1}{N}\right)\left(\frac{|\epsilon|}{T_c}\right)^2 \sum_{\substack{n=-\infty \\ n \ne 0}}^{\infty} \text{sinc}^2(n f_c|\epsilon|)\delta(f - n f_c)$$

$$+ \frac{N+1}{N^2}\left(\frac{|\epsilon|}{T_c}\right)^2 \sum_{\substack{m=-\infty \\ m \ne 0}}^{\infty} \text{sinc}^2\left(\frac{m f_c|\epsilon|}{N}\right)\delta\left(f - \frac{m f_c}{N}\right)$$

and for $T_c/2 \le |\epsilon| \le T_c$,

$$S_b(f,\epsilon) = \left(\frac{|\epsilon|}{T_c}\right)^2 \sum_{\substack{n=-\infty \\ n \ne 0}}^{\infty} \text{sinc}^2(n f_c|\epsilon|)\delta(f - n f_c) + \left(1 - \frac{|\epsilon|}{T_c}\right)^2 \delta(f)$$

$$- \frac{1}{N}\delta(f) + \frac{1}{N}\left(\frac{|\epsilon|}{T_c}\right)^2 \sum_{\substack{n=-\infty \\ n \ne 0}}^{\infty} \text{sinc}^2(n f_c|\epsilon|)\delta(f - n f_c)$$

$$+ \frac{1}{N}\left(1 - \frac{|\epsilon|}{T_c}\right)^2 \delta(f)$$

$$+ \frac{1}{N}\left(1 + \frac{1}{N}\right)\left(\frac{|\epsilon|}{T_c}\right)^2 \sum_{\substack{m=-\infty \\ m \neq 0}}^{\infty} \operatorname{sinc}^2\left(\frac{mf_c}{N}|\epsilon|\right)\delta\left(f - \frac{mf_c}{N}\right)$$

$$+ \frac{1}{N}\left(1 + \frac{1}{N}\right)\left(\frac{|\epsilon|}{T_c}\right)^2 \delta(f)$$

$$= \left[1 - \left(1 + \frac{1}{N}\right)\left(\frac{|\epsilon|}{T_c}\right)\right]^2 \delta(f) \tag{D-16}$$

$$+ \left(1 + \frac{1}{N}\right)\left(\frac{|\epsilon|}{T_c}\right)^2 \sum_{\substack{n=-\infty \\ n \neq 0}}^{\infty} \operatorname{sinc}^2(nf_c|\epsilon|)\delta(f - nf_c)$$

$$+ \left(\frac{N+1}{N^2}\right)\left(\frac{|\epsilon|}{T_c}\right)^2 \sum_{\substack{m=-\infty \\ m \neq 0}}^{\infty} \operatorname{sinc}^2\left(\frac{mf_c}{N}|\epsilon|\right)\delta\left(f - \frac{mf_c}{N}\right)$$

Before using (D-14) to simplify (D-13b) and (D-6b), the zero-frequency term must first be removed from the summation. Observe that (D-15) and (D-16) are identical, so that the desired power spectrum can be calculated from either for any $0 \leq |\epsilon| \leq T_c$.

Reference

[1] W. J. GILL, "Effect of Synchronization Error in Pseudo-random Carrier Communications," *Conf. Rec.*, First Annual IEEE Commun. Conf., pp. 187–191, June 1965.

APPENDIX **E**

Evaluation of Phase Discriminator Output Autocorrelation Functions and Power Spectra

E-1 Noncoherent Delay-Lock Tracking Loop

The discriminator output autocorrelation function is defined in (4-64) with $\epsilon(t,\delta)$ given in (4-63). To shorten the following equations, define

$$
\begin{aligned}
C_1 &= \frac{1}{2} K_1 P \left\{ R_c^2 \left[\left(\delta - \frac{\Delta}{2} \right) T_c \right] - R_c^2 \left[\left(\delta + \frac{\Delta}{2} \right) T_c \right] \right\} \\
C_2 &= \sqrt{2 K_1 P} \\
R_- &= R_c \left[\left(\delta - \frac{\Delta}{2} \right) T_c \right] \\
R_+ &= R_c \left[\left(\delta + \frac{\Delta}{2} \right) T_c \right]
\end{aligned}
\tag{E-1}
$$

Then

$$
\begin{aligned}
E[\epsilon(t,\delta)\epsilon(t + \tau, \delta)] = E\{&[C_1 + C_2\{R_-n_{2I}(t) - R_+n_{1I}(t)\} \cos[\phi - \phi' + \theta_d(t)] \\
&+ C_2\{R_-n_{2Q}(t) - R_+n_{1Q}(t)\} \sin[\phi - \phi' + \theta_d(t)] \\
&+ [n_{2I}(t)]^2 + [n_{2Q}(t)]^2 - [n_{1I}(t)]^2 - [n_{1Q}(t)]^2 \\
&\times [C_1 + C_2\{R_-n_{2I}(t + \tau) - R_+n_{1I}(t + \tau)\} \\
&\quad \times \cos[\phi - \phi' + \theta_d(t + \tau)] \\
&+ C_2\{R_-n_{2Q}(t + \tau) - R_+n_{1Q}(t + \tau)\} \\
&\quad \times \sin[\phi - \phi' + \theta_d(t + \tau)] \\
&+ [n_{2I}(t + \tau)]^2 + [n_{2Q}(t + \tau)]^2 \\
&- [n_{1I}(t + \tau)]^2 - [n_{1Q}(t + \tau)]^2]\}
\end{aligned}
\tag{E-2}
$$

Expansion of the products within this expected value yields the sum of a large number of terms. The expected value of each term can be separately evaluated. The random noise process is independent of the local oscillator phase so that the expectations can be factored. For example,

$$E[C_1 C_2 \{R_- n_{2I}(t + \tau) - R_+ n_{1I}(t + \tau)\} \cos[\phi - \phi' + \theta_d(t + \tau)]]$$

$$= C_1 C_2 E[R_- n_{2I}(t + \tau) - R_+ n_{1I}(t + \tau)] E[\cos[\phi - \phi' + \theta_d(t + \tau)]]$$

(E-3)

The cosine term can be further expanded yielding

$$E[\cos[\phi - \phi' + \theta_d(t + \tau)]] = E[\cos(\phi - \phi')] E[\cos(\theta_d(t + \tau))]$$

$$- E[\sin(\phi - \phi')] E[\sin(\theta_d(t + \tau))]$$

(E-4)

The received carrier phase and the local oscillator phase are uniformly distributed over $(0, 2\pi)$ so that

$$E[\cos(\phi - \phi')] = E[\sin(\phi - \phi')] = 0 \qquad \text{(E-5)}$$

Therefore, all expressions having the form of (E-3) may be set to zero in the expression. Expanding (E-2) yields

$$E[\epsilon(t, \delta)\epsilon(t + \tau, \delta)] = C_1^2 + C_1 E[\{n_{2I}(t + \tau)\}^2 + \{n_{2Q}(t + \tau)\}^2$$

$$- \{n_{1I}(t + \tau)\}^2 - \{n_{1Q}(t + \tau)\}^2]$$

$$+ C_2^2 E[\{R_- n_{2I}(t) - R_+ n_{1I}(t)\}$$

$$\times \{R_- n_{2I}(t + \tau) - R_+ n_{1I}(t + \tau)\}]$$

$$\times E[\cos\{\phi - \phi' + \theta_d(t)\} \cos\{\phi - \phi' + \theta_d(t + \tau)\}]$$

$$+ C_2^2 E[\{R_- n_{2I}(t) - R_+ n_{1I}(t)\}$$

$$\times \{R_- n_{2Q}(t + \tau) - R_+ n_{1Q}(t + \tau)\}]$$

$$\times E[\cos\{\phi - \phi' + \theta_d(t)\} \sin\{\phi - \phi' + \theta_d(t + \tau)\}]$$

$$+ C_2^2 E[\{R_- n_{2Q}(t) - R_+ n_{1Q}(t)\}$$

$$\times \{R_- n_{2I}(t + \tau) - R_+ n_{1I}(t + \tau)\}]$$

$$\times E[\sin\{\phi - \phi' + \theta_d(t)\} \cos\{\phi - \phi' + \theta_d(t + \tau)\}]$$

$$+ C_2^2 E[\{R_- n_{2Q}(t) - R_+ n_{1Q}(t)\}$$

(E-6)

$$\times \{R_- n_{2Q}(t + \tau) - R_+ n_{1Q}(t + \tau)\}]$$

$$\times E[\sin\{\phi - \phi' + \theta_d(t)\} \sin\{\phi - \phi' + \theta_d(t + \tau)\}]$$

$$+ C_1 E[\{n_{2I}(t)\}^2 + \{n_{2Q}(t)\}^2 - \{n_{1I}(t)\}^2 - \{n_{1Q}(t)\}^2]$$

$$+ E[\{[n_{2I}(t)]^2 + [n_{2Q}(t)]^2 - [n_{1I}(t)]^2 - [n_{1Q}(t)]^2\}$$

$$\times \{[n_{2I}(t + \tau)]^2 + [n_{2Q}(t + \tau)]^2$$

$$- [n_{1I}(t + \tau)]^2 - [n_{1Q}(t + \tau)]^2\}]$$

The baseband noise processes $n_{1I}(t)$, $n_{2I}(t)$, $n_{1Q}(t)$, and $n_{2Q}(t)$ have been assumed to be white Gaussian noise processes but have not been assumed to be independent. It was demonstrated earlier that, under the assumptions that $\Delta \geq 1.0$ and that the input noise have a significantly wider bandwidth than the received signal, these noise processes are uncorrelated and therefore independent. In order to complete this analysis, assume that the processes are independent so that the products of expected value factor. With this assumption and because

$$E[n_{1I}(t)] = E[n_{1Q}(t)] = E[n_{2I}(t)] = E[n_{2Q}(t)] = 0$$

Equation (E-6) simplifies to

$$
\begin{aligned}
E[\epsilon(t,\delta)\epsilon(t + \tau, \delta)] = C_1^2 &+ C_1 E[\{n_{2I}(t)\}^2 + \{n_{2Q}(t)\}^2 - \{n_{1I}(t)\}^2 - \{n_{1Q}(t)\}^2] \\
&+ \{C_2^2 R_-^2 E[n_{2I}(t)n_{2I}(t + \tau)] \\
&\quad + C_2^2 R_+^2 E[n_{1I}(t)n_{1I}(t + \tau)]\} \\
&\times E[\cos\{\phi - \phi' + \theta_d(t)\} \cos\{\phi - \phi' + \theta_d(t + \tau)\}] \\
&+ \{C_2^2 R_-^2 E[n_{2Q}(t)n_{2Q}(t + \tau)] \\
&\quad + C_2^2 R_+^2 E[n_{1Q}(t)n_{1Q}(t + \tau)]\} \\
&\times E[\sin\{\phi - \phi' + \theta_d(t)\} \sin\{\phi - \phi' + \theta_d(t + \tau)\}] \\
&+ C_1 E[\{n_{2I}(t + \tau)\}^2 + \{n_{2Q}(t + \tau)\}^2 \\
&\quad - \{n_{1I}(t + \tau)\}^2 - \{n_{1Q}(t + \tau)\}^2] \\
&+ E[\{[n_{2I}(t)]^2 + [n_{2Q}(t)]^2 - [n_{1I}(t)]^2 - [n_{1Q}(t)]^2\} \\
&\quad \times \{[n_{2I}(t + \tau)]^2 + [n_{2Q}(t + \tau)]^2 \\
&\quad - [n_{1I}(t + \tau)]^2 - [n_{1Q}(t + \tau)]^2\}]
\end{aligned}
$$

(E-7)

Further simplification of this result is achieved if it is noticed that all of the noise processes have identical statistics. Combining similar terms yields

$$
\begin{aligned}
E[\epsilon(t,\delta)\epsilon(t + \tau,\delta)] = C_1^2 &+ C_2^2(R_-^2 + R_+^2)E[n_b(t)n_b(t + \tau)] \\
&\times E[\cos\{\phi - \phi' + \theta_d(t)\} \cos\{\phi - \phi' + \theta_d(t + \tau)\}] \\
&+ C_2^2(R_-^2 + R_+^2)E[n_b(t)n_b(t + \tau)] \\
&\times E[\sin\{\phi - \phi' + \theta_d(t)\} \sin\{\phi - \phi' + \theta_d(t + \tau)\}] \\
&+ E[4\{n_b(t)n_b(t + \tau)\}^2 - 4\{n_b(t)\}^2\{n_b(t + \tau)\}^2] \\
= C_1^2 &+ C_2^2(R_-^2 + R_+^2)E[n_b(t)n_b(t + \tau)] \\
&\times E[\cos\{\theta_d(t) - \theta_d(t + \tau)\}] \\
&+ 4E[\{n_b(t)n_b(t + \tau)\}^2] - 4E^2[\{n_b(t)\}^2]
\end{aligned}
$$

(E-8)

where $n_b(t)$ is any one of the four baseband noise processes. Observe that this function is not independent of the data modulation because of the term $E[\cos\{\theta_d(t) - \theta_d(t + \tau)\}]$. Using (E-1), (E-8) becomes

$$R_\epsilon(\tau) = \tfrac{1}{4}K_1^2 P^2 \left\{ R_c^2\left[\left(\delta - \frac{\Delta}{2}\right)T_c\right] - R_c^2\left[\left(\delta + \frac{\Delta}{2}\right)T_c\right] \right\}^2$$

$$+ 2K_1 P \left\{ R_c^2 \left[\left(\delta - \frac{\Delta}{2} \right) T_c \right] + R_c^2 \left[\left(\delta + \frac{\Delta}{2} \right) T_c \right] \right\} E[n_b(t)n_b(t + \tau)] \tag{E-9}$$

$$\times E[\cos\{\theta_d(t) - \theta_d(t + \tau)\}]$$

$$+ 4E[\{n_b(t)n_b(t + \tau)\}^2] - 4E^2[\{n_b(t)\}^2]$$

The power spectrum at the discriminator output is found by taking the Fourier transform of (E-9). This Fourier transform will be considered one term at a time. The first term is not a function of τ so that its transform results in a delta function at dc. This dc component is the desired phase correction term.

The second term of (E-9) is a (signal $\times$ noise) term which is the result of the squaring operation. It is the product of two functions of τ so that the Fourier multiplication theorem can be used to calculate the Fourier transform. By definition

$$E[n_b(t)n_b(t + \tau)] = R_{n_b}(\tau) \tag{E-10}$$

and by the Wiener–Khintchine theorem

$$S_{n_b}(f) = \int_{-\infty}^{\infty} R_{n_b}(\tau)e^{-j2\pi f\tau} \, d\tau \tag{E-11}$$

The expected value of $\cos[\theta_d(t) - \theta_d(t - \tau)]$ is evaluated by considering the data modulated carrier

$$a(t) = \cos[\omega_{IF}t + \theta_d(t) + \beta] \tag{E-12}$$

with complex envelope

$$A(t) = \exp[j\theta_d(t) + j\beta] \tag{E-13}$$

The complex autocorrelation function is defined [1] by

$$R_A(\tau) = \tfrac{1}{2}E[A^*(t)A(t + \tau)] \tag{E-14}$$

where the complex conjugate arises from the fact that the difference frequency term of the product of $a(t)a(t + \tau)$ is the desired term in the real autocorrelation $R_a(\tau)$. Substituting (E-13) into (E-14) yields

$$R_A(\tau) = \tfrac{1}{2}E\{\exp[-j\theta_d(t) + j\theta_d(t + \tau)]\} \tag{E-15}$$

so that

$$E[\cos\{\theta_d(t) - \theta_d(t + \tau)\}] = 2\,\text{Re}[R_A(\tau)] \tag{E-16}$$

The Fourier transform of (E-16) is

$$S_{\theta_d}(f) = \int_{-\infty}^{\infty} 2\,\text{Re}[R_A(\tau)]e^{-j2\pi f\tau} \, d\tau \tag{E-17}$$

$$= \int_{-\infty}^{\infty} [R_A(\tau) + R_A^*(\tau)]e^{-j2\pi f\tau} \, d\tau$$

It is easily shown that $R_A^*(\tau) = R_A(-\tau)$ using a change of variable in (E-16) so that

$$S_{\theta_d}(f) = \int_{-\infty}^{\infty} R_A(\tau)e^{-j2\pi f\tau}\,d\tau + \int_{-\infty}^{\infty} R_A(-\tau)e^{-j2\pi f\tau}\,d\tau$$

$$= S_A(f) + S_A(-f) \tag{E-18}$$

Finally, the total contribution of the (signal $\times$ noise) term of (E-9) to the discriminator output power spectrum is

$$S_2(f) = 2K_1P\left\{R_c^2\left[\left(\delta - \frac{\Delta}{2}\right)T_c\right] + R_c^2\left[\left(\delta + \frac{\Delta}{2}\right)T_c\right]\right\}S_{n_b}(f) * S_{\theta_d}(f) \tag{E-19}$$

The third term of (E-9) is four times the autocorrelation function of the output of a square-law device with $n_b(t)$ as its input. The Fourier transform yields the power spectrum of the output of this same device. This output spectrum has been calculated in detail in Ref. 2. The square law device considered in this analysis is characterized by

$$y = ax^2 \tag{E-20}$$

where the input x is a bandlimited Gaussian noise process which has power spectrum $S_x(f)$ and total power σ_x^2. The power spectrum of y is shown to be

$$S_y(f) = a^2\sigma_x^4\delta(f) + 2a^2S_x(f) * S_x(f) \tag{E-21}$$

Using this result, the contribution of the third term in (E-9) is

$$S_3(f) = 4\sigma_{n_b}^4\delta(f) + 8S_{n_b}(f) * S_{n_b}(f) \tag{E-22}$$

where

$$\sigma_{n_b}^2 = \int_{-\infty}^{\infty} S_{n_b}(f)\,df \tag{E-23}$$

The fourth and last term of (E-9) is the square of the expected value of the square of $n_b(t)$. The expected value of the square of a random process is the power in that process. This term is not a function of τ so that its Fourier transform yields a delta function whose magnitude is

$$S_4(f) = 4\{\sigma_{n_b}^2\}^2\delta(f) \tag{E-24}$$

Observe that this dc component exactly cancels the dc component of $S_3(f)$. Combining all four terms of the discriminator output power spectrum yields

$$S_\epsilon(f) = \tfrac{1}{4}K_1^2P^2\left\{R_c^2\left[\left(\delta - \frac{\Delta}{2}\right)T_c\right] - R_c^2\left[\left(\delta + \frac{\Delta}{2}\right)T_c\right]\right\}^2$$

$$+ 2K_1P\left\{R_c^2\left[\left(\delta - \frac{\Delta}{2}\right)T_c\right] + R_c^2\left[\left(\delta + \frac{\Delta}{2}\right)T_c\right]\right\}S_{n_b}(f) * S_{\theta_d}(f) \tag{E-25}$$

$$+ 8S_{n_b}(f) * S_{n_b}(f)$$

E-2 Tau-Dither Noncoherent Tracking Loop

The discriminator output autocorrelation function is defined in (4-100) with $\epsilon(t,\delta)$ defined in (4-99). To shorten the following equations, define

$$C_1 = K_1 P$$

$$C_2 = 2\sqrt{K_1 P}$$

$$R_+ = R_c\left[\left(\delta + \frac{\Delta}{2}\right)T_c\right] \tag{E-26}$$

$$R_- = R_c\left[\left(\delta - \frac{\Delta}{2}\right)T_c\right]$$

Then

$$
\begin{aligned}
E[\epsilon(t,\delta)\epsilon(t+\tau,\delta)] = E\{&[C_1\{R_-^2 q_2(t) - R_+^2 q_1(t)\} \\
&+ C_2\cos[\phi - \phi' + \theta_d(t - T_d)]\{n_{2I}(t)R_- q_2(t) - n_{1I}(t)R_+ q_1(t)\} \\
&+ C_2\sin[\phi - \phi' + \theta_d(t - T_d)]\{n_{2Q}(t)R_- q_2(t) \\
&- n_{1Q}(t)R_+ q_1(t)\} + [n_{2I}(t)]^2 q_2(t) + [n_{2Q}(t)]^2 q_2(t) \\
&- [n_{1I}(t)]^2 q_1(t) - [n_{1Q}(t)]^2 q_1(t)] \\
\times &[C_1\{R_-^2 q_2(t + \tau) - R_+^2 q_1(t + \tau)\} \\
&+ C_2\cos[\phi - \phi' + \theta_d(t + \tau - T_d)] \\
\times &\{n_{2I}(t + \tau)R_- q_2(t + \tau) - n_{1I}(t + \tau)R_+ q_1(t + \tau)\} \\
&+ C_2\sin[\phi - \phi' + \theta_d(t + \tau - T_d)] \\
\times &\{n_{2Q}(t + \tau)R_- q_2(t + \tau) \\
&- n_{1Q}(t + \tau)R_+ q_1(t + \tau)\} \\
&+ [n_{2I}(t + \tau)]^2 q_2(t + \tau) \\
&+ [n_{2Q}(t + \tau)]^2 q_2(t + \tau) \\
&- [n_{1I}(t + \tau)]^2 q_1(t + \tau) - [n_{1Q}(t + \tau)]q_1(t + \tau)]\}
\end{aligned}
\tag{E-27}
$$

Observe that the function $q(t)$ and therefore $q_1(t)$ and $q_2(t)$ are stationary random processes only if a random delay or phase shift is included in their argument. All expected values involving $q_1(t)$ or $q_2(t)$ will be assumed to be over this implied random phase. When (E-27) is expanded, many terms can be immediately set equal to zero because of the independence of all random processes and because all of the noise processes have zero mean. In particular, any term including a factor of the form $E\{\cos[\phi - \phi' + \theta_d(t - T_d)]\}$ may be set to zero as shown earlier. After considerable but straightforward manipulation, it can be shown that

$$E[\epsilon(t,\delta)\epsilon(t + \tau,\delta) = E\{C_1^2[R_-^2 q_2(t) - R_+^2 q_1(t)][R_-^2 q_2(t + \tau) - R_+^2 q_1(t + \tau)]\}$$
$$+ E\{C_1[R_-^2 q_2(t) - R_+^2 q_1(t)][n_{2I}(t + \tau)]^2 q_2(t + \tau)\}$$
$$+ E\{C_1[R_-^2 q_2(t) - R_+^2 q_1(t)][n_{2Q}(t + \tau)]^2 q_2(t + \tau)\}$$
$$- E\{C_1[R_-^2 q_2(t) - R_+^2 q_1(t)][n_{1I}(t + \tau)]^2 q_1(t + \tau)\}$$
$$- E\{C_1[R_-^2 q_2(t) - R_+^2 q_1(t)][n_{1Q}(t + \tau)]^2 q_1(t + \tau)\}$$
$$+ E\{C_2^2 \cos[\phi - \phi' + \theta_d(t - T_d)]$$
$$\times \cos[\phi - \phi' + \theta_d(t + \tau - T_d)]\}$$
$$\times E\{[n_{2I}(t)R_- q_2(t) - n_{1I}(t)R_+ q_1(t)]$$
$$\times [n_{2I}(t + \tau)R_- q_2(t + \tau) - n_{1I}(t + \tau)R_+ q_1(t + \tau)]\}$$
$$+ E\{C_2^2 \sin[\phi - \phi' + \theta_d(t - T_d)]$$
$$\times \sin[\phi - \phi' + \theta_d(t + \tau - T_d)]\} \tag{E-28}$$
$$\times E\{[n_{2Q}(t)R_- q_2(t) - n_{1Q}(t)R_+ q_1(t)]$$
$$\times [n_{2Q}(t + \tau)R_- q_2(t + \tau) - n_{1Q}(t + \tau)R_+ q_1(t + \tau)]\}$$
$$+ E\{C_1[n_{2I}(t)]^2 q_2(t)[R_-^2 q_2(t + \tau) - R_+^2 q_1(t + \tau)]\}$$
$$+ E\{C_1[n_{2Q}(t)]^2 q_2(t)[R_-^2 q_2(t + \tau) - R_+^2 q_1(t + \tau)]\}$$
$$- E\{C_1[n_{1I}(t)]^2 q_1(t)[R_-^2 q_2(t + \tau) - R_+^2 q_1(t + \tau)]\}$$
$$- E\{C_1[n_{1Q}(t)]^2 q_1(t)[R_-^2 q_2(t + \tau) - R_+^2 q_1(t + \tau)]\}$$
$$+ E\{[[n_{2I}(t)]^2 q_2(t) + [n_{2Q}(t)]^2 q_2(t) - [n_{1I}(t)]^2 q_1(t)$$
$$- [n_{1Q}(t)]^2 q_1(t)]$$
$$\times [[n_{2I}(t + \tau)]^2 q_2(t + \tau) + [n_{2Q}(t + \tau)]^2 q_2(t + \tau)$$
$$- [n_{1I}(t + \tau)]^2 q_1(t + \tau) - [n_{1Q}(t + \tau)]^2 q_1(t + \tau)]\}$$

All of the baseband noise processes above have identical statistics so that a number of the terms of this equation can be combined to obtain

$$E[\epsilon(t,\delta)\epsilon(t + \tau,\delta)] = C_1^2 R_-^4 E[q_2(t)q_2(t + \tau)] - C_1^2 R_-^2 R_+^2 E[q_1(t)q_2(t + \tau)]$$
$$- C_1^2 R_-^2 R_+^2 E[q_2(t)q_1(t + \tau)] + C_1^2 R_+^4 E[q_1(t)q_1(t + \tau)]$$
$$+ 4C_1 \sigma_n^2 R_-^2 E[q_2(t)q_2(t + \tau)]$$
$$- 2C_1 \sigma_n^2 R_+^2 E[q_1(t)q_2(t + \tau)]$$
$$- 2C_1 \sigma_n^2 R_-^2 E[q_2(t)q_1(t + \tau)]$$
$$+ 4C_1 \sigma_n^2 R_+^2 E[q_1(t)q_1(t + \tau)]$$
$$- 2C_1 \sigma_n^2 R_+^2 E[q_2(t)q_1(t + \tau)]$$

$$- 2C_1\sigma_n^2 R_-^2 E[q_1(t)q_2(t + \tau)] \tag{E-29}$$

$$+ C_2^2 E[\cos\{\theta_d(t - T_d) - \theta_d(t + \tau - T_d)\}]$$

$$\times E[n_b(t)n_b(t + \tau)]$$

$$\times \{R_-^2 E[q_2(t)q_2(t + \tau)] + R_+^2 E[q_1(t)q_1(t + \tau)]\}$$

$$+ 2E[\{n_b(t)n_b(t + \tau)\}^2]E[q_2(t)q_2(t + \tau)]$$

$$+ 2\sigma_n^4 E[q_2(t)q_2(t + \tau)] - 4\sigma_n^4 E[q_2(t)q_1(t + \tau)]$$

$$+ 2\sigma_n^4 E[q_1(t)q_1(t + \tau)] - 4\sigma_n^4 E[q_1(t)q_2(t + \tau)]$$

$$+ 2E[\{n_b(t)n_b(t + \tau)\}^2]E[q_1(t)q_1(t + \tau)]$$

where

$$\sigma_n^2 = E[\{n_{1I}(t)\}^2] = E[\{n_{1Q}(t)\}^2] = E[\{n_{2I}(t)\}^2] = E[\{n_{2Q}(t)\}^2] \tag{E-30}$$

for any time t since these processes are stationary.

At this point it is convenient to define the crosscorrelation and autocorrelation functions involving $q_1(t)$ and $q_2(t)$ in terms of correlation functions of $q(t)$. These correlation functions can be calculated using the expected value over a random phase α, that is,

$$E[q(t)q(t + \tau)] = \int_{-\infty}^{\infty} q(t + \alpha)q(t + \tau + \alpha)p_T(\alpha) \, d\alpha \tag{E-31}$$

where $p_T(\alpha)$ is the probability density function of the random phase. The phase is assumed to be uniformly distributed over the period $2T_q$ of $q(t)$. Therefore,

$$E[q(t)q(t + \tau)] = \frac{1}{2T_q} \int_{-T_q}^{T_q} q(t + \alpha)q(t + \tau + \alpha) \, d\alpha = R_q(\tau) \tag{E-32}$$

The result of this integration is independent of t and is a periodic triangle function as illustrated in Figure E-1a. Using (4-90) it is easy to show that

$$E[q_1(t)q_1(t + \tau)] = E[q_2(t)q_2(t + \tau)] = \tfrac{1}{4} + \tfrac{1}{4}R_q(\tau) = R_{q_1}(\tau) \tag{E-33}$$

that

$$E[q_1(t)q_2(t + \tau)] = E[q_2(t)q_1(t + \tau)] = \tfrac{1}{4} - \tfrac{1}{4}R_q(\tau) \tag{E-34}$$

and that

$$E[q_1(t)q_2(t + \tau)] = \tfrac{1}{2} - E[q_1(t)q_1(t + \tau)] = \tfrac{1}{2} - R_{q_1}(\tau) \tag{E-35}$$

These functions are plotted in Figure E-1b and c.

Substituting the expressions of (E-33) through (E-35) into (E-29) and simplifying yields

$$E[\epsilon(t,\delta)\epsilon(t + \tau,\delta)] = C_1^2(R_-^2 + R_+^2)^2 R_{q_1}(\tau) - C_1^2 R_-^2 R_+^2$$

$$+ 8C_1\sigma_n^2(R_-^2 + R_+^2)R_{q_1}(\tau) - 2C_1\sigma_n^2(R_-^2 + R_+^2) \tag{E-36}$$

$$+ C_2^2 E[\cos\{\theta_d(t - T_d) - \theta_d(t + \tau - T_d)\}]$$

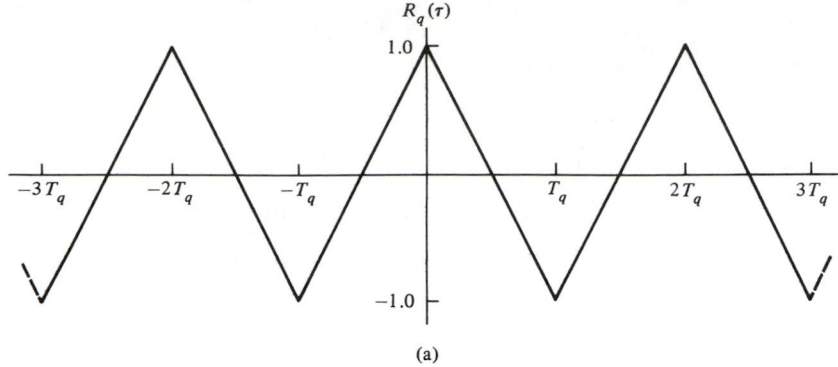

(a)

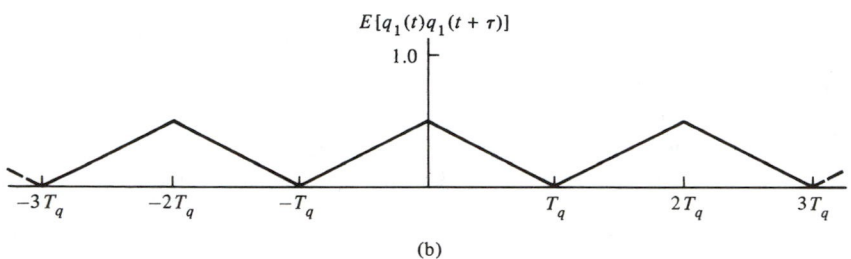

(b)

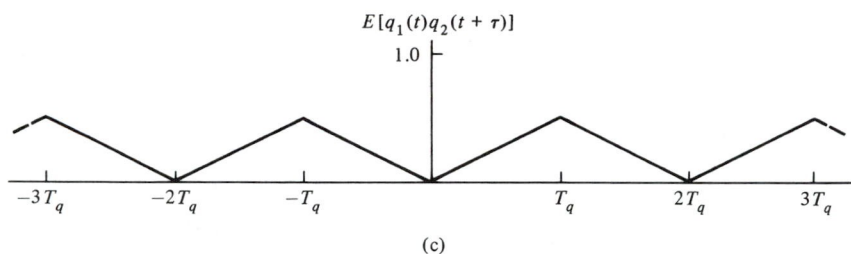

(c)

FIGURE E-1. Autocorrelation and cross-correlation function for tau-dither tracking loop switching functions: (a) $R_q(\tau)$; (b) $R_{q1}(\tau)$ or $R_{q2}(\tau)$; (c) $R_{q1q2}(\tau)$.

$$\times (R_-^2 + R_+^2)R_{n_b}(\tau)R_{q1}(\tau)$$
$$+ 4E[\{n_b(t)n_b(t + \tau)\}^2]R_{q1}(\tau)$$
$$+ 12\sigma_n^4 R_{q1}(\tau) - 4\sigma_n^4$$

Using (E-26) the desired autocorrelation becomes

$$R_\epsilon(\tau) = K_1^2 P^2 \left\{ R_c^2 \left[\left(\delta - \frac{\Delta}{2}\right)T_c \right] + R_c^2 \left[\left(\delta + \frac{\Delta}{2}\right)T_c \right] \right\}^2 R_{q1}(\tau)$$
$$- K_1^2 P^2 R_c^2 \left[\left(\delta - \frac{\Delta}{2}\right)T_c \right] R_c^2 \left[\left(\delta + \frac{\Delta}{2}\right)T_c \right]$$

$$+ 8K_1 P\sigma_n^2 \left\{ R_c^2 \left[\left(\delta - \frac{\Delta}{2} \right) T_c \right] + R_c^2 \left[\left(\delta + \frac{\Delta}{2} \right) T_c \right] \right\} R_{q1}(\tau)$$

$$- 2K_1 P\sigma_n^2 \left\{ R_c^2 \left[\left(\delta - \frac{\Delta}{2} \right) T_c \right] + R_c^2 \left[\left(\delta + \frac{\Delta}{2} \right) T_c \right] \right\}$$

$$+ 4K_1 P \left\{ R_c^2 \left[\left(\delta - \frac{\Delta}{2} \right) T_c \right] + R_c^2 \left[\left(\delta + \frac{\Delta}{2} \right) T_c \right] \right\} R_{n_b}(\tau) R_{q1}(\tau) \quad \text{(E-37)}$$

$$\times E[\cos\{\theta_d(t - T_d) - \theta_d(t + \tau - T_d)\}]$$

$$+ 4E[\{n_b(t)n_b(t + \tau)\}^2]R_{q1}(\tau)$$

$$+ 12\sigma_n^4 R_{q1}(\tau) - 4\sigma_n^4$$

The Fourier transform of $R_\epsilon(\tau)$ is the power spectrum needed to complete the analysis of the tau-dither tracking loop. The Fourier transform of the first term is a function of δ times the power spectrum $S_{q1}(f)$ of $q_1(t)$ which is given by

$$S_{q1}(f) = \sum_{n=-\infty}^{\infty} D_n \delta(f - nf_q) \quad \text{(E-38)}$$

where

$$D_n = \begin{cases} \dfrac{1}{4} & \text{for } n = 0 \\ 0 & \text{for } n \text{ even} \\ \left(\dfrac{1}{n\pi} \right)^2 & \text{for } n \text{ odd} \end{cases}$$

Thus

$$S_1(f) = K_1^2 P^2 \left\{ R_c^2 \left[\left(\delta - \frac{\Delta}{2} \right) T_c \right] + R_c^2 \left[\left(\delta + \frac{\Delta}{2} \right) T_c \right] \right\}^2 S_{q1}(f) \quad \text{(E-39)}$$

The second term is not a function of τ so that its Fourier transform is a delta function at dc,

$$S_2(f) = -K_1^2 P^2 R_c^2 \left[\left(\delta - \frac{\Delta}{2} \right) T_c \right] R_c^2 \left[\left(\delta + \frac{\Delta}{2} \right) T_c \right] \delta(f) \quad \text{(E-40)}$$

The third term is similar to the first in form, so that

$$S_3(f) = 8K_1 P\sigma_n^2 \left\{ R_c^2 \left[\left(\delta - \frac{\Delta}{2} \right) T_c \right] + R_c^2 \left[\left(\delta + \frac{\Delta}{2} \right) T_c \right] \right\} S_{q1}(f) \quad \text{(E-41)}$$

and the fourth term is again a delta function

$$S_4(f) = -2K_1 P\sigma_n^2 \left\{ R_c^2 \left[\left(\delta - \frac{\Delta}{2} \right) T_c \right] + R_c^2 \left[\left(\delta + \frac{\Delta}{2} \right) T_c \right] \right\} \delta(f) \quad \text{(E-42)}$$

The Fourier transform of the fifth term results in a convolution of three power spectra. The Fourier transform of $E[\cos\{\theta_d(t - T_d) - \theta_d(t + \tau - T_d)\}]$ is denoted by $S_{\theta_d}(f)$ and is defined in (E-13) through (E-16). Then

$$S_5(f) = 4K_1P\left\{R_c^2\left[\left(\delta - \frac{\Delta}{2}\right)T_c\right] + R_c^2\left[\left(\delta + \frac{\Delta}{2}\right)T_c\right]\right\}S_{n_b}(f) * S_{q1}(f) * S_{\theta_d}(f)$$

$$(\text{E-43})$$

The transform of the sixth term is the convolution of the power spectrum discussed in Section 4-3 for the output of a square-law device and the power spectrum of $q_1(t)$. From (E-21),

$$S_6(f) = \{4\sigma_n^4\delta(f) + 8S_{n_b}(f) * S_{n_b}(f)\} * S_{q1}(f)$$
$$= 4\sigma_n^4 S_{q1}(f) + 8S_{n_b}(f) * S_{n_b}(f) * S_{q1}(f)$$

$$(\text{E-44})$$

The transform of the seventh and eighth terms are

$$S_7(f) = 12\sigma_n^4 S_{q1}(f)$$

$$(\text{E-45})$$

and

$$S_8(f) = -4\sigma_n^4\delta(f)$$

$$(\text{E-46})$$

The desired discriminator output power spectrum is therefore

$$S_\epsilon(f) = \sum_{j=1}^{8} S_j(f)$$

$$(\text{E-47})$$

The complete discriminator output power spectrum can be calculated from (E-39) through (E-47).

In order to obtain results which are reasonably simple, the exact analysis of $S_\epsilon(f)$ must be abandoned at this point. Assume now that the dithering frequency is chosen such that the only components of $S_{q1}(f)$ which influence tracking performance are those at $f = 0$, f_q, and $3f_q$. Assume also that there is no data modulation so that $S_{\theta_d}(f) = \delta(f)$. With these assumptions

$$S_{q1}(f) = \frac{1}{9\pi^2}[\delta(f + 3f_q) + \delta(f - 3f_q)]$$

$$+ \frac{1}{\pi^2}[\delta(f + f_q) + \delta(f - f_q)] + \tfrac{1}{4}\delta(f)$$

$$(\text{E-48})$$

and after some simplification

$$S_\epsilon(f) = \tfrac{1}{4}K_1^2P^2\left\{R_c^2\left[\left(\delta - \frac{\Delta}{2}\right)T_c\right] - R_c^2\left[\left(\delta + \frac{\Delta}{2}\right)T_c\right]\right\}^2 \delta(f)$$

$$+ \left[K_1P\left\{R_c^2\left[\left(\delta - \frac{\Delta}{2}\right)T_c\right] + R_c^2\left[\left(\delta + \frac{\Delta}{2}\right)T_c\right]\right\} + 4\sigma_n^2\right]^2$$

$$\times \left[\frac{1}{\pi^2} \delta(f - f_q) + \frac{1}{\pi^2} \delta(f + f_q) + \frac{1}{9\pi^2} \delta(f + 3f_q) \right.$$

$$\left. + \frac{1}{9\pi^2} \delta(f - 3f_q) \right] \qquad \text{(E-49)}$$

$$+ \left[K_1 P \left\{ R_c^2 \left[\left(\delta - \frac{\Delta}{}\right) T_c \right] + R_c^2 \left[\left(\delta + \frac{\Delta}{}\right) T_c \right] \right\} \right]$$

$$\times \left[S_{n_b}(f) + \frac{4}{\pi^2} S_{n_b}(f - f_q) + \frac{4}{\pi^2} S_{n_b}(f + f_q) + \frac{4}{9\pi^2} S_{n_b}(f - 3f_q) \right.$$

$$\left. + \frac{4}{9\pi^2} S_{n_b}(f + 3f_q) \right]$$

$$+ 2S_T(f) + \frac{8}{\pi^2} S_T(f - f_q) + \frac{8}{\pi^2} S_T(f + f_q)$$

$$+ \frac{8}{9\pi^2} S_T(f - 3f_q) + \frac{8}{9\pi^2} S_T(f + 3f_q)$$

where

$$S_T(f) = S_{n_b}(f) * S_{n_b}(f)$$

References

[1] S. STEIN, and J. JONES, *Modern Communications Principles* (New York: McGraw-Hill, 1967).

[2] W. B. DAVENPORT, and W. L. ROOT, *An Introduction to the Theory of Random Signals and Noise* (New York: McGraw-Hill, 1958).

Numerical Approximations for the Chi-Squared Probability Distribution and Marcum's Q-Function

F-1 Introduction

This appendix includes information on the numerical computation of the central and noncentral chi-squared distributions and the generalized Marcum's Q-function, which is related to the noncentral chi-squared distribution. As mentioned in Chapter 5, C and Pascal programs for computing these distributions and their inverses are available on the Internet from Arthur Ross of Qualcomm, Inc., San Diego, California. These programs are on Qualcomm's FTP server. They can be obtained by anonymous FTP to: ftp.qualcomm.com in the directory/pub/Statlib. In the remainder of this appendix we discuss means for computing these functions and give a few Matlab [1] programs for computing a more limited set of them than available from Arthur Ross.

F-2 Computation of the (Central) Chi-Squared Distribution

Recall that the chi-squared random variable is the sum of the squares of ν independent Gaussian random variables each with zero mean and variance unity. Thus if the X_i's are unit-variance, zero-mean independent Gaussian random variables,

$$X^2 = \sum_{i=1}^{\nu} X_i^2 \tag{F-1}$$

is said to follow the chi-squared distribution with ν degrees of freedom. Following the notation of Abramowitz and Stegun [2], the probability that $X^2 \leq \chi^2$ is denoted as $P(\chi^2 \,|\, \nu)$ [i.e., the cumulative distribution function (cdf)] and the complementary cdf is denoted by $Q(\chi^2 \,|\, \nu) = 1 - P(\chi^2 \,|\, \nu)$. Series expressions given in Ref. 2 for $Q(\chi^2 \,|\, \nu)$ are

$$Q(\chi^2 \,|\, \nu) = 2Q(x) + 2Z(x) \sum_{r=1}^{(\nu-1)/2} \frac{\chi^{2r-1}}{1 \cdot 3 \cdot 5 \cdots (2r-1)} \qquad \nu \text{ odd} \tag{F-2}$$

$$Q(\chi^2 \,|\, \nu) = \sqrt{2\pi} Z(x) \left(1 + \sum_{r=1}^{(\nu-2)/2} \frac{\chi^{2r}}{2 \cdot 4 \cdots (2r)} \right) \qquad \nu \text{ even} \qquad \text{(F-3)}$$

where

$$Z(x) = \frac{e^{-x^2/2}}{\sqrt{2\pi}} \qquad \text{(F-4)}$$

An approximation for large ν is

$$Q(\chi^2 \,|\, \nu) = Q(x_2) \qquad \nu > 30 \qquad \text{(F-5)}$$

where

$$x_2 = \frac{(\chi^2/\nu)^{1/3} - [1 - 2/(9\nu)]}{\sqrt{2/(9\nu)}} \qquad \text{(F-6)}$$

and $Q(x)$ is the Gaussian Q-function, the approximation of which was discussed in Appendix B. A Matlab program for computing the chi-squared cdf based on these equations is given in Table F-1. Note that the Gaussian Q-function is computed with the Matlab function given in Table F-2.

TABLE F-1. Matlab Program for Computing the Chi-Square CDF

```
%       This program computes the cumulative chi-square
%       and complementary chi-square distributions from
%       Equations 26.4.4 and 26.4.5 of Abramowitz and Stegun.
%       Inputs are nu, the number of degrees of freedom, and
%       the upper limit of the cdf. A range of values can be
%       input for the upper limit. In addition to the arrays
%       of input and output values, a plot is also provided.
%
%       R. E. Ziemer
%       4/17/94
%
fprintf('This program computes the cumulative chi-square
    distribution \n')
X2P=input('Input starting value of ''upper limit'' parameter
    or chi^2 ');
N_X2=input('Number of steps for upper integral limit ');
if N_X2 >= 1
        X2inc=input('Step size for upper limit ');
end
nu=input('Input degrees of freedom - integer ');
iplot=input('Provide plot? Input 1; otherwise 0 ');
format long
dof=rem(nu,2);
if dof == 1
        NS=(nu-1)/2;
else
        NS=(nu-2)/2;
```

```
end
y=zeros(1,N_X2);
compl_chi=zeros(1,N_X2);
cum_chi=zeros(1,N_X2);
for n_step=1:1:N_X2
X2=X2P+(n_step-1)*X2inc;
chi=sqrt(X2);
y(n_step)=X2;
if nu < 30
        [Q,P]=qfn(chi);
        Z=exp(-X2/2)/sqrt(2*pi);
        if NS > 0
                T=zeros(1,NS);
                if dof == 1
                        T(1)=chi;
                        for r=1:NS-1
                                T(r+1)=X2*T(r)/(2*r+1);
                        end
                        QX2=2*Q+2*Z*sum(T);
                else
                        T(1)=chi^2/2;
                        for r=1:NS-1
                                T(r+1)=X2*T(r)/(2*(r+1));
                        end
                        QX2=sqrt(2*pi)*Z*(1+sum(T));
                end
        else
                if dof == 1
                        QX2=2*Q;
                else
                        QX2=sqrt(2*pi)*Z;
                end
        end
        PX2=1-QX2;
else
        K1=(X2/nu)^(1/3);
        K2=1-2/(9*nu);
        K3=sqrt(2/(9*nu));
        arg=(K1-K2)/K3;
        [A,B]=qfn(arg);
        QX2=A;
        PX2=1-QX2;
end
if X2 ==0
        QX2=1;
        PX2=0;
end
cum_chi(n_step)=PX2;
compl_chi(n_step)=QX2;
end
```

```
if iplot > 0
        plot(y,compl_chi),xlabel('chi-squared'),...
        ylabel('complement of the cumulative probability'),...
        title(['Complement of the chi-squared distribution for
            ',num2str(nu), ' degrees of freedom '])
end
fprintf('Cumulative central chi-square and its complement \n')
fprintf('Number of degrees of freedom = %f \n',nu)
disp(' Values for chi^2 ')
disp(y)
disp(' Cumulative probability ')
disp(cum_chi)
disp(' Complement of cumulative probability ')
disp(compl_chi)
```

TABLE F-2. Matlab Program for Gaussian *Q*-Function

```
%       This function computes the Gaussian Q-function
%       using the rational approximation 26.2.17 of
%       Abromowitz and Stegun. For sufficiently large
%       arguments, the asymptotic expansion 26.2.12 is used.
%       The rational approximation is accurate to within
%       7.5e-8. Q is the value of the Q-function (integral of
%       Gaussian pdf of zero mean and unit variance from the
%       input quantity to infinity, and P = 1 - Q is the
%       cumulative distribution function. The call is
%       [Q,P] = qfn(x).
%
%       R. E. Ziemer
%       4/17/94
%
function [Q,P]=qfn(x)
b(1)=0.31938153;
b(2)=-0.356563782;
b(3)=1.781477937;
b(4)=-1.821255978;
b(5)=1.330274429;
p=0.2316419;
y=abs(x);
T=1/(1+p*y);
TT=zeros(1,5);
for r=1:5
        TT(r)=T^r;
end
Z=exp(-y^2/2)/sqrt(2*pi);
if y<4
        Q1=Z*b*TT';
else
        Q1=(Z/y)*(1-1/y^2+3/y^4);
```

```
end
if x>0
        Q=Q1;
else
        Q=1-Q1;
end
P=1-Q;
```

F-3 Generalized Marcum's Q-Function

The generalization of Marcum's Q-function is

$$Q_M(a,b) = \frac{1}{a^{M-1}} \int_b^\infty x^M \exp\left(-\frac{x^2 + a^2}{2}\right) I_{M-1}(ax) \, dx \qquad \text{(F-7)}$$

where $I_N(x)$ is the modified Bessel function of order N. Parl [3] gives a recursive means for computing it. However, also given in [3] are the following series representations:

$$Q_M(a,b) = \exp\left[-\frac{(a-b)^2}{2}\right] \sum_{i=1-M}^\infty \left(\frac{a}{b}\right)^i \exp(-ab) I_i(ab) \qquad \text{(F-8)}$$

for $a < b$, and

$$Q_M(a,b) = 1 - \exp\left[-\frac{(a-b)^2}{2}\right] \sum_{i=M}^\infty \left(\frac{b}{a}\right)^i \exp(-ab) I_i(ab) \qquad \text{(F-9)}$$

for $a > b$. Marcum's Q-function is just $Q_1(a,b)$. For computational purposes note that $I_{-i}(x) = I_i(x)$. The series given above converge sufficiently rapidly so that they can be truncated at reasonable values of the summation indices.

Parl [3] notes that a way to compute the necessary modified Bessel functions is to use the backward recursion

$$I_{n-1}(z) = I_{n+1}(z) + \frac{2n}{z} I_n(z) \qquad \text{(F-10)}$$

by starting at a sufficiently large N with $KI_N = 1$ and $KI_{N-1} = 0$ and finding the constant K by the relation

$$1 = \sum_{n=-\infty}^\infty e^{-z} I_n(z) \qquad \text{(F-11)}$$

A Matlab program for computing the generalized Marcum's Q-function is given in Table F-3.

TABLE F-3. Matlab Program for Computing the Generalized Marcum *Q*-function

```
%       This program computes Marcum's Q-function, QM(a,b),
%       using Parl's series using modified Bessel functions
%       (IEEE Trans. on Inf. Theory, Vol. IT-26, pp. 121-124,
%       Jan. 1980).
%
%       R. E. Ziemer
%       4/6/94
%
fprintf('This program computes the Marcum Q-function,
   QM(a,b) \n')
al=input('Input ''a > 0'' parameter beginning ');
Na=input('Input ''a'' parameter number of steps ');
if Na > 1
        delal=input('Input ''a'' parameter step size ');
end
bl=input('Input ''b > 0'' parameter beginning ');
Nb=input('Input ''b'' parameter number of steps ');
if Nb > 1
        delbl=input('Input ''b'' parameter step size ');
end
M=input('Input ''M'' parameter - integer ');
n=1;
b2=bl+delbl*(Nb-1);
a2=al+delal*(Na-1);
A=zeros(1,Na);
B=zeros(1,Nb);
Q=zeros(Na,Nb);
for b = bl:delbl:b2
        B(n)=b;
        m=1;
        for a = al:delal:a2
                A(m)=a;
                z=a*b;
                %Modified Bessel function:
                K=101;
                Ix=zeros(1,K);
                if z >= 0.1
                        Ix(K)=1;
                        Ix(K-1)=0;
                        for k=K-1:-1:2
                                Ix(k-1)=Ix(k+1)+2*(k-1)*Ix(k)/z;
                        end
                        S=2*sum(Ix)-Ix(1);
                        I=Ix*exp(z)/S;
                else
                        Ix(1)=1;
                        for k=2:1:K
                                Ix(k)=(0.5*z)^(k-1)/(k-1);
```

```
                        end
                        I=Ix;
                end
                TP=zeros(size(K));
                if a<b
                        for k=1:K
                                TP(k)=(a/b)^(k-1)*exp(-z)*I(k);
                        end
                        TN=zeros(size(K));
                        if M>1
                                for j=1:M-1
                                        TN(j)=(b/a)^j*exp(-z)*I(j+1);
                                end
                        end
                        Q(m,n)=exp(-0.5*(a-b)^2)*(sum(TP) + sum(TN));
                else
                        for k=M+1:K
                                TP(k)=(b/a)^(k-1)*exp(-z)*I(k);
                        end
                        Q(m,n)=1-exp(-0.5*(a-b)^2)*sum(TP);
                end
                m=m+1;
        end
        n=n+1;
end
fprintf(' The values of variable a \n')
disp(A)
fprintf(' The values of variable b \n')
disp(B)
fprintf(' The QM(a, b) matrix for M = %f \n', M)
disp(Q)
if Na > 1 & Nb > 1
        mesh(Q)
end
```

F-4 Noncentral Chi-Squared Distribution

The noncentral chi-squared probability density function is

$$f(x) = \frac{1}{2}\left(\frac{x}{\lambda}\right)^{(\nu-2)/4} e^{-(x+\lambda)/2} I_{\nu/2-1}(\sqrt{\lambda x}) \quad x \geq 0 \tag{F-12}$$

where ν is the degrees of freedom and λ is the noncentrality parameter. Through a change of variables and suitable redefinition of parameters, it follows that the cdf for the noncentral chi-squared distribution can be expressed in terms of the generalized Marcum's Q-function as

$$P_{nc}(\chi^2 | \nu) = \int_0^{\chi^2} f(x) \, dx = 1 - Q_{\nu/2}(\sqrt{\lambda}, \chi) \tag{F-13}$$

where ν must be even in order to use this means of calculation. A Matlab program for computing the noncentral chi-squared cdf using the generalized Marcum's Q-function is given in Table F-4.

TABLE F-4. Matlab Program for Computing the Noncentral Chi-Square CDF

```
%       This program computes the cumulative noncentral
%       chi-square and complementary noncentral chi-square
%       distributions from Marcum's Q-function. Inputs are nu or
%       the number of degrees of freedom, the noncentrality
%       parameter, lambda, and the upper limit of the cdf. A
%       range of values can be input for the upper limit. In
%       addition to the arrays of input and output values, a
%       plot is also provided.
%
%       R. E. Ziemer
%       4/11/94
%
fprintf('This program computes the cumulative noncentral
    chi-square distribution \n')
lambda=input('Input ''non-centrality'' parameter ');
delta1=input('Input starting value of ''upper limit''
    parameter or chi^2 ');
no_steps=input('Enter number of values of the upper limit to
    step through ');
if no_steps >= 1
        ddelta=input('Step size for upper limit ');
end
N=input('Input ''degrees of freedom'' parameter - even integer ');
iplot=input('Provide plot? Input 1; otherwise 0 ');
format long
a=sqrt(lambda);
upper_limit=zeros(1,no_steps);
cumprob=zeros(1,no_steps);
comple_cumprob=zeros(1,no_steps);
for no = 1:1:no_steps
delta=delta1+(no-1)*ddelta;
upper_limit(no)=delta;
b=sqrt(delta);
M=N/2;
%Modified Bessel function:
z=a*b;
                K=101;
                Ix=zeros(1,K);
                if z >= 0.1
                        Ix(K)=1;
                        Ix(K-1)=0;
                        for k=K-1:-1:2
                                Ix(k-1)=Ix(k+1)+2*(k-1)*Ix(k)/z;
                        end
```

```
                              S=2*sum(Ix)-Ix(1);
                              I=Ix*exp(z)/S;
                 else
                              Ix(1)=1;
                              for k=2:1:K
                                       Ix(k)=(0.5*z)^(k-1)/(k-1);
                              end
                              I=Ix;
                 end
TP=zeros(1,K);
if a<b
         for k=1:K
                 TP(k)=(a/b)^(k-1)*exp(-z)*I(k);
         end
         TN=zeros(1,K);
         if M>1
                 for j=1:M-1
                          TN(j)=(b/a)^j*exp(-z)*I(j+1);
                 end
         end
         ccumchi=exp(-0.5*(a-b)^2)*(sum(TP) + sum(TN));
else
         for k=M+1:K
                 TP(k)=(b/a)^(k-1)*exp(-z)*I(k);
         end
         ccumchi=1-exp(-0.5*(a-b)^2)*sum(TP);
end
         cumchix=1-ccumchi;
cumprob(no)=cumchix;
comple_cumprob(no)=ccumchi;
end
if iplot > 0
         plot(upper_limit,comple_cumprob),xlabel('chi-squared'),...
         ylabel('complement of the cumlative probability'),...
         title(['Complement of noncentral chi^2 distribution for
            ',num2str(N),' dof and ',num2str(lambda),'
noncentrality parameter'])
end
fprintf('Cumulative chi-square and complement for %f
   non-centrality parameter \n',lambda)
fprintf('Number of degrees of freedom = %f \n',N)
disp(' Values for upper limit ')
disp(upper_limit)
disp(' Cumulative probability ')
disp(cumprob)
disp(' Complement of cumulative probability ')
disp(comple_cumprob)
```

References

[1] MATH WORKS, INC., *The Student Edition of Matlab*® (Englewood Cliffs, N.J.: Prentice Hall, 1992).

[2] M. ABRAMOWITZ and I. STEGUN, eds., *Handbook of Mathematical Functions* (New York: Dover, 1972).

[3] S. PARL, "A New Method of Calculating the Generalized Q-Function," *IEEE Trans. on Information Theory*, Vol. IT-26, pp. 121–124, January 1980.

Mathematical Tables

TABLE G-1. The Sinc Function

z	sinc z	sinc$^2 z$	z	sinc z	sinc$^2 z$
0.0	1.0	1.0	1.6	−0.18921	0.03580
0.1	0.98363	0.96753	1.7	−0.15148	0.02295
0.2	0.93549	0.87514	1.8	−0.10394	0.01080
0.3	0.85839	0.73684	1.9	−0.05177	0.00268
0.4	0.75683	0.57279	2.0	0	0
0.5	0.63662	0.40528	2.1	0.04684	0.00219
0.6	0.50455	0.25457	2.2	0.08504	0.00723
0.7	0.36788	0.13534	2.3	0.11196	0.01254
0.8	0.23387	0.05470	2.4	0.12614	0.01591
0.9	0.10929	0.01195	2.5	0.12732	0.01621
1.0	0	0	2.6	0.11643	0.01356
1.1	−0.08942	0.00800	2.7	0.09538	0.00910
1.2	−0.15591	0.02431	2.8	0.06682	0.00447
1.3	−0.19809	0.03924	2.9	0.03392	0.00115
1.4	−0.21624	0.04676	3.0	0	0
1.5	−0.21221	0.04503			

TABLE G-2. Trigonometric Identities

Euler's theorem: $e^{\pm ju} = \cos u \pm j \sin u$

$\cos u = \frac{1}{2}(e^{ju} + e^{-ju})$

$\sin u = \dfrac{(e^{ju} - e^{-ju})}{2j}$

$\sin^2 u + \cos^2 u = 1$

$\cos^2 u - \sin^2 u = \cos 2u$

$2 \sin u \cos u = \sin 2u$

$\cos^2 u = \frac{1}{2}(1 + \cos 2u)$

$\sin^2 u = \frac{1}{2}(1 - \cos 2u)$

$\sin(u \pm v) = \sin u \cos v \pm \cos u \sin v$

$\cos(u \pm v) = \cos u \cos v \mp \sin u \sin v$

$\sin u \sin v = \frac{1}{2}[\cos(u - v) - \cos(u + v)]$

$\cos u \cos v = \frac{1}{2}[\cos(u - v) + \cos(u + v)]$

$\sin u \cos v = \frac{1}{2}[\sin(u - v) + \sin(u + v)]$

$$\cos^{2n} u = \frac{\sum_{k=0}^{n-1} 2 \binom{2n}{k} \cos 2(n - k)u + \binom{2n}{n}}{2^{2n}}$$

$$\cos^{2n-1} u = \frac{\sum_{k=0}^{n-1} \binom{2n-1}{k} \cos (2n - 2k - 1)u}{2^{2n-2}}$$

$$\sin^{2n} u = \frac{\sum_{k=0}^{n-1} (-1)^{n-k} 2 \binom{2n}{k} \cos 2(n - k)u + \binom{2n}{n}}{2^{2n}}$$

$$\sin^{2n-1} u = \frac{\sum_{k=0}^{n-1} (-1)^{n+k-1} \binom{2n-1}{k} \sin (2n - 2k - 1)u}{2^{2n-2}}$$

where

$$\binom{n}{k} = \frac{n!}{(n - k)!k!}$$

TABLE G-3. Indefinite Integrals

$$\int \sin ax\, dx = -\frac{1}{a} \cos(ax)$$

$$\int \cos ax\, dx = \frac{1}{a} \sin(ax)$$

$$\int \sin^2 ax\, dx = \frac{x}{2} - \frac{\sin 2ax}{4a}$$

$$\int \cos^2 ax\, dx = \frac{x}{2} + \frac{\sin(2ax)}{4a}$$

$$\int x \sin ax\, dx = \frac{\sin ax - ax \cos ax}{a^2}$$

$$\int x \cos ax\, dx = \frac{\cos ax + ax \sin ax}{a^2}$$

$$\int x^m \sin x\, dx = -x^m \cos x + m \int x^{m-1} \cos x\, dx$$

$$\int x^m \cos x\, dx = x^m \sin x - m \int x^{m-1} \sin x\, dx$$

$$\int \sin ax \sin bx\, dx = \frac{\sin(a - b)x}{2(a - b)} - \frac{\sin(a + b)x}{2(a + b)} \qquad a^2 \neq b^2$$

$$\int \sin ax \cos bx \, dx = -\left[\frac{\cos(a - b)x}{2(a - b)} + \frac{\cos(a + b)x}{2(a + b)}\right], \qquad a^2 \neq b^2$$

$$\int \cos ax \cos bx \, dx = \frac{\sin(a - b)x}{2(a - b)} + \frac{\sin(a + b)x}{2(a + b)}, \qquad a^2 \neq b^2$$

$$\int e^{ax} \, dx = \frac{e^{ax}}{a}$$

$$\int x^m e^{ax} \, dx = \frac{x^m e^{ax}}{a} - \frac{m}{a} \int x^{m-1} e^{ax} \, dx$$

$$\int e^{ax} \sin bx \, dx = \frac{e^{ax}}{a^2 + b^2}(a \sin bx - b \cos bx)$$

$$\int e^{ax} \cos bx \, dx = \frac{e^{ax}}{a^2 + b^2}(a \cos bx + b \sin bx)$$

TABLE G-4. Definite Integrals

$$\int_0^\infty \frac{x^{m-1}}{1 + x^n} \, dx = \frac{\pi/n}{\sin(m\pi/n)}, \qquad n > m > 0$$

$$\int_0^{\pi/2} \sin^n x \, dx = \int_0^{\pi/2} \cos^n x \, dx = \begin{cases} \dfrac{1 \cdot 3 \cdot 5 \cdots (n - 1)}{2 \cdot 4 \cdot 6 \cdots (n)} \dfrac{\pi}{2} & n \text{ even, } n \text{ an integer} \\[2em] \dfrac{2 \cdot 4 \cdot 6 \cdots (n - 1)}{1 \cdot 3 \cdot 5 \cdots (n)}, & n \text{ odd} \end{cases}$$

$$\int_0^\pi \sin^2 nx \, dx = \int_0^\pi \cos^2 nx \, dx = \pi/2, \qquad n \text{ an integer}$$

$$\int_0^\pi \sin mx \sin nx \, dx = \int_0^\pi \cos mx \cos nx \, dx = 0, \qquad m \neq n, \, m \text{ and } n \text{ integer}$$

$$\int_0^\pi \sin mx \cos nx \, dx = \begin{cases} \dfrac{2m}{m^2 - n^2}, & m + n \text{ odd} \\[2em] 0 & m + n \text{ even} \end{cases}$$

$$\int_0^\infty e^{-a^2 x^2} \, dx = \frac{\sqrt{\pi}}{2a}, \qquad a > 0$$

$$\int_0^\infty x^n e^{-ax} \, dx = \frac{n!}{a^{n+1}}, \qquad n \text{ an integer and } a > 0$$

$$\int_0^\infty x^{2n} e^{-ax} \, dx = \frac{1 \cdot 3 \cdot 5 \cdots (2n - 1)}{2^{n+1} a^n} \sqrt{\frac{\pi}{a}}$$

$$\int_0^\infty e^{-ax} \cos bx \, dx = \frac{a}{a^2 + b^2}, \qquad a > 0$$

$$\int_0^\infty e^{-ax} \sin bx \, dx = \frac{b}{a^2 + b^2}, \qquad a > 0$$

$$\int_0^\infty e^{-a^2 x^2} \cos bx \, dx = \frac{\sqrt{\pi}}{2a} e^{-b^2/4a^2}$$

$$\int_0^\infty x^{a-1} \cos bx \, dx = \frac{\Gamma(\alpha)}{b^\alpha} \cos \pi\alpha/2, \qquad 0 < \alpha < 1, \, b > 0$$

$$\Gamma(\alpha) \triangleq \int_0^\infty x^{a-1} e^{-x} \, dx$$

$$\int_0^\infty x^{\alpha-1} \sin bx \, dx = \frac{\Gamma(\alpha)}{b^\alpha} \sin \pi\alpha/2, \qquad 0 < |\alpha| < 1, \, b > 0$$

$$\int_0^\infty x e^{-ax^2} I_k(bx) \, dx = \frac{1}{2a} e^{b^2/4a}$$

$$\int_0^\infty \text{sinc } x \, dx = \int_0^\infty \text{sinc}^2 x \, dx = \frac{1}{2}$$

$$\int_0^\infty \frac{\cos ax}{b^2 + x^2} \, dx = \frac{\pi}{2b} e^{-ab} \qquad a > 0, \, b > 0$$

$$\int_0^\infty \frac{x \sin ax}{b^2 + x^2} \, dx = \frac{\pi}{2} e^{-ab} \qquad a > 0, \, b > 0$$

TABLE G-5. Series Expansions

$$(u + v)^n = \sum_{k=0}^n \binom{n}{k} u^{n-k} v^k$$

where

$$\binom{n}{k} = \frac{n!}{(n-k)! \, k!}$$

Letting $u = 1$ and $v = x$, where $|x| \ll 1$, results in the following approximations:

$$(1 + x)^n \cong 1 + nx$$

$$(1 + x)^{1/2} \cong 1 + \tfrac{1}{2}x$$

$$(1 + x)^{-n} \cong 1 - nx$$

$$\ln(1 + u) = \sum_{k=1}^\infty (-1)^{k+1} \frac{u^k}{k}$$

$$\log_a u = \log_e u \log_a e; \; \log_e u = \ln u = \log_a u \log_e a$$

$$e^u = \sum_{k=0}^\infty \frac{u^k}{k!} \cong 1 + u, \qquad |u| \ll 1$$

$$a^u = e^{u \ln a}$$

$$\sin u = \sum_{k=0}^{\infty} \frac{(-1)^k u^{2k+1}}{(2k+1)!} \cong u - \frac{u^3}{3!}, \qquad |u| \ll 1$$

$$\cos u = \sum_{k=0}^{\infty} \frac{(-1)^k u^{2k}}{(2k)!} \cong 1 - \frac{u^2}{2!}, \qquad |u| \ll 1$$

$$\tan u = u + \tfrac{1}{3}u^3 + \tfrac{2}{15}u^5 + \cdots$$

$$\sin^{-1} u = u + \tfrac{1}{6}u^3 + \tfrac{3}{40}u^5 + \cdots$$

$$\tan^{-1} u = u - \tfrac{1}{3}u^3 + \tfrac{1}{5}u^5 - \cdots$$

$$\operatorname{sinc} u = 1 - \frac{(\pi u)^2}{3!} + \frac{(\pi u)}{5!} - \cdots$$

$$J_n(u) \cong \begin{cases} \dfrac{u^n}{2^n n!}\left[1 - \dfrac{u^2}{2^2(n+1)} + \dfrac{u^4}{2 \cdot 2^4(n+1)(n+2)} - \cdots\right], \\[2mm] \sqrt{\dfrac{2}{\pi u}} \cos\left(u - \dfrac{n\pi}{2} - \dfrac{\pi}{2}\right), \qquad u \gg 1 \end{cases}$$

$$I_0(u) \cong \begin{cases} 1 + \dfrac{u^2}{2^2} + \dfrac{u^4}{2^2 4^2} + \cdots \cong e^{u^2/4}, \qquad 0 \le u \ll 1 \\[2mm] \dfrac{e^u}{\sqrt{2\pi u}}, \qquad u \gg 1 \end{cases}$$

TABLE G-6. Fourier Transform Theorems

Name of Theorem	Signal	Transform
1. Superposition (a_1 and a_2 arbitrary constants)	$a_1 x_1(t) + a_2 x_2(t)$	$a_1 X_1(f) + a_2 X_2(f)$
2. Time delay	$x(t - t_0)$	$X(f)e^{-j2\pi f t_0}$
3a. Scale change	$x(at)$	$|a|^{-1} X\left(\dfrac{f}{a}\right)$
3b. Time reversal[a]	$x(-t)$	$X(-f) = X^*(f)$
4. Duality	$X(t)$	$x(-f)$
5a. Frequency translation	$x(t)e^{j\omega_0 t}$	$X(f - f_0)$

5b. Modulation	$x(t) \cos \omega_0 t$	$\frac{1}{2}X(f - f_0) + \frac{1}{2}X(f + f_0)$
6. Differentiation	$\dfrac{d^n x(t)}{dt^n}$	$(j2\pi f)^n X(f)$
7. Integration	$\displaystyle\int_{-\infty}^{t} x(t')\, dt'$	$(j2\pi f)^{-1}X(f) + \frac{1}{2}X(0)\delta(f)$
8. Convolution	$\displaystyle\int_{-\infty}^{\infty} x_1(t - t')x_2(t')\, dt'$ $= \displaystyle\int_{-\infty}^{\infty} x_1(t')x_2(t - t')\, dt'$	$X_1(f)X_2(f)$
9. Multiplication	$x_1(t)x_2(t)$	$\displaystyle\int_{-\infty}^{\infty} X_1(f - f')X_2(f')\, df'$ $= \displaystyle\int_{-\infty}^{\infty} X_1(f')X_2(f - f')\, df'$

[a] $\omega_0 = 2\pi f_0$; $x(t)$ is assumed to be real in 3b.

TABLE G-7. Fourier Transform Pairs

Pair Number	$x(t)$	$X(f)$
1.	$\Pi\left(\dfrac{t}{\tau}\right)$	$\tau \operatorname{sinc} \tau f$
2.	$2W \operatorname{sinc} 2Wt$	$\Pi\left(\dfrac{f}{2W}\right)$
3.	$\Lambda\left(\dfrac{t}{\tau}\right)$	$\tau \operatorname{sinc}^2 \tau f$
4.	$\exp(-\alpha t)u(t), \quad \alpha > 0$	$\dfrac{1}{\alpha + j2\pi f}$
5.	$t \exp(-\alpha t)u(t), \quad \alpha > 0$	$\dfrac{1}{(\alpha + j2\pi f)^2}$
6.	$\exp(-\alpha|t|), \quad \alpha > 0$	$\dfrac{2\alpha}{\alpha^2 + (2\pi f)^2}$
7.	$\delta(t)$	1
8.	1	$\delta(f)$
9.	$\delta(t - t_0)$	$\exp(-j2\pi f t_0)$
10.	$\exp(j2\pi f_0 t)$	$\delta(f - f_0)$
11.	$\cos 2\pi f_0 t$	$\frac{1}{2}\delta(f - f_0) + \frac{1}{2}\delta(f + f_0)$

12.	$\sin 2\pi f_0 t$	$\dfrac{1}{2j}\,\delta(f - f_0) - \dfrac{1}{2j}\,\delta(f + f_0)$
13.	$u(t)$	$(j2\pi f)^{-1} + \frac{1}{2}\delta(f)$
14.	$\mathrm{sgn}(t)$	$(j\pi f)^{-1}$
15.	$\dfrac{1}{\pi t}$	$-j\,\mathrm{sgn}(f)$
16.	$\hat{x}(t) = \dfrac{1}{\pi}\displaystyle\int_{-\infty}^{\infty} \dfrac{x(\lambda)}{t - \lambda}\,d\lambda$	$-j\,\mathrm{sgn}(f)X(f)$
17.	$\displaystyle\sum_{m=-\infty}^{\infty} \delta(t - mT_s)$	$f_s \displaystyle\sum_{m=-\infty}^{\infty} \delta(f - mf_s),\ f_s = T_s^{-1}$

Note: $\mathrm{sinc}\ u = \dfrac{\sin \pi u}{\pi u}$

$$\Pi(u) = \begin{cases} 1, & |u| \le \frac{1}{2} \\ 0, & \text{otherwise} \end{cases}$$

$$\Lambda(u) = \begin{cases} 1 - |u|, & |u| \le 1 \\ 0, & \text{otherwise} \end{cases}$$

Index